Fascinating Life Sciences

This interdisciplinary series brings together the most essential and captivating topics in the life sciences. They range from the plant sciences to zoology, from the microbiome to macrobiome, and from basic biology to biotechnology. The series not only highlights fascinating research; it also discusses major challenges associated with the life sciences and related disciplines and outlines future research directions. Individual volumes provide in-depth information, are richly illustrated with photographs, illustrations, and maps, and feature suggestions for further reading or glossaries where appropriate.

Interested researchers in all areas of the life sciences, as well as biology enthusiasts, will find the series' interdisciplinary focus and highly readable volumes especially appealing.

Gabriela-Alina Sauciuc
Adriano R. Lameira • Tomas Persson
Editors

Nature Beats

The What, How and Why of Rhythmic Animal Behaviours

Springer

Editors
Gabriela-Alina Sauciuc
Department of Philosophy/
Cognitive Science
Lund University
Lund, Sweden

Adriano R. Lameira
Department of Psychology
University of Warwick
Coventry, UK

Tomas Persson
Department of Philosophy/
Cognitive Science
Lund University
Lund, Sweden

ISSN 2509-6745 ISSN 2509-6753 (electronic)
Fascinating Life Sciences
ISBN 978-3-032-24249-5 ISBN 978-3-032-24250-1 (eBook)
https://doi.org/10.1007/978-3-032-24250-1

This Springer imprint is published by the registered company Springer Nature Switzerland AG
The registered company address is: Gewerbestrasse 11, 6330 Cham, Switzerland

Acknowledgments

We are grateful for the generous funding provided by the Bank of Sweden Tercentenary Foundations and Lund University, which has permitted the open access publication of all the contributions to this book. Editors Gabriela-Alina Sauciuc and Tomas Persson would like to acknowledge the contribution to science of the late chimpanzees Linda, Torsten, Manda, and Santino, who are all deeply missed.

Contents

Contributors

Lara S. Burchardt Leibniz Zentrum für Allgemeine Sprachwissenschaften Berlin, Berlin, Germany

Institute for Theoretical Biology, Humboldt-Universität zu Berlin, Berlin, Germany

Donders Institute for Brain, Cognition, and Behaviour, Radboud University, Nijmegen, Netherlands

Peter F. Cook Marine Mammal Science/Neuroscience, New College of Florida, Sarasota, FL, USA

Institute for Marine Sciences, University of California Santa Cruz, Santa Cruz, CA, USA

Chiara De Gregorio Department of Psychology, University of Warwick, Coventry, UK

Tecumseh W. Fitch Department of Behavioural and Cognitive Biology, University of Vienna, Vienna, Austria

Marco Gamba Department of Life Sciences and Systems Biology, University of Turin, Turin, Italy

Michael D. Greenfield ENES Bioacoustics Research Lab, CRNL, University of Saint-Etienne, CNRS, Inserm, Saint-Etienne, France

Department of Ecology and Evolutionary Biology, University of Kansas, Lawrence, KS, USA

Morgan L. Gustison Department of Psychology, Western University, London, ON, Canada

Yuko Hattori Institute for the Evolutionary Origins of Human Behavior (EHUB), Kyoto University, Inuyama, Japan

Denise Herzing Department of Biological Science, Florida Atlantic University, Boca Raton, FL, USA

Gerlinde Höbel Behavioural and Molecular Ecology Group, Department of Biological Sciences, University of Wisconsin–Milwaukee, Milwaukee, WI, USA

Marisa Hoeschele Acoustics Research Institute, Austrian Academy of Sciences, Vienna, Austria

Henkjan Honing Music Cognition Group, Institute for Logic, Language and Computation, University of Amsterdam, Amsterdam, The Netherlands

Adriano R. Lameira Department of Psychology, University of Warwick, Coventry, UK

Edward W. Large Department of Psychological Sciences, University of Connecticut, Storrs, CT, USA

Department of Physics, University of Connecticut, Storrs, CT, USA

Hugo Merchant Department of Neurophysiology, Instituto de Neurobiología, Universidad Nacional Autónoma de México, Campus Juriquilla, Santiago de Querétaro, Mexico

Bjorn Merker Independent Scholar, Kristianstad, Sweden

Kazuo Okanoya Advanced Comprehensive Research Organization (ACRO), Teikyo University, Tokyo, Japan

Nao Ota Museum of Nature and Human Activities, Hyogo, Sanda, Japan

Aniruddh Patel Department of Psychology, Tufts University, Medford, MA, USA

CIFAR Program in Brain Mind and Consciousness, Toronto, ON, Canada

Marcus Perlman Department of Linguistics and Communication, University of Birmingham, Birmingham, UK

Tomas Persson Department of Philosophy/Cognitive Science, Lund University, Lund, Sweden

Tudor Popescu Department of Cognition, Emotion, and Methods in Psychology, University of Vienna, Vienna, Austria

Shannon Proksch Department of Psychology, Augustana University, Sioux Falls, SD, USA

Program in Neuroscience, Center for Interdisciplinary Studies, Augustana University, Sioux Falls, SD, USA

Teresa Raimondi Department of Human Neurosciences, Sapienza University of Rome, Rome, Italy

Department of Life Sciences and Systems Biology, University of Turin, Rome, Italy

Andrea Ravignani Department of Human Neurosciences, Sapienza University of Rome, Rome, Italy

Center for Music in the Brain, Department of Clinical Medicine, Aarhus University, Aarhus, Denmark

Istituto di Scienze e Tecnologie della Cognizione, CNR, Rome, Italy

Stephan A. Reber Department of Philosophy/Cognitive Science, Lund University, Lund, Sweden

Colleen Reichmuth Institute for Marine Sciences, University of California Santa Cruz, Santa Cruz, CA, USA

Gabriela-Alina Sauciuc Department of Philosophy/Cognitive Science, Lund University, Lund, Sweden

Jennifer N. Schneider Department of Psychology, LCC International University, Klaipeda, Lithuania

What is Rhythm and How Can We Study It?

1

Lara S. Burchardt, Marisa Hoeschele, Henkjan Honing, Edward W. Large, Bjorn Merker, and Jennifer N. Schneider

Abstract

Rhythm. What is it? We all have an intuitive understanding of rhythm. The term is used widely to describe temporal structures of any kind, ranging from the change of tides and seasons, or the drumming of rain on our windowsill, to the movements of animals, heartbeats or breathing, or the captivating rhythm of a skilled drummer. As of yet, the term "rhythm" lacks established technical usage

L. S. Burchardt (✉)
Leibniz Zentrum für Allgemeine Sprachwissenschaften Berlin, Berlin, Germany

Institute for Theoretical Biology, Humboldt-Universität zu Berlin, Berlin, Germany

Donders Institute for Brain, Cognition, and Behaviour, Radboud University, Nijmegen, Netherlands
e-mail: lara.sophie.burchardt@hu-berlin.de

M. Hoeschele
Acoustics Research Institute, Austrian Academy of Sciences, Vienna, Austria
e-mail: marisa.hoeschele@oeaw.ac.at

H. Honing
Music Cognition Group, Institute for Logic, Language and Computation, University of Amsterdam, Amsterdam, The Netherlands
e-mail: H.J.Honing@uva.nl

E. W. Large
Department of Psychological Sciences, University of Connecticut, Storrs, CT, USA

Department of Physics, University of Connecticut, Storrs, CT, USA
e-mail: edward.large@uconn.edu

B. Merker
Independent Scholar, Kristianstad, Sweden

J. N. Schneider
Department of Psychology, LCC International University, Klaipeda, Lithuania
e-mail: jschneider@lcc.lt

G.-A. Sauciuc et al. (eds.), *Nature Beats*, Fascinating Life Sciences,
https://doi.org/10.1007/978-3-032-24250-1_1

in behavioral biology and other research fields and has therefore been applied to a wide range of behavioral phenomena in the temporal domain. This wide scope and lack of terminological specificity easily confound the study and discussion of the distribution of rhythmic behaviors across species. In this chapter, we tackle this issue across animal species (including humans) mostly in the context of cognitive and behavioral biology and neurobiology. We identify three general perspectives to study rhythm, starting with the physical domain, describing a purely physical phenomenon purposefully detached from any biological example. The second perspective, moving into biology, includes the perception of these physical signals. The third perspective is based on higher-order concepts and structures such as music and group interactions. Starting from the most general sense in which the term "rhythm" is used for phenomena in time, we add successive constraints to increase specificity or specialization. Thus, we proceed by stages from a sense of rhythm as present in all life forms, to one which, to our current knowledge, might only apply to humans. Lastly, we provide an extensive overview of issues and perspectives necessary to grasp the full concept of "rhythm" in modern rhythm research across all domains.

1.1 Introduction

Rhythm. What is it? We all have an intuitive understanding of rhythm: The term is used widely to describe temporal structures of any kind. These range from the rhythm of the tides or seasons and the random—yet rhythmic—drumming of rain on our windowsill, to the movements, heartbeats, and breathing of animals, or the captivating rhythm of a skilled drummer. Given that the term "rhythm" is used in research on music and dance, speech and animal communication, circadian organismic changes, neural and mechanical oscillators, and movement, as well as in research on acoustic, visual, chemical, or electrical signaling in general, it is unlikely that the term will find a single definition optimal for all its diverse uses.

In fact, the term "rhythm" lacks established technical usage in behavioral biology and other research fields, and has therefore been applied to a wide range of behavioral phenomena in the temporal domain. This wide scope and lack of terminological specificity easily confound the study and discussion of the distribution of rhythmic behaviors across species including humans. Behaviors may show rhythms from the millisecond to the year range. Rhythm is used in the description of individuals, dyads, or groups. Taking the term "rhythm" in its widest temporal sense of "periodicity," rhythmic behavior is exhibited by every living species, including plants [1]. From the rhythmic beating of a bacterial flagellum powered by a sophisticated molecular rotary motor [2] to critical oscillations of outer hair cells in the cochlea, to cortical oscillations in the mammalian brain, and the rhythmic timing of muscle activation by low-level neural pattern generators for heartbeat, breathing, suckling, chewing, and locomotion [3], periodicity is ubiquitous in nature.

From this set of examples, it already becomes clear what a vast field a single definition of rhythm would need to encompass. A lot of terminology is loaned from the well-established field of musicology, using terms and concepts that were thought of in a highly specific context of a cultural/societal phenomenon so far only known from the human species. The terminological, disciplinary, and subject-matter diversity often leads to misunderstandings between researchers, wrongful generalization from humans to animals, and ambiguity of definitions that should be unequivocal. As an informal exercise for orientation purposes, one may instead proceed by adopting different, increasingly specific, senses of the term, to explore the consequences of doing so for the distribution of rhythmic capacities across the animal kingdom. In this chapter, we do so across animal species, including humans, mostly in the context of cognitive and behavioral biology and neurobiology. First, we identify three general perspectives (Fig. 1.1), starting with the physical domain perspective, describing a purely physical phenomenon purposefully detached from any biological example. The second perspective, moving into biology, includes the perception of these physical signals. The third perspective is based on higher-order concepts and structures such as music and group interactions. Starting from the most general sense in which the term rhythm is used for phenomena in time, we add successive constraints to increase specificity or specialization. Thus, we proceed by stages from a sense of rhythm which is present in all life forms, to one which, to our current knowledge, might only apply

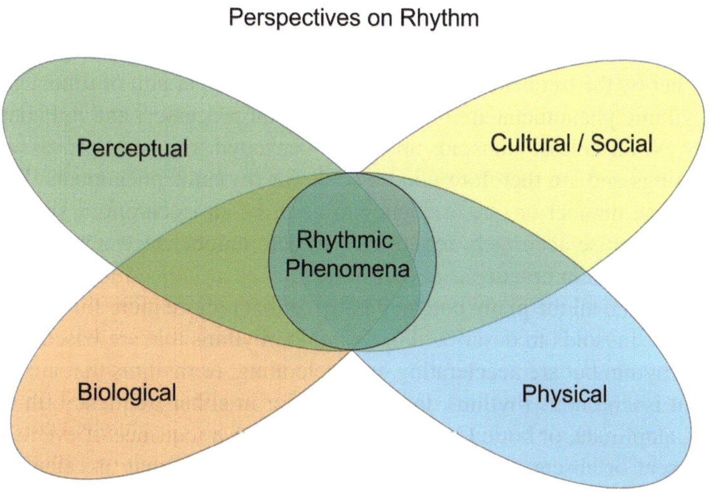

Fig. 1.1 Different approaches to studying rhythmic phenomena as discussed in this chapter. Rhythmic phenomena are abundant in nature and can be found to some extent in all life forms. Very different definitions and distinctions arise depending on the perspective from which rhythm is studied. Rhythmic phenomena could be studied from a purely physical point of view, from a biological point of view, from a perspective that is focused on the perception of rhythmic phenomena, and from a cultural and social perspective. Neither of these are mutually exclusive or all-encompassing

to humans. Lastly, we provide an extensive overview of issues and perspectives necessary to grasp the full concept of "rhythm" in modern rhythm research across all domains.

1.2 Rhythm from Three Perspectives

1.2.1 Physical Domain

An operationalizable definition of rhythms based on the physical domain of the signal needs to include repetitiveness, regularity, and predictability. The definition should be independent of but usable on timescales ranging from milliseconds to years or more. Different rhythm categories for completely different timescales (years vs. seconds, for example) could be established.

The aim is to be able to describe a physical signal, a series of events in time, quantitatively and qualitatively in a way that is universally applicable, precise, easy to understand in its precision, and suitable for all kinds of data. To achieve this goal, we first need to characterize all the patterns we might want to describe. Rhythm requires some kind of pattern that can be abstracted from the signal. Therefore, we should only consider signals to be rhythmic where this is possible. To illustrate this, we will look at acoustic signals as an example of rhythmic behaviors in general. Overall, acoustic signals range from white noise (random events across the spectrum) to a completely stereotyped event sequence of similar intervals, which we often call isochronous or metronome-like rhythm. This rhythm can be described by a single number, the duration or period between elements, and as such can be represented either by the frequency of a sinusoid or as events per unit of time. Of course, many rhythmic phenomena are discrete, even point processes, and not continuous sinusoids. Nevertheless, sinusoids are easily converted to a point process by their zero-crossings and are therefore good models for rhythmic phenomena. While we only need one number or one frequency to describe an isochronous signal, representing white noise accurately requires an infinite number of sinusoids across all frequencies, which in practice is not feasible.

In between lie all the many potential patterns that require more than one but less than infinite sinusoids to describe. For example, rhythms that are based on an isochronous rhythm but are accelerating or decelerating, or rhythms that are based on *n* different isochronous rhythms, that could differ in either frequency (in terms of cycles/s), amplitude, or both. Considering rhythms as a sequence of events that are either present or absent and have a certain amplitude, we can use sine waves or cosine waves as approximators to describe them. As we do not need negative values, we could, for example, simplify to

$$y(t) = |\cos(t)| \text{ or } y(t) = |\sin(t)|.$$

The period is the duration until the pattern repeats itself, and the frequency is the number of repetitions per unit of time, both of which also define the timescale of the rhythm. The amplitude of the sine wave can inform the investigator about different hierarchies between elements, for example, some elements might be emphasized by a different amplitude. This becomes especially important when considering perception (see Sect. 1.2.2). Furthermore, complex rhythms may be made up of multiple interacting rhythms with different frequencies, phases, and amplitudes. It is therefore important how many different oscillators make up the signal and how they are related.

A decelerating pattern could be described as:

$$T(t) = T_0 \times e^{-\alpha t}$$

where T_0 = the initial period and α = a constant determining the rate of decrease of the period.

The other way around, if we want to describe a linearly accelerating signal, the equation would be

$$T(t) = T_0 \times e^{-\alpha t}.$$

The use of mathematical equations as models for biological systems is typically idealizations or approximations, and as such may be more precise than is possible in practice, which is why it is also important to describe how well any model describes a given biological system. For example, to mark the difference between mathematically perfect isochrony and a regularly repeating pattern found in a biological system, the term "quasi-isochrony" may be used. Models of rhythm are more applicable to some species and behaviors than to others, which is an important topic of this chapter and the book as a whole.

The terminology used in this section was focused on the physical signal properties. These signals only obtain meaning once they are perceived. While we could say that the term "rhythm" can be based on the signal itself, we now move to rhythmicality or rhythmicity, a notion that we apply to any situation when attention (either conscious or subconscious) is placed on the production and/or perception of rhythms.

1.2.2 Perceptual Domain

As noted in the Introduction, rhythms are everywhere in the natural world. In many cases, rhythms occur simply because of efficiency or mechanical repetition. For example, a woodpecker pecking at its fastest speed would create a predictable isochronous rhythm (or quasi-isochronous) just by the nature of efficiency. A heartbeat also follows a repetitive pattern thereby creating a rhythm. But where the creation of rhythms becomes particularly interesting is where it interacts with perception. At this point, we move from talking about the physical properties of a signal—the rhythm itself—to the sensory/cognitive perception of rhythm—perceptual rhythmicity.

Even in the case where rhythm is produced for reasons of efficiency or mechanical repetition, these patterns may then be received and identified by a nearby animal. For example, an animal may detect the approach of a predator and the speed at which it approaches by the rhythm of its footsteps. At this point, it may be advantageous for the receiver to perceive the rhythmic elements of the signal. This is the beginning of perceptual rhythmicity.

Once a signal is perceived as rhythmic, perception may also, as a next step, begin to interact with production. If rhythms provide information to a perceiver, then signals may begin to be altered to provide specific rhythmic information. For example, vocal learning animals may display their quality by their ability to reproduce an acoustic signal accurately and steadily. The precise timing among notes and the predictability that they have can allow a listener to assess the skill of the singing individual, much like humans at a musical concert assess the skill of the performers.

Ultimately, perceptual rhythmicity depends on the physical domain and is also critical in developing intentional rhythmic behavior—a move from rhythm to rhythmicity and then to its use in a species' behavioral repertoire. Similarly, a perceptual/production feedback loop depends on perceptual rhythmicity and is necessary to allow animals to add meaning to the interpretation of rhythm. It is to this communicative and interactive signaling and its effect on species-specific cultures that we turn next.

1.2.3 Species-Specific Biology and Cultural Domain

1.2.3.1 Communicative Signaling

When adding the constraint that rhythmic behavior must be employed in communicative signaling, the pool of species under consideration contracts appreciably, because not all creatures engage in overt behavioral signaling. Signaling requires a receiver [4], but not every species has sensory means for receiving communicative signals. Thus, for example, the common vocal-acoustic mode of communicative signaling requires a sensory capacity for sound reception. That capacity is not merely a matter of being equipped with the peripheral sensory organ, but of possessing the considerable neural resources needed to use it. Such a resource is naturally missing in creatures without a brain and is even absent in some with a brain (e.g., *Amphioxus*: [5]). This still leaves a vast array of species that do have the sensory equipment needed to engage in communicative signaling, and which do so.

There is a particular reason for animals engaged in communicative signaling to do so rhythmically, and repetitively, in addition to the kind of ubiquitous periodicity mentioned earlier. Signaling typically takes place in an ambience of noise, which impedes reception, the very point of signaling. A basic strategy to increase the probability of reception under noise is to simply repeat the signal, which offers receivers the opportunity to extract the signal from noise by signal averaging [6] but also through attentional tuning and expectation [7–9]. Given repetitive signaling, the pacing of successive periods is again, as in repetitive behaviors, likely to be regular for reasons of economy of timing control.

So far, the signaling we considered consists of overt behavior employing any one of a range of production modalities: gestural, vocal, bioluminescent, or electric (for details see Chaps. 2 and 3). Whatever the behavior, its effects are simply broadcast from sender to receiver, as a unidirectional ("solo") performance. Its rhythmic form (the temporal patterning of its component periods, as well as their overall temporal arrangement) is under these assumptions a free variable, except for the stipulation that some kind of periodicity is in evidence to qualify as rhythmic (which, in turn, entails the passive, non-obligatory convenience of even pacing, as before). It is changes in behaviors that involve the temporal coordination of the signaling behavior of two or more individuals, which we consider next.

1.2.3.2 Interactive Coordination

By adding further the constraint that the communicative signaling be interactively coordinated in time, the pool of species of interest contracts to those that engage in behaviors such as turn-taking, countersinging, jamming, duetting, chorusing, and similar behavioral signal interactions. Even at this level of constraint, we are still a long way from being able to supply an exhaustive list of the species in question—they are many, indeed: for duetting alone, and birds alone, there are over 400 duetting species [10]—but at least we are approaching what might be made the topic of a concerted project of behavioral taxonomy.

Under the interactive constraint, temporal patterning of the communicative signal is no longer a free variable: the timing of the signal as a whole or of one or more of its components is now bound to that of the partner or partners in the interaction. In the simplest case of turn-taking, interindividual coordination of signal timing requires only the motive to respond and the rule to delay signal initiation until after a received signal has ended. The alternation that results requires no specialized mechanisms such as the predictive timing introduced in Sect. 1.2.3.3. It is achieved, rather, by responding to stimuli within the limitations of sensory-motor reaction time, and can be refined by incorporating learned accommodation of signal initiation to predictable perturbations in stimulus timing [11, 12].

Beyond the common case of simple alternation, a range of more complex patterns of interlocking individual signal contributions occur in some species, topped so far by the tightly coordinated interleaving of male and female syllable sequences in the duets of plain-tailed wrens [13]. The interleaving is so seamless that it gives the impression of a single bird singing (cf. hocket in human music). Its pace, moreover, is so rapid that it renders accounts reliant on syllable-by-syllable reactions implausible. It was therefore suggested that "each bird might learn the entire pattern as a single sequence of which it only contributes its own segment during performance" [14] (p. 5, fn. 1). This conjecture subsequently was confirmed by a study of patterns of neural activity recorded in plain-tailed wrens exposed to duet stimuli [15], a study which also suggested (on indirect behavioral grounds) that feedback from the bird's own, as well as its partner's voice, contributes to duet coordination. This case of complex and precise temporal coordination of joint signaling is noteworthy for the contrast between its suggested underlying mechanisms and those of another form of tight inter-individual coordination in time, to be considered next.

1.2.3.3 Interindividual Behavioral Synchrony

Among species whose members engage one another through interactively coordinated, communicative, periodic signals, some do so in sustained displays of temporally precise interindividual behavioral synchrony based on an underlying isochronous pulse, tactus, or beat. The regular recurrence of a temporal interval makes its next occurrence predictable. This predictability allows individuals to eliminate delays introduced by reaction time from their coordination through predictive timing [16]. The pulse or beat that underlies such inter-individual coordination is what we clap and nod to in rhythmic music. By adding this constraint of sustained, pulse-based interindividual synchrony, we finally arrive at a rather manageable assemblage of species found mostly among arthropods, but including also a few species of anurans and humans [17, 18]. Further examples of precise and sustained interindividual isochronous synchrony may of course emerge as research proceeds.

Note that sustained interindividual precision synchrony has few biological uses apart from the specialized circumstances under which some arthropods, anurans, and one primate have found their way to this specialized capacity. In all the nonhuman cases those circumstances center on the need for males to attract migrating females for mating (group living animals typically have no such need), and since a combination of female exogamy and male territoriality has been attributed to the common ancestor of humans and chimpanzees on cladistic grounds, that same need has been proposed to explain human membership of this select assemblage of rhythmic group synchronizers [14, 19, 20] (see also Chaps. 5 and 14, this volume).

Not only is the capacity for sustained precision group synchrony relatively rare in nature, but it is also dependent upon specialized sensory-motor mechanisms. As already noted, reaction time introduces an unavoidable delay between stimulus reception and response initiation. One therefore cannot get a rhythmic sequence of discrete signals to coincide with those of a partner by reacting, stimulus-by-stimulus, to the partner's output. Rather, to achieve lagless precision synchrony, it is necessary to "know" beforehand when the next stimulus is to occur to target it at the time it occurs. The enabling device for such predictive timing of behavior is the regularity of stimulus pacing, that is, isochrony, which makes the time of occurrence of the next stimulus in the sequence perfectly predictable.

The simple one, one, one, one, one of isochrony is therefore the underlying structural device on which all the species that engage in sustained interindividual precision synchrony rely, at times with an evenness of pacing whose coefficient of variation is smaller than what humans at their best produce [18]. The crucial role of predictive timing in lagless precision synchrony thus converts the nonobligatory, default mode of the various rhythmic ("periodic") behaviors covered in points 1 through 3 to an obligatory and fundamental feature of rhythmicity in the precision group synchrony sense covered here. With that conversion, the loose and local "default" mode of isochrony considered so far is enhanced (regularized) to a capacity for sustained production of precisely timed free-running isochronous output. Details on the various devices by which synchronizing species exploit this sustained isochrony to achieve synchrony are reviewed in Greenfield and Merker ([18], see also Chap. 5 this volume). In a manner of speaking, isochrony is that "same time" (here to be read in the sense of "shared time") to which the "syn-" (Greek for "with," "together") of "synchrony" alludes.

1.2.3.4 Role of Culture and Cultural Transmission

We can now add the further constraint of cultural transmission and differentiation of patterns of rhythmicity to the constraint of showing sustained and precise interindividual rhythmic synchrony. With these constraints and to our current knowledge we would end up with a single species, humans, because neither arthropod nor anuran synchronous chorusing carries cultural content. In sharp contrast to this, the rhythmic pattern content of human precision group synchrony, which, needless to say, is structurally complex far beyond the simple unison synchrony of nonhuman group synchronizers, is deeply embedded in cultural traditions. It is expressed primarily in music, dance, and drill, art forms whose rhythmic content is elaborated and subject to intergenerational transmission on a learned basis in culturally specific genres of rhythmic tradition [21–24]. This is only to say that humans maintain cultural traditions of rhythm in their own right, that is, rhythmicity as an independent dimension of musical content (as in Walz versus Tango versus Sinawi (Korean six-time)). We presumably arrived at these elaborate cultural traditions on paths similar to those of other species with cultural traditions of complex esthetic displays [25] (see also Chaps. 5 and 12).

1.3 Isochrony and Its Relation to Rhythm

The methods used to study underlying isochrony in biological systems are still subject to ongoing exploration and are undergoing further development. Especially for methods still in development, it remains unclear to what extent interpretations are consistent across vastly different datasets or which hidden assumptions must be met for these methods to function as intended. Therefore, the study of isochrony in animal behaviors, including human behavior, currently requires a multidimensional approach. This approach draws upon a diverse array of theoretical concepts and methodological tools described here, as well as those that may be developed in the future, to ensure interpretations are based on a solid foundation.

The pacing of many different repetitive behaviors tends to be regular, often approximating isochrony in the timing of their component movements, at least within the local temporal horizons of a few successive elements. No particular functional significance attaches to this even pacing: mere repetition simplifies the control of timing compared to having to determine the timing of each component behavior separately. The factor of control economy suffices to render isochrony the default mode of many such repetitive, self-timed movement sequences. As such, they are unstable, easily disrupted, or modulated by competing behaviors, but also easily regularized should functional demands require it. An example of the latter is the regular pacing necessary for interappendage coordination and maintenance of constant speed in locomotion. Certain physiological processes, such as breathing, which are quasi-isochronous, again by laws of efficiency, can be coupled to locomotion in locomotor-respiratory coupling and also to vocalizations in locomotor-respiratory-vocal coupling (see Chaps. 2 and 9 this volume). In all this, we must emphasize again that no biological system will be isochronous, but rather "quasi-isochronous" and well described by isochrony. The quantification of rhythms in biological systems will always be a model of the found patterns, and we need to acknowledge the importance of describing how well a given model fits the patterns found in the data.

Based on these assumptions and keeping necessary considerations in mind, it should be the start of any rhythmic analysis to look for isochrony, as it seems reasonable to find an underlying isochrony in many rhythmic systems on some level of signaling. This would be true across both time scales and modalities of signaling. In acoustic communication, this would mean finding the appropriate time frame for analysis, such as within a signal (a common example is pulsed sounds in fish [26]), between signals, or between phrases in a song, as well as between songs. There are complex rhythmic signals showing predictability and repetition that do not show an underlying isochronous rhythm, though, for which we need other methods.

In some rhythmic phenomena, such as in a lot of human rhythmic music, especially dance music, isochrony is a truly global phenomenon in the sense of being maintained consistently throughout a performance bout. Because occasional phase shifts beyond the minute ones of gradual tempo drift disrupt this sustained isochrony, local measures of isochrony are inherently incapable of capturing this attribute of human rhythmic music (see [21] p. 179 ff), because the occasional phase shifts will only marginally affect such local measures. More generally, various methods exist to check whether a signal could be modeled by an isochronous rhythm, complemented by subsequent analyses to quantify the frequencies able to model the raw signal. To be sure about isochrony being a global phenomenon, we need to find proof for that in global as well as local measures. Analyses such as interonset interval (IOI) analysis, assessments of variability parameters, visual inspection, and Fourier analysis provide means to dissect rhythmic patterns across different temporal scales [27]. These methods enable researchers to uncover underlying oscillatory dynamics, explore the predictability of temporal sequences, and discern the interplay between regularity and variability. As for local measures, the limitations of each method need to be checked carefully to render meaningful and comparable results.

In human music, rhythmic information is not always clearly marked but emerges from the entirety of the structure. With nonhuman animals, it can thus be difficult to know where to begin. One approach is to look at intervals between time points of interest. That is why the basis for most, if not all rhythm analyses, is intervals between such time points of interest. These could be intervals between starting points of elements, that is, individual syllables, calls, or clicks in a sound sequence, but also bioluminescent flashes, footfalls in stomping, or any other behavior of interest. It could also be intervals between pulses of one sound, or it could be intervals between the point of highest amplitude in a sound or the ends of sound elements (or elements of other modalities). The definition of interval is dependent on the dataset and question but must be specified in detail, to guarantee comparability [27].

Following the procedures described in Burchardt and Knörnschild [28] or Hersh et al. [27], we start by explaining the variability parameters and distributional measures, such as the coefficient of variation, normalized pairwise variability index and integer ratios. We will then proceed to short explanations of specific rhythm analysis approaches like the IOI approach and Fourier analysis and end with the important description of measures that indicate the goodness-of-fit of the models, that is, the isochronous rhythm rates.

To illustrate the need to distinguish between measures used to interpret isochrony, we apply them to four different hypothetical sequences. Each sequence of intervals consists of 1000 values. We show the interval distribution as histograms (Fig. 1.2) and the variability estimates produced by the different measures (Table 1.1). We have one isochronous sequence, where every interval is the same, in our case 1 (Fig. 1.2a). Sequence two is quasi-isochronous with an interval sequence mean of 1.0 and a standard deviation of 0.3 to mimic a biological signal sequence with an underlying isochronous beat (Fig. 1.2b). A third sequence illustrates a linear increase in interval duration: starting with an interval of 1 s, we add 0.01 s with every step so that the last interval is 10.99 s long (Fig. 1.2c). A last sequence is drawn randomly from values between 1 and 11, to show the differences in the linearly increasing sequence (Fig. 1.2d).

The coefficient of variation (CV) is calculated by dividing the standard deviation by the mean. It is independent of sample size and the mean, making it a powerful measure for comparing rhythmic variability across individuals, species, or studies (Eq. 1.1 [28]). For small sample sizes, an adjusted equation exists (Eq. 1.2 [29]).

$$C_V = \frac{S}{\overline{X}} \tag{1.1}$$

$$C_V^* = \left(1 + \frac{1}{4n}\right) C_V \tag{1.2}$$

C_V = coefficient of variation

C_V^* = unbiased coefficient of variation for very small sample sizes
s = standard deviation
$\overline{x}$ = sample mean
n = sample size

Established for speech quantification, the normalized Pairwise Variability Index (nPVI [28, 30, 31]) indicates the rhythmic structure of a sequence and how well each interval can be predicted by the preceding one, which makes it a local parameter (Eq. 1.3). In a mathematically isochronous sequence, the nPVI will be zero. In

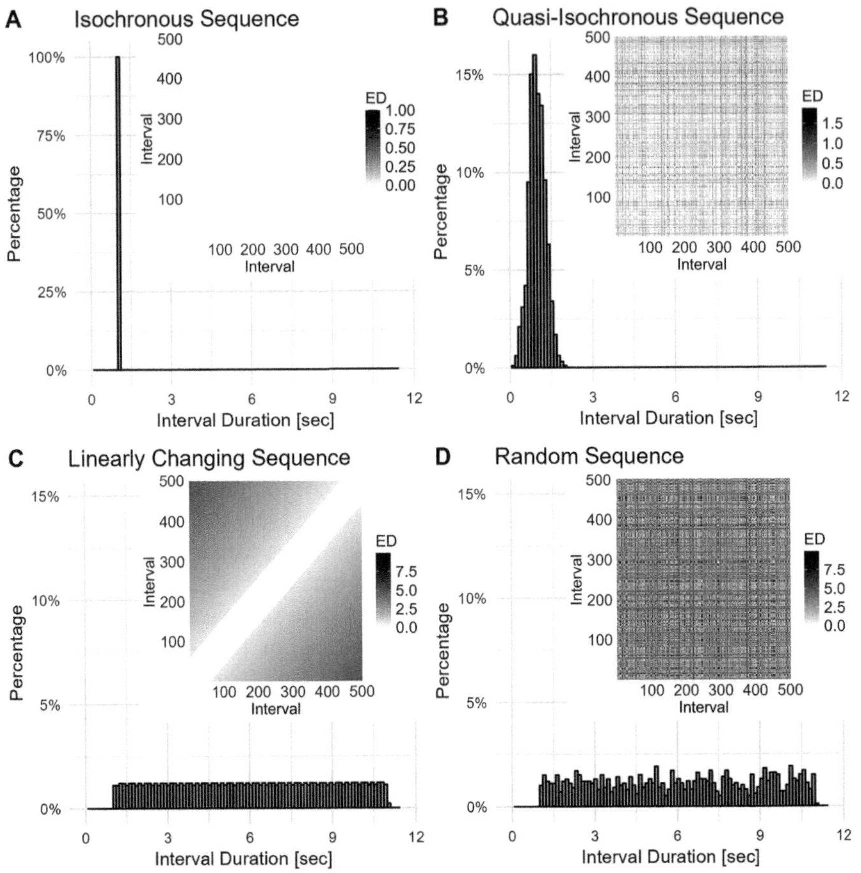

Fig. 1.2 Histogram of the distribution of interval duration in four different exemplary sequences of 1000 elements each. (**a**) In an isochronous sequence, every interval is exactly the same, which results in only one bin being filled in the histogram and a completely white recurrence plot. (**b**) In a biologically more realistic situation, we would have a quasi-isochronous sequence, with interval durations jittering around a mean interval, in this case of 1 s. The resulting histogram shows a steep unimodal distribution. This distribution is an absolute prerequisite for finding a global underlying isochronous rhythm able to model the sequence. The recurrence plot shows a jittered pattern with overall small Euclidean distances (up to 1.5). (**c**) In a sequence where intervals increase linearly by 0.01 s per interval, the histogram shows a uniform distribution, which does not hint at an underlying isochronous pattern. The recurrence plot clearly shows the linear transformation and string differences. Euclidean distances go up to 7.5. (**d**) The same is true for a sequence of random intervals of durations between 1 and 11 s. The recurrence plot shows no patterning, but higher Euclidean distances (up to 7.5). This is a novel analysis run for this chapter

most cases, very small numbers strongly indicate an underlying isochronous sequence. However, the nPVI needs to be combined with visual inspection and other variability parameters, to make sure to not falsely assume an underlying isochrony, as in the case illustrated in Fig. 1.2, where the results for a sequence with intervals linearly increasing in duration (Fig. 1.2c) yield an nPVI that is smaller than the one in Fig. 1.2b.

$$\mathrm{nPVI} = \sum_{k=1}^{m-1} \left| \frac{\mathrm{IOI}_k - \mathrm{IOI}_{k+1}}{\frac{\mathrm{IOI}_k + \mathrm{IOI}_{k+1}}{2}} \right| * \frac{100}{m-1} \tag{1.3}$$

nPVI = normalized Pairwise Variability Index k = index number m = total number of indices IOI = interonset interval

Another local parameter that needs to be handled very carefully is integer ratios—also called categorical rhythms, rhythmic categories, rhythmic ratios, or interval ratios—which in many studies are denoted as *rk*. This parameter was first described by Roeske et al. [32] (Eq. 1.4) and has subsequently been employed in other studies [33–37]. It describes the predominant ratio between adjacent intervals and, thus, this description is—similar to the nPVI—a very local parameter, as it analyzes only two adjacent intervals. All successive intervals in a quasi-isochronous sequence will have a ratio close to 1:1, but as in the already discussed case of the global isochrony that characterizes many forms of human rhythmic music, phase shifts that disrupt this global isochrony will be only marginally reflected in changes in the resulting rk of 0.5, which accordingly does not capture this crucial aspect of human musical rhythmicity (see rk entries in Table 1.1) and is therefore in this form not a suitable basis for comparison between human music and acoustic animal behavior. One way to avoid that issue could be to additionally analyze the ratios not only between adjacent intervals but also between random intervals in a sequence or all interval pairs. If the underlying isochrony is indeed global, both results should be similar.

Table 1.1 Calculated variability parameters for four different interval sequences

Sequence	CV	nPVI	Mean rk	Minimum rk	Maximum rk	Interval range
Isochronous sequence	0	0	0.50	0.50	0.50	1
Quasi-isochronous sequence	0.1	11.40	0.5	0.38	0.61	0.7–1.3
Linearly changing sequence	0.48	0.24	0.50	0.50	0.50	1–11
Random sequence	0.48	60.64	0.5	0.09	0.9	1–11

Each sequence consists of 1000 elements. The coefficient of variation (CV) normalized pairwise variability index (nPVI), and integer ratio (rk) are calculated. For the rk the mean, minimum, and maximum value is shown. Furthermore, the interval range of intervals in any of the four exemplary sequences is indicated

$$r_k = \frac{\mathrm{IOI}_n}{\mathrm{IOI}_n + \mathrm{IOI}_{n+1}} \tag{1.4}$$

IOI = interval
n = index

The results of the example sequences summarized in Fig. 1.2 and Table 1.1 nicely illustrate that the results of one single analysis can be very misleading when searching for underlying isochrony. The linearly changing sequence could be considered to have an underlying isochronous structure when looking at the variability parameters: the nPVI is very small and the r_k seems to indicate a perfectly isochronous 1:1 ratio of intervals. However, neither of these values can be considered as global parameters. As the local parameters they are, they only have limited power in describing the rhythm of a sequence, especially for longer sequences. This needs to be kept in mind when using them to make inferences about the global rhythmic nature of a sequence, especially when relating the analysis to musical rhythm, whose isochrony is conspicuously global.

To find these global rhythms, we advance to the next level of analysis. As a sequence of similar intervals, isochronous rhythms are structurally similar, but of course, these intervals can be of different duration between different sequences. To differentiate, we speak of different isochronous frequencies, or beat frequencies, which are reported in Hertz, which is beats per second. To calculate a best-fitting model of an isochronous frequency underlying a sequence, three methods have been described. The fastest and computationally easiest is the IOI approach (Eq. 1.5), where we divide 1 by the mean (or median) interval duration of the sequence [28]. Another approach, the generate-and-test approach, tests various frequencies and optimizes for the smallest root-mean-square deviation of elements to beats of the tested frequency [38]. The mathematically most challenging approach is based on a fast Fourier transform. Here we need to convert the interval sequence into a binary sequence, where points of interest are transcribed as 1 whereas everything else is transcribed as 0 in a timeseries. Then, the Fourier transformation decomposes the signal into its sinusoidal components. Fourier transformations are mostly calculated via computer programs, which is why we do not give the equation here. A discussion of advantages and disadvantages as well as datasets for which each method might be best suited and equations for the fast Fourier transform are given in Burchardt and Knörnschild [28].

$$\text{Frequency} = \frac{1}{\overline{\mathrm{IOI}}} \tag{1.5}$$

The three methods just described to calculate rhythm frequencies for isochronous sequences will result in a best-fitting frequency. The quality of this "best fit" must also be assessed, though, to facilitate comparisons of rhythmic and beat precision between different contexts, individuals, and studies. A universal goodness-of-fit value has been described that can be used for frequencies independent of the method they were calculated with, which indicates the beat precision (*bp*, Eq. 1.6 [39]). It is calculated per element in a sequence after a best-fitting isochronous frequency has been found. In such a case, we know the maximum possible deviation any element can have from the next beat, which is half the beat length (i.e., half the time between two adjacent beats of the corresponding frequency). Relating this with the absolute deviation an element has from the closest beat gives a number between zero and one. Zero would mean no deviation and thus a perfect fit of this particular element with the beat, while one would mean the maximum possible deviation of this particular element to the beat.

The best ways to utilize the *bp* are a topic of ongoing investigation and a couple of things need to be acknowledged when utilizing *bp* analysis. Currently, the first element in a sequence functions as an anchor for matching beats and elements and calculating deviations It would thus have a *bp* of 0, and should not be considered. The first element might be off, though, and not a good anchor point. Ideally, the beat position would be calculated globally by offsetting the isochronous beat starting time in increments up to a full beat to find the offset of the highest overlap with elements or minimum deviation of elements from the found beat, as it is, for example, also done in Norton and Scharff [38]. Furthermore, the *bp* values are analyzed over a sequence. It has been previously proposed to take an average of *bp* values over a sequence [39]. Nevertheless, when averaging over a sequence, this approach has an important limitation. Imagine an average *bp* of 0.1 and 0.9 for two sequences. A deviation of 0.1 would mean all elements are very close to the theoretical beat. A deviation of 0.9 would mean all elements are very close to the maximum possible deviation but in a very structured manner. This would indicate that they follow the same beat equally well but with a phase shift. Therefore, when summarizing and only when summarizing over a sequence, a symmetry around 0.5 exists, so summarized values should be transformed (Eq. 1.7, *work in development*). On the other hand, the raw distribution of *bp*s can give insights into finer structures. The *bp* can be a powerful measure when using its distributional information instead of a summarized value. This distributional information can, for example, be visualized with rose plots.

$$bp = \frac{|\Delta|}{\Delta_{max}} \tag{1.6}$$

$$bp_t = 0.5 - |bp - 0.5| \tag{1.7}$$

In conclusion, this short method excursion shows that further method development is necessary and needs to engage with rhythm definitions from different perspectives to enable cross-species comparison and meaningful interpretations. The tools' outputs should always be critically compared, and caution should be taken when different methodologies hint at different interpretations. Methods that reliably detect and quantify a rhythmic signal made up of several sinusoids are yet to be developed. By delving into the nuances of regularity, researchers will hopefully be able to uncover the fundamental principles governing rhythmic behavior across various species. This will pave the way for a deeper understanding of the intricate temporal dynamics that shape the natural world and will lay the foundation to study rhythmicity, rhythms in interaction, as well as higher-order concepts such as rhythm cultures in humans and other animals.

1.4 Perceptual Rhythmicity: Attention on Rhythm Production and Perception

1.4.1 Assessing Perceptual Rhythmicity Through Laboratory Experiments of Perception

As discussed in Sect. 1.2, the production of quasi-isochronous and other rhythmic patterns does not necessarily coincide with the use of these patterns as perceptual signals. However, the capacity to perceive these parameters is absolutely critical for the capacities that, for example, underlie the formation of rhythmic cultures. Therefore, to determine whether animals have rhythmicity, it is of particular importance to consider whether animals can perceive rhythmic patterns.

How can we assess whether animals truly perceive the rhythmic aspects of a signal and not other properties? In other words, how can we determine whether other species have rhythmicity? This is a non-trivial question. Take for example the coo vocalization of the collared dove. This vocalization contains five elements: short note, short silence, long note, long silence, and final note. Playback experiments revealed that the collared doves require the overall length and timing of the elements to be accurate for them to respond to it [40]. But is this because they are matching the song to a particular template? Or is it because they are identifying critical rhythmic properties within the signal?

Perceptual rhythmicity is ultimately an abstraction from the original signal. When an animal has perceptual rhythmicity, it can detect the presence of rhythmic features such as the ones we have discussed so far, for example, isochrony, synchrony, and beat frequency. Only if an animal can do this, may it also be able to detect higher-order patterns such as meter (one of the terms borrowed from musicology). Meter is a perceptual grouping of beats. In a waltz, for example, isochronous beats are grouped in threes. In typical popular music songs, the beats are grouped in fours. Another example of higher-order pattern recognition is tempo invariance. When we hear a familiar song performed slower or more quickly, we can recognize it because the relative spacing between events is the same.

With the collared dove example, we can see that even if rhythm is present in the physical signal, perceptual rhythmicity does not necessarily follow. The collared dove may instead be matching a template of absolute times to assess a received coo signal. The rhythmic properties and how they relate to other similar signals may not be processed; however, this is unclear from naturalistic experiments alone. By presenting artificial signals in laboratory experiments, researchers have begun to shed light on whether nonhuman species experience perceptual rhythmicity.

Many early experiments were conducted with small songbirds and parrots using operant conditioning. Both avian groups are known for their ability to sing "songs" and learn new vocalizations from experience, which made them interesting study groups for the evolution of music as a whole [41]. Overall, however, these studies showed that birds tend to pay attention to absolute features of sound rather than relative ones [42]. When it came to pitch, the birds struggled to learn to discriminate between simple relationships, such as sequences of notes that were rising vs. falling in pitch [43–49]. Instead, the birds tended to encode the absolute pitch of the notes [50, 51], which is rarely encoded in humans: only people with perfect pitch, which make up about as low as 1/60,000 of the population, can encode absolute pitch at a level that comes close to many of the tested birds [52].

Similarly, with rhythm, the birds also tended to pay attention to absolute rather than relative features [53]. Although the birds could easily learn to attend to specific time interval lengths, they struggled to identify overarching patterns in a sound, especially more complex ones like different kinds of meters. This was especially true for the one vocal nonlearning species that was tested, i.e., the pigeon was not able to encode arguably the simplest relative feature: isochrony [54]. For the birds that do vocally learn, the results were better. Older studies with European starlings [43] and jackdaws [55] suggest that these species can learn abstract information about arbitrary rhythms. A more recent study with zebra finches and budgerigars suggests that although the animals attend primarily to absolute features, some individuals of both species also attend to relative features [56]. Place preference experiments with budgerigars additionally showed that female (but not male) budgerigars prefer rhythmic to arrhythmic signals [57] and that isochrony is an important component of this preference (Haiduk et al., *in preparation*), suggesting that these species can attend to relative information in acoustic signals. However, relative information was more likely to be used in other contexts. While meter detection in nonhuman species appears to be a challenge, many species appear to be able to use rhythmic information in language to identify emphasized syllables in human speech (budgerigars [58], rats [59], zebra finches [60]). This suggests that perhaps animals are more likely to pay attention to relative information in vocal signals than in arbitrary, synthetic signals, which may have less biological relevance.

An argument could be made that vocal learning birds were more likely than vocal non-learners to be able to attend to the relative features necessary for perceptual rhythmicity given other recent research on beat synchronization. Of the animals that humans tend to choose as pets, parrots may be particularly interesting. Videos shared via social media indicate that parrots can develop the capacity to dance to music in captive environments. This ability was experimentally examined in a case study of a sulfur-crested cockatoo (*Cacatua galerita eleonora*) named "Snowball" [61] and later statistically evaluated by additional analyses applied to a larger corpus of videos of "dancing" pets, collected from social media. These analyses suggested that parrots (and one elephant) – unlike many other pets such as dogs – were stimulated to move to and track the beat in human music [61, 62]. This feat requires more than the ability to detect isochrony: depending on the style of music, the beat can be obscured/unaccented, yet the birds seem to spontaneously find it and move to it. It must be pointed out, however, that the synchronization observed in birds had much lower accuracy than that commonly observed in music and it also appeared to be highly intermittent, that is the parrots repeatedly drifted in and out of synchrony. Therefore, others have argued that any claims of rhythmic synchronization based on these parrot data are controversial (for more details and an extended discussion, see Chap. 14).

Patel has proposed that the reason parrots have this ability to synchronize with a beat is closely connected with their vocal flexibility (for more details on this hypothesis, see Chap. 7, Chap. 14, and Chap. 15). Specifically, he argued that because the birds need to coordinate their own vocalizations in time, which is a motor response to auditory information, they can also move other parts of their bodies in time with sound. Vocal learning animals, unlike vocal non-learning animals, also have direct connections between the auditory processing areas and motor areas of the brain, which Patel argues may be the mechanism underlying this ability [63]. However, others have criticized this hypothesis suggesting that the links between auditory and motor areas only serve as a feedback loop for the bird to hone its own vocalization, and not as a comparison to external stimuli. A more detailed discussion of this can be found in Chap. 14.

In addition, in recent years, there have also been efforts to show that other vocal non-learning animals are able to align bodily movements with a beat. Ronan the sea lion (*Zalophus californianus*) is probably the most striking example of an animal that was rigorously trained to move accurately to an isochronous stimulus for a food reward and then spontaneously was able to transfer this training to music in some instances [64]. However, because of the amount of training required, it is unclear to what degree this ability is comparable in its ontogenesis to the human capacity to synchronize to a beat (see Chaps. 10 and 14). There is also additional evidence from other species that auditory-motor synchronization is possible without a strong neural link between auditory and motor areas of the brain (for review see [65]). For example, Ai the chimpanzee was trained to tap two keys on a keyboard alternately. When provided with a metronome sound during this task, if the metronome was played at a tempo that was close to her natural tapping tempo Ai aligned her taps to the beat [66].

Considering our closest relatives when it comes to rhythmicity leads to additional insights, further expanded upon in Chap. 12. Suffice it to say here that, while other primates do not vocally learn to the extent that humans do (some adjust, e.g., the pitch of their vocalizations in response to other individuals, but they cannot learn new non-innate types of vocalizations), [65] that does not mean that rhythmicity is not important to them. A recent case study of two chimpanzees showed that these individuals were able to synchronize their bodily movements to one another in what appeared to be a ritualized dance [67]. It is therefore possible that at least in primates, dance and rhythmicity may have come before acoustic synchronization. However, it is unclear whether this behavior has to do with perceptual rhythmicity or whether it's simply a physical constraint of joint motor activity (Chaps. 9, 12 and 14). Neural and behavioral studies comparing rhesus macaques to humans have shown that this species has a subset of the rhythmicity found in humans. Most striking is perhaps the finding that human infants already directly after birth automatically detect missing beats in an isochronous pattern [68], while rhesus macaques do not [69].

Despite this, there are also many parallels between humans and other primates in terms of how temporal expectations are formed. The gradual audiomotor evolution (GAE) hypothesis (for more details, see Chap. 7) suggests that beat-based timing is particularly developed in humans relative to apes and monkeys, but that it stems from the interval-based timing found across the primate lineage [70–73]. There is evidence that there is only limited beat-based timing in other primates including chimpanzees [74], bonobos [75], macaques [71], and other nonhuman primates [76]. Because beat information comes from overarching structural information that involves more than one interval, comparing multiple intervals is critical to beat-based timing. The GAE hypothesis suggests that conserved neural architecture across the primate lineage supports single temporal interval processing, but the processing of multiple intervals varies across the lineage, peaking in humans and likely incorporating other areas of the brain critical for, for example, attention, spatial sense or relational processing ([70], as well as Chaps. 7 and 12 in this volume).

Here, we have been discussing situations where animals perceive rhythmic signals. In many cases, these signals were not the animal's species-specific vocalizations. However, by assessing whether the species can perceive these signals, we can determine whether and to what extent other species have perceptual rhythmicity. Of course, whether or not the animals then produce rhythmic information intentionally requires us also to assess whether we can find rhythmic information in the patterns they produce. Because, as discussed at length, rhythms may be produced for reasons other than perceptual rhythmicity, one needs data about both rhythmic production and perception to conclude that a species has the potential to produce rhythmic information intentionally. However, in many species with potentially interesting/meaningful rhythmic information in their signals, we know little about their perception of rhythm. We will turn to some of these species now.

1.4.2 Assessing Animal Vocalizations as an Indirect Measure of Rhythmicity

Despite the many recent advances, pursuit of the mystery of perceptual rhythmicity across species is still hindered by many obstacles. An obvious approach to understanding whether other animals have perceptual rhythmicity may be to simply turn to their vocalizations or other means of rhythmic production including both acoustic and non-acoustic means. However, as we touched on in Sect. 1.2.2, this is easier said than done. Rhythmic communication is ultimately about the perception and interpretation of signals; animals may not perceive rhythm in signals that we perceive as rhythmic, or vice versa. Thus, when trying to understand rhythmic communication, it is important to consider every aspect of an animal's Umwelt. Consider the vast difference in the temporal-spatial lifestyle of a vibrating insect and a blue whale. Such differences are reflected in the temporal scale at which species can produce and perceive rhythmic signals. For example, birds have better temporal resolution than humans [77], such that a signal may sound like it is pulsed to a bird, but continuous to a human. Likewise, animals may differ in the characteristics of the rhythmic signals that they find meaningful as well as in their tolerance for error.

It is challenging to determine what aspects of rhythm might be meaningful to the receiver. Variations that may appear to be noise to us may turn out to be of biological importance to animals, as has recently been discovered with sperm whale codas (patterned click trains). Scientists have historically categorized codas into distinct types based on relative temporal features (distinctive patterns in the interunit ratio) and absolute number of clicks. Differences in absolute temporal features were largely disregarded. However, recent research by Sharma and colleagues [78] on dyadic interactions between whales has led to a shift in the understanding of codas. During vocal exchanges, whales will smoothly vary the absolute duration of their codas, a phenomenon termed rubato by the researchers. Throughout the exchange, the whales respond to each other by matching each other's rubato, even when they are producing different coda types. Clearly, absolute temporal features are meaningful for the whales, though what that meaning is, and if it is perceived the same as rubato in human music, remains to be determined.

Alternatively, other species might have a high tolerance for jitter in terms of noise. As discussed above, all biological signals contain a certain amount of jitter. Despite this, signals are often described as rhythmic in the literature, leading to the question, how much jitter is too much jitter? Do the animals perceive those variations, and even if they do, are those variations biologically important? As discussed above, Snowball the dancing cockatoo speeds up or slows down his movements to match the tempo of the music to which he is dancing; however, very few of his movements are timed to the beat (see Fig. 1 in [61]). He lacks precision, yet the online comments sections of his videos are filled with exclamations of how he has much better rhythm than most humans. The average human viewer does not notice the variation, and we do not know what other cockatoos notice. We have similar reactions to seeing human babies "dance" [14]. Perhaps we are impacted more by the gestalt of the performance than the timing of the individual dance moves. Perhaps the same is true for a female songbird evaluating the song of a potential mate [79], in which case a little jitter may be offset by other elements of the performance.

1.4.3 Assessing Rhythmicity Through Neural Oscillations

In recent years, the study of neuronal oscillations has become a fascinating area of neuroscientific inquiry. This approach to brain function has the potential to establish links between cellular function, network dynamics, and overt behavior such as we describe throughout this book. Nonlinear dynamical systems provide a theoretical language to integrate information from neural and behavioral levels, providing novel predictions and testable hypotheses. In contrast to linear systems, nonlinear dynamical systems have pattern-forming properties that can account for rhythmic structures, such as those found in music. Neural resonance theory (NRT) uses tools of dynamical systems to derive predictions about rhythm performance and rhythm perception based on the general properties of oscillatory neurophysiology [80] (see also Chap. 8).

Neural rhythms are not symbolic representations of behavioral rhythms; they are real, physical rhythms, oscillations whose behaviors and interactions are governed by physical principles. The principles of NRT are based on generic properties of oscillating neural populations and are compatible with dynamical descriptions of perception-action at a behavioral level. The goal of NRT is to link rhythmicity at the neural and behavioral levels making predictions at both levels that are directly testable in human and non-human animals. The hypothesis of neural oscillation has been debated, however. Patel [81] claims that the ubiquity of neural oscillations should imply that all animals should synchronize to music, and Fitch [82] more colloquially suggested that this would suggest dogs (and other animals) should be able to dance. While such claims may seem somewhat contrived, they question a key premise of NRT on evolutionary grounds. In humans, however, rhythmicity develops over a time scale of many years [83]. Modeling work has shown that neural circuits can adapt and learn through interacting with the environment in a process called attunement [84, 85]. Hebbian learning via synaptic plasticity, which increases coupling strength [86], may explain perceptual narrowing to rhythms during the first year of life [85, 87] as well as the attunement of perceptual systems to motor-vestibular rhythms [88, 89]. Interestingly, recent modeling work suggests that oscillation itself is learnable by networks of excitatory and inhibitory neurons in nonhuman animals [90]. This could explain why nonvocal learning species can learn to synchronize to a musical beat [64, 73]. In Chap. 8, we will delve into these aspects of neural resonance theory and ask what these features of neural circuits mean for the study of rhythmicity in humans and non-human animals.

1.4.4 Assessing Musical Rhythm and Musicological Definitions

The temporal dimensions of music encompass a multitude of elements, including rhythmic patterns, metrical structure, tempo, and timing. These elements can manifest in diverse forms and are readily identifiable by most human listeners through the acoustic signal. Below, we offer concise descriptions of each aspect of musical rhythm (for a more elaborate discussion on the psychology of musical rhythm, see [91]; for an elaborate overview of the use of neuroscientific methods see [70]).

Rhythmic Pattern: A rhythmic pattern is the mental representation of rhythm on a discrete, symbolic scale, comparable to its notation in a musical score. Toussaint's cross-culturally wide-ranging categorization and mathematical (geometric) analysis of rhythmic patterns is a landmark achievement which has not received the attention it deserves [24]. This contrasts with a performed or expressively timed rhythm measured on a continuous scale from the audio signal. Rhythmic patterns involve the sequential pattern of interonset intervals (IOIs) and can be perceptually grouped at the surface level [92]. This grouping often leads to perceptual accents and hierarchical ordering [93, 94]. Rhythmic categorization, the process of deriving discrete rhythmic categories from a continuous rhythmic signal, is a key area of study in music cognition [95, 96]. This process underpins our ability to recognize and predict rhythmic patterns in music.

Pulse (or Beat): The pulse, or beat, is a fundamental component of musical time, representing regularly recurring, precisely equivalent psychological events [8, 21, 97]. Beat perception involves a cognitive process where listeners perceive a sensation of a regular pulse when listening to a rhythm [98]. The beat is the most salient regularity in the metrical structure of rhythmic music. It is closely related to the concept of tactus in music theory [21, 93]. Beat perception varies across individuals and can change with age and musical training, generally favoring a period of around 600 ms, equating to approximately 100 beats/min [99, 100]. And finally, beat perception is active early in life [68] and cannot be explained by statistical learning [101]. Hence, it is suggested to be a fundamental musical trait [102].

Meter: Meter refers to the organization of pulses into recurring patterns, such as, for example, the alternation of strong and weak beats that is common in Western music (for a cross-cultural discussion of meter, see [21]). This higher-order regularity involves at least two levels of pulse (or beat), creating a hierarchical framework known as the metrical structure. Common metrical patterns include the duple meter (march) and triple meter (waltz), with rates of meter and subdivisions related to each other at integer ratios. The metrical structure forms the backbone of musical rhythmic perception and is integral to the experience of musical rhythm. People typically tap to the level of the beat, reflecting their interaction with this hierarchical structure [100].

Tempo: Tempo is the impression of the speed or rate at which the music's pulse passes. While the exact nature of tempo perception is not fully understood, it is related to the cognitive notion of beat or tactus. Tempo can vary widely, influencing the emotional and aesthetic experience of music ([16, 103] and Chap. 11 this volume).

Expressive Timing: Expressive timing introduces nuances that can make a rhythm sound "mechanical," "laid-back," or "rushed" [104]. This phenomenon arises from a discrepancy between top-down, expectation-driven perception and bottom-up, data-driven perception. Some notes are played earlier or later than expected, adding expressiveness to the rhythm. Expressive timing is a crucial area of interest in music cognition, as it involves understanding how small deviations from expected timing contribute to the perception of musicality and emotional expression. It poses questions such as why a rhythm with a slightly shorter note isn't simply perceived as a different rhythm and how various components (rhythmic pattern, metrical structure, timing, tempo) interact to shape our perception of rhythm.

Integrative Perspectives: The study of pulse, meter, and musical rhythm integrates insights from psychology, neuroscience, anthropology, musicology, and evolutionary biology. As such, it promotes an integrative framework that encompasses the perceptual, cognitive, and neural dimensions of rhythm, contributing to a fuller understanding of the capacity for music (i.e., musicality) [105].

Advances in neuroimaging and electrophysiology have identified brain networks involved in rhythmic processing, highlighting the complexity of rhythm cognition. These findings emphasize the distributed nature of rhythm processing across various brain regions ([101, 106], as well as Chaps. 7, 9, this volume).

Cultural and ecological aspects of rhythm are also significant. Different musical traditions utilize diverse metrical patterns, reflecting cultural learning and preferences. Recently, more and more researchers have been examining how rhythmic abilities develop in various cultural contexts, emphasizing the role of the sociocultural environment in shaping rhythm cognition [107].

1.4.5 Future Directions in Animal Musicality Research

As noted by Doolittle et al. [108], Hoeschele et al. [41], and further explored by ten Cate and Honing [109], establishing whether other animals possess similar musical and/or cognitive abilities as humans requires a dual approach. The first approach involves observational and analytical methods, which can provide hints that an animal might have certain perceptual or productive capabilities. However, to conclusively demonstrate these abilities, a second experimental approach is necessary. This involves controlled experiments where animals are tested to see if they can learn, distinguish, and generalize artificially constructed sounds that vary in specific musical features. Together, these complementary approaches offer a more comprehensive understanding of the extent to which animals share cognitive capacities with humans.

In summary, to date, it is clear that at least some avian and primate species show perceptual rhythmicity, but it is still unclear what causes these abilities to arise. While we know very little about the perception of rhythmicity across species, especially non-avian and nonprimate species, we know a great deal more about the physical presence of rhythm and the use of these rhythms in signals across the animal kingdom. Our knowledge of rhythmic cultures and cultural transmission of rhythmic behaviors in animals other than humans is patchy at best and it will be an interdisciplinary effort to close the knowledge gap between what we inherently know about human music and possible counterparts in other animals.

1.5 Conclusions and Outlook in Rhythm Research

As shown in this chapter, rhythm research is a vast, interdisciplinary, and fascinating field in science. Rhythms have been studied for decades from very different angles and with various questions in mind, which led to a rag rug of rhythm-related definitions.

Musicology, linguistics, neurobiology, and ecology for a long time were studying rhythm without much overlap, doing important basic research in each respective field. Only now with a solid knowledge basis in all those fields have questions evolved to be more interdisciplinary. This resulted in attempts to connect the knowledge from different fields, for example, in search of evolutionary underpinnings of music [105]. In this more interdisciplinary approach, ambiguously used definitions and differing terminology became an issue. In an attempt to enable better communication about rhythm research across fields, we dove into definitions from all fields to find the common denominators and pitfalls.

We concluded that a "universal" definition of rhythm might not be achievable, or desirable. We can study rhythm from very different perspectives, that might even need—at least in part—divergent definitions to remain precise and clear. The three perspectives we discussed include studying rhythm a) as a purely physical signal, b) in a biological context (in terms of perception and cognition) as well as c) from a cultural and social perspective. When studying rhythm, researchers should be aware of these different perspectives and differing terminology. For example, adapting a term from the cultural perspective to a purely biological context can be problematic, especially when the term is stripped of its cultural aspects. However, the cultural perspective might sometimes be essential for answering questions rooted in biology. In such cases, it is crucial to fully understand the assumptions and scientific background of a definition from the cultural perspective. This is particularly true when applying musicological terms to animal examples without acknowledging the specific underlying assumptions. Biologically speaking, "rhythm" is not a unitary dimension, but a complex multidimensional construct that includes behaviors with little in common, either as regards the mechanisms that generate them, the ontogenies by which they develop, the selection pressures by which they evolve, or the functions they serve, to use Tinbergen's four keys to clarify in the analysis of behavior [110]. This heterogeneity of natural behaviors which all can be described as rhythmic can be ignored only at the price of drawing false analogies and spurious homologies, to the detriment of much-needed progress on the topic to which this book is dedicated.

Many open questions remain in the study of rhythms from all introduced perspectives. For example, it will be important to study whether rhythmic synchronization in humans utilizes direct auditory/motor connections in the brain and if that is the same in animals. In animals, we need to make sure to consider biologically relevant behaviors and stimuli in experiments and consider natural and lab behavior rather than one or the other. Only with a comparable basic knowledge about the physical signal, neurological processes, and cognitive perception of rhythms for humans and other animals will we be able to find common representations. Many of these topics will be discussed in more detail throughout the book. Our goal is that you, the reader, can answer the question we asked in the beginning (Rhythm: What is it?) with a shared set of representations and interpretations to another reader.

References

1. Bünning E (1957) Endogenous diurnal cycles of activity in plants. In: Rudnick D (ed) Rhythmic and synthetic processes in growth. Princeton University Press, Princeton, pp 111–126
2. Morimoto YV, Minamino T (2014) Structure and function of the bidirectional bacterial flagellar motor. Biomolecules 4:217–234
3. Bucher D, Haspel G, Golowasch J, Nadim F (2015) Central pattern generators. Encycl Life Sci 11:1–12
4. Freeberg TM, Book DL, Jung H, Kyle SC (2021) Communication, cues, and signals. In: Shackleford TK, Weekes-Shackleford VA (eds) Encyclopedia of evolutionary psychological science. Springer, Cham/Berlin, pp 1206–1214. https://doi.org/10.1007/978-3-319-19650-3_2728
5. Lacalli TC (2004) Sensory systems in amphioxus: a window on the ancestral chordate condition. Brain Behav Evol 64:148–162
6. Hassan U, Anwar MS (2010) Reducing noise by repetition: introduction to signal averaging. Eur J Phys 31:453–465
7. Barnes R, Jones MR (2000) Expectancy, attention and time. Cogn Psychol 41:254–311
8. Large EW, Jones MR (1999) The dynamics of attending: how people track time varying events. Psychol Rev 106:119–159
9. Shapiro KL, Arnell KM, Raymond JE (1997) The attentional blink. Trends Cogn Sci 1:291–295
10. Hall ML (2009) Chapter 3: A review of vocal duetting in birds. In: Advances in the study of behavior [Internet]. Academic Press, pp 67–121. https://www.sciencedirect.com/science/article/pii/S0065345409400032
11. Benichov JI, Benezra SE, Vallentin D, Globerson E, Long MA, Tchernichovski O (2016) The forebrain song system mediates predictive call timing in female and male zebra finches. Curr Biol 26:309–318
12. Benichov JI, Globerson E, Tchernichovski O (2016) Finding the beat: from socially coordinated vocalizations in songbirds to rhythmic entrainment in humans. Front Hum Neurosci 10:255
13. Mann NI, Dingess KA, Slater PJB (2006) Antiphonal four-part synchronized chorusing in a neotropical wren. Biol Lett 2:1–4
14. Merker B, Madison GM, Eckerdahl P (2009) On the role and origin of isochrony in human rhythmic entrainment. Cortex 45(1):4–17
15. Fortune ES, Rodríguez C, Li D, Ball GF, Coleman MJ (2011) Neural mechanisms for the coordination of duet singing in wrens. Science 334:666–670
16. Fraisse P (1982) Rhythm and tempo. In: Deutsch D (ed) Psychology of music. Academic Press, New York, pp 149–180
17. Greenfield MD, Honing H, Kotz SA, Ravignani A (1835) Synchrony and rhythm interaction: from the brain to behavioural ecology. Philos Trans R Soc B 2021(376):20200324
18. Greenfield M, Merker B (2023) Coordinated rhythms in animal species, including humans: entrainment from bushcricket chorusing to the philharmonic orchestra. Neurosci Biobehav Rev 153:105382
19. Merker B (1999) Synchronous chorusing and the origins of music. Music Sci 3(1_suppl):59–73
20. Merker B, Morley I, Zuidema W (2015) Five fundamental constraints on theories of the origins of music. Philos Trans R Soc B 370:20140095
21. Arom S (1991) African polyphony and polyrhythm: musical structure and methodology. Cambridge University Press, Cambridge. https://doi.org/10.1017/CBO9780511518317
22. Gottlieb RS (1993) Solo Tabla drumming of North India: its repertoire, styles, and performance practices. Motilal Banarsidass Publishers, Delhi
23. Kippen J (1988) The Tabla of Lucknow: a cultural analysis of a musical tradition. Cambridge University Press, Cambridge

24. Toussaint GT (2013) The geometry of musical rhythm: what makes a "good" rhythm good? CRC Press, Boca Raton
25. Merker B (2019) When extravagance impresses: recasting esthetics in evolutionary terms. In: Thaut M, Hodges D (eds) The Oxford handbook of music and the brain. Oxford University Press, Oxford, pp 66–86
26. Bass AH, McKibben JR (2003) Neural mechanisms and behaviors for acoustic communication in teleost fish. Prog Neurobiol 69(1):1–26
27. Hersh TA, Ravignani A, Burchardt LS (2023) Robust rhythm reporting will advance ecological and evolutionary research. Methods Ecol Evol 14(6):1398–1407. https://onlinelibrary.wiley.com/doi/abs/10.1111/2041-210X.14118
28. Burchardt LS, Knörnschild M (2020) Comparison of methods for rhythm analysis of complex animals' acoustic signals. PLoS Comput Biol 16(4):e1007755
29. Kello CT, Bella SD, Médé B, Balasubramaniam R (2017) Hierarchical temporal structure in music, speech and animal vocalizations: jazz is like a conversation, humpbacks sing like hermit thrushes. J R Soc Interface 14(135) http://rsif.royalsocietypublishing.org/content/14/135/20170231.abstract
30. Daniele JR, Patel AD (2013) An empirical study of historical patterns in musical RhythmAnalysis of German & Italian classical music using the nPVI equation. Music Percept 31(1):10–18
31. Grabe E, Low E (2002) Durational variability in speech and the rhythm class hypothesis. In: Warner N (ed) Laboratory phonology VII. Mouton de Gruyter, Berlin, pp 515–546
32. Roeske TC, Tchernichovski O, Poeppel D, Jacoby N (2020) Categorical rhythms are shared between songbirds and humans. Curr Biol 30(18):3544.e6–3555.e6
33. De Gregorio C, Valente D, Raimondi T, Torti V, Miaretsoa L, Friard O, Giacoma C, Ravignani A, Gamba M (2021) Categorical rhythms in a singing primate. Curr Biol 31(20):R1379–R1380
34. De Gregorio C, Maiolini M, Raimondi T, Carugati F, Miaretsoa L, Valente D, u. a. (2024) Isochrony as ancestral condition to call and song in a primate. Ann N Y Acad Sci 1537(1):41–50
35. Demartsev V, Manser MB, Tattersall GJ (2022) Vocalization-associated respiration patterns: thermography-based monitoring and detection of preparation for calling. J Exp Biol 225(5):jeb243474
36. Lameira AR, Hardus ME, Ravignani A, Raimondi T, Gamba M (2024) Recursive self-embedded vocal motifs in wild orangutans. eLife 12:RP88348
37. Osiecka AN, Fearey J, Ravignani A, Burchardt LS (2024) Isochrony in barks of Cape fur seal (*Arctocephalus pusillus pusillus*) pups and adults. Ecol Evol 14(3):e11085
38. Norton P, Scharff C (2016) "Bird song Metronomics": isochronous Organization of Zebra Finch Song Rhythm. Front Neurosci 10:309
39. Burchardt LS, Briefer EF, Knörnschild M (2021) Novel ideas to further expand the applicability of rhythm analysis. Ecol Evol 11(24):18229–18237
40. Slabbekoorn H, ten Cate C (1999) Collared dove responses to playback: slaves to the rhythm. Ethology 105:377–391
41. Hoeschele M, Merchant H, Kikuchi Y, Hattori Y, ten Cate C (2015) Searching for the origins of musicality across species. Philos Trans B 370:2140094. http://rstb.royalsocietypublishing.org/content/royptb/370/1664/20140094.full.pdf
42. Hoeschele M (2017) Animal pitch perception: melodies and harmonies. Comp Cogn Behav Rev 12:5–18
43. Hulse SH, Cynx J, Humpal J (1984) Absolute and relative pitch discrimination in serial pitch perception by birds. J Exp Psychol Gen 113:38–54
44. Hulse SH, Cynx J (1985) Relative pitch perception is constrained by absolute pitch in songbirds (*Mimus*, *Molothrus*, and *Sturnus*). J Comp Psychol 99:176–196
45. Hulse SH, Cynx J (1986) Interval and contour in serial pitch perception by a passerine bird, the European starling (*Sturnus vulgaris*). J Comp Psychol 100:215–228
46. MacDougall-Shackleton SA, Hulse SH (1996) Concurrent absolute and relative pitch processing by European starlings (*Sturnus vulgaris*). J Comp Psychol 110:139–146

47. Njegovan M, Weisman R (1997) Pitch discrimination in field- and isolation-reared black-capped chickadees (*Parus atricapillus*). J Comp Psychol 111:294–301
48. Page SC, Hulse SH, Cynx J (1989) Relative pitch perception in the European starling (*Sturnus vulgaris*): further evidence for an elusive phenomenon. J Exp Psychol Anim Behav Process 15:137–146
49. Weisman RG, Njegovan MG, Ito S (1994) Frequency ratio discrimintion by zebra finches (*Taeniopygia guttata*) and humans (*Homo sapiens*). J Comp Psychol 108:363–372
50. Friedrich A, Zentall T, Weisman R (2007) Absolute pitch: frequency-range discriminations in pigeons (*Columba livia*): comparisons with zebra finches (*Taeniopygia guttata*) and humans (*Homo sapiens*). J Comp Psychol 121:95–105
51. Lee TTY, Charrier I, Bloomfield LL, Weisman RG, Sturdy CB (2006) Frequency-range discriminations and absolute pitch in black-capped chickadees (*Poecile atricapillus*), mountain chickadees (*Poecile gambeli*), and zebra finches (*Taeniopygia guttata*). J Comp Psychol 120:217–228
52. Weisman RG, Balkwill LL, Hoeschele M, Moscicki MK, Bloomfield LL, Sturdy CB (2010) Absolute pitch in boreal chickadees and humans: exceptions that test a phylogenetic rule. Learn Motiv 41:156–173
53. Hoeschele M, Wagner B, Mann DC (2023) Lessons learned in animal acoustic cognition through comparisons with humans. Anim Cogn 26(1):97–116
54. Hagmann CE, Cook RG (2010) Testing meter, rhythm, and tempo discriminations in pigeons. Behav Process 85(2):99–110
55. Reinert J (1965) Takt- und Rhythmusunterscheidung bei Dohlen. Z Tierpsychol 22(6):623–671
56. ten Cate C, Spierings M, Hubert J, Honing H (2016) Can birds perceive rhythmic patterns? A review and experiments on a songbird and a parrot species. Front Psychol 7(May):1–14
57. Hoeschele M, Bowling DL (2016) Sex differences in rhythmic preferences in the Budgerigar (*Melopsittacus undulatus*): A comparative study with humans. Front Psychol 7(1543):1–10
58. Hoeschele M, Fitch WT (2016) Phonological perception by birds: budgerigars can perceive lexical stress. Anim Cogn 19(3):643–654
59. Toro JM, Hoeschele M (2017) Generalizing prosodic patterns by a non-vocal learning mammal. Anim Cogn 20(2):179–185
60. Spierings MJ, Cate C (2014) Zebra finches are sensitive to prosodic features of human speech. Proc R Soc B Biol Sci 281(May):20140480
61. Patel AD, Iversen JR, Bregman MR, Schulz I (2009) Experimental evidence for synchronization to a musical beat in a nonhuman animal. Curr Biol 19(10):827–830
62. Schachner A, Brady TF, Pepperberg IM, Hauser MD (2009) Spontaneous motor entrainment to music in multiple vocal mimicking species. Curr Biol 19(10):831–836
63. Patel AD (1835) Vocal learning as a preadaptation for the evolution of human beat perception and synchronization. Philos Trans R Soc B Biol Sci 2021(376):20200326
64. Cook P, Rouse A, Wilson M, Reichmuth C (2013) A California Sea lion (*Zalophus californianus*) can keep the beat: motor entrainment to rhythmic auditory stimuli in a non vocal mimic. J Comp Psychol 127(4):412–427
65. Tyack PL (2019) A taxonomy for vocal learning. Philos Trans R Soc B Biol Sci 375(1789):20180406
66. Hattori Y, Tomonaga M, Matsuzawa T (2013) Spontaneous synchronized tapping to an auditory rhythm in a chimpanzee. Sci Rep 3(1566):1–6
67. Lameira AR, Eerola T, Ravignani A (2019) Coupled whole-body rhythmic entrainment between two chimpanzees. Sci Rep 9(1):18914
68. Winkler I, Háden GP, Ladinig O, Sziller I, Honing H (2009) Newborn infants detect the beat in music. Proc Natl Acad Sci 106(7):2468–2471
69. Honing H, Merchant H, Háden GP, Prado L, Bartolo R (2012) Rhesus monkeys (*Macaca mulatta*) detect rhythmic groups in music, but not the beat. PLoS One 7(12):51369
70. Bouwer FL, Háden GP, Honing H (2024) Probing beat perception with event-related potentials (ERPs) in human adults, newborns, and nonhuman primates. In: Advances in experimental medicine and biology. Springer International Publishing, pp 227–256

71. Honing H, Bouwer FL, Prado L, Merchant H (2018) Rhesus monkeys (*Macaca mulatta*) sense isochrony in rhythm, but not the beat: additional support for the gradual audiomotor evolution hypothesis. Front Neurosci 12:475
72. Honing H, Merchant H (2014) Differences in auditory timing between human and non-human primates. Behav Brain Sci 27(6):557–558
73. Merchant H, Honing H (2014) Are non-human primates capable of rhythmic entrainment? Evidence for the gradual audiomotor evolution hypothesis. Front Neurosci 7(January):274
74. Hattori Y, Tomonaga M (2020) Rhythmic swaying induced by sound in chimpanzees (*Pan troglodytes*). Proc Natl Acad Sci USA 117(2):936–942
75. Large EW, Gray PM (2015) Spontaneous tempo and rhythmic entrainment in a bonobo (*Pan paniscus*). J Comp Psychol 129(4):317–328
76. Raimondi T, Di Panfilo G, Pasquali M, Zarantonello M, Favaro L, Savini T et al (1990) Isochrony and rhythmic interaction in ape duetting. Proc R Soc B Biol Sci 2023(290):20222244
77. Dooling RJ, Leek MR, Gleich O, Dent ML (2002) Auditory temporal resolution in birds: discrimination of harmonic complexes. J Acoust Soc Am 112(2):748–759
78. Sharma P, Gero S, Payne R, Gruber DF, Rus D, Torralba A et al (2024) Contextual and combinatorial structure in sperm whale vocalisations. Nat Commun 15(1):3617
79. Goller F (2022) Vocal athletics–from birdsong production mechanisms to sexy songs. Anim Behav 184:173–184
80. Large EW, Snyder JS (2009) Pulse and meter as neural resonance. Ann N Y Acad Sci 1169(1):46–57
81. Patel AD (2014) The evolutionary biology of musical rhythm: was Darwin wrong? PLOS Biol 12:e1001821. https://journals.plos.org/plosbiology/article?id=10.1371/journal.pbio.1001821
82. Fitch WT (2012) The biology and evolution of rhythm: unravelling a paradox. In: Language and music as cognitive systems. Oxford University Press, New York, pp 73–95
83. Snyder JS, Gordon RL, Hannon EE (2024) Theoretical and empirical advances in understanding musical rhythm, beat and metre. Nat Rev Psychol 3(7):449–462
84. Gibson JJ (1979) The ecological approach to visual perception. Houghton, Mifflin and Company, Boston. xiv, 332
85. Tichko P, Large EW (2019) Modeling infants' perceptual narrowing to musical rhythms: neural oscillation and Hebbian plasticity. Ann N Y Acad Sci 1453(1):125–139
86. Kim JC, Large EW (2021) Multifrequency Hebbian plasticity in coupled neural oscillators. Biol Cybern 115(1):43–57
87. Hannon EE, Trehub SE (2005) Tuning in to musical rhythms: infants learn more readily than adults. Proc Natl Acad Sci USA 102(35):12639–12643
88. Phillips-Silver J, Trainor LJ (2005) Feeling the beat: movement influences infant rhythm perception. Science 308(5727):1430
89. Tichko P, Kim JC, Large EW (2021) Bouncing the network: a dynamical systems model of auditory–vestibular interactions underlying infants' perception of musical rhythm. Dev Sci 24(5):e13103
90. Zemlianova K, Bose A, Rinzel J (2024) A recurrent neural network for rhythmic timing [Internet]. bioRxiv:2024.05.24.595797. https://www.biorxiv.org/content/10.1101/2024.05.24.595797v1
91. Honing H, Bouwer FL (2019) Rhythm. In: Rentfrow PJ, Levitin DJ (eds) Foundations in music psychology. The MIT Press, pp 33–69
92. Handel S (1993) Listening: an introduction to the perception of auditory events. MIT Press, Cambridge, MA
93. Lerdahl F, Jackendoff R (1983) A generative theory of tonal music. MIT Press, Cambridge, MA
94. Martin JG (1972) Rhythmic (hierarchical) versus serial structure in speech and other behavior. Psychol Rev 79(6):487–509
95. Desain PWM, Honing H (2003) The formation of rhythmic categories and metric priming. Perception 32(3):341–365
96. Jacoby N, McDermott JH (2017) Integer ratio priors on musical rhythm revealed cross-culturally by iterated reproduction. Curr Biol 27(3):359–370

97. Cooper G, Meyer LB (1960) The rhythmic structure of music. University of Chicago Press, Chicago, IL
98. Povel DJ, Essens P (1985) Perception of temporal patterns. Music Percept 2(4):411–440
99. Bouwer FL, Burgoyne JA, Odijk D, Honing H, Grahn JA (2018) What makes a rhythm complex? The influence of musical training and accent type on beat perception. PLoS One 13(1):1–26
100. London J (ed) (2012) Hearing in time: psychological aspects of musical meter, 2nd edn. Oxford University Press, Oxford. https://doi.org/10.1093/acprof:oso/9780199744374.002.0004
101. Háden GP, Bouwer FL, Honing H, Winkler I (2024) Beat processing in newborn infants cannot be explained by statistical learning based on transition probabilities. Cognition 243:105670
102. Honing H (2012) Without it no music: beat induction as a fundamental musical trait. Ann N Y Acad Sci 1252(1):85–91
103. Honing H (2013) Structure and interpretation of rhythm in music. In: Deutsch D (ed) Psychology of music, 3rd edn. Academic Press, London, pp 369–404
104. Clarke EF (1999) Rhythm and timing in music. In: Deutsch D (ed) The psychology of music, 2nd edn. Academic Press, New York, pp 473–500
105. Honing H, ten Cate C, Peretz I, Trehub SE (2015) Without it no music: cognition, biology and evolution of musicality. Philos Trans R Soc B 370(1664):20140088
106. Grahn JA, Brett M (2007) Rhythm and beat perception in motor areas of the brain. J Cogn Neurosci 19(5):893–906
107. Jacoby N, Polak R, Grahn JA, Cameron DJ, Lee KM, Godoy R, Undurraga EA, Huanca T, Thalwitzer T, Doumbia N, Goldberg D (2024) Commonality and variation in mental representations of music revealed by a cross-cultural comparison of rhythm priors in 15 countries. Nat Hum Behav. https://doi.org/10.1038/s41562-023-01800-9
108. Doolittle EL, Gingras B, Endres DM, Fitch WT (2014) Overtone-based pitch selection in hermit thrush song: unexpected convergence with scale construction in human music. Proc Natl Acad Sci 111(46):16616–16621
109. ten Cate C, Honing H (2022) Precursors of music and language in animals [Internet]. OSF. https://osf.io/4zxtr
110. Tinbergen N (1963) On aims and methods of ethology. Z Tierpsychol 20(4):410–433

How Animals Produce Rhythm: An Overview of Ethological Mechanisms for Rhythmic Communication

2

Gabriela-Alina Sauciuc and Jennifer N. Schneider

Abstract

Animals communicate with a staggering diversity of rhythmic signals, yet behind all this diversity lies a small number of ethological mechanisms. In this chapter, we provide a brief review of how animals produce rhythmic signals based on these mechanisms, which span tremulation, percussion, bodily displays, stridulation, tymbalation, aerodynamic sound production (including vocalization), as well as the production of luminescent, chromogenic, and electric rhythms. We point out that these ethological mechanisms which underlie the production of communicative rhythms appear to be derived from rhythmic 'praxic' behaviors that initially evolved to implement vital functions such as breathing, feeding, locomotion, defense, or procreation. Based on the findings reviewed by the chapter, we identify three important areas for future research for advancing our understanding of the evolution of rhythmic communication. First, there is a massive research gap when it comes to quantifying how rhythmic signals unfold and interact *across* perceptual and behavioral modalities. Second, there is a need to understand how behavioral rhythms relate to—and are constrained by—their underlying neuromuscular substrates. Third, very little is known about the relationship between communicative rhythms and the praxic behaviors in which they putatively originated.

Gabriela-Alina Sauciuc and Jennifer N. Schneider contributed equally with all other contributors.

G.-A. Sauciuc (✉)
Department of Philosophy/Cognitive Science, Lund University, Lund, Sweden
e-mail: gabriela-alina.sauciuc@lucs.lu.se

J. N. Schneider (✉)
Department of Psychology, LCC International University, Klaipėda, Lithuania
e-mail: jschneider@lcc.lt

G.-A. Sauciuc et al. (eds.), *Nature Beats*, Fascinating Life Sciences,
https://doi.org/10.1007/978-3-032-24250-1_2

2.1 Introduction

In this chapter, we preface more detailed characterizations of taxon-specific rhythmic signals (as in Chap. 3 [1]) or underlying mechanisms (as in Chaps. 6 [2], 7 [3], 8 [4], 9 [5] and 11 [6]), with a swift review of the behavior classes that actuate rhythmic communication. We begin by reviewing modes of rhythmic signaling that exploit locomotor appendages and postural body segments. Some of these, such as tremulation (Sect. 2.2), percussion (Sect. 2.3) and visual displays (Sect. 2.4), are based on specialized neuromotor programs that are mainly implemented in nonspecialized anatomical structures. Others, such as stridulation (Sect. 2.5) and membrane vibration (Sect. 2.6), also involve morphological adaptations that enable organisms to emit vibroacoustic signals. In Sect. 2.7, we discuss aerodynamic mechanisms of rhythmic communication (including vocalization), which exploit specialized respiratory and digestive neuromechanical substrates. In Sects. 2.8 and 2.9, we briefly discuss rhythmic signaling based on more unusual bioluminescent, chromogenic, and electric mechanisms. In Sect. 2.10, we conclude by suggesting avenues for future research targeting the ethological and physiological origins of rhythmic communication.

Overall, there is a general consensus among scholars of animal (including human) communication that communicative behaviors are the result of exaptation [7] of neuromechanical systems evolved to sustain vital biological functions (breathing, nutrition, propulsion, procreation, etc.) and the praxic behaviors associated with these (defense, foraging, wing flapping, etc.). A generally accepted idea is that, to serve a communicative function, certain praxic behaviors have undergone a process of ritualization toward becoming more unambiguous and resilient to interference by evolving exaggerated and repetitive characteristics [8]. Therefore, the borders between praxic and communicative function may sometimes be blurred and the signal value of certain behaviors contested. Here—and throughout the volume—we adopt the view that behavioral rhythms are communicative insofar as they exert effects on the states and behaviors of con- or heterospecific receivers.

2.2 Tremulation

Tremulation is a mode of vibroacoustic signaling that results from the bobbing, shaking or quivering movements of various body parts. Most commonly, tremulatory rhythms are generated by the contraction of thoracic or abdominal muscles, but other mechanisms, such as wing fanning and/or antennal oscillations, are also documented [9, 10]. Tremulatory vibrations propagate through the substrate, to which they are typically conducted by locomotor appendages. Tremulation is the predominant mode of rhythmic signaling in insects (being documented in 68% of signaling families), where it accomplishes functions ranging from courtship and sexual advertising, to defense, aggression, or encoding size, post-contest status, and territorial spacing (reviewed in [11]).

Whiteflies (*Aleyrodidae* spp.), for example, produce distinct vibroacoustic courtship and advertisement calls by rhythmically oscillating the thorax or abdomen up and down, through body thrusting or wing flicks [12]. In dipterans (true flies, mosquitos), rhythmically patterned wing and/or abdominal vibration serves in species recognition and mate attraction, with both pulse rate and the duration of interpulse intervals playing a biologically significant role [13]. As another example, honey bees produce warning piping signals by vibrating the thorax against the honeybee comb while contracting the wing muscles without flight activation [14–16].

Among vertebrates, this mode of rhythmic signaling has been described in ichthyic and amphibian species. In salmonids, rhythmic quivering powered by major locomotor muscles produces vibroacoustic signals during (pre-)spawning. In sculpins (Cottidae), gobies (Gobiidae), darters (Percidae), and cusk eels (Ophiidae), periodic contractions of the muscles that control the pectoral girdle produce pulsed vibroacoustic signals that, visually, may appear as rhythmic head or body bobbing (for recent reviews, see [17, 18]). As tremulation displays occur when fishes are close to solid substrate, they sometimes may take the form of—or be combined with—percussion (see Sect. 2.3), as, for example, described in the Padanian goby (*Padogobius martensii*, [19]). The tremulatory mechanisms of fish can generate a continuum of oscillatory frequencies that, at very short interpulse intervals, are no longer perceived by humans as pulsed signals, but as continuous tones [20]. An anuran example is the aggressive display of red-eyed tree frog (*Agalychnis callidryas*), conveyed by stereotyped tremulation at a 12 Hz rate, which propagates as vibration through the plant substrate [21]. Anurans also produce a number of visual displays using body or limb tremulation which can be rhythmic or not, depending on the species and context (for review see [22]).

2.3 Percussion

Percussive rhythms are generated by striking body parts—most commonly locomotor appendages—against a substrate or other body parts. The rhythmic pendling of drumming organs tightly mirrors the temporal patterning of underlying muscular contractions. Thus, percussive signals are well-suited to encode species identity, as they can capture temporal variations that reflect species-specific optimal rates and ranges of possibilities. They can also encode information about the body condition, sex and the motivational state of individuals, by capturing variations in how muscles render a species-specific pattern, which is known to be influenced by, for example, age or hormonal state (for details, see Chaps. 9 [5], 6 [6] and Chap. 11).

Percussion is mainly encountered in arthropods and mammals. Many species of insects, crustaceans, and arachnids have been reported to generate percussive rhythmic signals, variably labeled as tapping, stomping, thumping, or rapping. Termites, ants, wasps, heelwalkers, as well as bush crickets (*Meconema* spp.) and raspy crickets (Gryllacrididae) communicate with rhythmic signals produced by leg stamping on the substrate (reviewed by [23]). Whistling moths (*Hecatesia* spp.) generate percussive rhythms by colliding sexually dimorphic "castanets" located on the wings. Among crustaceans, fiddler crabs (*Uca* spp.) "rap" on the ground with their enlarged claw [24], while ghost crabs drum by hitting the ground with one or both enlarged claws, as well as by rapid stepping during "dance" displays [25]. Functionally, all these percussive rhythms have been documented in a mating context, either as advertising calls, courtship displays, and/or rival assessment. Spiders from multiple genera are known to drum with their pedipalps—appendages located anterior to legs, which serve as tactile organs and as secondary sexual organs—or their legs as part of complex courtship displays [26–29].

In mammals, fossorial and semifossorial moles and rodents drum their hindlimbs against the substrate, inside or outside the burrow, to advertise territorial occupancy or mating status, or to deter rivals or predators (for details, see Chap. 3 [1]). Just like humans, the nonhuman great apes exhibit manual drumming on various surfaces—their own and others' bodies, ground, trees, etc. This drumming is mainly produced during dominance display and in play contexts. Moreover, in captivity, all ape species use drumming as an attention getter (for details, see Chap. 12 [30]). Percussive communication has also been documented in pinnipeds, with front or rear flippers being repeatedly slapped against the water (see Chap. 3 [1]).

Percussion is rare in other vertebrate classes. Among birds, locomotor percussion has only been described in the red-cheeked (*Uraeginthus bengalus*) and blue-capped cordon-bleu (*U. cyanocephalus*), which rapidly drum their feet on the substrate (for a recent review, see [31]), and in palm cockatoos (*Probosciger aterrimus*), which drum using claw-held sticks or seed pods [32]. In all these avian species, the percussive rhythms are part of courtship displays. Among anurans, species of Dart Poison frogs (Dendrobatidae) and toads (Bufonidae) exhibit toe tapping, which stirs up prey species in the surrounding vegetation and aids hunting (reviewed in Chap. 3 [1]). Finally, a species of cyprinid fish (*Danionella cerebrum*) has a unique form of self-percussion: they produce vibrations by drumming their swim bladders with cartilage connected to a specialized rib and muscles [33].

Besides the most widespread percussion mode actuated by locomotor appendages, some vertebrate and invertebrate taxa have been documented to drum with other body parts. For example, woodpeckers produce rival assessment and territorial displays through bill drumming driven by rapid head movement (reviewed in Chap. 3 [1]), ruddy ducks produce multimodal displays that also include thumping sounds generated by drumming their bill onto their air sacs [34], while storks and albatrosses produce percussive signals with a courtship or aggressive function through bill clattering (reviewed in Chap. 4 [35]). Other drumming modes reported in the literature include head drumming in the fossorial mole rat *Spalax ehrenbergi* (Chap. 3 [1]) and some termite species, air sac drumming in anurans (reviewed in Chap. 3 [1]), and abdominal-tip drumming in termites [36] and stoneflies [37].

2.4 Rhythmic Visual Displays

Both invertebrates and vertebrates communicate with rhythmic movements that do not simultaneously generate acoustic or seismic vibrations, being solely broadcasted in the visual channel. The great majority of these visual signals are actuated by locomotor appendages (limb waving, flapping, twisting, etc.), but the involvement of other body parts, such as the head (e.g. head bobbing) or the trunk (e.g. swaying, rocking, lounging) is also common. The reliance on these body parts underscores the continuity between the purely 'choreographic' displays illustrated in this section and the two modes of rhythmic signaling discussed in the previous two sections, that is, tremulation and percussion.

Most visual displays are produced as part of elaborated multimodal displays, and may occur at the same time with other rhythmic or nonrhythmic signals or be intercalated with them. For example, in jumping spiders (*Salticidae*), the males perform temporally complex sequences of bodily motions that involve ornamented body parts, including leg waving, the furling and unfurling of abdominal flaps, trunk bobbing and rocking [38–40]. These motions can be combined with vibrational signaling, which is deployed intermittently during the performance of the visual displays and also as a function of male's distance to the female. Similarly, fiddler crabs (*Uca* spp.) are famous for their claw-waving displays, which can consist of simpler claw lifting or of complex combinations of vertical and horizontal claw movements that are also coupled with leg and body movements [41]. Some species further intercalate vibrational signals within this visual display [42] (and references therein). Crickets, which are famous for their vibroacoustic displays, exhibit silent rhythmic displays in the form of body rocking [43].

Multimodal displays that incorporate visual rhythms are also present in vertebrate taxa. Elaborated bird 'dances', for example, have already been noted by the founding fathers of evolutionary science more than a century ago [44, 45]. In fish taxa, visual displays take the form of head nodding [46] or elaborated swimming patterns [47–49]. In anurans, advertisement calling may occur in conjunction with rhythmic visual signals such as foot flagging, arm waving or vocal sac pumping [50]. Relevant examples are also found in mammals, including primates. For example, the courtship 'dance' of white-faced capuchin monkeys (*Cebus capucinus*) includes a behavioral suite of eye contact with protruded lips, rhythmic vocalizations, 'pirouetting' and rhythmic mutual chase in slow-motion [51]. The well-known dominance displays of the chimpanzees are also multi-modal, and besides vocalization, may also include foot or hand percussion, and are often preceded by bouts of whole-body swaying, which can be stationary or performed while locomoting (for details and more examples, see Chap. 12 [30]).

In contrast to the examples above, some rhythmic motions are unimodal. For example, in many fiddler crab species, the claw-waving display is silent. In vertebrates, a recently uncovered example is the rhythmic bodily jerk of female gibbons which are used as sexual solicitations [52, 53]. A highly relevant clade for the study of visual rhythms is Reptilia, given the relatively lower prevalence of acoustic communication as compared to other taxa. Thousands of lizard species exhibit rhythmic visual displays in the form of head bobbing and push-ups (for more details, see Chap. 4 [35]). In snakes, several kinds of rhythmic motions occur as part of courtship and male rivalry displays. Courtship behaviors, which in some species are also displayed by females, include staccato jerking of the head and/or body, sawing motions, tail waves, and more [54, 55]. Male rivalry display includes, among other rhythmic signals, the so-called 'snake dance', in which two rival males face each other while the head along with about one-third of the trunk is lifted off the ground and perform swaying movements [54–56]. Body swaying in snakes, however, may accomplish a defensive function, to distract a visually oriented predator [57] or to avoid detection by mimicking the motion of vegetation [58], as it has also been reported in as bats [59] and for several invertebrate species including mantices [60] and phasmid insects [61].

2.5 Stridulation

Stridulation generates rhythmic signals through friction, whereby a mobile body part (the *plectrum*) is rubbed against a stationary one (the *stridulitrum*) in a regular pattern [62]. Stridulation is widespread in arthropods, being found in tens of thousands of insect, chelicerate and crustacean species [63–65]. The phylum exhibits a staggering diversity of stridulatory mechanisms and organs, with recent surveys indicating that stridulation has evolved independently a great number of times: 84–94 times in Heteropteran insects ('true bugs'), 25 times in orthopteran insects (crickets, katydids, locusts), 57 times in spiders (Araneae), 24–27 times in other Arachnids, and 40 times in other Pancrustacea [64, 65].

The diversity of stridulatory mechanisms found in each arthropod class, however, can be reduced to a small number of types that account for the overwhelming majority of stridulatory rhythms produced by the respective class. Among heteropterans, for example, 82% of all stridulation cases are produced with the involvement of three mechanisms: leg rubbed on abdomen, leg rubbed on the wing, tergal plate rubbed against the hind wing [65]. Likewise, orthopterans stridulate mainly with wing-leg mechanisms, while crustaceans (e.g. marine hermit crabs, *Trizopagurus* spp. and mangrove crabs) stridulate by rubbing chelipeds (clawed legs) provided with ridges, spines or bristles against each other [66, 67]. Isopod crustaceans in the family Armadillidiidae (commonly known as woodlouse or pill bugs) stridulate using a ledge of scales situated on the terminal segments of their legs, but only during conglobation, that is, when they assume a ball shape [68].

Stridulation organs can also be found on body parts that are implicated in feeding and foraging. For example, in arachnids, stridulation is typically produced by structures located on the chelicerae [65]—foraging appendages situated in front of the mouth, that are used to hold and crush prey, or to deliver venom. Spiders have been reported to also stridulate using their pedipalps [69]. Stridulation powered by mouth muscles is additionally found in some insect species. For example, adults of some

Orthoptera species, as well as the larvae of some Coleoptera and Lepidoptera species generate rhythmic signals by rubbing together sclerotized mouth segments ([70] and references therein). The ghost crab (*Ocypode quadrata*) stridulates by the grinding of internal foregut teeth [71], which generates pulsed signals of aggression, which are thought to have evolved through the ritualization of incidental sounds of feeding. In the spiny lobster, stridulation is produced when a plectrum located at the base of each antenna rubs against a 'file' located near the eye, being thus driven by specific antenna movements.

Among vertebrates, several fish species stridulate using aspects of the pectoral fins that are powered by locomotor muscles. Catfish use abduction and/or adduction of locomotor muscles to rub the ridges on their pectoral spines over the rough surface of the shoulder girdle, producing pulsatile signals [72, 73]. In Indian catfish (*Scissor rhabdophorus*), stridulations can also be made with the dorsal fin [74]. Croaking gouramis (*Trichopsis vittatus*) stridulate by rubbing of hypertrophied tendons found on the pectoral fins [75], which results in a double-pulsed sound [76].

A smaller number of fish species stridulate using components of the feeding apparatus. For example, the pomacentrid fish *Dascyllus flavicaudatus* generates pulsed sounds through rapid jaw snapping. This rhythmic signal is thought to be an elaboration of the single jaw slam—an aggressive display that evolved through the ritualization of the biting motion, which occurs in several pomacentrid (clown fish) species. Grunts (*Haemulon* spp.) emit serial grating sounds in distress situations, which are generated by stridulating the teeth of the pharyngeal jaw apparatus [77]. Finally, seahorses (*Hippocampus* spp.) produce click series in a variety of contexts (stress, male–male competition, courtship, and feeding) via stridulation of the supraoccipital ridge against the grooved part of the coronet [78, 79].

In birds, limb-based stridulation has only been attested in the club-winged Manakin, where modified feathers are rubbed against one another, during male courtship displays, when the wings are joined, resulting in complex acoustic patterns that have been compared to song [80]. Among mammals, the only species known to stridulate are the tenrecs (Tenrecinae), which are able to produce repetitive sound patterns by rubbing specialized dorsal quills against each other. The stridulatory movements are created by alternate contractions of hypodermic muscles, which cause the quills to quickly turn clockwise and anticlockwise [81]. These stridulatory sounds have been shown to be species-specific and occur in contexts such as predator warning, mother-offspring localization or adult distal communication [81–83].

2.6 Membrane Pulsing

2.6.1 Tymbalation in Insects

Hemipterans (cicadas, leafhoppers, planthoppers) and lepidopterans (moths, butterflies) can emit pulsed trains of clicks by pulsating specialized vibroacoustic organs called *tymbals* [84–89]. The tymbal consists of bilateral membranous areas of the cuticle that can be cyclically buckled when the muscles attached to them contract. In many species that possess a tymbal, the membrane is ribbed or striated. These corrugations function as a frequency multiplier, where a train of vibroacoustic pulses—rather than a single pulse—is emitted with every muscle contraction. In addition, the specific configuration of ribs or striations determines the species-specific patterning of the pulses. Lepidopteran tymbals are located on metathoracic or wing structures, in the latter case being powered aero-elastically by wing fanning (moths: [86, 90]; butterflies: [91]). Hemipteran tymbals are typically located on the first or second abdominal segment [87, 89], being powered by hypertrophied muscles that are structurally and biochemically very similar to the wing muscles rather than being similar to other abdominal muscles [92].

2.6.2 Swim Bladder Pulsing in Teleost Fish

Many teleost fish produce sounds by pulsing their swim bladders via specialized sonic muscles. The size and shape of the swim bladder as well as the size, number, and configuration of the sonic muscles vary greatly between species (for review see [93]). Depending on the arrangement of these anatomical parts, the bladder may create a mixture of monopole, dipole, and quadrupole spreading, which may lead to directional sound transmission in some species [94]. The frequency of pulses is determined by the speed of the muscle contractions, with one pulse per contraction [93]. Pulses can be generated during the constriction of the bladder via fast muscles, or during the release from stretching by slow muscles [95]. Additionally, continued vibration of tendons and bones creates multicycled pulses in some species [94]. The role of the swim bladder as a resonance chamber is currently in question, as it seems to function more to dampen additional oscillations [94].

2.6.3 Laryngeal Pulsing in Anurans

While most anurans produce sounds via air-dynamic mechanism, aquatic frogs in the Pipidae family produce sounds in their larynx, but without vocal folds or airflow (see Sect. 2.7.1). They have long been thought to produce sounds in water via air bubble implosions [96]; however, more recent investigations provide evidence for mechanical excitation of the larynx [97]. Specialized arytenoid discs, or bony rods, are surrounded by elastic tissue that pulses when the discs are rapidly pulled apart creating clicks. Both the spectral and temporal properties of clicks differ between species [98]. Separate vibratory regions of the laryngeal tissue produce dyadic frequency bands in the calls [97] while neuronal pattern generators control rhythm production [99].

2.7 Rhythmic Signals Based on Aerodynamic Mechanisms

2.7.1 Pneumatic Sound Production

To date, the study of rhythmic communication in vertebrate species has focused primarily on tetrapod vocalizations—a signaling mode that implicates processes and structures originally evolved to control respiration and alimentation [100]. Typically, tetrapod vocal production can be characterized using the source-filter model of sound production [101, 102]. Sounds are created by modulating the pressure of air flow that traverses respiratory/alimentary cavities through the vibration of a valve-like organ (e.g. syrinx or larynx) powered by muscular oscillatory contractions. In addition to vocal organ musculature, pectoral and abdominal respiratory muscles regulate air expulsion, so that sound is usually generated during expiration [103], though some species produce sounds during inhalation (see Chap. 9 [5]). The vocal tract (larynx, trachea, nostrils, and/or adjustable mouth cavity) acts as a resonance filter or series of bandpass filters, emphasizing those frequencies that match the resonance frequency of the vocal tract and dampening other frequencies, creating formants. Finally, muscular oscillations of the upper vocal tract (e.g. mandibular oscillations) as well as lip or beak movements further structure vocal emissions.

The avian vocal system shares some important properties with the mammalian system but differs in keyways (for review see [104]). For example, the syrinx is divided such that sounds are produced separately in the right and left sides that can act unilaterally, independently, alternately, or in nonlinear combinations [105, 106]. Songbirds have adaptations that allow them to have greater control and variation in their vocal output, including vibratory labia in the syrinx and additional muscular control of the syrinx and labia that allow for precise frequency and amplitude modulation necessary for trill production [106].

Many aquatic animals have adaptations that allow them to internally recycle air when underwater. Physostomous fish gulp air at the surface before returning to normal swim depth and producing sounds [107], though not all species produce rhythmic sounds. Cetaceans produce series of sounds during long dives, requiring the internal circulation of air. In Mysticetes, singing is made possible due to the presence of a laryngeal sac, a multipurpose, flexible, muscular sac in the respiratory system [108]. The internal recycling of air between the lungs and the laryngeal sac/nasal passages constrains the duration of the sounds that can be produced and provides the temporal structure of the songs [109]. In addition, it is the vibration of tissue that transfers sound waves to the water, rather than the vibration of air [108]. In odontocetes, sounds are transferred to water via a fatty sac called the melon. Right and left phonic lips can operate independently [110].

Similar to terrestrial mammals, cetacean sounds have acoustic features consistent with a sound source-filter model of sound production [111]. Odontocetes generate sounds by using airflow to vibrate phonic tissue ('monkey lips') in the nasal cavity [112]; however, no filter has been identified though the melon has been suggested [113]. In mysticetes, the transverse arytenoid folds (U-fold) and the cricoid cushion have been identified as the sound sources for producing low frequency sounds (20-100 Hz) while humpbacks (*Megaptera novaeangliae*), right (*Eubalaena* spp.), and bowhead whales (*Balaena mysticetus*) likely produce high frequency sounds (>100 Hz) via vibrations of laryngeal tissue known as a U-fold [111], though the U is not complete in bowhead whales [114]. Both the laryngeal sac (low-pass filter) and the nasal cavity (formants) have been identified as possible 'filters' or resonance chambers [115, 116]. A variety of sounds (pulsed, mid-frequency, high-frequency), with and without formants, can be produced based on which valves of the respiratory system are opened or closed, as well as which parts of the U-fold/flaps are vibrating [116, 117].

2.7.1.1 Types of Pneumatically Produced Rhythmic Vocalizations

Multiunit signals can be produced in a variety of ways within this basic system. The simplest production mechanism is to produce a series of individual calls so that one call is being produced with each exhale. Call series produced in this way range from chaotic (interunit intervals are variable from unit to unit, but centre around a consistent mean) to nearly isochronic. Species across a variety of taxa produce contact calls in which the predictable timing of the individual calls may help recipients isolate the call in noise, but the rhythm itself is not meaningful so precision is not important; timing of individual calls can change with rapid changes in social context. In many bat species clicks during the search phase of echolocation are produced with wingbeat and breath [118].

Songs on the other hand are generally more organized and can fit into two categories: isochronic and categorical. Many mysticetes produce long repetitive sequences of sounds, with no clear beginning or end. The degree of isochrony differs between and within species. For example, even within the same humpback whale song rhythmic patterns change and different sections of song differ in the degree of isochrony [119]. Many other species, including most songbirds, produce categorical rhythms—signal components are organized in a fixed pattern with clear temporal rules [120]. Sperm whale codas also fit within this category [121].

Multiunit signals can also be created within a single exhalation via amplitude modulation either at the sound source or during the filtering phase of sound production. Air pressure at the vocal folds can be modulated to produce pulses of sounds in which the interpulse interval can be controlled. As hunting bats switch from the search phase to the terminal phase (i.e. buzz), interclick intervals get smaller and more clicks are produced per wingbeat/breath [118, 122]. As the interval between pulses decreases, pulses become buzzes or pulse-bursts, and as buzzes become faster, they can be perceived as tones, such as the squeals of sperm whales [123]. Since the vocal folds vibrate, they create changes in the sound pressure, such that all vocalizations are amplitude modulated. However, whether or not these sounds are perceived as pulsed, or tonal/continuous will depend on the temporal resolution of the auditory system, and is therefore species dependent.

Amplitude modulation at the filter phase occurs due to muscle contractions in the vocal tract as well as articulation movements of the jaw, tongue, lip, and/or beaks [124]. An example in humans would be the tongue movements that create the voiced rolled ‘r’ phoneme. Many species of songbirds produce single-breath, rapid repetition of song notes called trills or buzzes, which are produced by contractions of ‘superfast’ muscles in the syrinx [125].

2.7.2 Other Types of Sonification by Air Expulsion

Fish in the Clupeid family produce pulsed sounds by releasing air from their swim bladder out the anus, forming bubbles (reviewed in [93]). Wahlberg and Westerberg [126] propose that as the bubbles are forming externally, the addition of air to the bubble triggers oscillations, which in turn create sound waves in the water. The release of each bubble then produces a “chirp”.

While uncommon, sound generation through air expulsion is also encountered in insects and, similar to vertebrate vocalizations, it involves muscular mechanisms implicated in either respiration or alimentation. For example, adults and/or juveniles of Blattodea, Coleoptera, and Lepidoptera generate rhythmic sounds when body contractions cause air expulsion through spiracles—tracheal openings located on the abdomen or thorax [127, 128]. Death’s head hawkmoth (*Acherontia atropos*) adults produce single or serial defensive squeaks with a biphasic structure, which is imprinted by the alternating rhythm of pharyngeal cavity inflation and deflation, as air is sucked into and out of the proboscis [129]. In addition, serial rapid pulses are generated during the inflation phase by the vibrations of a valve-like structure [129]. In a similar vein, caterpillars of the Nexus sphinx hawkmoth (*Amphion floridensis*) generate stereotyped defensive squeaks by pulsating air jets through its two foregut chambers, which results in a rhythmic pattern that consists of the alternation of a longer unit (corresponding to the inflation phase) followed by multiple short-duration units, corresponding to the deflation phase [130].

2.7.3 Sonifications Generated by Air Cavitation

Snapping shrimp (*Alpheidae* spp.) also produce sounds via air bubbles, but via a substantially different mechanism than Clupeidae. Shrimp rapidly close their claws, creating a jet stream of water. As the water is pushed out, a cavernous air bubble is formed. The namesake “snap” is created when the air bubble collapses [131]. Air cavitation is also encountered as part of the multimodal displays of walruses (*Odobenidae* spp.), which generate loud sounds by bursting air bubbles through flipper clapping. This display, which may last from tens of seconds to minutes, has been confirmed to exhibit a highly isochronous rhythm [132].

2.8 Bioluminescent and Chromogenic Rhythms

The rhythmic and highly synchronous flashing of fireflies (beetles in the Lampyridae family) is one of the first rhythmic signals to have attracted substantial research attention. Fireflies produce these signals in a lantern organ that is located on the sixth and seventh abdominal segments, under a layer of translucent cuticle. This organ contains photocytes—specialized cells that hold the bioluminescent substance luciferin—and the tracheal and nervous circuits that contribute to light emission. Only the tracheoles, however, reach the photocytes. Fireflies generate luminescence through luciferin oxidation, which is catalysed by the enzyme luciferase [133, 134]. The latter determines the specific color (e.g. green, yellow, orange) of the signal [135]. The rhythmic patterning of flashing is regulated by central pattern generator circuits (CPGs), which, at lantern level, trigger the release of octopamine. This neurotransmitter sets in motion a chain of chemical processes leading up to the synthesis and diffusion of nitric oxide across cell membranes, allowing neural signals to reach the photocytes [136, 137]. Nitric oxide also underlies flash termination, as it is rapidly degraded by light.

Bioluminescence is also widely present among marine invertebrates and fish [138, 139], but it is currently unclear to what extent it is characterized by rhythmicity and whether it has a signal value. Blinking, pulsing, flashing or waving displays have been mentioned for several ostracod (Crustacea, see [140] and references therein), fish [141, 142] and cephalopod (for details, see Chap. 4 [35]) species, with luminescence being generated through different mechanisms. Cypridinid ostracods, for example, produce exogenous serial light pulses by ejecting luciferin, luciferase and mucus from specialized secretory cells located in the upper lip, which react with oxygen from the seawater [143]. Key variables that influence pattern parameters, such as duration and intensity, include the molecular structure of luciferase and the luciferin-luciferase ratio [144].

Species of flashlight fish (Anomalopidae) and ponyfish (Leiognathidae) generate bioluminescent blinking signals in subocular and circumesophageal light organs, respectively, that contain luminescent bacterial symbionts [145, 146]. In both families, the light-emitting system also contains reflective structures lined with guanine crystals, either within the light organ or in adjacent areas. In ponyfish, the light reflected by these areas is transmitted to the exterior via translucent skin patches, whose location, orientation, shape and size vary greatly across lineages, thereby generating a great diversity of blinking signals [145]. In both pony- and flashlight fish, blink patterns are generated by mechanisms that temporarily occlude the continual luminescence emitted by the bacteria. In flashlight fish, this is achieved by a downward rotation of the light organ, which exposes its darkly pigmented backside [147], while in ponyfish it is driven by a combination of muscular shutters of the light organ [148]. The temporal parameters of blinking signals are not fixed, but have been found to vary with the ecological [142] or social context [149]. Based on data from flashlight fish, this variability is driven by the flexibility of the inter-flash intervals (the duration of the occlusion time), rather than the duration of light exposure [142].

Among coleoid cephalopods (a subclass that includes squid, octopus and cuttlefish) there are more than 70 luminous genera [150] and several of these have been reported to produce bioluminescent rhythmic signals (detailed in Chap. 4 [35]). Several sepiolid and loliginid genera rely on bacterial symbionts for light generation, but many species have intrinsic light generation mechanisms based on luciferin and luciferase [138]. However, the most impressive and dramatic displays of coleoid cephalopods are those based on changes of skin color patterns, that is, chromogenic displays. These are famous for the ability to mimic the environment, but can also serve communicative purposes through, for example, rhythmic signaling (for details, see Chap. 3 [35]).

The chromogenic displays of coleoids are enabled by a unique, stratified dermal architecture. Directly under the epidermis, there are hundreds of thousands to millions of chromatophores, which are multicellular neuromuscular organs [151]. The core of the chromatophore is constituted by the chromatocyte—a cell that contains pigment granules within a highly flexible sacculus. Each chromatocyte is surrounded by dozens of radial muscle cells that are under monosynaptic brain control and can undergo contraction cycles as rapid as 200 ms that can also be produced serially, giving up to rhythmic displays. Each muscle contraction expands the chromatocyte into a pigmented disk whose area can be up to 14 times larger than in the relaxed state. The pigment granules within the chromatocyte reflect and transmit yellows, reds, and browns, as well as black in some species [152]. Granules of a given color predominate within a single chromatophore, but the other colors also occur. Underneath the chromatophore layer, coleids exhibit iridophores and leucophores that provide two types of structural coloration. The iridophores are composed of stacked platelet cells that contain reflectin (a reflective protein) and reflect light by thin-film interference, producing iridescent colors across the whole visible spectrum range [153]. This mechanism is similar to the reflectance that generates colored soap bubbles. Only cuttlefish and octopus species exhibit leucophores, which are composed of reflecting cells organized in microspheres that scatter full-spectrum light appearing white [154].

2.9 Electric Rhythms

Weakly electric fish in the superfamily Mormyridae and the order Gymnotiformes communicate with rhythmic electric discharges generated by specialized electric organs. These are composed of electrogenic cells—also known as electrocytes—that are derived from skeletal muscles. As an exception, species in the Apteronotidae family have neurogenic electrocytes, derived from the massive enlargement of axonal terminals of spinal motoneurons [155–157]. Electric organ discharges are produced by the synchronous depolarization of the electrocytes, which is elicited by command volleys from a medullary CPG (for details, see Chap. 9 [5]).

The rhythmic signals of weakly electric fishes are either wave- or pulse-type. Wave-type electric discharges have similar durations for the firing and interfiring intervals, which results in a pattern that resembles a quasi-sinusoidal waveform.

Pulse-type discharges are characterized by a short-to-long ratio between the firing period and the interfiring interval. The frequency and period of electric discharges are mainly controlled by the CPG, aiding in species, sex and individual recognition [155]. These variables, however, are flexibly adjusted by motivational states and functional contexts such as mating, intraspecific aggression, identification.

2.10 Conclusions and Future Directions

Animals communicate with a staggering diversity of rhythmic signals (detailed in Chaps. 3 [1] and 4 [35]). Yet, as we show in the present chapter, behind all this diversity lies a small number of ethological mechanisms. In turn, these are constrained by the main sensory and neuromotor systems that organisms exploit to interact with the world in order to fulfil vital functions. In most cases, the behaviors that actuate rhythmic communication are secondary characters of intrinsically rhythmic neuromotor systems whose original purpose is to enable breathing, feeding, and locomotion. Thus, the proclivity to encode biological information in rhythmic sequences can be accounted for by the intrinsic rhythmicity of the neuromechanical systems that underlie signaling. Moreover, the rhythmic activity of these systems has perceptual byproducts (sounds, vibrations, flickers, etc.) that can be transmitted in the vibrational, acoustic, or visual modality, as well as multimodally. Thus, even prior to evolving a signal function, such perceptual byproducts may accomplish a rudimentary communicative function, as cues that inform recipients about an individual's ongoing behavior, motivational state, and body fitness.

The relationship between the neuromuscular activity of the substrates that actuate rhythmic signals and the temporal structure of the signal is not fixed, but depends largely on the mode of signal production. Generally, rhythmic signals generated through tremulation, percussion or as visual displays, which do not rely on anatomical adaptations for signal production, exhibit a pulse frequency that closely reflects the contraction frequency of the actuating muscles. A similar relationship is found in electric and some bioluminescent signals. By contrast, the anatomical adaptations that enable stridulation, membrane vibration or vocalization may allow animals to generate signals with a faster pulse than that allowed by the physical capabilities of the underlying muscles.

The examples reviewed in this chapter also suggest several ways in which rhythmic displays can be multimodal. Frequently, signals are simultaneously broadcasted in at least two modalities. Vibrational modes of signaling can generate acoustic and visual byproducts, while visual displays can incidentally be broadcasted as acoustic and substrate-borne vibrations. Such multimodally broadcasted signals, however, may be picked up by intended and unintended receivers in distinct sensory modalities. For instance, many arthropod species produce signals that humans perceive as sounds, while conspecific bugs, which lack auditory organs, perceive them as substrate-borne vibrations [158]. As another example, water ripples—a byproduct of the advertisement calls of túngara frogs (*Physalaemus pustulosus*)—are picked up as substrate-born vibrations by male rivals and detected by predator bats through echolocation [159]. In other cases, signals in different modalities can be produced by the same set of structures, but using different ethological mechanisms. For example, the tergal plate of insects can generate vibrational signals through tremulation and vibro-acoustic signals through stridulation, while ghost crabs can use their claws to drum or to stridulate. Finally, rhythmic signals can co-occur simultaneously or sequentially with other rhythmic or nonrhythmic signals, which are produced in different modalities and using different body parts, as illustrated by some of the fish, bird or primate examples discussed in the chapter.

We identify the three aspects highlighted in this conclusive section of the chapter as important areas for future research. Future studies pursuing questions about (i) how rhythmic signals unfold and interact across perceptual and behavioral modalities, (ii) how they relate to their underlying neuromuscular substrates and (iii) how they relate to the praxic behaviors in which they putatively originated will substantially advance our understanding of the evolution of rhythmic communication.

Acknowledgements GA Sauciuc's contribution to this chapter has been funded by the Bank of Sweden Tercentenary foundation (grant nr. P17-0942:1) and the Swedish Research Council (grant 2019-03104).

References

1. Sauciuc G-A, Schneider JN, Burchardt LS et al (2025) Chapter 3: The functional and structural diversity of rhythmic communication across species. In: Sauciuc G-A, Lameira AR, Persson T (eds) Nature beats: the what, why and how of animal rhythmic behaviours. Springer Nature Switzerland, Cham
2. Schneider JN, Burchardt LS, Sauciuc G-A, Lameira AR (2025) Chapter 6: How environmental and social contexts shape and constrain rhythm structure and production. In: Sauciuc G-A, Lameira AR, Persson T (eds) Nature beats: the what, why and how of animal rhythmic behaviour. Springer Nature Switzerland, Cham
3. Proksch S, Honing H, Ravignani A, Merchant H (2025) Chapter 7: Converging notions on the main hypotheses regarding the origins of beat perception and synchronization: the key role of audiomotor interactions. In: Nature beats: the what, why and How of animal rhythmic behaviour. Springer Nature Switzerland, Cham
4. Large EW (2025) Chapter 8: Neural resonance and comparative rhythm cognition. In: Sauciuc G-A, Lameira AR, Persson T (eds) Nature beats: the what, why and how of animal rhythmic behaviours. Springer Nature Switzerland, Cham
5. Sauciuc G-A, Schneider JN, Hattori Y (2025) Chapter 9: The neuroethology of rhythmic communication: neural and mechanical building blocks and constraints on rhythm production and perception. In: Sauciuc G-A, Lameira AR, Persson T (eds) Nature beats: the what, why and how of animal rhythmic behaviours. Springer Nature Switzerland, Cham
6. Gustison ML, Sauciuc G-A (2025) Chapter 11: Moving the body and the mind—links between emotion and rhythmic communication and the underlying neurobiological mechanisms. In: Sauciuc G-A, Lameira AR, Persson T (eds) Nature beats: the what, why and how of animal rhythmic behaviours. Springer Nature Switzerland, Cham
7. Gould SJ, Vrba ES (1982) Exaptation—a missing term in the science of form. Paleobiology 8:4–15. https://doi.org/10.1017/S0094837300004310
8. Cullen JM (1966) E. Ritualization of animal activities in relation to phylogeny, speciation and ecology: reduction of ambiguity through ritualization. Philos Trans R Soc Lond Ser B Biol Sci 251:363–374. https://doi.org/10.1098/rstb.1966.0022

9. Virant-Doberlet M, Cokl A (2004) Vibrational communication in insects. Neotrop Entomol 33:121–134. https://doi.org/10.1590/S1519-566X2004000200001
10. Low ML, Naranjo M, Yack JE (2021) Survival sounds in insects: diversity, function, and evolution. Front Ecol Evol 9:641740. https://doi.org/10.3389/fevo.2021.641740
11. Virant-Doberlet M, Stritih-Peljhan N, Žunič-Kosi A, Polajnar J (2023) Functional diversity of vibrational signaling systems in insects. Annu Rev Entomol 68:191–210. https://doi.org/10.1146/annurev-ento-120220-095459
12. Kanmiya K (2005) Mating behaviour and vibratory signals in whiteflies (Hemiptera: Aleyrodidae). In: Drosopoulos S, Claridge M (eds) Insect sounds and communication. CRC Press, pp 365–379
13. Fabre CCG, Hedwig B, Conduit G et al (2012) Substrate-borne vibratory communication during courtship in *Drosophila melanogaster*. Curr Biol 22:2180–2185. https://doi.org/10.1016/j.cub.2012.09.042
14. Simpson J (1964) The mechanism of honey-bee queen piping. Z Vgl Physiol 48:277–282. https://doi.org/10.1007/BF00339456
15. Seeley T, Tautz J (2001) Worker piping in honey bee swarms and its role in preparing for liftoff. J Comp Physiol A 187:667–676. https://doi.org/10.1007/s00359-001-0243-0
16. Yamamoto T, Sugahara M, Okada R, Ikeno H (2021) Differences between queen piping temporal structures of two honeybee species, *Apis cerana* and *Apis mellifera*. Apidologie 52:524–534. https://doi.org/10.1007/s13592-021-00840-2
17. Zeyl JN, Malavasi S, Holt DE et al (2016) Convergent aspects of acoustic communication in darters, sculpins, and gobies. In: Sisneros JA (ed) Fish hearing and bioacoustics. Springer International Publishing, Cham, pp 93–120
18. Roberts L, Rice AN (2023) Vibrational and acoustic communication in fishes: the overlooked overlap between the underwater vibroscape and soundscape. J Acoust Soc Am 154:2708–2720. https://doi.org/10.1121/10.0021878
19. Lugli M, Torricelli P, Pavan G, Mainardi D (1997) Sound production during courtship and spawning among freshwater gobiids (Pisces, Gobiidae). Mar Freshw Behav Physiol 29:109–126. https://doi.org/10.1080/10236249709379003
20. Parmentier E, Kéver L, Boyle K et al (2013) Sound production mechanism in *Gobius paganellus* (Gobiidae). J Exp Biol 216:3189–3199. https://doi.org/10.1242/jeb.087205
21. Caldwell MS, Johnston GR, McDaniel JG, Warkentin KM (2010) Vibrational Signaling in the agonistic interactions of red-eyed treefrogs. Curr Biol 20:1012–1017. https://doi.org/10.1016/j.cub.2010.03.069
22. Hodl W, Amézquita A (2001) Visual signaling in anuran amphibians. In: Anuran communication. Smithsonian Inst. Press, Washington, DC, pp 121–141
23. Hill PSM (2014) Stretching the paradigm or building a new? Development of a cohesive language for vibrational communication. In: Cocroft RB, Gogala M, Hill PSM, Wessel A (eds) Studying vibrational communication. Springer, Berlin/Heidelberg, pp 13–30
24. Salmon M (1971) Signal characteristics and acoustic detection by the fiddler crabs, *Uca rapax* and *Uca pugilator*. Physiol Zool 44:210–224. https://doi.org/10.1086/physzool.44.4.30152493
25. Clayton D (2008) Singing and dancing in the ghost crab *Ocypode platytarsus* (crustacea, Decapoda, Ocypodidae). J Nat Hist 42:141–155. https://doi.org/10.1080/00222930701840530
26. Rinaldi I, Stropa A (1998) Sexual behaviour in *Loxosceles gaucho* Gertsch (Araneae, Sicariidae). Bull Br Arachnol Soc 11:57–61
27. Elias DO, Mason AC, Maddison WP, Hoy RR (2003) Seismic signals in a courting male jumping spider (Araneae:Salticidae). J Exp Biol 206:4029–4039. https://doi.org/10.1242/jeb.00634
28. Uetz GW, Clark DL, Roberts JA (2016) Multimodal communication in wolf spiders (Lycosidae)—an emerging model for study. In: Advances in the study of behavior. Elsevier, pp 117–159
29. Lallo MM, Uetz GW (2024) Multimodal courtship communication in a wolf spider. J Zool 324:214–223. https://doi.org/10.1111/jzo.13209

30. Sauciuc G-A, Persson T, Hattori Y, Lameira AR (2025) Chapter 12: The significance of the great ape heritage in the evolution of human rhythm in speech, music and dance. In: Sauciuc G-A, Lameira AR, Persson T (eds) Nature beats: the what, why and how of animal rhythmic behaviour. Springer Nature Switzerland, Cham
31. Ota N, Soma M (2022) Vibrational signals in multimodal courtship displays of birds. In: Hill PSM, Mazzoni V, Stritih-Peljhan N et al (eds) Biotremology: physiology, ecology, and evolution. Springer International Publishing, Cham, pp 237–259
32. Heinsohn R, Zdenek CN, Cunningham RB et al (2017) Tool-assisted rhythmic drumming in palm cockatoos shares key elements of human instrumental music. Sci Adv 3:e1602399. https://doi.org/10.1126/sciadv.1602399
33. Cook VANO, Groneberg AH, Hoffmann M et al (2024) Ultrafast sound production mechanism in one of the smallest vertebrates. Proc Natl Acad Sci 121:e2314017121. https://doi.org/10.1073/pnas.2314017121
34. Clark CJ, Brennan PLR (2024) Observations on the bill-drumming display of Ruddy duck (*Oxyura jamaicensis*). Wilson J Ornithol 136:471–479. https://doi.org/10.1676/23-00077
35. Lameira AR, De Gregorio C, Gamba M et al (2025) Chapter 4: uncharted rhythmic diversity across the animal world. In: Sauciuc G-A, Lameira AR, Persson T (eds) Nature beats: the what, why and how of animal rhythmic behaviour. Cham, Springer International Publishing
36. Howse PE (1964) The significance of the sound produced by the termite *Zootermopsis angusticollis* (Hagen). Anim Behav 12:284-IN8. https://doi.org/10.1016/0003-3472(64)90015-6
37. Sandberg JB, Stewart KW (2005) Life history of the stonefly *Isogenoides zionensis* (Plecoptera: Perlodidae) from the San Miguel River, Colorado. Illiesia 1:21–32. https://doi.org/10.5281/ZENODO.4754461
38. Elias DO, Land BR, Mason AC, Hoy RR (2006) Measuring and quantifying dynamic visual signals in jumping spiders. J Comp Physiol A 192:785–797. https://doi.org/10.1007/s00359-006-0116-7
39. Girard MB, Kasumovic MM, Elias DO (2011) Multi-modal courtship in the peacock spider, *Maratus volans* (O.P.-Cambridge, 1874). PLoS One 6:e25390. https://doi.org/10.1371/journal.pone.0025390
40. Girard MB, Elias DO, Kasumovic MM (2015) Female preference for multi-modal courtship: multiple signals are important for male mating success in peacock spiders. Proc R Soc B Biol Sci 282:20152222. https://doi.org/10.1098/rspb.2015.2222
41. How MJ, Zeil J, Hemmi JM (2009) Variability of a dynamic visual signal: the fiddler crab claw-waving display. J Comp Physiol A 195:55–67. https://doi.org/10.1007/s00359-008-0382-7
42. Mowles SL, Jennions M, Backwell PRY (2017) Multimodal communication in courting fiddler crabs reveals male performance capacities. R Soc Open Sci 4:161093. https://doi.org/10.1098/rsos.161093
43. Lin C-C, Hedwig B (2022) The central organisation of courtship and rivalry behaviour in *Gryllus bimaculatus* (deGeer) as revealed by lesions of abdominal connectives. Behav Brain Res 434:114005. https://doi.org/10.1016/j.bbr.2022.114005
44. Wallace AR (1869) The Malay archipelago: the land of the orang-utan, and the bird of paradise. A narrative of travel, with studies of man and nature. Harper and Brothers, New York
45. Darwin CR (1871) The descent of man, and selection in relation to sex. John Murray, London
46. Ord TJ, Tonia Hsieh S (2011) A highly social, Land-dwelling fish defends territories in a constantly fluctuating environment: behavior of a Land-dwelling fish. Ethology 117:918–927. https://doi.org/10.1111/j.1439-0310.2011.01949.x
47. Rowland WJ (1995) Do female stickleback care about male courtship vigour? Manipulation of display tempo using video playback. Behaviour 132:951–961. https://doi.org/10.1163/156853995X00388
48. Rosenthal GG, Evans CS, Miller WL (1996) Female preference for dynamic traits in the green swordtail, *Xiphophorus helleri*. Anim Behav 51:811–820. https://doi.org/10.1006/anbe.1996.0085

49. Sumarto BKA, Nofrianto AB, Mokodongan DF et al (2020) Variation in mating Behaviors between a tropical and a temperate species of medaka fishes. Zool Sci 38. https://doi.org/10.2108/zs200114
50. Grafe TU, Wanger TC (2007) Multimodal Signaling in male and female foot-flagging frogs *Staurois guttatus* (Ranidae): an alerting function of calling. Ethology 113:772–781. https://doi.org/10.1111/j.1439-0310.2007.01378.x
51. Manson JH, Perry S, Parish AR (1997) Nonconceptive sexual behavior in bonobos and capuchins. Int J Primatol 18:767–786. https://doi.org/10.1023/A:1026395829818
52. Fan P-F, Ma C-Y, Garber PA et al (2016) Rhythmic displays of female gibbons offer insight into the origin of dance. Sci Rep 6:34606. https://doi.org/10.1038/srep34606
53. Coye C, Caspar KR, Patel-Grosz P (2024) Dance displays in gibbons: biological and linguistic perspectives on structured, intentional, and rhythmic body movement. Primates. https://doi.org/10.1007/s10329-024-01154-4
54. Senter P, Harris SM, Kent DL (2014) Phylogeny of courtship and male-male combat behavior in snakes. PLoS One 9:e107528. https://doi.org/10.1371/journal.pone.0107528
55. Senter PJ (2022) Phylogeny of courtship and male-male combat behavior in snakes: an updated analysis. Curr Herpetol 41. https://doi.org/10.5358/hsj.41.35
56. Brattstrom BH (1974) The evolution of reptilian social behavior. Am Zool 14:35–49. https://doi.org/10.1093/icb/14.1.35
57. LeGros DL (2018) Side to side swaying as a defensive behaviour in the Dekay's Brownsnake (*Storeria dekayi*). Can Field-Nat 131:235–237. https://doi.org/10.22621/cfn.v131i3.1897
58. Fleishman LJ (1985) Cryptic movement in the vine Snake *Oxybelis aeneus*. Copeia 1985:242. https://doi.org/10.2307/1444822
59. Knörnschild M, Harview C, Moseley R, Von Helversen O (2009) Remaining cryptic during motion — behavioral synchrony in the proboscis bat (*Rhynchonycteris naso*). Acta Chiropterol 11:208–211. https://doi.org/10.3161/150811009X465839
60. Watanabe H, Yano E (2009) Behavioral response of mantid *Hierodula patellifera* to wind as an antipredator strategy. Ann Entomol Soc Am 102:517–522. https://doi.org/10.1603/008.102.0323
61. Tan EJ, Elgar MA, Bian X, Peters RA (2023) Interpreting animal behaviors—a cautionary note about swaying in phasmids. Front Ecol Evol 11:1065789. https://doi.org/10.3389/fevo.2023.1065789
62. Ewing AW (1989) Arthropod bioacoustics: neurobiology and behaviour. Comstock Publishing Associates, Ithaca
63. Matthews RW, Matthews JR (2009) Mechanocommunication. In: Insect behavior. Springer Netherlands, Dordrecht, pp 291–339
64. Song H, Béthoux O, Shin S et al (2020) Phylogenomic analysis sheds light on the evolutionary pathways towards acoustic communication in orthoptera. Nat Commun 11:4939. https://doi.org/10.1038/s41467-020-18739-4
65. Davranoglou L, Taylor GK, Mortimer B (2023) Sexual selection and predation drive the repeated evolution of stridulation in Heteroptera and other arthropods. Biol Rev 98:942–981. https://doi.org/10.1111/brv.12938
66. Field LH, Evans A, Macmillan DL (1987) Sound production and stridulatory structures in hermit crabs of the genus *Trizopagurus*. J Mar Biol Assoc U K 67:89–110. https://doi.org/10.1017/S0025315400026382
67. Boon P, Yeo D, Todd P (2009) Sound production and reception in mangrove crabs *Perisesarma* spp. (Brachyura: Sesarmidae). Aquat Biol 5:107–116. https://doi.org/10.3354/ab00136
68. Cividini S, Sfenthourakis S, Montesanto G (2020) Are terrestrial isopods able to use stridulation and vibrational communication as forms of intra and interspecific signaling and defense strategies as insects do? A preliminary study in *Armadillo officinalis*. Sci Nat 107:4. https://doi.org/10.1007/s00114-019-1656-3
69. Victor RT, Classen-Rodriguez L, Maloney TJ, Moore MK (2019) Stridulation by cosmetid harvestmen (Arachnida: Opiliones). J Arachnol 47:132. https://doi.org/10.1636/0161-8202-47.1.132

70. Bura VL, Kawahara AY, Yack JE (2016) A comparative analysis of sonic defences in Bombycoidea caterpillars. Sci Rep 6:31469. https://doi.org/10.1038/srep31469
71. Taylor JRA, de Vries MS, Elias DO (2019) Growling from the gut: co-option of the gastric mill for acoustic communication in ghost crabs. Proc R Soc B Biol Sci 286:20191161. https://doi.org/10.1098/rspb.2019.1161
72. Ladich F (1997) Comparative analysis of Swimbladder (drumming) and pectoral (stridulation) sounds in three families of catfishes. Bioacoustics 8:185–208. https://doi.org/10.1080/09524622.1997.9753362
73. Webb McAdams AL, Smith ME (2023) The relationship between body size and stridulatory sound production in loricariid catfishes. J Acoust Soc Am 154:3672–3683. https://doi.org/10.1121/10.0022575
74. Mahajan CL (1963) Sound producing apparatus in an Indian catfish (*Sisor rhapdophorus* Hamilton). J Linn Soc Lond Zool 44:721–724. https://doi.org/10.1111/j.1096-3642.1963.tb01638.x
75. Kratochvil H (1978) Der Bau des Lautapparates vom Knurrenden Gurami (*Trichopsis vittatus* Cuvier & Valenciennes) (Anabantidae, Belontiidae). Zoomorphologie 91:91–99. https://doi.org/10.1007/BF00994156
76. Daugherty JL, Marshall JA (1976) The sound-producing mechanism of the croaking gourami, *Trichopsis vittatus* (Pisces, Belontiidae). Physiol Zool 49:227–244. https://doi.org/10.1086/physzool.49.2.30152543
77. Bertucci F, Ruppé L, Van Wassenbergh S et al (2014) New insights into the role of the pharyngeal jaw apparatus in the sound-producing mechanism of *Haemulon flavolineatum* (Haemulidae). J Exp Biol 217:3862–3869. https://doi.org/10.1242/jeb.109025
78. Colson DJ, Patek SN, Brainerd EL, Lewis SM (1998) Sound production during feeding in *hippocampus* seahorses (Syngnathidae). Environ Biol Fish 51:221–229. https://doi.org/10.1023/A:1007434714122
79. Romanchek BAH, Uetz G, Scheifele PM (2024) Characterization of sound production by the pot-bellied seahorse (*Hippocampus abdominalis*) during feeding. J Fish Biol 105:124–128. https://doi.org/10.1111/jfb.15747
80. Bostwick KS, Prum RO (2005) Courting bird sings with stridulating wing feathers. Science 309:736–736. https://doi.org/10.1126/science.1111701
81. Zherebtsova OV (2006) Morphofunctional interpretation of the quills stridulating in tenrecs (Lipotyphla, Tenrecidae). Russ J Theriol 5:1–11. https://doi.org/10.15298/rusjtheriol.05.1.01
82. Gould E, Eisenberg JF (1966) Notes on the biology of the Tenrecidae. J Mammal 47:660–686. https://doi.org/10.2307/1377896
83. Marshall CD, Eisenberg JF (1996) Hemicentetes semispinosus. Mamm Species 1. https://doi.org/10.2307/3504327
84. Ossianilsson F (1949) Insect drummers, a study on the morphology and function Op the sound-producing organ of Swedish Auchenorrhyncha, with notes on their sound-production. Entomologiska Sällskapet i Lund, Lund
85. Corcoran AJ, Hristov NI (2014) Convergent evolution of anti-bat sounds. J Comp Physiol A 200:811–821. https://doi.org/10.1007/s00359-014-0924-0
86. Greenfield MD (2014) Acoustic communication in the nocturnal lepidoptera. In: Hedwig B (ed) Insect hearing and acoustic communication. Springer, Berlin/Heidelberg, pp 81–100
87. Wessel A, Mühlethaler R, Hartung V et al (2014) The tymbal: evolution of a complex vibration-producing organ in the Tymbalia (Hemiptera excl. Sternorrhyncha). In: Cocroft RB, Gogala M, Hill PSM, Wessel A (eds) Studying vibrational communication. Springer, Berlin/Heidelberg, pp 395–444
88. Dowdy NJ, Conner WE (2019) Characteristics of tiger moth (Erebidae: Arctiinae) anti-bat sounds can be predicted from tymbal morphology. Front Zool 16:45. https://doi.org/10.1186/s12983-019-0345-6
89. Davranoglou L-R, Mortimer B, Taylor GK, Malenovský I (2020) On the morphology and evolution of cicadomorphan tymbal organs. Arthropod Struct Dev 55:100918. https://doi.org/10.1016/j.asd.2020.100918

90. O'Reilly LJ, Agassiz DJL, Neil TR, Holderied MW (2019) Deaf moths employ acoustic Müllerian mimicry against bats using wingbeat-powered tymbals. Sci Rep 9:1444. https://doi.org/10.1038/s41598-018-37812-z
91. Yack JE, Otero LD, Dawson JW et al (2000) Sound production and hearing in the blue cracker butterfly *Hamadryas feronia* (lepidoptera, Nymphalidae) from Venezuela. J Exp Biol 203:3689–3702. https://doi.org/10.1242/jeb.203.24.3689
92. Iwamoto H (2017) The tymbal muscle of cicada has flight muscle-type sarcomeric architecture and protein expression. Zool Lett 3:15. https://doi.org/10.1186/s40851-017-0077-4
93. Kasumyan AO (2008) Sounds and sound production in fishes. J Ichthyol 48:981–1030. https://doi.org/10.1134/S0032945208110039
94. Fine ML, Parmentier E (2015) Mechanisms of fish sound production. In: Ladich F (ed) Sound communication in fishes. Springer, Vienna, pp 77–126
95. Boyle KS, Riepe S, Bolen G, Parmentier E (2015) Variation in swim bladder drumming sounds from three doradid catfish species with similar sonic morphologies. J Exp Biol jeb.123414. https://doi.org/10.1242/jeb.123414
96. Yager DD (1992) A unique sound production mechanism in the pipid anuran *Xenopus borealis*. Zool J Linnean Soc 104:351–375. https://doi.org/10.1111/j.1096-3642.1992.tb00927.x
97. Kwong-Brown U, Tobias ML, Elias DO, et al (2019) The return to water in ancestral *xenopus* was accompanied by a novel mechanism for producing and shaping vocal signals. eLife 8:e39946. doi:https://doi.org/10.7554/eLife.39946
98. Tobias ML, Evans BJ, Kelley DB (2011) Evolution of advertisement calls in African clawed frogs. Behaviour 148:519–549. https://doi.org/10.1163/000579511X569435
99. Barkan CL, Kelley DB, Zornik E (2018) Premotor neuron divergence reflects vocal evolution. J Neurosci 38:5325–5337. https://doi.org/10.1523/JNEUROSCI.0089-18.2018
100. Clark CJ (2016) Locomotion-induced sounds and Sonations: mechanisms, communication function, and relationship with behavior. In: Suthers RA, Fitch WT, Fay RR, Popper AN (eds) Vertebrate sound production and acoustic communication. Springer International Publishing, Cham, pp 83–117
101. Fant G (2012) Acoustic theory of speech production: with calculations based on X-ray studies of Russian articulations. De Gruyter Mouton, Tubingen
102. Fitch T (2006) Production of vocalizations in mammals. In: Encyclopedia of language & linguistics. Elsevier, pp 115–121
103. Bradbury JW, Vehrencamp SL (2011) Sound and sound signal production. In: Principles of animal communication, 2nd edn. Sinauer Associates, Sunderland, pp 19–63
104. Riede T, Goller F (2010) Peripheral mechanisms for vocal production in birds—differences and similarities to human speech and singing. Brain Lang 115:69–80. https://doi.org/10.1016/j.bandl.2009.11.003
105. Nowicki S, Capranica RR (1986) Bilateral syringeal interaction in vocal production of an oscine bird sound. Science 231:1297–1299. https://doi.org/10.1126/science.3945824
106. Suthers RA, Zollinger SA (2004) Producing Song: the vocal apparatus. Ann N Y Acad Sci 1016:109–129. https://doi.org/10.1196/annals.1298.041
107. Rountree RA, Juanes F, Bolgan M (2018) Air movement sound production by alewife, white sucker, and four salmonid fishes suggests the phenomenon is widespread among freshwater fishes. PLoS One 13:e0204247. https://doi.org/10.1371/journal.pone.0204247
108. Reidenberg JS, Laitman JT (2008) Sisters of the sinuses: cetacean air sacs. Anat Rec 291:1389–1396. https://doi.org/10.1002/ar.20792
109. Mercado E, III, Handel S (2012) Understanding the structure of humpback whale songs (L). J Acoust Soc Am 132:2947–2950. https://doi.org/10.1121/1.4757643
110. Cranford TW, Elsberry WR, Van Bonn WG et al (2011) Observation and analysis of sonar signal generation in the bottlenose dolphin (*Tursiops truncatus*): evidence for two sonar sources. J Exp Mar Biol Ecol 407:81–96. https://doi.org/10.1016/j.jembe.2011.07.010
111. Elemans CPH, Jiang W, Jensen MH et al (2024) Evolutionary novelties underlie sound production in baleen whales. Nature 627:123–129. https://doi.org/10.1038/s41586-024-07080-1

112. Cranford TW, Amundin M, Norris KS (1996) Functional morphology and homology in the odontocete nasal complex: implications for sound generation. J Morphol 228:223–285. https://doi.org/10.1002/(SICI)1097-4687(199606)228:3<223::AID-JMOR1>3.0.CO;2-3
113. Reidenberg JS, Laitman JT (2010) Generation of sound in marine mammals. In: Handbook of behavioral neuroscience. Elsevier, pp 451–465
114. Schoenfuss HL, Bragulla HH, Schumacher J et al (2014) The anatomy of the larynx of the bowhead whale, *Balaena mysticetus*, and its sound-producing functions. Anat Rec 297:1316–1330. https://doi.org/10.1002/ar.22907
115. Reidenberg JS, Laitman JT (2007) Discovery of a low frequency sound source in Mysticeti (baleen whales): anatomical establishment of a vocal fold homolog. Anat Rec 290:745–759. https://doi.org/10.1002/ar.20544
116. Cazau D, Adam O, Laitman JT, Reidenberg JS (2013) Understanding the intentional acoustic behavior of humpback whales: a production-based approach. J Acoust Soc Am 134:2268–2273. https://doi.org/10.1121/1.4816403
117. Damien J, Adam O, Cazau D et al (2019) Anatomy and functional morphology of the mysticete rorqual whale larynx: phonation positions of the U-fold. Anat Rec 302:703–717. https://doi.org/10.1002/ar.24034
118. Jones G (1999) Scaling of echolocation call parameters in bats. J Exp Biol 202:3359–3367. https://doi.org/10.1242/jeb.202.23.3359
119. Schneider JN, Mercado, III, E (2019) Characterizing the rhythm and tempo of sound production by singing whales. Bioacoustics, 28(3):239–256. (published online 2018, in print 2019) https://www.tandfonline.com/doi/pdf/10.1080/09524622.2018.1428827?casa_token=m3ugx_NEVvMAAAAA:gYYMdniFmiUHHXB1zKAGuNJsmkTYEm67Z9BgULQbInXekK4Z85ThYIsEzkvtTswIAm9veNThV120EFY
120. Roeske TC, Tchernichovski O, Poeppel D, Jacoby N (2020) Categorical rhythms are shared between songbirds and humans. Curr Biol 30:3699. https://doi.org/10.1016/j.cub.2020.08.026
121. Sharma P, Gero S, Payne R et al (2024) Contextual and combinatorial structure in sperm whale vocalisations. Nat Commun 15:3617. https://doi.org/10.1038/s41467-024-47221-8
122. Suthers RA, Fattu JM (1973) Mechanisms of sound production by echolocating bats. Am Zool 13:1215–1226. https://doi.org/10.1093/icb/13.4.1215
123. Weir CR, Frantzis A, Alexiadou P, Goold JC (2007) The burst-pulse nature of 'squeal' sounds emitted by sperm whales (*Physeter macrocephalus*). J Mar Biol Assoc U K 87:39–46. https://doi.org/10.1017/S0025315407054549
124. Goller F, Mallinckrodt MJ, Torti SD (2004) Beak gape dynamics during song in the zebra finch. J Neurobiol 59:289–303. https://doi.org/10.1002/neu.10327
125. Elemans CPH, Mead AF, Rome LC, Goller F (2008) Superfast vocal muscles control Song production in songbirds. PLoS One 3:e2581. https://doi.org/10.1371/journal.pone.0002581
126. Wahlberg M (2003) Sounds produced by herring (*Clupea harengus*) bubble release. Aquat Living Resour 16:271–275. https://doi.org/10.1016/S0990-7440(03)00017-2
127. Nelson MC (1979) Sound production in the cockroach, *Gromphadorhina portentosa*: the sound-producing apparatus. J Comp Physiol A 132:27–38. https://doi.org/10.1007/BF00617729
128. Bura VL, Rohwer VG, Martin PR, Yack JE (2011) Whistling in caterpillars (*Amorpha juglandis*, Bombycoidea): sound-producing mechanism and function. J Exp Biol 214:30–37. https://doi.org/10.1242/jeb.046805
129. Brehm G, Fischer M, Gorb S et al (2015) The unique sound production of the death's-head hawkmoth (*Acherontia atropos* (Linnaeus, 1758)) revisited. Sci Nat 102:43. https://doi.org/10.1007/s00114-015-1292-5
130. Rosi-Denadai CA, Scallion ML, Merrett CG, Yack JE (2018) Vocalization in caterpillars: a novel sound-producing mechanism for insects. J Exp Biol 221:jeb169466. https://doi.org/10.1242/jeb.169466
131. Versluis M, Schmitz B, Von Der Heydt A, Lohse D (2000) How snapping shrimp snap: through cavitating bubbles. Science 289:2114–2117. https://doi.org/10.1126/science.289.5487.2114

132. Larsen ON, Reichmuth C (2021) Walruses produce intense impulse sounds by clap-induced cavitation during breeding displays. R Soc Open Sci 8:210197. https://doi.org/10.1098/rsos.210197
133. Deluca M, McElroy WD (1978) [1] Purification and properties of firefly luciferase. In: Methods in enzymology. Elsevier, pp 3–15
134. Krafsur ES, Moon RD, Albajes R et al (2008) Fireflies: control of flashing. In: Capinera JL (ed) Encyclopedia of entomology. Springer Netherlands, Dordrecht, pp 1452–1463
135. Oba Y, Konishi K, Yano D et al (2020) Resurrecting the ancient glow of the fireflies. Sci Adv 6:eabc5705. https://doi.org/10.1126/sciadv.abc5705
136. Trimmer BA, Aprille JR, Dudzinski DM et al (2001) Nitric oxide and the control of firefly flashing. Science 292:2486–2488. https://doi.org/10.1126/science.1059833
137. Aprille JR (2004) Role of nitric oxide and mitochondria in control of firefly flash. Integr Comp Biol 44:213–219. https://doi.org/10.1093/icb/44.3.213
138. Haddock SHD, Moline MA, Case JF (2010) Bioluminescence in the sea. Annu Rev Mar Sci 2:443–493. https://doi.org/10.1146/annurev-marine-120308-081028
139. Widder EA (2010) Bioluminescence in the ocean: origins of biological, chemical, and ecological diversity. Science 328:704–708. https://doi.org/10.1126/science.1174269
140. Morin JG (2019) Luminaries of the reef: the history of luminescent ostracods and their courtship displays in the Caribbean. J Crustac Biol 39:227–243. https://doi.org/10.1093/jcbiol/ruz009
141. Davis MP, Holcroft NI, Wiley EO et al (2014) Species-specific bioluminescence facilitates speciation in the deep sea. Mar Biol 161:1139–1148. https://doi.org/10.1007/s00227-014-2406-x
142. Jägers P, Wagner L, Schütz R et al (2021) Social signaling via bioluminescent blinks determines nearest neighbor distance in schools of flashlight fish *Anomalops katoptron*. Sci Rep 11:6431. https://doi.org/10.1038/s41598-021-85770-w
143. Huvard AL (1993) Ultrastructure of the light organ and immunocytochemical localization of luciferase in luminescent marine ostracods (crustacea: Ostracoda: Cypridinidae). J Morphol 218:181–193. https://doi.org/10.1002/jmor.1052180207
144. Hensley NM, Ellis EA, Gerrish GA et al (2019) Phenotypic evolution shaped by current enzyme function in the bioluminescent courtship signals of sea fireflies. Proc R Soc B Biol Sci 286:20182621. https://doi.org/10.1098/rspb.2018.2621
145. Chakrabarty P, Davis MP, WmL S et al (2011) Evolution of the light organ system in ponyfishes (Teleostei: Leiognathidae). J Morphol 272:704–721. https://doi.org/10.1002/jmor.10941
146. Gruber DF, Phillips BT, O'Brien R et al (2019) Bioluminescent flashes drive nighttime schooling behavior and synchronized swimming dynamics in flashlight fish. PLoS One 14:e0219852. https://doi.org/10.1371/journal.pone.0219852
147. Johnson GD, Rosenblatt RH (1988) Mechanisms of light organ occlusion in flashlight fishes, family Anomalopidae (Teleostei: Beryciformes), and the evolution of the group. Zool J Linnean Soc 94:65–96. https://doi.org/10.1111/j.1096-3642.1988.tb00882.x
148. Sparks JS, Dunlap PV, Smith WL (2005) Evolution and diversification of a sexually dimorphic luminescent system in ponyfishes (Teleostei: Leiognathidae), including diagnoses for two new genera. Cladistics 21:305–327. https://doi.org/10.1111/j.1096-0031.2005.00067.x
149. Sasaki A, Ikejima K, Aoki S et al (2003) Field evidence for bioluminescent Signaling in the pony fish, *Leiognathus elongatus*. Environ Biol Fish 66:307–311. https://doi.org/10.1023/A:1023959123422
150. Herring PJ (1977) Bioluminescence of marine organisms. Nature 267:788–793. https://doi.org/10.1038/267788a0
151. Messenger JB (2001) Cephalopod chromatophores: neurobiology and natural history. Biol Rev 76:473–528. https://doi.org/10.1017/S1464793101005772
152. Bower DQ, Senft SL, Hanlon RT, Deravi LF (2024) Pigment granule architecture varies across yellow, red, and brown chromatophores in squid *Doryteuthis pealeii*. Sci Rep 14:31417. https://doi.org/10.1038/s41598-024-83021-2

153. Mäthger LM, Hanlon RT (2007) Malleable skin coloration in cephalopods: selective reflectance, transmission and absorbance of light by chromatophores and iridophores. Cell Tissue Res 329:179–186. https://doi.org/10.1007/s00441-007-0384-8
154. Mäthger LM, Senft SL, Gao M et al (2013) Bright white scattering from protein spheres in color changing, flexible cuttlefish skin. Adv Funct Mater 23:3980–3989. https://doi.org/10.1002/adfm.201203705
155. Kramer B (2009) Electric communication and electrolocation. In: Binder MD, Hirokawa N, Windhorst U (eds) Encyclopedia of neuroscience. Springer, Berlin/Heidelberg, pp 1038–1045
156. Dunlap KD, Silva AC, Smith GT, Zakon HH (2017) Weakly electric fish: behavior, neurobiology, and neuroendocrinology. In: Hormones, brain and behavior. Elsevier, pp 69–98
157. Ilieş I, Zupanc GKH (2023) Computational modeling predicts regulation of central pattern generator oscillations by size and density of the underlying heterogenous network. J Comput Neurosci 51:87–105. https://doi.org/10.1007/s10827-022-00835-7
158. Virant-Doberlet M, Kuhelj A, Polajnar J, Šturm R (2019) Predator-prey interactions and eavesdropping in vibrational communication networks. Front Ecol Evol 7:203. https://doi.org/10.3389/fevo.2019.00203
159. Halfwerk W, Jones PL, Taylor RC et al (2014) Risky ripples allow bats and frogs to eavesdrop on a multisensory sexual display. Science 343:413–416. https://doi.org/10.1126/science.1244812

The Functional and Structural Diversity of Rhythmic Communication Across Species

3

Gabriela-Alina Sauciuc, Jennifer N. Schneider, Lara S. Burchardt, Michael D. Greenfield, Gerlinde Höbel, and Nao Ota

Abstract

Research on the rhythmic communication (RC) of nonhuman animals is undergoing a period of great effervescence and expansion. Here, we synthesize this rapidly growing literature with the aim of mapping RC forms and usage contexts in an unprecedented diversity of taxa, spanning insects, fish, frogs, birds and

G.-A. Sauciuc (✉)
Department of Philosophy/Cognitive Science, Lund University, Lund, Sweden
e-mail: Gabriela-Alina.Sauciuc@lucs.lu.se

J. N. Schneider
Department of Psychology, LCC International University, Klaipėda, Lithuania
e-mail: jschneider@lcc.lt

L. S. Burchardt
Leibniz Zentrum für Allgemeine Sprachwissenschaften Berlin, Berlin, Germany

Institute for Theoretical Biology, Humboldt-Universität zu Berlin, Berlin, Germany

Donders Institute for Brain, Cognition, and Behaviour, Radboud University, Nijmegen, Netherlands
e-mail: lara.sophie.burchardt@hu-berlin.de

M. D. Greenfield
ENES Bioacoustics Research Lab, CRNL, University of Saint-Etienne, CNRS, Inserm, Saint-Etienne, France

Department of Ecology and Evolutionary Biology, University of Kansas, Lawrence, KS, USA
e-mail: michael.greenfield@univ-tours.fr

G. Höbel
Behavioural and Molecular Ecology Group, Department of Biological Sciences, University of Wisconsin–Milwaukee, Milwaukee, WI, USA
e-mail: hoebel@uwm.edu

N. Ota
Museum of Nature and Human Activities, Hyogo, Sanda, Japan
e-mail: n-ota@hitohaku.jp

G.-A. Sauciuc et al. (eds.), *Nature Beats*, Fascinating Life Sciences,
https://doi.org/10.1007/978-3-032-24250-1_3

mammals. We see this as a first large-scale effort toward charting the functional and structural traits of RC to identify taxon-specific and shared motifs and, thus, evolutionary trends. Our synthesis shows that all the covered taxa use RC in reproductive contexts and suggest that RC undergoes an increasing functional complexification as a function of sociocognitive complexification. By contrast, structural complexity fluctuates less at cross-taxonomic level, such that, at the production level, within each studied class or order, some species exhibit highly complex RC forms and rhythmic coordination, while others exhibit the simplest forms of RC. However, based on the current data, it remains unclear if species with complex RC structures also possess sophisticated RC perception. This chapter also reveals considerable theoretical and methodological heterogeneity in the study of animal RC and the need for cross-talks between taxon-specialized researchers toward harmonising key terms, concepts and measures to improve data comparability.

3.1 Introduction

Research on the rhythmic communication (RC) of nonhuman animals has proliferated in recent years, spurring a period of great effervescence in the field. In this chapter, we synthesize the major findings of this rapidly expanding literature, focusing on the forms that RC takes on across taxa and the functional contexts in which it occurs. We regard this endeavour as a necessary first step toward larger-scale efforts to comprehensively map structural and functional traits in order to identify cross-taxon shared motifs, and taxon- or species-specific traits. The outcomes of such comparisons are bound to provide an empirical foundation of unprecedented span and depth for addressing questions of RC evolution.

As the study of RC has not proceeded at a similar pace across the various taxa, our criteria for inclusion in this review vary to some extent across the focal taxa. The highest priority was given to studies that provide statistical information on rhythmic parameters. As a second best choice, we included studies that provide quantitative analyses of the temporal structure of signals, if these were suggestive of rhythmic patterning (as defined in Chap. 1 [1]). Finally, we also included behaviours described—but not yet quantified—as rhythmic.

The research on animal RC is characterized by considerable conceptual and terminological heterogeneity, which partly reflects distinct traditions regarding levels of analysis and focal topics solidified in taxon- or species-specific research. For this reason, whenever rhythm statistics were absent, we decided to use the original terminology of the source studies when referring to aspects of structural and temporal organisation (e.g. syllable, pulse, strike, phrase, drumroll, etc.). For the same reason, we also decided to organize this review according to a taxonomic, rather than thematic, criterion. Our review, thus, covers major findings concerning RC forms and usage contexts in insects (Sect. 3.2), fish (Sect. 3.3), anurans (Sect. 3.4), birds (Sect. 3.5) and mammals (Sect. 3.6). Due to extreme data scarcity, potential RC in additional taxa (e.g. cephalopods, reptilians, arachnids) is provided in Chap. 4 [2]—which identifies new relevant species and behaviours for RC research.

To address questions related to the potential functions of rhythmic patterning in animal communication, in this chapter we focused on data about (i) the functional contexts in which rhythmic signals are used by animals; (ii) the role of temporal patterning in context discrimination; (iii) animals' sensitivity to manipulations of the rhythmic parameters that constitute the species-specific norm; (iv) data concerning the distribution of rhythmic signals across sexes. To address questions related to rhythmic structure, we primarily considered data on (i) pattern complexity (e.g. singleton vs. syllables vs. motifs); (ii) levels of organisation (flat vs. hierarchical); (iii) periodicity (e.g. isochrony, acceleration, deceleration); (iv) the presence and forms of rhythmic interactions (e.g. duet, chorus, etc.).

3.2 Arthropod Rhythm

Given the vast diversity of the earth's arthropod species, notably the range of morphology, ecology and behaviour in this group, it is not surprising that arthropod rhythms cover an extremely broad spectrum of (temporal) structure and function. But in keeping with the focus of other chapters in this volume and of other taxa in this chapter, we concentrate on relatively fast (ultradian; period generally <10 s) rhythms that are largely expressed in an interactive behavioural context, that is, communication. Thus, we only cover circadian and other long-term rhythms insofar as they describe when or how individuals display their communicative behaviour over the course of a day or their daily activity period. Likewise, purely physiological (e.g. cardiac rate) and locomotor rhythms (e.g. gait and wingbeat) are omitted unless they are associated with some form of interindividual messaging. This concentration on RC ensures that the rhythmic activity can be approached from both motor (production) and sensory (perception; behavioural response) perspectives, a hallmark of the volume.

Rather than offering an encyclopaedia of RC across the phylum, we dwell on key examples that have been studied in detail and that demonstrate how arthropods generate rhythmic signals, how receiving individuals perceive and respond to the rhythms, and how signal rhythms vary among related species. Most forms of arthropod RC entail the repetition of mechanical (acoustic or substrate vibration) and optical (bioluminescent or reflected light) signals. Of these, airborne acoustic and bioluminescent signals are more easily studied. They are generally conspicuous to human observers or are detectable with conventional instrumentation and can therefore be readily analysed. Moreover, signals in the acoustic and bioluminescent modalities can be reproduced synthetically and transmitted with precision to test receiver responses in the laboratory and field. On the other hand, simulation of substrate vibration and optical signals based on reflected light remains difficult in many cases, and careful testing is often beyond our reach.

3.2.1 Acoustic Signals and Rhythms

The most thoroughly studied arthropod acoustic signals are the chirps, trills, buzzes, etc. of ensiferan orthopterans (crickets and katydids). Other than those species that have secondarily lost their songs over the course of evolution, adult male ensiferans broadcast 'advertisement' calls that announce their presence to potentially receptive females (and rival males) within a given radius [3]. That distance may extend to as long as 500 m in larger species with intense songs (e.g. https://orthsoc.org/sina/195a.htm). In many crickets and katydids males produce different calls [4, 5], as well as substrate vibrations [6], during close-range courtship or aggressive encounters, and females too call in response to preferred male advertisements in some katydid species [7]. These additional signals notwithstanding, the male advertisement calls are the signposts by which ensiferan species are recognized by human observers—and by conspecifics [8].

3.2.1.1 Ensiferan Stridulation

Ensiferans rely on specialized forewing-forewing stridulation to generate their advertisement calls. A transverse row of a great many miniscule teeth (the file) on the ventral side of the base of one wing is rapidly moved against a raised longitudinal vein (the scraper, or plectrum) on the dorsal side of the base of the other wing, and with each toothstrike of the file against the plectrum one or both forewings are deflected and generate one or more sound waves [9, 10]. A membranous area at the forewing base serves to amplify the vibration and radiate the sound. Crickets have a complete acoustic apparatus (file, scraper, triangular membranous area termed the 'harp') on both forewings, whereas katydids bear the file on the left wing and the scraper and membranous area on the right (see [9, 11] for details on minor asymmetry between the left and right forewings in crickets). During stridulation ensiferans elevate their forewings and move them laterally, opening and closing them through a small angular arc [12]. Notably, their hindwings are kept folded against the body and are not directly involved in singing, which they perform while on the ground or vegetation. Although the same muscles are used to move the forewings during stridulation and flight, the latter entails stroking the hindwings as well, extending both pairs of wings perpendicular to the body and moving them vertically and through a much greater angular arc, and typically at a slower rhythm (cycles of wing movement per s) than in stridulation [13].

3.2.1.1.1 Nested Rhythms

While some Neotropical rainforest species of katydids call only sporadically and do so without regular repetition of song or of song elements [14], most ensiferans call for extended intervals during their daily activity period [15] and their advertisements often feature a set of nested rhythms. The first rhythm, the rate at which sound waves are produced—the carrier frequency of the song—is ubiquitous and almost trivial, but it forms the basis for all higher-level rhythms, so we begin here. In crickets, the carrier frequency—mostly between 4 and 7 kHz, but higher than 20 kHz in some species [16]—is identical to the rate at which teeth on the file strike the plectrum [9]. This 'toothstrike rate' can be very regular, a feature contributing to the relatively pure (narrow-band; high Q) frequencies—their 'pitch'—that make cricket song pleasant to human ears. A regular toothstrike rate arises because adjacent teeth along the file are evenly spaced, the file advances steadily and incrementally against the plectrum during forewing movement, and—most critically—a morphological 'escapement mechanism' ensures that the file jumps from one tooth to the next over a constant time interval at each increment [17] (but see [18] for complications).

The toothstrike rate is typically very close to the natural resonant frequency of the membranous harp, a correspondence that suppresses radiation of other frequencies and thereby retains the narrow-band frequency character of the song [9]. The forewing structure responsible for amplification and radiation of sound in katydids is different, and more complex than in crickets, and a toothstrike often generates a highly damped pulse of sound rather than a single sound wave. Consequently, katydid song is generally broadband and perceived as 'buzzy' by the human ear. Owing to their small size and material properties, the natural resonant frequency of the forewing acoustic apparatus in crickets as well as katydids is high (>2–3 kHz), well beyond the rate of muscle contraction. Thus, the file and plectrum device is essentially a 'frequency multiplier', converting the rate at which wings can be opened and closed (<250 Hz), which is limited by muscular constraints, to a toothstrike rate and carrier frequency that are 50–100× faster [9]. Katydid song is typically much higher in carrier frequency than cricket song, and many katydid species sing partly or entirely in the ultrasonic (>20 kHz) band.

The next rhythm in ensiferan stridulation is the number of cycles of lateral forewing movement per second. In many crickets, file-plectrum toothstrikes are typically made only during the closing phase of the wing movement cycle, leaving a pronounced silent interval during the opening phase. The sound produced during the closing phase of a given cycle of wing movement may be designated a 'pulse' (although the literature is replete with a wealth of onomatopoetic terms for sound units in acoustic insects), and the pulse rate corresponds with cycle rate. For example, in the well-studied European field cricket *Gryllus bimaculatus*, whose advertisement call has a carrier frequency ≈4.7 kHz, a wing closure generates a regular sequence of 105 toothstrikes over a 22-ms interval, which form a sound pulse. The cycle rate, and hence pulse rate, is ≈30 Hz, and an 11-ms silent interval separates consecutive pulses [12, 19].

Rhythms whose periods are longer than a cycle of wing movement occur in a great many crickets and katydids. Continuing with *G. bimaculatus* song, a calling male typically produces 4 pulses at the 30-Hz rhythm, pauses for ≈200 ms, and then repeats. The group of 4 pulses is termed a 'chirp', with the chirp rhythm averaging ≈3 Hz [19]. Variations on this theme are found in many ensiferans, as well as in other arthropods broadcasting various mechanical and optical signals. Both pulse and chirp rhythms often continue over many hundreds of cycles with moderate to high isochrony (CV < 0.10; recordings archived on the internet site Singing Insects of North America (SINA): orthsoc.org). In both crickets and katydids human observers generally note the chirp rhythm, whereas the much faster pulse rhythm is often the more critical parameter for conspecifics [20, 21]. Rhythms may also include components shorter than a cycle of wing movement, particularly in species that produce sound during both the wing closure and opening, albeit with a brief silent gap when the wings reverse direction at the onsets of closure and opening [22, 23]. Thus, a complete cycle of wing movement comprises two intervals, each with a characteristic sound. Generally, the closing interval is longer and higher in amplitude than the opening one. In addition to songs generated by a rhythmic series of a given type of wing movement cycles, some katydid species produce a series of several rhythms over the course of a song, each one comprising the repetition of a different wing movement and its corresponding sound [24, 25]. These different wing movements can vary in their repetition rate, the angular arc of wing opening and the force with which the file teeth strike the scraper. All three factors may influence the sound broadcast during a particular rhythm.

Most ensiferans sing during a specific daily activity period—morning, evening, beginning of the night [3, 15, 26, 27]—and during this period males within a local population may temporally organize their song in bouts separated by intervals of silence, or relative silence [15]. But bout lengths are not uniform (10 s–15 min), bouts are not repeated with strict regularity, and isochronous rhythms at this temporal scale are not evident at an individual or collective level. Singing in bouts may be an effective means for an individual to apportion limited energy to sexual advertisement throughout a lengthy daily activity period [28]. At a collective level, loosely organized group bouts may occur because some individuals are more 'motivated' to initiate song than others, and those less motivated individuals then respond and sing because doing so offers them their main chance of achieving any mating success. That is, they are forced, if possible, to at least match the calls of nearby rivals [29].

Are the mechanisms controlling the circadian activity pattern of song linked to the control of song rhythm itself? We do not yet have any direct evidence from crickets and katydids but work on several 'model organisms' among arthropods indicates a general control over timing at these very different period lengths. Beginning with the landmark investigation of *Drosophila melanogaster* by [30], studies have shown that the per gene, which regulates the length of the endogenous, circadian period of general activity, also influences period length of ultradian rhythms in courtship behaviour. These ultradian rhythms range from roughly 1-min long periods during which courtship song pulse rate, reflecting cycles of wing movement, fluctuates from fast to slow and then back to fast (*Drosophila melanogaster* [30, 31]); to repetitions of approx. 350-ms 'pulse trains' in courtship song (*Drosophila montana* [32]); to an approx. 250-ms inter pulse-train interval in courtship song (the tephritid fly *Bactrocera cucurbitae* [33]); and to an approx. 10-s train of courtship head movements (the pteromalid wasp *Nasonia vitripennis* [34]). Changes in circadian and ultradian periods are parallel, that is, 'long day' per strains exhibit longer periods during courtship behaviour in all cases except *N. vitripennis*. Moreover, in *D. melanogaster* the interstrain changes in circadian activity are proportional to those in courtship behaviour [30, 31].

3.2.1.2 Intraspecific Features and Variation

Whereas collective singing bouts and other song interactions [35, 36] indicate clearly that ensiferans can modify their songs and song rhythms upon perceiving the songs of conspecific neighbours, the regular expression of song elements such as pulses and chirps by isolated individuals, upon reaching maturity, demonstrates that song rhythms are endogenous and not learned [35, 37]. But song constancy applies to a given individual singing under a particular ambient condition. Ensiferans being primarily ectothermic, the critical ambient condition influencing the rate of muscle contraction, and hence stridulatory rhythm, is temperature. Crickets and katydids generally accelerate their song rhythms as temperature rises within a 'middle' range, from 15 to 30 °C [38, 39]. The slope of song rhythm vs. temperature is roughly linear within this range (see [39] for discussion of linearity), and in the crickets *Gryllus integer* and *Teleogryllus oceanicus* it was found to be more significant for pulse rate [40, 41]—which is directly related to repeated muscle contraction—than for chirp rate or carrier frequency, the latter being partly controlled by the material properties of the stridulatory apparatus [42]. Similar thermal effects on communication rhythms are reported elsewhere in ectotherms (other acoustic insects, such as cicadas [43], anurans [44] and in the bioluminescent flash rhythms of lampyrid beetles [45–47], see also Chap. 6 [48] this volume). The firefly rhythms are controlled biochemically [49, 50] rather than directly by muscle contractions, indicating that the repetition rate of non-muscular activity may also be temperature dependent. A major exception to thermal dependence is reported for the approx. 1-min-long period during which courtship song cycles between a fast and slow pulse rhythm in *Drosophila melanogaster* (see previous section). This ultradian period, influenced by the per gene that notably controls the daily activity pattern [30], thus follows the general 'rule' of circadian activity being largely unaffected by ambient temperature.

At a given temperature, pulse and chirp rhythms, as well as chirp lengths (number of pulses per chirp), present different levels of intraspecific variation across ensiferans and other arthropods. In some species, male pulse rhythms exhibit little variation and are centred around a mean value. The narrow range of rhythms likely reflects stabilising selection imposed by females that exhibit a 'closed' preference function for song: they walk or jump toward rhythms—a 'phonotactic' response—close to that mean value and ignore those much faster or slower [51]. Closed female preference functions and limited intraspecific variation in male pulse rhythm may be particular to taxa where sympatric species broadcast similar signals differing primarily in that rhythm character. By virtue of closed preference functions and limited rhythm variation, species recognition might be enhanced, and cross-species attraction and hybrid mating may be reduced [52]. But a species' characteristic pulse rhythm would shift up and down with changes in temperature, and it would only ensure conspecific pairing if female preference functions for pulse rhythm exhibit commensurate shifts. Some evidence for such parallel shifts has been found in ensiferans [53–55] and other ectotherms [56, 57], a phenomenon known as 'temperature coupling', but the extent to which signaller and receiver thermal shifts have coevolved—and expressly favour conspecific attraction—remains controversial

[58, 59]. Possibly, coevolution of signaller and receiver thermal shifts has occurred mostly in those taxa where pulse rhythm is a very reliable species indicator.

In marked contrast to constancy in pulse rhythm are the great many cases where a rhythm character varies substantially among individuals within a population and the female preference function is 'open-ended' rather than centred around a given value. That is, females show an increasing level of response for faster rhythms or longer chirps, and this increase continues for values beyond those broadcasted in the male population [37, 51]. Here, selection is directional rather than stabilising, and it favours extreme values—which only a few males can approach. Such directional selection may be expected where there are no sympatric species broadcasting similar signals that might confuse and beckon females in the focal population. By choosing males displaying faster rhythms or longer chirps, both of which demand more energy investment, females may be opting for mates with superior ability, possibly genetically based, to acquire energy or to use it efficiently. That relatively few males display these preferred rhythm and chirp values may signify the inability of most individuals to have acquired the requisite energy, or their obligation to devote their acquired energy resources to general body maintenance and defence.

Before leaving the topic of intraspecific variation, the notion of open-ended preference functions should be clarified to preclude a common misunderstanding. Contrary to the literal meaning of the term, these functions do not encompass an increasing preference level without limit. Rather, the increased preference level continues to the highest character values in the (male) population and then slightly beyond, but at some value preference reaches an asymptote and then eventually declines [60]. For pulse rhythm, the decline may occur when the rhythm of successive pulses exceeds neural constraints. For example, the (synthetic) pulse rhythm as presented in experimental bioassays, is faster than the maximum value of the organism's sustainable rate of action potentials ($\approx$250 Hz), causing many pulses to be 'unregistered' by the peripheral nervous system [61]. For chirp length, longer and longer chirps may begin to be evaluated as continuous trills, as the essential 'chirp character' of the song is no longer recognized, and a positive behavioural response is not elicited [62, 63].

3.2.1.3 Interspecific Interference: Predator Avoidance

In addition to the problem of potential hybrid mating presented by related sympatric species broadcasting similar songs, acoustic insect species may confront other rhythmic signals that present more dire consequences. Such threats arise specifically from foraging insectivorous bats, who emit high-carrier frequency rhythmic echolocation signals to navigate and hunt in darkness. Thus, females in any potential prey species that broadcasts rhythmic song must be able to discriminate conspecific signals from those of local bats. To accomplish this obviously essential task, an acoustic species must overcome at least three difficulties. First, bat diversity is high in most temperate and tropical regions of the world [64], and an acoustic insect would have to distinguish its conspecific call from the echolocation signals of multiple bat species, which may collectively encompass a diverse range of rhythms. Second, for acoustic insects who themselves communicate with high-frequency calls, discrimination cannot be achieved by evaluating the carrier frequency profile of perceived signals. With the exception of some cicada species, frequency tuning in insect hearing is too crude to distinguish all but the most major differences; for example, below 20 kHz vs. above [37]. Thus, any effective discrimination would be restricted to temporal features such as rhythm and the length of rhythmically repeated elements (chirps, pulses, etc.). Third, discrimination must be accomplished rapidly, as any delay due to excessively fine evaluation could very well lead to death. But overly coarse evaluation, immediately categorising nearly all rhythmic high-frequency signals as predator ones, would run the risk of missing many legitimate, conspecific calls. These are the general expectations for discrimination in signal detection theory and are normally analysed with a receiver operating curve to predict the optimal strategy [65].

The acoustic insects that may be most susceptible to confusion by bat echolocation signals are nocturnal lepidopterans (moths). More than 55% of the world's moth species hear in the ultrasound (>20 kHz) range [66, 67], and many perform various behaviours that evade bats, mostly while in flight [68] but also when on the substrate [69]. Moreover, in several moth families, some species broadcast pulsed male advertisement or courtship songs whose carrier frequencies and rhythms crudely resemble bat echolocation signals [67].

Anti-bat behaviour has been intensively studied in one of these acoustic species, the crambid moth *Achroia grisella*, and the findings show how conspecific and predator signals are distinguished. The male advertisement calls in *A. grisella* are a rhythmic sequence of high-frequency (70–130 kHz) pulses that are delivered at ≈100 Hz at 25 °C. Pulses are 100–500 μs long and generated by wing fanning (≈50 Hz) while remaining stationary on the substrate, which causes a pair of tymbals at the left and right wing bases to resonate at each wing upstroke and downstroke [61]. For comparison, 'typical' echolocation signals by foraging bats are 30–120 kHz in carrier frequency, 2–10 ms long, and delivered at a 15–30 Hz rhythm [70, 71].

A. grisella shows little response to conspecific song while in flight, and their anti-bat behaviour in the air is simple. Following a 30-ms latency beginning at the onset of a single high-frequency test pulse >1 ms in length and above 75 dB in peak sound pressure level, both males and females stop beating their wings and drop to the ground. A bat foraging in the vicinity of a flying moth can arrive within 1–2 s of detecting it [72], and the simple flight interruption response, employing minimum signal evaluation, can rapidly put an *A. grisella* out of harm's way. False alarms are few because the *A. grisella* song pulses are extremely brief and generally do not trigger flight interruption, which requires the integration of the threshold sound energy over an interval many times longer than a typical moth pulse.

Anti-bat behaviour on the substrate, where mating activities take place, is another matter. A substantial percentage of bat species forage (glean) for prey that remain on the substrate as well as those that are in flight [73, 74]. The approach of a bat while gleaning is much slower than in aerial hunting, and it relies on listening to the inadvertent sounds made by moving prey, and to singing ones. This lengthy approach means that a targeted moth has sufficient time to evaluate more than a single sound pulse, which would be critical given that singing males—potential mates or rivals—are likely to be adjacent. Behavioural assays of *A. grisella* males and females on the substrate show that conspecific vs. predator discrimination is done by categorical perception of the test pulse rhythm they hear: conspecific if >35–40 Hz, and predator if slower [71]. That is, a singing male becomes silent upon hearing a slow rhythm, and a female in the process of moving toward a singing male stops running. Thus, the *A. grisella* female preference function, which is open-ended at high (male) pulse rhythm, is sharply truncated at the low end in order to avoid predators [75].

3.2.2 Interspecific Variation and Rhythm Evolution

How do arthropod rhythms vary across species, and how has interspecific diversity in rhythm evolved? This is an intractable question to address globally among arthropods because a great many factors are likely involved at that scale and data are lacking for many species, but we can acquire some meaningful glimpses of variation and evolution by focusing on a few exemplary taxa. We have chosen to look at rhythm diversity in several genera of acoustic and bioluminescent arthropods that meet a range of criteria. First, the taxon would include a relatively large number of species, which would afford the opportunity to make many inferences about rhythm diversity and evolution. Second, species in the taxon all use an acoustic or bioluminescent male advertisement signal and do not appear to rely on other signal modalities for long-distance pair formation. This stipulation would rule out strictly diurnal arthropods, where communication using reflected light may be used. Third, the taxon would be monophyletic and have a robust molecular phylogeny—features necessary for making evolutionary inferences. Using these criteria, we have selected three North American clades: Nearctic members of the field cricket genus *Gryllus*, the coneheaded katydid genus *Neoconocephalus*, and the Nearctic firefly genus *Photinus*. All three clades have over 20 species, acoustic or bioluminescent signals are precisely described in each species within the clade, and recent molecular phylogenies exist. Several Palearctic genera/clades meet some of the criteria, but they either have fewer species or many species whose rhythms have yet to be fully described.

The Nearctic *Gryllus* presently includes 44 crickets whose male advertisement songs can be categorized according to whether they are a series of discrete ‘chirps’ or are continuous ‘trills’, and if chirped, by the chirp rhythm and the number of pulses per chirp [76]. Accordingly, the following scheme has been proposed (Fig. 3.1): trillers (6 species with an uninterrupted pulse rhythm); fast, short chirpers (4 species with chirp rates from 5 to 8 Hz (at approx. 25 °C), each chirp including 3 or fewer pulses); fast, long chirpers (3 species with chirp rates from 5 to 8 Hz, each chirp including 4 or more pulses); slow chirpers (≈30 species with chirp rates slower than 4 Hz, each chirp including 4 or more pulses); no advertisement song (2 species; note that one of these species, *G. ovisopis* does broadcast close-range courtship and aggressive songs [39]). Fast, long chirpers are a monophyletic group, but all other categories are distributed across the *Gryllus* phylogeny [77]. The main

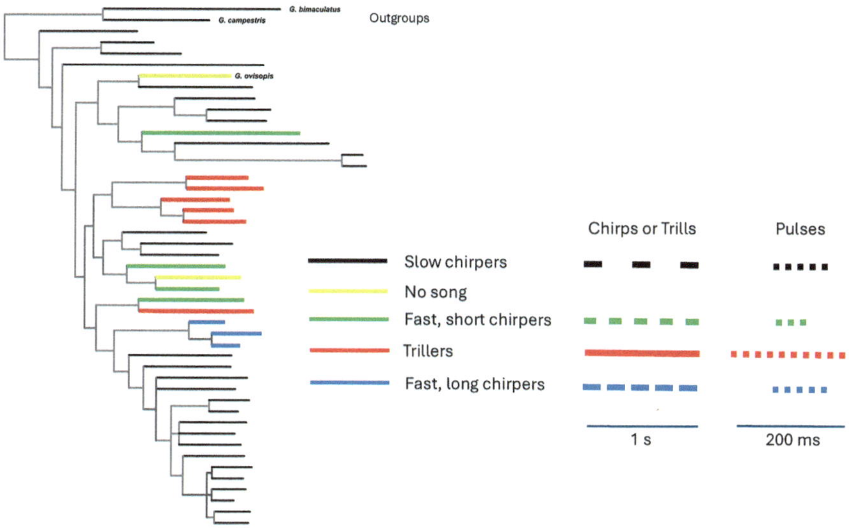

Fig. 3.1 Phylogeny of North American species of the field cricket (Gryllinae) genus *Gryllus*, showing male advertisement song rhythm. A pictorial key to rhythm timelines is shown to the right. The phylogenetic tree was generated by anchored hybrid enrichment using RAxML analysis of concatenated data (Fig. 12 in [77]). Outgroups are two European species of *Gryllus*, both of which are slow chirpers. Rhythms are from songs recorded at 25 °C or corrected to values expected at that temperature

inference that can be made from the phylogenetic information is that slow chirping is the ancestral character state, and song absence, fast chirping (both short and long), and trilling are all derived and have each appeared independently on several occasions. The transitions from slow to fast chirping, and to trilling, may have entailed increased energy expenditure, as expected from a higher overall number of pulses per s (= chirp rate X pulses per chirp X pulse rate). Possibly, sexual selection drove these transitions. Predation risk and demography may also have played a role—in retaining slow chirping in most species—as this rhythmic pattern, with its long silent intervals, might afford a singer more opportunity to perceive natural enemies and male rivals. Presumably, predation risk is also responsible for the several cases of advertisement song absence.

The genus *Neoconocephalus* includes 25 katydids in North and Central America and the Caribbean islands [78] (and species information with recordings archived on the internet site Singing Insects of North America (SINA), orthsoc.org). Songs are known for nearly all species, but only 17 species were included in a molecular phylogeny [79]. An unknown number of additional species are found in South America. *Neoconocephalus* songs can be categorized by a scheme similar to that used for *Gryllus* crickets but with more complications (Fig. 3.2): continuous buzzes of indeterminate length and delivered at a fast pulse rhythm (100–250 Hz; 7 species); continuous buzzes of indeterminate length and delivered at a slow pulse rhythm (< 50 Hz; 2 species); buzzes (verses of multiple pulses) of determinate

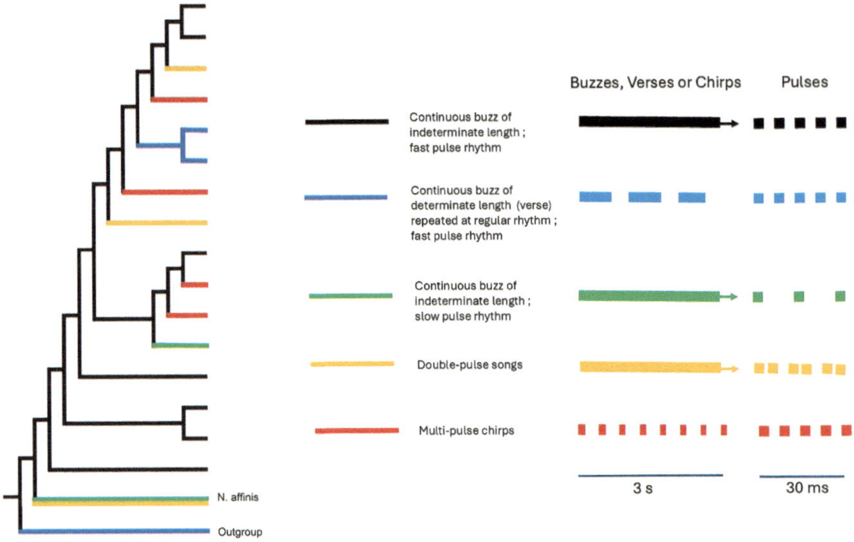

Fig. 3.2 Phylogeny of North and Central American and Caribbean species of the conehead (Copiphorini) katydid genus *Neoconocephalus*, showing male advertisement song rhythm. A pictorial key to rhythm timelines is shown to the right. The phylogenetic tree was generated by a total evidence protocol, with branch lengths drawn unscaled for simplification (Fig. 5 in [79]). Only 17 of the 25 known species from the region were included in the molecular phylogeny. *N. affinis*, at the base of the tree, may be more closely related to unknown South American species. Outgroups are species in the copiphorine genera *Belocephalus* and *Bucrates*

length (0.1–1.0 s) repeated at a regular rhythm (0.5–3 Hz: 4 species); chirps (song units comprising 1–6 pulses; 20–60 ms long; repeated at 2–10 Hz; 2 species); double-pulse songs (continuous buzzes in which 2 'pulse types' alternate, with females evaluating the paired double pulses; 3 species; note that one of these species, *N. affinis*, sings with a slow pulse rhythm and is therefore in two different song categories). The phylogeny indicates several independent transitions from continuous fast-pulse buzzes toward chirps, toward continuous double-pulse buzzes and toward continuous slow-pulse buzzes [79]. Chirps may afford singers improved listening for natural enemies and conspecifics [80], and both double-pulse buzzes and slow-pulse buzzes may offer conspecific females more accurate evaluation of the conspecific pulse rhythm, potentially a challenge where pulse rhythms are very fast (>150 Hz) [20, 81]. The phylogeny also shows that continuous fast-pulse buzzers predominate at the root of the tree [79], but the species at the very root is the slow-pulse buzzer with double pulses (*N. affinis*, from Central America and the Caribbean Islands [78]). This anomaly might be explained by the incomplete coverage of New World *Neoconocephalus*: an unknown number of South American species are missing from analysis, and some could represent a clade of fast-pulse buzzers in which *N. affinis* is an offshoot with derived song characters.

Diversification of signal rhythms in the Nearctic firefly genus *Photinus* exhibits several close parallels with that in *Gryllus* crickets and *Neoconocephalus* katydids,

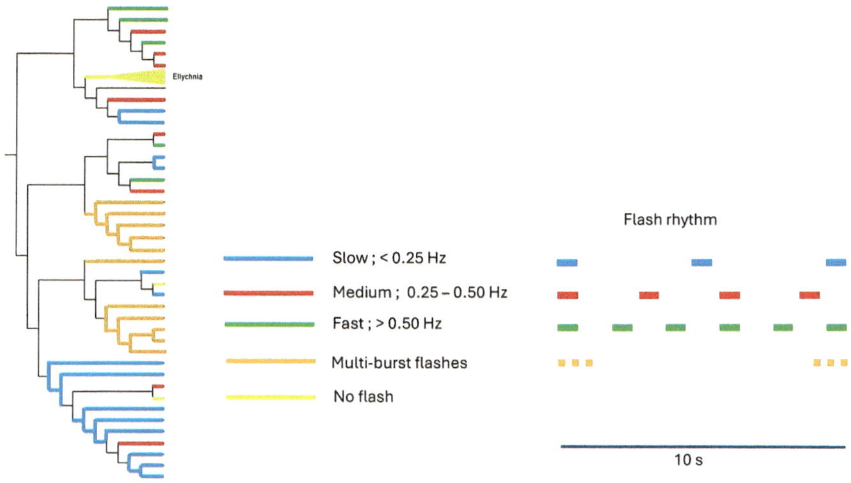

Fig. 3.3 Phylogeny of North American species of the firefly (Lampyridae) genus *Photinus*, showing male advertisement flash rhythm. A pictorial key to rhythm timelines is shown to the right. The phylogenetic tree was constructed using a best maximum likelihood phylogeny and is based on data from three protein-coding genes (Fig. 2 in [84]). *Ellychnia* fireflies were formerly considered as a separate genus but are currently assigned to Photinus; the triangular shape of its branch reflects the several species in the taxon. Black colour for the taxon immediately below *Ellychnia* indicates that flash rhythm has not been determined. See Fig. 3 in [84] for details on ancestral character state reconstruction of flash rhythm. Flash rhythms have been corrected to values expected at 20 °C

remarkable similarities given the different signal modality and the date of the most recent common ancestor (MRCA) of Coleoptera and Orthoptera (≈380 mya [82]). Male *Photinus* produce rhythmic, bioluminescent flashes while flying slowly several meters above the ground. A male typically hovers in place at the instant when emitting a flash, resumes horizontal flight, and then emits his next flash in the same manner [83]. Females generally remain on the ground or vegetation, and they reply to preferred conspecific flashes with their own characteristic flash following a species-specific delay measured from the ending of the male flash. Fireflies have excellent vision, and a male typically responds to a female reply flash by descending and approaching the female. In each *Photinus* species, the male flashes are characterized by a specific length (200–500 ms) and are repeated at a specific rhythm (0.15–0.70 Hz; Fig. 3.3). Flashes are further characterized by whether they are only a single burst of light vs. two or more bursts that are repeated at a fast 'burst rhythm' (0.50–3.00 Hz); that is, the bioluminescent equivalent of a multipulse chirp in *Gryllus* crickets (Fig. 3.1). When female replies are given with a specific rhythm, it generally matches the 'flash rhythm' in the male advertisement. The current *Photinus* phylogeny and flash descriptions are complete, and the information indicates a monophyletic taxon of at least 41 species, with single-burst flashes as the ancestral character state [84]. Transitions to multiple-burst flashes may have occurred independently on several occasions. Flash rhythms vary among both single- and multiple-burst flash species, and transitions toward faster and toward slower rhythms are found.

Transitions toward faster and slower rhythms may be accompanied by shortening and lengthening, respectively, of flash length, suggesting a conservation of energy expenditure. Sexual selection may have driven transitions to multiple-burst flashes and to faster rhythms, whereas predator avoidance could have been responsible for transitions to slower rhythms, and to the three cases where flashes have been lost entirely [84]. Analyses of sympatric distributions indicate that reproductive character displacement may explain many differences in flash rhythm between *Photinus* species [84], an inference also made for *Gryllus* crickets [85–87] (see [88] for a historical perspective) and *Neoconocephalus* katydids [79]. There is no evidence that habitat differences between species—the 'environmental adaptation hypothesis' [89]—explain their specific flash rhythms [84].

Overall, analyses of the selected arthropod groups show how rhythm structure functions in communication, how it varies, and how that variation may have originated. These analyses have been possible because of the relative ease with which many arthropods can be studied in the laboratory and field, the ease of studying acoustic and bioluminescent communication, and the simplicity of communication in the selected groups. That is, the critical parameters were temporal—and thus very relevant to rhythm—rather than spectral, and secondary signalling modalities were either absent or were unimportant.

3.3 Rhythmic Communication in Fish

3.3.1 Acoustic Communication in Fish

Fish sounds differ from other species' sounds in that there are various sound production mechanisms in the same species group, with some fish species not being able to produce sounds at all. Sound production mechanisms can be grouped into three main mechanisms: stridulation, hydrodynamic sound production and sounds produced via the swim bladder (for more details on these production mechanisms, see Chap. 2 [90] this volume. Fish sounds mostly consist of broadband pulsed sounds with a distinct temporal patterning [91]. These highly stereotyped, pulsatile sounds are for example generated by rhythmic contractions of muscles, acting under a rhythmically active vocal motor network [92]. As such, the periodicity within the fish sound elements can be indicative of the periodical activity of neurophysiological processes [93, 94]. Scientific evidence suggests that fish generate, perceive, and recognize acoustic regularity in time. Fish auditory systems are well suited for temporal processing [95], as teleost temporal resolution abilities are defined in the order of a few milliseconds. Although fish lack a cochlea, it has been suggested that they rely on alternative auditory mechanisms to efficiently distinguish temporal patterns and process signals in the presence of noise [96]. For some species, it has been proven that they can discriminate each pulse within a sound [97, 98]. Further, sound temporal patterns, as well as amplitude fluctuations and frequency content, are reflected at the level of the brainstem, so that fish auditory systems can accurately encode the temporal patterns of conspecific sounds [99–101].

Given this, correlations found between within-sound rhythms and environmental parameters or certain behaviours are likely caused by the different rhythms and not merely a by-product of something we can't measure or are not aware of. Therefore, it is not surprising that we find various such correlations for different species and parameters.

3.3.1.1 Within-Sound Rhythms and Their Relevance

The measurement and analysis of temporal features within single fish sound elements are routinely included in fish bioacoustics studies. Parameters often included are the duration of the whole element, pulse/burst number per element as well as pulse/burst duration. Those within-sound temporal features can code for species identity [102] as well as for specific behavioural patterns or motivational states [103–105]. Furthermore, within-sound temporal features can be influenced by environmental conditions such as water temperature, as this can impact the performance of the morpho-physiological processes underlying sound production [106, 107] (for more details, see Chap. 6 [48]). It can also be affected by man-made acoustic disturbances [108]. We will go into more detail in the following sections.

In Sciaenidae, temporal features are known to be relevant during reproduction, namely the formation of seasonal reproductive aggregations at sea and call rates appear to be closely related [109]. Furthermore, the time in which the number of calls peak (i.e. time of day) is positively related to the timing and number of eggs collected at sea [110]. Captivity studies proved that call temporal features reflect spawning activity in many species. In *Scienops ocellatus*, an increase in the number of calls, with longer calls and more pulses, has been recorded when spawning occurs [111] and *Cynoscion nebulosus* was more likely to spawn when male call rate was higher [112]. Calling rate, call temporal features (i.e. sound duration, number of pulses, and pulse period) and call sound pressure levels are indicative of breeding conditions and of spawning events in *Argyrosomus regius*, *Sciaenops ocellatus*, and *Umbrina cirrose*. Longer sounds, made up of a higher number of pulses (i.e. repeated faster), were found during *Sciaena umbra* spawning choruses recorded at sea [105].

A role of temperature was found in different species of croaking gourami, including *Trichopsis schalleri*, *Trichopsis pumilus*, and *Trichopsis vittatus* [113]. This observation underscores the role of temperature in the acoustic communication of this genus, with temporal characteristics like sound interval, burst period, and pulse period showing temperature-related fluctuations. For instance, under different temperature conditions, sound intervals diminished from 492 to 259 ms, burst periods decreased from 51 to 35 ms, and pulse periods shifted from 5.8 to 5.1 ms. Intriguingly, the number of sounds and bursts within a sound remained impervious to temperature variations. This comprehensive analysis sheds light on how temperature uniquely influences the spectral and temporal features of sounds in the croaking gourami, highlighting the impact of ambient temperature on acoustic communication within this group [114]. For more examples on the temperature influence on fish (for details, see Chap. 6 [48]).

In other species, the combination of pulse duration and pulse period provides insights into the communication context. For example, in *Dascyllus albisella*, agonistic chirps exhibit shorter pulse periods and longer pulse durations than courtship chirps [91]. The role of temporal patterns in mate attraction was studied in the plainfin midshipman fish (*Porichthys notatus*). Males of this species produce a long, repetitive hum to attract females to their nesting sites, which vary in duration, repetition rate and intensity. Females show increased phonotaxis as hum durations increase and gap duration between repetitive hums decreases. So even though the temporal pattern is important here, in the end, it seems to be more about sound duration than rhythm [115, 116].

A consistent pattern found across various fish species, influencing temporal characteristics such as sound duration, the number of bursts or pulses within sounds, pulse duration, and burst/pulse periods is growth-related (for details, see Chap. 9 this volume). Numerous studies [117–122] consistently highlight the increase in these temporal parameters with the growth or size of the fish, which can make temporal features a developmental indicator. Nevertheless, exceptions to this trend exist, as observed in the case of the toadfish *Halobatrachus didactylus*, where the number of pulses within a sound, and consequently sound duration, decreases during ontogenetic growth [123]. The size of the sound-generating mechanism also plays a pivotal role [124, 125], whereby there is a close association between sound duration and the size of these mechanisms, specifically the length of the pectoral spine across a selection of catfish species spanning different families. These observations illuminate the intricate temporal dynamics inherent in fish sounds connected to sound production mechanisms and their physical boundaries, as well as environmental conditions, and provide valuable insights into acoustic RC.

While all the above examples rely on within-sound rhythms, much less is known about between-sound rhythms and potential functions. Considering some strong biomechanical and physical constraints on within-fish sound rhythms, between-sound rhythms might offer more degrees of freedom and therefore add important nuances to the picture of temporal structures in acoustic fish communication.

3.3.1.2 Between-Sound Rhythms

Between-sound rhythms are studied far less than within-sound rhythms in fish. In three of the most common fish in the Mediterranean basin, they have been studied comparatively, though. *Sciaena umbra* sounds have a simple, isochronous, metronome-like rhythm of around 0.3–0.4 Hz in addition to within-sound regularity. This between-sound rhythm was found to be quite stable across time (in the same habitat recorded 12 years apart) and space (across four different habitats in the Mediterranean basin: Mallorca, Venice, Trieste, and Crete). Interestingly, the rhythms are not dependent on temperature but seem to be influenced by season and anthropogenic noise [126, 127]. The/Kwa/sound type emitted by *Scorpaena* spp. has a more complex rhythm than *Sciaena umbra*, still presenting an underlying isochronous pattern. Calls of *Ophidion rochei* males present no rhythm, but a random temporal succession of sounds [126]. It is to be highlighted that the within-sound temporal structure of *O. rochei* was found to be very rhythmic [128].

3.3.2 Rhythmic Body Movements and Other Visual Signals

Delving into the world of rhythmic phenomena in fish behaviour, we even find hints at the origins of gestures. Research by Bass and Chagnaud revealed a captivating connection: the rhythmic acoustic signals produced by pectoral fins in ray-finned (teleost) fish may hold the key to understanding the link between vocalisation and upper-limb control [115]. Their comprehensive review encompasses the role of the caudal hindbrain in teleost fish, shedding light on how it orchestrates both muscle control for movement and the rhythms found in acoustic signalling. This prompts us to consider that the neural coregulation of the pectoral system with rhythmic acoustic signals dates to ancient times, possibly even preceding the emergence of terrestrial quadrupeds. It is within this rhythmic connection that we uncover the intriguing interplay between the rhythmic signals of fish and their corresponding rhythmic body movements, which will be discussed in this section.

Many species of cichlid fish are known for their intricate courtship rituals, which often involve rhythmic body movements. Males of different cichlid species for example perform a courtship dance that includes rapid, synchronized lateral body movements and fin displays to attract females. Similar movements are produced in male-male agonistic interactions. These movements are believed to signal readiness for mating in the first context and to establish dominance in the second. Involved here is a sensory system called the lateral line, which detects water movements and pressure changes, so the rhythmic movements are not only visual signals but potentially also mechanosensory signals and therefore multimodal [129, 130].

Apart from rhythmic courtship dances, we find other highly rhythmic and stereotyped movements in the schooling behaviour of many fish species. One example is the Firehead Tetra (*Hemigrammus bleheri*). In this species, synchronisation with the nearest neighbour is crucial for the emergence of the collective synchronisation of a school. Tail beating frequencies (rhythms) are very similar in neighbouring pairs of fish and are either in-cycle or out-of-cycle, which means movements are synchronized in both modes, but either in-phase or 180° phase shifted [131].

Different mating behaviours are also described as rhythmic for example in Whitefishes (Coregonidae) in the Baikal and Ob basin. In a mating act, two fish swim side by side, conducting a rhythmic parallel movement, during which the males strike the back of the female's abdomen. These strikes are coordinated rhythmically and different striking frequencies have been reported for different Whitefish species, averaging 17 Hz in *Coregonus migratorius* and *Coregonus pidschian*, or 25–27 Hz for *Coregonus tugun* and *Coregonus peled* [132].

In some blenny species, a rhythmic head nodding can be observed during different contexts, but this rhythmic nodding has not been quantified. *Parablennius rouxi* uses nodding during courtship rituals, while it can be observed in agonistic interactions in *Parablennius pilicornis*. The nodding has generally been described as a central part of the behavioural repertoire in social interactions in intertidal blenny species like *Coryphoblennius galerita*, *Lipophrys pholis* or *Salaria pavo* [133].

Another schooling example involving behavioural rhythms is found in *Anomalops katoptron*. This flashlight fish is exhibiting another multimodal behaviour:

synchronized swimming normally guided by visual cues is achieved during nighttime via synchronized and therefore rhythmic bioluminescent flashing displays [134]. This is only one form of rhythmic bioluminescent light pattern found in deep-sea fish, though. Many deep-sea fish species, such as lanternfish (Myctophidae) or flashlightfish (Anomalopidae), employ rhythmic bioluminescent light patterns as a form of communication. The specialized light-producing organs called photophores (for details on this production mechanisms, see Chap. 2 this volume) emit flashes or pulses of light in various patterns of which some have been described as rhythmic. Various purposes have been suggested for these displays including attracting prey, confusing predators, and communicating with potential mates or conspecifics without an ambient light source. The rhythmic flashing display has been quantified in *Photoblepharon*. In undisturbed fish kept in constant dark, blink rates changed diurnally, being higher during the day than during nighttime. Moreover, isolated fish showed individual 'basal' blink rates, for example, one fish showed 10 blinks per minute and a second 50 blinks per minute, and they increased to 40 and 60 blinks per minute, respectively, when the two fish could see each other indicating a role in interindividual communication [135].

3.4 Rhythmic Communication in Anurans

With over 7500 species, anurans (frogs and toads) are the largest and most vocal group of amphibians [136]. Most species communicate primarily or exclusively using acoustic signals [3, 137], and the raucous choruses consisting of hundreds or thousands of calling males are one of the great spectacles of nature. Some species have relatively short calls that are repeated at a high rate (60 or more calls per minute), creating the impression of a stereotyped rhythm. The stereotyped calling behaviour of the European Treefrog, *Hyla arborea* has enticed scientists to let males interact with a metronome and test whether they can adjust their calling to different beats of the metronome [138]. Yet, not all species show this stereotyped rhythmic calling behaviour—the acoustic structure of anuran vocalisations is diverse, as are the rhythms at which they are produced.

3.4.1 Sound Production

There are two sound production mechanisms known in anurans—air-driven sound production and contraction of bony rods [139]. Most species produce sound by passing air over vocal cords that then passively vibrate as a result [140, 141]. The airflow goes from the lungs through the larynx into the oral cavity. Unlike the vocalisation process of many other vertebrates, the air does not exit the mouth but cycles back and forth between the buccal cavity and the lungs. This back and forth is facilitated by the vocal sac, a flexible membrane of skin present in most male frogs and toads [142]. Some species call without a vocal sac, suggesting that the vocal sac is not required for making sound [143]. Sound production in anurans mainly occurs during expiration, that is, by air passing from the lungs into the vocal sac [140], but fire-bellied toads of the genus *Bombina* produce calls during inspiration [144], and *Discoglossus* produce intermittently inspiratory and expiratory notes [145]. Pulmonary pressure is correlated with call frequency (pitch), and frogs may vary the frequency of their calls by varying the pressure in their lungs [141]. Sound production is powered by cyclical contraction of trunk muscles [146], and in *Hyla versicolor* and *H. chrysoscelis,* the operating frequencies of the trunk muscles match the pulse frequency within a call [147].

A completely different mechanism of sound production evolved in the fully aquatic family Pipidae. These frogs lack vocal cords but have a box-like enlarged larynx with paired, movable bony rods that end in discs [148]. How these structures produce sound is still under debate [139], but they allow these frogs to call underwater for hours, and do so seemingly motionless because no air is shuttled between the lungs and the throat [149].

3.4.2 Rhythm in Frog Sounds

The acoustic structure of anuran vocalisations is diverse, ranging from simple to complex, and stereotyped to highly variable [150]. Calls can be simple, that is, composed of a single note of various lengths, can have various degrees of amplitude modulation ('pulses'), combine several similar simple or pulsed notes, or combine notes that differ in their temporal and spectral composition (complex calls [150]). The species-specific nature of frog vocalisation is well recognized now, and call descriptions are available for many species. Data on rhythmic behaviour relating to the pulse or note components of calls are available for a wide range of species. Data on rhythmic behaviour relating to the repetition rate of calls, call groups, or call bouts, on the other hand, show a dichotomous pattern. While data for call rate (number of calls per unit time) or its reciprocal measure, call period (time from onset of one call to onset of following call) is generally reported for species with high call rates (above 30–60 calls/min), species that call at slower rates or that are considered 'sporadic' callers generally do not have call rate data included as part of their call descriptions.

The variation in the temporal structure of anuran vocalisations can give rise to several rhythmic patterns. At one extreme are simple calls, where rhythm is only present in the repetition pattern of the call itself (i.e. call rate). Pulsed calls add the rhythm of the pulses (pulse rate) to the rhythm of the call. Calls that are composed of multiple pulsed notes may add a third or fourth rhythm. Sometimes calling is not continuous, but concentrated during calling bouts, time periods of varying durations that are separated by periods of silence, adding yet another potential temporal pattern.

How regular/stereotyped the rhythms are can also vary. Within-vocalisation rhythms (i.e. at the level of pulse rate) tend to be highly stereotyped, likely because it is important for species recognition in many species [3]. How quickly and regularly vocalisations are repeated (i.e. rhythm at the level of call rate) on the other hand, can vary greatly between species. Some species have relatively short calls that are repeated at high rates of 60 or more calls per minute, giving the impression of a highly stereotyped rhythm. The North American Green Treefrog (*Hyla cinerea*) has simple, short one-note calls (150 ms) that are repeated quite regularly every 800 ms for many hours (Höbel, personal observations). Other frogs, like the Neotropical Gliding Frog (*Agalchnis spurelli*), have short one-note calls (200 ms) that are repeated at relatively regular but long intervals (every 16 ± 1.5 s) [151]. North American Green Frogs (*Rana clamitans*) repeat their calls every 60–90 s [152], time frames over which the impression of rhythmic calling is lost, at least to the human observer. Finally, some anurans, like the Chinese Giant Spiny Frog, *Paa spinosa* [153] or the Neotropical Rainfrog *Eleutherodactylus fitzingeri* [154] vocalize so sporadically that the call rate, although readily available from the original recordings, is not even reported in the call descriptions. An interesting question for future studies will be to expand call descriptions beyond the vocalisation itself and incorporate information on the rate and consistency with which the vocalisations are produced. Seemingly a-rhythmic callers might in fact show rhythmic behaviour, but at a rate that is too slow for listening humans to pick up on.

3.4.3 Rhythmic Interactions Between Calling Males

The conspicuous nature of the large frog choruses formed by many temperate anurans may suggest that this is the predominant form of social organisation in anurans. Yet, many species are territorial and call from within their territories, like the Poison Dart Frogs [155], or have developmental modes (i.e. direct development in terrestrial eggs) that emancipate them from aggregating at ponds and allow them to call dispersed in their habitat [156, 157]. Vocal interactions between males are nevertheless common. Many male anurans tend to space their calls so that the vocalisations of one individual do not overlap those of nearby individuals [158, 159]. Depending on the signal timing rules of the respective species, this can lead to antiphonal alternations between two calling males (duets), as well as to trios, or quartets in a chorus [160, 161]. Call responses leading to alternation can come about because the internal pacemaker that generates an individual's calling rhythm can be altered by interactions with other males. A rival's call inhibits call production until sometime after the end of the rival's call [162].

While some type of behaviourally mediated, rather precise call timing is likely the norm for chorus-forming anurans with stereotyped call repetition, avoidance of call overlap in species where males call at irregular intervals is likely due to the low signal-to-noise ratio [163]. There is also a small number of frogs that have been reported to actively overlap their vocalisations. Some species with simple one-note calls show actual synchronisation of their calls: *Kassina fusca* [164], *K. senegalensis* [165], *Buergeria japonica* [166]. Some species that produce longer pulsed calls (*Assa darlingtonia* [167]), call groups or call bouts (*Diasporus diastema* [157]), do overlap the call/call groups/bouts, but the pulses/calls themselves are alternated. There are also some species that have complex calls consisting of an introductory note and variable number of secondary notes. In *Smilisca sila*, the secondary notes are so short and fast that they cannot be alternated with those of other males, resulting in overlap of all call portions. But in *Cochranella granulosa* [168] and *Hyla ebraccata* [169], the first pulse overlaps but secondary pulses may alternate. For the latter two species, it has also been reported that the lagging male may add on pulses to the end of his call, resulting in his last pulses being free of overlap [170].

3.4.4 Other Rhythmic Behaviours: Toe Tapping

A number of anuran species exhibit rhythmic gestural displays, either involving their entire arms and legs (arm waving, foot flagging) [171–173] or behaviours involving only fingers and toes. Toe tapping has been documented in a number of species across different families but is best documented for diurnal Poison Dart Frogs (family Dendrobatidae) and toads (family Bufonidae) (reviewed in [174–176]). Although a potential role in courtship, aggression or simply being a 'displacement behaviour' has been discussed in the literature [177–179], the most likely role of toe tapping seems to be in prey capture [174, 176, 180]. Several studies have demonstrated that toe tapping increases up to ten-fold in the presence of prey [180–183]. Slower tap rhythms of 300 taps/min (in cane toads, *Rhinella marina* [180] may function as visual lures for specific prey items (particularly other toads). Faster rhythms of 500 taps/min (in Dyeing poison frogs, *Dendrobates tinctorius* [183]) seem to serve more to agitate prey. Movement is essential for the detection of prey items in anurans [184], and tapping may stir small arthropods via substrate vibrations and thereby facilitate prey detection by toe-tapping frogs [183].

3.5 Rhythmic Communication in Avian Species

If asked to identify animal RC, birdsongs and elaborate courtship displays are likely the first things to come to mind. There are over 11,000 identified species (though some estimate the number to be closer to 18,000), divided into 41 orders and 250 families. Of these, the passerines are the most well studied in terms of acoustic communication, often with detailed descriptions of song structure, as well as dance. Here, we give a general overview of the function of rhythm in both birdsong and dance. Additionally, we review nonvocal acoustic signals in other families. Rhythm research in avian species also spans experimental work on rhythm perception and synchronisation in parrot species. As this chapter focuses on functional and structural aspects of RC, this literature is not reviewed here—for discussions regarding this research, we refer the reader to Chaps. 7 [185] and 14 [186].

3.5.1 Rhythmic Communication in Birdsong

Birdsong, perhaps more than songs of other taxa, tends to share many features of music, such as categorical rhythms [187], prosody [188], and *accelerando* [189]. With over 4800 species of songbirds and a far greater number of song types, birdsongs are diverse. They range from simple to complex, and from stereotyped to highly variable. A song might be composed from one or more rhythmic patterns. Some songs contain sections with rapid, rhythmic repetition of notes or phrases, known as trills or buzzes. Some songbirds are closed-ended learners, while others are open-ended, with varying degrees of vocal plasticity. Some species sing singly, while others sing in duets. This structural and behavioural diversity makes singing behaviour an important focal point for RC research.

3.5.1.1 Rhythmicality in Solo Male Birdsong

Many species from temperate zones exhibit sexual dimorphism with respect to size, colour, and/or calling, with males doing most of the singing. The form of song depends heavily on the type of mating system and the degree of territoriality of the species. In species with polygynous or serially monogamous mating systems, mate attraction is one of the main functions of song; therefore, female choice is the driving force behind song evolution. In polygamous species, where females have a short time to evaluate males, songs tend to be highly extravagant and stereotyped [190], features that are regulated by steroid sex hormones seasonally (testosterone) as well as daily (oestradiol) ([191], as well as Chaps. 9 and 11, this volume).

Additionally, in territorial species, resident birds sing at dawn to broadly advertise to neighbours that they still occupy a given area [192]; therefore, song can act as an aggressive signal [193]. Some species perform song 'matching', whereby an individual produces the same signal type in response to a rival's signal [194]. The degree of matching has been associated with the motivational level of the signallers [194–196], and may allow for more direct comparison of singing ability. Alternatively, yellowhammers (*Emberiza citrinella*) show rhythmic differentiation between neighbours [197].

The importance of sexual selection on song structure has led to the 'performance hypothesis', the idea that a male's quality can be assessed by his singing ability. Rhythm perception is flexible with regard to rate [198]; therefore, the production rate of rhythms (segments or categorical rhythm patterns) can be flexible. Rate plays a role in maximising the performance of elements that are performed close to their physical limits. These often involve a production cost or a trade-off, such as between repetition rate and frequency bandwidth in trills [199] or the internote interval and frequency shift in nontrilled songs or song segments [200] (see Chap. 9 [201] for a discussion of production constraints).

Increased production rates of trilled and nontrilled songs have been observed in intruder playback studies [202–204] and male recipients discriminate between singers with different performance levels [205, 206]. Females show preferences for impossible trills in playback studies [207] and solicit males with faster song production in natural situations [208]. Song performance improves with age, size, and body condition ([203, 209, 210], as well as Chap. 9 this volume). Together, these indicate that rhythm is important for inter- and intrasexual communication.

Rhythm may also play a role in performance evaluation through stereotypy, how reliably a rhythm is produced, and/or by contributing to Gestalt aesthetics. The degree of consistency of notes within songs changes with age [211, 212], with high consistency associated with greater response of territorial males and prospective mates to playbacks [213, 214] and greater reproductive success [215]. Studies that include temporal consistency show similar results [216, 217] though more research is needed. Goller proposed that females judge song quality based on aesthetic processing, which he called the song's 'acoustic Gestalt' [218]. Song production is complicated, with many trade-offs; therefore, it is unlikely that any male will be good at producing all elements well [218]. Rather, females may be choosing mates they perceive to have the 'sexiest' voice based on individual aesthetic preferences rather than a comparison of a specific element [218, 219].

3.5.1.2 Rhythmicality in Solo Female Birdsong

Studies of solo female singing are relatively rare, though female song is common in the tropics and females sing in 64% of all songbird species with sufficient data to determine which sexes sing [220]. The functions of solo female song are still under preliminary investigation [221]. Elaboration of song may depend on function, with songs used for interpair communication being less elaborate [222] than those used for intrasexual competition [223, 224] or for mate attraction and maintenance [225, 226]. However, there is little to no data on how rhythm functions in female song.

3.5.1.3 Rhythmicality in Avian Duets

Duetting is common in socially monogamous tropical passerine species, members of the Ralidae family (rails) that hold year-round territories [227], and group-living passeriforms [228]. Duets are formed when one member of the pair initiates and the other chooses to respond. The simplest duets are uncoordinated, overlapping singing bouts. Rhythmicity emerges when duets are sung simultaneously or antiphonally, with high degrees of coordination between singers [229]. The degree of coordination differs by species. For example, coordination in Neotropical wren species with longer breeding seasons shows higher coordination during duetting than species with shorter breeding seasons [230]. Both the timing and type of syllable may be produced according to a 'duet code' which may be pair specific or shared with neighbours [229, 230], though many species can produce content [231] and vary timing flexibly [228].

In many species, duetting occurs in the form of rapid turns, on a syllable-to-syllable basis, with sub-second latencies [232–234]. The mechanisms of producing these antiphonal duets range from entrainment—where the quality of the duet depends on the temporal precision of each individual's syllable production rather than mate familiarity [235]—to experiential learning—where the quality of the duet improves as each member learns to adjust to the other over time [230]. Neuronal activity in the HVC (proper name of an avian song hub) alternates in an excitatory/inhibitory pattern between the brains of mated pairs, providing a brain-coupling mechanism that supports precise timing in dyadic interactions [236, 237]. Likewise, many species can distinguish between coordinated and uncoordinated duets [238]. Extracellular recordings in the premotor cortex (i.e. HVC) of anesthetized plain-tailed wrens (*Pheugopedius euophrys*) exposed to typical and manipulated song stimuli show heightened sensitivity to intersyllable intervals [234]. Some wren species can adjust the timing and the phrase type they produce in response to their mates [239, 240]. Hence, for many species rhythmicity matters, being central to eliciting behavioural/neural responses and thus to fulfilling the putative function ascribed to the respective signal.

Recent trends favour a multifunctional view of tropical bird duetting that encompasses cooperative functions (joint resource defence, signalling commitment, maintaining contact) as well as mate guarding [241–244]. The multifunctionality framework is supported by the flexible production of duets observed in many species. While there are clear structural differences between interpair and intrapair communication, whereby interpair duets tend to be high amplitude and often contain sex specific content [228], the role of rhythm in either is not clear. For example, Rufous hornero (*Furnarius rufus*) pairs can alter the degree of temporal coordination in response to intruders [245]. However, coordinated duets do not necessarily evoke a stronger behavioural response in listening birds [238, 245]. It is possible that temporal coordination is serving an intrapair function even within an interpair context. Intrapair duetting is often low amplitude and likely under-reported [246], leaving a dearth of data.

While most duets happen when pairs are within 10 m of each other, duets do happen at distances at which timing is affected by speed of sound delays [239, 243]. There is some evidence that birds adapt the timing of their duets to each other [234] and therefore might be able to compensate for timing delays resulting from distance (see Chap. 6 [48] for further discussion of synchrony constraints). Multimodal displays are important in this context. In 84% of 20 species reviewed in [228], duetting was accompanied by simultaneous bodily displays. Visual displays help pairs coordinate duets when distance is a factor [247].

3.5.2 Rhythms in Nonvocal and Multimodal Courtship Displays in Birds

RC is not necessarily restricted to vocalisations. Conspicuous body movements that are mainly performed in courtship contexts often produce multimodal signals (i.e. visual, acoustic, and tactile signals). Several studies have reported that birds produce nonvocal sounds through feathers (e.g. greater sage-grouse [248], hummingbirds [249], snipes [250], manakins [251]), bills (e.g. storks [252], Java sparrows [253], woodpeckers [254]), and feet (e.g. cordon-bleus [255], palm cockatoos [256]). Vibrations via substrate can also serve as a communicative signal [257]. Here, we introduce several examples of multimodal RC in birds and discuss their possible functions.

3.5.2.1 Drumming Behaviour in Woodpeckers

The most well-known example of a nonvocal rhythmical signal in birds is probably the drumming display by woodpeckers (family Picidae). Woodpeckers produce rhythmic nonvocal sounds by rapidly hammering their bill on trees. This behaviour primarily contributes to nest excavation and foraging but also functions as a social signal, called the drumming display. The speed of the drumming display for communication is generally faster than the drilling speed of foraging and nest excavation [254], reaching around 20 beats per second [258] and varies within and across genera [254]. These facts suggest that the rhythms of drumming sounds play a key role in woodpecker communication. Many species have been reported to increase drumming behaviour when an individual's breeding territory is invaded by another (reviewed in [254]). Moreover, playback studies carried out with multiple species show that playback of drumming readily elicits aggressive behaviours—calls, attack flights, drumming—from both male and female resident birds. Besides such close agonistic encounters, individuals also advertise territory at a specific time of the day, usually at dawn [254, 259–263].

In recent studies, there has been a focus on the effects of drum speed, length, and rhythm during territorial contests, showing that parameters such as speed and length vary considerably within populations and among species [258, 261, 263, 264]. For example, spontaneous dawn drumming are shorter than aggressive drumming elicited by territorial intrusion [265]. Playback of drum bouts longer than the typical speed of a species elicits heightened aggressive displays in both males and females, suggesting that drumming is adjusted as a function of context to regulate signal effectiveness or potency [261].

Fast rates elicit more vigorous aggressive displays, even when manipulations are in the order of 8 ms [263] and receivers attempt to match such rapid drums by speeding up their own drumming [254]. Such data suggest high sensitivity to drum tempo and at least some degree of motor control/flexibility of drumming output. With respect to tempo functionality, it has been proposed that a higher drum tempo signals a more potent threat from an intruder.

There is also evidence that woodpeckers are sensitive to the rhythmic patterning of the signal, which appears to be functional for recognising the species. For example, downy woodpeckers (*Dryobates pubescens*) cease to respond when a downy woodpecker signal is manipulated so that its typical pattern of linear rate deceleration is replaced by a pattern of exponential rate acceleration [263]. They respond aggressively when the accelerating pattern is heard as part of the drumming of sympatric red-bellied woodpeckers (*Melanerpes carolinus*). Lower behavioural responses are obtained in other species as well, when drumming speed, cadence and acceleration patterns are manipulated [260]. Large-scale comparative analyses suggest that rhythmic pattern encodes species identity in woodpeckers [266], where sympatric sister taxa are more likely to have distinct cadence and, to a lower extent, distinct acceleration patterns compared to allopatric sister taxa.

3.5.2.2 Drumming Behaviour in Other Birds

Some birds also exhibit behaviours that produce nonvocal sounds, called 'drumming', in a manner different from woodpeckers. Albatrosses and storks are known to produce drumming-like sounds by clattering their bill, as described in Chap. 4 [2]. Male ruffed grouses (*Bonasa umbellus*) drum by beating their wings 40–50 times in rapid succession over 8–11 s and produce a low-frequency sound [267, 268]. The drumming behaviour is assumed to attract females and defend a territory, just like in woodpeckers. Drumming speed and duration vary not only among individuals but also within individuals over the years [267]. Ruffed grouses increase their drumming display when drumming sounds are played [267]. The drumming sound has a different directionality compared to vocalisations. It radiates primarily along the front and back of the displaying bird, which differs from vocalisations that radiate primarily toward the front. It is still unclear what the function of this directionality of drumming sounds is, but they may more accurately demonstrate their location and territory by emitting highly directional sounds in multiple directions at multiple locations in their territory [268].

Palm cockatoos (*Probosciger aterrimus*) perform a unique courtship display called drumming. The male palm cockatoos drum with a modified stick or seedpod on hollow trees/tree branches as intersexual displays, sometimes as part of more elaborated multimodal displays that may also feature calling, darkening of red cheek patch, crest erection [256, 269, 270]. This drumming exhibits a regular pulse, with intertap interval varying from 0.09 to 2.77 s, both within and between individuals. Drumming occurs predominantly when a female is present. Palm cockatoo males exhibit individual-specific drumming patterns, especially with respect to intertap interval [256]. These facts suggest that the male palm cockatoos advertise individual information such as physical and cognitive abilities toward surrounding potential mates and rivals via drumming.

3.5.2.3 Rhythmic Dance Displays in Songbirds

Conspicuous dance displays by birds are often comprised of rhythmic and repetitive body movements like bobbing. Estrildid finches are a well-studied songbird family in terms of singing behaviour, from neural mechanisms to communicative functions. They sing and dance simultaneously as a courtship behaviour in males and some females [271]. At least 35 species of Estrildid finches are known to perform repetitive bobbing [257], and the bobbing sometimes produces multimodal signals in addition to visual signals. In blue-capped cordon-bleus (*Uraeginthus cyanocephalus*), for instance, both males and females sing songs while performing rhythmic bobbing, which produces distinct nonvocal sounds. By recording these displays with a high-speed video camera, we previously discovered that in addition to bobbing, their visual courtship display includes quite rapid step-dancing, which is assumed to produce vibrations and nonvocal sounds [255, 272]. One bobbing spans one to six steps which are almost invisible to the naked eye. One step takes about 20–40 ms at an estimated rate of 25 Hz to 50 Hz. The bobbing speed is on average 1.39 bobs per second. The dance performances do not differ between sexes but vary among individuals. The bobbing speed gets faster when an opposite-sex individual is nearby [255]. Further, the dance display is promoted in the presence of an audience in addition to a potential mate [273]. The effective range of multimodal signals by dancing (i.e. visual, nonvocal sounds and vibrations) is assumed to be narrower than songs [272, 274]. The dance display might play an important role in close-range communication on the same perch or at least within sight.

Simultaneous song and dance displays can influence each other's rhythm. In blue-capped cordon-bleus, songs with dance displays were longer and more stereotyped than songs without dance displays [275]. Bobbing speed gets faster when accompanying songs. In addition to these, there are several reports on song and dance interactions in songbirds. In male brown-headed cowbirds (*Molothrus ater*), the movement of the wings is inserted in the silent parts of the courtship song [276]. Dance displays (swing and wing spread) by male montezuma oropendolas (*Psarocolius montezuma*) are each choreographed to coincide with the loudest note and lowest peak frequency of the song, respectively [277]. Male superb lyrebirds (*Menura novaehollandiae*) match a specific dance display to each of the different songs [278]. Male Java sparrows (*Padda oryzivora*) produce bill-click sounds several times during the song, which is coordinated with the song-note sequences [253]. Although these dance behaviours may not be in themselves rhythmic, they may contribute to the rhythmical production of vocalisations. Much is still unknown about how multimodal and multicomponent signals in courtship behaviour interact and what their communicative function is. Investigating them would provide new insights into how and why birds generate rhythmical signals.

3.6 Rhythmic Communication in Mammalian Species

Mammalian species are diverse, making broad generalisations difficult. Vocalisations and other rhythmic behaviours are produced in a variety of forms and contexts, and the state of knowledge also differs greatly between taxa. For many taxa, temporal measurements and perceptual studies have not been done. For these taxa, we review vocalisations and behaviours that have been described as repetitive or rhythmic in the literature for major taxa divisions. Therefore, each taxonomic group is presented differently.

3.6.1 Rhythmic Communication in Chiroptera

3.6.1.1 Acoustic Rhythms

Chiropterans, encompassing more than 1450 known species today, are one of the biggest mammalian orders surpassed in species richness only by rodents. They are characterized by being the only mammals with the ability to fly, furthermore, a large sub-group (Yangochiroptera) also possesses the ability to echolocate [279]. We know a lot about the temporal structure of echolocation calls. Same as in odontocetes, echolocation clicks or pulses are produced in a rhythmic pattern that depends on the context. Generally speaking, during foraging, we can observe a search phase, with very consistent interonset intervals and a final buzz phase with much faster intervals. Concrete Inter-Onset Intervals (IOI) depend on the species and environment. In many bat species, it is known that—if sensory needs allow—echolocation production is often coupled to wingbeat frequencies [280–284]. Echolocation frequencies also depend on species size and feeding ecology, for example, we see differences between bats in open vs. closed habitats (for details, see Chaps. 6 [48] and 9 [201]).

Less is known about the temporal structures of other acoustic signals, though many bat species produce a diversity of social communication sounds in addition to echolocation calls. Concrete rhythms were analysed in at least three different social contexts – territorial song, isolation calls and babbling of pups – in two different species. The greater sac-winged bat *Saccopteryx bilineata* is a vocal learning species and rhythms were analysed in the learned territorial song [285] produced by adult males. We find rhythm between 10 and 20 Hz, corresponding to IOIs between 100 and 50 ms, respectively [286]. On a larger timescale, the number of long buzz syllables in a territorial song, a parameter potentially influencing the rhythm, is positively correlated to reproductive success in this species [287].

Similar rhythms and IOI ranges can be found during the vocal ontogeny in a babbling phase, which is reminiscent of human infant babbling [288, 289]. Even in the innate isolation calls of this species, roughly the same intervals between sound elements were found [286]. All the rhythms match the rhythms of echolocation in the search phase, which is coupled to the wingbeat as in many other bat species (for details see Chap. 9 [201]). This is intriguing, because all other three contexts are produced, while animals are not flying.

In another species, such as *Carollia perspicillata*, isochronous rhythms were found again in innate isolation calls, produced with frequencies of 20–30 Hz and, thus, IOIs of 50 ms to 30 ms respectively. In a small dataset of echolocation calls, rhythmic frequency ranged between ~10 and 43 Hz, potentially indicating a less strict coupling of rhythms between different acoustic signal types [290, 291]. Exciting findings from adjacent areas of research suggest high temporal control of vocal output in bats [292–294], coordinated neural activity between individuals in social and communication contexts [295, 296] as well as strong individuality in acoustic signals of individuals [297, 298]. Together with a solid knowledge of underlying neural circuitry [299, 300] this makes bats an even more interesting group to study rhythmic communication in.

3.6.1.2 Bodily Rhythms

The proboscis bat *(Rhynchonycteris naso)* shows a peculiar group behaviour that was described as synchronized. It is to be noted that the term 'synchrony' is used very loosely here, describing a coordinated co-occurring behaviour in a group of bats. The bats roost on tree trunks, in broad daylight. Sometimes, many or all the group members can be seen participating in a behaviour that can be described as gentle rocking or swinging. The behaviour occurs more frequently with higher wind velocity and is therefore thought to resemble parts of the surrounding vegetation swaying in the wind and to function as a concealing mechanism [285]. Accompanying the rocking, sometimes also synchronous urinating and grooming can be observed; this might serve the function to hide otherwise conspicuous movements, necessary for grooming or urinating [285].

3.6.2 Rhythmic Communication in Rodentia

Rodents form the largest mammalian order, counting more than 2000 species in over 400 genera, which represents 40% of all mammalian species. They live in a diversity of habitats on all continents (but Antarctica) and exhibit a diversity of morphological traits and forms of social organisation. The evolutionary success of rodents is generally attributed to large litter size, short gestation period and early onset of sexual maturity—traits that also make them popular species in laboratory research, including research on vocal communication and auditory perception. While the latter has generated a vast literature, few studies have focused specifically on RC. To address this topic, in this brief review, we prioritize studies that (i) provide quantifications of rhythmic patterning in rodent communication; (ii) assess rhythm perception and/or the biological significance of temporal patterning with discrimination or playback experiments; (iii) describe rhythmic patterning of higher complexity; (iv) illustrate nonvocal rhythmic signals.

3.6.2.1 Rhythm Production and Perception in Laboratory Mice and Rats

Rhythmic ultrasonic vocalisations (USV) have been mentioned for several rodent taxa, for example, *Meriones*, *Peromyscus*, *Rattus*, *Baiomys*, *Scotinomys*, *Spermophilus*, *Tamiasciurus* ([301] and references therein). These have been noted to occur in a variety of contexts, including pup isolation calls, courtship and mating, social play or antipredator defence. The characterisation of rodent ultrasonic vocalisations as rhythmic is primarily based on subjective assessments rather than objective rhythm quantification, although there are also exceptions. For example, it has been recently confirmed that the courtship calls of mice males exhibit a hierarchically organized rhythmic structure, where single USVs are produced in groups, which, in turn, are combined into longer series [301]. Inter-USV intervals are shorter than 100 ms and correspond to silent inhalations, while intergroup intervals range between 100 and 275 ms, corresponding to a silent sniff. Two USV-types have been identified depending on temporal parameters, which parallels results from acoustic analyses of rat USV [302].

The stereotyped rhythms of mice courtship arise from USV-type collocation within a series, whereby consecutive short USVs predominate at series onset and offset, while consecutive long USVs predominate within a series. Modelling data suggests that this patterning has biological significance, as it is sufficient for discriminating male courtship calls from pup or female calls [301]. This corroborates previous findings that intervocalisation intervals in male mouse courtship calls are stable within- and between males [303] and extends findings from playback experiments showing that the temporal structure of pup calls is both sufficient and necessary for eliciting approach behaviour in lactating females ([304, 305] and references therein).

Rodent capacities for rhythm perception have been investigated experimentally in rats, using discrimination tasks. These studies show that rats can distinguish rhythmic isochrony from sound strings with irregular IOI [306] and may exhibit temporal pattern recognition. Rats could, for example, identify the rhythmic pattern of a habituated melody in versions with distorted spectral features and distinguish it from nonrhythmic versions [307]. In addition, there is evidence that rats can distinguish between metrical and nonmetrical rhythmic sequences [308]. Rhythm perception by rats is also supported by findings that rats can synchronize bodily movements (head bopping, licking) to audio-visual rhythms [309–311] and that neural activity in the rat auditory cortex entrains to the rhythm of musical excerpts [310].

3.6.2.2 Song Rhythms in Baiomyine Mice

Mice in the genera *Baiomys* (*B. taylori* and *B. musculus*) and *Scotinomys* (*S. teguina* and *S. xerampelinus*) are known to sing, that is, to produce highly stereotyped and spectro-temporally complex long calls [312]. Baiomyine songs are characterized by a progressive modulation of spectro-temporal features, with note duration, IOI, bandwidth and amplitude increasing as the song progresses [312, 313]. Both males and females sing, and singing is accompanied by a characteristic posture thought to amplify sound propagation, where the snout is oriented upward, and the individual often stands in an erected position on the hindpaws.

Scotinomys species produce songs that contain both sonic and ultrasonic notes, while *Baiomys* species emit predominantly ultrasonic songs. *S. teguina* exhibits the longest and most elaborated display, which is characterized by sexual divocalism in the temporal domain: male song lasts in average 9 s, comprising an average of 113 pulses, while female song lasts in average 6 s with an average of 84 pulses [312]. By comparison, the song of *S. xerampelinus* and the song of the *Baiomys* species last about 2–2.5 s, exhibiting minimal or no sexual dimorphism. Duty cycle is, however, higher in *S. xerampelinus*, averaging 35 pulses/s compared to *Baiomys* species where pulse/s ranges between 17 (*B. taylori*) and 20 (*B. musculus*).

The functional significance of the baiomyine song is not entirely resolved. The song exhibits robust propagation in the wild, which supports an *overall* function of conspecific localisation [312]. Data from *S. teguina* also suggest a role in reproductive behaviour, as the removal of a previously present female intensifies male singing, while female song is not influenced by male behaviour [314]. The empirical evidence is consistent with interpretations of sexual selection both by female choice and by intermale competition. Consistent with the former, signals subjected to sexual selection by female choice are typically highly conspicuous in size (i.e. larger or longer) and elaborated [315], as it is also the case in *S. teguina*. As vocal effort in this species is predicted by levels of leptin—a metabolic hormone that tracks long-term variation in adiposity and influences investment in costly traits—male song quality may broadcast male fitness quality [316]. This includes RC variables such as the ability to sustain the species-specific rhythmic pattern for a longer time or to produce it more often. Experimental data show that females prefer high-performance songs: they approach faster-rate songs more often, more readily and for a longer duration than slow-rate songs [317].

The intermale competition interpretation is supported by two lines of evidence: song effort and dominance are strongly influenced by androgen levels (further detailed in Chap. 11 [318]) and exposure to a male's song elicits counter-song by other males. Countersinging is a prolonged vocal interaction with accurate interindividual coordination [319], during which mice males time their contribution to avoid acoustic overlap. Experimental manipulations (based on, e.g. intracranial microstimulation, cooling, pharmacological treatment) indicate that rhythmic patterning and accurate temporal coordination is associated with motor cortex activity that emerges only in contexts that require dynamic temporal adaptation, that is, during countersinging [319].

3.6.2.3 Seismic Rhythms in Fossorial and Semifossorial Rodents

More than 250 rodent species spend a significant portion of their lives underground, exhibiting either a strictly fossorial or semifossorial lifestyle [320]. The underground environment presents unique constraints to mammalian life, given the lack of sunlight, low temperature, high humidity, reduced levels of ventilation and solid matter surroundings. In such environmental conditions, (semi)fossorial rodents have evolved a range of adaptations, including adaptations for communication through seismic waves, which is found in unrelated rodent lineages (for reviews, see [321–323]), including mole-rat species (family Spalacidae and Bathyergidae), gerbils (genera *Meriones, Psammomys, Rhombomys, Tatera, Gerbillus*), voles (genus *Microtus*), squirrels (genus *Cynomys*), agoutis (genera *Dasyprocta, Myoprocta*), jerboa (*Jaculus jaculus*) and kangaroo rats (genus *Dipodomys*). This trait is shared with nonrodent fossorial mammals, such as true moles (Talpidae) and marsupial moles (Notoryctidae), afrosoricid moles (Chrysochloridae), and Elephant shrews (*Elephantulus* spp.). The most widespread behavioural mechanism of mammalian vibrational communication is foot drumming, but (semi)fossorial mammals can also drum using their head (e.g. mole rat species [324, 325]), teeth (European mole vole, *Microtus pyrenaicus* [326]) or chest (giant mole rat, *Fukomys mechowii* [327]).

The seismic signals of (semi)fossorial mammals are delivered as either singleton foot- (head, chest) drums that are repeated (the desert kangaroo rat, *Dipodomys deserti* [328]; gerbils: [329]; mole rats [330]) or as drum rolls composed of a variable number of pulses, which are given in repeated bouts, forming a series. Temporal parameters are available for most of these signals, including total drumroll duration, number of pulses in a drumroll, IOI, number of drumrolls in a series, and interdrumroll duration. Generally, seismic signals are described in terms suggestive of rhythmic structure, that is, as exhibiting regular temporal patterning, stable periodicity, constant acceleration or deceleration, or combinations of these [325, 330, 331]. However, proper rhythm quantifications are currently lacking.

The drumming bouts of subterranean mammals are generally brief, spanning a few seconds, although longer drumrolls have been also documented, for example, in the giant kangaroo rat, *D. ingens* [331]. By contrast, duty-cycle tends to be high, averaging 20 foot-strikes/s in the giant kangaroo rat [331] or 23 head strikes/s in the demon African mole rat (*Tachyoryctes daemon*) [325]. IOI durations are, consequently, very short, and appear to exhibit low variability, which is suggestive of rhythmicality. By contrast, interdrumroll intervals are more variable and tend to have suprasecond durations.

Whether and how socioecological variables influence vibrational communication in subterranean mammals remains to be established. Seismic communication has been reported in rodents that occupy habitats of great diversity with respect to soil composition, hardness, humidity, temperature, and food resources. On the sociality spectrum, seismic signals are primarily found in solitary species, but have also been reported in highly social species such as gerbils, the giant mole rat and eusocial mole rats [322, 323, 332]. Within-genus comparisons indicate that the temporal parameters of drumming encode species identity in some gerbil and kangaroo rat species [329, 331, 333]. For similar findings in a nonrodent genus, that is, African elephant shrews, see [334]. Moreover, in mole- and kangaroo rat species drumming also encodes individual identity, which is based on temporal/rhythmic plasticity [325, 331]. The banner-tailed kangaroo rat (*Dipodomys spectabilis*), for example, maintains life-long drumming signatures which can, however, be updated when the individuals move into new areas or when population density increases. Statistical analyses show that males develop drumming signatures that are distinct from their mothers' drumming patterns, which has been interpreted as a mechanism of inbreeding avoidance [335]. Moreover, playback experiments indicate that kangaroo rats discriminate between the drumming patterns of familiar and unfamiliar neighbours (reviewed in [332]).

Seismic signals occur primarily in intraspecific communication. Solitary mole-rats [324, 325, 330] and kangaroo rats [321] drum to advertise territorial occupancy, which may play a spacing function to minimize the costs of direct encounters, given the high social intolerance of many fossorial species. In blind mole rats [336] and banner-tailed kangaroo rats [335], males also engage in drumming exchanges as part of territorial challenges, while in gerbils and kangaroo rats males engage in intense drumming competitions within the home area of an oestrous female [332]. These interactions are energetically demanding: a single male kangaroo rat may drum for hours, producing thousands of foot rolls [321]. Drumming duration, however, does not seem to be related to reproductive success. Mole- and kangaroo rats, as well as gerbils of both sexes may drum to coordinate mating, in sexually dimorphic ways [330, 335]. In gerbils and kangaroo rats, male precopulatory drumming is described as 'softer' than drumming produced in other contexts [332], and may also occur after mating, as a mate-guarding signal [335]. Interestingly, in the Cape mole rat (*Georychus capensis*), males and females engage in drumming interactions, during which the sexual dimorphism of drumming patterns is attenuated through interindividual temporal adaptation and mutual entrainment [330].

Less commonly, drumming also occurs in response to disturbance (for gerbils, kangaroo rats see [332]; for mole rats see [325]) or in response to specific predators. In the social great gerbil (*Rhombomys opimus*), both males and females produce alarm signals through drumming, with temporal patterning that varies depending on degree of risk and response urgency [337]. Gerbils, prairie dogs and kangaroo rats are known for drumming when detecting snakes, which may play a predator-deterrent role [321]. Finally, drumming has been suggested to function as a kind of subterranean echolocation in blind mole rats. This species produces digging knocks that exhibit a distinct temporal patterning compared to the knocks used in intraspecific communication and are given at increased rates when the rats dig around obstacles [338].

Given the diversity of production mechanisms, contexts of usage, and variable socioecology of rodent species that engage in seismic communication, it is generally agreed that this communication form is not phylogenetically determined but has evolved by convergence in unrelated lineages. In addition, it is commonly speculated that mammalian seismic signals originated in the ritualisation of instrumental behaviours (e.g. stress response, chasing rivals, digging), which have been moulded into stereotyped, patterned emissions for the purposes of communication [332].

3.6.3 Rhythmic Communication in Cetacea

Cetaceans live in an ecosystem where they must breathe voluntarily, adjust to strong sound propagation effects, locate prey, predators, and conspecifics in three-dimensional, low-visibility space, and coordinate actions in temporary or long-term social groups. As a result, cetaceans have developed multiple sound production mechanisms, use sound flexibly to hunt prey (e.g. via echolocation), and demonstrate vocal convergence, imitation, and innovation. Such vocal plasticity makes them an ideal taxon for studying the use of rhythm in vocal communication. However, the same constraints that likely have led to the evolution of their vocal communication signals also make them difficult to study - i.e. most of their behaviour occurs underwater.

3.6.3.1 Acoustic Rhythms in Mysticetes (Baleen Whales)

Much of what is known about mysticetes acoustic signals comes from analyses of autonomous remote recordings. Contextual observations are limited and, therefore, the function of calls is difficult to determine. Contexts can roughly be divided into three categories: short-range contexts such as during social/sexual behaviour, foraging, and within surface-active social groups, medium-range contexts where reciprocal calling occurs between separated individuals and mother/calf pairs, and long-range communication where the signal travels many kilometres.

Short-range and medium-range calls are often described as nonsong vocalisations and are identified by their shorter duration and often singleton nature. However, some nonsong signals are described as pulses and occur as singletons or in short series (e.g. fin whales, *Balaenoptera physalus* [339]). For example, gray whales (*Eschrichtius robustus*) produce a pulsed call, called a knock, that has been observed during social/sexual behaviour [340]. Humpback whales (*Megaptera novaeangliae*) produce a cry, a stereotyped tonal call series, during coordinated [341] or solitary feeding on herring [342]. Additionally, seven types of repetitive calls were identified in active social groups of migrating humpbacks [343], as least one of which was heard mostly after two groups joined [344].

Contact call exchanges involve the repetition of a single call type, often alternating between individuals, ending when the individuals are reunited (for review see [345]). Intercall intervals are rarely reported in the literature. Contact situations are presumably contextual and therefore may be repetitive without being rhythmic. However, differences in mean interunit intervals are one of the primary characteristics differentiating contact call types in Bryde's whales (*Balaenoptera edeni*) [346] and the contact calls of a Bryde's whale calf consist of a series of broadband pulses [347]. Further research into the temporal nature of these interactions is needed across mysticete species.

Songs, long stereotyped sequences of sounds, have been described in most mysticetes, except for gray whales, Bryde's whales, right whales (*Eubalaena* sp.), and pygmy right whales (*Caperea marginata*). The structure of songs differs greatly between species from the single repeated call of Sei whales (*Balaenoptera borealis*) [348] to the complex songs of humpback whales and bowheads (*Balaena mysticetus*) [349, 350]. Unlike the songs of most birds, most whale songs have no definitive start and end point and may be sung without breaks for minutes to hours, though for more complex songs, motifs are repeated in a predictable pattern. Little research has been done on the temporal patterning of whale song. The temporal patterns of rorqual songs have been assumed to be fairly stable; however, recent findings indicate fin whales change the internote intervals of their songs across breeding seasons [351] and pygmy blue whales (*Balaenoptera musculus brevicauda*) show temporal variability between and within individuals [352]. Humpback whales flexibly use a variety of temporal patterns including several types of stable patterns, accelerating and decelerating patterns, and less structured patterns; changes in temporal structure do not always align with changes in acoustic structure [353].

Song is presumed to have a reproductive function as it is most common during the winter breeding season, often produced by single individuals, and (almost) all sexed singers have been male (for a review see [345]). However, it has also been suggested that song could act as sonar [354, 355]. Recently, fin whale song has been used by humans to determine the crustal structure of the ocean floor [356]. How song functions within the reproductive context is largely unknown, with most of the evidence coming from observations of humpback whales. Observations from the breeding grounds, where singers are predominantly solitary, largely lean toward an intrasexual function, in which song serves to space singers from one another [357], as well as mediate male-male interactions between singers and nonsingers [358, 359]. Evidence from migrating whales, on the other hand, suggests an intersexual function, in that singers escort females during migration [360] and males escorting females are less likely to sing in the presence of potential competitors [361].

3.6.3.2 Acoustic Rhythms in Odontocetes (Toothed Whales)

Odontocetes flexibly produce and use a variety of tonal sounds and broadband clicks. Like mysticetes, several odontocete species produce call series that are thought to function as contact calls and for maintaining group cohesion and coordination during traveling and foraging [362–364]. The degree of rhythmicality of these calls has rarely been measured. Both northern right whale dolphins (*Lissodelphis borealis*) and narwhals (*Monodon monoceros*) produce stereotyped, rhythmic burst-pulse series, the function of which is still unclear [365, 366]. Long-finned pilot whales (*Globicephala melas*) produce relatively evenly spaced call sequences when socialising, particularly in large groups, with call frequency increasing with group size (up to 100 s of individuals) [367].

Odontocetes also produce click trains, which can vary in rate and patterns (periodic accelerating, decelerating, or aperiodic) both within and between bouts. The most well-known functions of click trains are echolocation and navigation. Similar to bats, echolocation during foraging can be divided into search, approach, and terminal phases. Blainville's beaked whales (*Mesoplodon densirostris*) produce clicks with relatively consistent interclick intervals (ICIs; 300–400 ms) during the first two stages and buzzes (ICIs = 10 ms) during the terminal phase, with an acceleration of click rate as they approach the prey [368, 369]. Observations of both captive and wild animals reveal similar behaviour across odontocetes. Therefore, the rhythmic patterns produced are often a by-product of context and may provide useful information for eavesdroppers as well as the foraging individual.

Click trains can also be used for communication, particularly in species that do not whistle [370]. Communication click trains tend to be shorter in duration than foraging buzzes and are often termed burst pulses. They can have either shorter or longer mean ICIs than foraging buzzes, depending on the species [371–373]. Pulsed calls are not necessarily rhythmic; for example, spotted dolphins produce genital buzzes that range in repetition rate from 8 to 2000 clicks per second [374]. Many species produce burst pulses with short, consistent ICIs, which in some cases, are perceived by humans as a tonal 'cry' [375, 376]. For example, the burst pulses of White-beaked dolphins (*Lagenorhynchus albirostris*) have ICIs of only 1.39 (SD = 0.41) ms [377].

In general, pulsed calls are associated with high levels of activity such as fast swimming [376], aerial behaviours [378], play [379], aggression and courtship [374, 380], and sexual behaviour [381]. Click rates and rhythmic patterning can differ between contexts. For example, Risso's dolphins (*Grampus griseus*) use accelerating click trains for echolocation and decelerating pulsed calls (presumably) for communication [372]. Likewise, captive harbour porpoises produced fast, accelerating click trains during aggressive interactions, slower accelerating click trains during mother-calf separation, and variable interval click trains in other contexts [382]. Clicks are high intensity and occur in aggressive and sexual interactions, indicating that they may also serve as a form of tactile communication [383].

In social contexts, female sperm whales emit categorical rhythmic patterns of clicks (called codas) which can be produced at different tempos but fixed interval ratios [384]. The codas used differ between clans, with regional dialects [385, 386], and remain stable within groups over several years [384]. Individuals may differ in the timing of clicks within the most shared coda [387]. Hence, rhythm may be important for group and/or individual identification. However, recent analyses have revealed that tempo is important in dyadic interactions as lagging whales match the changing tempos of lead whales, opening up possibilities for more nuanced communication [388].

3.6.3.3 Synchrony in Cetaceans

Many cetaceans have been observed doing synchronized movements in a number of behavioural contexts such as traveling in social groups [389], mother-calf interactions [390], group foraging [391, 392], anti-predator (or threat) responses [393], displays by males within an alliance [394], and interspecific aggression [395]. Thus far, descriptions of these synchronized movements do not clearly indicate rhythmic movement, warranting further investigation. However, in a few cases, synchronized movements are accompanied by rhythmic vocalisations. For example, spinner dolphins (*Stenella longirostris*) produce search phase-like clicks when changing formation during synchronized foraging, though the clicks are not at the correct interval for detecting prey. They may be detecting the location of or communicating with other dolphins in the group to coordinate movements [396].

Observations of male cooperative alliances of dolphins reveal that dolphins from different locations coordinate their sound production. Indo-Pacific bottlenose dolphins (*Tursiops truncatus*) that have formed cooperative alliances coordinate pop-train production (low-frequency clicks used for threat) when corralling females. Coordination includes simultaneous production of pop-trains, with alternating pops, and adjustment of tempo [391]. Herzing [397] found that male pairs of Atlantic bottlenose dolphins produce synchronized or overlapping buzzes, often in combination with other pulsed sounds (brays) and whistles, when fighting other dyads or chasing females. Similarly, groups of 3–8 Atlantic spotted dolphins (*Stenella frontalis*) produce synchronized squawks and screams during intra- and interspecific aggression [397]. Thus, synchronized rhythmic vocal displays may reinforce cooperative bonds and/or communicate increased threat levels to out-groups.

3.6.4 Rhythmic Communication in Pinnipedia

The suborder Pinnipedia comprises 34 species [398] grouped in three families: Phocidae (earless, or true, seals; 18 species in 10 genera), Otariidae (sea lions and fur seals; 15 species in 7 genera) and Odobenidae (one walrus species). Molecular and genetic evidence supports a monophyletic evolution from a common musteloid ancestor, with Phocids being the first to diverge [399]. The pinnipeds have recently drawn attention as potential model species for research on the evolution of complex rhythmic capabilities, such as those deployed in human speech and music [400]. First, phocids and walruses exhibit flexible learning abilities (reviewed in [401]) and, according to the popular 'vocal learning-beat perception and synchronisation' hypothesis rhythmic synchronisation abilities are a by-product of flexible vocal learning, owing to enhanced audiomotor couplings required for the latter [402, 403] (see also Chaps. 7 [185] and 14 [186]). Second, and contrary to this hypothesis, one of the few nonhuman animals ever shown to accurately synchronize bodily movement to the beat of complex musical stimuli is a California sea lion (*Zalophus californianus*) named Ronan, that is, an individual of a species that is not considered to have flexible vocal learning abilities [404, 405] (for further discussions, see also Chap. 14 [186]). Remarkably, after being trained on metronomic stimuli, Ronan was able to predictively synchronise head bobs to rich musical stimuli at various tempos. Moreover, cross-species comparisons revealed that Ronan's synchronisation skills were just as good or even superior to those of American college undergraduates asked to entrain an arm movement of similar amplitude to the pinniped's head bobbing [404, 405].

Besides such trained rhythmicality, the pinnipeds exhibit rhythmic aspects in their naturally occurring communicative signals, and rhythm-focused studies have been conducted on both nonvocal [406, 407] and vocal signals [408–416]. Most knowledge about pinniped communication comes from studies conducted during the breeding season and concerns signals produced in two broad contexts: mother-pup localisation and male reproductive signals.

Mother- and pup localisation calls are most elaborated in otariids [417, 418], where mothers forage at sea during the lactation period and, thus, the stereotypic signature of the call aids mother-pup recognition and reunion. Generally, pinniped mother- and pup-localisation calls consist of serially repeated units, which, in some species—but not in others [419] (and references therein)—gives rise to a basic rhythmic pattern. It is currently unclear, however, if the temporal patterning of call-units plays any role in individual signatures and recognition.

Male reproductive signals with a rhythmic structure include bodily displays (found in members of all three pinniped families), stereotyped vocal displays (common in otariids), and elaborated vocal or mixed 'song' displays in odobenids and some phocid species [414, 418]. Typically, otariid vocalisations are airborne and given as staccato series of grunts, barks, etc., which may co-occur and synchronize with (seemingly) rhythmic bodily movements, such as forward lounging toward rival males or head-bobbing (California sea lion, the Steller sea lion, *Eumetopias jubatus*). The latter two species, as well as the northern fur seal (*Callorhinus ursinus*) also produce a range of underwater pulsed vocalisations of variable frequency and amplitude [418] (and references therein). The rhythmic structure seems to encode levels of fitness and/or arousal, thereby informing the audience on a male's energetic and/or motivational capacity to defend resources. Indeed, in sea lions and cape fur seals (*Arctocephalus pusillus pusillus*) within-call frequency is highest in high-arousal situations, such as female herding and male confrontation [415, 420]. A playback experiment further demonstrated a capacity for tempo perception in cape fur seals, whereby conspecifics tailored their vigilance behaviours depending on the broadcasted bark rate: high-rate and high-arousal confrontation barks vs. low-rate territorial advertising barks [415].

Odobenid males exhibit some of the longest and most complex reproductive displays known in mammals. Their highly stereotyped and multimodal song lasts 3–7 min and is repeated continuously for several hours and even days. This display consists of airborne and underwater vocalisations, and percussive signals, including bell-like sounds that are produced by striking throat patches with the front flipper and loud impulse sounds produced by bursting air bubbles produced by flipper clapping. Rhythm quantification has been carried out for clapping-induced cavitation sounds in a Pacific walrus (*Odobenus rosmarus divergens*) male, revealing remarkable isochrony, whereby bubble bursts occurred at a steady 1.2-s IOI in sequences that lasted from tens of seconds to minutes [406]. Extant data suggest, moreover, that rhythmic parameters—pulse number, pattern and rate—of the overall display exhibit individual variation and may encode information about individual fitness [406, 421].

The structure of the walrus song is reminiscent of that found in whales, in that basic elements (tap, knock, double-knock, triple-knock, bell-knock, coda) are assembled into patterned sequences that combine into larger sequences which are given in a specific order. Two basic elements are given as pulsed sequences, potentially giving rise to rhythms: taps and knocks. The knock is louder and slower (1–3 Hz), while the tap is less intense, but faster (10 Hz). Depending on the patterning of tap and knock sounds, walruses produce three types of song: diving, coda and strum cycles [421]. The song type—coda or diving—that predominates in a season may vary, with data suggesting that the preferred song-type of the female-attending male is adopted by nonattending males [422]. Certain sequences exhibit individual variation, whereas others are fixed. For example, the underwater portion of a diving song always begins with a so-called diving vocalisation: a rapid knock-series, followed by 100–300 slower pulses, which is concluded with a bell-knock. Song types have both a long, standard version and a short version where certain middle sequences are omitted [422]. The long versions tend to occur when males sing alone, while short versions are frequent when the males sing in proximity to attended females. It is believed that short songs minimize the time that female-attending males spend underwater, thereby minimising the risk that attended females interact with other males. Moreover, short versions entail an increased frequency of sequences that encode individual identity (surface coda, diving vocalisation), thereby suggesting that short song versions also entail a more intense schedule of identity and occupancy advertising [422].

The reproductive signals of male phocids include nonvocal displays (e.g. flipper slapping in the harbour seal [423], flipper clapping in grey seals [407, 424]), as well as airborne and underwater vocalisations, which, in some species (elephant seals, the hooded seal, *Cystophora cristata*) are amplified by resonating organs. The airborne calls of phocids are generally less elaborated with respect to repertoire size and structure (reviewed in [425]. By contrast, underwater displays exhibit high interspecific variation, whereby some species only produce a small inventory of simple, pulsed calls (e.g. spotted seal, *Phoca largha*; crabeater seal, *Lobodon carcinophagus*), while others have large inventories, including several song types (e.g. harp seals, *Pagophilus groenlandicus* and Weddell seals, *Leptonychotes weddellii*). This variability appears to be accounted for by socioecological variables rather than phylogeny (for more details, see Chap. 6 [48]).

In species that call underwater while holding their breath for the entire duration of the dive, the quality of vocal rhythms may function as an honest fitness signal, that is, of stamina and breath-holding capacity. In support of this hypothesis, in the leopard seal, the temporal parameters of rhythmic calls are individually distinctive and associated with male size [412]. Specifically, larger males produce more robust and stable rhythmic calling across the breeding season (which lasts for several months), while rhythmic stability in smaller males decreases progressively throughout the breeding season.

In other species, the males can emit a range of rhythmically structured call types, which may be assembled into longer song displays (harp seal, [409]; Weddell seals [410]). Harp seals exhibit three distinct rhythmic patterns, whereby the most common pattern is characterized by sub-second IOI, while the less common patterns exhibit either supra-second IOI or a patterned 'short-long' combination of sub- and supra-second IOI [409]. The sub-second type presents a variation with a short-long contrast, thereby exhibiting a similar structure to the mixed (sub–/supra-second) type, albeit at a faster rate.

Rhythmic patterns have been documented for 10 (of 34) underwater call types produced by Weddell seals, including an isochronous, an accelerating and a decelerating pattern, as well as several stable combinations of these basic patterns. The functional significance of these rhythmic patterns is unresolved, although it has been speculated that they may serve multiple functions, such as encoding contextual and emotional information and/or supporting species recognition [409, 410]. A recent case study of unprecedented detail suggests that the underwater vocalisations of Weddell seals primarily play a role in intermale competition, and secondarily in female choice [426]. It is, thus, plausible, that rhythmic patterning may encode information related to male fitness and motivational states. With respect to species recognition, call repetitiveness is hypothesized to provide an anti-masking effect in a noisy environment, while call predictability is hypothesized to optimize attentional and perceptual processes, in a way consistent with the dynamic attending theory (for details, see Chaps. 7 [185] and 8 [427]).

Finally, in species that emit airborne vocalisations (e.g. the northern elephant seal), rhythmic parameters—call pulse rate and number of pulses per call—function as individual signatures [411, 428]. These rythmic parameters guide the social navigation of males, minimising costly direct encounters as subordinate males promptly halt approaching females upon hearing the rhythmic pattern of a dominant male. Such decisions, however, are not simply based on cues that encode fighting potential or motivation. They are, instead, based on the memorisation of individual rhythmic signatures and social ranks that are re-negotiated on a yearly basis, through fighting, at the start of the breeding season [411, 428]. This suggests not only a capacity for rhythm (in particular, tempo) perception, but also a capacity for flexible associations between a rhythmic signature and its variable social significance.

As a final point in this brief review, we return to the vocal learning and rhythmic synchronisation hypothesis, whereby flexible vocal learning species are predicted to exhibit advanced rhythmic abilities, in particular beat perception and accurate synchronisation. Currently, aspects of rhythmic coordination have been investigated in the mother-pup contact vocalisations of two pinniped species—a flexible vocal learner (the harbour seal, *Phoca vitulina* [416]) and a vocal nonlearning species (the cape fur seal [419]). The outcomes of these investigations are relatively similar, that is, irrespective of vocal learning status, neither species exhibits rhythmic isochrony—considered to be the stepping stone of rhythmic synchronisation—in their spontaneous contact calls. Moreover, rhythmic synchronisation seems to be absent in naturalistic vocal interactions between harbour seal pups. The call of both species, however, exhibits features of asynchronous rhythmic coordination, whereby individuals flexibly time their vocal contributions to avoid overlapping with other conspecifics' calls. Such timing plasticity is reminiscent of the loose rhythmic structuring of human speech (as opposed to the remarkably tight rhythmic structuring characteristic of human music). This phenomenon is likely to be widespread in pinnipeds, considering demands for anti-masking mechanisms in noisy environments, especially when the main source of noise is represented by conspecifics, as it was the case in both studies. By contrast, in environments with predominantly abiotic—and thus temporally irregular—a noise sources, isochrony may be a better anti-masking mechanism, as exemplified by the stereotyped rhythms of leopard seals.

3.6.5 Rhythmic Communication in Terrestrial Carnivores

While terrestrial carnivores do produce some rhythmic or semi-rhythmic behaviours, research is lacking on the temporal qualities and differential usage of these behaviours. Below we highlight a few behaviours that merit further investigation. The discussion of potential acoustic rhythmic behaviour is arranged by broad vocal categories, generalising across species with the intent to encourage cross-species comparisons of rhythm and rhythm usage. In addition, tail-wagging is addressed. While tail-wagging occurs in a broad range of taxa, this review focuses only on the domestic dog (*Canis familiaris*) as they are easy to study in laboratory situations.

3.6.5.1 Terrestrial Carnivore Pulsed Calls

There are two types of pulsed calls that occur across several carnivore families: growls and purr-like vocalisations, growls being higher frequency (e.g. cheetah: ≈40 Hz [429]) and often higher amplitude. Growls are generally used for threat in a wide variety of species and contexts (e.g. canids [430]; jaguars and otters [431]), and contain information about body size (e.g. domestic dog [432]). Further research into the variability of growls within and across species is needed.

One of the distinguishing vocalisations of felids is their purrs, which likely are produced by all but six felid species as well as juvenile large-spotted genets (*Genetta tigrina*). Purring is a pulsed vocalisation (15–25 Hz) that is produced during inhalation and exhalation, and can continue for seconds to minutes. Purrs are produced during close contact social interactions and may communicate contentment, appeasement, or might be self-soothing [433, 434]. Purr-like vocalisations have been reported in a variety of species, including giant otters (*Pteronura brasiliensis*) [435], most Ursidae, beech (*Martes foina*) and pine (*M. martes*) martens, raccoons (*Procyon lotor*), binturong (*Arctictis binturong*), and possibly aardwolf (*Proteles cristatus*) [434]. These vocalisations differ from true purrs in that they are only produced during exhalation [434]. However, they are sometimes called purrs because they occur in similar contexts [435], such as after nursing, but may be sustained into adulthood (e.g. Ursidae [434]).

Other pulsed calls may be specific to more narrowly defined taxa. For example, oestrous females of the six nonpurring felids produce a precopulation pulsed call that has sometimes been described as purr-like [434]. Additionally, both dholes [436] and New Guinea singing dogs [437] produce a trill-like pulsed call, though the function of these calls is uncertain.

3.6.5.2 Canidae Repeated Calls

Canids produce several types of tonal sounds that can occur as singletons or in series with varying degrees of regularity that have not yet been quantified. For example, yelps, which are generally thought to be defensive or warning calls in antagonistic interactions [430], are produced in series by a number of species. The most common repeated call type in foxes is the bark, which can function as a contact call [438] or be used in mate and/or territorial defence [439, 440]. Barks can take many acoustic forms [441] and can vary within individuals in different contexts. Arousal can increase breathing rate which in turn affects the pace, amplitude, and number of repetitions [430]. On the other hand, swift foxes (*Vulpes velox*) can be individually identified using mean IOIs [442].

Several canid species produce short-duration, low-amplitude, tonal sounds in series, though no research has been done on the degree of isochrony of these calls. These vocalisations differ in fundamental frequency within and between species and are sometimes biphonated [436, 444]. While published spectrograms show that these sounds often occur in series, the series are rarely described. As a result, the categorisation, nomenclature, and proposed function are inconsistent within and between species (examples of names include squeaks, whines, whimpers, and wuhs).

In general, these vocalisations occur in prosocial contexts (e.g. adult-pup interactions, hanging out, greeting, and play), and are considered to be contact soliciting or of friendly intent [441, 443–445]. They may also indicate appeasement in status interactions [445, 446], alert others of one's presence (e.g. at food), and serve a reconciliation function after aggressive interactions [445]. However, other studies found that tonal call series are associated with frustration and/or arousal in captive animals [446, 447]. It is unclear if these differences in behavioural observations are due to species differences, categorisation of tonal sound types not aligning with function, or differences in the constitution and captive situations of the studied packs. Adult and juvenile swift foxes use chuckle vocalisations differently, with adults using them solely with friendly intent and juveniles using them either in friendly or antagonistic contexts [440]. Therefore, tonal vocal series may be used differently based on species, individual, and context.

Other repeated call types may be species specific. For example, female New Guinea singing dogs (*Canis dingo hallstromi*) produce a 3-min series of yelp-screams during copulation [437]. African wild dogs have perhaps the richest repertoire with a variety of 'staccato' vocalisations. Arctic foxes produce a repeated cooing sound when interacting with pups [448]. They produce four types of twitters that are used in pro-social (greeting and play) and aggressive contexts (mobbing and attacking), begging cries that occur during distress, and *hoos* that are used in long-distance communication [443].

3.6.5.3 Tail Wagging

An animal's tail position and movement can convey a variety of different messages to both conspecifics and allospecifics. Of particular interest to the understanding of rhythm is tail wagging, the back-and-forth movement of the tail. Of the terrestrial carnivores, the most in-depth research has been done with domestic dogs. A combination of tail position, wagging rate, and degree and direction of symmetry may convey information about the emotional state of the animal. Wagging rate may be associated with arousal level, while direction and degree of asymmetry are associated with different types of stimuli in laboratory conditions. Dogs produce larger amplitude strokes to the left when alone or encountering an unfamiliar conspecific [449] and show higher heart rates when watching videos of dogs with left asymmetry [450]. Dogs produce right-sided asymmetry when encountering their owner (largest degree of asymmetry), unfamiliar humans, or a cat (slight asymmetry) [449]. The degree of regularity of tail-wagging has yet to be quantified. If speed is relatively constant in both directions, then asymmetrical wagging should produce a doublet pattern.

3.6.6 Rhythmic Qualities in Nonhuman Primate Communication

Characterising primate communication presents many challenges. Primates exhibit graded vocal repertoires, where call types have intermediate variants that shift into one another. Thus, the identification of unique calls and subunits is difficult and requires large samples to overcome issues related to individual and contextual variability. Moreover, the functional classification of primate signals poses challenges related to usage complexity and the intricacies of primate sociality [451]. A signal type may be used in multiple behavioural contexts, which makes it difficult to pinpoint its function. Contextual bivalence is also a complicating factor, e.g. signals of friendly intent may occur within the antagonistic framework of dominance relationships. To cover the structural and functional complexity of primate RC in a succinct and coherent way, we start by reviewing long-range signals (Sect. 3.6.6.1), followed by close-range signals (Sect. 3.6.6.2). Special sections are then dedicated to rhythmic copulation signals (Sect. 3.6.6.3) and to temporal coordination in primate RC (Sect. 3.6.6.4). By necessity, this review will only capture a glimpse of all relevant signals, prioritising signals with quantified rhythmic qualities and signal classes that are commonly described as rhythmic and are widespread across primate taxa.

3.6.6.1 Long Range Rhythmic Communication

Primates produce a diversity of species-specific long-range calls, which ensure communication with more or less dispersed conspecifics in various kinds of environments where visual access may be obstructed. Call parameters, such as communication flow (who signals to whom in which situations), flexibility and complexity vary greatly across primate species, with novel data and findings continuously enriching our understanding of both functional and structural aspects. Case in point, a series of recently published studies have uncovered rhythmic qualities in the long-distance calls of species that pertain to nearly all the superfamilies in the order, including lemurs (Indri lemur, *Indri indri*), tarsiers (genus *Tarsius*), New World monkeys (genus *Callicebus* and *Plecturocebus*) and apes (Hylobatidae; Pongidae). Given their spectro-temporal elaboration, such long-range calls (with the exception of the orangutan long call) have been designated as song.

The typical structural template of these calls is a hierarchical nested structure, where basic sound types (roars, melodic notes, pants, etc.) combine into groups (or phrases) based on principles of alternation or repetition, which are assembled into larger sequences, following a fixed, stereotyped order (indri [452], titi monkeys [453], tarsiers [454], gibbons [455], orangutans [456, 457]). Rhythmic isochrony has been reported for call components, including the final 'descending' phrase of the indri song cycle, tarsier male notes, titi monkey and gibbon duetting, the long call of orangutan males and the alarm call of orangutan females [454–459]. In addition, categorical rhythms (i.e. based on intervals related by small integer ratios such as 1:1 or 1:2) have been described for portions of Indri song [460]. Interestingly, in orangutans both the male long call and the female alarm call exhibit recursive self-embedded isochrony with two or three levels of rhythmic hierarchy [456, 457]. This suggests that recursion—a trait long believed to be uniquely human—has evolved gradually in the hominid lineage (see also Chap. 12 [461]). Notably, for some of the calls mentioned here, developmental research is also underway (for details, see Chap. 10 [462]).

While the received view is that primate 'singing' is associated with monogamy, being shaped by sexual selection [463, 464], this has been recently questioned. Nearly half of the 'singing' species exhibit other forms of social organisation than the assumed default of pair-living, while for the other half, the details of social organisation are still to be mapped [465]. By contrast to monogamy, 'territoriality' and 'social bonds' are traits shared by all 'singing' species, suggesting that primate song primarily fulfils group-related functions. Consistent with this, in the few singing species where song function has been studied comparatively (e.g. Black-fronted titi monkeys (*Callicebus nigrifrons*), territorial gibbon species), the rate of coordinated loud calls increases significantly during periods of food abundance but not during periods of female oestrus [466, 467]. Moreover, song is emitted in contexts where there is a need to reaffirm territorial occupancy. For example, tarsiers 'sing' at dawn when returning to their sleeping tree [468], while gibbon species 'sing' often after a nightlong sleep. 'Singing' is also frequent in intergroup encounters in tarsiers [468], indri lemurs [469] and titi monkeys [466]. Overall, this aligns primate songs with the broader class of coordinated loud calls, which are common within the order and independent of specific forms of social organisation or mating system.

Some 'singing' species exhibit spectro-temporally distinct calls for territorial display and other long-range functions, such as contact calls (e.g. black-fronted titi monkeys [466, 470]). In other species, however, contextual information is encoded by temporal patterning variations within the song template. For example, in Indri lemurs, the intergroup display song has shorter IOI and is five times longer than occupancy advertising song which, in turn, is more stereotyped and longer than contact song [469]. Likewise, in gibbon song, the number and arrangement of note types, as well as the timing of the female great call and male coda reliably discriminate between predatory and occupancy advertising [471].

Temporal patterning encodes contextual information in nonsinging species as well. The loud call of male Thomas langur (*Presbytis thomasi*) exhibits distinct temporal patterning (i.e. number, rate and acceleration of tonal and nontonal units) if emitted as a territorial occupancy call at predawn, during intergroup display or in the presence of predators [472]. In guereza monkeys (*Colobus guereza*), the duration and number of roaring phrases distinguish between eagle-, leopard, and predawn calls, while repetition rates within the initial call discriminate between alarm and nonalarm calls [473, 474]. In the orangutan loud call, differences in the number and rate of pulses, number of pulse types, and duration of ascending frequency discriminate between spontaneous calls, calls given in response to other males' loud calls and calls given in response to environmental disturbances [475]. In the chimpanzee (*Pan troglodytes*) pant hoot, the duration, rate, acceleration and number of elements given in the various phases of the call (build-up, climax and letdown phase) vary systematically between pant hoots given in the context of food, travel, and group displays [476–479]. Similarly, temporal features discriminate between the snake and the hunt bark, with snake barks having longer units compared to hunt barks [480]. The bark can also be combined with other call types or with drumming, and such combinatorial variants discriminate between additional contexts, such as travel, aggression, and responding to calls from neighbouring communities. Playback experiments demonstrate that such subtle variations of temporal structure are perceived by conspecifics, which react with the expected adaptive behaviours (langurs [481], orangutans [475], guereza monkeys [473], gibbons [482]).

3.6.6.2 Rhythmic Vocalisations and Bodily Displays in Close-Range Interactions

Although considered an important component of primate vocal behaviour, close-range calls have received far less research attention. Rhythm-focused analyses come primarily from a handful of studies on the rate of acoustic signals generated by fast mandibular movements. These have been described in several primate species, including voiced lip-smacking (so-called wobbles) in gelada monkeys (*Theropithecus gelada*) [483], unvoiced lip-smacking and/or teeth chattering in long-tailed macaques (*Macaca fascicularis*) [484], stump-tailed macaques (*Macaca arctoides*) [485] and chimpanzees [486], as part of the male coda in gibbon song [487] or as species-atypical sounds in orangutans [488]. The behavioural contexts of these signals vary from affiliation (chimpanzees, gelada monkeys, long-tailed macaques), mating (stump-tailed macaques—teeth chattering), agonism (stump-tailed macaques—panting), or territorial display (gibbons). Despite considerable mechanistic and functional heterogeneity, these signals are often grouped together because their periodicity rate matches that of human speech, which averages around 5 Hz (range: 3–8 Hz [489]). This rhythm is a fundamental component of human speech and its disruption affects comprehension [490].

Based on available spectrograms and behavioural descriptions, the inventory of rhythmic close-range signals could be expanded with serially produced vocalisation types that are widespread across primate taxa, such as grunts and pants. Generally, grunts signal affiliative intentions (vervet monkeys [491], mangabeys [492], baboons [493, 494], chimpanzees [495]), with contextual nuances that may be species-specific. A relevant example is the so-called rhythmic grunt of baboons (*Papio* species), which consists of a series of evenly spaced tonal syllables with an approximate IOI of 450 ms [493]. 'Rhythmic grunts' are given by both sexes in affiliative contexts [493, 494], to signal friendly intent upon approaching subordinates or infant carrying females, to initiate interaction with infants, to appease screaming infants, or during adult hugging. Male-specific contexts include solicitation of copulation, postcopulation and infant-directed vocalisation, infant-holding and infant-bouncing. Current analyses suggest that interspecies discrimination in papionins is primarily driven by spectral features and overall call duration [494], while some contextual information is encoded by temporal features. For example, in Guinea Baboons (*Papio papio*), rhythmic grunts with an appeasement function exhibit faster IOI [493].

Male baboons also produce rhythmic combinatorial variants of grunts that have nonaffiliative functions, including the 'roar-grunts' (given in response to disturbances) and two types of panted grunts, given in agonistic contexts. The 'roar-grunt' has the same IOI as the rhythmic grunt described above, but exhibits progressively increasing amplitude and is often encased between an initial 'hum' and a final longer grunt or roar. The two agonistic pant types consist of rapid series of panted grunts, which are given either as monosyllabic series during dominance displays, or as 'two-phase grunts' during intermale agonism. The 'two-phase' grunt consists of 11–18 pairs of panted grunts, where each pair exhibits a fixed durational (long-short) and spectral (high-low) contrast [493].

The chimpanzee grunt is another noteworthy example, where the sub-class of 'food grunt' has potential referential function, with both spectral and temporal features shown to encode food preference and food type irrespective of arousal. In affiliative contexts, chimpanzees give grunts and pant-grunts as 'greeting' calls in the presence of dominants, especially when individuals reunite after a prolonged separation [495] or prior to feeding [496].

'Pant' vocalisations exhibit functional heterogeneity across primate species, and may be produced in both affiliative and agonistic contexts (e.g. the 'pant-threats' of drill [497, 498] and macaque [499] species). As reviewed above, the loud call of titi monkeys exhibits pant elements, and panted variants of baboon grunts (which otherwise are affiliative vocalisations) occur during dominance displays and intermale competition. Drill pants consist of monosyllabic series lasting up to 10 s [497]. Macaque pant-threats are, similarly, given as series of monosyllabic sounds, which dominant individuals direct toward subordinates in aggressive situations [499]. Interestingly, in captive macaques, pant-threats precede or follow another rhythmic signal, that is, display drumming, which is produced by slapping surfaces or shaking cage mesh, and is exhibited by dominant individuals [500]. Macaque drumming bouts are short, lasting on average 1.5 s, and are delivered with an average rate of 5/s. Agonistic drumming is also present in great ape species, although the apes also exhibit drumming in nonagonistic contexts such as play, mindful exploration, attention-getting or nut-cracking (reviewed in more detail in Chap. 12 [461]). The great apes, moreover, exhibit a range of nonvocal rhythmic behaviours that occur in a variety of contexts ranging from courtship to dominance display and play (see Chaps. 12 [461] and 13 [501]).

3.6.6.3 Rhythmic Copulation Signals

In a mating context, primates exhibit an array of copulatory calls and bodily displays, whose temporal structure has been consistently described as rhythmic [502–505]. These can occur prior, during or after copulation, and may be given by males, females or by both sexes. Across species, there is considerable variability with respect to which sex produces them in which moment of the mating act. The functional significance of primate copulation signals is still debated. Most research attention, thus far, has been given to female postcopulatory calls, while precopulatory signals are understudied. Generally, the functional hypotheses that receive the strongest empirical support vary across species, as a function of contextual details (who gives the signal when) and the social organisation and mating system of the species.

Current data suggest that precopulatory primate signals serve to solicit copulation and, potentially, signal female fertility [505]. Male precopulation signals have been reported in distant phylogenetic taxa such as ruffed lemurs (*Varecia* spp., [506]) and great apes (gorillas: [507, 508]); chimpanzees: [509, 510]). In form, they may vary from 'whine' calls in ruffed lemurs [506] to 'sex grunts' in mountain gorillas (*Gorilla beringei beringei*), to pulsed 'horse-like sounds' in Western gorillas (*Gorilla gorilla*) [507, 508] and to rhythmic bodily movements, including multifunctional (rocking, bouncing, head bobbing) or context-specific signals, such as pelvic thrust [509, 510]. Female precopulatory signals may take the form of rhythmic calls (e.g. gelada monkeys [505, 511]) or rhythmic bodily movements, such as head bobbing, squatting and jerky limb movements (polygynous gibbon species [512, 513]).

Intracopulatory calls given by both sexes have been, for example, reported in gelada monkeys and Western gorillas, in the form of acoustically distinct 'copulation grunts'. In gelada monkeys, copulation grunts are emitted by males for the entire duration of the mating act, and only intermittently and optionally by females [514]. The copulation grunt of Western gorillas consists of staccato series of more than 80 'huh' sounds that, in both sexes, are produced at a rate of 9 pulses/s [508]. In other species, intracopulatory calls are the prerogative of either males (e.g. rhesus macaques (*Macaca mulatta)* [499]) or females (e.g. Barbary macaques (*Macaca sylvanus*) [515, 516], bonobos (*Pan paniscus*) [517], long-tailed macaques [518]). Acoustically, these sex-specific vocalisations may range from staccato pant-grunts (e.g. Barbary macaque females, long-tail macaque females) to high-pitch squeaks and screams (e.g. rhesus macaque males, bonobo females).

Postcopulatory calls (i.e. produced toward the end of and after copulation) consist of pulsed sounds that may vary spectrally from low-pitch grunts (e.g. macaques, baboons) to high-pitch screams (e.g. chimpanzees), while sharing the temporal feature of rhythmicality [502, 503, 505]. They are primarily produced by females and are common across primate taxa, but otherwise rare in the animal kingdom. Although female postcopulatory calls have received much theoretical and empirical attention, their functional significance remains poorly understood. Currently, the most viable hypotheses are that such calls serve to enhance female fitness, either indirectly through male-male competition (i.e. through sperm competition) or as a mechanism of female mate choice.

According to the 'sperm competition' hypothesis, females call towards the end of or after copulation to attract and mate with as many males as possible and, thus, increase the probability of obtaining the 'best' sperm. Consistent with this explanation, in several species the females promptly increase distance from the male after the mating act (reviewed in [502]). According to the 'female mate choice' hypothesis, postcopulatory calls are given to motivate consorting behaviours and mate guarding from high-ranking males, which results in increased paternity certainty and paternal offspring protection. Consistent with this interpretation, in some species, the occurrence of female postcopulation calls is positively associated with male rank (Guinea baboons [502], chimpanzees [519]). There are, however, indications that these hypotheses are neither mutually exclusive, nor exhaustive. A recent study aimed at concomitantly testing the above hypotheses in olive baboons (*Papio anubis*) found some level of support for both, while also reporting considerable unexplained variability [504]. The alternative perspective suggested by such findings is that female postcopulatory calls are multifunctional and may vary with social (e.g. hierarchy stability, personal history with the consorting male) or ecological (e.g. ingestion of plants that affect levels of reproductive hormones) factors.

3.6.6.4 Temporal Coordination in Primate Rhythmic Communication

Primate loud calls are often communal, that is, emitted by two or more individuals, either sequentially, as 'duets' or simultaneously as 'chorus'. Structural analyses suggest that duets exhibit accurate coordination, whereby individual contributions are swiftly delivered with a maximum latency of 200 ms [228]. Often, the precise timing of primate duets is compared to the temporal coordination of turn-taking in human dialogue, which exhibits a similar latency. Arguments have been made, however, that human speech is essentially an offset-cued—and therefore reactive—type of coordination, as opposed to rhythmic coordination in musical behaviours, which is onset-cued and, thus, predictive [520]. In reactive coordination, turn-taking is timed based on local cues, such as, for example, a descending intonational contour. In predictive coordination, the timing of a contribution is cued by—and adapted to—the onset on the next-last unit in a rhythmic string, which requires the formation of a rhythm percept.

Experimental and emerging observational research shows that primates can exhibit predictive coordination to an exogenous rhythm (auditory, visual, tactile), either spontaneously [521–524] or after intensive training [525–527] (for more details, see Chaps. 7 [185], 10 [461] and Chap. 12 [461]). However, the exact type of coordination found in primate duets, *in naturalistic contexts*, remains to be established. An intriguing case is represented by the loud call of white-handed gibbons, where statistical analyses indicate the presence of both reactive and predictive timing. Specifically, [487] found that interindividual coordination was locally cued by acoustic features of incumbent phrase ending. By contrast, [455] found evidence that individual vocal onsets were causally related, being predicted by previous vocal onsets of the partner. Besides sequential coordination, this species also produces overlapping contributions with higher than chance synchrony. A combination of sequential and overlapped calling is also found in tarsiers and Indri song and in both species the males exhibit higher levels of temporal adaptation relative to the female contributions. Tarsier males have been reported to be capable of adjusting note timing relative to the female using both acceleration and deceleration [454, 528, 529]. In Indri lemurs, overlaps occur selectively between the reproducing couple (30% in average), being rare between adults and offspring [530]. While contribution timing is mutually influenced among callers, temporal adaptation is stronger in males.

By contrast to the cases presented above, participant contributions exhibit high levels of overlap in titi monkeys, gelada monkeys and chimpanzees. In titi monkeys, both males and females often produce the same inventory of syllable or phrase types, but with a temporal shift [453, 466, 531, 532]. Thus, when the male emits loud phrases, the female pants and vice versa. When duetting, titi monkeys show the capacity to adjust rhythmic features (pulse rate, phrase duration) to one another, which results in call convergence [454].

Reportedly, accurate temporal coordination is also present in gelada monkeys, where segments of the so-called Long Series that consist of egressive and ingressive cycles with 'steady and stable' rhythm and tempo show in-phase and anti-phase synchronisation [533]. Mean asynchrony in synchronized bouts was found to range between ±130 ms, which overlaps with the asynchrony interval reported for human rhythmic synchronisation. Gelada vocal synchronisation is described as flexible, with individuals being capable to adjust to one-another's call frequencies after short latencies averaging 2.37 cycles.

While chimpanzee pant hoot chorusing is often described as chaotic, statistical analyses demonstrate interindividual temporal adaptation between participating callers in the build-up and climax phases of the call. The build-up serves as a preamble to the culminating part of the call, during which chimpanzees can recruit other callers. Accordingly, the duration of the build-up phase is correlated with the joining latency of co-callers, suggesting that chorus initiators prolong the build-up phase until additional participants are recruited [534]. The duration and number of elements of the climax phase show a significant increase during chorus compared to solo calling, consistent with the idea that chimpanzees prolong the loudest phase of the call when calling with others. When chorus partners join early (i.e. already in the introduction phase), the number of climax elements correlates between callers, thereby suggesting that callers mirror each other's temporal structure [534]. Overall, it has been proposed that the chimpanzee pant hoot exhibits design features for facilitating effective chorus displays, which promote interindividual coordination.

3.7 Concluding Remarks

The research reviewed in this chapter reveals that rhythmic structure is pervasive in animal communication. At the same time, rhythmic signals tend to be costly, either directly through energy expenditure or indirectly, due to increased risk of detection by predators and/or rivals, which follows from emitting highly conspicuous signals. This begs the obvious question of what makes such rhythmic signals worthwhile. What adaptive benefits do rhythmic parameters (e.g. repetitiveness, regularity, patterning) provide to animals, so that the costs of conspicuous and, often, also metabolically demanding signals are exceeded by profits in the great book-keeping of life? To channel future research on fruitful avenues for answering such evolutionary questions, we conclude this chapter with a thematic summary of the wealth of data reviewed above and outline several research objectives for the short and long term.

3.7.1 Diversity of Contexts of Use and Rhythmic Structure

Our synthesis of RC data suggests that, across the surveyed taxa, RC exhibits an increasing functional diversification as a function of sociocognitive complexity and that the most prominent context for RC use, which is shared by all surveyed taxa, is reproduction. In the latter context, RC roles span mate attraction, courtship displaying and copulation signals. Mate attraction rhythms—emitted to capture the interest of the opposite sex—are mainly actuated by long-range acoustic or vibrational signals, which can be highly stereotypical or flexible. High stereotypy maximizes the unambiguous broadcasting of species identity. In addition, through rhythmic flexibility, individuals can also broadcast information about body condition and/or motivational states through variations in the execution of rhythmic parameters, which are readily picked up by choosy receivers.

A basic setting in the mate attraction context is that of males emitting rhythmic signals that broadcast the males' location—and potentially fitness or motivational state—to roaming females, with examples found in all the surveyed zoological classes. This may give rise to *leks,* that is, large aggregations of displaying males, which favours the emergence of male-male rhythmic interactions with various degrees of temporal coordination. In the reverse setting, where males are the 'searching' sex and engage in rhythmic displays within the home ranges of receptive females (e.g. fossorial rodents; pinnipeds), the function of such behaviours can more accurately be described as courtship, rather than mere mate attraction. This function is also found in contexts where both sexes are stationary (e.g. walrus). Finally, another variation is represented by (more or less) coordinated rhythmic displays with the participation of both sexes, typically reported in (serially) monogamous species (e.g. bird and primate duets). Such displays can attain high levels of complexity with the implication of several signalling modalities (birds), thereby incorporating both long-range (i.e. acoustic) and close-range (e.g. acoustic, visual, vibrational) rhythmic signals. Besides being functional in courtship, these displays also contribute to pair bonding and evaluation, often through the execution of pair-unique sequences of rhythmic—and other—behaviours.

A potentially distinct class of close-range rhythmic signals given in a reproductive context is represented by copulation signals, that is, signals given prior, during or after the mating act. This class is extremely rich in primates, with functions that vary depending on the emitting sex, species and timing relative to the mating act. In short, precopulation calls function as mating solicitations and may also signal fertility and/or sexual arousal. Intracopulatory calls may serve to express or enhance sexual arousal and, thus, mating success, while postcopulatory calls may serve as mate guarding signals or to enhance female reproductive fitness through mating with high-fitness or high-ranking males. Interestingly, while rhythmic copulation signals are very common in primates, sexual advertising and courtship displays seem to be less usual than in the other taxa reviewed here.

Across the studied taxa, rhythmic signals are also commonly encountered in inter-individual or intergroup interactions related to territorial regulation. Generically defined, animal territories are areas that provide resources to the individual or group, by providing access to food, shelter, breeding grounds, etc. Some territories persist across the year(s), being vacated when residents die or are displaced through competition, while others persist for shorter periods, typically the breeding season. The most basic setting in rhythmic territorial communication is that in which single individuals advertise occupancy using long-range signals, such as vocal (e.g. anurans) or percussive (woodpeckers, fossorial rodents) signals. Prominent examples of communal occupancy advertisement are the dawn and dusk choruses of birds and primates. Rhythmic signals are also emitted in territorial defence displays, when intruders approach or enter territorial borders. In this context, rhythmic parameters of execution provide cues about the physical and motivational potential of individuals or groups to defend resources. As reported for some species (e.g. pinnipeds, primates), occupancy advertisements and territorial defence displays may be highly similar, being mainly differentiated by temporal/rhythmic parameters. Altogether, by enforcing interindividual or intergroup spacing and preventing territorial intrusions, such rhythms minimize the risks (and costs) entailed by direct aggression. In a communal setting, a long-term effect of RC is the strengthening of pair/group bonding and the emergence of group-specific rhythms (e.g. birds, cetaceans, primates).

Rhythmic signals with a group regulatory function span long-range signals that coordinate group movement (contact calls, alarm calls), as well as close- or long-range signals that moderate social relationships (dominance, alliances, affiliation, etc.). Contact calls coordinate group movement, by broadcasting the location of separated members. Our review provides examples of rhythmically structured contact calls within the mother-infant dyad (bats, cetaceans, pinnipeds), within larger social groupings composed of kin- and/or nonkin members, emitted during travel or foraging (e.g. cetaceans, primates) or as a signal to regroup for resting (e.g. primates). Alarm calls coordinate group movement in relation to conspecific intruders or predators. In primates, there are several examples where temporal structure parameters encode predator type, with evidence that conspecifics readily pick up such fine-tuned differences.

Affiliation rhythms span various forms of vocalisations and gestures that occur during socialising (e.g. cetaceans, terrestrial carnivores, primates), contact initiation (e.g. terrestrial carnivores, primates), grooming (primates) or social play (terrestrial carnivores, primates). Given such contexts of use and indications that affiliative interactions last longer if accompanied by RC, the latter seems to contribute to the formation and consolidation of social bonds. Antagonistic rhythms occur as displays of dominance, aggression, appeasement and submission (terrestrial carnivores, primates), or as individual signatures that guide social navigation within the hierarchy (e.g. elephant seals).

Finally—and less commonly—rhythmic signals have been also reported in contexts related to prey-predator interspecific interactions and instrumental behaviours. Certain species (e.g. bioluminescent fish, anurans) use RC-forms to attract or confuse prey. Conversely, prey species may direct rhythmic signals to predators, either to confuse or deter them (e.g. insects, fossorial rodents). Instrumental rhythmic signals arise when the intended receiver is also the emitting individual, as found in echolocation (bats; fossorial mammals; cetaceans) or as a by-product of extractive foraging activities (e.g. woodpeckers, primates) where they may result in individual signatures.

Across the contexts summarized above, there is great structural diversity, expressed as species-specific rhythmic patterns and, sometimes, also as individual rhythmic signatures. Rhythmic patterns vary from serial singleton units that form isochronous, accelerating or decelerating sequences to various levels of elaboration and complexity. 'Horizontally' elaborated signals encompass patterned combinations of basic units (e.g. notes, flashes, movements), which are concatenated following various rules (repetition, alternation, addition, subtraction etc.). 'Vertically' elaborated signals exhibit hierarchical structuring of sequences nested within sequences, where rhythmic patterning may occur at one or several hierarchical levels.

The data synthesized in this chapter shows that metronomic-like isochrony and nested rhythms—as well as many variations in between—occur within each of the reviewed zoological classes, from insects to primates. Variable levels of structural complexity are also recorded within each of the contexts discussed above. Based on the currently available data, the greatest complexity seems to be found in rhythmic signals with a reproductive function, followed by communal territorial defence signals.

Structurally, rhythmic signals also vary along a spectrum of stereotypy and flexibility, in a manner that, again, is not taxon dependent. Notably, even species with highly stereotypical signals exhibit certain forms of flexibility, owing to environmental constraints or social influences exerted by conspecifics during rhythmic interactions (for details, see Chap. 6 [48]). The latter may result in various forms of rhythmic coordination, including precisely timed turn-taking in bird, rodents or primate duets or accurate interindividual synchronisation of phase and period, as exemplified by the stridulatory and bioluminescent signals of some insect species. Besides these obvious examples, statistical analyses reveal more subtle forms of rhythmic flexibility and coordination. For example, interindividual coordination may be more pronounced for certain segments of the signal (e.g. primate song), shown within a single modal channel of a multimodal signal (e.g. birds) or occur in the form of temporal alignment where co-signalling individuals adjust their habitual pace of rhythmic signals to match that of their signalling partners (e.g. chimpanzees).

3.7.2 Evolutionary Trends, Outstanding Questions and Avenues for Future Research

Several evolutionary patterns transpire from the research reviewed here. By mapping the functional contexts and forms of rhythmic communication, we find that rhythmic properties in the physical signal are pervasive in the animal kingdom, with a constant presence across taxa that are separated by more than 500 million years of evolution. This indicates that rhythm is an ancestral and highly resilient trait of animal signalling. Within each zoological class, however, there appears to be considerable structural diversity, whereby some genera/species within that class exhibit simple, metronome-like signals and others exhibit highly elaborated 'song' or multimodal displays. Signal modality seems to constrain signal complexity, whereby the highest levels of complexity are achieved in the acoustic channel, with the caveat that rhythmic quantification in nonacoustic modalities is currently scarce.

The observations made above—considerable gradients of structural complexity and channel diversity manifested within each zoological class—suggest that rhythmic signals have evolved several times independently. This hypothesis can be verified in some insect clades, where detailed information is available (Sect. 3.2.2) for both molecular phylogenies and signal structure. In other taxa, however, this is an endeavour that will have to wait for studies of rhythmic quantification in a representative number of species.

The taxonomic distribution of rhythmic complexity at the physical level of signal production shows that rhythmic complexity is not the prerogative of cognitively complex taxa, nor of advanced sociocognitive abilities. By contrast, few conclusions can be currently drawn concerning the perceptual level of explanation (for details, see Chap. [1]), that is, if conspecifics perceive this rhythmic complexity or even a rhythmic pattern at all. While rhythm perception studies are missing in many taxa, the available data indicate that, within each taxonomic class considered in this chapter, at least some species perceive isochrony and/or make functional discriminations based on temporal patterning. Studies with insects further suggest that, despite the presence of nested rhythmic structure in the physical signal, rhythm perception may be restricted to a specific level and temporal scale, which transmits biologically significant information to conspecifics. The question of which other species than humans perceive hierarchically structured rhythms at multiple timescales is a topic for future research.

In humans, rhythm perception has been conceptualized as a relational processing ability that allows individuals to match relations between relations in the temporal domain (Chaps. 1 [1] and 7 [185]). In this perspective, isochrony perception is the perception of similar durations across successive interonset intervals. Traditionally, being able to perceive the similarity of relations-between-relations (also called 'high-level' relational processing) has been showcased as a human-unique ability that, among other things, enables the processing of hierarchical structure and recursion in human language and music (as discussed in Chap. 1 [1]). This would suggest that the ability to perceive hierarchically structured rhythms is constrained by cognitive ability. Research on rhythm perception and relational processing in various domains (social, spatial, etc.) carried out with human infants and nonhuman species (detailed in Chap. 12 [461]) suggests a more complex picture, where relational processing can be implicit or explicit. This distinction is certainly consistent with findings that neonates perceive isochrony and even meter (see Chap. 10 [462]), while the ability to synchronize to exogenous rhythms emerges at the same age as explicit relational processing. It is also consistent with findings that, with hard training that increases the salience of interonset intervals, animals learn to flexibly—and *volitionally*—synchronize motor behaviour with exogenous rhythms, including complex musical stimuli (see Chaps. 7 [185] and 14 [186]). These parallels are speculative but raise fascinating questions about the underlying cognitive mechanisms of rhythm perception and rhythmic synchronisation across taxa and from an evolutionary perspective.

Important evolutionary questions are also raised by multimodal rhythms. The general assumption is that multimodal signals are perceived in all the involved modalities and that multimodality is adaptive by providing intersensory redundancy, which supports signal conservation. In some species, however, rhythms have evolved to signal to different receivers in distinct modalities. This is, for example, the case for vibroacoustic rhythms produced by arthropod species that have not evolved auditory perception (for details, see Chap. 2 [90]). In these cases, only the vibrational rhythms have biological significance to conspecifics, while the conspicuous acoustic rhythms actuate anti-predator strategies, thus, being the result of selective pressures related to predation.

The ubiquity of rhythmic structure in animal communication, across taxa and contexts of use, urges a central question to this volume: *why rhythm*? As shown in this chapter, rhythmic signals mediate communication in a range of social contexts, which suggests that they could contribute to a whole range of survival functions, including reproduction, resource protection, social regulation, food procurement or escape from predation. But how do rhythms benefit animals in these communicative contexts? The literature reviewed in this chapter suggests several possible answers to be followed up in future research. At the most general level, temporal patterning offers tremendous margins of flexibility for signal structuring. As shown throughout this review, rhythmic signals and the way in which they are executed is varied across taxa and species in a myriad of patternings and combinations that can encode species identity, individual identity, individual fitness, functional context, levels of arousal and motivational states. This is an evolutionarily efficient solution, as a single mechanism can be flexibly and variably implemented across modalities and species.

Rhythmic features, such as stereotypy and repetition, enhance signal identity and conservation. These have, therefore, been discussed as effective means for both conspecific localisation and signal propagation. These two functions are certainly crucial to the range of survival functions discussed above. Plausibly, in environments where the signalling space is agglomerated (due to high levels of sympatry or high population density) repetition supports signal resilience through redundancy, while the high stereotypy of species-specific rhythms provides an unambiguous signalling niche for reliable species and/or individual recognition. Environments with lower degrees of sympatry arguably entail less stringent pressures for strict stereotypy, thereby allowing for more signal variability. Thus, in such environments, besides fulfilling a localisation function, rhythmic structure may encode additional layers of information about individual quality and motivation. Thus, rhythmic variations contribute to mate choice, intermale competition, territorial and social regulation.

Rhythm is ideally suited to encode the types of information mentioned above, owing to the intrinsic rhythmicality of the neuromechanical systems—respiratory and locomotor—that, across evolutionary times, have been recruited into signalling. As it will be detailed in Chaps. 7 [185], 9 [201] and 8 [427], physiological rhythms that determine and modulate RC are found at both the mechanical (muscular) and neural level of motor output. As already discussed in Chap. 2 [90], the noncommunicative functions of these underlying systems have rhythmic perceptual by-products (sounds, vibrations, etc.) that may inherently provide cues about body condition, action tendencies and levels of arousal (see Chap. 11 [318]). Putatively, the basic rhythmicality inherited from the noncommunicative functions of underlying neuromechanical systems has been further shaped through ritualisation by sexual and socioecological selection pressures derived from mate choice, intermale competition, habitat, population density, etc. (as detailed in Chap. 6 [48]).

Owing to the physiological link between body condition, level of motor control and deployed motor output, rhythmic signals are suited to transmit information about signallers' condition and motivational states (see Chaps. 9 [201] and 11 [318] this volume). Given high stereotypy and repetition rates, rhythmic signals can be difficult and costly to produce. High stereotypy places high demands for accurate execution and offers receivers a succession of behavioural samples to evaluate signallers against the standard (i.e. stereotypy). It also entails levels of exertion that are higher than those typically deployed in noncommunicative behaviours. Consider, for example, the levels of exertion and demands for motor precision found in human dancing, as opposed to walking. During competitive display—whether for mates, territorial resources or social dominance—rhythmic signalling may push individuals to the physiological limit of their motor capabilities, so that 'top quality' individuals are able to produce faster rate signals and to sustain such demanding signalling for longer periods of time. In turn, as shown in some of the examples reviewed in this chapter, such rhythmic parameters can serve as indicators of stamina and energetic reserves.

To conclude, the ubiquity of rhythmic signalling can be accounted for by inherent constraints in the neuromechanical systems that actuate them and by sensory biases to attend to rhythmic cues as indicators of biologically significant states.

Moreover, rhythmic signalling is associated with reproductive, social and ecological benefits, which likely accounts for their evolutionary resilience and diversification, despite considerable costs incurred from metabolic demands and increased risks of predation or competition. The handful of evolutionary reflections summarized in this section certainly constitute but a glimpse of a more complex discourse that will be elaborated by the other chapters of the volume, as well as by future research. The inevitable conclusion of the review provided in this chapter is that it is up to future research to see through the two critical first steps that are required for understanding the evolutionary and functional significance of RC: detailed analyses of the temporal structure of the displays and accurate mapping of the behavioural contexts in which such displays spontaneously occur.

While form-function mappings based on detailed structural and contextual analyses are generally lacking, there are also some notable exceptions. This reflects the uneven development of the field. In insect, anuran and avian research, for example, there is a longer tradition of investigating how rhythmic signals are produced, structured, perceived, or interact with other rhythmic signals and what effects they have on con- and heterospecifics. For the great majority of taxa, however, the study of rhythm provides an entirely new dimension to investigate in the context of animal communication. To date, the field has been dominated by the study of acoustic signals, which provides a biased picture of the rhythmic landscape of animal communication. More research efforts, thus, are needed to catch up on rhythmic signalling in other modalities of display. No doubt, a major challenge to these future endeavours is the conceptual, terminological and methodological heterogeneity of the field. A higher degree of harmonisation is needed to ensure data comparability across taxa, which can only be achieved through dialogue and collaborative works of synthesis carried out by researchers with diverse backgrounds (as also attempted here).

Acknowledgements GA Sauciuc's contribution to this chapter has been funded by the Bank of Sweden Tercentenary foundation (grant nr. P17-0942:1) and the Swedish Research Council (grant 2019-03104).

References

1. Burchardt LS, Hoeschele M, Honing H et al (2025) Chapter 1: What is rhythm and how can we study it? In: Sauciuc G-A, Lameira AR, Persson T (eds) Nature beats: the what, why and how of animal rhythmic behaviour. Springer
2. Lameira AR, De Gregorio C, Gamba M et al (2025) Chapter 4: Uncharted rhythmic diversity across the animal world. In: Sauciuc G-A, Lameira AR, Persson T (eds) Nature beats: the what, why and how of animal rhythmic behaviour. Springer
3. Gerhardt HC, Huber F (2002) Acoustic communication in insects and anurans: common problems and diverse solutions. University of Chicago Press, Chicago
4. Balakrishnan R, Pollack GS (1996) Recognition of courtship song in the field cricket, *Teleogryllus oceanicus*. Anim Behav 51:353–366. https://doi.org/10.1006/anbe.1996.0034
5. Rebar D, Bailey NW, Zuk M (2009) Courtship song's role during female mate choice in the field cricket *Teleogryllus oceanicus*. Behav Ecol 20:1307–1314. https://doi.org/10.1093/beheco/arp143

6. Velilla E, Muñoz M, Quiroga N et al (2020) Gone with the wind: is signal timing in a neotropical katydid an adaptive response to variation in wind-induced vibratory noise? Behav Ecol Sociobiol 74:59. https://doi.org/10.1007/s00265-020-02842-z
7. Heller K-G (1992) Risk shift between males and females in the pair-forming behavior of bushcrickets. Naturwissenschaften 79:89–91. https://doi.org/10.1007/BF01131812
8. Walker TJ (1964) Cryptic species among sound-producing Ensiferan Orthoptera (Gryllidae and Tettigoniidae). Q Rev Biol 39:345–355
9. Ewing AW (1989) Arthropod bioacoustics: neurobiology and behaviour. Comstock Publishing Associates, Ithaca
10. Dethier VG (1992) Crickets and katydids, concerts and solos. Harvard University Press, Cambridge, MA
11. Montealegre-Z F, Jonsson T, Robert D (2011) Sound radiation and wing mechanics in stridulating field crickets (Orthoptera: Gryllidae). J Exp Biol 214:2105–2117. https://doi.org/10.1242/jeb.056283
12. Lin C-C, Hedwig B (2021) Wing movements underlying sound production in calling, rivalry, and courtship songs of the cricket *Gryllus bimaculatus* (DeGeer). J Insect Physiol 134:104299. https://doi.org/10.1016/j.jinsphys.2021.104299
13. Dudley R (2002) The biomechanics of insect flight: form, function, evolution. Princeton University Press, Princeton
14. Ter Hofstede HM, Symes LB, Martinson SJ et al (2020) Calling songs of Neotropical katydids (Orthoptera: Tettigoniidae) from Panama. J Orthoptera Res 29:137–201. https://doi.org/10.3897/jor.29.46371
15. Walker TJ (1983) Diel patterns of calling in nocturnal Orthoptera. In: Gwynne DT, Morris GK (eds) Orthopteran mating systems: sexual competition in a diverse group of insects. Westview Press, Boulder, pp 45–72
16. Robillard T, Desutter-Grandcolas L (2004) High-frequency calling in Eneopterinae crickets (Orthoptera, Grylloidea, Eneopteridae): adaptive radiation revealed by phylogenetic analysis. Biol J Linn Soc 83:577–584. https://doi.org/10.1111/j.1095-8312.2004.00417.x
17. Elliott CJH, Koch UT (1985) The clockwork cricket. Naturwissenschaften 72:150–153. https://doi.org/10.1007/BF00490404
18. Bennet-Clark HC, Bailey WJ (2002) Ticking of the clockwork cricket: the role of the escapement mechanism. J Exp Biol 205:613–625. https://doi.org/10.1242/jeb.205.5.613
19. Jacob PF, Hedwig B (2020) Modular timer networks: abdominal interneurons controlling the chirp and pulse pattern in a cricket calling song. J Comp Physiol A 206:921–938. https://doi.org/10.1007/s00359-020-01448-0
20. Deily JA, Schul J (2004) Recognition of calls with exceptionally fast pulse rates: female phonotaxis in the genus *Neoconocephalus* (Orthoptera:Tettigoniidae). J Exp Biol 207:3523–3529. https://doi.org/10.1242/jeb.01179
21. Schöneich S, Kostarakos K, Hedwig B (2015) An auditory feature detection circuit for sound pattern recognition. Sci Adv 1:e1500325. https://doi.org/10.1126/sciadv.1500325
22. Walker TJ (1975) Stridulatory movements in eight species of *Neoconocephalus* (Tettigoniidae). J Insect Physiol 21:595–603. https://doi.org/10.1016/0022-1910(75)90163-8
23. Gwynne DT (2001) Katydids and bush-crickets: reproductive behavior and evolution of the Tettigoniidae. Comstock Pub. Associates, Ithaca
24. Walker TJ, Dew D (1972) Wing movements of calling katydids: fiddling finesse. Science 178:174–176. https://doi.org/10.1126/science.178.4057.174
25. Nityananda V, Balakrishnan R (2021) Synchrony of complex signals in an acoustically communicating katydid. J Exp Biol:jeb.241877. https://doi.org/10.1242/jeb.241877
26. Greenfield MD (1988) Interspecific acoustic interactions among katydids *Neoconocephalus*: inhibition-induced shifts in diel periodicity. Anim Behav 36:684–695. https://doi.org/10.1016/S0003-3472(88)80151-9
27. Greenfield MD (1992) The evening chorus of the desert clicker, *Ligurotettix coquilletti* (Orthoptera: Acrididae): mating investment with delayed returns. Ethology 91:265–278. https://doi.org/10.1111/j.1439-0310.1992.tb00868.x

28. Schwartz JJ (1991) Why stop calling? A study of unison bout singing in a Neotropical treefrog. Anim Behav 42:565–577. https://doi.org/10.1016/S0003-3472(05)80240-4
29. Greenfield MD (2015) Signal interactions and interference in insect choruses: singing and listening in the social environment. J Comp Physiol A 201:143–154. https://doi.org/10.1007/s00359-014-0938-7
30. Kyriacou CP, Hall JC (1980) Circadian rhythm mutations in *Drosophila melanogaster* affect short-term fluctuations in the male's courtship song. Proc Natl Acad Sci 77:6729–6733. https://doi.org/10.1073/pnas.77.11.6729
31. Wheeler DA, Kyriacou CP, Greenacre ML et al (1991) Molecular transfer of a species-specific behavior from *Drosophila simulans* to *Drosophila melanogaster*. Science 251:1082–1085. https://doi.org/10.1126/science.1900131
32. Lagisz M, Wen S-Y, Routtu J et al (2012) Two distinct genomic regions, harbouring the period and fruitless genes, affect male courtship song in *Drosophila montana*. Heredity 108:602–608. https://doi.org/10.1038/hdy.2011.129
33. Miyatake T, Kanmiya K (2004) Male courtship song in circadian rhythm mutants of *Bactrocera cucurbitae* (Tephritidae: Diptera). J Insect Physiol 50:85–91. https://doi.org/10.1016/j.jinsphys.2003.10.002
34. Benetta ED, van de Zande L, Beukeboom LW (2021) Courtship rhythm in *Nasonia vitripennis* is affected by the clock gene period. Behaviour 158:685–704. https://doi.org/10.1163/1568539X-bja10087
35. Greenfield MD, Aihara I, Amichay G et al (2021) Rhythm interaction in animal groups: selective attention in communication networks. Philos Trans R Soc B Biol Sci 376:20200338. https://doi.org/10.1098/rstb.2020.0338
36. Greenfield MD, Merker B (2023) Coordinated rhythms in animal species, including humans: entrainment from bushcricket chorusing to the philharmonic orchestra. Neurosci Biobehav Rev 153:105382. https://doi.org/10.1016/j.neubiorev.2023.105382
37. Greenfield MD (2002) Signalers and receivers: mechanisms and evolution of arthropod communication. Oxford University Press, Oxford/New York
38. Walker TJ (1962) Factors responsible for intraspecific variation in the calling songs of crickets. Evolution 16:407–428. https://doi.org/10.1111/j.1558-5646.1962.tb03234.x
39. Walker TJ (1975) Effects of temperature, humidity, and age on stridulatory rates in *Atlanticus* spp. (Orthoptera: Tettigoniidae: Decticinae). Ann Entomol Soc Am 68:607–611. https://doi.org/10.1093/aesa/68.3.607
40. Martin SD, Gray DA, Cade WH (2000) Fine-scale temperature effects on cricket calling song. Can J Zool 78:706–712. https://doi.org/10.1139/z99-262
41. Walker SE, Cade WH (2003) The effects of temperature and age on calling song in a field cricket with a complex calling song, *Teleogryllus oceanicus* (Orthoptera: Gryllidae). Can J Zool 81:1414–1420. https://doi.org/10.1139/z03-106
42. Fletcher NH (1992) Acoustic systems in biology. Oxford University Press, New York
43. Fonseca P, Allen Revez M (2002) Temperature dependence of cicada songs (Homoptera, Cicadoidea). J Comp Physiol A 187:971–976. https://doi.org/10.1007/s00359-001-0267-5
44. Gayou DC (1984) Effects of temperature on the mating call of *Hyla versicolor*. Copeia 1984:733–738. https://doi.org/10.2307/1445157
45. Lloyd JE (1973) Fireflies of Melanesia: bioluminescence, mating behavior, and synchronous flashing (Coleoptera: Lampyridae). Environ Entomol 2:991–1008. https://doi.org/10.1093/ee/2.6.991
46. Branham MA, Greenfield MD (1996) Flashing males win mate success. Nature 381:745–746. https://doi.org/10.1038/381745b0
47. Iguchi Y (2002) The influence of temperature on flash interval in the Genji firefly *Luciola cruciata* (Coleoptera: Lampyridae). Entomol Rev Jpn 57:119–122
48. Schneider JN, Burchardt LS, Sauciuc G-A, Lameira AR (2025) Chapter 6: How environmental and social contexts shape and constrain rhythm structure and production. In: Sauciuc G-A, Lameira AR, Persson T (eds) Nature beats: the what, why and how of animal rhythmic behaviour. Springer

49. Greenfield MD (2001) Missing link in firefly bioluminescence revealed: NO regulation of photocyte respiration. BioEssays 23:992–995. https://doi.org/10.1002/bies.1144
50. Trimmer BA, Aprille JR, Dudzinski DM et al (2001) Nitric oxide and the control of firefly flashing. Science 292:2486–2488. https://doi.org/10.1126/science.1059833
51. Rodríguez RL, Boughman JW, Gray DA et al (2013) Diversification under sexual selection: the relative roles of mate preference strength and the degree of divergence in mate preferences. Ecol Lett 16:964–974. https://doi.org/10.1111/ele.12142
52. Gerhardt HC (1991) Female mate choice in treefrogs: static and dynamic acoustic criteria. Anim Behav 42:615–635. https://doi.org/10.1016/S0003-3472(05)80245-3
53. Walker TJ Jr (1957) Specificity in the response of female tree crickets (Orthoptera, Gryllidae, Oecanthinae) to calling songs of the Males1. Ann Entomol Soc Am 50:626–636. https://doi.org/10.1093/aesa/50.6.626
54. Doherty JA (1985) Temperature coupling and 'trade-off' phenomena in the acoustic communication system of the cricket, *Gryllus bimaculatus* de Geer (Gryllidae). J Exp Biol 114:17–35. https://doi.org/10.1242/jeb.114.1.17
55. Pires A, Hoy RR (1992) Temperature coupling in cricket acoustic communication. J Comp Physiol A 171:79–92. https://doi.org/10.1007/BF00195963
56. Gerhardt HC (1978) Temperature coupling in the vocal communication system of the gray tree frog, Hyla versicolor. Science 199:992–994. https://doi.org/10.1126/science.199.4332.992
57. von Helversen D, Von Helversen O (1981) Korrespondenz zwischen Gesang und auslösendem Schema bei Feldheuschrecken. Nova Acta Leopoldina 54:449–462
58. Ritchie MG, Saarikettu M, Livingstone S, Hoikkala A (2001) Characterization of female preference functions for *Drosophila montana* courtship song and a test of the temperature coupling hypothesis. Evolution 55:721–727. https://doi.org/10.1111/j.0014-3820.2001.tb00808.x
59. Greenfield MD, Medlock C (2007) Temperature coupling as an emergent property: parallel thermal effects on male song and female response do not contribute to species recognition in an acoustic moth. Evolution 61:1590–1599. https://doi.org/10.1111/j.1558-5646.2007.00140.x
60. Kilmer JT, Fowler-Finn KD, Gray DA et al (2017) Describing mate preference functions and other function-valued traits. J Evol Biol 30:1658–1673. https://doi.org/10.1111/jeb.13122
61. Jang Y, Greenfield MD (1996) Ultrasonic communication and sexual selection in wax moths: female choice based on energy and asynchrony of male signals. Anim Behav 51:1095–1106. https://doi.org/10.1006/anbe.1996.0111
62. Greenfield MD, Schul J (2008) Mechanisms and evolution of synchronous chorusing: emergent properties and adaptive functions in *Neoconocephalus* katydids (Orthoptera: Tettigoniidae). J Comp Psychol 122:289–297. https://doi.org/10.1037/0735-7036.122.3.289
63. Greenfield MD, Marin-Cudraz T, Party V (2017) Evolution of synchronies in insect choruses. Biol J Linn Soc 122:487–504. https://doi.org/10.1093/biolinnean/blx096
64. Neuweiler G (2000) The biology of bats. Oxford University Press, New York
65. Eckhartt GM, Ruxton GD (2022) Artificial light at night may decrease predation risk for terrestrial insects. Biol Lett 18:20220281. https://doi.org/10.1098/rsbl.2022.0281
66. Scoble MJ (2002) The lepidoptera: form, function and diversity, Reprinted. The Natural History Museum in association with Oxford University Press, London/Oxford
67. Greenfield MD (2014) Acoustic communication in the nocturnal Lepidoptera. In: Hedwig B (ed) Insect hearing and acoustic communication. Springer, Berlin/Heidelberg, pp 81–100
68. Roeder KD (1998) Nerve cells and insect behavior, 1st rev edn. Harvard University Press/Harvard University Press, Cambridge, MA
69. Greenfield MD, Baker M (2003) Bat avoidance in non-aerial insects: the silence response of signaling males in an acoustic moth. Ethology 109:427–442. https://doi.org/10.1046/j.1439-0310.2003.00886.x
70. Greenfield MD, Weber T (2000) Evolution of ultrasonic signalling in wax moths: discrimination of ultrasonic mating calls from bat echolocation signals and the exploitation of an antipredator receiver bias by sexual advertisement. Ethol Ecol Evol 12:259–279. https://doi.org/10.1080/08927014.2000.9522800

71. Greig E, Greenfield M (2004) Sexual selection and predator avoidance in an acoustic moth: discriminating females take fewer risks. Behaviour 141:799–815. https://doi.org/10.1163/1568539042265626
72. Rodríguez RL, Greenfield MD (2004) Behavioural context regulates dual function of ultrasonic hearing in lesser waxmoths: bat avoidance and pair formation. Physiol Entomol 29:159–168. https://doi.org/10.1111/j.1365-3032.2004.00380.x
73. Schnitzler H-U, Kalko EKV (2001) Echolocation by insect-eating bats: we define four distinct functional groups of bats and find differences in signal structure that correlate with the typical echolocation tasks faced by each group. Bioscience 51:557–569. https://doi.org/10.1641/0006-3568(2001)051[0557:EBIEB]2.0.CO;2
74. Schnitzler H-U, Moss CF, Denzinger A (2003) From spatial orientation to food acquisition in echolocating bats. Trends Ecol Evol 18:386–394. https://doi.org/10.1016/S0169-5347(03)00185-X
75. Zhou Y, Kelly JK, Greenfield MD (2011) Testing the fisherian mechanism: examining the genetic correlation between male song and female response in waxmoths. Evol Ecol 25:307–329. https://doi.org/10.1007/s10682-010-9421-2
76. Weissman DB, Gray DA (2019) Crickets of the genus Gryllus in the United States (Orthoptera: Gryllidae: Gryllinae). Zootaxa 4705. https://doi.org/10.11646/zootaxa.4705.1.1
77. Gray DA, Weissman DB, Cole JA, Lemmon EM (2020) Multilocus phylogeny of *Gryllus* field crickets (Orthoptera: Gryllidae: Gryllinae) utilizing anchored hybrid enrichment. Zootaxa 4750:zootaxa.4750.3.2. https://doi.org/10.11646/zootaxa.4750.3.2
78. Walker TJ, Greenfield MD (1983) Songs and systematics of Caribbean *Neoconocephalus* (Orthoptera: Tettigoniidae). Trans Am Entomol Soc 1890–109:357–389
79. Snyder RL, Frederick-Hudson KH, Schul J (2009) Molecular phylogenetics of the genus *Neoconocephalus* (Orthoptera, Tettigoniidae) and the evolution of temperate life histories. PLoS One 4:e7203. https://doi.org/10.1371/journal.pone.0007203
80. Greenfield MD (1990) Evolution of acoustic communication in the genus *Neoconocephalus*: discontinuous songs, synchrony, and interspecific interactions. In: Bailey WJ, DCF R (eds). Springer, Berlin/Heidelberg, pp 71–97
81. Bush SL, Beckers OM, Schul J (2009) A complex mechanism of call recognition in the katydid *Neoconocephalus affinis* (Orthoptera: Tettigoniidae). J Exp Biol 212:648–655. https://doi.org/10.1242/jeb.024786
82. Misof B, Liu S, Meusemann K et al (2014) Phylogenomics resolves the timing and pattern of insect evolution. Science 346:763–767. https://doi.org/10.1126/science.1257570
83. Lloyd JE (1997) Firefly mating ecology, selection and evolution. In: Crespi BJ, Choe JC (eds) The evolution of mating systems in insects and arachnids. Cambridge University Press, Cambridge, pp 184–192
84. Stanger-Hall KF, Lloyd JE (2015) Flash signal evolution in Photinus fireflies: character displacement and signal exploitation in a visual communication system. Evolution 69:666–682. https://doi.org/10.1111/evo.12606
85. Jang Y, Gerhardt HC (2006) Divergence in the calling songs between sympatric and allopatric populations of the southern wood cricket *Gryllus fultoni* (Orthoptera: Gryllidae). J Evol Biol 19:459–472. https://doi.org/10.1111/j.1420-9101.2005.01014.x
86. Jang Y, Gerhardt HC (2006) Divergence in female calling song discrimination between sym patric and allopatric populations of the southern wood cricket *Gryllus fultoni* (Orthoptera: Gryllidae). Behav Ecol Sociobiol 60:150–158. https://doi.org/10.1007/s00265-005-0151-3
87. Jang Y, Gerhardt HC (2007) Temperature effects on the temporal properties of calling songs in the crickets *Gryllus fultoni* and *G. vernalis*: implications for reproductive isolation in sympatric populations. J Insect Behav 20:33–52. https://doi.org/10.1007/s10905-006-9061-0
88. Walker TJ (1974) Character displacement and acoustic insects. Am Zool 14:1137–1150. https://doi.org/10.1093/icb/14.4.1137
89. Wiley RH, Richards DG (1978) Physical constraints on acoustic communication in the atmosphere: implications for the evolution of animal vocalizations. Behav Ecol Sociobiol 3:69–94

90. Sauciuc G-A, Schneider JN (2025) Chapter 2: How animals produce rhythm – an overview of ethological mechanisms for rhythmic communication. In: Nature beats: the what, why and how of animal rhythmic behaviours. Springer
91. Amorim MCP (2006) Diversity of sound production in fish. In: Ladich F, Collin SP, Moller P, Kapoor BG (eds) Communication in fishes. Science Publishers, Enfield, pp 71–105
92. Bass AH, McKibben JR (2003) Neural mechanisms and behaviors for acoustic communication in teleost fish. Prog Neurobiol 69:1–26. https://doi.org/10.1016/S0301-0082(03)00004-2
93. Sprague MW (2000) The single sonic muscle twitch model for the sound-production mechanism in the weakfish, *Cynoscion regalis*. J Acoust Soc Am 108:2430–2437. https://doi.org/10.1121/1.1315296
94. Fine ML, Parmentier E (2015) Mechanisms of fish sound production. In: Ladich F (ed) Sound communication in fishes. Springer, Vienna, pp 77–126
95. Vasconcelos RO, Fonseca PJ, Amorim MCP, Ladich F (2011) Representation of complex vocalizations in the Lusitanian toadfish auditory system: evidence of fine temporal, frequency and amplitude discrimination. Proc R Soc B Biol Sci 278:826–834. https://doi.org/10.1098/rspb.2010.1376
96. Wysocki LE, Ladich F (2005) Effects of noise exposure on click detection and the temporal resolution ability of the goldfish auditory system. Hear Res 201:27–36. https://doi.org/10.1016/j.heares.2004.08.015
97. Popper AN (1972) Auditory threshold in the goldfish (*Carassius auratus*) as a function of signal duration. J Acoust Soc Am 52:596–602. https://doi.org/10.1121/1.1913150
98. Wysocki LE, Ladich F (2002) Can fishes resolve temporal characteristics of sounds? New insights using auditory brainstem responses. Hear Res 169:36–46. https://doi.org/10.1016/S0378-5955(02)00336-2
99. Wysocki LE, Ladich F (2003) The representation of conspecific sounds in the auditory brainstem of teleost fishes. J Exp Biol 206:2229–2240. https://doi.org/10.1242/jeb.00417
100. Wysocki LE (2006) Detection of communication sounds. In: Ladich F, Collin SP, Moller P, Kapoor BG (eds) Communication in fishes. Science Publishers, Enfield, pp 175–203
101. Bass AH, Ladich F (2008) Vocal–acoustic communication: from neurons to behavior. In: Webb JF, Fay RR, Popper AN (eds) Fish bioacoustics. Springer, New York, pp 253–278
102. Parmentier E, Lecchini D, Frederich B et al (2009) Sound production in four damselfish (*Dascyllus*) species: phyletic relationships? Biol J Linn Soc 97:928–940. https://doi.org/10.1111/j.1095-8312.2009.01260.x
103. Hawkins AD, Amorim MCP (2000) Spawning sounds of the male haddock, *Melanogrammus aeglefinus*. Environ Biol Fish 59:29–41. https://doi.org/10.1023/A:1007615517287
104. Bolgan M, Crucianelli A, Mylonas CC et al (2020) Calling activity and calls' temporal features inform about fish reproductive condition and spawning in three cultured Sciaenidae species. Aquaculture 524:735243. https://doi.org/10.1016/j.aquaculture.2020.735243
105. Picciulin M, Fiorin R, Facca C, Malavasi S (2020) Sound features and vocal rhythms as a proxy for locating the spawning ground of Sciaena umbra in the wild. Aquat Conserv Mar Freshw Ecosyst 30:1299–1312. https://doi.org/10.1002/aqc.3340
106. Ladich F (2018) Acoustic communication in fishes: temperature plays a role. Fish Fish 19:598–612. https://doi.org/10.1111/faf.12277
107. Kéver L, Boyle KS, Parmentier E (2015) Effects of seawater temperature on sound characteristics in *Ophidion rochei* (Ophidiidae). J Fish Biol 87:502–509. https://doi.org/10.1111/jfb.12730
108. Picciulin M, Sebastianutto L, Codarin A et al (2012) Brown meagre vocalization rate increases during repetitive boat noise exposures: a possible case of vocal compensation. J Acoust Soc Am 132:3118–3124. https://doi.org/10.1121/1.4756928
109. Ramcharitar J, Gannon DP, Popper AN (2006) Bioacoustics of fishes of the family Sciaenidae (croakers and drums). Trans Am Fish Soc 135:1409–1431. https://doi.org/10.1577/T05-207.1
110. Luczkovich JJ, Sprague MW, Johnson SE, Pullinger RC (1999) Delimiting spawning areas of weakfish *Cynoscion regalis* (family Sciaenidae) in Pamlico sound, North Carolina using

Passive Hydroacoustic Surveys. Bioacoustics 10:143–160. https://doi.org/10.1080/09524622.1999.9753427

111. Montie EW, Kehrer C, Yost J et al (2016) Long-term monitoring of captive red drum *Sciaenops ocellatus* reveals that calling incidence and structure correlate with egg deposition. J Fish Biol 88:1776–1795. https://doi.org/10.1111/jfb.12938
112. Montie EW, Hoover M, Kehrer C et al (2016) Acoustic monitoring indicates a positive relationship between calling frequency and spawning in captive spotted seatrout (*Cynoscion nebulosus*). Peer J Inc
113. Ladich F, Bischof C, Schleinzer G, Fuchs A (1992) Intra- and interspecific differences in agonistic vocalization in croaking Gouramis (genus: *Trichopsis*, Anabantoidei, TELeostei). Bioacoustics 4:131–141. https://doi.org/10.1080/09524622.1992.9753212
114. Ladich F, Schleinzer G (2015) Effect of temperature on acoustic communication: sound production in the croaking gourami (labyrinth fishes). Comp Biochem Physiol A Mol Integr Physiol 182:8–13. https://doi.org/10.1016/j.cbpa.2014.11.013
115. Bass AH, Chagnaud BP (2012) Shared developmental and evolutionary origins for neural basis of vocal–acoustic and pectoral–gestural signaling. Proc Natl Acad Sci 109:10677–10684. https://doi.org/10.1073/pnas.1201886109
116. McIver EL, Marchaterre MA, Rice AN, Bass AH (2014) Novel underwater soundscape: acoustic repertoire of plainfin midshipman fish. J Exp Biol:jeb.102772. https://doi.org/10.1242/jeb.102772
117. Connaughton MA, Taylor MH, Fine ML (2000) Effects of fish size and temperature on weakfish disturbance calls: implications for the mechanism of sound generation. J Exp Biol 203:1503–1512. https://doi.org/10.1242/jeb.203.9.1503
118. Amorim MCP, Hawkins AD (2005) Ontogeny of acoustic and feeding behaviour in the Grey gurnard, *Eutrigla gurnardus*. Ethology 111:255–269. https://doi.org/10.1111/j.1439-0310.2004.01061.x
119. Colleye O, Frederich B, Vandewalle P et al (2009) Agonistic sounds in the skunk clownfish *Amphiprion akallopisos*: size-related variation in acoustic features. J Fish Biol 75:908–916. https://doi.org/10.1111/j.1095-8649.2009.02316.x
120. Tellechea JS, Martinez C, Fine ML, Norbis W (2010) Sound production in the whitemouth croaker and relationship between fish size and disturbance call characteristics. Environ Biol Fish 89:163–172. https://doi.org/10.1007/s10641-010-9709-7
121. Knight L, Ladich F (2014) Distress sounds of thorny catfishes emitted underwater and in air: characteristics and potential significance. J Exp Biol:jeb.110957. https://doi.org/10.1242/jeb.110957
122. Hadjiaghai O, Ladich F (2015) Sex-specific differences in agonistic behaviour, sound production and auditory sensitivity in the callichthyid armoured catfish *Megalechis thoracata*. PLoS One 10:e0121219. https://doi.org/10.1371/journal.pone.0121219
123. Vasconcelos RO, Ladich F (2008) Development of vocalization, auditory sensitivity and acoustic communication in the Lusitanian toadfish *Halobatrachus didactylus*. J Exp Biol 211:502–509. https://doi.org/10.1242/jeb.008474
124. Ladich F (1997) Comparative analysis of Swimbladder (drumming) and pectoral (stridulation) sounds in three families of catfishes. Bioacoustics 8:185–208. https://doi.org/10.1080/09524622.1997.9753362
125. Pruzsinszky I, Ladich F (1998) Sound production and reproductive behaviour of the armoured catfish *Corydoras paleatus* (Callichthyidae). Environ Biol Fish 53:183–191. https://doi.org/10.1023/A:1007413108550
126. Burchardt LS, Picciulin M, Parmentier E, Bolgan M (2021) A primer on rhythm quantification for fish sounds: a Mediterranean case study. R Soc Open Sci 8:210494. https://doi.org/10.1098/rsos.210494
127. Picciulin M, Bolgan M, Burchardt LS (2024) Rhythmic properties of Sciaena umbra calls across space and time in the Mediterranean Sea. PLoS One 19:e0295589. https://doi.org/10.1371/journal.pone.0295589

128. Parmentier E, Bouillac G, Dragičević B et al (2010) Call properties and morphology of the sound-producing organ in *Ophidion rochei* (Ophidiidae). J Exp Biol 213:3230–3236. https://doi.org/10.1242/jeb.044701
129. Butler JM, Maruska KP (2015) The mechanosensory lateral line is used to assess opponents and mediate aggressive behaviors during territorial interactions in an African cichlid fish. J Exp Biol 218:3284–3294. https://doi.org/10.1242/jeb.125948
130. Webb JF (2020) Morphology of the mechanosensory lateral line system of fishes. In: The senses: a comprehensive reference. Elsevier, pp 29–46
131. Ashraf I, Godoy-Diana R, Halloy J et al (2016) Synchronization and collective swimming patterns in fish (*Hemigrammus bleheri*). J R Soc Interface 13:20160734. https://doi.org/10.1098/rsif.2016.0734
132. Semenchenko S, Smeshlivaya N (2021) Spawning behaviour of whitefishes (Coregonidae). Ann Zool Fenn 58:129–140. https://doi.org/10.5735/086.058.0402
133. Almada VC, Santos RS (1995) Parental care in the rocky intertidal: a case study of adaptation and exaptation in Mediterranean and Atlantic blennies. Rev Fish Biol Fish 5:23–37. https://doi.org/10.1007/BF01103364
134. Gruber DF, Phillips BT, O'Brien R et al (2019) Bioluminescent flashes drive nighttime schooling behavior and synchronized swimming dynamics in flashlight fish. PLoS One 14:e0219852. https://doi.org/10.1371/journal.pone.0219852
135. Morin JG, Harrington A, Nealson K et al (1975) Light for all reasons: versatility in the behavioral repertoire of the flashlight fish. Science 190:74–76. https://doi.org/10.1126/science.190.4209.74
136. AmphibiaWeb (2024) AmphibiaWeb
137. Wells KD (2007) The ecology and behavior of amphibians. University of Chicago Press, Chicago
138. Busnel R-G, Dumortier B (1955) Phonoreactions de male d' *Hyla arborea* a des signaux acoustiques artificiels. Bull Soc Zoo 80:66–69
139. Kwong-Brown U, Tobias ML, Elias DO et al (2019) The return to water in ancestral *Xenopus* was accompanied by a novel mechanism for producing and shaping vocal signals. eLife 8:e39946. https://doi.org/10.7554/eLife.39946
140. Martin WF (1971) Mechanics of sound production in toads of the genus *Bufo*: Passive elements. J Exp Zool 176:273–293. https://doi.org/10.1002/jez.1401760304
141. Gridi-Papp M (2014) Is the frequency content of the calls in North American Treefrogs limited by their larynges? Int J Evol Biol 2014:198069. https://doi.org/10.1155/2014/198069
142. Liu C (1935) Types of vocal sac in the Salientia. Boston Society of Natural History, Boston
143. Boistel R, Aubin T, Cloetens P et al (2011) Whispering to the deaf: communication by a frog without external vocal sac or tympanum in noisy environments. PLoS One 6:e22080. https://doi.org/10.1371/journal.pone.0022080
144. Zweifel RG (1959) Effect of temperature on call of the frog, *Bombina variegata*. Copeia 1959:322–327. https://doi.org/10.2307/1439890
145. Weber E (1974) Comparative bio-acoustic investigations in the *Discoglossus pictus*, Otth 1837 and the *Discoglossus sardus* Tschudi 1837 (Discoglossidae, Anura). Zool Jahrbucher Physiol 78:40–84
146. Girgenrath M, Marsh RL (1997) In vivo performance of trunk muscles in tree frogs during calling. J Exp Biol 200:3101–3108. https://doi.org/10.1242/jeb.200.24.3101
147. Girgenrath M, Marsh RL (1999) Power output of sound-producing muscles in the tree frogs *Hyla versicolor* and *Hyla chrysoscelis*. J Exp Biol 202:3225–3237. https://doi.org/10.1242/jeb.202.22.3225
148. Yager DD (1992) A unique sound production mechanism in the pipid anuran *Xenopus borealis*. Zool J Linnean Soc 104:351–375. https://doi.org/10.1111/j.1096-3642.1992.tb00927.x
149. Tobias ML, Evans BJ, Kelley DB (2011) Evolution of advertisement calls in African clawed frogs. Behaviour 148:519–549. https://doi.org/10.1163/000579511X569435

150. Köhler J, Jansen M, Rodríguez A et al (2017) The use of bioacoustics in anuran taxonomy: theory, terminology, methods and recommendations for best practice. Zootaxa 4251:1–124. https://doi.org/10.11646/zootaxa.4251.1.1
151. Cossio R, Medina-Barcenas E (2020) Acoustic repertoire and calling behavior of the gliding Treefrog, *Agalychnis spurrelli* (Anura: Hylidae). South Am J Herpetol 17:71–78. https://doi.org/10.2994/SAJH-D-18-00040.1
152. Bee MA, Perrill SA (1996) Responses to conspecific advertisement calls in the green frog (*Rana clamitans*) and their role in male-male communication. Behaviour 133:283–301. https://doi.org/10.1163/156853996X00152
153. Yu B, Zheng R (2009) The advertisement call of the giant spiny frog *Paa spinosa*. Curr Zool 55:411–415. https://doi.org/10.1093/czoolo/55.6.411
154. Höbel G (2005) On the acoustic communication system of *Eleutherodactylus fitzingeri* (Anura: Leptodactylidae). Herpetol Rev 36:241–243
155. Pröhl H (2005) Territorial behavior in Dendrobatid frogs. J Herpetol 39:354–365
156. Duellman WE, Trueb L (1994) Biology of amphibians. Johns Hopkins University Press, Baltimore
157. Capshaw G, Foss-Grant AP, Hartmann K et al (2020) Timing of the advertisement call of the common tink frog (*Diasporus diastema*) shifts with the acoustic behaviour of local conspecifics. Bioacoustics 29:79–96. https://doi.org/10.1080/09524622.2018.1555715
158. Rosen M, Lemon RE (1974) The vocal behavior of spring peepers, *Hyla crucifer*. Copeia 1974:940–950. https://doi.org/10.2307/1442594
159. Höbel G (2011) Variation in signal timing behavior: implications for male attractiveness and sexual selection. Behav Ecol Sociobiol 65:1283–1294. https://doi.org/10.1007/s00265-011-1142-1
160. Goin CJ (1948) The peep order in peepers; a swamp water serenade. Q J Fla Acad Sci 11:59–61
161. Hardy DF (1959) Chorus structure in the striped chorus frog, *Pseudacris nigrita*. Herpetologica 15:14–16
162. Loftus-Hills JJ (1974) Analysis of an acoustic pacemaker in Strecker's chorus frog, *Pseudacris streckeri* (Anura: Hylidae). J Comp Physiol 90:75–87. https://doi.org/10.1007/BF00698369
163. Wells KD (1977) The social behaviour of anuran amphibians. Anim Behav 25:666–693. https://doi.org/10.1016/0003-3472(77)90118-X
164. Grafe TU (1999) A function of synchronous chorusing and a novel female preference shift in an anuran. Proc R Soc Lond B Biol Sci 266:2331–2336. https://doi.org/10.1098/rspb.1999.0927
165. Wickler W, Seibt U (1974) Rufen und Antworten bei *Kassina senegalensis*, *Bufo regularis* und anderen Anuren. Z Fuer Tierpsychol 34:524–537
166. Legett HD, Aihara I, Bernal XE (2020) Signal synchrony and alternation among neighbor males in a Japanese stream breeding Treefrog, *Buergeria japonica*. Curr Herpetol 39:80–85. https://doi.org/10.5358/hsj.39.80
167. Clulow S, Mahony M, Elliott L et al (2017) Near-synchronous calling in the hip-pocket frog *Assa darlingtoni*. Bioacoustics 26:249–258. https://doi.org/10.1080/09524622.2016.1260054
168. Ibáñez R (1991) Synchronized Calling in *Centrolenella granulosa* and *Smilisca sila* (Amphibia, Anura). University of Connecticut
169. Wells KD, Schwartz JJ (1984) Vocal communication in a neotropical treefrog, *Hyla ebraccata*: advertisement calls. Anim Behav 32:405–420. https://doi.org/10.1016/S0003-3472(84)80277-8
170. Reichert MS (2012) Call timing is determined by response call type, but not by stimulus properties, in the treefrog *Dendropsophus ebraccatus*. Behav Ecol Sociobiol 66:433–444. https://doi.org/10.1007/s00265-011-1289-9
171. Murphy JB (1976) Pedal luring in the Leptodactylid frog, *Ceratophrys calcarata* Boulenger. Herpetologica 32:339–341

172. Hodl W, Amézquita A (2001) Visual signaling in anuran amphibians. In: Anuran Communication, pp 121–141
173. Anderson NK, Gururaja KV, Mangiamele LA et al (2021) Insight into the evolution of anuran foot flag displays: a comparative study of color and kinematics. Ichthyol Herpetol 109:1047–1059. https://doi.org/10.1643/h2020160
174. Sloggett JJ, Zeilstra I (2008) Waving or tapping? Vibrational stimuli and the general function of toe twitching in frogs and toads (Amphibia: Anura). Anim Behav 76:e1–e4. https://doi.org/10.1016/j.anbehav.2008.08.005
175. Erdmann J (2017) The function of toe movement in feeding by The Gulf Coast Toad (*Incilius nebulifer*). Southeastern Louisiana University
176. Claessens LSA, Ganchev NO, Kukk MM et al (2020) An investigation of toe-tapping behaviour in anurans by analysis of online video resources. J Zool 312:158–162. https://doi.org/10.1111/jzo.12815
177. Furtado R, Nomura F (2014) Visual signals or displacement activities? The function of visual displays in agonistic interactions in nocturnal tree frogs. acta ethologica 17:9–14. https://doi.org/10.1007/s10211-013-0160-6
178. Furtado R, Márquez R, Hartz SM (2017) In front of a mirror: visual displays may not be aggressive signals in nocturnal tree frogs. J Nat Hist 51:443–454. https://doi.org/10.1080/00222933.2016.1262078
179. Starnberger I, Maier PM, Hödl W, Preininger D (2018) Multimodal signal testing reveals gestural tapping behavior in spotted reed frogs. Herpetologica 74:127–134. https://doi.org/10.1655/Herpetologica-D-17-00053.1
180. Hagman M, Shine R (2008) Deceptive digits: the functional significance of toe waving by cannibalistic cane toads, *Chaunus marinus*. Anim Behav 75:123–131. https://doi.org/10.1016/j.anbehav.2007.04.020
181. Schulte LM, König Y (2023) Experimental evidence that toe-tapping behavior in the green-and-black poison frog (*Dendrobates auratus*) is related to prey detection. acta ethologica 26:139–143. https://doi.org/10.1007/s10211-023-00422-8
182. Vergara-Herrera N, Cocroft R, Rueda-Solano LA (2023) Eating to the beat of the drum: vibrational parameters of toe tapping behavior in *Dendrobates truncatus* (Anura: Dendrobatidae). Evol Ecol. https://doi.org/10.1007/s10682-023-10277-x
183. Parrish TQ, Fischer EK (2024) Tap dancing frogs: posterior toe tapping and feeding in *Dendrobates tinctorius*. Ethology 130:e13465. https://doi.org/10.1111/eth.13465
184. Ewert J-P (1987) Neuroethology of releasing mechanisms: prey-catching in toads. Behav Brain Sci 10:337–368. https://doi.org/10.1017/S0140525X00023128
185. Proksch S, Honing H, Ravignani A, Merchant H (2025) Chapter 7: Converging notions on the main hypotheses regarding the origins of beat perception and synchronization: the key role of audiomotor interactions. In: Nature beats: the what, why and how of animal rhythmic behaviour. Springer
186. Merker B, Cook PF, Reichmuth C et al (2025) Chapter 14: Human group synchrony in light of the comparative method: what we know and would like to know. A peer commentary treatment. In: Sauciuc G-A, Lameira AR, Persson T (eds) Nature beats: the what, why and how of animal rhythmic behaviour. Springer
187. Roeske T, Larrouy-Maestri P, Sakamoto Y, Poeppel D (2020) Listening to birdsong reveals basic features of rate perception and aesthetic judgements. Proc R Soc B Biol Sci 287:20193010. https://doi.org/10.1098/rspb.2019.3010
188. Mol C, Chen A, Kager RWJ, Ter Haar SM (2017) Prosody in birdsong: a review and perspective. Neurosci Biobehav Rev 81:167–180. https://doi.org/10.1016/j.neubiorev.2017.02.016
189. Rothenberg D, Roeske TC, Voss HU et al (2014) Investigation of musicality in birdsong. Hear Res 308:71–83. https://doi.org/10.1016/j.heares.2013.08.016
190. Price JJ (2013) Why is birdsong so repetitive? Signal detection and the evolution of avian singing modes. Behaviour 150:995–1013. https://doi.org/10.1163/1568539X-00003051

191. Alward BA, Cornil CA, Balthazart J, Ball GF (2018) The regulation of birdsong by testosterone: multiple time-scales and multiple sites of action. Horm Behav 104:32–40. https://doi.org/10.1016/j.yhbeh.2018.04.010
192. Burt JM, Vehrencamp SL (2005) Dawn chorus as an interactive communication network. In: McGregor PK (ed) Animal communication networks. Cambridge University Press, Cambridge, pp 320–343
193. Searcy WA, Beecher MD (2009) Song as an aggressive signal in songbirds. Anim Behav 78:1281–1292. https://doi.org/10.1016/j.anbehav.2009.08.011
194. Krebs JR, Ashcroft R, Orsdol KV (1981) Song matching in the great tit *Parus major* L. Anim Behav 29:918–923. https://doi.org/10.1016/S0003-3472(81)80029-2
195. Vehrencamp SL (2001) Is song-type matching a conventional signal of aggressive intentions? Proc Biol Sci 268:1637–1642
196. Burt JM, Campbell SE, Beecher MD (2001) Song type matching as threat: a test using interactive playback. Anim Behav 62:1163–1170. https://doi.org/10.1006/anbe.2001.1847
197. Osiecka AN, Oliva MQ, Kouřil J et al (2025) Yellowhammer (*Emberiza citrinella*) males sing using individual isochronous rhythms and maximise rhythmic dissimilarity with neighbours
198. Rouse AA, Patel AD, Kao MH (2021) Vocal learning and flexible rhythm pattern perception are linked: evidence from songbirds. Proc Natl Acad Sci 118:e2026130118. https://doi.org/10.1073/pnas.2026130118
199. Podos J (1997) A performance constraint on the evolution of trilled vocalizations in a songbird family (Passeriformes: Emberizidae). Evolution 51:537–551. https://doi.org/10.1111/j.1558-5646.1997.tb02441.x
200. Geberzahn N, Aubin T (2014) Assessing vocal performance in complex birdsong: a novel approach. BMC Biol 12:58. https://doi.org/10.1186/s12915-014-0058-4
201. Sauciuc G-A, Schneider JN, Hattori Y (2025) Chapter 9: The neuroethology of rhythmic communication: neural and mechanical building blocks and constraints on rhythm production and perception. In: Sauciuc G-A, Lameira AR, Persson T (eds) Nature beats: the what, why and how of animal rhythmic behaviours. Springer
202. DuBois AL, Nowicki S, Searcy WA (2009) Swamp sparrows modulate vocal performance in an aggressive context. Biol Lett 5:163–165. https://doi.org/10.1098/rsbl.2008.0626
203. Vehrencamp SL, Yantachka J, Hall ML, De Kort SR (2013) Trill performance components vary with age, season, and motivation in the banded wren. Behav Ecol Sociobiol 67:409–419. https://doi.org/10.1007/s00265-012-1461-x
204. Linhart P, Jaška P, Petrusková T et al (2013) Being angry, singing fast? Signalling of aggressive motivation by syllable rate in a songbird with slow song. Behav Process 100:139–145. https://doi.org/10.1016/j.beproc.2013.06.012
205. DuBois AL, Nowicki S, Searcy WA (2011) Discrimination of vocal performance by male swamp sparrows. Behav Ecol Sociobiol 65:717–726
206. Phillips JN, Derryberry EP (2017) Equivalent effects of bandwidth and trill rate: support for a performance constraint as a competitive signal. Anim Behav 132:209–215. https://doi.org/10.1016/j.anbehav.2017.08.012
207. Drăgănoiu TI, Nagle L, Kreutzer M (2002) Directional female preference for an exaggerated male trait in canary (*Serinusanaria*) song. Proc R Soc Lond B Biol Sci 269:2525–2531. https://doi.org/10.1098/rspb.2002.2192
208. Ballentine B (2004) Vocal performance influences female response to male bird song: an experimental test. Behav Ecol 15:163–168. https://doi.org/10.1093/beheco/arg090
209. Ballentine B (2009) The ability to perform physically challenging songs predicts age and size in male swamp sparrows, *Melospiza georgiana*. Anim Behav 77:973–978. https://doi.org/10.1016/j.anbehav.2008.12.027
210. Juola FA, Searcy WA (2011) Vocalizations reveal body condition and are associated with visual display traits in great frigatebirds (*Fregata minor*). Behav Ecol Sociobiol 65:2297–2303. https://doi.org/10.1007/s00265-011-1240-0

211. De Kort SR, Eldermire ERB, Valderrama S et al (2009) Trill consistency is an age-related assessment signal in banded wrens. Proc R Soc B Biol Sci 276:2315–2321. https://doi.org/10.1098/rspb.2009.0127
212. Rivera-Gutierrez HF, Pinxten R, Eens M (2012) Tuning and fading voices in songbirds: age-dependent changes in two acoustic traits across the life span. Anim Behav 83:1279–1283. https://doi.org/10.1016/j.anbehav.2012.03.001
213. Rivera-Gutierrez HF, Pinxten R, Eens M (2011) Songs differing in consistency elicit differential aggressive response in territorial birds. Biol Lett 7:339–342. https://doi.org/10.1098/rsbl.2010.0962
214. Sierro J, De Kort SR, Hartley IR (2023) Sexual selection for both diversity and repetition in birdsong. Nat Commun 14:3600. https://doi.org/10.1038/s41467-023-39308-5
215. Botero CA, Rossman RJ, Caro LM et al (2009) Syllable type consistency is related to age, social status, and reproductive success in the tropical mockingbird. Anim Behav 77:701–706. https://doi.org/10.1016/j.anbehav.2008.11.020
216. Byers BE (2007) Extrapair paternity in chestnut-sided warblers is correlated with consistent vocal performance. Behav Ecol 18:130–136. https://doi.org/10.1093/beheco/arl058
217. Byers BE, Akresh ME, King DI (2016) Song and male quality in prairie warblers. Ethology 122:660–670. https://doi.org/10.1111/eth.12513
218. Goller F (2022) Vocal athletics—from birdsong production mechanisms to sexy songs. Anim Behav 184:173–184. https://doi.org/10.1016/j.anbehav.2021.04.009
219. Prum RO (2012) Aesthetic evolution by mate choice: Darwin's *really* dangerous idea. Philos Trans R Soc B Biol Sci 367:2253–2265. https://doi.org/10.1098/rstb.2011.0285
220. Webb WH, Brunton DH, Aguirre JD et al (2016) Female song occurs in songbirds with more elaborate female coloration and reduced sexual dichromatism. Front Ecol Evol 4. https://doi.org/10.3389/fevo.2016.00022
221. Riebel K, Odom KJ, Langmore NE, Hall ML (2019) New insights from female bird song: towards an integrated approach to studying male and female communication roles. Biol Lett 15:20190059. https://doi.org/10.1098/rsbl.2019.0059
222. Hall ML, Rittenbach MRD, Vehrencamp SL (2015) Female song and vocal interactions with males in a neotropical wren. Front Ecol Evol 3. https://doi.org/10.3389/fevo.2015.00012
223. Illes AE, Yunes-Jimenez L (2009) A female songbird out-sings male conspecifics during simulated territorial intrusions. Proc R Soc B Biol Sci 276:981–986. https://doi.org/10.1098/rspb.2008.1445
224. Pilowsky JA, Rubenstein DR (2013) Social context and the lack of sexual dimorphism in song in an avian cooperative breeder. Anim Behav 85:709–714. https://doi.org/10.1016/j.anbehav.2013.01.012
225. Langmore NE, Davies NB, Hatchwell BJ, Hartley IR (1996) Female song attracts males in the alpine accentor *Prunella collaris*. Proc Biol Sci 263:141–146
226. Price JJ, Yunes-Jiménez L, Osorio-Beristain M et al (2008) Sex-role reversal in song? Females sing more frequently THAN males in the streak-backed oriole. Condor 110:387–392. https://doi.org/10.1525/cond.2008.8430
227. Goldberg DL, Sadd BM, Capparella AP (2023) A rallid ballad: communal signaling is correlated with year-round territoriality in the most duet-rich family of birds (Gruiformes: Rallidae). Ornithology 140:ukac054. https://doi.org/10.1093/ornithology/ukac054
228. Dahlin CR, Benedict L (2014) Angry birds need not apply: a perspective on the flexible form and multifunctionality of avian vocal duets. Ethology 120:1–10. https://doi.org/10.1111/eth.12182
229. Hall ML (2009) Chapter 3: A review of vocal duetting in birds. In: Advances in the study of behavior. Academic, pp 67–121
230. Keenan EL, Odom KJ, Araya-Salas M et al (2020) Breeding season length predicts duet coordination and consistency in Neotropical wrens (Troglodytidae). Proc R Soc B Biol Sci 287:20202482. https://doi.org/10.1098/rspb.2020.2482

231. Lemazina A, Trost L, Gahr M, Hoffmann S (2021) The multifaceted vocal duets of white-browed sparrow weavers are based on complex duetting rules. J Avian Biol 52:jav.02703. https://doi.org/10.1111/jav.02703
232. Mann NI, Marshall-Ball L, Slater PJB (2003) The complex song duet of the plain wren. Condor 105:672–682
233. Mann NI, Dingess KA, Barker FK et al (2009) A comparative study of song form and duetting in NEOTROPICAL Thryothorus wrens. Behaviour 146:1–43
234. Fortune ES, Rodríguez C, Li D et al (2011) Neural mechanisms for the coordination of duet singing in wrens. Science 334:666–670. https://doi.org/10.1126/science.1209867
235. Jedlikowski J, Ręk P (2025) Rhythmic coordination in rapid duets of little crakes: cooperative signalling without shared experience. Proc R Soc B Biol Sci 292:20251100. https://doi.org/10.1098/rspb.2025.1100
236. Hoffmann S, Trost L, Voigt C et al (2019) Duets recorded in the wild reveal that interindividually coordinated motor control enables cooperative behavior. Nat Commun 10:2577. https://doi.org/10.1038/s41467-019-10593-3
237. Coleman MJ, Day NF, Rivera-Parra P, Fortune ES (2021) Neurophysiological coordination of duet singing. Proc Natl Acad Sci USA 118:e2018188118. https://doi.org/10.1073/pnas.2018188118
238. Kovach KA, Hall ML, Vehrencamp SL, Mennill DJ (2014) Timing isn't everything: responses of tropical wrens to coordinated duets, uncoordinated duets and alternating solos. Anim Behav 95:101–109. https://doi.org/10.1016/j.anbehav.2014.06.012
239. Templeton CN, Mann NI, Ríos-Chelén AA et al (2013) An experimental study of duet integration in the happy wren, *Pheugopedius felix*. Anim Behav 86:821–827. https://doi.org/10.1016/j.anbehav.2013.07.022
240. Rivera-Cáceres KD (2015) Plain wrens *Cantorchilus modestus zeledoni* adjust their singing tempo based on self and partner's cues to perform precisely coordinated duets. J Avian Biol 46:361–368
241. Hall ML (2004) A review of hypotheses for the functions of avian duetting. Behav Ecol Sociobiol 55:415–430. https://doi.org/10.1007/s00265-003-0741-x
242. Mennill DJ (2006) Aggressive responses of male and female rufous-and-white wrens to stereo duet playback. Anim Behav 71:219–226. https://doi.org/10.1016/j.anbehav.2005.05.006
243. Mennill DJ, Vehrencamp SL (2008) Context-dependent functions of avian duets revealed by microphone-Array recordings and multispeaker playback. Curr Biol 18:1314–1319. https://doi.org/10.1016/j.cub.2008.07.073
244. Sandoval L, Juárez R, Villarreal M (2018) Different messages are transmitted by individual duet contributions and complete duets in a species with highly overlapped duets. Open Ornithol J 11:56–67. https://doi.org/10.2174/1874453201811010056
245. Diniz P, Ramos DM, Webster MS, Macedo RH (2021) Rufous horneros perceive and alter temporal coordination of duets during territorial interactions. Anim Behav 174:175–185. https://doi.org/10.1016/j.anbehav.2021.02.007
246. Elie JE, Mariette MM, Soula HA et al (2010) Vocal communication at the nest between mates in wild zebra finches: a private vocal duet? Anim Behav 80:597–605. https://doi.org/10.1016/j.anbehav.2010.06.003
247. Ręk P, Magrath RD (2020) Visual displays enhance vocal duet production and the perception of coordination despite spatial separation of partners. Anim Behav 168:231–241. https://doi.org/10.1016/j.anbehav.2020.08.002
248. Koch RE, Krakauer AH, Patricelli GL (2015) Investigating female mate choice for mechanical sounds in the male greater sage-grouse. Auk 132:349–358. https://doi.org/10.1642/AUK-14-63.1
249. Clark CJ, Elias DO, Prum RO (2013) Hummingbird feather sounds are produced by aeroelastic flutter, not vortex-induced vibration. J Exp Biol:jeb.080317. https://doi.org/10.1242/jeb.080317

250. Miller EH, Areta JI, Jaramillo A et al (2020) Snipe taxonomy based on vocal and non-vocal sound displays: the South American Snipe is two species. Ibis 162:968–990. https://doi.org/10.1111/ibi.12795
251. Prum RO (1998) Sexual selection and the evolution of mechanical sound production in manakins (Aves: Pipridae). Anim Behav 55:977–994. https://doi.org/10.1006/anbe.1997.0647
252. Eda-Fujiwara H, Yamamoto A, Sugita H et al (2004) Sexual dimorphism of acoustic signals in the oriental White Stork: non-invasive identification of sex in birds. Zool Sci 21:817–821. https://doi.org/10.2108/zsj.21.817
253. Soma M, Mori C (2015) The songbird as a percussionist: syntactic rules for non-vocal sound and song production in Java sparrows. PLoS One 10:e0124876. https://doi.org/10.1371/journal.pone.0124876
254. Schuppe ER, Rutter AR, Roberts TJ, Fuxjager MJ (2021) Evolutionary and biomechanical basis of drumming behavior in woodpeckers. Front Ecol Evol 9:649146. https://doi.org/10.3389/fevo.2021.649146
255. Ota N, Gahr M, Soma M (2015) Tap dancing birds: the multimodal mutual courtship display of males and females in a socially monogamous songbird. Sci Rep 5:16614. https://doi.org/10.1038/srep16614
256. Heinsohn R, Zdenek CN, Cunningham RB et al (2017) Tool-assisted rhythmic drumming in palm cockatoos shares key elements of human instrumental music. Sci Adv 3:e1602399. https://doi.org/10.1126/sciadv.1602399
257. Ota N, Soma M (2022) Vibrational signals in multimodal courtship displays of birds. In: Hill PSM, Mazzoni V, Stritih-Peljhan N et al (eds) Biotremology: physiology, ecology, and evolution. Springer, Cham, pp 237–259
258. Miles MC, Schuppe ER, Ligon RM, Fuxjager MJ (2018) Macroevolutionary patterning of woodpecker drums reveals how sexual selection elaborates signals under constraint. Proc R Soc B Biol Sci 285:20172628. https://doi.org/10.1098/rspb.2017.2628
259. Winkler H, Short LL (1978) A comparative analysis of acoustical signals in pied woodpeckers (Aves, *Picoides*). Bulletin of the AMNH 160:Article 1
260. Dodenhoff DJ, Stark RD, Johnson EV (2001) Do woodpecker drums encode information for species recognition? Condor 103:143–150. https://doi.org/10.1093/condor/103.1.143
261. Schuppe ER, Sanin GD, Fuxjager MJ (2016) The social context of a territorial dispute differentially influences the way individuals in breeding pairs coordinate their aggressive tactics. Behav Ecol Sociobiol 70:673–682. https://doi.org/10.1007/s00265-016-2088-0
262. Figarski T (2017) Contrasting seasonal reactions of two sibling woodpeckers to playback stimulation in urban areas – implications for inventory and monitoring of the Syrian woodpecker. Behaviour 154:981–996. https://doi.org/10.1163/1568539X-00003452
263. Schuppe ER, Fuxjager MJ (2018) High-speed displays encoding motor skill trigger elevated territorial aggression in downy woodpeckers. Funct Ecol 32:450–460. https://doi.org/10.1111/1365-2435.13010
264. Stark RD, Dodenhoff DJ, Johnson EV (1998) A quantitative analysis of woodpecker drumming. Condor 100:350–356. https://doi.org/10.2307/1370276
265. Wilkins HD, Ritchison G (1999) Drumming and tapping by red-bellied woodpeckers: description and possible causation (Descripción y Posible causa del Tamborileo y Picoteo lento en *Melanerpes carolinus*). J Field Ornithol 70:578–586
266. Garcia M, Theunissen F, Sèbe F et al (2020) Evolution of communication signals and information during species radiation. Nat Commun 11:4970. https://doi.org/10.1038/s41467-020-18772-3
267. Aubin AE (1972) Aural communication in ruffed grouse. Can J Zool 50:1225–1229. https://doi.org/10.1139/z72-165
268. Garcia M (2012) Directionality of the drumming display of the ruffed grouse. Condor 114:500–506. https://doi.org/10.1525/cond.2012.110158
269. Murphy S, Legge S, Heinsohn R (2003) The breeding biology of palm cockatoos (*Probosciger aterrimus*): a case of a slow life history. J Zool 261:327–339. https://doi.org/10.1017/S0952836903004175

270. Zdenek CN, Heinsohn R, Langmore NE (2015) Vocal complexity in the palm cockatoo (*Probosciger aterrimus*). Bioacoustics 24:253–267. https://doi.org/10.1080/09524622.2015.1070281
271. Soma M, Garamszegi LZ (2015) Evolution of courtship display in Estrildid finches: dance in relation to female song and plumage ornamentation. Front Ecol Evol 3. https://doi.org/10.3389/fevo.2015.00004
272. Ota N (2020) Tap dancers in the wild: field observations of multimodal courtship displays in socially monogamous songbirds. Sci Nat 107:30. https://doi.org/10.1007/s00114-020-01686-x
273. Ota N, Gahr M, Soma M (2018) Couples showing off: audience promotes both male and female multimodal courtship display in a songbird. Sci Adv 4:eaat4779. https://doi.org/10.1126/sciadv.aat4779
274. Ota N, Gahr M, Soma M (2017) Songbird tap dancing produces non-vocal sounds. Bioacoustics 26:161–168. https://doi.org/10.1080/09524622.2016.1231080
275. Ota N, Gahr M (2022) Context-sensitive dance–vocal displays affect song patterns and partner responses in a socially monogamous songbird. Ethology 128:61–69. https://doi.org/10.1111/eth.13240
276. Cooper BG, Goller F (2004) Multimodal signals: enhancement and constraint of song motor patterns by visual display. Science 303:544–546. https://doi.org/10.1126/science.1091099
277. Miles MC, Fuxjager MJ (2018) Animal choreography of song and dance: a case study in the Montezuma oropendola, *Psarocolius montezuma*. Anim Behav 140:99–107. https://doi.org/10.1016/j.anbehav.2018.04.006
278. Dalziell AH, Peters RA, Cockburn A et al (2013) Dance choreography is coordinated with song repertoire in a complex avian display. Curr Biol 23:1132–1135. https://doi.org/10.1016/j.cub.2013.05.018
279. Jones G, Teeling E (2006) The evolution of echolocation in bats. Trends Ecol Evol 21:149–156. https://doi.org/10.1016/j.tree.2006.01.001
280. Schnitzler H-U (1971) Fledermäuse im Windkanal. Zeitschrift für Vergleichende Physiologie 73:209–221. https://doi.org/10.1007/BF00304133
281. Suthers RA, Thomas SP, Suthers BJ (1972) Respiration, wing-beat and ultrasonic pulse emission in an echo-locating bat. J Exp Biol 56:37–48. https://doi.org/10.1242/jeb.56.1.37
282. Kalko E (1994) Coupling of sound emission and wingbeat in naturally foraging European Pipistrelle bats (Microchiroptera: Vespertilionidae). Folia Zool
283. Wong JG, Waters DA (2001) The synchronisation of signal emission with wingbeat during the approach phase in soprano pipistrelles (Pipistrellus pygmaeus). J Exp Biol 204:575–583. https://doi.org/10.1242/jeb.204.3.575
284. Moss CF, Bohn K, Gilkenson H, Surlykke A (2006) Active listening for spatial orientation in a complex auditory scene. PLoS Biol 4:e79. https://doi.org/10.1371/journal.pbio.0040079
285. Knörnschild M, Harview C, Moseley R, Von Helversen O (2009) Remaining cryptic during motion – behavioral synchrony in the proboscis bat (*Rhynchonycteris naso*). Acta Chiropterologica 11:208–211. https://doi.org/10.3161/150811009X465839
286. Burchardt LS, Norton P, Behr O et al (2019) General isochronous rhythm in echolocation calls and social vocalizations of the bat *Saccopteryx bilineata*. R Soc Open Sci 6:181076. https://doi.org/10.1098/rsos.181076
287. Behr O, Von Helversen O, Heckel G et al (2006) Territorial songs indicate male quality in the sac-winged bat *Saccopteryx bilineata* (Chiroptera, Emballonuridae). Behav Ecol 17:810–817. https://doi.org/10.1093/beheco/arl013
288. Knörnschild M, Behr O, von Helversen O (2006) Babbling behavior in the sac-winged bat (*Saccopteryx bilineata*). Naturwissenschaften 93:451–454. https://doi.org/10.1007/s00114-006-0127-9
289. Fernandez AA, Burchardt LS, Nagy M, Knörnschild M (2021) Babbling in a vocal learning bat resembles human infant babbling. Science 373:923–926. https://doi.org/10.1126/science.abf9279

290. Burchardt LS, Knörnschild M (2020) Comparison of methods for rhythm analysis of complex animals' acoustic signals. PLoS Comput Biol 16:e1007755. https://doi.org/10.1371/journal.pcbi.1007755
291. Burchardt LS, Briefer EF, Knörnschild M (2021) Novel ideas to further expand the applicability of rhythm analysis. Ecol Evol 11:18229–18237. https://doi.org/10.1002/ece3.8417
292. Kiai A, Clemens J, Kössl M et al (2023) Flexible control of vocal timing in *Carollia perspicillata* bats enables escape from acoustic interference. Commun Biol 6:1153. https://doi.org/10.1038/s42003-023-05507-5
293. Moss CF, Chiu C, Surlykke A (2011) Adaptive vocal behavior drives perception by echolocation in bats. Curr Opin Neurobiol 21:645–652. https://doi.org/10.1016/j.conb.2011.05.028
294. Hörpel SG, Baier AL, Peremans H et al (2021) Communication breakdown: limits of spectro-temporal resolution for the perception of bat communication calls. Sci Rep 11:13708. https://doi.org/10.1038/s41598-021-92842-4
295. Zhang W, Yartsev MM (2019) Correlated neural activity across the brains of socially interacting bats. Cell 178:413–428.e22. https://doi.org/10.1016/j.cell.2019.05.023
296. Rose MC, Styr B, Schmid TA et al (2021) Cortical representation of group social communication in bats. Science 374:eaba9584. https://doi.org/10.1126/science.aba9584
297. Prat Y, Taub M, Yovel Y (2016) Everyday bat vocalizations contain information about emitter, addressee, context, and behavior. Sci Rep 6:39419. https://doi.org/10.1038/srep39419
298. Knörnschild M, Nagy M, Russo D (2025) Bats resolve conflicting sensory information for individual recognition. Curr Biol 35:1883–1889.e3. https://doi.org/10.1016/j.cub.2025.02.060
299. Babl SS, Kiai A, García-Rosales F, Hechavarría JC (2025) Neuronal activity underlying vocal production in bats. Ann N Y Acad Sci 1550:37–54. https://doi.org/10.1111/nyas.15410
300. Hoerpel SG, Vernes SC, Firzlaff U (2025) Representation of vocalizations in the frontal auditory field and the dorsal auditory cortex of bats. Ann N Y Acad Sci 1547:116–130. https://doi.org/10.1111/nyas.15336
301. Castellucci GA, Calbick D, McCormick D (2018) The temporal organization of mouse ultrasonic vocalizations. PLoS One 13:e0199929. https://doi.org/10.1371/journal.pone.0199929
302. Brudzynski SM, Bihari F, Ociepa D, Fu X-W (1993) Analysis of 22 kHz ultrasonic vocalization in laboratory rats: long and short calls. Physiol Behav 54:215–221. https://doi.org/10.1016/0031-9384(93)90102-L
303. Mahrt E, Agarwal A, Perkel D et al (2016) Mice produce ultrasonic vocalizations by intra-laryngeal planar impinging jets. Curr Biol 26:R880–R881. https://doi.org/10.1016/j.cub.2016.08.032
304. Gaub S, Ehret G (2005) Grouping in auditory temporal perception and vocal production is mutually adapted: the case of wriggling calls of mice. J Comp Physiol A 191:1131–1135. https://doi.org/10.1007/s00359-005-0036-y
305. Uematsu A, Kikusui T, Kihara T et al (2007) Maternal approaches to pup ultrasonic vocalizations produced by a nanocrystalline silicon thermo-acoustic emitter. Brain Res 1163:91–99. https://doi.org/10.1016/j.brainres.2007.05.056
306. Celma-Miralles A, Toro JM (2020) Discrimination of temporal regularity in rats (*Rattus norvegicus*) and humans (Homo sapiens). J Comp Psychol 134:3–10. https://doi.org/10.1037/com0000202
307. Celma-Miralles A, Toro JM (2020) Non-human animals detect the rhythmic structure of a familiar tune. Psychon Bull Rev 27:694–699. https://doi.org/10.3758/s13423-020-01739-2
308. Mayayo F, Toro JM (2025) Changes in temporal cues elicit rhythmic discrimination in rats (*Rattus norvegicus*). J Comp Psychol. https://doi.org/10.1037/com0000412
309. Katsu N, Yuki S, Okanoya K (2021) Production of regular rhythm induced by external stimuli in rats. Anim Cogn 24:1133–1141. https://doi.org/10.1007/s10071-021-01505-4
310. Ito Y, Shiramatsu TI, Ishida N et al (2022) Spontaneous beat synchronization in rats: neural dynamics and motor entrainment. Sci Adv 8:eabo7019. https://doi.org/10.1126/sciadv.abo7019
311. Rajendran VG, Tsdaka Y, Keung TY et al (2024) Rats synchronize predictively to metronomes. iScience 27:111053. https://doi.org/10.1016/j.isci.2024.111053

312. Miller JR, Engstrom MD (2007) Vocal stereotypy and singing behavior in Baiomyine mice. J Mammal 88:1447–1465. https://doi.org/10.1644/06-MAMM-A-386R.1
313. Campbell P, Pasch B, Pino JL et al (2010) Geographic variation in the songs of neotropical singing mice: testing the relative importance of drift and local adaptation. Evolution. https://doi.org/10.1111/j.1558-5646.2010.00962.x
314. Fernández-Vargas M, Tang-Martínez Z, Phelps SM (2011) Singing, allogrooming, and allomarking behaviour during inter- and intra-sexual encounters in the Neotropical short-tailed singing mouse (*Scotinomys teguina*). Behaviour 148:945–965
315. Ryan MJ, Keddy-Hector A (1992) Directional patterns of female mate choice and the role of sensory biases. Am Nat 139:S4–S35
316. Burkhard TT, Westwick RR, Phelps SM (2018) Adiposity signals predict vocal effort in Alston's singing mice. Proc R Soc B Biol Sci 285:20180090. https://doi.org/10.1098/rspb.2018.0090
317. Pasch B, George AS, Campbell P, Phelps SM (2011) Androgen-dependent male vocal performance influences female preference in Neotropical singing mice. Anim Behav 82:177–183. https://doi.org/10.1016/j.anbehav.2011.04.018
318. Gustison ML, Sauciuc G-A (2025) Chapter 11: Moving the body and the mind – links between emotion and rhythmic communication and the underlying neurobiological mechanisms. In: Sauciuc G-A, Lameira AR, Persson T (eds) Nature beats: the what, why and how of animal rhythmic behaviours. Springer
319. Okobi DE, Banerjee A, Matheson AMM et al (2019) Motor cortical control of vocal interaction in neotropical singing mice. Science 363:983–988. https://doi.org/10.1126/science.aau9480
320. Begall S, Burda H, Schleich CE (2007) Subterranean rodents: news from underground. In: Begall S, Burda H, Schleich CE (eds) Subterranean rodents. Springer, Berlin/Heidelberg, pp 3–9
321. Randall JA (2014) Vibrational communication: spiders to kangaroo rats. In: Witzany G (ed) Biocommunication of animals. Springer, Dordrecht, pp 103–133
322. Mason MJ, Narins PM (2001) Seismic signal use by fossorial mammals. Am Zool 41:1171–1184. https://doi.org/10.1093/icb/41.5.1171
323. Schleich C, Francescoli G (2018) Three decades of subterranean acoustic communication studies. In: Dent ML, Fay RR, Popper AN (eds) Rodent bioacoustics. Springer, Cham, pp 43–69
324. Rado R, Levi N, Hauser H et al (1987) Seismic signalling as a means of communication in a subterranean mammal. Anim Behav 35:1249–1251. https://doi.org/10.1016/S0003-3472(87)80183-5
325. Hrouzková E, Dvořáková V, Jedlička P, Šumbera R (2013) Seismic communication in demon African mole rat *Tachyoryctes daemon* from Tanzania. J Ethol 31:255–259. https://doi.org/10.1007/s10164-013-0374-0
326. Giannoni SM, Márquez R, Borghi CE (1997) Airborne and substrate-borne communications of *Microtus (Terricola) gerbei* and *M. (T.) duodecimcostatus*. Acta Theriol (Warsz) 42:123–141. https://doi.org/10.4098/AT.arch.97-16
327. Bednářová R, Hrouzková-Knotková E, Burda H et al (2013) Vocalizations of the giant mole-rat (*Fukomys mechowii*), a subterranean rodent with the richest vocal repertoire. Bioacoustics 22:87–107. https://doi.org/10.1080/09524622.2012.712749
328. Randall JA (1987) Field observations of male competition and mating in Merriam's and Bannertail kangaroo rats. Am Midl Nat 117:211. https://doi.org/10.2307/2425723
329. Bridelance P (1989) Communication acoustique entre individus adultes des genres *Meriones, Pachyuromys, Psammomys* (Gerbillidae, Rodentia). I – Analyse des répertoires. Mammalia 53. https://doi.org/10.1515/mamm.1989.53.1.3
330. Narins PM, Reichman OJ, Jarvis JUM, Lewis ER (1992) Seismic signal transmission between burrows of the Cape mole-rat, *Georychus capensis*. J Comp Physiol A 170:10.1007/BF00190397

331. Randall JA (1997) Species-specific footdrumming in kangaroo rats: *Dipodomys ingens, D. deserti, D spectabilis*. Anim Behav 54:1167–1175. https://doi.org/10.1006/anbe.1997.0560
332. Randall JA (2001) Evolution and function of drumming as communication in mammals. Am Zool 41:1143–1156
333. Bridelance P, Hemim Y (1989) Communication acoustique entre individus adultes des genres *Meriones, Pachyuromys* (Gerbillidae, Rodentia). II. Analyse phénétique du comportement acoustique de *M. shawi grandis, M. shawi shawi, M. libycus, M. crassus, M. tristrami, M. persicus*. Mammalia 53. https://doi.org/10.1515/mamm.1989.53.3.337
334. Faurie AS, Dempster ER, Perrin MR (1996) Footdrumming patterns of southern African elephant-shrews. Mammalia 60. https://doi.org/10.1515/mamm.1996.60.4.567
335. Randall JA (2010) Drummers and stompers: vibrational communication in mammals. In: O'Connel-Rodwell CE (ed) The use of vibrations in communication: properties, mechanisms and function across taxa. Transworld Research Network, Kerala, pp 99–120
336. Heth G, Frankenberg E, Pratt H, Nevo E (1991) Seismic communication in the blind subterranean mole-rat: patterns of head thumping and of their detection in the *Spalax ehrenbergi superspecies* in Israel. J Zool 224:633–638. https://doi.org/10.1111/j.1469-7998.1991.tb03791.x
337. Randall JA, Rogovin KA, Shier DM (2000) Antipredator behavior of a social desert rodent: footdrumming and alarm calling in the great gerbil, *Rhombomys opiums*. Behav Ecol Sociobiol 48:110–118. https://doi.org/10.1007/s002650000199
338. Kimchi T, Reshef M, Terkel J (2005) Evidence for the use of reflected self-generated seismic waves for spatial orientation in a blind subterranean mammal. J Exp Biol 208:647–659. https://doi.org/10.1242/jeb.01396
339. Watkins WA (1981) Activities and underwater sounds of fin whales. Sci Rep Whales Res Inst 33:83–117
340. Youngson BT, Darling JD (2016) The occurrence of pulse, "knock" sounds amidst social/sexual behavior of gray whales (*Eschrichtius robustus*) off Vancouver Island. Mar Mamm Sci 32:1482–1490. https://doi.org/10.1111/mms.12325
341. Cerchio S, Dahlheim M (2001) Variation in feeding vocalizations of humpback whales *Megaptera novaeangliae* from Southeast Alaska. Bioacoustics 11:277–295. https://doi.org/10.1080/09524622.2001.9753468
342. Fournet MEH, Gabriele CM, Sharpe F et al (2018) Feeding calls produced by solitary humpback whales. Mar Mamm Sci 34:851–865. https://doi.org/10.1111/mms.12485
343. Dunlop RA, Noad MJ, Cato DH, Stokes D (2007) The social vocalization repertoire of east Australian migrating humpback whales (*Megaptera novaeangliae*). J Acoust Soc Am 122:2893–2905. https://doi.org/10.1121/1.2783115
344. Dunlop RA, Cato DH, Noad MJ (2008) Non-song acoustic communication in migrating humpback whales (*Megaptera novaeangliae*). Mar Mamm Sci 24:613–629. https://doi.org/10.1111/j.1748-7692.2008.00208.x
345. Edds-Walton PL (1997) Acoustic communication signals of Mysticete whales. Bioacoustics 8:47–60. https://doi.org/10.1080/09524622.1997.9753353
346. Wang Z, Duan P, Chen M et al (2022) Vocalization of Bryde's whales (*Balaenoptera edeni*) in the Beibu Gulf, China. Mar Mamm Sci 38:1118–1139. https://doi.org/10.1111/mms.12917
347. Edds PL, Odell DK, Tershy BR (1993) Vocalizations of a captive juvenile and free-ranging adult-calf pairs of Bryde's whales, *Balaenoptera edeni*. Mar Mamm Sci 9:269–284. https://doi.org/10.1111/j.1748-7692.1993.tb00455.x
348. Cerchio S, Weir CR (2022) Mid-frequency song and low-frequency calls of sei whales in The Falkland Islands. R Soc Open Sci 9:220738. https://doi.org/10.1098/rsos.220738
349. Tervo OM, Parks SE, Miller LA (2009) Seasonal changes in the vocal behavior of bowhead whales (*Balaena mysticetus*) in Disko Bay, Western-Greenland. J Acoust Soc Am 126:1570–1580. https://doi.org/10.1121/1.3158941
350. Payne RS, McVay S (1971) Songs of humpback whales: humpbacks emit sounds in long, predictable patterns ranging over frequencies audible to humans. Science 173:585–597. https://doi.org/10.1126/science.173.3997.585

351. Romagosa M, Nieukirk S, Cascão I et al (2024) Fin whale song evolution in the North Atlantic. eLife 13:e83750. https://doi.org/10.7554/eLife.83750
352. Jolliffe CD, McCauley Robert D, Gavrilov AN (2023) Variability in temporal characteristics of the South Eastern Indian Ocean pygmy blue whale song. Anim Behav Cogn 10:211–231. https://doi.org/10.26451/abc.10.03.02.2023
353. Schneider JN, Mercado E, III (2019) Characterizing the rhythm and tempo of sound production by singing whales. Bioacoustics 28:239–256. https://doi.org/10.1080/09524622.2018.1428827
354. Frazer LN, Mercado E, III (2000) A sonar model for humpback whale song. IEEE J Ocean Eng 25:160–182. https://doi.org/10.1109/48.820748
355. Mercado E, III (2021) Spectral interleaving by singing humpback whales: signs of sonar. J Acoust Soc Am 149:800–806. https://doi.org/10.1121/10.0003443
356. Kuna VM, Nábělek JL (2021) Seismic crustal imaging using fin whale songs. Science 371:731–735. https://doi.org/10.1126/science.abf3962
357. Frankel AS, Clark CW, Herman LM, Gabriele CM (1995) Spatial distribution, habitat utilization, and social interactions of humpback whales, Megaptera novaeangliae, off Hawai'i, determined using acoustic and visual techniques. Can J Zool 73:1134–1146. https://doi.org/10.1139/z95-135
358. Darling JD, Bérubé M (2001) Interactions of singing humpback whales with other males. Mar Mamm Sci 17:570–584. https://doi.org/10.1111/j.1748-7692.2001.tb01005.x
359. Darling JD, Jones ME, Nicklin CP (2006) Humpback whale songs: do they organize males during the breeding season? Behaviour 143:1051–1101
360. Smith JN, Goldizen AW, Dunlop RA, Noad MJ (2008) Songs of male humpback whales, *Megaptera novaeangliae*, are involved in intersexual interactions. Anim Behav 76:467–477. https://doi.org/10.1016/j.anbehav.2008.02.013
361. Dunlop RA, Noad MJ (2016) The "risky" business of singing: tactical use of song during joining by male humpback whales. Behav Ecol Sociobiol 70:2149–2160. https://doi.org/10.1007/s00265-016-2218-8
362. Janik VM, King SL, Sayigh LS, Wells RS (2013) Identifying signature whistles from recordings of groups of unrestrained bottlenose dolphins (*Tursiops truncatus*). Mar Mamm Sci 29:109–122. https://doi.org/10.1111/j.1748-7692.2011.00549.x
363. Sayigh L, Quick N, Hastie G, Tyack P (2013) Repeated call types in short-finned pilot whales, *Globicephala macrorhynchus*. Mar Mamm Sci 29:312–324. https://doi.org/10.1111/j.1748-7692.2012.00577.x
364. Kaplan MB, Aran Mooney T, Sayigh LS, Baird RW (2014) Repeated call types in Hawaiian melon-headed whales (*Peponocephala electra*). J Acoust Soc Am 136:1394–1401. https://doi.org/10.1121/1.4892759
365. Rankin S, Oswald J, Barlow J, Lammers M (2007) Patterned burst-pulse vocalizations of the northern right whale dolphin, *Lissodelphis borealis*. J Acoust Soc Am 121:1213–1218. https://doi.org/10.1121/1.2404919
366. Walmsley SF, Rendell L, Hussey NE, Marcoux M (2020) Vocal sequences in narwhals (*Monodon monoceros*). J Acoust Soc Am 147:1078–1091. https://doi.org/10.1121/10.0000671
367. Zwamborn EMJ, Whitehead H (2017) Repeated call sequences and behavioural context in long-finned pilot whales off Cape Breton, Nova Scotia, Canada. Bioacoustics 26:169–183. https://doi.org/10.1080/09524622.2016.1233457
368. Madsen PT, Johnson M, De Soto NA et al (2005) Biosonar performance of foraging beaked whales (*Mesoplodon densirostris*). J Exp Biol 208:181–194. https://doi.org/10.1242/jeb.01327
369. Johnson M, Hickmott LS, Aguilar Soto N, Madsen PT (2008) Echolocation behaviour adapted to prey in foraging Blainville's beaked whale (*Mesoplodon densirostris*). Proc R Soc B Biol Sci 275:133–139. https://doi.org/10.1098/rspb.2007.1190
370. Morisaka T (2012) Evolution of communication sounds in odontocetes: a review. Int J Comp Psychol 25. https://doi.org/10.46867/IJCP.2012.25.01.04

371. Lammers MO, Au WWL, Herzing DL (2003) The broadband social acoustic signaling behavior of spinner and spotted dolphins. J Acoust Soc Am 114:1629–1639. https://doi.org/10.1121/1.1596173
372. Arranz P, DeRuiter SL, Stimpert AK et al (2016) Discrimination of fast click series produced by tagged Risso's dolphins (*Grampus griseus*) for echolocation or communication. J Exp Biol:jeb.144295. https://doi.org/10.1242/jeb.144295
373. Sørensen PM, Wisniewska DM, Jensen FH et al (2018) Click communication in wild harbour porpoises (Phocoena phocoena). Sci Rep 8:9702. https://doi.org/10.1038/s41598-018-28022-8
374. Herzing DL (1996) Vocalizations and associated underwater behavior of free-ranging Atlantic spotted dolphins, *Stenella frontalis* and bottlenose dolphins, *Tursiops truncatus*. Aquat Mamm 22:61–80
375. Dawson SM, Thorpe CW (1990) A quantitative analysis of the sounds of Hector's dolphin. Ethology 86:131–145. https://doi.org/10.1111/j.1439-0310.1990.tb00424.x
376. Henderson EE, Hildebrand JA, Smith MH, Falcone EA (2012) The behavioral context of common dolphin (*Delphinus* sp.) vocalizations. Mar Mamm Sci 28:439–460. https://doi.org/10.1111/j.1748-7692.2011.00498.x
377. Simard P, Mann DA, Gowans S (2008) Burst-pulse sounds recorded from white-beaked dolphins (*Lagenorhynchus albirostris*). Aquat Mamm 34:464–470. https://doi.org/10.1578/AM.34.4.2008.464
378. Dawson SM (1991) Clicks and communication: the Behavioural and social contexts of Hector's dolphin vocalizations. Ethology 88:265–276. https://doi.org/10.1111/j.1439-0310.1991.tb00281.x
379. Dudzinski KM (1996) Communication and behavior in the Atlantic spotted dolphins (*Stenella frontalis*): relationships between vocal and behavioral activities. Texas A&M University
380. Blomqvist C, Amundin M (2004) High-frequency burst-pulse sounds in agonistic/aggresive interactions in bottlenose dolphins, *Tursiops truncatus*. In: Thomas J, Moss CF, Vater M (eds) Echolocation in bats and dolphins. University of Chicago Press, Chicago, pp 425–431
381. Terada T, Morisaka T, Wakabayashi I, Yoshioka M (2022) Communication sounds produced by captive narrow-ridged finless porpoises (*Neophocaena asiaeorientalis*). J Ethol 40:245–256. https://doi.org/10.1007/s10164-022-00755-0
382. Clausen KT, Wahlberg M, Beedholm K et al (2011) Click communication in harbour porpoises *phocoena phocoena*. Bioacoustics 20:1–28. https://doi.org/10.1080/09524622.2011.9753630
383. Herzing DL, dos Santos ME (2004) Functional aspects of echolocation in dolphins. In: Thomas TA, Moss CF, Vater M (eds) Echolocation in bats and dolphins. University of Chicago Press, Chicago, pp 386–392
384. Moore KE, Watkins WA, Tyack PL (1993) Pattern similarity in shared codas from sperm whales (*Physeter catodon*). Mar Mamm Sci 9:1–9. https://doi.org/10.1111/j.1748-7692.1993.tb00421.x
385. Rendell L, Whitehead H (2005) Spatial and temporal variation in sperm whale coda vocalizations: stable usage and local dialects. Anim Behav 70:191–198. https://doi.org/10.1016/j.anbehav.2005.03.001
386. Rendell LE, Whitehead H (2003) Vocal clans in sperm whales (*Physeter macrocephalus*). Proc R Soc Lond B Biol Sci 270:225–231. https://doi.org/10.1098/rspb.2002.2239
387. Antunes R, Schulz T, Gero S et al (2011) Individually distinctive acoustic features in sperm whale codas. Anim Behav 81:723–730. https://doi.org/10.1016/j.anbehav.2010.12.019
388. Sharma P, Gero S, Payne R et al (2024) Contextual and combinatorial structure in sperm whale vocalisations. Nat Commun 15:3617. https://doi.org/10.1038/s41467-024-47221-8
389. Boileau A, Blais J, Mercier L et al (2023) Synchronous swimming and diving behaviour in a group of fin whales (*Balaenoptera physalus*). Aquat Mamm 49:87–93. https://doi.org/10.1578/AM.49.1.2023.87
390. Fellner W, Bauer GB, Stamper SA et al (2013) The development of synchronous movement by bottlenose dolphins (*Tursiops truncatus*). Mar Mamm Sci 29. https://doi.org/10.1111/j.1748-7692.2012.00609.x

391. Moore BL, Connor RC, Allen SJ et al (2020) Acoustic coordination by allied male dolphins in a cooperative context. Proc R Soc B Biol Sci 287:20192944. https://doi.org/10.1098/rspb.2019.2944
392. Allen JA, Cade DE, Casey CB et al (2024) Evidence of sociality and group foraging in Antarctic minke whales (*Balaenoptera bonaerensis*). Behav Ecol Sociobiol 78:61. https://doi.org/10.1007/s00265-024-03481-4
393. Aguilar De Soto N, Visser F, Tyack PL et al (2020) Fear of killer whales drives extreme synchrony in deep diving beaked whales. Sci Rep 10:13. https://doi.org/10.1038/s41598-019-55911-3
394. Connor RC, Smolker R, Bejder L (2006) Synchrony, social behaviour and alliance affiliation in Indian Ocean bottlenose dolphins, *Tursiops aduncus*. Anim Behav 72:1371–1378. https://doi.org/10.1016/j.anbehav.2006.03.014
395. Myers AJ, Herzing DL, Bjorklund DF (2017) Synchrony during aggression in adult male Atlantic spotted dolphins (*Stenella frontalis*). acta ethologica 20:175–185. https://doi.org/10.1007/s10211-017-0262-7
396. Benoit-Bird KJ, Au WWL (2009) Phonation behavior of cooperatively foraging spinner dolphins. J Acoust Soc Am 125:539–546. https://doi.org/10.1121/1.2967477
397. Herzing D (2015) Synchronous and rhythmic vocalizations and correlated underwater behavior of free-ranging Atlantic spotted dolphins (*Stenella frontalis*) and bottlenose dolphins (*Tursiops truncatus*) in the Bahamas. Anim Behav Cogn 2:14–29. https://doi.org/10.12966/abc.02.02.2015
398. Wilson DE, Mittermeier RA (2014) Handbook of the mammals of the world. Lynx Ediciones, Barcelona
399. Paterson RS, Rybczynski N, Kohno N, Maddin HC (2020) A Total evidence phylogenetic analysis of pinniped phylogeny and the possibility of parallel evolution within a monophyletic framework. Front Ecol Evol 7:457. https://doi.org/10.3389/fevo.2019.00457
400. Ravignani A, Fitch WT, Hanke FD et al (2016) What pinnipeds have to say about human speech, music, and the evolution of rhythm. Front Neurosci 10. https://doi.org/10.3389/fnins.2016.00274
401. Reichmuth C, Casey C (2014) Vocal learning in seals, sea lions, and walruses. Curr Opin Neurobiol 28:66–71. https://doi.org/10.1016/j.conb.2014.06.011
402. Patel AD, Iversen JR, Bregman MR, Schulz I (2009) Experimental evidence for synchronization to a musical beat in a nonhuman animal. Curr Biol 19:827–830. https://doi.org/10.1016/j.cub.2009.03.038
403. Patel AD (2024) Beat-based dancing to music has evolutionary foundations in advanced vocal learning. BMC Neurosci 25. https://doi.org/10.1186/s12868-024-00843-6
404. Cook P, Rouse A, Wilson M, Reichmuth C (2013) A California Sea lion (*Zalophus californianus*) can keep the beat: motor entrainment to rhythmic auditory stimuli in a non vocal mimic. J Comp Psychol 127:412–427. https://doi.org/10.1037/a0032345
405. Cook PF, Hood C, Rouse A, Reichmuth C (2025) Sensorimotor synchronization to rhythm in an experienced sea lion rivals that of humans. Sci Rep 15. https://doi.org/10.1038/s41598-025-95279-1
406. Larsen ON, Reichmuth C (2021) Walruses produce intense impulse sounds by clap-induced cavitation during breeding displays. R Soc Open Sci 8:210197. https://doi.org/10.1098/rsos.210197
407. Kocsis K, Duengen D, Jadoul Y, Ravignani A (2024) Harbour seals use rhythmic percussive signalling in interaction and display. Anim Behav 207:223–234. https://doi.org/10.1016/j.anbehav.2023.09.014
408. Serrano A (2001) New underwater and aerial vocalizations of captive harp seals (*Pagophilus groenlandicus*). Can J Zool 79:75–81. https://doi.org/10.1139/z00-182
409. Moors HB, Terhune JM (2003) Repetition patterns within harp seal (Pagophilus groenlandicus) underwater calls. Aquat Mamm 29:278–288. https://doi.org/10.1578/016754203101024211
410. Moors HB, Terhune JM (2004) Repetition patterns in Weddell seal (*Leptonychotes weddellii*) underwater multiple element calls. J Acoust Soc Am 116:1261–1270. https://doi.org/10.1121/1.1763956

411. Mathevon N, Casey C, Reichmuth C, Charrier I (2017) Northern elephant seals memorize the rhythm and timbre of their rivals' voices. Curr Biol 27:2352–2356.e2. https://doi.org/10.1016/j.cub.2017.06.035
412. Rogers TL (2017) Calling underwater is a costly signal: size-related differences in the call rates of Antarctic leopard seals. Curr Zool 63:433–443. https://doi.org/10.1093/cz/zox028
413. Ravignani A, Kello CT, De Reus K et al (2019) Ontogeny of vocal rhythms in harbor seal pups: an exploratory study. Curr Zool 65:107–120. https://doi.org/10.1093/cz/zoy055
414. Terhune JM (2019) The underwater vocal complexity of seals (Phocidae) is not related to their phylogeny. Can J Zool 97:232–240. https://doi.org/10.1139/cjz-2018-0190
415. Martin M, Gridley T, Elwen SH, Charrier I (2022) Feel the beat: cape fur seal males encode their arousal state in their bark rate. Sci Nat 109:5. https://doi.org/10.1007/s00114-021-01778-2
416. Anichini M, De Reus K, Hersh TA et al (2023) Measuring rhythms of vocal interactions: a proof of principle in harbour seal pups. Philos Trans R Soc B Biol Sci 378:20210477. https://doi.org/10.1098/rstb.2021.0477
417. Charrier I (2020) Mother–offspring vocal recognition and social system in pinnipeds. In: Aubin T, Mathevon N (eds) Coding strategies in vertebrate acoustic communication. Springer, Cham, pp 231–246
418. Charrier I (2021) Vocal communication in otariids and odobenids. In: Campagna C, Harcourt R (eds) Ethology and behavioral ecology of otariids and the odobenid. Springer, Cham, pp 265–289
419. Osiecka AN, Gridley T, Fearey J (2022) Temporal patterns in Cape fur seal (*Arctocephalus pusillus pusillus*) mother and pup attraction calls. Belg J Zool 152. https://doi.org/10.26496/bjz.2022.103
420. Schusterman RJ (1977) Temporal patterning in sea lion barking (*Zalophus californianus*). Behav Biol 20:404–408. https://doi.org/10.1016/S0091-6773(77)90964-6
421. Stirling I, Calvert W, Spencer C (1987) Evidence of stereotyped underwater vocalizations of male Atlantic walruses (*Odobenus rosmarus rosmarus*). Can J Zool 65:2311–2321. https://doi.org/10.1139/z87-348
422. Sjare B, Stirling I, Spencer C (2003) Structural variation in the songs of Atlantic walruses breeding in the Canadian High Arctic. Aquat Mamm 29:297–318. https://doi.org/10.1578/016754203101024121
423. Hayes SA, Costa DP, Harvey JT, Le Boeuf BJ (2004) Aquatic mating strategies of the male pacific harbor seal (*Phoca vitulina richardii*): are males defending the hotspot? Mar Mamm Sci 20:639–656. https://doi.org/10.1111/j.1748-7692.2004.tb01184.x
424. Hocking DP, Burville B, Parker WMG et al (2020) Percussive underwater signaling in wild gray seals. Mar Mamm Sci 36:728–732. https://doi.org/10.1111/mms.12666
425. Charrier I, Casey C (2022) Social communication in phocids. In: Costa DP, McHuron EA (eds) Ethology and behavioral ecology of phocids. Springer, Cham, pp 69–100
426. Russell L, Purdy J, Davis R (2016) Social context predicts vocalization use in the courtship behaviors of Weddell seals (*Leptonychotes weddellii*): a case study. Anim Behav Cogn 3:95–119. https://doi.org/10.12966/abc.04.05.2016
427. Large EW (2025) Chapter 8: Neural resonance and comparative rhythm cognition. In: Sauciuc G-A, Lameira AR, Persson T (eds) Nature beats: the what, why and how of animal rhythmic behaviours. Springer
428. Casey C, Charrier I, Mathevon N, Reichmuth C (2015) Rival assessment among northern elephant seals: evidence of associative learning during male–male contests. R Soc Open Sci 2:150228. https://doi.org/10.1098/rsos.150228
429. Smirnova DS, Volodin IA, Demina TS, Volodina EV (2016) Acoustic structure and contextual use of calls by captive male and female cheetahs (*Acinonyx jubatus*). PLoS One 11:e0158546. https://doi.org/10.1371/journal.pone.0158546
430. Tembrock G (1976) Canid vocalizations. Behav Process 1:57–75. https://doi.org/10.1016/0376-6357(76)90007-3
431. Leuchtenberger C, Almeida SB, Andriolo A, Crawshaw PG (2016) Jaguar mobbing by giant otter groups. Acta Ethologica 19:143–146. https://doi.org/10.1007/s10211-016-0233-4

432. Taylor AM, Reby D, McComb K (2010) Size communication in domestic dog, *Canis familiaris*, growls. Anim Behav 79:205–210. https://doi.org/10.1016/j.anbehav.2009.10.030
433. Peters G (1987) Acoustic communication in the genus Lynx (Mammalia: Felidae)—comparative survey and phylogenetic interpretation. Bonn Zool Beitr 38:315–330
434. Peters G (2002) Purring and similar vocalizations in mammals. Mammal Rev 32:245–271. https://doi.org/10.1046/j.1365-2907.2002.00113.x
435. Leuchtenberger C, Sousa-Lima R, Duplaix N et al (2014) Vocal repertoire of the social giant otter. J Acoust Soc Am 136:2861–2875. https://doi.org/10.1121/1.4896518
436. Volodin IA, Volodina EV, Isaeva IV (2001) Vocal repertoire in the dhole *Cuon alpinus* (Carnivora, Canidae) in captivity. Entomol Rev 81:S346
437. Koler-Matznick J, Brisbin IL, Feinstein M, Bulmer S (2003) An updated description of the New Guinea singing dog (*Canis hallstromi*, Troughton 1957). J Zool 261:109–118. https://doi.org/10.1017/S0952836903004060
438. Murdoch JD, Ralls K, Cypher B, Reading R (2008) Social interactions among San Joaquin kit foxes before, during, and after the mating season. J Mammal 89:1087–1093. https://doi.org/10.1644/07-MAMM-A-152.1
439. Frommolt K-H, Goltsman ME, Macdonald DW (2003) Barking foxes, *Alopex lagopus*: field experiments in individual recognition in a territorial mammal. Anim Behav 65:509–518. https://doi.org/10.1006/anbe.2003.2064
440. Darden SK, Dabelsteen T (2006) Ontogeny of swift fox Vulpes velox vocalizations: production, usage and response. Behaviour 143:659–681
441. Newton-Fisher N, Harris S, White P, Jones G (1993) Structure and function of red fox *Vulpes vulpes* vocalisations. Bioacoustics 5:1–31. https://doi.org/10.1080/09524622.1993.9753228
442. Darden SK, Dabelsteen T, Pedersen SB (2003) A potential tool for swift fox (*Vulpes velox*) conservation: individuality of long-range barking sequences. J Mammal 84:1417–1427. https://doi.org/10.1644/BEM-031
443. Robbins R (2000) Vocal communication in free-ranging african wild dogs (*Lycaon pictus*). Behaviour 137:1271–1298. https://doi.org/10.1163/156853900501926
444. Schneider JN, Anderson RE (2011) Tonal vocalizations in the red wolf (*Canis rufus*): potential functions of nonlinear sound production. J Acoust Soc Am 130:2275–2284. https://doi.org/10.1121/1.3628331
445. Weir JN, Schneider JN, Anderson RE (2021) The squeaking vocalization of grey wolves (*Canis lupus*): individuality in a close-range affiliative mammalian vocalisation. Behaviour 159:501–535. https://doi.org/10.1163/1568539X-bja10140
446. Déaux ÉC, Clarke JA (2013) Dingo (*Canis lupus dingo*) acoustic repertoire: form and contexts. Behaviour 150:75–101
447. Schassburger RM (1993) Vocal communication in the timber wolf, *Canis lupus*, Linnaeus: structure, motivation, and ontogeny; with 6 tables. Parey, Berlin/Hamburg
448. Buhler KJ, Samelius G, Alisauskas R, Jenkins E (2024) What does the fox say? Arctic fox vocalization and associated den behaviours. Polar Res 43. https://doi.org/10.33265/polar.v43.9430
449. Quaranta A, Siniscalchi M, Vallortigara G (2007) Asymmetric tail-wagging responses by dogs to different emotive stimuli. Curr Biol 17:R199–R201. https://doi.org/10.1016/j.cub.2007.02.008
450. Siniscalchi M, Lusito R, Vallortigara G, Quaranta A (2013) Seeing left- or right-asymmetric tail wagging produces different emotional responses in dogs. Curr Biol 23:2279–2282. https://doi.org/10.1016/j.cub.2013.09.027
451. Shultz S, Dunbar RIM (2007) The evolution of the social brain: anthropoid primates contrast with other vertebrates. Proc R Soc B Biol Sci 274:2429–2436. https://doi.org/10.1098/rspb.2007.0693
452. Sorrentino V, Gamba M, Giacoma C (2012) A quantitative description of the vocal types emitted in the indri's song. In: Masters J, Gamba M, Génin F (eds) Leaping ahead. Springer, New York, pp 315–322

453. Adret P, Dingess K, Caselli C et al (2018) Duetting patterns of Titi monkeys (primates, Pitheciidae: Callicebinae) and relationships with phylogeny. Animals 8:178. https://doi.org/10.3390/ani8100178
454. Clink DJ, Tasirin JS, Klinck H (2020) Vocal individuality and rhythm in male and female duet contributions of a nonhuman primate. Curr Zool 66:173–186. https://doi.org/10.1093/cz/zoz035
455. Raimondi T, Di Panfilo G, Pasquali M et al (2023) Isochrony and rhythmic interaction in ape duetting. Proc R Soc B Biol Sci 290:20222244. https://doi.org/10.1098/rspb.2022.2244
456. Lameira AR, Hardus ME, Ravignani A et al (2024) Recursive self-embedded vocal motifs in wild orangutans. eLife 12:RP88348. https://doi.org/10.7554/eLife.88348
457. De Gregorio C, Gamba M, Lameira AR (2025) Third-order self-embedded vocal motifs in wild orangutans, and the selective evolution of recursion. Ann N Y Acad Sci 1549:219–229. https://doi.org/10.1111/nyas.15373
458. De Gregorio C, Raimondi T, Bevilacqua V et al (2024) Isochronous singing in 3 crested gibbon species (*Nomascus* spp.). Curr Zool 70:291–297. https://doi.org/10.1093/cz/zoad029
459. De Gregorio C, Atonini P, Heymann E, Gamba M (2025) Isochrony in titi monkeys duets: social context as a proximate cause of duets' rhythm and regularity. Proc R Soc B Biol Sci 292:20242805. https://doi.org/10.1098/rspb.2024.2805
460. De Gregorio C, Valente D, Raimondi T et al (2021) Categorical rhythms in a singing primate. Curr Biol 31:R1379–R1380. https://doi.org/10.1016/j.cub.2021.09.032
461. Sauciuc G-A, Persson T, Hattori Y, Lameira AR (2025) Chapter 12: The significance of the great ape heritage in the evolution of human rhythm in speech, music and dance. In: Sauciuc G-A, Lameira AR, Persson T (eds) Nature beats: the what, why and how of animal rhythmic behaviour. Springer
462. Sauciuc G-A, Raimondi T, De Gregorio C (2025) Chapter 10: The developmental trajectory of rhythmic abilities in humans and singing primates. In: Sauciuc G-A, Lameira AR, Persson T (eds) Nature beats: the what, why and how of animal rhythmic behaviour. Springer
463. Oliveira DAG, Ades C (2004) Long-distance calls in Neotropical primates. An Acad Bras Ciênc 76:393–398. https://doi.org/10.1590/S0001-37652004000200031
464. Wich S, Nunn C (2002) Do male "long-distance calls" function in mate defense? A comparative study of long-distance calls in primates. Behav Ecol Sociobiol 52:474–484. https://doi.org/10.1007/s00265-002-0541-8
465. De Gregorio C, Valente D, Ferrario V et al (2024) Who you live with and what you duet for: a review of the function of primate duets in relation to their social organization. J Comp Physiol A 210:281–294. https://doi.org/10.1007/s00359-023-01689-9
466. Caselli CB, Mennill DJ, Bicca-Marques JC, Setz EZF (2014) Vocal behavior of black-fronted titi monkeys (*Callicebus nigrifrons*): acoustic properties and behavioral contexts of loud calls. Am J Primatol 76:788–800. https://doi.org/10.1002/ajp.22270
467. Bartlett TQ (2007) The Hylobatidae: small apes of Asia. In: Campbell CJ, Fuentes A, MacKinnon KC, Bearder SK (eds) Primates in perspective. Oxford University Press, New York, pp 274–289
468. MacKinnon J, MacKinnon K (1980) The behavior of wild spectral tarsiers. Int J Primatol 1:361–379. https://doi.org/10.1007/BF02692280
469. Torti V, Gamba M, Rabemananjara ZH, Giacoma C (2013) The songs of the indris (Mammalia: primates: Indridae): contextual variation in the long-distance calls of a lemur. Ital J Zool 80:596–607. https://doi.org/10.1080/11250003.2013.845261
470. Caselli CB, Mennill DJ, Gestich CC et al (2015) Playback responses of socially monogamous black-fronted titi monkeys to simulated solitary and paired intruders. Am J Primatol 77:1135–1142. https://doi.org/10.1002/ajp.22447
471. Clarke E, Reichard UH, Zuberbühler K (2006) The syntax and meaning of wild gibbon songs. PLoS One 1:e73. https://doi.org/10.1371/journal.pone.0000073
472. Wich SA, Koski S, De Vries H, Van Schaik CP (2003) Individual and contextual variation in Thomas Langur male loud calls. Ethology 109:1–13. https://doi.org/10.1046/j.1439-0310.2003.00837.x

473. Schel AM, Zuberbühler K (2012) Predator and non-predator long-distance calls in Guereza colobus monkeys. Behav Process 91:41–49. https://doi.org/10.1016/j.beproc.2012.05.004
474. Schel AM, Zuberbühler K (2009) Responses to leopards are independent of experience in Guereza colobus monkeys. Behaviour 146:1709–1737. https://doi.org/10.1163/000579509X12483520922007
475. Spillmann B, Dunkel LP, Van Noordwijk MA et al (2010) Acoustic properties of long calls given by flanged male orang-Utans (*Pongo pygmaeus wurmbii*) reflect both individual identity and context. Ethology 116:385–395. https://doi.org/10.1111/j.1439-0310.2010.01744.x
476. Clark AP, Wrangham RW (1993) Acoustic analysis of wild chimpanzee pant hoots: do Kibale Forest chimpanzees have an acoustically distinct food arrival pant hoot? Am J Primatol 31:99–109. https://doi.org/10.1002/ajp.1350310203
477. Notman H, Rendall D (2005) Contextual variation in chimpanzee pant hoots and its implications for referential communication. Anim Behav 70:177–190. https://doi.org/10.1016/j.anbehav.2004.08.024
478. Fedurek P, Zuberbühler K, Dahl CD (2016) Sequential information in a great ape utterance. Sci Rep 6:38226. https://doi.org/10.1038/srep38226
479. Desai NP, Fedurek P, Slocombe KE, Wilson ML (2022) Chimpanzee pant-hoots encode individual information more reliably than group differences. Am J Primatol 84:e23430. https://doi.org/10.1002/ajp.23430
480. Crockford C, Boesch C (2003) Context-specific calls in wild chimpanzees, *Pan troglodytes verus*: analysis of barks. Anim Behav 66:115–125. https://doi.org/10.1006/anbe.2003.2166
481. Wich SA, Sterck EHM (2003) Possible audience effect in Thomas langurs (primates; *Presbytis thomasi*): an experimental study on male loud calls in response to a tiger model. Am J Primatol 60:155–159. https://doi.org/10.1002/ajp.10102
482. Andrieu J, Penny SG, Bouchet H et al (2020) White-handed gibbons discriminate context-specific song compositions. PeerJ 8:e9477. https://doi.org/10.7717/peerj.9477
483. Bergman TJ (2013) Speech-like vocalized lip-smacking in geladas. Curr Biol 23:R268–R269. https://doi.org/10.1016/j.cub.2013.02.038
484. Ghazanfar AA, Chandrasekaran C, Morrill RJ (2010) Dynamic, rhythmic facial expressions and the superior temporal sulcus of macaque monkeys: implications for the evolution of audiovisual speech. Eur J Neurosci 31:1807–1817. https://doi.org/10.1111/j.1460-9568.2010.07209.x
485. Toyoda A, Maruhashi T, Malaivijitnond S, Koda H (2017) Speech-like orofacial oscillations in stump-tailed macaque (*Macaca arctoides*) facial and vocal signals. Am J Phys Anthropol 164:435–439. https://doi.org/10.1002/ajpa.23276
486. Pereira AS, Kavanagh E, Hobaiter C et al (2020) Chimpanzee lip-smacks confirm primate continuity for speech-rhythm evolution. Biol Lett 16:20200232. https://doi.org/10.1098/rsbl.2020.0232
487. Terleph TA, Malaivijitnond S, Reichard UH (2018) An analysis of white-handed gibbon male song reveals speech-like phrases. Am J Phys Anthropol 166:649–660. https://doi.org/10.1002/ajpa.23451
488. Lameira AR, Hardus ME, Bartlett AM et al (2015) Speech-like rhythm in a voiced and voiceless orangutan call. PLoS One 10:e116136. https://doi.org/10.1371/journal.pone.0116136
489. Chandrasekaran C, Trubanova A, Stillittano S et al (2009) The natural statistics of audiovisual speech. PLoS Comput Biol 5:e1000436. https://doi.org/10.1371/journal.pcbi.1000436
490. Shannon RV, Zeng F-G, Kamath V et al (1995) Speech recognition with primarily temporal cues. Science 270:303–304. https://doi.org/10.1126/science.270.5234.303
491. Cheney DL, Seyfarth RM (1992) Précis of how monkeys see the world. Behav Brain Sci 15:135–147. https://doi.org/10.1017/S0140525X00067911
492. Range F, Fischer J (2004) Vocal repertoire of sooty mangabeys (*Cercocebus torquatus atys*) in the Taï National Park. Ethology 110:301–321. https://doi.org/10.1111/j.1439-0310.2004.00973.x
493. Kemp C, Legou T, Boë L-J, Berthommier F (2017) Vocal repertoire of captive Guinea baboons (*Papio papio*). In: Boë L-J, Fagot J, Perrier P, Schwartz J-L (eds) Origins of human language: continuities and discontinuities with nonhuman primates. Peter Lang D, pp 15–58

494. Hammerschmidt K, Fischer J (2019) Baboon vocal repertoires and the evolution of primate vocal diversity. J Hum Evol 126:1–13. https://doi.org/10.1016/j.jhevol.2018.10.010
495. Fedurek P, Tkaczynski PJ, Hobaiter C et al (2021) The function of chimpanzee greeting calls is modulated by their acoustic variation. Anim Behav 174:279–289. https://doi.org/10.1016/j.anbehav.2021.02.002
496. De Waal FBM (1989) Food sharing and reciprocal obligations among chimpanzees. J Hum Evol 18:433–459. https://doi.org/10.1016/0047-2484(89)90074-2
497. Kudo H (1987) The study of vocal communication of wild mandrills in Cameroon in relation to their social structure. Primates 28:289–308. https://doi.org/10.1007/BF02381013
498. Gartlan JS (1970) Preliminary notes on the ecology and behaviour of the drill, *Mandrillus leucophaeus*. Ritgen 1824. In: Napier J, Napier PH (eds) Old world monkeys. Academic, New York, pp 445–480
499. Hauser MD, Marler P (1993) Food-associated calls in rhesus macaques (*Macaca mulatta*): II. Costs and benefits of call production and suppression. Behav Ecol 4:206–212. https://doi.org/10.1093/beheco/4.3.206
500. Remedios R, Logothetis NK, Kayser C (2009) Monkey drumming reveals common networks for perceiving vocal and nonvocal communication sounds. Proc Natl Acad Sci 106:18010–18015. https://doi.org/10.1073/pnas.0909756106
501. Lameira AR, Perlman M, Sauciuc G-A, Persson T (Forthcoming) Chapter 13: Rhythm-induced altered states and the evolution of human dance and music. In: Sauciuc G-A, Lameira AR, Persson T (eds) Nature beats: the what, why and how of animal rhythmic behaviour. Springer
502. Maestripieri D, Roney JR (2005) Primate copulation calls and postcopulatory female choice. Behav Ecol 16:106–113. https://doi.org/10.1093/beheco/arh120
503. Dixson AF (2012) Primate sexuality: comparative studies of the prosimians, monkeys, apes, and humans, 2nd edn. Oxford University Press, Oxford
504. Bouquet Y, Stephan C, Johnson CA et al (2018) Comparing functions of copulation calls in wild olive baboons, Papio anubis, using multimodel inference. Anim Behav 135:187–197. https://doi.org/10.1016/j.anbehav.2017.11.019
505. Zanoli A, Gamba M, Lemasson A et al (2022) Inter-sexual multimodal communication during mating in wild geladas: the leading role of females. Curr Zool 68:570–580. https://doi.org/10.1093/cz/zoab093
506. Gamba M, Giacoma C (2005) Vocal production mechanisms in ruffed lemurs: a Prosimian model for the basis of primate phonation. In: Manfredi C (ed) Models and analysis of vocal emissions for biomedical applications. 4th international workshop, October 29–31, 2005 – Firenze, Italy. Firenze University Press, Firenze, pp 207–210
507. Perlman M, Salmi R (2017) Gorillas may use their laryngeal air sacs for whinny-type vocalizations and male display. J Lang Evol 2:126–140. https://doi.org/10.1093/jole/lzx012
508. Salmi R, Hammerschmidt K, Doran-Sheehy DM (2013) Western gorilla vocal repertoire and contextual use of vocalizations. Ethology 119:831–847. https://doi.org/10.1111/eth.12122
509. Van Lawick-Goodall J (1968) The behaviour of free-living Chimpanzees in the Gombe Stream Reserve. Anim Behav Monogr 1:161–IN12. https://doi.org/10.1016/S0066-1856(68)80003-2
510. Sauciuc G-A, Zlakowska J, Klang K-J et al (2021) Rhythmic behaviors in chimpanzees: range, functional contexts, sex differences and emotional correlates. In: Proceedings of the annual meeting of the Cognitive Science Society, 43, Vienna, p 1007
511. Gustison ML, Le Roux A, Bergman TJ (2012) Derived vocalizations of geladas (*Theropithecus gelada*) and the evolution of vocal complexity in primates. Philos Trans R Soc B Biol Sci 367:1847–1859. https://doi.org/10.1098/rstb.2011.0218
512. Zhou J, Wei F, Li M et al (2008) Reproductive characters and mating behaviour of wild *Nomascus hainanus*. Int J Primatol 29:1037–1046. https://doi.org/10.1007/s10764-008-9272-7
513. Fan P-F, Ma C-Y, Garber PA et al (2016) Rhythmic displays of female gibbons offer insight into the origin of dance. Sci Rep 6:34606. https://doi.org/10.1038/srep34606

514. Aich H, Moos-Heilen R, Zimmermann E (1990) Vocalizations of adult gelada baboons (*Theropithecus gelada*): acoustic structure and behavioural context. Folia Primatol (Basel) 55:109–132. https://doi.org/10.1159/000156508
515. Fischer J, Hammerschmidt K (2002) An overview of the barbary macaque, *Macaca sylvanus*, vocal repertoire. Folia Primatol (Basel) 73:32–45. https://doi.org/10.1159/000060417
516. Fischer J, Hammerschmidt K, Todt D (1998) Local variation in barbary macaque shrill barks. Anim Behav 56:623–629. https://doi.org/10.1006/anbe.1998.0796
517. Clay Z, Pika S, Gruber T, Zuberbühler K (2011) Female bonobos use copulation calls as social signals. Biol Lett 7:513–516. https://doi.org/10.1098/rsbl.2010.1227
518. Engelhardt A, Fischer J, Neumann C et al (2012) Information content of female copulation calls in wild long-tailed macaques (*Macaca fascicularis*). Behav Ecol Sociobiol 66:121–134. https://doi.org/10.1007/s00265-011-1260-9
519. Townsend SW, Deschner T, Zuberbühler K (2008) Female chimpanzees use copulation calls flexibly to prevent social competition. PLoS One 3:e2431. https://doi.org/10.1371/journal.pone.0002431
520. Brown S (2007) Contagious heterophony: a new theory about the origins of music. Music Sci 11:3–26. https://doi.org/10.1177/102986490701100101
521. Hattori Y, Tomonaga M, Matsuzawa T (2013) Spontaneous synchronized tapping to an auditory rhythm in a chimpanzee. Sci Rep 3:1566. https://doi.org/10.1038/srep01566
522. Large EW, Gray PM (2015) Spontaneous tempo and rhythmic entrainment in a bonobo (*Pan paniscus*). J Comp Psychol 129:317–328. https://doi.org/10.1037/com0000011
523. Lameira AR, Eerola T, Ravignani A (2019) Coupled whole-body rhythmic entrainment between two chimpanzees. Sci Rep 9:18914. https://doi.org/10.1038/s41598-019-55360-y
524. Schweinfurth MK, Baldridge DB, Finnerty K et al (2022) Inter-individual coordination in walking chimpanzees. Curr Biol 32:5138–5143.e3. https://doi.org/10.1016/j.cub.2022.09.059
525. Takeya R, Kameda M, Patel AD, Tanaka M (2017) Predictive and tempo-flexible synchronization to a visual metronome in monkeys. Sci Rep 7:6127. https://doi.org/10.1038/s41598-017-06417-3
526. Gámez J, Yc K, Ayala YA et al (2018) Predictive rhythmic tapping to isochronous and tempo changing metronomes in the nonhuman primate. Ann N Y Acad Sci 1423:396–414. https://doi.org/10.1111/nyas.13671
527. Castillo-Almazán A, Pérez O, Prado L et al (2025) Flexible tapping synchronization in macaques: dynamic switching of timing strategies within rhythmic sequences. J Neurophysiol 134:580–590. https://doi.org/10.1152/jn.00158.2025
528. Nietsch A (1999) Duet vocalizations among different populations of Sulawesi tarsiers. Int J Primatol 20:567–583. https://doi.org/10.1023/A:1020342807709
529. Clink DJ, Lau AR, Bales KL (2019) Age-related changes and vocal convergence in titi monkey duet pulses. Behaviour 156:1471–1494. https://doi.org/10.1163/1568539X-00003575
530. Gamba M, Torti V, Estienne V et al (2016) The indris have got rhythm! Timing and pitch variation of a primate song examined between sexes and age classes. Front Neurosci 10. https://doi.org/10.3389/fnins.2016.00249
531. Robinson JG (2010) An analysis of the Organization of Vocal Communication in the Titi monkey Callicebus moloch. Z Für Tierpsychol 49:381–405. https://doi.org/10.1111/j.1439-0310.1979.tb00300.x
532. Müller AE, Anzenberger G (2002) Duetting in the Titi monkey *Callicebus cupreus*: structure, pair specificity and development of duets. Folia Primatol (Basel) 73:104–115. https://doi.org/10.1159/000064788
533. Richman B (1987) Rhythm and melody in gelada vocal exchanges. Primates 28:199–223. https://doi.org/10.1007/BF02382570
534. Fedurek P, Schel AM, Slocombe KE (2013) The acoustic structure of chimpanzee pant-hooting facilitates chorusing. Behav Ecol Sociobiol 67:1781–1789. https://doi.org/10.1007/s00265-013-1585-7

Uncharted Rhythmic Diversity Across the Animal World

4

Adriano R. Lameira, Chiara De Gregorio, Marco Gamba, Nao Ota, Teresa Raimondi, Stephan A. Reber, Jennifer N. Schneider, and Gabriela-Alina Sauciuc

Abstract

Historically, rhythmic capacities in animals have been systematically studied in few taxa, like fireflies and crickets, while other behaviours have been considered rhythmic but lack de facto rigorous analysis, such as chimpanzee "rain dance". Emerging fields like animal rhythm often suffer from data scarcity, leading to premature theories that overemphasize limited data and confuse the absence of evidence with evidence of absence. Species are assumed to have negligible or no rhythmic capacities when in fact they have only received negligible or no empirical effort. To understand rhythm behaviour and its evolution across lineages and under different conditions (including

A. R. Lameira (✉) · C. De Gregorio
Department of Psychology, University of Warwick, Coventry, UK
e-mail: Adriano.Lameira@warwick.ac.uk; Chiara.De-Gregorio@warwick.ac.uk

M. Gamba
Department of Life Sciences and Systems Biology, University of Turin, Turin, Italy
e-mail: marco.gamba@unito.it

N. Ota
Museum of Nature and Human Activities, Sanda, Hyogo, Japan
e-mail: n-ota@hitohaku.jp

T. Raimondi
Department of Life Sciences and Systems Biology, University of Turin, Torino, Italy

Department of Human Neurosciences, Sapienza University of Rome, Rome, Italy
e-mail: teresa.raimondi@uniroma1.it

S. A. Reber · G.-A. Sauciuc
Department of Philosophy/Cognitive Science, Lund University, Lund, Sweden
e-mail: stephan.reber@lucs.lu.se; Gabriela-Alina.Sauciuc@lucs.lu.se

J. N. Schneider
Department of Psychology, LCC International University, Klaipėda, Lithuania
e-mail: jschneider@lcc.lt

G.-A. Sauciuc et al. (eds.), *Nature Beats*, Fascinating Life Sciences,
https://doi.org/10.1007/978-3-032-24250-1_4

along the human lineage), it is therefore crucial that rhythm is comprehensively described, characterized, and analysed across the widest taxonomic span possible. It is necessary to proactively seek out new, unnoticed and overlooked animal rhythms. This chapter presents a list of everyday animal behaviours that offer intriguing and promising opportunities to study rhythm and synchronization across taxa. Our aim is to invite and guide renewed efforts towards new insights about rhythm and its evolution in the animal world.

4.1 Introduction

Scientific studies attempt to maximize data collection and assemble large datasets to create a picture of the natural world with better resolution, literally. A detailed image *shows* how various variables may affect the phenomena one is seeking to understand. This helps researchers to avoid too many inferences, and consequently, prevents the creation of spurious or bogus theories that attempt to connect a few solitary data points.

Rhythm is an old theme among anthropological, historical, sociological, cultural and ethnographic records; however, among the biological and psychological sciences, the study of rhythm is relatively new, with more open questions than conclusive answers [1]. For decades, the role of rhythm and the rhythmic capacities in animals has remained restricted to a few classic taxa (e.g. fireflies, Family: Lampyridae, crickets, Family: Gryllidae). Other cases of animal behaviour have been described as rhythmic but have never undergone rigorous analyses to conclude so, or to accurately characterize the rhythm(s) in point (e.g. chimpanzee "rain dance", *Pan troglodytes*). Without basic knowledge across species, genera, families, orders and classes about animal behavioural tempi and rhythmic capacities, it will remain difficult, *if possible at all*, to understand the various possible paths of rhythm evolution, including in the human lineage. Snowball, the dancing cockatoo (*Cacatua galerita eleonora*), who paced to human music [2], is a good example of how a new data point can "snowball" a renewed interest in the systematic study of animal rhythmicity—including ensuing criticisms that Snowball's synchrony to music tempo is not as impressive or revealing as originally claimed (see Chap. 5). This is exactly how science progresses.

Data-poor situations are characteristic of emerging fields and this applies to the expanding field of animal rhythm and rhythm evolution. However, this has not prevented scholars from conceiving new but premature theories of rhythm in animals and humans, overplaying the availability of a few data points at the expense of the many still missing. Absence of evidence is confused with evidence of absence (see for example the ill-considered "vocal learning and rhythmic synchronization hypothesis" [3, 4] and the "perfect synchrony in philopatric males on patchy resources" in Chap. 14 [5]). By making broad and risky inferences, these theories often steer research effort along the trendlines apparent in available data, that is, towards confirmatory data. Their prematurity would often require, however, that research focus on the lacunas and towards contradictory data.

To fill these gaps in our understanding of rhythm, its various components and possible underpinning mechanisms, this chapter presents a list of "everyday" animal behaviours that exhibit striking opportunities to explore new aspects of rhythm and expedite an ever more detailed and comprehensive picture of rhythm in the natural world. Without proactively seeking new or unpredicted rhythmic capacities in the most unlikely and unsuspecting of places and species, one can never reject the possibility that missing data points are missing due to lack of effort, other than missing due to species' true own rhythmic limitations.

The list in this chapter is meant to provide an accessible entry point to students of rhythm and a stimulant to hone one's observation skills and research questions, seek circumstances where the role of rhythm has remained unnoticed, ignored or hidden in plain sight. These new observations hold outstanding heuristic potential, for a single observation can be sufficient to prove an established theory false. Scientific revolutions oftentimes have anecdotal beginnings.

We organize these behaviours into intra- and interindividual behaviours, as these may demand distinct observational or experimental setups and be potentially motivated by slightly different questions or interests. The suggestions that follow can be implemented through standard video analyses, using freeware (e.g. BORIS) and code available in various open-access papers.

4.2 Intra-individual Rhythms

4.2.1 Spaniels' Tail Wag

When a scent hound (*Canis lupus familiaris*) picks and follows a smell, its tail paces vigorously, sometimes in circle wags. The social, motivational and emotional underpinning for tail-wagging are relatively well understood [6–8], but tail rhythm has never received empirical interest. How does tail rhythm relate to the dog's front and hind leg movement? While the rhythm of front and hind legs will probably approach antiphase synchrony between each other, is tail wag an independent rhythm altogether? There are no obvious bodily constraints that could limit the movement of the tail in relation to the body. Nevertheless, the tail may play a potential role in body balance, suggesting that direct physical coupling effects between leg and tail rhythms should not be excluded a priori. A shift between left- and right-asymmetry in a dog's tail wag is triggered by and triggers different emotional states [6–8], but does a change in side predominance alter a tail's rhythm, or vice-versa? If a dog's tail movement is coupled with its legs while searching for a scent, should one expect 1:1 ratio between leg and tail rhythms? Or does tail rhythm better predict a dog's respiratory rate instead, for example? What possible mechanisms should be invoked if leg and tail rhythms are associated in higher multiples, say 1:7? Should different underlying neuro-motoric oscillators be invoked to explain such phenomena?

4.2.2 Spider Interlimb Coordination

Spiders (Class: Arachnida) have four pairs of legs and a pair of fangs or appendices by the mouth used to seize and immobilize prey or to wipe clean their multiple eyes.

For adequate locomotion, hunting, communication and mating, flawless integration of movement is required between leg and fang pairs. What could orb-weavers' vibrations or peacock spiders' mating rituals reveal about rhythmic coordination and synchrony in spiders? Given that spiders have exoskeletons, the movement of each limb can be expected to be virtually independent from one another. This provides a very amicable biological system to investigate how coordination is achieved in the near absence of anatomical constraints. Coordination between legs and/or fangs, where present, could thus be expected to be strictly based on neurological pacemakers or oscillators, even though the nervous system of these animals is simpler than that of vertebrates. Are there effects of limb position, shape, or functional specialization on the degrees with which a limb can move in sync with others? In behaviours that require the simultaneous rhythmic movement of different limb and body parts, what is their rhythmic relationship?

4.2.3 Seagull Worm-Charming

Gulls (Genus: *Larus*) exhibit a "feeding technique" that consists of quickly stomping their webbed feet on the ground, a behaviour proposed to mimic the sound of raindrops or the revolving sounds of a mole advancing a new underground tunnel, which entices worms to the surface and which the gull then eats. What is the tempo of this rhythm and what is its relation to other rhythmic behaviours of seagulls? Does this rhythm represent a novel behaviour, or is it the same rhythm at which gull walk, but then performed on the same spot? Or is it the same rhythm at which gull beat their wings? Can seagulls adjust their feet rhythm depending on the number of worms that surface and on their feeding success?

4.2.4 A Dance of Slaps

Animals living in aquatic habitats can make a surprising variety of movements with their limbs other than simply using them for locomotion or propulsion. Whales (Order: Cetacea), dolphins, seals (Suborder: Pinnipedia), orcas, and even turtles (Order: Testudines) are masters in rhythmically slapping their tails, feet, and flippers, in close parallel to how humans do so during synchronized swimming.

If you have ever wandered near a pond with turtles, you might have seen them perform a very peculiar behaviour. At first glance, you may have thought to witness an aggressive showdown that might resemble a slapping contest between two turtles aiming at each other's faces. In fact, you would have witnessed two animals interacting positively. This behaviour is called *titillation* and can be performed either as a sexual display or as an element of social play [9]. During titillation, one animal rapidly vibrates the digits of the front feet above the eyes of the other one. How does titillation rhythm relate to the outcome of this social interaction? Could the tempo of feet movements influence the behaviour of the animal receiving the titillation? Could it advertise the quality of the titillating animal? Is the tempo distinct from those used in locomotion? Titillation consists of a complex and stereotyped series of movements, and therefore, we might hypothesize that only the strongest individuals would be able to produce a longer and faster series of movements. Could this behaviour influence turtles' social interactions?

Another participant in the animal slapping contest is certainly the humpback whale (*Megaptera novaeangliae*). This time, the slaps fall on the water surface rather than on another whale's face. Humpback whales can often be seen swimming on their backs just under the surface, lifting a pectoral flipper out of the water and then repeatedly slapping it on the water's surface, creating a large splash with a loud noise. This behaviour is known as *pectoral slap* or *pec slap*, and despite the substantial interest in this spectacular behaviour (imagine the splash created by a pectoral flipper that can be 5 m long!), its function(s) is not fully understood. Research on the topic indicates that both sexes perform pec slapping, but it seems that while females use it to attract mates, males practice pec slap to maintain affiliation with each other [10]. Another hypothesis is that the powerful series of slaps on the surface removes parasites. Social media is filled with videos of this spectacular behaviour, often taken during whale-watching tours, and pectoral slaps can be incredibly diverse depending on the whale that is performing them. Some pec slap series can be long, others are short, and some animals can even slap both flippers. Moreover, these splashes can be performed in various social contexts. What is the rhythm of this behaviour? Are the series of splashes performed regularly? The sound of pectoral slaps can travel several kilometres, and if whales use it to communicate with distant individuals, one could expect that the structure of the signal might carry some information about the animal performing the slaps, such as its rhythm. Would different clans perform pectoral slaps with different rhythms? Would younger whales slap with the same rhythm as older ones?

4.2.5 Courtship of Birds of Paradise

Multimodal displays potentially allow us to understand how separate communicative channels rhythmically interact with each other. Does the rhythmic pattern of a specific communicative channel (e.g. vocal) temporally correlate with that of another one (e.g. visual), in composite, multimodal behavioural displays? In other words, are concomitant behavioural outputs the result of the same endogenous oscillatory pattern?

The animal models that could lay the foundations for the investigation in this direction are potentially numerous, as multimodality is per se common and adaptive—it enhances information transfer, reliability, flexibility and memorability. One emblematic example of multimodal communication that offers a promising model to address these open questions is the courtship displays of male birds of paradise (Family: Paradisaeidae).

The birds of paradise are a monophyletic group, endemic to New Guinea and its satellite islands, eastern Australia, and the Moluccan Islands of Indonesia. The majority of species of this taxon show a polygynous mating system and sexual dimorphism, with males displaying remarkable, multiple ornamental phenotypes that play a crucial role in striking courtship displays. This behaviour is characterized by whimsical, composite, multimodal outputs, dominated by colours, movements, and sounds that leave any ethologist in wonder. The visual channel is, for instance, based on mobile specialized nuptial feathers, organized in shields, collars, and skirts. These ornaments are then often accentuated by motor patterns consisting of head bobs, bows, hops, steps, turns, and wing movements. These displays are then often accompanied by acoustic signals, both vocal (e.g. calls) and nonvocal (e.g. wings and bill snapping).

Quantifying the temporal features and correlations between display components may follow the following steps. First, the quantification of the tempo of different components of the display. Second, the quantification of the lag/synchrony between two components. Third, the verification of intra-component (i.e. does a specific component follow specific integer ratios?) and intercomponent rhythmic categories (i.e. does the temporal relationship between two or more components follow specific integer ratios?). Finally, causality metrics may be used to provide overall indication of whether the onset of one component accurately predicts or determines that of another component.

If the vocal and motor components of such displays are derived from the same endogenous oscillator, we then should expect one of the following scenarios: the vocal and motor components of the display occur at the exact same time (i.e. in synchrony); the display components are temporally separated from each other by intervals that are constant through time (e.g. the same endogenous oscillatory motion generates the different outputs separated by a constant lag for mechanical/physiological reasons); the tempo of a component is equal to an integer multiple of the tempo of another component (e.g. one component's tempo serves as a carrier to other component's rhythm). If the vocal and motor components of these displays do not derive from the same endogenous oscillators, we should then expect tempi that are not related by any small integers between components.

4.2.6 Solo Female Birdsong

While a plethora of research has been conducted on male birdsong (Suborder: Passeri) and pair dueting, studies of solo female singing are relatively rare. However, female song is common in the tropics and females sing in 64% of all songbird species with sufficient data to determine which sexes sing [11]. Even when female song has been investigated, it is often in comparison to male song and with little detailed analysis of song structure reported in the literature. Neurological and behaviour research of song production areas has viewed females as recipients of songs, rather than as singers. The rhythm features of such songs remain largely uncharacterized.

4.2.7 Growling

Growling is so ubiquitous across mammals that it is almost taken for granted. However, there are potentially important variations in growls. For example, red wolf (*Canis rufus*) growls range from noisy to harmonic, with the former being more common during aggressive interactions and the latter during playful interactions (Jennifer Schneider, *unpublished data*). Similarly, yellow mongooses (*Cynictis penicillata*) use different growls for alarm/threat than during food competition [12]. Possible variations in pulse rate need to be investigated. These examples suggest that rhythmic properties of growls, including absolute pulse rate and pulse rate variability may indicate subtle differences in the growling animal's mood or intent, and therefore may have important communicative value. Such possibilities warrant systematic cross-species studies to determine if these observations extend to additional species and if so, if there are universal versus species-specific temporal patterns.

4.2.8 Rattle Snake Tail Rattle

Among nonavian reptiles (Class: Sauropsida), and maybe even generally among tetrapods, rattle snakes might be the only animals which carry a quintessential rhythmic instrument on their body. To study sensitivity to rhythm, the tail rattling of rattle snakes (Subfamily: Crotalinae) could be an interesting behaviour, yet with rather unusual circumstances. Most importantly, the rattle snake rattling appears to be an exclusively interspecific communication behaviour. The rattling produces an acoustic signal, despite snakes being barely able to hear airborne vibrations. Snakes are specialized lizards, which have lost their external ears. Still, they can perceive vibrations traveling through the ground as they are transmitted via the lower jaw (that usually rests on the ground) to the inner ear [13]. However, a typical rattling display involves a body posture completely unsuitable for such bone conduction hearing: the rattle snake raises its head high off the ground, holds it horizontally and lays its neck into an S-shape, which would allow for a fast strike forward. Simultaneously, the rattle, plus most of the entire tail, is lifted fully off the ground. Hence, the jawbone is not in contact with the ground and any vibrations reaching it are negligible. The rattling signal is a warning to other species and the intended receivers are animals that might actively try to predate the snake or could accidentally step on it. If the warning is ignored, the rattle snake will strike and, in certain cases, envenomate the forewarned attacker or disturber.

Any perception of the rhythm of the rattle would therefore come from the receivers. However, the rattle snakes may adjust the rhythm of the rattle to the specific animal in front of them and to the specific context. There is evidence that the rattling behaviour is very plastic and/or under voluntary control. The western diamond back rattle snake (*Crotalus atrox*) habituates to external disturbance by reducing the rate of rattling but can recover fast if the type of disturbance changes [14]. The same species also modifies the rate of rattling in response to an approaching human. It increases the rattling frequency with the reduction of the distance to the perceived threat [15]. If a human is fairly close to the diamond back, it changes the sustained rattle of around 40 Hz abruptly to a much higher frequency of 60–100 Hz. Experiments with humans showed that this shift in frequency causes the sensation that the snake is closer than it actually is [15]. This interspecies communication system offers interesting research opportunities. For instance, one could study which rattling rates have high salience for certain species or whether rattle snakes adjust the rate depending on the species approaching. And this all with the peculiar caveat that the producer of the signal cannot actually hear it.

4.2.9 Mating Calls in Tortoises

Turtles were for the longest time considered quiet, even silent, animals. However, new evidence suggests that most species do actually vocalize on occasion [16]. A clear exception is the mating calls of tortoises (terrestrial turtles). Males of many different species produce loud calls during mounting and these appear to have multiple communicative functions. They are also often, at least anecdotally, described as rhythmical. These mount vocalizations occur in at least 35 species and phylogenetic bracketing suggests that they are an ancestral trait within Cryptodira (one of the two major lineages of living turtle species) [17]. Interestingly, mount calls do not appear to be predictable by-products of heavy breathing during copulation; the extent of amplitude and frequency modulation is inconsistent with such an explanation [17]. Some observations could be of interest for rhythm research. For instance, the number of mounting successes in male marginated tortoises (*Testudo marginata*) correlates positively with the number of mount calls, but negatively with the duration of them [18]. In the same study, behaviours related to "courtship intensity" also predicted mounting success. This suggests that the vocal display during mounting probably indicates mate quality and is used in female choice. Indeed, in Hermann's tortoises (*Testudo hermanni*), females preferred playbacks of mount calls with shorter duration, higher pitch, and higher call rate [19]. Particularly, the high call rate was interesting because, in males of this species, it correlates strongly with haematocrit values [20], that is, the percentage of red blood cells in the total blood volume. The females might hence be able to use the call rate for choosing mates of high physical quality. Precisely the same pattern was later also found in Greek tortoises (*Testudo graeca*) [21].

Tortoises do not live in stable groups and might find mates only infrequently. Therefore, being able to judge the quality of a potential mate out of sight could be advantageous. Plus, there is the added aspect of "mate choice copying": a male already involved in courtship must have been judged to be a suitable mate by another female and hence is, more likely than not, an acceptable mate. Given that high call rate predicts male quality and that these calls are produced during a rhythmic activity, females might indeed also be sensitive to rhythm. Healthy males might be able to maintain a more consistent rhythm during physically straining activities. Playback experiments using synthesized mount call series could deduce whether females prefer more rhythmic displays and maybe also whether there is an upper limit for the calling rate, that is, whether a faster call rate keeps getting more attractive even if it reaches unnatural levels and hence whether this is a case of runaway selection.

4.2.10 Gecko Vocal Knocking

The infraorder Gekkota, which contains all living gecko species, belongs to the oldest branch of the Squamata (all lizards, including snakes) family tree. That is, they are thought to have split from the common ancestor of all other living lizards before they diversified. Interestingly, they are by far the most vocal lizards, which suggests that Squamata evolution in general steered away from relying on vocal communication.

Geckos call to attract mates and to repel territorial rivals. Their call sequences contain idiosyncratic patterns that appear to be used for species recognition and that consequently inherently tend to pre-empt hybridization [22]. Call sequences may vary slightly between individuals, particularly in the number of small call elements ("rattle units") and the duration of pauses between them [23]. These differences probably reflect small variations in vocal anatomy and body condition, offering advertising cues about mate quality or rival strength. Some species also have the ability of flexibly modifying call sequences. If loud background noise threatens to mask the calls of the tokay gecko (*Gekko gecko*), males can prolong the duration of their shorter call notes and choose to produce more syllables of higher amplitude. In doing so, they specifically do not increase the amplitude of elements from their repertoire but compose their call sequence out of elements that will be loud enough for the current conditions [24].

Given that there are close to 2000 gecko species alive today, it is well possible that certain species attend to rhythm in call sequences to assess rivals and mates, particularly as the producers seem to exercise a certain level of control over the composition of their call bouts.

4.2.11 Head Bobbing in Lizards

Outside of geckos, lizards mainly use visual displays for communication [25]. In the Iguania (one of the youngest branches of the Squamata phylogeny), many species produce the head bobbing display, in which the head makes a fast "nodding" motion and the lizard simultaneously performs an incomplete push up with all four, or at least the two front, legs [26]. Many head bobs may be produced in fast succession, which creates sequences of informative value to a perceiver [25]. In a dyad of lizards, either both parties head bob or one of them will appease the "head bobber" by standing still, lifting one arm off the ground and waving in a circular motion. Head bobbing is used in territorial defence as well as in courtship, and the sequence can be indicative of species identity and mate quality. It is clearly a rhythmic behaviour. However, it is also very stereotypical and does not appear to be particularly flexible. This behaviour is often studied with robotic lizards, which are used to entice head bobbing and which can imitate the head bobbing display [27]. One of these studies found that lizards might be aware of the syntactic rules of their own species-specific bobbing behaviour: Sagebrush lizards (*Sceloporus graciosus*) produced more aggressive displays to a robot, which they perceived as a conspecific, that violated their syntactic rules [28]. However, the same study also tested whether sagebrush lizards would adjust their own displays to match that of this artificial opponent, that is, they investigated whether signal matching (or even "song" sharing like in songbirds) could occur in lizards. The subjects did not alter their display and hence showed no evidence of perceiving a rhythm different from their own idiosyncratic one.

Despite the lack of evidence so far, lizard head bobbing could become an interesting area for studying rhythmic behaviour. Iguania also contains around 2000 species inhabiting very diverse habitats ranging from deserts and tropical rain forest to even marine habitats [29]. In this diverse group, some understudied species might indeed be sensitive to rhythm.

4.2.12 Bill Clattering in Storks and Albatross

Some socially monogamous waterbirds, such as storks (Family: Ciconiidae) and albatrosses (Family: Diomedeidae), rapidly clap their mandibles to generate sounds for communication. This behaviour is sometimes described by different expressions, such as bill clattering, rattling, drumming, clappering, et cetera, but here we will refer to it as bill clattering.

Albatrosses exhibit bill clattering at different developmental stages [30]. Juveniles perform bill clattering when facing an intruder, such as carnivorous birds, which is assumed to be a defensive behaviour. Adult birds perform bill clattering in the same context (i.e. toward intruders), but they also perform it as one of the various behavioural elements in the mutual courtship display of males and females. During the courtship, the clattering speed is around 15–30 times per second. Bill clattering during courtship takes place in various postures and at specific times. For instance, Laysan albatross (*Phoebastria immutabilis*) and black-footed albatross (*Phoebastria nigripes*) perform bill clattering while standing erect and while bowing.

Storks perform bill-clattering behaviours in the context of courtship during the breeding season and in aggressive context throughout the year [31, 32]. In oriental storks (*Ciconia boyciana*), it has been reported that the acoustic structures of bill clattering convey information about sex. While the storks did not change the acoustic structure of bill clattering depending on social contexts (courtship and noncourtship), the fundamental frequency and the second harmonic frequency are lower in males than in females [32]. The fundamental frequency correlated positively with bill length, suggesting that bill morphology contributes to the sexual dimorphism of clattering sounds.

The functions of bill clattering still need to be clarified, as there is almost no quantitative study except for Eda-Fujiwara and colleagues [31, 32]. There are several open questions regarding the behavioural mechanisms and functions of bill clattering. For instance, it would be fruitful to investigate what kinds of information are included in bill clattering and its acoustic structure at within-individual and between-individual levels. Rhythmic features and acoustic traits (e.g. duration, frequency, amplitude) of bill clattering might reflect breeding status, age, context and physical condition, which can be useful information for potential mates and rivals. Investigating how they coordinate bill clattering with other body movements and postures would provide significant insights into multimodal communication (e.g. bill clattering while the fluttering of the wings). Since bill clattering is performed in various postures, there might be behavioural constraints and/or synergistic interactions with the momentum and/or position of other body parts [33, 34]. Additionally, how signal receivers respond to the clattering behaviour would be a critical aspect in elucidating the functions of clattering behaviour as a communicative signal.

Bill clattering behaviour is also intriguing in that both males and females perform it, often in a duet-like manner [30, 31]. Although it must be confirmed by empirical research, bill clattering might contribute to the formation and maintenance of pair bonding [35] and territory defence, like song duets [36]. The degree of the coordination of bill clattering could potentially reflect the coalition quality of the pair [37].

4.2.13 Wriggling Caterpillars

Caterpillars of various butterfly and moth species (Order: Lepidoptera) show an interesting defence mechanism when undergoing metamorphosis, namely during the pupal stage of development—they wiggle when touched. The behaviour is often referred to as pupal wriggling or pupal response. The putative function is defence again potential predators or parasites, startling or dislodging them or making it difficult to grasp and handle the pupa, namely, against ants or parasitoid wasps (Suborder: Apocrita). The rhythm characterization of this behaviour has not been done. Given that the caterpillar is undergoing one of the most spectacular and complete physiological reorganizations in the animal kingdom, it would be particularly interesting to verify how pupal wriggling relates to other rhythmic behaviours in the species, before (e.g. peristalsis' rhythm of the caterpillar gait during locomotion) and after the metamorphosis (e.g. the rhythm of the nominal butterfly or moth wing beat).

4.3 Inter-individual Rhythms

4.3.1 Flashing Squids and Wavy Cuttlefish

Cephalopods exhibit a unique dermal architecture that enables rapid changes of body colour patterning, effected by neuromuscular dermal organs that contain elastic vesicles with pigment (chromatophores) or light-emitting substances (photophores) (for details, see Chap. 2 this volume). The shape and aperture of these vesicles can be modified through muscular contraction, which generates changes in opacity, translucency and reflectivity [38, 39]. Besides enabling well-known and impressive camouflaging in cephalopods, rapid skin changes are also implicated in intra- and interspecific communication, including rhythmic displays in the form of *flashing* and *rhythmic passing waves* (for a review see [40]).

Flashing is produced by the synchronous and repetitive opening and closing of chromatophore/photophore fields across the body, which results in a rhythmic transition between two skin patterns. The ultimate and proximate mechanisms of this signal are yet to be uncovered. Currently, the best documented species is the Humboldt squid (*Dosidicus gigas*), whose flashing occurs frequently and exclusively in the presence of conspecifics, thereby suggesting a role in intraspecies communication [41]. The scarce observations available to date place Humboldt squid flashing in two contexts: group spawning—where it may function as an antagonistic intermale display—and collective foraging, where it may function as a signal of foraging intent, thereby mitigating foraging competition and potential cannibalization [41, 42]. Structurally, the flashing pattern of the Humbold squid is characterized by a global oscillation in the 2–4 Hz range, where the entire body changes colour between white and red. Anecdotal observations suggest the presence of similar flashing in other ommastrephid species, such as the purple-back squid (*Sthenoteuthis oualaniensis*, [41]).

Interestingly, Humboldt squids have been observed to often make physical contact after initiating a flashing bout and to co-flash. When coflashing, they seem to adjust the amplitude, frequency and phase of their displays. Currently available reports indicate that the phase- and period-alteration during coflashing may yield various forms of temporal alignment [41]. Thus, rhythmicity seems to play an important role in Humboldt squid communication and, potentially, more broadly in the ommastrephid family.

Bioluminescent flashing has been described in deep ocean species, such as the Dana squid (*Taningia danae*), the Vampire squid (*Vampyrotheuthis infernalis*) and *Octopoteuthis deletron* [43–46]. In all these species, flashing is emitted by photophores located on the arm-tip, with species-specific shapes that result from the number of engaged arm-tips. Bioluminescent flashing has potentially rhythmic qualities as it has been described as pulsed [45] or as repeated at subsecond intervals [46]. Reportedly, bioluminescent flashing plays a role in both intra- and interspecific communication. Illustrating the former, Dana squids have been observed to respond with flashing when approaching and circling a blue light torch that resembled conspecific flashing [45], while a vampire squid male has been reported to respond with flashing to the flashing of a nearby located female. With respect to interspecific communication, flashing has been linked to both predatory and anti-predatory contexts. In the vampire squid, for example, flashing may mislead predators to attack the arm tips, which are regenerative and, thus, can be sacrificed [46]. By contrast, in

O. deletron such deceptive flashing has a predatory function, whereby the squid mimics the food its prey [47].

The *rhythmic passing wave* pattern consists of contrast (i.e. light-dark) bands that 'move' across the body as a result of highly coordinated neuromuscular activity across multiple chromatophore arrays. This display has been documented in cuttlefish species of the genera *Sepia* and *Metasepia*, as well as the octopus species *Wunderpus photogenicus* [40, 47–51]. The contrast bands of a rhythmic wave can cover the whole body of the individual (cuttlefish species) or only areas surrounding the eye stalks, as in *W. photogenicus* [39]. Some cuttlefish species exhibit unidirectional waves, where the bands move in a single direction (along the anterior-posterior axis), as described in the common cuttlefish (*S. officinalis*), the giant cuttlefish (*S. apama*), the dwarf cuttlefish (*S. bandensis*) and the broadclub cuttlefish (*S. latimanus*) *photogenicus* [40, 47–49]. Other species exhibit coordinated band motion in two (*S. apama*, the flamboyant cuttlefish *M. pfefferi*, the Paintpot cuttlefish, *M. tullbergi* [40, 50, 51]) or up to five motion fields (the mourning cuttlefish *S. plangon*, the reaper cuttlefish, *S. mestus* [40]). Generally, the frequency range of the rhythmic wave display is similar to that of squid flashing (1–2 Hz in *S. officinalis*, *S. mestus and W. photogenicus*), although exceptions have also been reported (e.g. 0.3–0.5 Hz in *S. plangon*).

The rhythmic passing wave display appears to be a defensive strategy in some species (*S. officinalis*, *M. tullbergi*, *S mestus*), being emitted to approaching predators and scuba-divers either as an intimidation signal or as an aposematic signal of toxicity [40, 50]. The display has also been reported during hunting in juveniles of *S. officinalis* [49], as well as in *S. latimanus*, and *W. photogenicus* [40]. Finally, in *S. apama*, the rhythmic wave display may be linked to reproductive function, as it has been reported to occur during agonistic interactions between males [48].

Together, these examples position cephalopods as key taxa to examine for studying the evolution of rhythmic communication. Unfortunately, however, this is all we have about their fascinating rhythmic signals—a handful of scattered observations accompanied by anecdotal interpretations of the documented behaviour. The mechanisms used by cephalopods to emit rhythmic signals are unique among the animal species that inhabit planet Earth, and very little is currently known about how their intra- and interspecific sociality compares to that of other taxa. We cannot, thus, get inspiration from other model species to get insights into when, why and to whom cephalopods signal rhythmically. The knowledge gap is just as abysmal as the oceanic environments that cephalopods inhabit. Obviously, such habitats constitute a major challenge for obtaining suitable empirical data with respect to sample size, video quality and relevant socioecological variables. Technological progress may help improve data collection. More systematic data collection will allow us to answer questions concerning the ultimate and proximate mechanisms of cephalopod rhythmic communication and, whether they engage in rhythmic interactions with conspecifics.

4.3.2 Geese V Flight Formation

Geese, swans (Family: Anatidae), and many other bird species fly and migrate in social flocks, where individual birds position themselves along a V-shaped formation in the direction of travel. This interindividual positioning minimizes air resistance for the individuals travelling behind, increasing their glide, and reducing energy expenditure. For this reason, birds take turns in the front position and birds flying alone tend to beat their wings faster than those flying in formation. Beyond relative position, air movement across the flock and between adjacent individuals predicts that individuals should also seek to synchronize their wingbeat. Across large distances, individuals who beat their wings at the best timing to ride the whirl created by the individuals in front would be expected to reap large energetic benefits versus those who do not. Wingbeat probably approximates isochrony, but optimal synchronization between individuals should probably be expected to be slightly out of phase and potentially correlated with interindividual distance. Causality analyses could also shed light on possible coordination mechanisms between individuals. Moreover, birds produce vocalizations during flight formations. It would be interesting to examine when birds opt to vocalize in the wingbeat cycle, given biomechanical constraints imposed by the wing muscles that surround the respiratory systems and chest, and how potential vocal synchronization may occur as an emergent property of the wingbeat. This behaviour is observable across countries, continents, and bird classes. Comparative research across these sets could offer insight into the role of air temperature and pressure, as well as bird body shape and weight, on rhythm and synchrony, and inform whether different bird lineages exhibit distinct synchronization mechanisms or not.

4.3.3 Spider Courtship and Mating

In some spider families, sexual dimorphism can be impressive, with males being considerably smaller than females. In some species of the infraorder Mygalomorphae, males are at risk of being eaten by females during copulatory behaviours to the point that they may have tibial spurs on the first pair of legs to block the female's chelicerae during courtship. A pilot study at the University of Turin identified three rhythmic behaviours. A single bout of female drumming and male leg and body vibrations occurred cyclically throughout the courtship. The drumming frequency of the female was 30% slower than the leg vibrating of the male. The male's vibratory behaviours repeatedly occurred during courtship, suggesting crucial importance for the success of copulation, but it is also questionable whether these behaviours (or their disruption) play a role in the likelihood of a male being consumed by the female prior to the copulatory act. Comparative research across families could offer new insight into courtship in these spiders, explaining the role of certain rhythmic features in the reproductive success of males. Going further in this perspective and knowing how females in some families commonly react to simple drumming, one might imagine that synthetic stimuli could be developed to understand whether females react differently to different rhythmic structures and, in the future, even to think about building decoys with which to carry out manipulative experiments.

Acknowledgements G-A Sauciuc's contribution to this chapter has been funded by the Bank of Sweden Tercentenary foundation (grant nr. P17-0942:1) and the Swedish Research Council (grant 2019-03104). The contributions by AR Lameira to this chapter have been funded by UK Research & Innovation, Future Leaders Fellowship (grant nr. MR/T04229X/1).

References

1. Greenfield MD, Honing H, Kotz SA, Ravignani A (2021) Synchrony and rhythm interaction: from the brain to behavioural ecology. Philos Trans R Soc B Biol Sci 376:20200324. https://doi.org/10.1098/rstb.2020.0324
2. Patel AD, Iversen JR, Bregman MR, Schulz I (2009) Experimental evidence for synchronization to a musical beat in a nonhuman animal. Curr Biol 19:827–830. https://doi.org/10.1016/j.cub.2009.03.038
3. Patel AD, Iversen JR, Bregman MR, Schulz I (2009) Studying synchronization to a musical beat in nonhuman animals. Ann N Y Acad Sci 1169:459–469. https://doi.org/10.1111/j.1749-6632.2009.04581.x
4. Patel AD (2021) Vocal learning as a preadaptation for the evolution of human beat perception and synchronization. Philos Trans R Soc B Biol Sci 376. https://doi.org/10.1098/rstb.2020.0326
5. Merker B, Cook PF, Reichmuth C et al (2025) Chapter 14: Human group synchrony in light of the comparative method: what we know and would like to know. A peer commentary treatment. In: Sauciuc G-A, Lameira AR, Persson T (eds) Nature beats: the what, why and how of animal rhythmic behaviour. Springer
6. Siniscalchi M, Lusito R, Vallortigara G, Quaranta A (2013) Seeing left- or right-asymmetric tail wagging produces different emotional responses in dogs. Curr Biol 23:2279–2282. https://doi.org/10.1016/j.cub.2013.09.027
7. Artelle KA, Dumoulin LK, Reimchen TE (2011) Behavioural responses of dogs to asymmetrical tail wagging of a robotic dog replica. Later Asymmetries Body Brain Cogn 16:129–135. https://doi.org/10.1080/13576500903386700
8. Ren W, Wei P, Yu S, Zhang YQ (2022) Left-right asymmetry and attractor-like dynamics of dog's tail wagging during dog-human interactions. iScience 25:104747. https://doi.org/10.1016/j.isci.2022.104747
9. Kramer M, Burghardt GM (1998) Precocious courtship and play in emydid turtles. Ethology 104:38–56. https://doi.org/10.1111/j.1439-0310.1998.tb00028.x
10. Deakos MH (2002) Humpback whale (*Megaptera novaeangliae*) communication: The context and potential functions of pec-slapping behavior on the Hawaiian wintering grounds. Master thesis. University of Hawaii at Manoa
11. Webb WH, Brunton DH, Aguirre JD et al (2016) Female song occurs in songbirds with more elaborate female coloration and reduced sexual dichromatism. Front Ecol Evol 4. https://doi.org/10.3389/fevo.2016.00022
12. Le Roux A, Cherry MI, Manser MB (2009) The vocal repertoire in a solitary foraging carnivore, *Cynictis penicillata*, may reflect facultative sociality. Naturwissenschaften 96:575–584. https://doi.org/10.1007/s00114-008-0506-5
13. Friedel P, Young BA, Van Hemmen JL (2008) Auditory localization of ground-borne vibrations in snakes. Phys Rev Lett 100:048701. https://doi.org/10.1103/PhysRevLett.100.048701
14. Place AJ, Abramson CI (2008) Habituation of the rattle response in western diamondback rattlesnakes, *Crotalus atrox*. Copeia 2008:835–843. https://doi.org/10.1643/CE-06-246

15. Forsthofer M, Schutte M, Luksch H et al (2021) Frequency modulation of rattlesnake acoustic display affects acoustic distance perception in humans. Curr Biol 31:4367–4372.e4. https://doi.org/10.1016/j.cub.2021.07.018
16. Jorgewich-Cohen G, Townsend SW, Padovese LR et al (2022) Common evolutionary origin of acoustic communication in choanate vertebrates. Nat Commun 13:6089. https://doi.org/10.1038/s41467-022-33741-8
17. Galeotti P, Sacchi R, Fasola M, Ballasina D (2005) Do mounting vocalisations in tortoises have a communication function? A comparative analysis. Herpetol J 15:61–71
18. Sacchi R, Galeotti P, Fasola M, Ballasina D (2003) Vocalizations and courtship intensity correlate with mounting success in marginated tortoises *Testudo marginata*. Behav Ecol Sociobiol 55:95–102. https://doi.org/10.1007/s00265-003-0685-1
19. Galeotti P, Sacchi R, Rosa DP, Fasola M (2005) Female preference for fast-rate, high-pitched calls in Hermann's tortoises *Testudo hermanni*. Behav Ecol 16:301–308. https://doi.org/10.1093/beheco/arh165
20. Pellitteri Rosa D (2003) Selezione sessuale in *Testudo hermanni* (Reptilia, Testudinidae): relazioni tra condizioni morfo-fisiologiche, corteggiamento, canto e successo di accoppiamento. Msc thesis. University of Pavia
21. Pellitteri-Rosa D, Sacchi R, Galeotti P et al (2011) Courtship displays are condition-dependent signals that reliably reflect male quality in greek tortoises, *Testudo graeca*. Chelonian Conserv Biol 10:10–17. https://doi.org/10.2744/CCB-0840.1
22. Jono T (2016) Absence of temporal pattern in courtship signals suggests loss of species recognition in gecko lizards. Evol Ecol 30:583–600. https://doi.org/10.1007/s10682-016-9829-4
23. Yu X, Peng Y, Aowphol A et al (2011) Geographic variation in the advertisement calls of *Gekko gecko* in relation to variations in morphological features: implications for regional population differentiation. Ethol Ecol Evol 23:211–228. https://doi.org/10.1080/03949370.2011.566581
24. Brumm H, Zollinger SA (2017) Vocal plasticity in a reptile. Proc R Soc B Biol Sci 284:20170451. https://doi.org/10.1098/rspb.2017.0451
25. Carpenter CC, Ferguson GW (1977) Variation and evolution of stereotyped behavior in reptiles. In: Gans C, Tinkle DW (eds) Biology of the reptilia. Academic, London, pp 335–554
26. Macedonia JM, Clark DL, Fonley MR et al (2019) Analysis of bobbing displays in four species of Galápagos lava lizards using conventional and novel quantitative methods. Herpetologica 75:290. https://doi.org/10.1655/Herpetologica-D-19-00015.1
27. Frohnwieser A, Murray JC, Pike TW, Wilkinson A (2016) Using robots to understand animal cognition: robots in animal cognition. J Exp Anal Behav 105:14–22. https://doi.org/10.1002/jeab.193
28. Brian Smith C, Martins EP (2006) Display plasticity in response to a robotic lizard: signal matching or song sharing in lizards? Ethology 112:955–962. https://doi.org/10.1111/j.1439-0310.2006.01253.x
29. Gauthier JA, Kearney M, Maisano JA et al (2012) Assembling the squamate tree of life: perspectives from the phenotype and the fossil record. Bull Peabody Mus Nat Hist 53:3–308. https://doi.org/10.3374/014.053.0101
30. Tickell WLN (2000) Albatrosses. Yale University Press, New Haven
31. Kahl MP (1972) A revision of the family Ciconiidae (Aves). J Zool 167:451–461. https://doi.org/10.1111/j.1469-7998.1972.tb01736.x
32. Eda-Fujiwara H, Yamamoto A, Sugita H et al (2004) Sexual dimorphism of acoustic signals in the oriental white stork: non-invasive identification of sex in birds. Zoolog Sci 21:817–821. https://doi.org/10.2108/zsj.21.817
33. Cooper BG, Goller F (2004) Multimodal signals: enhancement and constraint of song motor patterns by visual display. Science 303:544–546. https://doi.org/10.1126/science.1091099
34. Ota N, Gahr M, Soma M (2015) Tap dancing birds: the multimodal mutual courtship display of males and females in a socially monogamous songbird. Sci Rep 5:16614. https://doi.org/10.1038/srep16614
35. Wachtmeister C-A (2001) Display in monogamous pairs: a review of empirical data and evolutionary explanations. Anim Behav 61:861–868. https://doi.org/10.1006/anbe.2001.1684

36. Hall ML (2004) A review of hypotheses for the functions of avian duetting. Behav Ecol Sociobiol 55:415–430. https://doi.org/10.1007/s00265-003-0741-x
37. Hall ML, Magrath RD (2007) Temporal coordination signals coalition quality. Curr Biol 17:R406–R407. https://doi.org/10.1016/j.cub.2007.04.022
38. Florey E (1969) Ultrastructure and function of cephalopod chromatophores. Am Zool 9:429–442. https://doi.org/10.1093/icb/9.2.429
39. Messenger JB (2001) Cephalopod chromatophores: neurobiology and natural history. Biol Rev 76:473–528. https://doi.org/10.1017/S1464793101005772
40. How MJ, Norman MD, Finn J et al (2017) Dynamic skin patterns in cephalopods. Front Physiol 8:393. https://doi.org/10.3389/fphys.2017.00393
41. Rosen H, Gilly W, Bell L et al (2015) Chromogenic behaviors of the Humboldt squid (*Dosidicus gigas*) studied *in situ* with an animal-borne video package. J Exp Biol 218:265–275. https://doi.org/10.1242/jeb.114157
42. Burford BP, Robison BH (2020) Bioluminescent backlighting illuminates the complex visual signals of a social squid in the deep sea. Proc Natl Acad Sci 117:8524–8531. https://doi.org/10.1073/pnas.1920875117
43. Roper CF, Vecchione M (1993) A geographic and taxonomic review of *Taningia danae* Joubin, 1931 (Cephalopoda: Octopoteuthidae), with new records and observations on bioluminescence. In: Okutani T, O'Dor RK, Kubodera T (eds) Recent advances in cephalopod fisheries biology. Tokai University Press, Tokyo, pp 441–456
44. Kubodera T, Koyama Y, Mori K (2007) Observations of wild hunting behaviour and bioluminescence of a large deep-sea, eight-armed squid, *Taningia danae*. Proc R Soc B Biol Sci 274:1029–1034. https://doi.org/10.1098/rspb.2006.0236
45. Robison BH, Reisenbichler KR, Hunt JC, Haddock SHD (2003) Light production by the arm tips of the deep-sea cephalopod *Vampyroteuthis infernalis*. Biol Bull 205:102–109. https://doi.org/10.2307/1543231
46. Bush SL, Robison H, Caldwell RL (2009) Behaving in the dark: locomotor, chromatic, postural, and bioluminescent behaviors of the deep-sea squid *Octopoteuthis deletron* Young 1972. Biol Bull 216:7–22
47. Hanlon RT, Messenger JB (2018) Cephalopod behaviour, 2nd edn. Cambridge University Press, Cambridge
48. Hall K, Hanlon RT (2002) Principal features of the mating system of a large spawning aggregation of the giant Australian cuttlefish *Sepia apama* (Mollusca: Cephalopoda). Mar Biol 140:533–545. https://doi.org/10.1007/s00227-001-0718-0
49. Adamo SA, Ehgoetz K, Sangster C, Whitehorne I (2006) Signaling to the enemy? Body pattern expression and its response to external cues during hunting in the cuttlefish *Sepia officinalis* (Cephalopoda). Biol Bull 210:192–200. https://doi.org/10.2307/4134557
50. Laan A, Gutnick T, Kuba MJ, Laurent G (2014) Behavioral analysis of cuttlefish traveling waves and its implications for neural control. Curr Biol 24:1737–1742. https://doi.org/10.1016/j.cub.2014.06.027
51. Thomas A, MacDonald C (2016) Investigating body patterning in aquarium-raised flamboyant cuttlefish (*Metasepia pfefferi*). PeerJ 4:e2035. https://doi.org/10.7717/peerj.2035

Endogenous Rhythm, Isochrony, and the Path to Perfect Synchrony

5

Michael D. Greenfield and Bjorn Merker

Abstract

Various species of crickets and katydids, one cicada species, various frogs, and one bird species (a rail) display striking coordination of acoustic signal rhythms with conspecific neighbors. A counterpart is seen in various firefly and crab species where optical signal rhythms are coordinated. The signals being coordinated function in sexual advertisement and courtship, their rhythms are ultradian (period <10 s) and endogenous (free-running), and the expressed coordination is either synchrony (~0° phase alignment) or alternation (anti-synchrony; ~180° alignment). Rhythm alignment is a sensorimotor process that is achieved via mutual phase adjustment or the combined adjustment of both phase and endogenous tempo (signal repetition rate). In the latter case, alignment can be extremely precise, continuous, and involve neighbors whose rhythm tempos are initially quite different. A key feature of coordination is rhythm isochrony, a relatively constant tempo, measured by the coefficient of variation, over many periods. Isochrony allows alignment to be anticipatory as opposed to reactive. Thus, displays of synchrony and anti-synchrony are forms of behavioral entrainment, a process in which rhythms in their entirety are coordinated as opposed to a simple series of one-off responses to a neighbor's separate signals. While some cases of rhythmic entrainment among animal species are artifacts of interneighbor competition, the more precise alignments, involving both phase and tempo adjustment, are adaptations that retain essential characters in broadcast signals or

M. D. Greenfield (✉)
ENES Bioacoustics Research Lab, CRNL, University of Saint-Etienne, CNRS, Inserm, Saint-Etienne, France

Department of Ecology and Evolutionary Biology, University of Kansas, Lawrence, KS, USA
e-mail: michael.greenfield@univ-tours.fr

B. Merker
Independent Scholar, Kristianstad, Sweden

G.-A. Sauciuc et al. (eds.), *Nature Beats*, Fascinating Life Sciences,
https://doi.org/10.1007/978-3-032-24250-1_5

enhance the collective broadcast of a local group. Rhythmic entrainment is not yet reported from natural populations of non-human mammals, notably non-human primates, indicating that human entrainment, which appears to be a universal trait and is an essential ingredient in music, dance, and related cultural activities, arose de novo in the hominid lineage.

5.1 Preamble

The title of our volume, *Nature Beats: The What, Why and How of Animal Rhythmic Behaviours*, brings together two domains of rhythmicity, namely, *musical rhythm* announced by the musical term "beat" in the title, and the vast domain of rhythmic behaviors found in the animal kingdom. That does not mean, of course, that animal rhythmicity is relevant to musical rhythmicity by the mere fact of being rhythmic. All motile animals, from unicellular flagellates to humans, propel themselves by rhythmic movements, and when they move at constant speed, the pacing of those movements approximates isochrony (cf. → Chap. 1 [1]).

More generally, when a given task cannot be accomplished in a single unitary act, repetition is necessary. In the absence of additional constraints on performance, equidistant repetition in time, i.e., isochrony, suggests itself as the default timing option, as in driving in a nail with a hammer, or in the rhythms of locomotion, breathing, and chewing. Vocal behavior is not exempt from this logic, as shown by untold instances of repetitive solo calling in mammals and birds (among which South-East Asian barbets and African tinkerbirds stand out for the regularity of their isochronous calling).

Given the compelling nonmusical reasons for animals to engage in evenly paced rhythmic behaviors, there is no reason to assign particular significance to such behavior in the comparative study of human music (cf. → Chap. 1 [1]). This is all the more so given that evenly paced rhythmicity by no means is a necessary or inherent structural characteristic of human music as such. Many forms of human music, widely spread across cultures, do without evenly paced rhythmicity. They have been given the generic label "unmeasured music" (see detailed discussion in [2], pp. 179 ff), exemplified by plain chant (*cantus planus*), melodic recitation of the Qur'an, introductory (alap) sections of North Indian ragas, many a solo guitar prelude to flamenco dance, and some versions of so-called freeform jazz in which the use of a steady beat is deliberately avoided.

Unmeasured music does not lack temporal dynamics or patterning in the sense of phrasing, complex timing, breathing cycles, and the like. That patterning is not, however, structured by the *unbroken iteration of a unit duration*, as it is in all rhythmic music, called "measured music" on account of this iterated unit duration ([2], pp. 179 ff). The unbroken nature of the iteration of that unit duration[1] makes it a global structural characteristic of rhythmic music, which, as its isochronous pulse, beat or tactus, provides the underlying temporal context for the timing of its constituent parts and determines their temporal relations by multiples and subdivisions of the unit duration ([1], pp. 179 ff). It is this unit duration that underlies the temporal structure of all rhythmic music that enables us to clap to such music, however complex its temporal patterning might be (for further details, see p. 8 in [8]).

It is by *discretizing time* through the unit duration of its sustained pulse, beat, or tactus that rhythmic music conquers all its distinctive temporal characteristics, principally the following three:

1. The elementary fact that rhythmic music has a determinate tempo [8, 9].
2. The temporal precision with which several participants can synchronize their contributions to a joint performance ([10], Table 1), whether by voice, by instrument, or by dancing to the music, through recourse to its shared isochronous musical pulse; and finally:
3. A profusion of small whole integer ratios in the temporal patterns of rhythmic music based on multiples or subdivisions of the iterated unit duration of the musical pulse [11–13].

These, then, are "cardinal characteristics of human musical rhythmicity" for which we would like to find parallels in the vast space of animal rhythmic behaviors, whether by analogy or homology, to serve as targets for comparative analysis. Searches are shortened by the extent to which they are conducted under relevant constraints. No facet of musical rhythmicity provides greater leverage in that regard than the way in which its structure allows several individuals to synchronize their behaviors to a joint performance with high temporal precision (point 2 above). At a

[1] "Unbroken" here designates sustained phase coherence, not the absence of gaps or pauses such as occur when one or more beats in a sequence are "skipped." As long as the phase before and after the pause are aligned with one another, iteration of the unit duration has not been broken. Phase discontinuities, on the other hand, violate the very point of iteration of a unit duration, which is to enable precise behavioral synchrony by making the time of occurrence of the next beat in the sequence perfectly predictable, and therefore "targetable" by predictive timing. A corollary of this timing principle is that no merely local measures over a few successive elements of a temporal sequence suffice to establish the musically crucial property of unbroken iterative continuity that define the musical pulse or beat. Nevertheless, claims of musical relevance based on such local measures have been made by numerous studies ever since such measures were introduced in 2020 by Roeske and colleagues [3]. The lack of unbroken isochrony in many of these studies—typically apparent on mere inspection of sonagrams—is hardly surprising: they analyze solo songs, that is, behavior without the need to establish inter-individual behavioral synchrony (see main text that follows). Even when coordinated with a partner in duetting, as in gibbon song and the duets of many bird species, the point of entry of one partner's contributions to the joint performance is timed, not by recourse to the temporal scaffolding of a sustained isochrony, but by contextual cues provided in the unfolding of the other partner's performance. Among primates, the most complex organization of such contextually cued duet organization occurs in siamangs [4, 5]. For the ontogeny of cued timing in a young crested gibbon female, see [6], pp. 101, 105, and Figure 6, with a telling instance shown in Fig. 5.2b of that paper.

stroke, this constraint narrows the vast search space of animal rhythmicity to a small subset of species that exhibit temporally precise interindividual behavioral synchrony.

The utility for comparative purposes of this structural feature of musical rhythmicity is further enhanced by its having conceivable biological functions, functions that would provide selection pressures for its evolution [14–16]. The same cannot be said for either keeping signaling behavior at a determinate tempo or adorning it with figures whose components exhibit small whole integer ratios on grounds other than multiplication or subdivision of the unit duration of unbroken isochrony (footnote 1), rendering their relevance to the comparative study of musical rhythmicity moot. Both the tempo and integer ratio features of rhythmic music come "for free," of course, once the signaling behavior is structured in time by unbroken iteration of a unit duration supplied by the musical pulse, beat, or tactus.

We have almost—but not quite—arrived at a criterion for finding candidate species for the comparative study of human musical rhythmicity, namely, "look for species capable of sustaining behavioral synchrony between individuals with high temporal precision." There is, however, a way to achieve a joint performance with precise inter-individual temporal synchrony without relying on unbroken iteration of a unit duration.

In footnote 1 of [8], one of us predicted that the complex, tightly timed, four-part "duetting" of the neotropical plain-tailed wren [17] is achieved by each bird learning the entire stereotyped joint pattern as a single, whole sequence of which it then only contributes its own segment during performance in a hocket-like fashion. This prediction, subsequently confirmed by direct recording of neural activity in wrens exposed to duet stimuli [18], was motivated by the fact that no shared isochronous time-giver is evident in the joint, complexly timed, bird chorus. The birds must therefore rely on some other means to align the timing of their parts with precision, for which a holistic, real-time, memory template to guide performance lies near at hand. The birds do not start their chorus in unison, which means that if a bird, once any wren has started to sing, simply "joins in" (perhaps in initial silence), guided by its template, its part of the performance will be temporally aligned with those of the others.

This, then, allows us to complete the comparative search strategy as follows: "First, find species whose natural behavior includes bouts of inter-individual behavioral synchrony sustained with high temporal precision. Then determine whether that inter-individual synchrony is achieved by the timing predictability offered by a shared isochronous time-giver (footnote 1 and [8])." This second desideratum is the only reason isochrony is of interest to the comparative study of musical rhythmicity, namely, to capture the distinctive way in which rhythmic music achieves precise temporal synchrony between individuals, and to distinguish it from other means of doing so.[2]

On coming across a species that meets both desiderata of the proposed search strategy, you have found a candidate species whose behavior deserves further scrutiny for its potential bearing on our understanding of the biology of human musical rhythmicity. Its natural behavior instantiates the core cardinal characteristic of human musical rhythmicity, of which its other two principal characteristics noted above are an automatic consequence in one case, and follow with necessity from signal complexity under time discretization in the other.

The proposed comparative strategy is no mere hypothetical exercise, because there are members of the animal kingdom who not only synchronize their behavior to a sustained and unbroken isochronous pulse, but do so spontaneously in the wild. Moreover, among them are some that do so by means that are analogous to what allows us to do the same, namely, by adjusting the phase and the period of their endogenous isochronous rhythmicity to end up in behavioral group synchrony with a precision that can surpass our own. These "natural synchronizers" are found in unexpectedly remote domains of the animal kingdom, to which we now turn.

5.2 Entrainment Across Animal Species

Among the noteworthy attributes of rhythmic animal behaviors, the coordination of rhythms between individuals is arguably the most remarkable [22]. However striking animal rhythms may appear in solo is generally compounded when neighbors join, and particularly when they entrain to one another. Such coordination can dazzle us by the sheer number of rhythmists and the collective intensity that results, by its precise timing, and by its complexity. Examples include synchronous fireflies in the Indo-Malayan region [23, 24], some of the periodical cicadas in central North America [25], and, of course, human music, dance, and related activities [14, 26]. But beyond appealing to our esthetic sensibilities, rhythm coordination poses major questions in neurophysiology and evolution. How do animals control their rhythmic behavior vis-à-vis their conspecific neighbors, and why do they bother to effect such control in the first place? In this chapter, we address these questions in non-human

[2] It should go without saying that this is not to be taken to mean that memorized musical content plays no role in human rhythmic music. It means only that the role of such material in inter-individual timing precision is subsidiary to the primary role of the shared isochronous time-giver in that regard, and is framed by it, if the music is rhythmic. Once its means to temporal synchrony is in place, various conventions for what is called expressive timing may be developed and learned in a given genre of music [8]. Similarly, the human capacity to "keep together in time" by means of a shared isochronous time-giver enables us to synchronize with sequences featuring gradual change in tempo, such as accelerating or decelerating ones [19]. The mechanism that enables us to latch on to an isochronous beat is ensured against being derailed by biologically inevitable temporal noise and drift by means of phase and period correction [8, 20, 21]. Jointly those two devices suffice to maintain synchrony with such formally non-isochronous sequences.

animal species where rhythm coordination appears regularly in natural populations and has been analyzed and modeled.

Whereas rhythms are pervasive in living organisms and involve most aspects of behavior and physiology [27], we focus specifically on rhythms evident in communication between conspecific individuals broadcasting and receiving sexual and social signals. It is in the signal exchanges between neighbors that rhythm coordination attains its fullest expression. This expression is largely a matter of relative timing, but an active spatial positioning—coordinated movement—of participating neighbors accompanies the temporal interaction in some species. A second focus is that the coordinated signals are mechanical or optical and apparently not chemical. The absence of chemical signals reflects the inability of odors, whether transmitted by diffusion or convection, to be precisely timed in the manner of pulses of sound or flashes of light except within a very short distance from the source [28]. Coordinated mechanical signals are generally acoustic, although the relative lack of these phenomena in the realm of substrate vibration may simply indicate that scientists are just now beginning to study vibrational communication in earnest. Optical signals entail either reflected light or bioluminescence, the latter being some of the most thoroughly analyzed cases of rhythm coordination. A third focus, following directly from the very notion of rhythm—repetition of events—is that the signals are discontinuous, e.g., chirps of sound or flashes of light, as opposed to unbroken whines, buzzes, or displays of color that continue for an indefinite time interval. Signal repetition rates—designated as "tempo"—range from 0.1 to 12 Hz (periods from 10 s down to 80 ms), meaning that the rhythms are "ultradian" as opposed to circadian ($\approx$ 24-h periods) or controlled by other long-term astronomical phenomena, e.g., lunar or solar time-givers (zeitgebers).

The format of rhythm coordination in animal species is typically some variant of synchrony, an approximately 0° phase alignment between neighbors, or alternation (anti-synchrony), an approximately 180° alignment. Synchrony can involve a great multitude of individuals in a local population, and the signalers can maintain their mutual timing with very high precision. Alternation is normally restricted to a few signalers, and their adherence to a given phase relationship is usually less strict. Beyond these two formats, a wide range of "signal interactions" are found in which animals modify their broadcasts when perceiving conspecific signals: they may change their tempo and temporal or spectral characteristics of their individual signals, or they might suddenly begin signaling after a long interval of inactivity [29]. These interactions usually do not entail actual rhythm coordination, and even when coordination is seemingly present, the phase alignment of individuals' rhythms may be an artifact. In the following sections, we explain what rhythm coordination is and is not, and we describe the analyses with which different levels of coordination can be identified. The criteria we present are not merely semantic, as it is only through the coordination of both signal tempo and phase that rhythm alignment can be precise, robust, and stable over a great many periods. Because rhythm alignment is critical for effective communication—and ultimately fitness—in some animal species, it befits us to pay attention to its parameters and how it is attained.

We approach the issue of precise alignment and coordination of animal rhythms from the perspective of "entrainment," the synchronization, or more broadly the phase alignment, of endogenous (= free-running) rhythms to an external time-giver. Using the circadian sleep patterns in humans as an example, the onset of sleep is roughly synchronized with nightfall, and the continuation of an approximately 24-h sleep cycle (normally closer to 25 h) when the astronomical time-giver is experimentally removed indicates that the rhythm is endogenous. Proceeding to animal communications and their ultradian rhythms, we replace the astronomical time-giver with a conspecific, signaling neighbor, or a synthetic stimulus representing that neighbor. For an animal to entrain to a synthetic neighbor, it would have to adjust its phase—provided that the two rhythmists have the same tempo—and come into alignment, or both its phase and tempo if the two do not. Entrainment of two live rhythmists would be similar except that the process is mutual. Rhythm alignment may rely on mutual evaluation of a neighbor's phase and tempo at a particular instant during the signal period, e.g., the onset or end of a 50-ms bush cricket chirp, treefrog call, or firefly flash during a 0.5-s call or flash period, or on continuous mutual evaluation throughout the period. The latter is conceivable in some optical signals entailing reflected light, e.g., rhythmic waving of an enlarged front leg as performed by males in some fiddler crab species [30]. Acoustic and optical signaling in arthropods and anurans often entails "nested rhythms," as where a 50-ms bush cricket chirp, delivered at 2 Hz, itself comprises five 6-ms pulses of sound, delivered at 90 Hz. Rhythm coordination usually involves only the longer (or longest) period, i.e., the slower (or slowest) rhythm, although secondary rhythms are sometimes coordinated independently [31]. This generalization may result from neurophysiological limitations that prevent alignment of very rapid rhythms, but selective factors specifically favoring alignment of the slower rhythms may also play a role. Lastly, it is easy to overlook the requirement that signal rhythms must be isochronous, i.e., maintain a rather constant period for a great many cycles, for entrainment to occur. Bluntly put, rhythm alignment is essentially the ability to anticipate when a neighbor's next signal will occur and adjust one's own next signal accordingly, an unlikely process should consecutive free-running period lengths fluctuate or drift, or when regular signaling is intermittent.

Readers may have noted that the examples given thus far, aside from human music and dance, are taken from arthropods and anurans, and for good reason. Most of the solid cases of rhythm coordination satisfying the criteria outlined above are in these taxa, and a basic model describing the neuroethology of solo and interactive rhythmic behavior was developed from an Indo-Malayan firefly, *Pteroptyx cribellata* [32, 33] (Fig. 5.1). This model assumes an underlying central nervous system (CNS) oscillator that steadily rises to a peak level at which point the behavior, an optical signal in *P. cribellata* but an acoustic one in other arthropods and in anurans, is triggered. A brief "effector delay" ensues between triggering and actual signal onset, and it comprises the time necessary for transmission of the neural message to the motor apparatus that creates the signal and the time for activation of that apparatus [34]. Meanwhile, the CNS oscillator rapidly descends to the basal level and begins its next cycle. Thus, a graphical depiction of the model presents a "sawtooth profile" as viewed over multiple cycles (Fig. 5.1a).

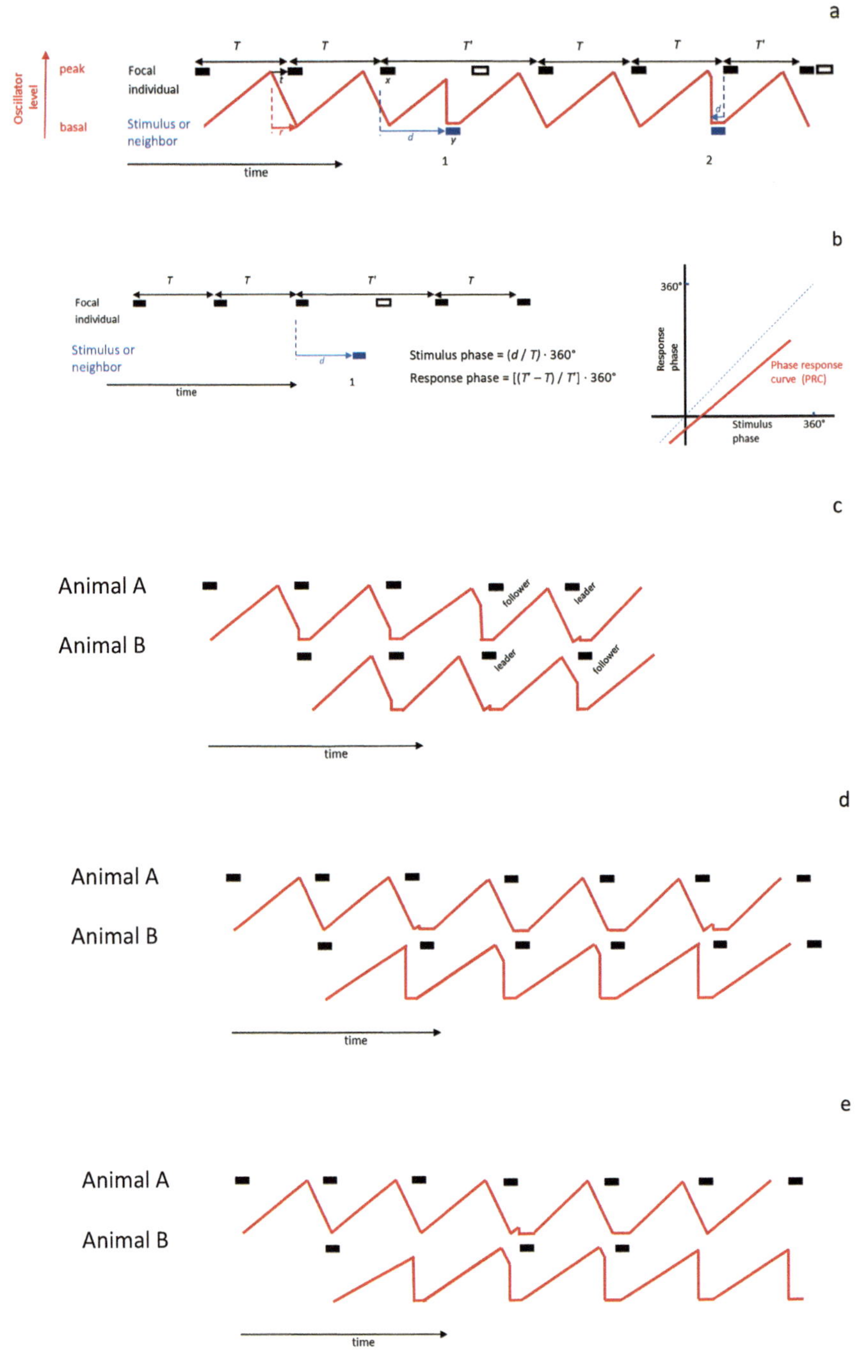

Fig. 5.1 Phase delay synchrony. (**a**) Sawtooth model depicting the rise and fall of a central nervous system (CNS) oscillator (red line) generating a periodic signal. Actual signal broadcast (horizontal black bar) occurs following a short interval *t*, an 'effector delay,' measured from the point when the oscillator attains its peak (trigger) level. A stimulus (horizontal blue bar) causes the oscillator to descend immediately to its basal level and remain inhibited until the stimulus ends. Thus, a stimulus perceived during the oscillator's ascent (point 1) yields a lengthened modified

5.2.1 Coupled Oscillators: Phase Delay

When an individual perceives an external conspecific signal, in some species, the CNS oscillator is fully reset, descending immediately to the basal level. For signals that are not momentary, i.e., whose length exceeds more than several milliseconds, the generator is typically inhibited at the basal level until the end of the signal, when the generator re-ascends to the peak level [35]. This ascent may be faster than in the free-running mode, a sort of "post-inhibitory rebound." Often, the individual returns to its free-running rhythm immediately following the resetting, with rhythm modification lasting for one and only one period. Playback tests broadcasting synthetic acoustic or optical stimuli have been useful in revealing this mechanism, termed "phase delay," and in identifying the effector delay component [36]. When a single, isolated stimulus signal is broadcast immediately before the tested individual is expected to call or flash, the individual will go ahead and call or flash anyway. This call or flash occurs because it had already been triggered by the CNS oscillator, and the earliest point in the signal onset—signal onset period at which the stimulus can be broadcast without resetting the tested individual indicates the length of the effector delay. It is the interval between that earliest point and signal onset, and measurements in different arthropod species give values from 50 to 200 ms, while 30 ms is given for an anuran [37]. For comparison, effector delays in humans of 95–120 ms are reported [24, 38]. We will later rely on the effector delay for distinguishing rhythm coordination, notably in synchrony, from other interactions in which an individual simply times a given signal with respect to a given signal of a neighbor.

Overall, the phase delay process is characterized by a lengthening of the concurrent signal period when a stimulus begins during the ascent of the oscillator, and a shortening of the next period—actually a "phase advance"—when the stimulus begins during the oscillator descent (Fig. 5.1a). These effects can be modeled algebraically [37] and graphed with a phase response curve (PRC), a quantitative tool well known in studies of biological periodicity (Fig. 5.1b). The abscissa of the PRC is the phase during the signal period when a stimulus begins (stimulus phase), and the ordinate is the proportional increase or decrease of the modified period (response phase). PRCs generally ascend diagonally, though not necessarily in a linear fashion, and cross the y-axis just below the origin. This crossing at a negative ordinate value reflects the post-inhibitory rebound often seen during the cycle immediately after resetting by a stimulus.

Fig. 5.1 (continued) period (T´), a phase delay, whereas a stimulus during the oscillator's descent yields a shortened period (point 2), a phase advance. The open horizontal bar indicates the expected timing of a signal in the absence of the stimulus. (**b**) Oscillator resetting depicted with a phase response curve (PRC). The modified period is determined as $T' = s \cdot d + (T + \varepsilon)$, where s is the slope of the PRC and ε is a stochastic element. (**c**) Two animals with identical free-running rhythms and CNS oscillators. Phase delay synchrony results when the PRC slope is ≈ 1, but stochastic variation in the oscillators is expected to yield an imperfect coordination in which the leading and following roles often switch between the animals. (**d**) Imperfect phase delay synchrony when the free-running rhythm of one animal (B) is approximately 10% slower. (**e**) Frequent interruptions of phase delay synchrony when the free-running rhythm of one animal (B) is approximately 30% slower; i.e., B's oscillator is often reset before attaining its peak level and triggering a signal. (Modified from Greenfield and Merker [7])

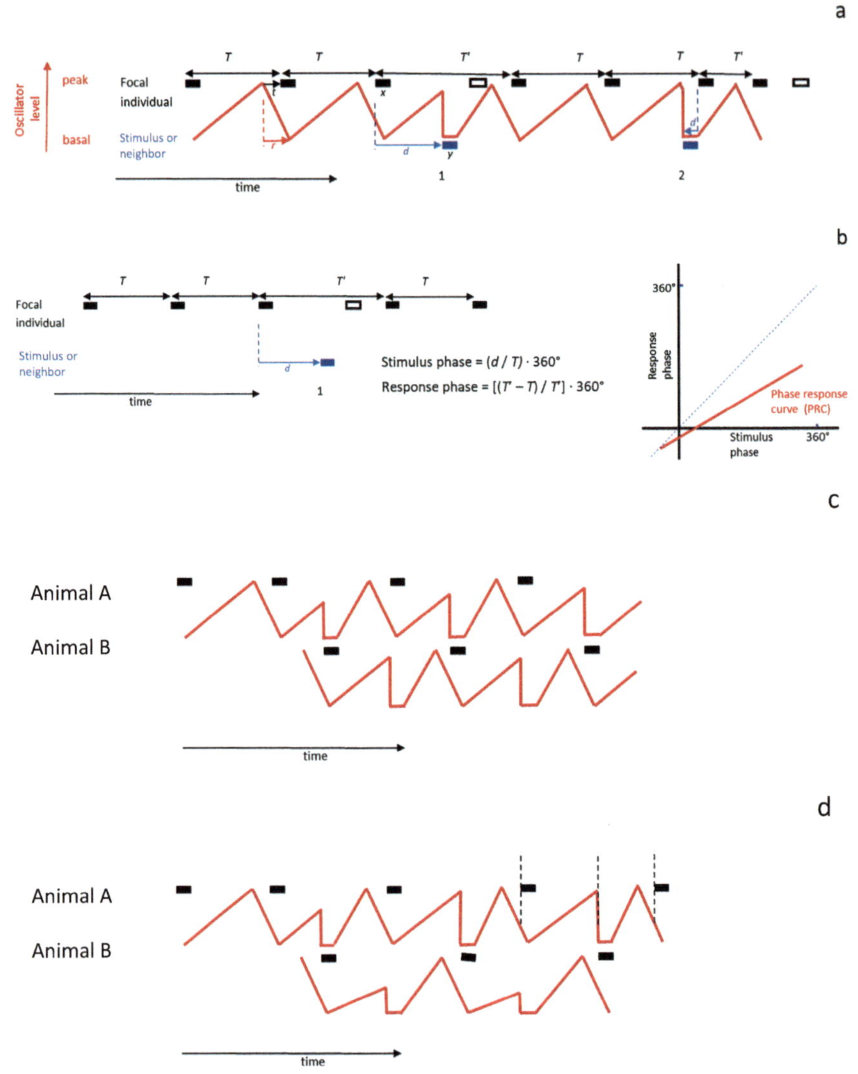

Fig. 5.2 Phase delay alternation. (**a**) Similar to the sawtooth model shown in Fig. 5.1a, except that oscillator ascent following inhibition by a stimulus is much faster, a 'post-inhibitory rebound.' (**b**) PRC determined as in Fig. 5.1b; owing to postinhibitory rebound, PRC slope is $<< 1$. (**c**) When 2 animals with identical free-running rhythms and PRC slopes $<< 1$ initiate joint signaling with a phase delay $\approx 180°$ (first signal broadcast by B), as will typically occur, alternation results. (**d**) When the free-running rhythm of one animal (B) is approximately 30% slower, alternation still results, but the slower animal will lag on each cycle; i.e., A → B intervals are consistently longer than B → A intervals. (Modified from Greenfield and Merker [7])

When two live rhythmists signal together, they form a "coupled oscillator," and each one influences its neighbor. If the two have similar rhythms—including the signals, the tempo and the phase delay mechanism—either a synchrony or an alternation arises. A synchrony occurs when the PRC is relatively steep, having a slope

≈ 1 (Fig. 5.1b, c), while an alternation occurs under shallower PRCs (slope < 0.7) (Fig. 5.2). An additional tendency is that synchrony typically involves relatively fast rhythms, periods being <1 s, whereas alternation predominates among slower rhythms [37]. The connection of alternation with slower rhythms may occur because a substantial post-inhibitory rebound is central to alternation, and fast rhythmists, already signaling close to their physiological limit, may be unable to accelerate for even a single period.

Phase delay can be considered the simplest form of entrainment of signal rhythm in animal species. The process only entails phase adjustment and yields an imperfect synchrony or alternation that is entirely dependent on equivalent tempos among the rhythmists. Nonetheless, once a focal individual's rhythm is reset by its neighbor, that focal individual's rhythm continues in its reset phase even in the absence of its neighbor's signals. This continuation shows that the rhythms themselves were coordinated and that the interaction is not simply a series of separate responses to each successive signal.

5.2.2 Emergence and Adaptation

Phase-delay entrainment is often an "emergent property" [35, 39], but it may also be a specific adaptation. Phase-delay synchrony can emerge by "default" from simple preferences for the relative timing of signals. When female receivers prefer a male's advertisement signal that "leads" his neighbor's signal by a brief time interval, which may result from a "precedence effect" or simply "forward masking" in cognition, male timing mechanisms that reduce the broadcast of ineffective "following" signals are favored. Phase delay happens to be an effective means of both foregoing the broadcast of following signals and increasing the likelihood of producing leading ones in subsequent cycles. But when two equivalent, rapid signalers coordinate their rhythms with a phase-delay mechanism, they reset each other mutually and an "imperfect synchrony" emerges. The synchrony is designated as imperfect because the signalers, not being metronomes, are subject to stochastic variation in their periods and their resetting, and a slight leading role typically fluctuates from one individual to the other over successive cycles (Fig. 5.1c). Specifically, however, these fluctuations may arise from the phase advance that an individual experiences after having been a follower. Imperfect synchrony as such, which can include numerous neighbors, is known among several arthropod taxa [39, 40]. The temporally structured group display is considered emergent as the signalers do not benefit from their synchrony per se, the receivers neither require nor prefer it, and the display depends entirely on the signalers having equivalent free-running tempos. When tempos are not equivalent, the faster signaler will generally lead the slower by a small phase angle each period (Fig. 5.1d). As tempo differences increase further, the slower signaler will miss its broadcast during more and more cycles while the faster one continues relatively unaffected (Fig. 5.1e).

When slow signalers employ a phase-delay mechanism, they usually generate an imperfect alternation in a manner analogous to the imperfect synchrony of rapid

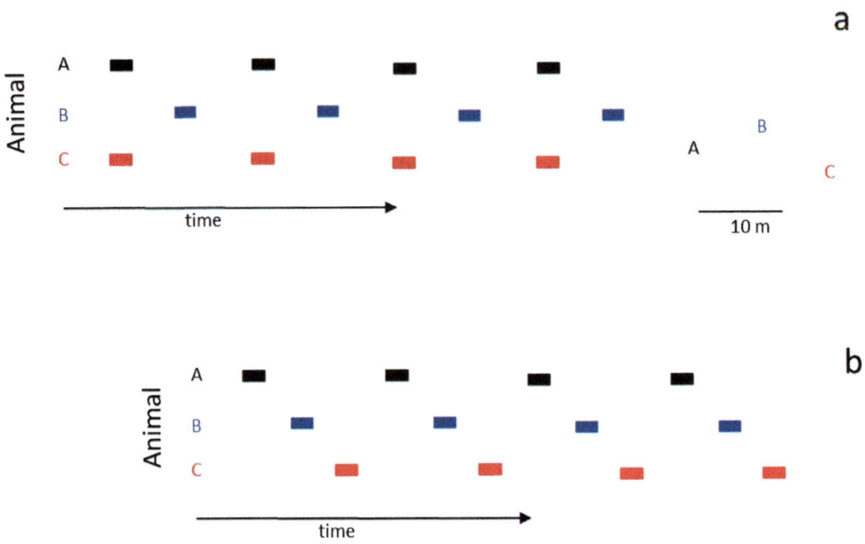

Fig. 5.3 Rhythm alternation with 3 signalers. (**a**) Clustered anti-phase synchronization in which each animal alternates with its nearest neighbor(s); A alternates with B, C alternates with B, A and C synchronize by default, and B alternates with the synchronized signals of A + C. (**b**) N-way (tri-phase) alternation in which each animal exhibits a phase delay of 120° with respect to one of its neighbors. (Modified from Greenfield and Merker [7])

signalers described above (Fig. 5.2). But there are two major differences. First, when more than two signalers are present, their expected alternation represents a conundrum. This problem, which is commonplace in natural populations, is overcome in several ways. (a) The signalers may spread out in the landscape provided that suitable habitat is available [41]. But the likelihood of this movement could also depend on the behavior and dispersion of receivers. (b) Signalers may selectively attend to their nearest neighbor while ignoring more distant ones whose perceived intensities fall below a threshold set by sensory adaptation, e.g., the level of a background chorus [42, 43]. Here, a signaler effectively alternates with its major competitor for receivers, i.e., avoids the broadcast of calls immediately following that competitor, while synchronizing, by default, with its lesser competitor. This tactic would establish what has been termed "clustered anti-phase synchrony," which can extend in a chain reaction fashion to involve more distant neighbors [44] (Fig. 5.3a). The phenomenon is reported in various acoustic insects and anurans. (c) Particularly in anurans, when local density increases such that clustered anti-phase synchrony is no longer possible, signalers may wait until a short lull in the chorus intensity appears and then quickly insert a call [45]. This tactic possibly entails a modification of the phase-delay process in which the CNS oscillator is kept "on hold" slightly below peak level until a quiet window occurs, when the oscillator is released and rapidly completes its ascent.

The second major difference between phase-delay synchrony and alternation is that the latter may often be a specific adaptation in addition to, or instead of, an

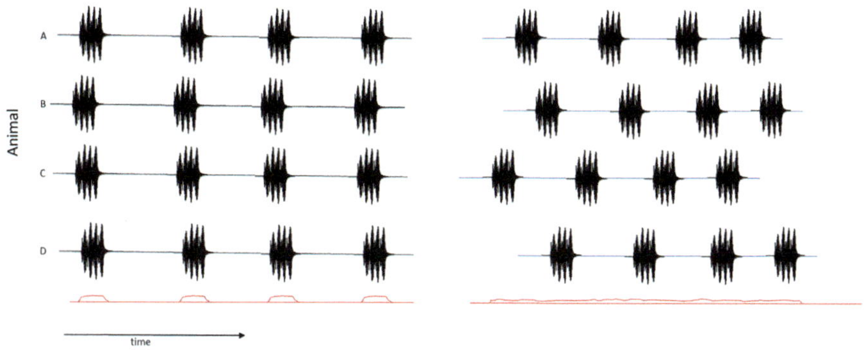

Fig. 5.4 Each panel shows signals broadcast by 4 individuals, A-D. Synchronization of the 4 animals in the left panel retains the characteristic signal envelope and signal rhythm, as depicted in the red trace below animal D's signals, whereas random signal timing in the right panel blurs these features. In addition, the peak collective intensity, indicated by the red trace, is substantially higher for synchronized signals relative to random ones. (Modified from Greenfield and Merker [7])

emergent property derived from receiver preferences for leading signals. When signalers alternate their rhythms, they can assess their neighbors more clearly and respond accordingly, if necessary [35]. This clarity issues from the general incompatibility of simultaneous signaling and perception, a particular problem in acoustic communication because an animal's own broadcasts can physically mask incoming sounds. But the problem may also occur in optical communication because an animal's fundamental attention to external stimuli can be attenuated during its own signaling [46]. And conversely, alternation can allow receivers to evaluate the signalers more accurately. The specific advantages of alternation do not require a precise 180° phase angle between neighbors' rhythms, and the imperfect entrainment afforded by a phase-delay mechanism (Fig. 5.2d) is more than adequate in most cases.

Like alternation, synchrony can be a specific adaptation under several conditions. The most likely is where the profile of amplitude over the course of a signal, the signal's "envelope," is a critical feature that receivers must perceive (Fig. 5.4). For example, female receivers might have to recognize the conspecific male signal envelope before responding to any one male, and this recognition could be impaired if local males do not align and instead collectively create a continuous broadcast as perceived by local females. There is experimental evidence for this situation in several synchronizing acoustic insects where successive male signals must be separated by quiet gaps of minimum duration and amplitude depth (10–12 dB) for females to initiate orientation [47, 48]. Thus, when males are grouped, all are forced to cooperate—synchronize with each other—for anyone to obtain even a chance of mating success. This bind may be especially severe in species whose signaling has a high "duty cycle" (= signal length / signal period), as failure to synchronize would certainly eliminate many of the requisite gaps. Tellingly, the synchronizing acoustic insects in which gaps were found to be requisite for female orientation feature high duty cycle [47, 48]. A closely related condition that could favor synchrony would occur where signal tempo—at a given temperature, as tempo is commonly influenced thermally in arthropods and anurans—would be necessary for conspecific recognition.

Another situation conducive to the bind forcing cooperative synchrony may occur in species where local density can be so high that nonaligned males would "clutter the landscape" and prevent females from cognitively focusing on any one signaler. The clutter effect has been proposed for bioluminescent synchrony in some firefly species [49], and it may hold even when the signal duty cycle is low. As above, when one or more males do not synchronize, females may not respond to any male in the group.

When local signalers align their rhythms in synchrony, they not only retain the signal envelope and tempo and reduce signal clutter, but they also increase the peak signal intensity of the group owing to the summation of simultaneous individual signals (Fig. 5.4). Groups that do not synchronize would not broadcast these enhanced peak intensities, and their effective signaling range in the landscape could be considerably lower. Thus, groups of male signalers that synchronize may benefit from higher attractiveness to female receivers in the general area. They may also make a stronger "impression" on rival male groups during competitive encounters. This peak intensity factor has been proposed as selectively favoring rhythmic synchrony in both humans [7] and non-human animals [50], but the effect awaits empirical study.

5.2.3 Coupled Oscillators: Phase Differential Mechanisms

Much of the thinking about adaptive synchrony was inspired by observations and tests with fireflies, notably the synchronous species in the Indo-Malayan region [51]. Their collective displays had been noted by Western observers as early as the sixteenth century, and more systematic study began with several scientific expeditions during the mid-twentieth century. These studies paid special attention to a remarkable species, *Pteroptyx malaccae*, where large numbers of males congregate on trees along riverbanks and flash 100-ms signals in unison at ca. 1.8 Hz. Measurements showed that the male flash rhythm was extremely isochronous in solo, that a male could align his flashes with an isochronous, synthetic flash signal and do so with a negligible phase angle, i.e., asynchrony, and that males attained similar alignment with other males [32, 52, 53]. This alignment was seen to continue for hundreds of cycles, and in attaining and keeping the synchrony, individuals modified their tempos by as much as 20%, slower ones speeding up while faster ones slowed down.

It was recognized that phase delay could not be responsible for this level of strict, nonemergent entrainment, but it was noted that a PRC with a very shallow slope could nonetheless roughly describe rhythm coordination in *P. malaccae*. However, it took the theoretical work of Yoshiki Kuramoto, a physical chemist—not a neuroscientist,

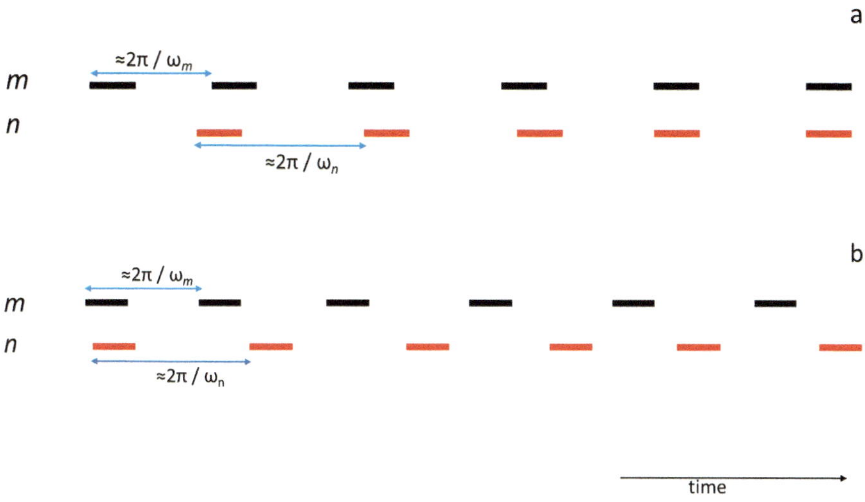

Fig. 5.5 Phase differential (or phase oscillator) synchrony and alternation generated by the Kuramoto model $d\theta_n / dt = \omega_n + \sum_{m \neq n} \Gamma nm(\theta m - \theta n)$, where θ is the phase of an oscillator, ω is its free-running frequency measured in radians, m and n are 2 different oscillators, and Γnm is the interaction function coupling the oscillators. Using a simple sinusoidal function, a positive coupling yields synchrony, whereas a negative coupling yields alternation. a) Synchronous signaling generated by a phase oscillator model with the interaction function $\Gamma nm(\theta m - \theta n)K \cdot \sin(\theta m - \theta n)$; K is the coefficient of coupling strength. The slower oscillator, n, becomes steadily faster, while the faster oscillator, m, slows until both converge and are aligned at a phase = 0°. b) Alternating signaling generated with the interaction function $\Gamma nm(\theta m - \theta n) = -K \cdot \sin(\theta m - \theta n)$. Again, the slower oscillator, n, becomes steadily faster, while the faster oscillator, m, slows until both converge, but here they are aligned at a phase of 180°. (Modified from Greenfield and Merker [7])

or even a biologist—to model a general "phase differential"[3] mechanism [54], which when applied to animal rhythmists explained how individuals might adjust their tempos in a way generating convergence on a common signal rate. In the basic Kuramoto model, two rhythms are coupled, e.g., by an optical signal and vision in our context, such that a given rhythm's incremental or decremental change in phase during a cycle is proportional to the (endogenous) phase difference between the two rhythms (Fig. 5.5). This link between phase difference and phase increment (or decrement) inevitably leads to rhythm convergence, and an interaction function in the model regulates the speed of that convergence (Fig. 5.5a). Changing the sign of the function from positive to negative yields divergence, i.e., rhythm alternation (Fig. 5.5b).

The model readily accommodates any number of rhythmists beyond a pair, and it would seem to account for many of the cases of strict, and probably adaptive, synchrony reported among arthropods. But would it? No sooner than its introduction to the world of animal rhythm, biologists and modelers realized that the basic Kuramoto model was inadequate for describing the "perfect synchrony" in *P. malaccae*, as well as in related firefly species and some species of acoustic insects with

[3] Termed "phase oscillator" models or mechanisms in applied mathematics and physics [32].

similar attributes. The main problem was that simulations using the Kuramoto model indicate than any stochastic variation in a rhythmist's tempo—inevitable in living rhythmists—initiate slowly damped, multi-cycle asynchrony [55]. Consequently, the strict alignment regularly seen [52, 53] should seldom occur.

To reconcile theory with observation, the applied mathematician George Bard Ermentrout proposed a refinement wherein rhythmists do not retain their endogenous tempos, as in the classical Kuramoto model, and adapt tempo—as well as phase—dynamics to their neighbors' rhythms [56].[4] This refinement, along with related ones since proposed by other modelers [57], do seem to yield a stricter and more robust synchrony as observed in *P. malaccae*. Our current issue is that no one has yet conducted the necessary neuroethological tests on live fireflies, or their equals in strict synchrony among acoustic insects (but see [58]), to confirm and/or fine-tune the models further. But some observations from less strict acoustic synchronizers show how simple adjustments can modify tempo and yield rhythm alignment. One method is lengthening or shortening signals by adding or subtracting pulses from multi-pulse chirps [59]. Another is forgoing the initiation of a signal during a given time interval measured from the end of a neighbor's signal [48]. Both approaches are seen in species signaling with high duty cycles, where the primary objective is retention of silent gaps between successive signals.

Before concluding our coverage of perfect synchrony, two groups in addition to fireflies and acoustic insects deserve attention. The main rivals of arthropods in synchrony are humans, whether they are trained musicians or not. Measurements of humans while performing or in cognitive psychology laboratories show that asynchronies less than 30 ms between rhythmists are regularly attained [7, 10], approaching—but not matching—*P. malaccae* and some acoustic insects. In humans neuroethological tests have been done, and these reveal that both phase and tempo correction are responsible for the ability to coordinate rhythm in synchrony [7], agreeing with Bard Ermentrout's model.

Anurans enter the conversation in a different way, as rhythm coordination is almost always alternation, and in small groups, i.e., pairs, the alternation appears to be a phase-delay process [37, 39], but with aspects that indicate entrainment based on tempo adjustment. Recent studies on alternation in the Japanese treefrog *Hyla japonica* that focused on the responsible mechanisms, however, point specifically toward control by a phase-differential process that can be described by the basic

---

[4] Ermentrout [56] refined the Kuramoto model by including tempo dynamics as follows:

$$d\,\omega_n / dt = \epsilon(\acute{\omega}_n - \omega_n) + \textstyle\sum_{m \neq n} P_{nm}(\theta_m) \bullet G_n(\omega_n, \theta_n),$$

where $\acute{\omega}_n$ is the intrinsic or preferred frequency of oscillator n, $\epsilon(\acute{\omega}_n - \omega_n)$ measures the "natural relaxation rate" of oscillator n to $\acute{\omega}_n$, Pnm is the strength of a "pulse"of oscillator m observed by oscillator n, and G_n measures the effect of a neighbor's pulse on the intrinsic frequency of oscillator n when that pulse occurs at phase θ_n. Effectively, G_n represents the phase response curve (PRC) of oscillator n, an increase in the tempo ω_n when the pulse phase is early (negative) and a decrease in the tempo when the pulse phase is late (positive). The specific form of G_n is chosen to match what the live oscillators, for example, fireflies, actually do.

Kuramoto model [60]. Assuming that alternation in *H. japonica* is adaptive, its rapid signaling and high duty cycle probably explain the need for a precise mechanism of rhythm coordination that incorporates both tempo and phase corrections. The males broadcast 100-ms notes at ca. 3 Hz [44], and neighbors would overlap their calls extensively without strict control.

Beyond the pairwise arrangement, *H. japonica* exhibits two coordination modes when a third male joins the chorus. They exhibit either clustered anti-phase synchrony, as already discussed for phase delay coordination, or n-way (tri-phase) alternation, in which each male signals at a 120° phase angle with respect to a given neighbor while retaining their fundamental 3-Hz tempos (Fig. 5.3b). Both "pacts" are predicted by simulations of the basic Kuramoto model [55]—with a negative interaction function and three rhythmists—although the conditions that favor one coordination mode over the other are not clear. The live frogs and the simulations indicate fast changes, within several cycles at most, from one mode to the other. N-way alternation, which can include up to 5 rhythmists with each male signaling at a 72° phase angle relative to a given neighbor—an impressive feat of coordination—has been seen in other anurans [22], but it is unknown in arthropods.

Our knowledge of rhythm coordination among animal species and its mechanisms is obviously incomplete, but certain patterns are beginning to emerge. One is actually the absence of pattern, that is a lack of distinct association between mechanism and phylogeny. This independence is clearly seen in *Pteroptyx* fireflies and *Neoconocephalus* bush crickets, where both phase-delay and phase-differential mechanisms function in different species within the same genus [7]. The variation in mechanism points to an evolutionary lability where the animals are fitted with a set of elementary neural components that may be summoned and assembled according to the selection pressures experienced during diversification. Alternatively, novel components may evolve rapidly under these pressures.

5.2.4 Where Are the "Higher Vertebrates?"

Perhaps influenced by the superior, and near universal, ability with which we humans can coordinate rhythm, not a few scientists have probed the occurrence of entrainment in other primates, in other mammals, and in birds. The notion that our specialized entrainment ability could be traced back to a vertebrate ancestor rather than having evolved de novo within the hominid lineage was implicit in the search. Many of the investigations involved captive animals that had undergone elaborate and specific training or that had lived with humans for extended intervals, although some work was conducted in the field or in large enclosures [61, 62]. To date, no investigation has reported entrainment, as we describe in this chapter, in natural populations of a non-human mammal. This assessment may change in the future as more animals are thoroughly tested, particularly species in marine and tropical rainforest environments or that use signal modalities for which we must rely on instrumentation to detect, e.g., ultrasound. Thus, promising groups for study include cetaceans and microchiropterans, both of which are also distinguished by signaling that is more isochronous than that observed in other mammals.

In birds, entrainment has been reported in one species of rail, the little crake (*Zapornia parva*), in which the male and female in a duetting pair alternate very rapid (12 Hz) calls with precision [63]. The alternation relies on both phase and tempo adjustment, continues for a great many cycles, and may function in clearly announcing the pair's shared territorial defense.

Research on captive mammals and birds has been similarly problematic [61, 62]. With the exception of the sea lion Ronan [64, 65], none of the captive animals met the basic requirements of entrainment. In general, the animals did not adjust their tempo to both a faster and a slower rhythmic stimulus, and they did not maintain an alignment uninterruptedly for a great many cycles. When synchrony was observed, positive asynchronies with the stimulus, i.e., the animal signaled just afterward, typically exceeded the expected effector delay and were likely independent, rapid responses to each signal and not coordination of the entire rhythm per se. And in the case of Ronan, one might ask whether her success at synchronous entrainment pointed as much to the prowess of the trainer and/or to the particular amenability of the trainee to training. Perhaps the most appropriate way of viewing these observations of trained animals is that they represent "incipient" or "latent synchrony." Some underlying sensorimotor "apparatus" is present, but the species was probably never subjected to critical selection pressure that could have favored full expression of entrainment in the natural milieu.

5.3 A Primer on Testing Entrainment

What is to be done? The fundamental aspect of entrainment is an endogenous rhythm, and without establishing that basis, it does not pay to proceed further. A hypothetical example illustrates the point. Two animals interact acoustically, each one initiating a brief call 450 ms following the end of its neighbor. There is no endogenous call rhythm in solo, but the two exhibit a regular alternation with a call rhythm of 2 Hz for the pair and of 1 Hz for each individual. Given that there was no rhythm to begin with, the observed alignment in duo is an artifact best described as interactive "turn-taking," a widespread process in mammals [66, 67] and well known to students of human conversation [68].

Once an endogenous rhythm is confirmed, we ask whether it is isochronous. But what exactly is isochrony? The attribute has two components, the degree of constancy of tempo and the continuity of that constancy, and a biologically relevant, minimum standard for each component cannot be made just by setting general, probably arbitrary, thresholds, e.g., a coefficient of variation (cv = standard deviation/mean) of signal period <0.20 and a continuity ≥ 10 "regular" periods. Ideally, these standards are set such that signaling deemed isochronous would allow an animal to predict the next signal(s) of its neighbor and align appropriately. This predictive ability could depend on the animal's cognition, e.g., the number of signal periods it must perceive in order to judge tempo, and the type of synchronous alignment to be made, e.g., merely retaining silent gaps or attaining a strict, $\approx 0°$ phase angle.

We do not anticipate these estimates to come easily, but rather expect to shift thinking about the role of isochrony in entrainment. At the very least, isochrony cannot be judged and designated from a small number of periods, intermittent signaling, and wide fluctuations in period length. The constancy should be measured by cv, not by the often applied "rhythm ratio" in which the lengths of 2 consecutive periods are compared, then advancing to the next 2 consecutive periods and repeating, and potentially continuing the calculation for several additional period pairs. The rhythm ratio can be a useful local measure for some temporal features of animal signaling, but as a global index of isochrony, notably in the context of entrainment, it is insensitive to accelerating or decelerating drift in period length and can give a false indication. For the record, detailed surveys of call periods in the two acoustic groups highlighted in this chapter, insects and anurans, showed that cv ranges from 0.04 to 0.25, with approximately one-half of the species below 0.10 [69, 70]. This level of constancy is uncommon in birds (cf. spectrograms in Xeno Canto [71]; see exceptions noted in the Preamble above) and is seldom seen in non-human mammals outside of cetaceans and bats [72], a factor that may be partly responsible for the apparent absence, or at least rarity, of entrainment.

The litmus test for entrainment is the ability to "latch onto" an external rhythm without "significant asynchronies" and to do so whether initially faster or slower than that rhythm. The primary tool for conducting this test is the PRC, as this will reveal the necessary phase advances and phase delays when alignment is initiated as well as maintained (Figs. 5.1 and 5.2). Moreover, the PRC will function equally well for phase delay and the more precise and adaptive phase differential mechanisms. It is unfortunate that PRCs have seldom been applied in studies of rhythm in birds and non-human mammals. An absence of significant asynchronies is an aspect particular to synchronous alignment. It pertains specifically to the length of delays (positive asynchronies) between the external stimulus and the focal animal's signals. Should these positive asynchronies regularly exceed the animal's effector delay, the animal's rhythm is unlikely to be latching onto the external one. Rather, the animal is probably making separate, rapid responses to each signal in the stimulus train. The occurrence of one rhythm latching onto another without meaningful time lags was first emphasized by Bard Ermentrout in the firefly *Pteroptyx malaccae*, and he coined the term "perfect synchrony" to describe the strict alignment [56]. It has since been confirmed in other fireflies and some bush crickets [73], adding arthropods to the list, which had already included humans, with their well-appreciated musical and related abilities.

5.4 Synopsis

This review has concentrated on arthropods and anurans among non-human animals because of the authors' collective knowledge of rhythm in invertebrate, vertebrate, and human behavior. Our overall familiarity has shown us that the non-human animals most similar to humans in the ability to coordinate rhythm are phylogenetically very distant from us. The similarities are most unlikely to be homologous, and we are forced to consider analogous selection pressures leading to the observed convergence. In so doing, we may be reminded that rhythm coordination is not the only complex behavioral trait exhibiting such improbable convergence. Students of communication might recognize that a code of abstract symbols providing information on events distant in space and time shows an equally disjunct pattern: recruitment "dances" in honeybees (and some of their close relatives) and language in humans. We suggest that refocusing away from homology may lead to an improved understanding of the fascinating world of rhythm coordination.

Acknowledgments Many of the ideas presented here owe their synthesis to two conferences, "Synchrony and rhythm interaction: from neurons to ecology" held at the Lorentz Center, Leiden Univ., the Netherlands, in July 2019, and "Nature beats: the what, why and how of rhythmic animal behaviours" held in Furuvik, Sweden, in May 2022. We greatly appreciate the administrative work and intellectual contributions of organizers at the Leiden Conference, Patricia Backwell, Henkjan Honing, Sonja Kotz, and Andrea Ravignani, and at the Furuvik conference, Elainie Madsen, Tomas Persson, and Gabriela-Alina Sauciuc. We are also grateful to the many participants at the conferences for having shared their knowledge and insights and to an anonymous referee who helped us to hone the final version of our chapter.

References

1. Burchardt LS, Hoeschele M, Honing H, et al (2025) Chapter 1: what is rhythm and how can we study it?
2. Arom S, Thom M, Tuckett B, et al (2004) African polyphony and polyrhythm: musical structure and methodology, first paperback edition 2004. Cambridge University Press, Cambridge
3. Roeske TC, Tchernichovski O, Poeppel D, Jacoby N (2020) Categorical rhythms are shared between songbirds and humans. Curr Biol 30:3699. https://doi.org/10.1016/j.cub.2020.08.026
4. Haimoff EH (1981) Video analysis of Siamang (*Hylobates syndactylus*) songs. Behaviour 76:128–151. https://doi.org/10.1163/156853981X00040
5. Geissmann T (2000) Duet songs of the siamang (*Hylobates syndactylus*): I. Structure and organisation. Primate Rep 56:33–60
6. Merker B, Cox C (1999) Development of the female great call in *Hylobates gabriellae*: a case study. Folia Primatol (Basel) 70:97–106. https://doi.org/10.1159/000021680
7. Greenfield MD, Merker B (2023) Coordinated rhythms in animal species, including humans: entrainment from bushcricket chorusing to the philharmonic orchestra. Neurosci Biobehav Rev 153:105382. https://doi.org/10.1016/j.neubiorev.2023.105382
8. Merker BH, Madison GS, Eckerdal P (2009) On the role and origin of isochrony in human rhythmic entrainment. Cortex 45:4–17. https://doi.org/10.1016/j.cortex.2008.06.011
9. Fraisse P (1982) Rhythm and tempo. In: Psychology of music. Elsevier, pp 149–180
10. Jacoby N, Polak R, London J (2021) Extreme precision in rhythmic interaction is enabled by role-optimized sensorimotor coupling: analysis and modelling of West African drum ensemble music. Philos Trans R Soc B Biol Sci 376:20200331. https://doi.org/10.1098/rstb.2020.0331
11. Merker B (2002) Music: the missing Humboldt system. Music Sci 6:3–21. https://doi.org/10.1177/102986490200600101
12. Desain P, Honing H (2003) The formation of rhythmic categories and metric priming. Perception 32:341–365. https://doi.org/10.1068/p3370

13. Merker B (2014) Groove or swing as distributed rhythmic consonance: introducing the groove matrix. Front Hum Neurosci 8. https://doi.org/10.3389/fnhum.2014.00454
14. Merker B (1999) Synchronous chorusing and the origins of music. Music Sci 3:59–73. https://doi.org/10.1177/10298649000030S105
15. Mehr SA, Krasnow MM, Bryant GA, Hagen EH (2021) Origins of music in credible signaling. Behav Brain Sci 44. https://doi.org/10.1017/s0140525x20000345
16. Savage PE, Loui P, Tarr B et al (2021) Music as a coevolved system for social bonding. Behav Brain Sci 44. https://doi.org/10.1017/s0140525x20000333
17. Mann NI, Dingess KA, Slater PJB (2006) Antiphonal four-part synchronized chorusing in a Neotropical wren. Biol Lett 2:1–4. https://doi.org/10.1098/rsbl.2005.0373
18. Fortune ES, Rodríguez C, Li D et al (2011) Neural mechanisms for the coordination of duet singing in wrens. Science 334:666–670. https://doi.org/10.1126/science.1209867
19. Madison G, Merker B (2005) Timing of action during and after synchronization with linearly changing intervals. Music Percept 22:441–459. https://doi.org/10.1525/mp.2005.22.3.441
20. Repp BH (2005) Sensorimotor synchronization: a review of the tapping literature. Psychon Bull Rev 12:969–992. https://doi.org/10.3758/bf03206433
21. Repp BH, Su Y-H (2013) Sensorimotor synchronization: a review of recent research (2006–2012). Psychon Bull Rev 20:403–452. https://doi.org/10.3758/s13423-012-0371-2
22. Greenfield MD (2005) Mechanisms and evolution of communal sexual displays in arthropods and anurans. In: Advances in the study of behavior. Elsevier, pp 1–62
23. Buck J, Buck E (1966) Biology of synchronous flashing of fireflies. Nature 211:562–564. https://doi.org/10.1038/211562a0
24. Buck J, Buck E (1968) Mechanism of rhythmic synchronous flashing of fireflies: fireflies of Southeast Asia may use anticipatory time-measuring in synchronizing their flashing. Science 159:1319–1327. https://doi.org/10.1126/science.159.3821.1319
25. Sheppard LW, Mechtley B, Walter JA, Reuman DC (2020) Self-organizing cicada choruses respond to the local sound and light environment. Ecol Evol 10:4471–4482. https://doi.org/10.1002/ece3.6213
26. Fraisse P (1980) Les synchronizations sensori-motrices aux rhythmes. In: Anticipation et Comportement. Editions de C.N.R.S, Paris, pp 233–257
27. Aschoff J (1981) Biological rhythms. Springer US, Boston
28. Dusenbery DB (1992) Sensory ecology: how organisms acquire and respond to information. W. H. Freeman, New York
29. Tuttle MD, Ryan MJ (1982) The role of synchronized calling, ambient light, and ambient noise, in anti-bat-predator behavior of a treefrog. Behav Ecol Sociobiol 11:125–131. https://doi.org/10.1007/BF00300101
30. Backwell P, Jennions M, Passmore N, Christy J (1998) Synchronized courtship in fiddler crabs. Nature 391:31–32. https://doi.org/10.1038/34076
31. Nityananda V, Balakrishnan R (2021) Synchrony of complex signals in an acoustically communicating katydid. J Exp Biol:jeb241877. https://doi.org/10.1242/jeb.241877
32. Hanson FE, Case JF, Buck E, Buck J (1971) Synchrony and flash entrainment in a New Guinea firefly. Science 174:161–164. https://doi.org/10.1126/science.174.4005.161
33. Buck J, Buck E, Case JF, Hanson FE (1981) Control of flashing in fireflies: V. Pacemaker synchronization in *Pteroptyx cribellata*. J Comp Physiol A 144:287–298. https://doi.org/10.1007/BF00612560
34. Buck J, Buck E, Hanson FE et al (1981) Control of flashing in fireflies: IV. Free run pacemaking in a synchronic *Pteroptyx*. J Comp Physiol A 144:277–286. https://doi.org/10.1007/BF00612559
35. Greenfield MD (1994) Cooperation and conflict in the evolution of signal interactions. Annu Rev Ecol Syst 25:97–126. https://doi.org/10.1146/annurev.es.25.110194.000525
36. Greenfield MD, Roizen I (1993) Katydid synchronous chorusing is an evolutionarily stable outcome of female choice. Nature 364:618–620. https://doi.org/10.1038/364618a0

37. Greenfield MD, Tourtellot MK, Snedden WA (1997) Precedence effects and the evolution of chorusing. Proc R Soc Lond B Biol Sci 264:1355–1361. https://doi.org/10.1098/rspb.1997.0188
38. Hubel KA, Reed B, Yund EW et al (2013) Computerized measures of finger tapping: effects of hand dominance, age, and sex. Percept Mot Skills 116:929–952. https://doi.org/10.2466/25.29.PMS.116.3.929-952
39. Greenfield MD (1994) Synchronous and alternating choruses in insects and anurans: common mechanisms and diverse functions. Am Zool 34:605–615. https://doi.org/10.1093/icb/34.6.605
40. Perez DM, Klunk CL, Araujo SBL (2021) Imperfect synchrony in animal displays: why does it occur and what is the true role of leadership? Philos Trans R Soc B Biol Sci 376:20200339. https://doi.org/10.1098/rstb.2020.0339
41. Nityananda V, Stradner J, Balakrishnan R, Römer H (2007) Selective attention in a synchronising bushcricket: physiology, behaviour and ecology. J Comp Physiol A 193:983–991. https://doi.org/10.1007/s00359-007-0251-9
42. Greenfield M, Snedden WA (2003) Selective attention and the spatio-temporal structure of orthopteran choruses. Behaviour 140:1–26. https://doi.org/10.1163/156853903763999863
43. Marin-Cudraz T, Greenfield MD (2016) Finely tuned choruses: bush crickets adjust attention to neighboring singers in relation to the acoustic environment they create. Behav Ecol Sociobiol 70:1581–1589. https://doi.org/10.1007/s00265-016-2166-3
44. Aihara I, Takeda R, Mizumoto T et al (2011) Complex and transitive synchronization in a frustrated system of calling frogs. Phys Rev E 83:031913
45. Höbel G (2014) Effect of temporal and spectral noise features on gap detection behavior by calling green treefrogs. Behav Process 108:43–49. https://doi.org/10.1016/j.beproc.2014.09.013
46. Nityananda V (2016) Attention-like processes in insects. Proc R Soc B Biol Sci 283:20161986. https://doi.org/10.1098/rspb.2016.1986
47. Greenfield MD, Schul J (2008) Mechanisms and evolution of synchronous chorusing: emergent properties and adaptive functions in *Neoconocephalus* katydids (Orthoptera: Tettigoniidae). J Comp Psychol 122:289–297. https://doi.org/10.1037/0735-7036.122.3.289
48. Greenfield MD, Marin-Cudraz T, Party V (2017) Evolution of synchronies in insect choruses. Biol J Linn Soc 122:487–504. https://doi.org/10.1093/biolinnean/blx096
49. Moiseff A, Copeland J (2010) Firefly synchrony: a behavioral strategy to minimize visual clutter. Science 329:181–181. https://doi.org/10.1126/science.1190421
50. Nityananda V, Balakrishnan R (2009) Modeling the role of competition and cooperation in the evolution of katydid acoustic synchrony. Behav Ecol 20:484–489. https://doi.org/10.1093/beheco/arp022
51. Buck J (1988) Synchronous rhythmic flashing of fireflies. II. Q Rev Biol 63:265–289. https://doi.org/10.1086/415929
52. Hanson FE (1978) Comparative studies of firefly pacemakers. Fed Proc 37:2158–2164
53. Hanson FE (1982) Pacemaker control of rhythmic flashing of fireflies. In: Cellular pacemakers. Wiley, New York, pp 81–100
54. Kuramoto Y (1984) Chemical oscillations, waves, and turbulence. Springer Berlin Heidelberg, Berlin, Heidelberg
55. Kuramoto oscillation. https//www.ruthfong.com/kuramoto/
56. Ermentrout B (1991) An adaptive model for synchrony in the firefly *Pteroptyx malaccae*. J Math Biol 29:571–585. https://doi.org/10.1007/BF00164052
57. Ott E, Antonsen TM (2017) Frequency and phase synchronization in large groups: low dimensional description of synchronized clapping, firefly flashing, and cricket chirping. Chaos Interdiscip J Nonlinear Sci 27:051101. https://doi.org/10.1063/1.4983470
58. Nityananda V, Balakrishnan R (2007) Synchrony during acoustic interactions in the bushcricket *Mecopoda* 'chirper' (Tettigoniidae: Orthoptera) is generated by a combination of chirp-by-chirp resetting and change in intrinsic chirp rate. J Comp Physiol A 193:51–65. https://doi.org/10.1007/s00359-006-0170-1
59. Walker TJ (1969) Acoustic synchrony: two mechanisms in the snowy tree cricket. Science 166:891–894. https://doi.org/10.1126/science.166.3907.891

60. Ota K, Aihara I, Aoyagi T (2020) Interaction mechanisms quantified from dynamical features of frog choruses. R Soc Open Sci 7:191693. https://doi.org/10.1098/rsos.191693
61. Merker B, Morley I, Zuidema W (2015) Five fundamental constraints on theories of the origins of music. Philos Trans R Soc B Biol Sci 370:20140095. https://doi.org/10.1098/rstb.2014.0095
62. Merker B, Morley I, Zuidema W (2018) Five fundamental constraints on theories of the origins of music. In: Honing H (ed) The origins of musicality. The MIT Press, pp 49–80
63. Jedlikowski J, Ręk P (2025) Rhythmic coordination in rapid duets of little crakes: cooperative signalling without shared experience. Proc R Soc B Biol Sci 292:20251100. https://doi.org/10.1098/rspb.2025.1100
64. Cook P, Rouse A, Wilson M, Reichmuth C (2013) A California Sea lion (Zalophus californianus) can keep the beat: motor entrainment to rhythmic auditory stimuli in a non vocal mimic. J Comp Psychol 127:412–427. https://doi.org/10.1037/a0032345
65. Cook PF, Hood C, Rouse A, Reichmuth C (2025) Sensorimotor synchronization to rhythm in an experienced sea lion rivals that of humans. Sci Rep 15. https://doi.org/10.1038/s41598-025-95279-1
66. Demartsev V, Strandburg-Peshkin A, Ruffner M, Manser M (2018) Vocal turn-taking in Meerkat group calling sessions. Curr Biol 28:3661–3666.e3. https://doi.org/10.1016/j.cub.2018.09.065
67. Pougnault L, Levréro F, Leroux M et al (2022) Social pressure drives "conversational rules" in great apes. Biol Rev 97:749–765. https://doi.org/10.1111/brv.12821
68. Schegloff EA (2007) Sequence Organization in Interaction: a primer in conversation analysis, 1st edn. Cambridge University Press
69. Gerhardt HC, Huber F (2002) Acoustic communication in insects and anurans: common problems and diverse solutions. University of Chicago Press, Chicago
70. SINA (Singing insects of North America). https://www.orthsoc.org/sina
71. Xeno Canto (Sharing wildlife sounds from around the world). https://www.xeno-canto
72. Burchardt LS, Norton P, Behr O et al (2019) General isochronous rhythm in echolocation calls and social vocalizations of the bat *Saccopteryx bilineata*. R Soc Open Sci 6:181076. https://doi.org/10.1098/rsos.181076
73. Murphy MA, Thompson NL, Schul J (2016) Keeping up with the neighbor: a novel mechanism of call synchrony in *Neoconocephalus ensiger* katydids. J Comp Physiol A 202:225–234. https://doi.org/10.1007/s00359-016-1068-1

How Environmental and Social Contexts Shape and Constrain Rhythm Structure and Production

6

Jennifer N. Schneider, Lara S. Burchardt, Gabriela-Alina Sauciuc, and Adriano R. Lameira

Abstract

For animals to communicate effectively, communication signals must fit the environmental and social contexts in which they are produced. Hence, in this chapter we review a wide range of constraints that, in combination, shape the structure of rhythmic signals both proximately and ultimately. These constraints include broad environmental factors (cyclic patterns of light and temperature, ecological niche, and indirect natural selection); immediate environmental factors (habitat structure and soundscapes); and social context. Additionally, we discuss the implications of these constraints in terms of anthropogenic effects on the environment, the development of culture, and speciation. We emphasize the importance of considering multiple constraints when designing studies and interpreting data. Most of the research we review in the chapter involves acoustic communication; the near absence of other modalities highlights the major gap in our understanding of socio-environmental constraints. Finally, we emphasize the importance of further research into the effects of anthropological disturbances of natural systems on rhythmic behavior.

J. N. Schneider (✉)
Department of Psychology, LCC International University, Klaipėda, Lithuania
e-mail: jschneider@lcc.lt

L. S. Burchardt
Leibniz Zentrum für Allgemeine Sprachwissenschaften Berlin, Berlin, Germany

Institute for Theoretical Biology, Humboldt-Universität zu Berlin, Berlin, Germany

Donders Institute for Brain, Cognition, and Behaviour, Radboud University, Nijmegen, The Netherlands

G.-A. Sauciuc
Department of Philosophy/Cognitive Science, Lund University, Lund, Sweden

A. R. Lameira
Department of Psychology, University of Warwick, Coventry, UK

G.-A. Sauciuc et al. (eds.), *Nature Beats*, Fascinating Life Sciences,
https://doi.org/10.1007/978-3-032-24250-1_6

6.1 Introduction

To survive, animals have tuned their behaviors to the socio-environmental conditions offered by their habitats and communities. This also applies to communicative behaviors, which can be simultaneously constrained by a number of external factors. Hence, an animal's surroundings narrow the optimal parameters for effective communication to well within the limits of what it is capable of producing. In other words, an animal utilizes its realized signal niche (limited by external constraints), rather than its fundamental signal niche (limited only by internal constraints), in terms of rhythmic production. Here, we review various environmental and social factors that constrain and shape rhythmic communication signals.

In this chapter, we review how the cycles and rhythms of major physical phenomena funnel rhythmic production (Sects. 6.1 and 6.2), how the physical and acoustic structures of the environment affect the production (Sect. 6.3) and propagation of rhythmic signals (Sects. 6.4 and 6.5), and how rhythms may differ structurally in relation to the audience (Sect. 6.6), as well as interactions between these factors (Sect. 6.7). Finally, we look at the implications of these constraints in terms of anthropogenic effects on the environment and the development of culture (Sects. 6.8 and 6.9).

6.2 Light Intensity

The amount, chromaticity, and intensity of natural light vary depending on time, location, climate, and atmospheric conditions. Light variation across time occurs periodically at multiple timescales. The revolution of the Earth around the sun causes seasonal variations in day length, which affects the amount of light exposure (photoperiod) on an annual basis. The Earth's rotation on its own axis causes the circadian cycle of light and dark periods. And light intensity during nighttime varies across the lunar cycle, being highest during the full moon phase and lowest during the new moon phase. These predictable photoperiod variations provide crucial cues for the daily and seasonal regulation of animal physiology and behavior, including (rhythmic) communication (as will be detailed further down). Light cycles, however, are further affected by location, being more pronounced as the distance from the equator increases, and by climatic and atmospheric conditions. For example, light intensity is enhanced by intermittent clouds and severely reduced by overcast and rain conditions ([1] and references therein).

To date, most research on how light variation affects animal communication has focused on the generic effects of photoperiod on the timing and overall duration of signals. For example, the seasonal fluctuation of light intensity upregulates (or downregulates) physiological mechanisms, such as melatonin, which affects the overall activity budgets of the animals (as detailed further down), and sex steroid hormones (as detailed in Chap. 11 [2] this volume), which activate (or inhibit) breeding behaviors, including communication. Rhythmic signaling, which features heavily in reproductive contexts (for a comprehensive review, see Chap. 3 [3] this volume), is, thus, inherently subjected to such effects.

Light changes across the day cycle dictate the onset of distinct kinds of behaviors (foraging, resting, etc.) in patterns that depend on whether animals are diurnal or nocturnal. These are further influenced by moonlight variation across the lunar cycle, which increases or decreases the time budget of certain activities in species-specific ways. As reviewed in Chap. 3 [3], many species perform daily rhythmic displays that are cued by light intensity thresholds, such as the dawn and dusk choruses found in insects, coral reef communities, frogs, birds, and some mammalian species. A frequent finding is that, during full-moon nights, bird and primate dawn choruses start earlier compared to other lunar phases, while crepuscular calling may shift activity to the nighttime [1, 4]. Moonlight may even induce diurnal species to display during the night [5]. Conversely, rainy days and diminished sunlight may cause chorus delays or other disruptions. For example, Yucatán black howler monkeys (*Alouatta pigra*) reduce chorusing activity to a single midday session (instead of both dawn and dusk sessions) when sunlight hours decrease during the rainy season [6].

During the nighttime, communication—and overall activity budgets—are influenced by the effects of light levels on the ability to detect food, conspecifics, and/or predators and on the risk of being detected by predators. For instance, in frog species, the overall duration of nocturnal vocal activity is inhibited or enhanced in species-specific ways across the lunar cycle, depending on moonlight brightness [7, 8]. Small, olfaction-oriented mammals (e.g., rodents, insectivores) tend to be lunaphobic, exhibiting lower levels of activity and calling during the full-moon phase. By contrast, more visually oriented species, such as nocturnal primates, tend to be luna-philic, increasing overall levels of activity, including calling [9, 10]. This variation in vocal activity appears to be modulated by species-specific strategies for predator defense or sexual advertisement. Moonlit nights, for example, facilitate early predator detection in visually oriented primates, which have also been found to increase co-travel on such nights, as a way of boosting vigilance and predator mobbing [9].

In coral reef communities, the acoustic activity of both fish and invertebrates varies with both daily and lunar cycles [11, 12]. The high-frequency pulse trains of fish (0.5–1.5 kHz) increase during moonlight hours, while low-frequency fish sounds (0.1–0.3 kHz) and invertebrate sounds decrease in the same period [11]. Across the lunar cycle, the low-frequency sounds of Floridian coral-reef communities have been found to attain maximum complexity and output during the low-light lunar phases [12].

Effects on the total duration of rhythmic signals (as reviewed above) can be the result of changes in the temporal structure of calls, for example, changes in the total number of sub-units (pulses, notes, etc.) or interunit intervals. To date, only a handful of studies have addressed the effects of illumination on these aspects of rhythmic structure. This (currently limited) research shows that at least two porpoise species (Phocoenidae) exhibit diel changes in the temporal parameters of echolocation click trains. For example, harbor porpoises (*Phocoena phocoena*) emit a higher proportion of click trains with very short interclick intervals during the night (compared to during the day) [13], while evidence from East Asian finless porpoises (*Neophocaena asiaeorientalis sunameri*) shows that both click duration and interclick intervals are significantly shorter at night compared to during the day [14]. These changes may facilitate information processing at night by increasing sonar resolution as a way to compensate for the lack of visual information [14].

In a nocturnal bird species—the vermilion flycatcher (*Pyrocephalus rubinus*)—males sing longer songs, with a higher number of introductory elements and lower entropy as the sunrise approaches [15, 16]. Longer songs with more elements have also been recorded at brighter compared to darker sites and, within the same site, during the full moon phase [17]. By contrast, in diurnal (e.g., zebra finch, *Taeniopygia guttata*) and cathemeral (e.g., Japanese quail, *Coturnix japonica*) birds, a radical enhancement of the photoperiod was found to be associated with shorter song motifs, shorter syllables, and intersyllable intervals, corresponding to an overall increase in the temporal frequency of vocal output [18]. In some cases, such changes resulted in the loss of rhythmic patterning at the song level. Moreover, in the zebra finch, which is a vocal learning species, excessive exposure to light across the diel cycle disrupted the acquisition of species-specific rhythmic patterns [18]. A similar effect on the subsecond scale of call patterning was found in the nocturnal midshipman fish (*Porichthys notatus*), where an excessively extended photoperiod was associated with shorter *hum* calls and an overall decrease in the total duration of calling activity.

Taken together, the findings reviewed above show that light changes affect acoustic rhythms at the seasonal, lunar, and circadian timescales by inducing changes in the timing and overall duration of calling periods and, at the sub- and supra-second levels, by inducing changes in the rhythmic patterning of individual calls. Mechanistically, both types of changes can be explained by the inhibitory effects of prolonged photoperiod on melatonin synthesis. In diurnal species, melatonin suppression increases arousal and prolongs the wake period of the circadian cycle, which, in turn, affects the onset and time budget of behaviors. By contrast, in nocturnal species, elevated melatonin levels are associated with elevated activity levels.

Crucially, melatonin exerts regulatory effects that are specific to acoustic signaling, as melatonin receptors are abundantly expressed within neural circuits that control acoustic output across taxa, such as in fish [19], frogs [20], and birds [21]. Melatonin is also involved in neuroendocrine processes underlying reproductive function. Thus, light variation is translated into an endocrine (melatonin) signal that relays information about the appropriate timing of vocalization directly to the vocal control areas of the brain. This has both stimulating and inhibitory effects on the timing of vocal emission, in species-specific ways. In turn, this is consistent with the longer duration of calls and call units found in nocturnal flycatchers and midshipman fish (reviewed above), as a function of exposure to darkness. Experimental data show that, in zebra finches, the surgical removal of the pineal gland (which synthesizes melatonin) has similar effects to those exerted by extended photoperiod on adult song performance and on the acquisition of song rhythm in young birds [18].

In midshipman fish, treatment with a melatonin analog counters the effects of extended photoperiod on overall call activity and single call duration [19].

Finally, melatonin can also influence rhythmic output indirectly, through influences exerted on the neuroendocrine reproductive axis, whereby fluctuations in melatonin levels influence the frequency and amount of hypothalamic and gonadotropic hormones (for a recent review, see [22]). Besides having a crucial role in physiological changes that trigger the onset of birdsong and other seasonal rhythmic vocalizations (for details, see Chap. 11 [2] this volume), androgen levels may also affect the rhythmic parameters of calls, as in singing mice [23] (for more details, see also Chap. 9 [24] this volume).

6.3 Temperature

Temperature, as an abiotic ecological factor, is an important modulator of a wide range of physiological, morphological, and behavioral traits. Impacting traits on an individual, population, and global taxonomic level, it also exhibits substantial temporal variation throughout a day, interseasonally, or between seasons.

6.3.1 Ectotherms

Especially in ectotherms, animals that rely on external sources to regulate their body temperature, temperature plays a crucial role in signal production and perception in general, and has a high impact on rhythmic structures in different species. Temperature could affect the rhythmic structure of communication signals in three ways (as suggested in [25]): (1) the signal as produced by the signaler changes, but the receiver is not able to perceive the change; (2) the signal stays the same, but the perception changes because of temperature, and (3) both are influenced by temperature as was first suggested under the term "temperature coupling" in 1978 in the gray treefrog (*Dryophytes versicolor*) [26]. To better help explain these mechanisms, in the following, we will summarize examples from different ectotherm classes: Osteichthyes (bony fish), Amphibia, Insecta, and Arachnida.

6.3.1.1 Osteichthyes (Bony Fish)

Sound production is present in about two-thirds of ray-finned fish (Actinopterygii), and probably evolved independently over 30 times [27, 28]. All examples listed here belong to the ray-finned fish. The sounds are produced in various ways (see Chap. 2 [29] this volume). Different sound production mechanisms in fish are differently influenced by temperature. A detailed review of the influence of temperature on acoustic communication in fish—both on the production as well as on the perception side—is available from Ladich [30]. Several studies have been carried out asking about the effects of ambient temperature on sound production, while fewer studies have looked at how this impacts hearing sensitivity and the coupling between the two. Many fish produce pulsatile sounds, where one sound consists of multiple "pulses," each the result of a distinct muscle contraction [31]. Many fish then produce sounds in sequence. We therefore have to differentiate between within-sound temporal structures and between-sound temporal structures that could be influenced by temperature. Ladich [30] summarizes that calling activity in relation to temperature changes has been described in approximately a dozen fish families, where most studies found a positive correlation, whereby there is an increase in calling activity (sound production per unit of time) with an increase in temperature.

Other factors, most importantly light levels, are likely playing a salient role in many of these studies in changing calling activity. Examples, where illumination most likely did not cause the correlation between calling activity and temperature, include, for example, the sounds produced by the croaking gourami (*Trichopsis vittata*) under lab conditions. The sound characteristics were analyzed for three temperatures: 25, 30, and 35 °C. In the croaking gourami, a sound consists of multiple bursts and a burst consists of two pulses (see [32], Figure 1); therefore, the temporal structure can change on three levels. Next to spectral parameters that changed depending on temperature, the intervals changed on all three levels, decreasing between sounds, between bursts within sounds, and between pulses in bursts. Other species for which an increase in calling activity was reported with increasing temperatures are the Lusitanian toadfish (*Halobatrachus didactylus*) [33], the spotted weakfish (*Cynoscion nebulosus*) [34–36], or the Raphael (*Platydoras armatulus*) [37]. A recent study of between-call rhythms in one of the most common vocal fish in the Mediterranean Sea, the brown meager (*Sciaena umbra*), looked at rhythmic properties of fish across space and time, expecting to see an effect of temperature. This effect could not be found, though. The consistency in rhythms across four different locations with different temperatures and other varying environmental factors highlights the potential fitness relevance of the relationship between sound rhythms in this specific species [38].

Much less is known about the effect of temperature on hearing and signal perception in fish. Hearing in fish is connected to so-called otoliths in the inner ear structure. The group of otophysines is characterized by the presence of a specialized hearing structure called the Weberian apparatus. Temperature increase enhances hearing sensitivity in several otophysine species, such as the Eurasian minnow (*Phoxinus laevis*) [39], the goldfish (*Carassius auratus*) [40], the channel catfish (*Ictalurus punctatus*) and pictus cat (*Pimelodus pictus*) [41], the Alaska pollock (*Gadus chalcogrammus*), the striped Raphael [37], and the common carp (*Cyprinus carpio*) and Wels catfish (*Silurus glanis*) [42]. Unfortunately, close to nothing is known yet on whether these changes in sensitivity or sound characteristics are coupled in the sense of "temperature coupling." Only one study has been conducted analyzing this in fish, but on the spectral characteristics. In the plainfin midshipman, the "hum" produced by males shows a temperature-dependent change in fundamental frequency, which is mirrored in the temperature-dependent frequency preference found in females [43]. In other ectothermic taxa, this temperature coupling has to selectively affect only temporal features. The assumption is that such a coupling exists to make sound communication more meaningful. Assuming that sound characteristics are used to assess the fitness of a conspecific, the animal must be able to separate changes in physical properties of sounds due to temperature from those due to fitness, for example, size.

6.3.1.2 Amphibia

The phenomenon of "temperature coupling," was suggested based on work on female preference in the gray tree frog (*Hyla versicol*). Specifically, the preference of female gravid frogs was tested at different temperatures on synthetic calls. Between the different temperature regimes, different pulse rates of calls were tested, as the calls of this frog species are pulsed, as described for many fish sounds above. Females preferred a rhythm frequency of call pulse rate of 15 Hz at 16 °C and of 24 Hz at 24 °C. This preference corresponds to the temporal characteristics of male calls produced at the respective temperature [26]. Hoy et al. [44] suggested that this is the case because of a shared genetic basis, with similar neural pathways being involved in both pattern generation and pattern recognition.

Cases where either production or perception is influenced by temperature have been found in additional species. The effect of temperature on the hearing sensitivity of frogs, as with fish, has only been tested on spectral parameters. In the grass frog (*Rana t. temporaria* L.) and the fire-bellied toad (*Bombina bombina* L.), hearing sensitivity is enhanced with increasing temperature [45]. Vocal production in anurans has been shown to be temperature-dependent in a number of species. The Brazilian torrent frog (*Hylodes heyeri*), for example, changes its call duration with temperature changes, and even though the number of pulses per call (terminology used in this study: notes per call) was highly correlated with the call duration, the number of pulses per call was not itself correlated to temperature. This analysis shows how intricate certain biological correlations are, and that we always have to keep statistical limitations in mind when comparing such results. The intervals between calls also do not show any correlation with air temperature in the given data [46].

Another example for the complex relations between temporal structure, temperature, and other factors stems from the Andean frog *Hyla labialis*. These frogs live at very different elevation levels across the Andes and, thus, their habitats have different average temperatures with substantial variations. Here, temperature sensitivity was found to be influenced by the average temperature; frogs that lived in colder habitats showed less temperature sensitivity. However, on average, pulse repetition rates increased with temperature while call duration decreased [47]. This phenomenon of decreased temperature sensitivity in eurythermic species has been shown in *Colostethus subpunctatus* [48] as well, where frogs living at higher altitudes with stronger temperature changes within the habitat exhibit less temperature-dependent changes in temporal features of calls than frogs living at lower altitudes. The temporal features affected are call duration and calling rate. Pulse rate is not affected in this species. Therefore, the authors argue that species with short, unmodulated calls (either frequency or pulse-modulated), such as *C. subpunctatus*, are potentially less affected by temperature than species with longer or more frequency-modulated calls. The temperature likely affects the contractile properties of muscles in both the trunk and the larynx, which is another important degree of freedom to keep in mind when studying the temperature dependency of temporal patterns in acoustic communication in general.

Independent of that, there are various other frog species that show a temperature dependency on temporal call parameters. In 13 additional species of frog, a temperature-dependent change in at least one of the parameters, call duration, pulse rate, or call rate was found: *Acris crepitans blanchardi*, *Bombina variegate scabra*, *Bufo alvarius*, *B. americanus*, *B. microscaphus microscaphus*, *B. microscaphus californicus*, *B. viridis*, *Gastrotheca marsupiata*, *Hyla savignyi*, *Rana balcanica*, *R. lessonae*, *R. perezi*, and *R. ridibunda* (summarized in [48], Table 5). Yet another example that shows the intricacy of temperature dependency was reported from the frog *Crinia signifera*. Temperature and the social environment interact in this species: calling rate increases with temperature in individuals calling without conspecifics present, but this relationship disappears when frogs call in a duet or in a chorus [49]. To further illustrate the effect of the social environment on temporal parameters of frog calls, in two competing Australian tree frog species, calling rate increased when the competitor was present [50]. Whenever such effects of parameters on the behavior of animals are investigated, we need to be aware of the complex net of relationships between variables and encourage greater consideration of how physical and social environments might interact to influence the signal content and perception of acoustic and other behavioral displays.

6.3.1.3 Insecta

Temperature coupling, where production and perception of temporal structures seem to be dependent on temperature in a similar way, has also been reported in insects, for example, in *Gryllus bimaculatus*. This species produces sounds through stridulation. The repetition rates of various element types exhibit a linear increase with rising temperatures within the range of 15–24 °C, with no discernible effect observed at higher temperatures. Females presented with different repetition rates at different temperatures preferred the rhythmic characteristics of male calls that mirrored the specific temperature under examination (15, 24, or 30 °C). Similar to frogs and fish, this observation suggests a coupling between the physiological mechanisms governing pattern generation and those involved in pattern recognition among crickets [51]. In *G. integer*, a similar trend was found, male crickets singing at higher temperatures had higher production rates, evident in shorter pauses between chirps and shorter intervals between pulses in a chirp (called syllable period in the original paper). The overall number of syllables did not change. Interestingly, here the authors did not study the female preference for chirps' characteristics, but the temperature preference of males, where males preferred warmer micro-habitats to sing [52].

Another factor interacting with the temperature dependence of communication signals in insects, similar to that described above in frogs, is the social environment. Signals in a community of *Oecanthus* tree crickets increase in pulse rate with increasing temperature. The lower the temperature, the more similar the signals between the different species. The preference of females followed the temperature-dependent pulse rates, but at lower temperatures, signal similarity increased more strongly than the discrimination of females. This leads to potential signal interference and increases the likelihood of mismating at lower temperatures in *Oecanthus* tree cricket communities, as the general phonotaxis of females is not affected by temperature [53]. Temperature dependency is also found in the visual, rhythmic communication of fireflies, which—in contrast to many of the described acoustic dependencies—is not linear, but shows a curvilinear relationship across the tested temperature range [54].

6.3.1.4 Arachnida

So far, we have only talked about unimodal, primarily acoustic displays and their temperature dependency. But a multimodal example has recently been described in the jumping spider, *Habronattus clypeatus*. In this species, various aspects of visual and vibratory male signals changed with temperature. Generally, with increasing temperature, signal components of both modalities decreased in duration and increased in speed (and on the spectral level in frequency). As reported for the visual signals of fireflies, the relationships broadly follow a curvilinear shape. Production rates and frequencies increased to a point, then flattened out or slightly decreased at extremely high temperatures. The authors suggest that a temperature threshold exists—in this species, around 40 °C—beyond which males are not able to increase their performance [55].

Furthermore, in the courtship display of the wolf spider, *Schizocosa floridana*, the complex song is composed of vibrational sounds produced by three different body parts that might be differentially affected by temperature changes [56]. Vibrations are produced by the stridulation of the pedipalps, by abdominal tremulation, and by foreleg percussion tapping of the forelegs [57]. Certain components of the courtship behavior of *S. floridana* increase in production rate as the temperature rises, while others decrease or show no response to temperature changes. Female preference, and therefore potential selection pressures, also changes with temperature. While some components of the courtship display experience directional selection (where individuals with certain traits are favored over others) at higher temperatures, they do not at lower temperatures. This suggests that temperature influences which traits are favored during mate selection. These complex interactions highlight how a detailed understanding of the effect of environmental variation on communication is critical to understanding mate choice and the evolution of complex signals [56].

6.3.2 Endotherms

In ectotherms, we can see strong, direct effects of temperature on the whole body and all processes involved in communication (both perception and production). In endotherms, which maintain a roughly stable body temperature, these effects are not to be expected in the same strength, nevertheless there are multiple ways in which more extreme ambient temperatures can influence communication. Heat stress, for example, can increase respiratory and metabolic rates, which can lead to vocal fatigue and a change in the temporal structure of vocal communication. Another effect of heat stress can be dehydration, which in extreme cases could impair vocal communication due to dehydrated mucous membranes and vocal cords [58]. As the effects are expected to be much smaller, and potentially indirect, not many studies in endotherms focus on the effects of temperature on communication, let alone on the temporal structures of communication. The examples we have stem from highly vocal species, with strong locomotor-respiratory coupling, like bats, where an effect is to be expected. We have livestock examples, where a possible effect of heat stress on vocal communication is studied as a potentially easy, noninvasive way to monitor well-being, and we have examples of classical model organisms such as zebra finches.

6.3.2.1 Mammalia

In two bat species of the genus *Molossus*, the duration of sound elements during echolocation increases with higher temperature and humidity, while peak frequency also changes. A change in temperature leads to higher atmospheric attenuation. So here, temperature is likely only indirectly influencing the temporal parameters, as bats adjust their calls to still enable precise orientation under higher atmospheric attenuation [59].

In many livestock animals, such as pigs, various parameters are tested as potential indicators of welfare, which is important to monitor in animal husbandry to ensure animal well-being. In piglets, an increase in air temperature increased the overall number and length of vocalizations, which potentially changes the temporal structure of communication [60]. It is unclear how well-defined communication sequences are and what an underlying rhythm might be, so these temperature effects are to be considered with caution. However, these observations serve as good indicators for temperature effects on communication in mammals.

6.3.2.2 Aves

In turkeys, the number of turkey vocalizations increased by 43% under heat stress compared to control conditions (34 °C vs. 21 °C control). Turkeys responded to heat stress within 30 min by producing more vocalizations [61]. This is again just an indication that rhythm and temporal parameters might be affected by heat stress, but not a direct measurement of rhythmic changes.

The direct impact of temperature on song production was studied in zebra finches. In the attempt to find the brain regions responsible for the temporal structure of zebra finch song production, different areas of the brain were cooled to see if and how cooling would alter the temporal structure of song. Temperature can have effects on both higher- and lower-level mechanisms of vocal control, with top-down and bottom-up effects. Here, the known top-down effects of temperature on neural mechanisms, such as synaptic transmission, were used. The strongest effect was found for the so-called HVC region. When HVC is cooled down, song production slows down as well, which is why it is believed to be crucial in maintaining the temporal structure of song in zebra finches [62]. As a byproduct, these experiments show a direct effect of extreme temperature on sound production in these birds.

6.4 Indirect Natural Selection

Besides influencing directly when and how animals behave and deploy rhythmic signals through variations in physical factors (such as light and temperature), the environment imposes many other features, forces, and fluctuations on the lives of individuals and communities, such as variations in food production, the presence of competitor species, or the size and type of a predator guild. Such ecological factors exert considerable effects on survival and fitness, playing a significant role as selective drivers and impetus toward local adaptation, molding not only a species' behavior but also, in the long term, its biology. These ensuing "bodily" changes (e.g., referred to as "ecological character displacement" in the context of interspecific competition) automatically and inherently shift, release, or levy new constraints on an animal's capacity to produce rhythm.

6.4.1 Feeding Ecology

Ecological character displacement can have an indirect effect on sexual signals by constraining the morphology used in rhythm production. Avian beak morphology is a classic example. Beak morphology is largely shaped by food availability and resource partitioning [63]; however, given that the beak is also part of the vocal system, its size and shape constrain song output, especially frequency properties [64]. Along with the trachea and the larynx, the beak acts as a resonance filter, reducing the wideband harmonic sounds produced in the syrinx to pure tones or whistles [65]. The degree of beak gape determines the final frequency content of each note, making beak movement essential for proper song performance [66]. The biomechanics of large beaks constrain vocal flexibility, including production rates of song elements such as trills, that is, rapidly repeated units or sequences of units. The stronger musculature required to crack larger seeds is associated with larger beak depth [67], whereby the velocity necessary for a bird to rapidly open and close its beak is diminished in favor of seed-cracking force [68]. Additionally, birds with large beaks might have higher moments of inertia [69].

These effects are present in a variety of songbird clades, including Darwin's finches [Geospizinae] [70, 71], woodcreepers [Dendrocolaptinae] [72], and other oven birds [Furnariidae] [69], tanagers [Thraupidae] [73], and sparrows [Passerellideae] [74, 75]. For example, white-crown sparrow (*Zonotrichia leucophrys*) trill rates have decreased along with beak depth over several decades [75]. However, for song structure to be constrained by beak morphology, a species' song must be near maximum performance level (see Chap. 9 [24] this volume), which is not the case in all species. Not all of Darwin's finches show a relationship between beak size/shape and trill rate [76]. Other social and environmental factors affect production rates, such that not all species sing near threshold [72, 77].

6.4.2 Bipedal Walking

An interesting case of indirect environmental effect has also shaped and led to, perhaps, the most prevalent and frequent rhythmic behavior in modern humans—bipedal walking. Although classically treated as an "adaptation to the savannah," this belief has now been discredited by various strands of data from great apes [78] and humans [79]. The new view that has gained convergent empirical support is that bipedal walking derived from the bipedal stance of an arboreal ancestor at the tips of thin tree branches for the collection of fruit [80, 81]. Because hominids are the heaviest primate family, and because fruit comprises a large portion of the hominid diet whenever available, this required ancestral hominids to balance and manage their weight when trying to reach for fruit, which typically grows on the tips of branches, where support is the weakest and most prone to breaking. This posed a serious challenge for hominids: being heavier than most arboreal primates, a fall due to breaking a support branch could prove lethal, but being heavier also meant that hominids had higher nutritional requirements that energy-rich fruit could fulfill. The solution was for individuals to spread their weight across many branches by taking on a bipedal posture, spreading both legs to stand on a supporting branch, using one hand to grab on higher branches for balance and partial weight offload, and using the free hand to reach out for the fruit.

Once paleoclimate change started shifting world forests into gradually more open and patchy landscapes, primarily across Africa and Eurasia, inviting ancestral hominids to descend from the trees, they would have "hit the ground walking." Only then, on an open and stable substrate, could hominids sustain their bipedal stance and start using it for locomotion, shifting their weight from one leg to the other cyclically.

Eventually, bipedal walking became an evolutionary force in its own right, reshaping our bodies and, notably, our skulls and faces. For example, the foramen magnum, the hole at the base of our skulls through which the spinal cord passes and where the human column connects to the head, migrated anteriorly. The head became positioned above the spine to help maintain balance and stability in a vertical stance, unlike quadrupeds, whose foramen magnum is characteristically in the back of the skull [82]. The repositioning of the foramen magnum and a new upright position meant that the human vocal tract was also reconfigured. It could no longer operate as a straight tube. Instead, it had to make a right angle at the intersection of a horizontal mouth cavity and a vertical pharynx. This shift repositioned the basis of the tongue, allowing for new tongue constrictions and maneuvers that redirected airflow coming from the lungs and the vocal folds in new ways. This increased articulatory leverage meant that new types of shapes could be created inside the vocal tract, and with those, new resonant frequencies, or formants, that "cast" the sound waves coming from the vocal folds [83]. Because formants are the basis of vowel recognition, this chain of events culminated in an increased vowel space in upright hominids, which together with increased lingual capacity now endowed the capacity to produce, and quickly shift between, a larger range of sounds, setting up the scene to the evolution of what would one day become our species' most diagnostic rhythmic behavior—speech.

6.5 Habitat Structure

The distance at which a signal can be detected and decoded, its *active space*, is delimited by how a signal degrades during transmission and propagation. These effects are imposed by habitat features that reflect, refract, or absorb energy from a signal during its broadcast, such as the physical structure of the habitat (e.g., leaves and branches that obstruct a signal and through which a signal must transfer through or rebound from), or features that mask and conflict with the signal, such as ambient noise. Different habitats impose different constraints on spectral and temporal characteristics of communication signals [84, 85], such that the same signals will have different active spaces in different environments under different conditions. The desired active space of a signal will depend on the distance to the intended recipients and the nature of the message. A message about a signaler's presence only needs to be detected and will tend to have a larger active space than a more specific or elaborated message, such as information about a signaler's fitness or context, which requires decoding subtle aspects of the signal. Therefore, the same signal may have a different active space for each category of recipients, which can be differentially affected by the habitat.

Several rhythmic signals may have evolved to ensure clearer communication over long distances and/or instances when signals tend to be significantly degraded or masked before reaching the intended receiver(s). This may include, for example, seasonally unstable environments, where individuals need to ensure that a signal will have good chances of being received under highly variable conditions, say, both during wet and cold vs. dry and hot seasons. Additionally, animals with large home ranges and/or migrating animals may encounter multiple habitat types and, hence, will be required to produce the same signal across highly disparate conditions that affect transmission differently.

The evolutionary response and solution to these communicative challenges has often been the emergence of repeating and rhythmic signals. The production of signal units in strings or sequences functions, in most of these conditions, as "functional backups" of each other. If some units become too degraded or unreliable, or are lost during transmission and propagation, other units may be able to reach the receivers as intended (efficacy backup hypothesis [86]). Different types of signal units may also be stringed along the same sequence, where each unit type is reliable under different environmental conditions (fluctuating environments hypothesis [87]). For example, one unit type may have a larger active space than another that is easier to localize (efficacy trade-off hypothesis [86]). The same information may also be coded in different parameters of the same signal for detection at different distances, as is the case, for example, with individual identification cues in zebra finches [88].

Because habitat types can grade into or be hugely different from others, different physical constraints are to be expected on different signals and rhythms in different habitats. For example, aquatic vs. terrestrial, 2D (terrestrial open) vs. 3D (terrestrial arboreal, sub-aquatic, aerial), arctic vs. equatorial, low land vs. mountainous, and subterranean vs. above ground. Regardless, the constraints for signal and rhythm production will always invariably be directly shaped by the physical properties of, and physical conditions experienced in, the different habitats, as they "constrain" behavior into specific moments and timings. The more attuned and "coupled" behaviors become to their environment, the more efficient they will be.

6.5.1 Terrestrial Habitats

Sound transmission in air is constrained by a number of factors: attenuation of low-frequency sounds due to ground impedance, high-frequency attenuation (>3 kHz) due to scattering caused by vegetation, temporal "blurring" due to reverberations created by sound scattering off of objects, micrometeorological heterogeneities in the atmosphere, and amplitude modulations due to atmospheric turbulence [89].

Habitats with higher vegetation cover, such as forests, create more reverberations than open habitats such as savannahs, potentially masking temporal and frequency features of signals. For example, fast-paced signals are often unsuited to forested habitats because sound degradation due to scattering, reflections, and reverberations quickly masks the short silent gaps that divide signal units from each other. The signal becomes a single continuous, protracted unit. On the other hand, in open habitats, atmospheric turbulence is more severe (e.g., wind gusts, air movement due to temperature shifts), but fast-paced signals can have greater chances of reaching the receiver with minimal variability between components if well timed by the producer. The entirety of the message can be quickly transmitted and reach the receiver within relatively short windows of stable air conditions. Indeed, many species of songbirds produce more notes and faster trills or faster-paced songs in open rather than densely vegetated habitats [72, 75].

These constraints also affect primate calls and are presumed to have shaped in potent ways the evolution of speech and language. For example, similar to any human language, great ape repertoires are composed of both consonant-like and vowel-like calls [90]. The articulation of the two is very distinct. The sound source of consonant-like calls is characteristically in the mouth, such as clicks, smacks, raspberries, and sizzles produced by the articulation of the tongue, lips, and jaw, in coordination with airflow control. Conversely, the sound source of vowel-like calls is characteristically in the larynx, such as grunts and hoots produced by the oscillation of the vocal folds [91, 92]. These two mechanisms generate very distinct acoustics, endowing consonant-like and vowel-like calls with distinct propagation attributes. In playback experiments conducted in a forest, both orangutan consonant-like and vowel-like calls could be detected up to 100 m distance [93]. Signals degraded with distance from the source, but the information they encoded did not. That is, information about an individual's identity, the context of call production, and the individual's natal population could all be extracted and assessed efficiently from signals that were either close by or far away, up to 100 m.

Different results were obtained, however, when the same experiments were replicated in a savannah [94]. Both call types could equally be detected up to 100 m, as in the forest, but their broadcast performance diverged significantly from that mark onwards. Vowel-like calls quickly degraded to the extent that at 400 m away, less than 20% of the calls could still be detected. This made any effort to extract their encoded information very challenging, if possible at all. Conversely, more than 80% of the consonant-like calls could be detected at 400 m, significantly increasing the chances for detection and decoding. In evolutionary terms, these results show that ecological shifts can transform the active space of certain signals or call types or categories, enhancing the functional advantages relative to others in the species repertoire. This alerts one to the notion that paleo-climate change likely played an important role in shaping the vocal system of human ancestors. Where consonant-like and vowel-like calls were functionally evenly matched in forested habitats, consonant-like calls started to gain a functional edge as habitats gradually became more open and drier. Such advantages could help explain why various mechanisms of speech and language perception and processing rely in modern humans on the acoustic saliency of consonants [94], even if consonants are relatively late innovations along the primate lineage [95].

6.5.2 Aquatic Habitats

Water is better suited for long-distance acoustic communication than air in that it has a low absorption coefficient, allowing sounds to travel farther [96]. However, sound propagation becomes increasingly complex when the distance to the intended recipient is greater than the water depth. Both the surface and the bottom (depending on the substrate) act as reflective surfaces. The reflected sounds merge with the original sound wave, creating complex convergent and divergent interference patterns. Additionally, sound velocity changes as a function of water temperature, salinity, and depth. If these factors change dramatically, such as the temperature changes in a thermocline, a reflective boundary will occur within the water column, resulting in an underwater sound corridor (horizontal) or wall (vertical) [97]. Bottom composition, topography, and flora also affect sound propagation [98]. Water depth and bottom composition create a lower frequency cut-off in shallow waters, while sandy bottoms scatter high-frequency sounds more than silt-clay bottoms [99]. How sounds degrade is going to depend on the depth of both the signaler and the receiver [100].

Rorqual whales take advantage of the deep sound channel created by oceanic thermoclines, which in turn constrains the temporal organization of their songs. As reviewed by Bass and Clark [100], upper and lower barriers are created by refracting thermoclines, rather than reflecting boundaries (i.e., the surface or bottom), leading to minimal transmission loss, but potentially giving rise to "echoes" that conflict with the original signal. The depth of the deep channel changes with bottom topography and latitude, such that whales need only go to depths of 100–500 m to use the channel that will carry sounds across ocean basins at depths greater than 2000 m. Not only are the songs of rorquals infrasonic to maximize transmission distance, but they also have longer internote intervals than their coastal counterparts, who communicate over shorter distances.

Coastal waters (100–200 m deep) in which humpback whales (*Megaptera novaeangliae*) sing are relatively shallow compared to the deep waters in which rorquals sing. The constructive and destructive interference patterns created by boundary reflections are frequency dependent, such that sounds in different frequency bands differentially oscillate in amplitude with distance from the source, greatly altering the spectral content of the sound [96, 101, 102], as well as creating reverberations. Reverberations can extend the duration of a sound, and if overlapping additional units with the same frequency bands, amplify sounds, as has been seen in humpback whale songs. The spectral and temporal structure of the songs may be arranged in such a way as to maximize signal strength via regularly spaced units that create reverb, and localizability with sounds that produce less reverb [103].

In contrast, delphinids are able to flexibly change the interclick intervals of their click trains to avoid overlap of echoes. Hence, click trains produced when foraging (close targets) are faster than click trains produced when traveling (distance targets) [104]. However, delphinids also increase their click rates with depth [105, 106], indicating that features of the aquatic environment shape rhythm production separately from target distance. Socializing dolphins might modify their click rates so that group members can track their movements. Researchers have been able to use the timing of echoes of click trains and burst pulses reflecting off the substrate and surface to localize and track spinner dolphins (*Stenella longirostris*) [107]. Such ability requires some familiarity with the environment (e.g., known water depth), and, for researchers, ideal conditions such as smooth bottom substrate and calm sea state. Since dolphin perceptual systems are honed to calculating minute differences between echoes, it is probable that they also have the perceptual capabilities of tracking their group members in more diverse conditions. Further research is needed to determine if social groups of dolphins moderate their click rates with changes in the environmental context, such as water depth, substrate, and sea state.

The shallow water environment (depths <5 m) in which many teleost fish live is even more restrictive, with limited communication ranges. This restrictive range constrains the evolution of sound production and detection mechanisms, as well as acoustic behavior. Sound production and detection are often limited to the near field, relying on particle motion rather than sound pressure propagation, except for species with a connection between the swimbladder (or other gas bubble) and inner ear that allows for far-field sound detection [108]. Additionally, propagation distance changes with tide level. Lusitanian toadfish make less acoustic effort in low tide, when sound propagation is minimal [109]. Most teleost fishes produce signals that do not match the propagation in their environment [100]. In some situations, it may be advantageous for signalers to reduce or minimize active space, to avoid eavesdropping by competitors or predators (eavesdropping avoidance hypothesis), demonstrate quality, or engage in auto-communication [110]. Reduced active space in fishes may preserve important temporal qualities of the signals [100].

6.5.3 Subterranean Habitats

The subterranean habitat presents unique challenges to animal communication, given the high-density medium (e.g., sand, soil) that creates a dark and hypoxic ecotope. The visual and acoustic modalities, which are extensively exploited by terrestrial and aquatic animals, are severely restricted in the underground. Visual signals are difficult to decipher in the absence of light, and sounds are readily attenuated by tunnel walls. At reduced oxygen levels, intense and energetically costly rhythmic displays may be difficult to sustain. Research in fossorial mammals, spanning rodent, insectivore, and marsupial species (for details, see Chap. 3 [3] this volume) shows that, in the face of such constraints, underground-dwelling species have adapted to exploit the seismic channel of communication [111]. This places selective demands on the mechanisms for signal production, selecting for behaviors that can produce signals that propagate well as substrate-borne mechanical vibrations, and on mechanisms for information encoding, selecting for signal parameters that are most resilient in the given medium.

Consistent with this, among underground mammals, the main behavioral mechanism for seismic signal production is percussion, whereby animals drum their feet, and less commonly other body parts (teeth, chest, head, etc.), on the ground or the surrounding tunnel walls (as reviewed in Chap. 2 [29] and Chap. 3 [3] this volume). While drumming generates both air- and substrate-borne vibrations, only the latter can propagate within the solid medium to sufficiently long distances for ensuring interindividual communication. For example, in the Cape mole-rat (*Georychus capensis*), the acoustic output of foot drumming becomes indistinguishable from the background noise within 1 m of the source, while the seismic component is detectable at a distance of 3–4 m, which corresponds to the natural interindividual spacing of the species [112]. In a similar vein, the peak signal amplitude generated by banner-tailed kangaroo rats (*Dipodomys spectabilis*) through foot-drumming is 40 dB greater in seismic waves compared to airborne sound [113].

Such propagation differences between the acoustic and seismic output of percussive signals in the underground habitat are accounted for by differences in the physical properties of the carrier wave in terms of direction and speed of energy propagation, dispersion, and attenuation. Propagation experiments show that, in the underground, seismic signals are primarily conveyed by bending waves of the Rayleigh type, which propagate on the substrate with a rolling elliptical motion that moves ground particles up and down, as well as backward and forward in the direction in which the waves propagate. Compared to sound waves, bending waves travel much slower (i.e., about 40 m/s compared to about 100 m/s), but they also attenuate more slowly (i.e., 3 dB for every doubling of distance compared to 6 dB for sound), thereby conserving their salience and integrity at biologically significant distances (for reviews and discussions, see [114, 115]). Across and within-species, however, the propagation distance of the seismic signals varies with source amplitude, which, in turn, is positively associated with source size.

The aspects discussed above (propagation medium, transmission channel, and signal production mechanisms) impose constraints on information encoding.

Biotremologists reiterate the idea that temporal patterning provides an efficient way for information encoding in percussion-based seismic signals. First, in the underground, the heterogeneity and unpredictability of the medium generate highly variable conditions for signal production and propagation. This variability significantly affects the reliability of spectral parameters. For example, the radiated power of the signal (force × velocity) is modulated by substrate impedance, whereby power is attenuated in soft- compared to hard soil. In such conditions, where the spectral parameters of signals cannot be trusted, only the temporal properties constitute a reliable source of information. The resilience of rhythmic patterns in bio-seismic waves is clearly demonstrated by research in the emergent field of urban seismology, showing that the beat of musical pieces performed at concerts can be decoded in the seismic waves created by the dancing crowds [116, 117].

Second, the vibrational signals produced by fossorial mammals are typically brief (lasting a few seconds to a minute), but very fast, averaging in some species more than 20 foot- or head-strikes per second (for details, see Chap. 3 [3] this volume). These patterns may be constrained by the molecular and energetic profile of the muscles implicated in drumming, which, in turn, are constrained by the environment. Indeed, foot drumming muscles are dominated by type IIb muscle fibers, which can generate high-force and high-speed contractions, but only for a short period, in the order of seconds (as detailed in Chap. 9 [24] this volume). Since type IIb fibers rely on glycolytic (rather than oxidative) bioenergetic pathways, this muscular composition could be an adaptation to the hypoxic environment of the burrows, by relying more heavily on anaerobic muscle metabolism [118, 119].

6.6 Soundscapes

Soundscapes are made up of all the biophonies, geophonies (e.g., waterfalls), and anthrophonies that an organism is exposed to in its natural habitat. Combined together, these sounds can be described by an ambient noise profile, which often changes in predictable diurnal and seasonal cycles. Here, we examine the effects of this ambient noise profile on signal structure and signaling behavior. In addition, we focus specifically on how biophonies, the sounds produced by conspecifics and sympatric species, shape rhythm production.

6.6.1 Ambient Noise

Ambient noise has had a profound effect on acoustic signal production and perception at the level of taxa, population, and individual, which includes rhythmic signals. Like physical features of a signaler's habitat, the presence of other noises (including anthropogenic noise) in a habitat effectively constrains the distance at which a signal can be detected and interpreted [120]. When the spectral content of the signal overlaps with the spectral content of ambient noise, the signal may be masked. Hence, the same signal will have a different active space when produced next to a babbling brook, in a quiet field, or in a group of other signaling animals. For example, tropical katydids (*Mecopoda elongata*) desynchronize in high natural noise conditions, as the noise masks conspecific signals, making it difficult to achieve entrainment [121]. Synchronizing treefrogs (*Smilisca sila*), on the other hand, seem to take advantage of the masking effects of waterfalls to avoid predation [122].

In environments where the ambient noise profile is fairly predictable, animals may evolve mechanisms to increase the probability of detection [123]. The frequency bandwidth of animal signals is often matched to ambient noise profiles (baleen whales: [100], primates: [124], birds: [125]). In cases where sound production is constrained so that evolutionary pressure does not allow for the production of high-frequency sounds outside the ambient noise profile (see Chap. 2 [29] this volume), taxa may develop hearing sensitivity in tune with conspecific sounds. For example, hearing sensitivity and sound production in gobies (family Gobiidae) match the quietest band of the low-frequency ambient sound profile [126]. Some crickets and katydids have narrowband hearing ranges that filter background noise in favor of conspecific signals (for a review, see [127]), while the Australian bush cricket can switch between broadband and narrowband hearing [128].

Noisy environments may also constrain the structure of communication signals. Rhythm discrimination ability aids humans in detecting speech in noise [129], thus temporally organized signals may be well suited for communication in noisy environments. Another way to increase detectability in noise is with multi-modal signals such as wing waving in song sparrows [130] or foot flagging in anurans [131]. Anuran species that live near noisy water environments are more likely to use foot flagging as part of their courtship displays than other species.

The same physical environment may show large variations in the range and intensity of ambient sounds at any given time, leading to behavioral plasticity. Common signal modifications include spectral changes such as increases in frequency and amplitude, and temporal changes such as increased duration, calling rates, and/or redundancy, all of which can increase signal detection in noise (for reviews, see [132–136]). While changes in temporal features do not necessarily imply changes in rhythm, such changes will affect signals that already contain rhythmic components. Examples include increased pulse rate in a marine fish [137], increased call rate of multi-component calls of a tree frog [138], and increased singing effort of categorical rhythms in a suburban bird [139]. In contrast, some species of frogs reduce their call rates in response to traffic noise playback, possibly to compensate for the higher energy cost of producing higher frequency calls [140]. Whether call rate increases or decreases may be situationally dependent, especially if the noise is perceived as threatening or difficult to overcome [133].

Species differ in their degree of plasticity in response to changes in noise levels. Chaffinches (*Fringilla coelebs*) produced higher song repetition rates of one song before switching songs in a naturally noisy environment (i.e., near a noisy stream) [141], but not in an anthropomorphically noisy environment (i.e., noisy town) [142]. Temporal features of blackbird (*Turdus merula*) songs differ at the population level, but not the individual level, in relation to ambient noise [143]. The degree of plasticity may depend on the degree of natural noise variation in the environment, with highly variable environments leading to short-term adjustments in signal production [144]. Additionally, the type of modification may depend on the purpose of the signal and the relative costs and benefits of ensuring detection on the part of the signaler and the recipient [123].

6.6.2 Conspecifics

The signals of conspecifics have a density-dependent effect on the soundscape, which can constrain total output, structure, and coordination of rhythmic sounds. Competitors often need to be heard over the noise, while group members need to be recognized. Many animals increase their vocalization rates when competitor numbers are higher or neighbors are closer [145–147]. Others alter the structure of their signals. For example, in laboratory experiments, echolocating bats [148], dolphins [149], and electrolocating fish [150] alter temporal characteristics of their signals presumably to avoid jamming. Blackbirds dedicate higher portions of their songs to twitter, a difficult element to produce, when territories are denser [143]. In this case, the goal is not just to be heard, but to demonstrate fitness in the face of competition.

6.6.2.1 Chorusing

When multiple individuals are signaling in the same time period, the group behavior is often termed chorusing. The timing of signal production may be synchronous (e.g., flashing fireflies), alternating (e.g., dueting birds), or chaotic (e.g., howling wolf packs), with signals from one individual influencing the signals of others. Synchronization in insects and anurans likely developed due to female choice. In some species, females are not attracted to overlapping, unsynchronized signals, leading to selective pressure to synchronize. In other species, females prefer the leading male, and synchrony is therefore a byproduct of males competing to be first (for a review, see [151]). Adjusting vocal timing to avoid overlap with conspecific competitors has been described in some insect and frog species. Additionally, synchrony and turn-taking can play a part in reinforcing social/pair bonds and in territory defense (siamangs [152], birds [153]).

While many species produce rhythmic signals, only a few synchronize their signals. Both crickets and whales produce long monotonous series of sounds, yet crickets form synchronized choruses and whales do not. The problem may be threefold. First, true synchrony, as seen in insects, requires coupling of endogenous oscillators [154]. Solo performance of humpback whale song does not show as precise isochrony as the mating signals of insects [155] and does not suggest an endogenous oscillator. However, some anurans and dueting species are able to roughly time their signals in response to other conspecifics, producing the effect of synchrony or alternation [156]. Some parts of humpback whale songs as well as some rorqual songs are regular enough that other whales should be able to match the timing.

Second, the number of individuals in a signaling group also plays a role in coordinated timing. Individuals pay attention to a select number of neighbors when adjusting the timing of their signals, not the whole group [157, 158]. Because of this, the farther away you get from a focal individual, the more likely the second individual will be out of sync. Hence, even in groups of fireflies, which have the highest degree of synchronization, sometimes flashing will occur as a wave from one end of the group to the other [159]. Similarly, alternation of signals is limited to 3–4 individuals, as there is a limit on the space between calls in which new individuals can signal [154]. During the peak of the breeding season, humpback whales will sing within hearing distance of up to 20–25 other singing whales (assuming 5 km spacing between singers) [160], producing a cacophony of sound that is often difficult to parse.

A final constraint in producing synchronous rhythms is the social distance between signalers. The further apart two individuals are, the longer the timing delay between the production of the signal and the receipt of the signal. In general, species that synchronize produce sound at rates of 2–14 Hz [154], avoiding significant timing delays to their nearest neighbors. Whales, on the other hand, produce sounds at much slower rates (0.2–1 Hz) [155] and much greater distances. Singing humpback whales tend to be regularly spaced on the breeding grounds, maintaining distances of approximately 5 km [160], meaning a time delay of over 3 s between when the sound is produced and when it is heard by the nearest neighbor.

The combination of poor isochrony, large group size, and distance-based timing delays makes it nearly impossible for whales to synchronize their songs. Rather, what we observe is entrainment on a much longer time scale. There is some evidence that humpbacks may match the phrase type to their nearest neighbor [161], they do not do so all the time, and all whales within hearing distance do not match phrases in real time. However, whales do match phrase types and order as songs evolve throughout the year, showing population wide entrainment.

6.6.2.2 Signature Vocalizations

In species that live in high-density colonies, individual recognition may depend on having distinct vocal patterns, as can be seen in pinniped mother-pup communication localization calls (i.e., vocal signals that aid in mother-pup reunion after periods of separation). Call parameters are constrained by demands for mother-pup vocal recognition, which, in turn, are determined by socio-ecological selection pressures related to breeding substrate (floating ice, fast ice, or land), population density, the duration of maternal attendance (weeks, months, or years), and the type of maternal foraging during the lactation period [162, 163]. Specifically, high population density, frequent mother-pup separation (due to maternal foraging), and extended maternal attendance—as encountered in sea lions and walruses—create demands for vocal signals that are malleable enough to allow for a great variety of individual signatures and can withstand high levels of background noise. Rhythmic patterning offers a means for achieving both. On the one hand, rhythmic calls can be actuated by highly stereotypical sequences that encode individual signatures, which can be based on individual-specific inter-onset intervals, call-unit duration, call-type combinations, etc. On the other hand, rhythmic signaling entails the regular recurrence of temporal patterns, which provides redundancy and, thus, ways to ensure signal conservation in a noisy environment.

Extant data show that species that fit the socio-ecological profile outlined above have high levels of individual stereotypy in their localization calls and multiparametric vocal recognition at a call-unit level (reviewed in [162, 163]). As noted in Chap. 3 [3], it has been mathematically determined that the localization calls of several species are rhythmically patterned ([164] and references therein). Whether or which rhythmic (temporal) parameters play a role in individual signatures and mother-pup recognition remains to be studied. However, the presence of call stereotypy and its variable patterning between successive pups of a single mother suggest highly flexible capacities for call analysis and vocal recognition in pinniped females, in particular in otariid mothers that need to regularly update pup signatures at 1–2-year intervals.

6.6.3 Sympatric Species

The final important component of soundscapes that constrains rhythmic signals is the presence of sympatric species. Animals living in the same area compete over acoustic space in much the same way as they compete for other resources. As a result, sympatric species may evolve to partition the available habitat into sub-niches, with each species shifting the acoustic properties of their signals to avoid interference (acoustic niche partitioning hypothesis [165]). Such changes in signal structure fit under the broader umbrella of divergent agonistic character displacement, which occurs as a result of interspecific interference competition [166]. Partitioning the acoustic space often involves signaling at different times of the day and/or in different locations [167], or constraining sounds to specific, nonoverlapping frequency bands [168]. However, partitioning can also be achieved by adjusting the fine scale temporal features of signals, as has been reported across taxa, including insects [169], amphibians [170], fish [171], birds [172], and mammals [173].

Convergent signal evolution has been observed in some birds. While signal divergence is thought to aid species recognition and avoid hybridization (reproductive character displacement [166]), female antbirds can distinguish between species and between individuals based on subtle voice cues despite signal convergence [174]. Thus, evolutionary pressure is focused on perception rather than production. Selective pressure to converge signals may occur between species that compete for the same territory, as evidenced by convergence of territory defense but not other signals (convergent agonistic character displacement [175]). Signal divergence may be more important for avoiding interference and/or hybridization during mate attraction, while convergence may be more important for threat recognition during territorial defense. When a song serves more than one function, selective pressure may act differently on different aspects of birdsong, as song rhythm is under strong biomechanical constraints, while syllable morphology is often learned [176].

Several factors go into shaping rhythmic signals, so that character displacement due to interspecies competition may not be the strongest factor. For example, divergent character displacement of drumming patterns only occurs in sympatric woodpeckers that are closely related, while ecological factors play a much bigger role [177]. Similarly, the calls of sympatric whistling rats differ in the number of components as well as spectral features, some of which fit the acoustic adaptation hypothesis and some of which do not and might be due to interference avoidance [173].

6.7 Social Context

Just as the physical environment shapes the signals that animals produce, the social environment can also constrain rhythm. An animal's social environment is often complex, being composed of a combination of intended (e.g., offspring, mates, group members) and unintended (e.g., predators, parasites, competitors) recipients. The signals that an animal produces depend on who is watching or listening. Each type of recipient has its own resource needs and social role, often necessitating the need to tailor information for different types of intended recipients and concealing or falsifying information for unintended recipients. Some rhythmic signals have evolved or are flexibly produced in such a way as to address these different audiences.

6.7.1 Mating Systems in Pinnipeds

During the mating season, pinniped males produce a whole range of rhythmic signals (as detailed in Chap. 3 [3] this volume). In phocid species, the structural complexity of vocal calls correlates nearly perfectly with rhythmic diversity and, in turn, is strongly shaped by socio-ecological variables, including the mating system, population density, and type of breeding grounds [178]. For example, polygynous species, where the females form large aggregations on stable ice during the breeding seasons (e.g., the harp seal, *Pagophilus groenlandicus*) [179] and the Weddell seal, *Leptonychotes weddellii* [180], exhibit the largest call repertoires that include multiple rhythmic patterns, and highly complex underwater vocalizations classified as "songs."

By contrast, species characterized by serial monogamy, pack-ice breeding, and low population density (the ribbon seal [*Histriophoca fasciata*], the leopard seal [*Hydrynga leptonyx*], the Ross seal [*Ommatophoca rossii*], the bearded seal [*Erignathus barbatus*]), exhibit lower vocal complexity and a single rhythmic pattern. It has been argued that this type of mating and breeding conditions raises demands for call features that are adapted for advertising at a distance [181]. Such demands are, for example, well met by the rhythmic advertisements of leopard seals, which are characterized by high stereotypy and persistent repetitiveness [182].

Finally, a third socio-ecological niche, characterized by polygynous breeding where males and females form large colonies on land, is found in the northern elephant seal (*Mirounga angustirostris*), being also shared with sea lions (family Otariidae). This socio-ecological niche entails selection pressures for short-range signals that are adapted to a terrestrial environment (e.g., airborne vocalizations, visual displays) and used for rival assessment. In otariid males, rhythmic calls that encode levels of male fitness and/or arousal are used to regulate territorial disputes without direct engagement [183, 184]. In elephant seals, these feats are combined with a system where breeding hierarchies are first established through direct competition and then maintained through the recognition of individual rhythmic signatures [185, 186].

6.7.2 Audience Effects

The composition of an individual's social group to a large extent constrains the likelihood of different behaviors that he or she may engage in. These effects are generally known as "audience effects," highlighting the notion that animals are always inserted within a social arena with other conspecifics, whose presence, identity, ranking, sex, familiarity, and/or actions interplay with what an individual might decide to do or avoid doing [187]. These effects can have powerful evolutionary consequences.

For instance, take the example of a primate individual intending to make a friendly approach to another with whom he *infrequently* socializes. The congruent behavior will be a slow, gentle, and peripheral approach to signal benign intent. A fast and sudden approach could, alternatively, be likely perceived as signaling confrontational intent. On the other hand, if he or she is approaching a sibling or friend, then a fast, sudden, and direct approach may signal playful intent. If the approach is toward an alpha animal, the safest option for an individual will be to move in slowly and softly. A jolt or quick move could result in a reprimanding lash by the alpha. Over developmental and evolutionary time, social interactions such as these tend to become ritualized. That is, if across the social interactions over an individual's lifetime and across generations, a particular behavior or rhythm predictably results in a (more) successful outcome, then this rhythm goes through a selective process against other possible alternative behaviors and rhythms. For instance, if approaching a distant social peer or alpha individual at a walking pace of 0.5 Hz repeatedly elicits a tolerant response, this pace of approach will be preferred in similar future contexts. With time, this rhythm can become so constant and obvious that a dominant or alpha individual begins to *expect* it, triggering a reinforcement feedback loop between the approacher and approached. A shared "convention" emerges between the two—a 0.5 Hz pace signals benign intent and will be tolerated. Anything significantly deviant will not. Stereotypical behaviors, such as "mating dances," are a classic example of selective run-aways of these types of reinforcement dynamics. Ritualization can, in this fashion, narrow the range of behaviors and rhythms that an individual is likely to perform under a specific situation.

Species whose rhythm production is flexible may change the temporal properties of their signals in response to differences in the status or social roles of conspecific recipients. In a dyadic interaction between individuals of different status, only one individual may make adjustments while the other does not. For example, weakly electric fish that produce pulse-type electric organ discharges have been observed to alter their interpulse intervals to either avoid or intentionally jam the electrolocation signals of conspecifics. These changes have been posited to play a role in reinforcing school cohesion and establishing social hierarchies (for a review, see [188]). Likewise, adjustments may be audience dependent, as in the case of Peruvian warbling-antbird (*Hypocnemis perviana*) duets [189]. Bonded pairs produce precisely timed duets in response to simulated rival pairs, as a form of territory defense. However, when the perceived audience is a rival female, focal females engage in mate guarding by changing the timing of their duet contributions to interfere with their mate's contributions. Meanwhile, males adjust their timing to try to avoid jamming.

6.7.3 Eavesdroppers

The presence of unintended recipients, or eavesdroppers, can also constrain signal structure. As alluded to above, according to the eavesdropping avoidance hypothesis [122], signals may evolve or be adjusted to avoid detection by interspecific (predators, parasites, and prey) or intraspecific (competitors, consorts) eavesdroppers. In the former case, there is a natural selection vs. sexual selection trade-off where the change in the signal to avoid detection by predators may also decrease the likelihood of being detected by potential mates. Furthermore, Peake [190] distinguishes between interceptive eavesdropping (a single signal) and social eavesdropping (interactions), each of which could constrain signal structure differentially. While interceptive eavesdropping avoidance often leads to synchrony, social eavesdropping can lead to changes in the rhythmic structure. For example, male katydids (*Scudderia pistillata*) will produce a female-like click in the presence of satellite males during the time window when potential mates should respond, possibly drawing rival males' attention away from the female [191].

Synchrony may decrease predator risk in insects and anurans. When sounds from two locations arrive at the ears simultaneously, the recipient may perceive the sound as coming from a phantom central location [192], potentially confusing both intended and unintended recipients. Hence, both eavesdroppers [193] and prospective mates [158, 194, 195] often show a precedence effect for leading males when signals alternate or partially overlap. From a sexual selection perspective, synchrony may be advantageous at longer distances, acting as a long-range attraction signal for females (beacon effect hypothesis [196]), but disadvantageous at shorter distances. For example, female Japanese treefrogs (*Buergeria japonica*) show preferences for synchronization when away from a chorus, but no preferences at the chorus [197]. The degree to which the precedence effect exists in intended and unintended recipients differs between call types and species. For example, synchronized frog calls attract fewer frog-eating bats, in the case of pug-nosed tree frogs (*Smilisca sila*), and frog-biting midges, in the case of túngara frogs (*Engystomops pustulosus*), but not vice versa [198]. Despite reduced parasite risk, túngara frogs produce alternating rather than synchronized calls in natural conditions possibly due to the decreased benefits of synchronization in enemy avoidance and stronger precedence effects in females. Hence, broad generalizations cannot be made across species.

Alternatively, there are many arthropods that produce rhythmic signals as a self-defense strategy against potential predators or parasitoids, aimed at startling the predator (deimatic display), jamming the echolocation signals of bats, imitating the enemy's enemy, warning potential enemies of additional defensive mechanisms such as toxins, and/or producing vibrations that overstimulate the predator (for a review, see [199]). An estimated 20% of moth species produced ultrasonic signals either to jam bat echolocation or as acoustic aposematism (for a review, see [200]). For example, moths from the genus *Yponomeuta* produce sounds via tymbalation with their wing beats, even though they cannot detect the sounds themselves [201]. However, these signals are not always rhythmic. For example, acoustic defense signals in some tettigoniid species are temporally more variable than mating songs, and in *Poecilimon ornatus* differ between sexes [202].

6.7.4 Social Learning and Cultural Constraints

Across many taxa, conspecifics are also relied on or exploited as external sources of information and knowledge. That is, individuals use others' behaviors, choices, and strategies to guide their own [203], learning alternatives or improved means of doing things in their context. Reliance on social learning ranges from great apes [204] to fruit flies [205]; however, the greater the behavioral flexibility of a species, the larger the consequences of social learning. Behavioral flexibility increases the kind and degree of behaviors that individuals can perform in a controlled and purposeful way and, hence, it increases the rate and diversity of novel behaviors and innovations occurring at any one time in a population, that is, behaviors hitherto unknown or unseen but found(ed) by an "inventor" individual. Once proven effective or popular to others, novel behaviors can spread across a community or population very rapidly, within one or a few generations. When an innovation becomes prevalent across a community, it is said to have become a local tradition, which is typically assumed to be absent in other distant or independent communities given that original seed innovations are always local-specific (unless another individual in another community makes the same discovery and conceives the same innovation).

When a population exhibits a collection of two or more traditions, it is said to have a local culture. Social learning and cultural means of acquiring information from others can become so potent that it has been described as the second inheritance system, besides the first DNA-based system [206]. For an individual who is born into a community with a local culture, this means that he or she will only be exposed to the behaviors that have been locally established and conventionalized and to which most of its family, peers, and conspecifics conform. In this way, even if traditions are based on behaviors that were originally novel, once they become established, traditions also impose a restriction on individuals' development and a curb on their behavioral repertoire. An individual will behave and act in a certain way only because that is what it has learned to do by observing role-model individuals. Though clearly advantageous because an individual can learn many things that he or she does not have to discover or invent itself, traditions and cultures also ultimately silo individuals, keeping them bound to the local rules and costumes that could (have already) become ineffective, cumbersome, outdated or plainly obsolete, even if better alternatives are available in neighboring communities. Humans are the richest cultural species; therefore, these effects have permeated virtually all aspects of life, habits, and routines, including human rhythm. Speech rhythm varies across populations [207] as well as musical rhythms [208] and dance styles. Other rhythmic innovations have proven so successful that they have gained global status, from hammering, screwing, sawing, sewing, riding bicycles, using mortars and pestles to process foods, writing, and typing on a computer keyboard. These rhythms have taken up such a central role in our species globally that they have gained industrial status, governing and constraining what any one person may set out to invent or innovate.

6.8 Anthropogenic Effects

As detailed in scientific studies across various fields, we are facing severe changes to ecosystems due to anthropogenic effects. This ranges from the large-scale climate crisis [209] we are all facing, to physical changes of habitats due to agriculture or human infrastructures, as well as noise and light introduced into various soundscapes and environments across the globe. These anthropogenic effects act on rhythm production and perception on different levels, ranging from changes in species composition, to changes in physiology through fluctuating temperatures or the need to adjust communication signals in response to noise.

6.8.1 Light Pollution

In recent times, the natural cycles of light variation have been drastically affected by artificial illumination. The excessive use of artificial light during nighttime has given rise to the phenomenon of "light pollution," with significant effects on wildlife behavior and physiology (for recent reviews, see [1, 210]), as well as human health [211–214] and development [215]. During nighttime, artificial illumination can cause a 10 to 30-fold increase in light intensity. Current estimates suggest that at least 83% of the world's human population is affected by light pollution, and this area increases by about 2% annually [216].

Light pollution disrupts circadian, lunar, and seasonal endocrine cycles, which can lead to short-term physiological and behavioral changes that may be reversed with a return to more natural lighting conditions [217, 218]. However, prolonged exposure can result in long-term impacts such as disrupted reproductive cycles. For example, artificial light can induce birdsong irrespective of the reproductive status of the signaler, leading to mistimed mating calls [217, 218]. It also disrupts the learning of species-specific signals, primarily affecting rhythm acquisition [18]. Moreover, light pollution affects song rhythm in ways that could be detrimental to reproductive success or homeostasis, leading to physiologically unsustainable activation [218, 219]. It disrupts time-activity budgets, affecting the timing of signaling and the allocation of time to other activities [220]. Additionally, brighter nights may increase predatory activity, which can diminish signaling duration and affect functions that depend on signaling [221]. Alternatively, animals might continue to sing despite higher predation risks.

6.8.2 Temperature

Many studies highlight the potential effect of anthropogenic climate change, where more extreme temperatures or temperature changes throughout the day have the potential to heavily impact a species' mating and communication system. Different temperatures favor different selection pressures; for example, in wolf spiders [56] and jumping spiders [55], variations in temperature can influence the effectiveness of courtship displays. In various frog species, rising temperatures can uncouple sound production and detection systems, preventing females from perceiving male frequencies [222]. Similar issues can be speculated to occur in various vocal and ectothermic fish species. The importance of temperature dependencies on temporal and spectral aspects of signal production and perception cannot be overstated. This calls for further detailed analysis to uncover the intricate interactions between signal characteristics, social environment, and environmental factors such as temperature.

6.8.3 Soundscapes

The negative effects of anthropogenic noise are becoming an increasing concern for a number of taxa and have been problematized for marine and terrestrial environments similarly [223, 224]. The necessary adjustments are often energetically costly, and the inability to adjust can be reproductively costly. Species that rely on acoustic signals for mating or territory defense, such as birds and amphibians, are particularly vulnerable. For instance, birds may alter the frequency or timing of their songs to avoid overlap with human-made noises, such as traffic or industrial sounds [225]. This can lead to reduced mating success if the altered songs are less attractive to potential mates or if they interfere with the intended message.

6.8.4 Interference on Species Composition

Anthropogenic effects also interfere with species composition in various ecosystems. Changes in habitat due to deforestation or urbanization can lead to a decrease in species diversity and a shift in community structures. These changes can disrupt the acoustic environment, affecting the ability of species to communicate effectively, or might lead to acoustic competition between species. Species that rely on specific acoustic niches may find it harder to compete or may be forced to alter their signaling behavior, potentially leading to decreased reproductive success and population declines. These detrimental effects have been seen in cases where an invasive species' calls have overlapped the acoustic niche of native species (for a review, see [226]). One specific example, where the species composition is altered due to local disturbances to the nutrient load and pH of water bodies, can be found in the Wallum areas in Australia. The Wallum sedgefrog (*Litoria olongburensis*) and the eastern sedgefrog (*Litoria fallax*), formerly separated from each other, now compete for the same acoustic space, leading to acoustic competition altering the rhythm of calls [50]. Similarly, swamp frogs (*Limnodynastes convexiusculus*) altered their call rate and intercall interval variability in response to playbacks of invasive cane toads [227].

6.9 Conclusions

Because the natural world organizes itself by major cyclic events (e.g., the seasons, the diurnal cycle), animals behave in a "timely" manner with regard to such cycles to meet their needs and wants in an efficient and successful way. Effectively, their behavioral rhythms can be considered as "coupled" with, or "constrained" by, the physical rhythms that the environment imposes. Light and temperature, in particular, have a profound effect on the temporal aspects of signal production. Additionally, rhythm production can be indirectly affected by environmental adaptations unrelated to communication, but which alter anatomy and/or physiology, inducing physical constraints on the signal.

The efficacy of communication also depends on the efficacy of signals to propagate, which is related to how suitable the properties of the signals are to the socio-ecological conditions. For instance, signals can only function within the physical bandwidth available in nature for their transmission, that is, along the channels that are available between a sender and a receiver for the effective transmission from the former to the latter. Hence, the environment imposes very specific routes through which signals can work and perform biological functions on behalf of the individuals, populations, and species that produce and receive them. Finally, signals may be constrained or adapted to both the intended and nonintended audience, either to "be heard," to minimize the chance of being detected, or to communicate flexibly to different categories of recipients.

We have examined a variety of environmental and social constraints to better understand the significance of each to rhythm production. However, in nature, these factors are never working in isolation, but rather rhythmic signals are shaped by a variety of constraints that interplay with each other. Research designs need to be constructed in such a way as to consider multiple constraints as well as their interactions. Exclusive focus on one constraint can lead to a failure to consider alternative explanations for complex rhythmic signals. For example, humpback whale song structure has long been thought to be a product of runaway sexual selection, though it does not meet the typical criteria for neuromechanical fitness indicators outlined in Chap. 9 [24]. Alternatively, the controversial sonar hypothesis explains song structure as a product of environmental constraints, while taking into account competition but not sexual selection [228].

On an evolutionary time scale, socio-environmental interactions may lead to speciation. As habitats and/or resources change, selection on complex signals, such as rhythmic signals, may be bidirectional such that one signal becomes two. Assortative mating may lead some females to prefer one over the other, resulting in behavioral speciation. Podos and Nowicki [67] have posited that bidirectional changes in beak morphology should lead to song divergence and speciation (see Sect. 6.4.1). As an intermediary step, territorial males respond differently to songs from matching morphs than nonmatching morphs, demonstrating song-based discrimination within the same species [229]. Similarly, differences in vegetation density correlate with the temporal divergence of gray-breasted wood-wren (*Henicorhina leucophrys*) songs [230]. While song learning has been shown to influence speciation of songbirds [231, 232], future research should also investigate the role of socio-environmental factors.

Two additional avenues of future research become apparent: what is missing, and what is essential. The literature is mostly constrained to a single modality, acoustic rhythmic signals. This focus is not surprising given human proclivity for vocal production in combination with the increased challenges of studying visual and tactile rhythms. Nonetheless, a deeper understanding of the socio-environmental constraints on other modalities is needed. Finally, the profound anthropogenic impacts described in Sect. 6.8 underscore the urgent need for further research into their effects on natural rhythms and communication systems. Understanding these complex interactions is crucial, as the perturbations caused by climate change, noise, light pollution, habitat alteration, and other anthropogenic effects can significantly affect the survival of species and the health of ecosystems. Future research must delve deeper into these mechanisms in order to develop effective strategies for mitigating the far-reaching effects of human activities on the natural world.

Acknowledgments GA Sauciuc's contribution to this chapter has been funded by the Bank of Sweden Tercentenary foundation (grant nr. P17-0942:1) and the Swedish Research Council (grant nr. 2019-03104). AR Lameira's contributions to this chapter has been funded by UK Research & Innovation, Future Leaders Fellowship (grant nr. MR/T04229X/1).

References

1. Dickerson AL, Hall ML, Jones TM (2023) Effects of variation in natural and artificial light at night on acoustic communication: a review and prospectus. Anim Behav 198:93–105. https://doi.org/10.1016/j.anbehav.2023.01.018
2. Gustison ML, Sauciuc G-A (2026) Chapter 11: Moving the body and the mind—links between emotion and rhythmic communication and the underlying neurobiological mecha-

nisms. In: Sauciuc G-A, Lameira AR, Persson T (eds) Nature beats: the what, why and how of animal rhythmic behaviour. Springer Nature
3. Sauciuc G-A, Schneider JN, Burchardt LS et al (2026) Chapter 3: The functional and structural diversity of rhythmic communication across species. In: Sauciuc G-A, Lameira AR, Persson T (eds) Nature beats: the what, why and how of animal rhythmic behaviour. Springer Nature
4. York JE, Young AJ, Radford AN (2014) Singing in the moonlight: dawn song performance of a diurnal bird varies with lunar phase. Biol Lett 10:20130970. https://doi.org/10.1098/rsbl.2013.0970
5. Alonso JC, Colón A, Inmaculada, Palacín C (2021) Moonlight triggers nocturnal display in a diurnal bird. Anim Behav 171:87–98. https://doi.org/10.1016/j.anbehav.2020.11.011
6. Horwich RH, Gebhard K (1983) Roaring rhythms in black howler monkeys (*Alouatta pigra*) of Belize. Primates 24:290–296. https://doi.org/10.1007/BF02381093
7. Grant RA, Halliday T, Chadwick EA (2012) Amphibians' response to the lunar synodic cycle—a review of current knowledge, recommendations, and implications for conservation. Behav Ecol 24:53–62. https://doi.org/10.1093/BEHECO/ARS135
8. Lima MSCS, Pederassi J, Caramaschi U et al (2021) Frog vocalization is influenced by moon phases: Brazilian frogs tend to prefer low-albedo phases. Web Ecol 21:1–13. https://doi.org/10.5194/we-21-1-2021
9. Gursky S (2003) Lunar philia in a nocturnal primate. Int J Primatol 24:351–367. https://doi.org/10.1023/A:1023053301059
10. Nash LT (2007) Moonlight and behavior in nocturnal and cathemeral primates, especially *Lepilemur leucopus*: illuminating possible anti-predator efforts. In: Gursky SL, Nekaris KAI (eds) Primate anti-predator strategies. Developments in primatology: progress and prospects. Springer, Boston
11. Duane D, Freeman S, Freeman L (2024) Moonlight-driven biological choruses in Hawaiian coral reefs. PLoS One 19:0299916. https://doi.org/10.1371/journal.pone.0299916
12. Staaterman E, Paris-Limouzy CB, Deferrari HA et al (2014) Celestial patterns in marine soundscapes. Mar Ecol Prog Ser 508:17–32. https://doi.org/10.3354/MEPS10911
13. Carlström J (2005) Diel variation in echolocation behavior of wild harbor porpoises. Mar Mamm Sci 21:1–12
14. Ogawa M, Kimura SS (2023) Variations in echolocation click characteristics of finless porpoise in response to day/night and absence/presence of vessel noise. PLoS One 18:e0288513. https://doi.org/10.1371/journal.pone.0288513
15. Ríos-Chelén AA, Quirós-Guerrero E, Gil D (2013) Dealing with urban noise: vermilion flycatchers sing longer songs in noisier territories. Behav Ecol Sociobiol 67:145–152. https://doi.org/10.1007/s00265-012-1434-0
16. Ríos-Chelén AA, Cuatianquiz-Lima C, Bautista A (2018) No reliable evidence for immediate noise-induced song flexibility in a suboscine. Urban Ecosyst 21:15–25. https://doi.org/10.1007/s11252-017-0690-1
17. Nakamura-Garcia MT, Ríos-Chelén AA (2022) More than noise: light, moon phase, and singing behavior in a passerine. Urban Ecosyst 25:291–303. https://doi.org/10.1007/s11252-021-01142-2
18. Derégnaucourt S, Saar S, Gahr M (2012) Melatonin affects the temporal pattern of vocal signatures in birds. J Pineal Res 53:245–258. https://doi.org/10.1111/j.1600-079X.2012.00993.x
19. Feng NY, Bass AH (2016) "Singing" fish rely on circadian rhythm and melatonin for the timing of nocturnal courtship vocalization. Curr Biol 26:2681–2689. https://doi.org/10.1016/j.cub.2016.07.079
20. Howard CM, Lutterschmidt DI (2015) The effects of melatonin on brain arginine Vasotocin: relationship with sex and seasonal differences in melatonin receptor type 1 in Green treefrogs (*Hyla cinerea*). J Neuroendocrinol 27:670–679. https://doi.org/10.1111/jne.12292
21. Fusani L, Gahr M (2015) Differential expression of melatonin receptor subtypes MelIa, MelIb and MelIc in relation to melatonin binding in the male songbird brain. Brain Behav Evol 85:4–14. https://doi.org/10.1159/000367984

22. viviD D, Bentley GE (2018) Seasonal reproduction in vertebrates: melatonin synthesis, binding, and functionality using Tinbergen's four questions. Molecules 23:652
23. Pasch B, George AS, Campbell P, Phelps SM (2011) Androgen-dependent male vocal performance influences female preference in neotropical singing mice. Anim Behav 82:177–183
24. Sauciuc G-A, Schneider JN, Hattori Y (2026) Chapter 9: The neuroethology of rhythm communication: neuromechanical building blocks and constraints on rhythm production and perception. In: Sauciuc G-A, Lameira AR, Persson T (eds) Nature beats: the what, why and how of animal rhythmic behaviour. Springer Nature
25. Broder ED, Elias DO, Rodríguez RL et al (2021) Evolutionary novelty in communication between the sexes. Biol Lett 17:20200733
26. Gerhardt HC (1978) Temperature coupling in the vocal communication system of the gray tree frog, *Hyla versicolor*. Science 199:992–994
27. Parmentier E, Diogo R, Fine ML (2017) Multiple exaptations leading to fish sound production. Aust Fish 18:958–966
28. Rice AN, Farina SC, Makowski AJ et al (2022) Evolutionary patterns in sound production across fishes. Ichthyol Herpetol 110:1–12
29. Sauciuc G-A, Schneider JN (2026) Chapter 2: How animals produce rhythm—an overview of ethological mechanisms for rhythmic communication. In: Sauciuc G-A, Lameira AR, Persson T (eds) Nature beats: the what, why and how of animal rhythmic behaviour. Springer Nature
30. Ladich F (2018) Acoustic communication in fishes: temperature plays a role. Aust Fish 19:598–612
31. Bass AH, McKibben JR (2003) Neural mechanisms and behaviors for acoustic communication in teleost fish. Prog Neurobiol 69:1–26
32. Ladich F, Schleinzer G (2015) Effect of temperature on acoustic communication: sound production in the croaking gourami (labyrinth fishes). Comp Biochem Physiol A Mol Integr Physiol 182:8–13
33. Amorim MCP, Vasconcelos RO, Marques JF, Almada F (2006) Seasonal variation of sound production in the Lusitanian toadfish *Halobatrachus didactylus*. J Fish Biol 69:1892–1899
34. Montie EW, Kehrer C, Yost J et al (2016) Long-term monitoring of captive red drum *Sciaenops ocellatus* reveals that calling incidence and structure correlate with egg deposition. J Fish Biol 88:1776–1795
35. Monczak A, Berry A, Kehrer C, Montie EW (2017) Long-term acoustic monitoring of fish calling provides baseline estimates of reproductive timelines in the May River estuary, southeastern USA. Mar Ecol Prog Ser 581:1–19
36. Montie EW, Hoover M, Kehrer C et al (2017) Acoustic monitoring indicates a positive relationship between calling frequency and spawning in captive spotted seatrout (*Cynoscion nebulosus*). PeerJ 5:e2944
37. Papes S, Ladich F (2011) Effects of temperature on sound production and auditory abilities in the striped Raphael catfish *Platydoras armatulus* (family Doradidae). PLoS One 6:26479
38. Picciulin M, Bolgan M, Burchardt LS (2024) Rhythmic properties of *Sciaena umbra* calls across space and time in the Mediterranean Sea. PLoS One 19:0295589
39. van Heel WH (1956) Pitch discrimination in the minnow (*Phoxinus laevis*) at different temperature levels. Experientia 12:75–77
40. Fay RR, Ream TJ (1992) The effects of temperature change and transient hypoxia on auditory nerve fiber response in the goldfish (*Carassius auratus*). Hear Res 58:9–18
41. Wysocki LE, Montey K, Popper AN (2009) The influence of ambient temperature and thermal acclimation on hearing in a eurythermal and a stenothermal otophysan fish. J Exp Biol 212:3091–3099
42. Maiditsch IP, Ladich F (2014) Effects of temperature on auditory sensitivity in eurythermal fishes: common carp Cyprinus carpio (family Cyprinidae) versus Wels catfish *Silurus glanis* (family Siluridae). PLoS One 9:108583
43. McKibben JR, Bass AH (1998) Behavioral assessment of acoustic parameters relevant to signal recognition and preference in a vocal fish. J Acoust Soc Am 104:3520–3533

44. Hoy RR, Hahn J, Paul RC (1977) Hybrid cricket auditory behavior: evidence for genetic coupling in animal communication. Science 195:82–83. https://doi.org/10.1126/science.831260
45. Walkowiak W (1980) Sensitivity, range and temperature dependence of hearing in the grass frog and fire-bellied toad. Behav Process 5:363–372
46. Lingnau R, Bastos RP (2007) Vocalizations of the Brazilian torrent frog *Hylodes heyeri* (Anura: Hylodidae): repertoire and influence of air temperature on advertisement call variation. J Nat Hist 41:1227–1235
47. Lüddecke H, Sánchez OR (2002) Are tropical Highland frog calls cold–adapted? The case of the Andean frog Hyla labialis. Biotropica 34:281–288
48. Navas CA, Bevier CR (2001) Thermal dependency of calling performance in the Eurythermic frog *Colostethus subpunctatus*. Herpetologica 57:384–395
49. Wong BBM, Cowling ANN, Cunningham RB, Donnelly CF (2004) Do temperature and social environment interact to affect call rate in frogs (*Crinia signifera*)? Austral Ecol 29:209–214
50. Filer A, Burchardt LS, Rensburg BJ (2021) Assessing acoustic competition between sibling frog species using rhythm analysis. Ecol Evol 11:8814–8830
51. Doherty JA (1985) Temperature coupling and "trade-off" phenomena in the acoustic Communication system of the cricket, *Gryllus bimaculates* De Geer (Gryllidae). J Exp Biol 114:17–35
52. Hedrick A, Perez D, Lichti N (2002) Temperature preferences of male field crickets (*Gryllus integer*) alter their mating calls. J Comp Physiol A 188:799–805
53. Symes LB, Rodríguez RL, Höbel G (2017) Beyond temperature coupling: effects of temperature on ectotherm signaling and mate choice and the implications for communication in multispecies assemblages. Ecol Evol 7:5992–6002
54. Edmunds LN (1963) The relation between temperature and flashing intervals in adult male fireflies, *Photinus pyralis*. Ann Entomol Soc Am 56:716–718
55. Brandt EE, Kelley JP, Elias DO (2018) Temperature alters multimodal signaling and mating success in an ectotherm. Behav Ecol Sociobiol 72:191
56. Rosenthal MF, Elias DO (2019) Nonlinear changes in selection on a mating display across a continuous thermal gradient. Proc R Soc B Biol Sci 286:20191450. https://doi.org/10.1098/rspb.2019.1450
57. Rosenthal MF, Wilkins MR, Shizuka D, Hebets EA (2018) Dynamic changes in display architecture and function across environments revealed by a systems approach to animal communication. Evolution 72:1134–1145
58. Lenis Sanin Y, Zuluaga Cabrera AM, Tarazona Morales AM (2016) Adaptive responses to thermal stress in mammals. Rev Med Vet 31:121–135
59. Chaverri G, Quirós OE (2017) Variation in echolocation call frequencies in two species of free-tailed bats according to temperature and humidity. J Acoust Soc Am 142:146–150
60. Ferrari S, Costa A, Guarino M (2013) Heat stress assessment by swine-related vocalizations. Livest Sci 151:29–34
61. Liu L, Ni JQ, Li Y et al (2018) Assessment of heat stress in turkeys using animal vocalization analysis. In: 2018 ASABE annual international meeting. American Society of Agricultural and Biological Engineers, p 1
62. Long M, Fee M (2008) Using temperature to analyse temporal dynamics in the songbird motor pathway. Nature 456:189–194
63. Grant PR, Grant BR (2007) How and why species multiply: the radiation of Darwin's finches. Princeton University Press
64. Giraudeau M, Nolan PM, Black CE et al (2014) Song characteristics track bill morphology along a gradient of urbanization in house finches (*Haemorhous mexicanus*). Front Zool 11:1–8
65. Nowicki S, Marler P (1988) How do birds sing? Music Percept 5:391–426
66. Hoese WJ, Podos J, Boetticher NC, Nowicki S (2000) Vocal tract function in birdsong production: experimental manipulation of beak movements. J Exp Biol 203:1845–1855
67. Podos J, Nowicki S (2004) Beaks, adaptation, and vocal evolution in Darwin's finches. Bioscience 54:501–510

68. Herrel A, Podos J, Vanhooydonck B, Hendry AP (2009) Force–velocity trade-off in Darwin's finch jaw function: a biomechanical basis for ecological speciation? Funct Ecol 23:119–125
69. Derryberry EP, Seddon N, Derryberry GE et al (2018) Ecological drivers of song evolution in birds: disentangling the effects of habitat and morphology. Ecol Evol 8:1890–1905
70. Podos J (2001) Correlated evolution of morphology and vocal signal structure in Darwin's finches. Nature 409:185–188
71. Christensen R, Kleindorfer S, Robertson J (2006) Song is a reliable signal of bill morphology in Darwin's small tree finch *Camarhynchus parvulus*, and vocal performance predicts male pairing success. J Avian Biol 37:617–624
72. Derryberry EP, Seddon N, Claramunt S et al (2012) Correlated evolution of beak morphology and song in the neotropical woodcreeper radiation. Evolution 66:2784–2797
73. Demery A-JC, Burns KJ, Mason NA (2021) Bill size, bill shape, and body size constrain bird song evolution on a macroevolutionary scale. Ornithology 138:ukab011. https://doi.org/10.1093/ornithology/ukab011
74. Ballentine B (2006) Morphological adaptation influences the evolution of a mating signal. Evolution 60:1936–1944
75. Derryberry EP (2009) Ecology shapes birdsong evolution: variation in morphology and habitat explains variation in white-crowned sparrow song. Am Nat 174:24–33
76. Huber SK, Podos J (2006) Beak morphology and song features covary in a population of Darwin's finches (*Geospiza fortis*). Biol J Linn Soc 88:489–498
77. Porzio NS, Repenning M, Fontana CS (2019) Do beak volume and bite force influence the song structure of sympatric species of seedeaters (Thraupidae: Sporophila)? Emu-Austral Ornithol 119:71–78
78. Drummond-Clarke RC, Kivell TL, Sarringhaus L et al (2022) Wild chimpanzee behavior suggests that a savanna-mosaic habitat did not support the emergence of hominin terrestrial bipedalism. Sci Adv 8:eadd9752. https://doi.org/10.1126/sciadv.add9752
79. Venkataraman VV, Kraft TS, Dominy NJ (2012) Tree climbing and human evolution. In: Proceedings of the National Academy of Sciences
80. Thorpe S, Holder R, Crompton R (2007) Origin of human bipedalism as an adaptation for locomotion on flexible branches. Science 316:1328–1331. https://doi.org/10.1126/science.1140799
81. Crompton RH, Sellers WI, Thorpe SKS (2010) Arboreality, terrestriality and bipedalism. Philos Trans R Soc B Biol Sci 365:3301–3314. https://doi.org/10.1098/rstb.2010.0035
82. Lieberman D (2011) The evolution of the human head. Harvard University Press. https://www.hup.harvard.edu/books/9780674046368. Accessed 5 Nov 2024
83. Ekström AG, Edlund J (2023) Evolution of the human tongue and emergence of speech biomechanics. Front Psychol 14:1150778. https://doi.org/10.3389/fpsyg.2023.1150778
84. Marten K, Marler P (1977) Sound transmission and its significance for animal vocalization: I. Temp HABITATS. Behav Ecol Sociobiol 2:271–290
85. Nemeth E, Dabelsteen T, Pedersen SB, Winkler H (2006) Rainforests as concert halls for birds: are reverberations improving sound transmission of long song elements? J Acoust Soc Am 119:620–626
86. Hebets EA, Papaj DR (2005) Complex signal function: developing a framework of testable hypotheses. Behav Ecol Sociobiol 57:197–214
87. Bro-Jørgensen J (2010) Dynamics of multiple signalling systems: animal communication in a world in flux. Trends Ecol Evol 25:292–300
88. Mouterde SC, Theunissen FE, Elie JE et al (2014) Acoustic communication and sound degradation: how do the individual signatures of male and female zebra finch calls transmit over distance? PLoS One 9:102842
89. Wiley RH, Richards DG (1978) Physical constraints on acoustic communication in the atmosphere: implications for the evolution of animal vocalizations. Behav Ecol Sociobiol 3:69–94. https://doi.org/10.1007/BF00300047
90. Lameira AR, Maddison I, Zuberbühler K (2014) Primate feedstock for the evolution of consonants. Trends Cogn Sci 18:60–62

91. Ekström AG, Moran S, Sundberg J, Lameira A (2023) PREQUEL: supervised phonetic approaches to analyses of great ape quasi-vowels. In: Skarnitzl R, Volín J (eds) Proceedings of the 20th International Congress of Phonetic Sciences. Guarant International, pp 3076–3080
92. Ekström AG, Nellissen L, Bortalato T et al (2024) Phonetic properties of chimpanzee, gorilla, and orangutan hoots tell a uniform story and point to new frontiers. PsyArXiv. https://doi.org/10.31234/osf.io/cuxjv
93. Lameira AR, Alexandre A, Gamba M et al (2021) Orangutan information broadcast via consonant-like and vowel-like calls breaches mathematical models of linguistic evolution. Biol Lett 17:20210302. https://doi.org/10.1098/rsbl.2021.0302
94. Gannon C, Hill RA, Lameira AR (2023) Open plains are not a level playing field for hominid consonant-like versus vowel-like calls. Sci Rep 13:21138. https://doi.org/10.1038/s41598-023-48165-7
95. Lameira AR (2023) Arboreal origin of consonants and thus, ultimately, speech. Trends Cogn Sci 27:122–124. https://doi.org/10.1016/j.tics.2022.11.012
96. Urick RJ (1983) Principles of underwater sound. McGraw-Hill, New York
97. Mackenzie KV (1981) Nine-term equation for sound speed in the oceans. J Acoust Soc Am 70:807–812
98. Miksis-Olds JL, Miller JH (2006) Transmission loss in manatee habitats. J Acoust Soc Am 120:2320–2327
99. Desharnais F, Ellis DD (1997) Data-model comparisons of reverberation at three shallow-water sites. IEEE J Ocean Eng 22:309–316
100. Bass AH, Clark CW (2003) The physical acoustics of underwater sound communication. In: Acoustic communication. Springer, New York, pp 15–64
101. Mercado E, III, Frazer LN (1999) Environmental constraints on sound transmission by humpback whales. J Acoust Soc Am 106:3004–3016
102. Mercado E, III, Schneider JN, Green SR et al (2007) Acoustic cues available for ranging by humpback whales. J Acoust Soc Am 121:2499–2502
103. Mercado E, III (2016) Acoustic signaling by singing humpback whales (*Megaptera novaeangliae*): what role does reverberation play? PLoS One 11:0167277
104. Tellechea J (2020) Echolocation inter-click interval variation among specific behaviours in free-ranging bottlenose dolphins from the coast of Uruguay. J Cetacean Res Manag 21:141–149. https://doi.org/10.47536/jcrm.v21i1.192
105. Simard P, Hibbard AL, McCallister KA et al (2010) Depth dependent variation of the echolocation pulse rate of bottlenose dolphins (*Tursiops truncatus*). J Acoust Soc Am 127:568–578. https://doi.org/10.1121/1.3257202
106. Marques CS, Marques DA, Blackwell SB et al (2024) Narwhal (*Monodon monoceros*) echolocation click rates to support cue counting passive acoustic density estimation. J Acoust Soc Am 155:891–900. https://doi.org/10.1121/10.0024723
107. Aubauer R, Lammers MO, Au WWL (2000) One-hydrophone method of estimating distance and depth of phonating dolphins in shallow water. J Acoust Soc Am 107:2744–2749. https://doi.org/10.1121/1.428660
108. Ladich F (1999) Did auditory sensitivity and vocalization evolve independently in otophysan fishes? Brain Behav Evol 53:288–304. https://doi.org/10.1159/000006600
109. Amorim MCP, Simões JM, Almada VC, Fonseca PJ (2011) Stereotypy and variation of the mating call in the Lusitanian toadfish, *Halobatrachus didactylus*. Behav Ecol Sociobiol 65:707–716
110. Reichard DG, Anderson RC (2015) Why signal softly? The structure, function and evolutionary significance of low-amplitude signals. Anim Behav 105:253–265
111. Nevo E (2022) Evolution of communication systems underground in a blind mammal, *Spalax*. In: Hill PSM, Mazzoni V, Stritih-Peljhan N, Virant-Doberlet M, Wessel A (eds) Biotremology: physiology, ecology, and evolution. Animal signals and communication, vol 8. Springer, Cham

112. Narins PM, Reichman OJ, Jarvis JUM (1992) Seismic signal transmission between burrows of the cape mole-rat, *Georychus capensis*. J Comp Physiol A 170:13–21. https://doi.org/10.1007/BF00190397
113. Randall J, Lewis E (1997) Seismic communication between the burrows of kangaroo rats, *Dipodomys spectabilis*. J Comp Physiol A 181:525–531. https://doi.org/10.1007/s003590050136
114. Elias DO, Mason AC (2014) The role of wave and substrate heterogeneity in vibratory communication: practical issues in studying the effect of vibratory environments in communication. In: Studying vibrational communication, vol 3. Springer Nature. https://doi.org/10.1007/978-3-662-43607-3_12
115. Narins PM, Stoeger AS, O'Connell-Rodwell C (2016) Infrasonic and seismic communication in the vertebrates with special emphasis on the Afrotheria: an update and future directions. In: Suthers R, Fitch W, Fay R, Popper A (eds) Vertebrate sound production and acoustic communication. Springer handbook of auditory research. Springer, Cham
116. Díaz J, Ruiz M, Sánchez-Pastor PS (2017) Urban seismology: on the origin of earth vibrations within a city. Sci Rep 7:15296. https://doi.org/10.1038/s41598-017-15499-y
117. Tepp G, Stubailo I, Kohler M et al (2024) Shake to the beat: exploring the seismic signals and stadium response of concerts and music fans. Seismol Res Lett 95:2179–2194. https://doi.org/10.1785/0220230385
118. Bennett NC, Faulkes CG (2000) African mole-rats: ecology and eusociality. Cambridge University Press
119. Sahd L, Doubell N, Bennett NC, Kotzé SH (2022) Hind foot drumming: myosin heavy chain muscle fiber distribution in the hind limb muscles of three African mole-rat species (Bathyergidae). Anat Rec 305:170–183
120. Putland RL, Merchant ND, Farcas A, Radford CA (2018) Vessel noise cuts down communication space for vocalizing fish and marine mammals. Glob Chang Biol 24:1708–1721
121. Hartbauer M, Siegert ME, Fertschai I, Römer H (2012) Acoustic signal perception in a noisy habitat: lessons from riangulumng insects. J Comp Physiol A 198:397–409
122. Tuttle MD, Ryan MJ (1982) The role of synchronized calling, ambient light, and ambient noise, in anti-bat-predator behavior of a treefrog. Behav Ecol Sociobiol 11:125–131. https://doi.org/10.1007/BF00300101
123. Wiley RH (2017) How noise determines the evolution of communication. Anim Behav 124:307–313
124. Waser PM, Waser MS (1977) Experimental studies of primate vocalization: specializations for long-distance propagation. Z Für Tierpsychol 43:239–263
125. Ryan MJ, Brenowitz EA (1985) The role of body size, phylogeny, and ambient noise in the evolution of bird song. Am Nat 126:87–100
126. Lugli M, Yan HY, Fine ML (2003) Acoustic communication in two freshwater gobies: the relationship between ambient noise, hearing thresholds and sound spectrum. J Comp Physiol A 189:309–320
127. Morley EL, Jones G, Radford AN (2014) The importance of invertebrates when considering the impacts of anthropogenic noise. Proc R Soc B Biol Sci 281:20132683
128. Römer H, Bailey W (1998) Strategies for hearing in noise: peripheral control over auditory sensitivity in the bushcricket *Sciarasaga quadrata* (Austrosaginae: Tettigoniidae). J Exp Biol 201:1023–1033
129. Slater J, Kraus N (2016) The role of rhythm in perceiving speech in noise: a comparison of percussionists, vocalists and non-musicians. Cogn Process 17:79–87
130. Akçay Ç, Beecher MD (2019) Multi-modal communication: song sparrows increase signal redundancy in noise. Biol Lett 15:20190513
131. Hödl W, Amézquita A (2001) Visual signaling in anuran amphibians. In: Ryan MJ (ed) Anuran communication. Smithsonian Institution Press, Washington, DC, pp 121–141
132. Brumm H, Slabbekoorn H (2005) Acoustic communication in noise. Adv Study Behav 35:151–209

133. Weilgart LS (2007) A brief review of known effects of noise on marine mammals. Int J Comp Psychol 20:159–168
134. Brumm H, Zollinger SA (2013) Avian vocal production in noise. In: Brumm H (ed) Animal communication and noise. Springer-Verlag, Berlin Heidelberg, pp 187–227
135. Shannon G, McKenna MF, Angeloni LM et al (2016) A synthesis of two decades of research documenting the effects of noise on wildlife. Biol Rev 91:982–1005
136. Duquette CA, Loss SR, Hovick TJ (2021) A meta-analysis of the influence of anthropogenic noise on terrestrial wildlife communication strategies. J Appl Ecol 58:1112–1121
137. Picciulin M, Sebastianutto L, Codarin A et al (2012) Brown meagre vocalization rate increases during repetitive boat noise exposures: a possible case of vocal compensation. J Acoust Soc Am 132:3118–3124. https://doi.org/10.1121/1.4756928
138. Kaiser K, Hammers J (2009) The effect of anthropogenic noise on male advertisement call rate in the neotropical treefrog, *Dendropsophus riangulum*. Behaviour 146:1053–1069
139. Díaz M, Parra A, Gallardo C (2011) Serins respond to anthropogenic noise by increasing vocal activity. Behav Ecol 22:332–336
140. Cunnington GM, Fahrig L (2010) Plasticity in the vocalizations of anurans in response to traffic noise. Acta Oecol 36:463–470
141. Brumm H, Slater PJ (2006) Ambient noise, motor fatigue, and serial redundancy in chaffinch song. Behav Ecol Sociobiol 60:475–481
142. Deoniziak K, Osiejuk TS (2016) Disentangling relations among repertoire size, song rate, signal redundancy and ambient noise level in European songbird. Ethology 122:734–744
143. Ripmeester EA, Kok JS, Rijssel JC, Slabbekoorn H (2010) Habitat-related birdsong divergence: a multi-level study on the influence of territory density and ambient noise in European blackbirds. Behav Ecol Sociobiol 64:409–418
144. Luther D, Gentry K (2013) Sources of background noise and their influence on vertebrate acoustic communication. Behaviour 150:1045–1068
145. Bastos RP, Alcantara MB, Morais AR et al (2011) Vocal behaviour and conspecific call response in *Scinax centralis*. Herpetol J 21:43–50
146. Yoon J, Sillett TS, Morrison SA, Ghalambor CK (2012) Breeding density, not life history, predicts interpopulation differences in territorial aggression in a passerine bird. Anim Behav 84:515–521
147. Barrero A, Gómez-Catasús J, Pérez-Granados C et al (2023) Conspecific density and habitat quality drive the defence and vocal behaviour of a territorial passerine. Ibis 166:1–18
148. Amichai E, Blumrosen G, Yovel Y (2015) Calling louder and longer: how bats use biosonar under severe acoustic interference from other bats. Proc R Soc B Biol Sci 282:20152064. https://doi.org/10.1098/rspb.2015.2064
149. Kloepper LN, Branstetter BK (2019) The effect of jamming stimuli on the echolocation behavior of the bottlenose dolphin, *Tursiops truncatus*. J Acoust Soc Am 145:1341–1352. https://doi.org/10.1121/1.5093636
150. Heiligenberg W (1974) Electrolocation and jamming avoidance in a *Hypopygus* (Rhamphichthyidae, Gymnotoidei), an electric fish with pulse-type discharges. J Comp Physiol 91:223–240. https://doi.org/10.1007/BF00698054
151. Greenfield MD, Marin-Cudraz T, Party V (2017) Evolution of synchronies in insect choruses. Biol J Linn Soc 122:487–504
152. Geissmann T, Orgeldinger M (2000) The relationship between duet songs and pair bonds in siamangs, *Hylobates syndactylus*. Anim Behav 60:805–809
153. Hall ML (2009) A review of vocal duetting in birds. Adv Study Behav 40:67–121
154. Greenfield MD, Merker B (2023) Coordinated rhythms in animal species, including humans: entrainment from bushcricket chorusing to the philharmonic orchestra. Neurosci Biobehav Rev 153:105382
155. Schneider JN, Mercado E, III (2019) Characterizing the rhythm and tempo of sound production by singing whales. Bioacoustics 28:239–256
156. Ravignani A, Bowling DL, Fitch W (2014) Chorusing, synchrony, and the evolutionary functions of rhythm. Front Psychol 5:104182

157. Greenfield MD, Rand AS (2000) Frogs have rules: selective attention algorithms regulate chorusing in *Physalaemus pustulosus* (Leptodactylidae). Ethology 106:331–347
158. Greenfield MD, Siegfreid E, Snedden WA (2004) Variation and repeatability of female choice in a chorusing katydid, *Ephippiger ephippiger*: an experimental exploration of the precedence effect. Ethology 110:287–299
159. Buck J (1988) Synchronous rhythmic flashing of fireflies. II Q Rev Biol 63:265–289. https://doi.org/10.1086/415929
160. Frankel AS, Clark CW, Herman L, Gabriele CM (1995) Spatial distribution, habitat utilization, and social interactions of humpback whales, *Megaptera novaeangliae*, off Hawai'i, determined using acoustic and visual techniques. Can J Zool 73:1134–1146
161. Cholewiak DM, Cerchio S, Jacobsen JK et al (2018) Songbird dynamics under the sea: acoustic interactions between humpback whales suggest song mediates male interactions. R Soc Open Sci 5:171298
162. Charrier I (2020) Mother–offspring vocal recognition and social system in Pinnipeds. In: Aubin T, Mathevon N (eds) Coding strategies in vertebrate acoustic communication. Animal signals and communication. Springer, Cham
163. Charrier I (2021) Vocal communication in otariids and odobenids. In: Ethology and behavioral ecology of otariids and the odobenid. Springer, Cham, pp 265–289. https://doi.org/10.1007/978-3-030-59184-7_14
164. Osiecka AN, Gridley T, Fearey J (2022) Temporal patterns in cape fur seal (*Arctocephalus pusillus pusillus*) mother and pup attraction calls. Belg J Zool 152. https://doi.org/10.26496/bjz.2022.103
165. Krause B (1987) The niche hypothesis: how animals taught us to dance and sing. Whole Earth Rev 57:14–16
166. Grether GF, Losin N, Anderson CN, Okamoto K (2009) The role of interspecific interference competition in character displacement and the evolution of competitor recognition. Biol Rev 84:617–635
167. Staniewicz A, Sokołowska E, Muszyńska A, Budka M (2023) Competition for acoustic space in a temperate-forest bird community. Behav Ecol 34:1043–1054
168. Lima MSCS, Pederassi J, Pineschi RB, Barbosa DBS (2018) Acoustic niche partitioning in an anuran community from the municipality of Floriano, Piauí, Brazil. Braz J Biol 79:566–576
169. Ancillotto L, Labadessa R (2023) A noisy neighbour: acoustic competition drives changes in song traits and spatiotemporal behaviour of orthopterans. Anim Behav 201:117–124
170. Prasad VK, Chuang MF, Das A et al (2022) Coexisting good neighbours: acoustic and calling microhabitat niche partitioning in two elusive syntopic species of balloon frogs, *Uperodon systoma* and *U. Globulosus* (Anura: Microhylidae) and potential of individual vocal signatures. BMC Zool 7:27
171. Weimann SR, Black AN, Leese J et al (2018) Territorial vocalization in sympatric damselfish: acoustic characteristics and intruder discrimination. Bioacoustics 27:87–102
172. Hart PJ, Paxton K, Ibanez T et al (2020) Acoustic niche partitioning in two tropical wet forest bird communities. BioRxiv. https://doi.org/10.1101/2020.08.17.253674
173. Le Roux A, Jackson TP, Cherry MI (2002) Differences in alarm vocalizations of sympatric populations of the whistling rats, *Parotomys brantsii* and *P. littledalei* (Rodentia: Muridae). J Zool 257:189–194
174. Seddon N, Tobias JA (2010) Character displacement from the receiver's perspective: species and mate recognition despite convergent signals in suboscine birds. Proc R Soc B Biol Sci 277:2475–2483
175. Tobias JA, Seddon N (2009) Signal design and perception in *Hypocnemis* antbirds: evidence for convergent evolution via social selection. Evolution 63:3168–3189
176. Love J, Goller F (2021) Processes underlying complex patterns of song trait evolution in a *Setophaga* hybrid zone. Ecol Evol 11:7264–7277
177. Garcia M, Theunissen F, Sèbe F et al (2020) Evolution of communication signals and information during species radiation. Nat Commun 11:4970

178. Terhune JM (2019) The underwater vocal complexity of seals (Phocidae) is not related to their phylogeny. Can J Zool 97:232–240. https://doi.org/10.1139/cjz-2018-0190
179. Moors HB, Terhune JM (2003) Repetition patterns within harp seal (*Pagophilus groenlandicus*) underwater calls. Aquat Mamm 29:278–288. https://doi.org/10.1578/016754203101024211
180. Moors HB, Terhune JM (2004) Repetition patterns in Weddell seal (*Leptonychotes weddellii*) underwater multiple element calls. J Acoust Soc Am 116:1261–1270. https://doi.org/10.1121/1.1763956
181. Rogers T (2005) Influences of female pupping habitat and maternal care on the vocal repertoire size of male phocid seals. Aquat Mamm 31:96–103
182. Rogers TL (2017) Calling underwater is a costly signal: size-related differences in the call rates of Antarctic leopard seals. Curr Zool 63:433–443
183. Schusterman RJ (1977) Temporal patterning in sea lion barking (*Zalophus californianus*). Behav Biol 20:404–408. https://doi.org/10.1016/S0091-6773(77)90964-6
184. Martin M, Gridley T, Elwen SH, Charrier I (2021) Feel the beat: cape fur seal males encode their arousal state in their bark rate. Naturwissenschaften 109:5. https://doi.org/10.1007/s00114-021-01778-2
185. Casey C, Charrier I, Mathevon N, Reichmuth C (2015) Rival assessment among northern elephant seals: evidence of associative learning during male–male contests. R Soc Open Sci 2:150228. https://doi.org/10.1098/rsos.150228
186. Mathevon N, Casey C, Reichmuth C, Charrier I (2017) Northern elephant seals memorize the rhythm and timbre of their rivals' voices. Curr Biol 27:2352–2356.e2. https://doi.org/10.1016/j.cub.2017.06.035
187. Zuberbühler K (2008) Audience effects. Curr Biol 18:R189–R190. https://doi.org/10.1016/j.cub.2007.12.041
188. Jones TK, Allen KM, Moss CF (2021) Communication with self, friends and foes in active-sensing animals. J Exp Biol 224:jeb242637. https://doi.org/10.1242/jeb.242637
189. Tobias JA, Seddon N (2009) Signal jamming mediates sexual conflict in a duetting bird. Curr Biol 19:577–582. https://doi.org/10.1016/j.cub.2009.02.036
190. Peake TM (2005) Eavesdropping in Communication networks. In: McGregor PK (ed) Animal communication networks. Cambridge University Press, Cambridge, pp 13–37
191. Villarreal SM, Gilbert C (2014) Male *Scudderia pistillata* katydids defend their acoustic duet against eavesdroppers. Behav Ecol Sociobiol 68:1669–1675
192. Litovsky RY, Colburn HS, Yost WA, Guzman SJ (1999) The precedence effect. J Acoust Soc Am 106:1633–1654
193. Lee N, Elias DO, Mason AC (2009) A precedence effect resolves phantom sound source illusions in the parasitoid fly *Ormia ochracea*. Proc Natl Acad Sci 106:6357–6362
194. Bosch J, Márquez R (2002) Female preference function related to precedence effect in an amphibian anuran (*Alytes cisternasii*): tests with non-overlapping calls. Behav Ecol 13:149–153
195. Höbel G (2010) Interaction between signal timing and signal feature preferences: causes and implications for sexual selection. Anim Behav 79:1257–1266. https://doi.org/10.1016/j.anbehav.2010.02.026
196. Buck J, Buck E (1978) Toward a functional interpretation of synchronous flashing by fireflies. Am Nat 112:471–492
197. Legett HD, Aihara I, Bernal XE (2021) The dual benefits of synchronized mating signals in a Japanese treefrog: attracting mates and manipulating predators. Philos Trans R Soc B 376:20200340
198. Legett HD, Page RA, Bernal XE (2019) Synchronized mating signals in a communication network: the challenge of avoiding predators while attracting mates. Proc R Soc B 286:20191067
199. Davranoglou LR, Taylor GK, Mortimer B (2023) Sexual selection and predation drive the repeated evolution of stridulation in Heteroptera and other arthropods. Biol Rev 98:942–981
200. Barber JR, Plotkin D, Rubin JJ et al (2022) Anti-bat ultrasound production in moths is globally and phylogenetically widespread. Proc Natl Acad Sci 119:2117485119

201. O'Reilly LJ, Agassiz DJ, Neil TR, Holderied MW (2019) Deaf moths employ acoustic Müllerian mimicry against bats using wingbeat-powered tymbals. Sci Rep 9:1444
202. Kowalski KN, Lakes-Harlan R, Lehmann GU, Strauß J (2014) Acoustic defence in an insect: characteristics of defensive stridulation and differences between the sexes in the tettigoniid *Poecilimon ornatus* (Schmidt 1850). Zoology 117:329–336
203. Hoppitt W, Lala KN (2013) Social learning: an introduction to mechanisms, methods, and models. Princeton University Press
204. Whiten A, van Schaik CP (2007) The evolution of animal 'cultures' and social intelligence. Philos Trans R Soc B Biol Sci 362:603–620. https://doi.org/10.1098/rstb.2006.1998
205. Danchin E, Nöbel S, Pocheville A et al (2018) Cultural flies: conformist social learning in fruitflies predicts long-lasting mate-choice traditions. Science 362:1025–1030. https://doi.org/10.1126/science.aat1590
206. Whiten A (2005) The second inheritance system of chimpanzees and humans. Nature 437:52–55. https://doi.org/10.1038/nature04023
207. Coupé C, Oh YM, Dediu D, Pellegrino F (2019) Different languages, similar encoding efficiency: comparable information rates across the human communicative niche. Sci Adv 5:eaaw2594. https://doi.org/10.1126/sciadv.aaw2594
208. Mehr SA, Singh M, Knox D et al (2019) Universality and diversity in human song. Science 366:eaax0868. https://doi.org/10.1126/science.aax0868
209. Lee H, Calvin K, Dasgupta D et al (2023) IPCC, 2023: climate change 2023: synthesis report, summary for policymakers. Contribution of Working Groups I, II and III to the sixth assessment report of the Intergovernmental Panel on Climate Change [Core Writing Team, Lee H, Romero J (eds)]. IPCC, Geneva. https://doi.org/10.59327/IPCC/AR6-9789291691647.001. Accessed 30 Oct 2024
210. Sanders D, Frago E, Kehoe R et al (2021) A meta-analysis of biological impacts of artificial light at night. Nat Ecol Evol 5:74–81. https://doi.org/10.1038/s41559-020-01322-x
211. Dominoni DM, Borniger JC, Nelson RJ (2016) Light at night, clocks and health: from humans to wild organisms. Biol Lett 12:20160015. https://doi.org/10.1098/rsbl.2016.0015
212. Walker WH, Borniger JC, Gaudier-Diaz MM et al (2020) Acute exposure to low-level light at night is sufficient to induce neurological changes and depressive-like behavior. Mol Psychiatry 25:1080–1093. https://doi.org/10.1038/s41380-019-0430-4
213. Münzel T, Hahad O, Daiber A (2021) The dark side of nocturnal light pollution. Outdoor light at night increases risk of coronary heart disease. Eur Heart J 42:831–834. https://doi.org/10.1093/eurheartj/ehaa866
214. Karska J, Kowalski S, Gładka A et al (2024) Artificial light and neurodegeneration: does light pollution impact the development of Alzheimer's disease? GeroScience 46:87–97. https://doi.org/10.1007/s11357-023-00932-0
215. Fonken LK, Nelson RJ (2016) Effects of light exposure at night during development. Curr Opin Behav Sci 7:33–39. https://doi.org/10.1016/j.cobeha.2015.10.008
216. Falchi F, Cinzano P, Duriscoe D et al (2016) The new world atlas of artificial night sky brightness. Sci Adv 2:1600377. https://doi.org/10.1126/sciadv.1600377
217. Da Silva A, Valcu M, Kempenaers B (2016) Behavioural plasticity in the onset of dawn song under intermittent experimental night lighting. Anim Behav 117:155–165. https://doi.org/10.1016/j.anbehav.2016.05.001
218. Dias KS, Dosso ES, Hall AS et al (2019) Ecological light pollution affects anuran calling season, daily calling period, and sensitivity to light in natural Brazilian wetlands. Naturwissenschaften 106:46. https://doi.org/10.1007/s00114-019-1640-y
219. Da Silva A, Samplonius JM, Schlicht E, Valcu M, Kempenaers B (2014) Artificial night lighting rather than traffic noise affects the daily timing of dawn and dusk singing in common European songbirds. Behav Ecol 25:1037–1047. https://doi.org/10.1093/beheco/aru103
220. Da Silva A, Kempenaers B (2017) Singing from north to south: latitudinal variation in timing of dawn singing under natural and artificial light conditions. J Anim Ecol 86:1286–1297. https://doi.org/10.1111/1365-2656.12739

221. Hall AS (2016) Acute artificial light diminishes Central Texas anuran calling behavior. Am Midl Nat 175:183–193
222. Narins PM, Meenderink SWF (2014) Climate change and frog calls: long-term correlations along a tropical altitudinal gradient. Proc R Soc B Biol Sci 281:20140401
223. Slabbekoorn H, Bouten N, Opzeeland I et al (2010) A noisy spring: the impact of globally rising underwater sound levels on fish. Trends Ecol Evol 25:419–427
224. Francis CD, Barber JR (2013) A framework for understanding noise impacts on wildlife: an urgent conservation priority. Front Ecol Environ 11:305–313
225. Gomes L, Solé M, Sousa-Lima RS, Baumgarten JE (2022) Influence of anthropogenic sounds on insect, anuran and bird acoustic signals: a meta-analysis. Front Ecol Evol 10. https://doi.org/10.3389/fevo.2022.827440
226. Hopkins JM, Edwards W, Schwarzkopf L (2022) Invading the soundscape: exploring the effects of invasive species' calls on acoustic signals of native wildlife. Biol Invasions 24:3381–3393. https://doi.org/10.1007/s10530-022-02856-w
227. Bleach IT, Beckmann C, Both C et al (2015) Noisy neighbours at the frog pond: effects of invasive cane toads on the calling behaviour of native Australian frogs. Behav Ecol Sociobiol 69:675–683. https://doi.org/10.1007/s00265-015-1879-z
228. Mercado E, III (2018) The sonar model for humpback whale song revised. Front Psychol 9:1156. https://doi.org/10.3389/fpsyg.2018.01156
229. Podos J (2010) Acoustic discrimination of sympatric morphs in Darwin's finches: a behavioural mechanism for assortative mating? Philos Trans R Soc B Biol Sci 365:1031–1039. https://doi.org/10.1098/rstb.2009.0289
230. Dingle C, Halfwerk W, Slabbekoorn H (2008) Habitat-dependent song divergence at subspecies level in the grey-breasted wood-wren. J Evol Biol 21:1079–1089. https://doi.org/10.1111/j.1420-9101.2008.01536.x
231. Lachlan RF, Servedio MR (2004) Song learning accelerates allopatric speciation. Evolution 58:2049–2063. https://doi.org/10.1111/j.0014-3820.2004.tb00489.x
232. Mason NA, Burns KJ, Tobias JA et al (2017) Song evolution, speciation, and vocal learning in passerine birds. Evolution 71:786–796. https://doi.org/10.1111/evo.13159

Converging Notions on the Main Hypotheses Regarding the Origins of Beat Perception and Synchronization: The Key Role of Audiomotor Interactions

7

Shannon Proksch, Henkjan Honing, Andrea Ravignani, and Hugo Merchant

Abstract

Beat perception and synchronization (BPS) describes a suite of cognitive and physical abilities that allow humans and some nonhuman animals to perceive and move to a rhythmic auditory stimulus. This chapter reviews two hypotheses that seek to explain the emergence of BPS in humans and a select few other animals (primarily nonhuman primates): the vocal learning hypothesis (VLH) and the gradual audiomotor evolution (GAE) hypothesis. VLH argues that vocal learning is a prerequisite to complex BPS, while GAE argues that BPS abilities may have been built upon extant interval timing mechanisms. Both hypotheses highlight the importance of sensorimotor networks in the brain, including dorsal pathways

S. Proksch (✉)
Department of Psychology, Augustana University, Sioux Falls, SD, USA

Program in Neuroscience, Center for Interdisciplinary Studies, Augustana University, Sioux Falls, SD, USA
e-mail: Shannon.Proksch@augie.edu

H. Honing
Music Cognition Group, Institute for Logic, Language and Computation, University of Amsterdam, Amsterdam, The Netherlands
e-mail: H.J.Honing@uva.nl

A. Ravignani
Department of Human Neurosciences, Sapienza University of Rome, Rome, Italy

Center for Music in the Brain, Department of Clinical Medicine, Aarhus University, Aarhus, Denmark

Istituto di Scienze e Tecnologie della Cognizione, CNR, Rome, Italy
e-mail: andrea.ravignani@uniroma1.it

H. Merchant
Department of Neurophysiology, Instituto de Neurobiología, Universidad Nacional Autónoma de México, Campus Juriquilla, Santiago de Querétaro, Mexico
e-mail: hugomerchant@unam.mx

G.-A. Sauciuc et al. (eds.), *Nature Beats*, Fascinating Life Sciences,
https://doi.org/10.1007/978-3-032-24250-1_7

and basal-ganglia-thalamocortical loops, but they differ in the importance of vocal learning abilities in developing these pathways. Recent revisions to each hypothesis have been made to address an apparent continuum of BPS abilities across both vocal and nonvocal learning species. We suggest that there may be convergence between these revised hypotheses toward a more complete story of the emergence of rhythm production and perception.

7.1 Introduction

The ability to generate rhythmic motor output relies on strong coupling between motor and sensory processes. The motor systems of primates, other mammals, and birds are particularly well suited for managing the rhythmic movements of manual gestures, complex reach and grasp behaviors, gait patterns, wingbeats, and the actions behind carefully timed and coordinated vocal calls. Primates have a sophisticated cortical premotor system engaged in controlling a large repertoire of voluntary movements of the proximal appendages, the neck, and the vocal musculature [1–3]. The specific sensory modality coupled to rhythmic motor processes among rhythmic species varies based on the relative salience and evolutionary functions/ecological importance of that modality. Nevertheless, motor processes play a particularly important role in the emergence of rhythmic output (production) and perception. The importance of motor processes arises from the particular strength of sensory-motor and motor-motor connections both in the brain and through coupled rhythmic activity in sensorimotor/motor-motor behavior. At its most basic, rhythm production and perception involve the generation and tracking of temporal regularities within a specific modality. In music, the generation of temporal regularities in sound is realized through the patterning of beats into larger rhythmic patterns; temporal regularities in speech are organized through the patterning of phonemes into the rhythmic patterns of sentences. In nature, examples of temporal regularities are observed through rhythmic timing of pinniped calls, bird songs, some aspects of nonhuman primate drumming behavior, the beating of bat wings, and the timing of gait patterns (for details and examples, see Chaps. 2 [4], 3 [5], and 4 [6]). Neural processes that serve sensorimotor control are intricately linked with the neural processes that enable intricate timing perception [7]. Rhythms of locomotion, such as walking gait or wing beats, are coupled with respiratory rhythms (motor-motor couplings, see Chap. 9 [8]) and rhythmic production of sounds is generated by rhythmic motor-effector processes, which in turn stimulate rhythmic auditory sensations (sensorimotor couplings).

Beyond the incidental coupling of motor and sensory process, the process of sensorimotor synchronization (SMS) occurs when an individual synchronizes their movement with an external perceptual signal [9–11]. In humans, SMS is strongest when synchronizing to an auditory rhythm [12], although humans are also capable of synchronizing to visual or tactile rhythms to a lesser and a similar degree, respectively [13]. SMS ability in nonhuman primates, such as the Rhesus macaque (*Macaca mulatta*), does not show a dominance for the auditory domain [12, 14]. However, when trained properly Rhesus monkeys perform tempo flexible and predictive SMS with auditory metronomes [14, 15]. SMS abilities in chimpanzees (*Pan troglodytes*) are still subject of debate (but see Eleuteri et al. [16]). Humans and a select few other species are further capable of *musical* beat perception and synchronization (BPS)—which includes the ability to predictively, constructively, and flexibly synchronize to hierarchically patterned periodic rhythms within complex sound sequences across a range of tempi [17]. While human evidence is mechanistic and neural, the few animal cases of BPS are mostly based on behavioral data and include both vocal learning animals— Snowball, the Sulfur-Crested Cockatoo (*Cacatua galerita eleonora*) [18, 19]—and nonvocal learning animals—Ronan, the California Sea Lion (*Zalophus californianus*) [20–22] and a pair of Rhesus macaques (c.f. [23])—that demonstrate BPS ability and can flexibly synchronize to music.

Two hypotheses have received significant attention in their efforts to explain the emergence of rhythmic output and audiomotor coordination, and why a subset of these abilities (namely, SMS and BPS) is present in humans and a select few other animals, but are weak or nonexistent in other species. These are the vocal learning hypothesis (VLH) and the gradual audiomotor evolution (GAE) Hypothesis. VLH argues that vocal learning is a prerequisite to complex BPS, while GAE argues that beat perception and synchronization abilities may have been built upon extant interval timing mechanisms. Both hypotheses highlight the importance of sensorimotor networks in the brain, including dorsal pathways and basal-ganglia-thalamocortical loops [24, 25], but they differ in the importance of vocal learning abilities in developing these pathways. Recently, there have been substantial revisions to each hypothesis to address an apparent continuum of BPS abilities across both vocal and nonvocal learning species [23, 26]. In this chapter, we provide background on the neural and behavioral mechanisms underlying SMS and BPS. We then review the different commitments of VLH and GAE and the recent revisions to each hypothesis. We suggest that there may be convergence between these revised hypotheses toward a more complete story of the emergence of rhythm production and perception.

7.2 Neural and Behavioral Mechanisms Underlying SMS and BPS

7.2.1 Synchronizing Movements to an External Stimulus

7.2.1.1 Sensorimotor Synchronization

What is it? Sensorimotor synchronization refers to the intentional process of coordinating a rhythmic movement to an external perceptual signal [9, 10]. Behaviorally, SMS is frequently studied in finger tapping paradigms, where a participant is asked to tap in response to a rhythmic auditory stimulus, such as a metronome. Human SMS is characterized by negative mean asynchronies in finger tapping, that is, the tap precedes the onset of the auditory tone, indicating the prediction or anticipation of upcoming rhythmic stimuli. Human SMS is not static, however, as humans tend to exhibit positive asynchronies for shorter intervals and negative asynchronies for longer intervals. SMS is also characterized by intentional error correction when the tapping and external stimuli do not align [27, 28], and error correction has been observed in monkeys during isochronous tapping synchronization [14, 29]. Variability in intertap intervals during SMS tasks generally correlates with skill and is lower in individuals with more musical training. Similar sensorimotor synchronization paradigms have been implemented in nonhuman primate research. When the reward contingencies emphasize small asynchronies, the monkeys can flexibly and predictably synchronize their tapping to auditory and visual metronomes [15]. In addition, in saccadic instead of tapping tasks, recent research has indicated that macaques also show both predictive timing as well as tempo-flexible synchronization in response to periodic *visual* stimuli and saccadic movements with appropriate training [30].

In humans, stronger SMS is observed for auditory as opposed to visual rhythmic stimuli. That is, humans have an auditory advantage in SMS. Nonhuman primates, such as the macaque and chimpanzee, are able to more strongly couple to visual rhythms than auditory rhythms (though see Sect. 7.5.2 for a more thorough description of SMS variability for monkeys). The only experiment in pinnipeds has shown that a California sea lion can be trained to synchronize its head bobbing to both an auditory metronome and to complex musical stimuli [20–22].

7.2.2 Extracting a Beat from Complex Stimuli

Human musicality involves more than merely synchronizing movements to an external metronome. Within the rhythmic domain, our auditory and motor systems extract a regular pulse from a complex, multilayered, and timbrically diverse auditory stream.

7.2.2.1 Beat Perception and Synchronization

What is it? BPS refers to the sensorimotor and cognitive functions that enable an individual to synchronize their movements with their acoustic environment and is more specific than sensorimotor synchronization [31]. BPS abilities include that rhythmic timing movements are predictive, tempo flexible, constructive, hierarchical, modality biased, and engage the motor system ([17] summarized in Box 7.1). Original descriptions of BPS characterize it as a unique feature of human music [31]. Although foundational to human musicality, some building blocks of BPS abilities have been identified in nonhuman animal species. These include vocal learning species, such as birds [18, 32], pinnipeds [20], and nonhuman primates [33, 34].

Box 7.1 Principles of beat perception and synchronization
(Summarized from Patel and Iversen [17])

Predictive—when tapping to an auditory beat, humans (and animals with BPS ability) will often exhibit negative-mean asynchronies, tapping slightly earlier than the arrival of a beat.

Tempo flexible—an individual is able to entrain their movements to a range of tempi within the preferred tempo range for that species.

Constructive—entrainment to a beat is determined by top-down attentional mechanisms, and not merely bottom-up synchronization to the physical stimulus. This allows us to hear a downbeat in a syncopated rhythm.

Hierarchical—beats can be arranged in patterns that consist of combinations smaller beats, and shaped into stronger and weaker beats. Different metrical patterns can be perceived by tapping or attending to certain beats within a phrase (i.e., duple vs. triple meter).

Modality-biased—humans are best able to entrain to and perceive an auditory beat. In contrast, primates are best able to entrain to visual beats.

Engages the motor system—perceiving a beat in the absence of overt movement engages neural resources throughout the motor system, and, in humans, is associated with enhanced connectivity between motor and auditory regions of the brain

Where simple sensorimotor synchronization addresses how an organism synchronizes their actions to the patterns of a physical stimulus, beat perception involves a *psychological* perception of a beat which may or may not be present in the stimulus itself [35, 36]. In humans, the psychological phenomenon of "feeling the beat" consistently recruits basal ganglia and cerebellar activity [37, 38]. BPS is also associated with stronger coupling between auditory and motor regions [39]. Specifically, the activation of motor planning regions (e.g., supplementary motor area (SMA); primary motor cortex (PMC)) is thought to provide an internal predictive model of a (musical) beat that aids the auditory prediction of upcoming beats (a concept expanded upon in more detail in Sect. 7.3.1.1). Communication between basal ganglia structures, thalamus, and auditory-motor cortical structures further facilitates beat perception including the ability to synchronize to a beat (expanded upon in more detail in Sect. 7.4).

7.3 Is Vocal Learning Necessary for BPS? Views from VLH/ASAP and GAE Hypotheses

There are two competing hypotheses for the emergence of rhythmic output and audiomotor coordination. Both claim that strong sensorimotor coupling (between auditory and motor regions) is necessary for the emergence of auditory beat perception and synchronization. However, they differ as to the role of vocal learning in the emergence of BPS, and the extent to which BPS may be observable across species. These hypotheses are the vocal learning hypothesis and the gradual audio-motor evolution hypothesis, emphasizing the role of dorsal pathways and the motor cortico-basal-ganglia-thalamocortical loop, respectively [40]. Nevertheless, in the case of vocal learner birds, the audiomotor circuit is quite different from the primate audiomotor system, emphasizing convergent evolution for BPS. Since their original proposals, there have been significant updates to each hypothesis.

Below, we will explain the general problem space for each hypothesis, namely, what component of BPS is explained by each hypothesis, and which neural structures are implicated. We will also explain recent revisions to each hypothesis. In Sect. 7.6, we will highlight where these two general approaches to BPS converge or are still in disagreement. We will end with a look toward how the pathways implicated in each hypothesis may have been jointly implicated in the emergence of rhythm production in the hominin line due to their roles in increasingly hierarchical action planning.

7.3.1 Vocal Learning and Rhythmic Synchronization Hypothesis

The vocal learning and rhythmic synchronization hypothesis (frequently referred to as just the vocal learning hypothesis, or VLH) explains the capacity for BPS as it is related to vocal learning and was put forth by Patel [31] as a potential necessary evolutionary precursor for beat perception and synchronization. According to VLH, the strong auditory-motor connections observed in BPS were likely co-opted from strong auditory-motor connections needed to learn complex, rhythmic vocal calls from conspecifics. Further evidence for this hypothesis was found in early research in nonvocal learning primates [24]. In this early work, Rhesus macaques showed difficulty synchronizing taps to auditory metronomes: reacting rather than anticipating an auditory beat and demonstrating similar variability when tapping to auditory versus visual metronomes (compared to the auditory advantage observed in humans) [12]. Chimpanzees, also nonvocal learners, showed deficits in tempo flexibility [33], and EEG work in monkeys showed that the neural markers of beat detection were not present in the monkeys [41] (although see updated research in these nonhuman primates in Sect. 7.5.2). Tight connections within a premotor-basal ganglia-auditory network and a dorsal auditory stream in vocal learners may have been co-opted for rhythm production, enhancing rhythm perception and production in humans compared to nonhuman primates [24]. The VLH was expanded upon and strengthened with the proposal of the auditory simulation for action prediction hypothesis [17]. ASAP provides an explanation for motor activity observed in the absence of movement when listening to an auditory stimulus. In ASAP, the dorsal auditory stream, in conjunction with the cortical premotor system, is elaborated on as the substrate for BPS. Hence, an important gap between the VLH and ASAP is the comparison in the degree of functional and anatomical homology between the brain pathways for vocal learning in birds [42] and the cortical audiomotor circuit in humans [43].

7.3.1.1 Action Simulation for Auditory Prediction

ASAP explains the role of motor processes in beat perception even in the absence of movements [17]. While the activation of motor systems is expected for physical entrainment to rhythmic stimuli, a prevalent observation of musical beat and rhythm perception in humans is the activation of motor regions even during passive music listening [40]. Nozaradan [44] used frequency tagging to show that the amplitude of cortical responses is increased at the location of the physical beat as well as at the location of an imagined metrical beat in rhythmic auditory stimuli. The relative amplitude at beat frequency is reduced in patients with cerebellar and basal ganglia lesions when compared with healthy controls, implicating these regions as neural substrates for beat-based timing [45]. fRMI experiments have shown increased basal ganglia activity (specifically the putamen) in beat-detection tasks during attentive listening [46] and in attentive listening tasks in response to beat rhythms compared with nonbeat control rhythms [38], with increased connectivity between putamen and supplementary motor area, and between putamen, supplementary motor area, primary motor cortex, and auditory cortex, respectively. Teki and colleagues [47] observed increased activity in a striato-thalamo-cortical network (basal ganglia (putamen, caudate nucleus), thalamus, SMA, PMC, and dorsolateral prefrontal cortex) after beat-based rhythmic priming when judging the duration between the final successive tones of a sequence. In individuals with more perceptual sensitivity to a beat, SMA and left PMC show higher activation [48]. Chen and colleagues [49] observed higher activity of SMA, PMC, and cerebellum when participants listened in anticipation of tapping to a beat and during passive listening, and an ALE meta-analysis by Gordon and colleagues [50] found activation of left and right lateral PMC, right PMC, and left cerebellum during passive listening. Accordingly, the same striato-thalamo-cortical network and the cerebellum are engaged during the Synchronization-Continuation tapping task [51].

According to ASAP, the strong predictive timing ability of the motor system is linked to auditory beat-based timing perception via the parietal cortex and the dorsal auditory pathway, whereby motor planning regions simulate periodic actions and thus facilitate the prediction of upcoming beats [17]. Specifically, the parietal cortex is a locus of sensorimotor integration [52], active during beat perception [53, 54] and via connection with the premotor cortex is thought to facilitate temporal expectation [55] and predictability [56]. Causal interventions using transcranial magnetic stimulation demonstrate that downregulating parietal activity causes deficits in beat-based timing perception, providing support for the role of the parietal cortex as a hub between motor and auditory cortices along the dorsal auditory pathway for beat-based timing perception [57]. Further, the dorsal auditory pathway is also much more developed in humans than nonhuman primates. Under the VLH, these findings are expected if the dorsal auditory pathway and connections between auditory and motor cortices happen to be strengthened in vocal learning species, enabling strong BPS capacity. Recent experiments and modeling work have led to important updates of VLH and ASAP, which we explain in more detail in Sects. 7.5.1 and 7.5.2.

7.4 Gradual Audiomotor Evolution

The gradual audiomotor evolution hypothesis explains the capacity for BPS as orthogonal or unrelated to vocal learning. In the same year as the VLH and ASAP hypotheses were published, Merchant and Honing [25] published a contrasting explanation for the difference in the degree of BPS abilities between humans and primates. The Gradual Audiomotor Evolution hypothesis similarly points to an important role of the dorsal auditory stream in beat perception abilities [25]. However, this proposal highlights subcortical structures involved in beat perception—specifically, strong connections within the motor cortico-basal ganglia thalamo-cortical (mCBGT) circuit—and their important role in the evolution of beat-based timing in the primate lineage.

The basal ganglia are involved in voluntary motor control and, in humans, in beat-based timing perception [37, 58, 59]. The basal ganglia project to cortical motor areas via the mCBGT loop, with the putamen and SMA found to be active in both perception and production of simple metrical rhythms [37, 47, 51, 60, 61]. The basal ganglia (and putamen in particular) are thought to underlie the prediction of sub-second timing information that can then be used to anticipate future beat onsets. Indeed, during a synchronization-continuation tapping task, the oscillations in the primate putamen switch from gamma-band during sensory-driven synchronization to beta-band during interval-driven continuation [62–64]. Dysfunction of these regions leads to the movement impairments observed in Parkinson's disease due to loss of dopamine-producing neurons in the substantia nigra [65] and includes deficits in timing [66], a large increase in tapping variability during a synchronization-continuation task [67], and in perceptual beat-based rhythmic discrimination tasks [68]. Since the basic mCBGT neural circuitry is shared among primates, GAE proposes that gradual adaptation of this pathway to include greater connection with auditory areas and increased development of the dorsal auditory stream may have eventually led to human-level beat perception ability, including both interval- and beat-based timing abilities [25, 69, 70].

7.5 Continuums: Recent Updates to VLH and GAE

7.5.1 Updates to VLH: The Vocal Learning Continuum

As originally presented, the vocal learning and rhythmic synchronization hypothesis suggested a binary distinction between animals that either have the capacity for vocal learning or do not. However, the synchronization ability of the California sea lion—Ronan—strongly brought the VLH into question [20]. Sea lions are not a vocal learning species, and yet, Ronan was able to bob its head to a metronome and to complex musical stimuli, albeit with substantial operant training [20, 71]. This calls into question whether complex vocal learning is a *necessary* precursor to beat perception, or, as suggested by Patel [26], is merely a helpful preadaptation for BPS. Patel argues that Ronan's synchronization abilities do not completely refute the VLH as sea lions do exhibit *vocal flexibility* [72] and notes that the specific strategies and neural connections used by Ronan to move to the beat may differ in important ways from humans and other vocal learning animals [26]. Specifically, connections within the dorsal and ventral auditory pathway in humans [73] and an analogous, recently identified dual pathway system in parrots [74, 75] support complex vocal learning and provide strong auditory-motor forebrain connections which facilitate synchronization to an external auditory stimulus [76]. This connection is observed to be weaker in nonhuman primates whose BPS abilities seem more strongly related to subcortical motor processes [77] (which Patel [26] argues could similarly be the case with Ronan, the sea lion) and stronger in songbirds, parrots, and humans. Accordingly, the revised vocal learning hypothesis has expanded to accommodate predictions of the vocal learning continuum hypothesis—that is, the possibility that vocal learning ability may exist on a continuum extending from vocal nonlearners, to moderate and up to high vocal learners [42, 78, 79]—and predicts that BPS ability should correlate with vocal learning capacity.

While BPS abilities may exist along a continuum of vocal learning capacity, the case of Ronan the sea lion suggests the possibility that BPS abilities may also depend on whether or not synchronizing with a rhythmic stimulus is *intrinsically rewarding*. The intrinsic reward and rhythmic synchronization hypothesis [30, 80] proposes that there may be capacity for both vocal and nonvocal learners to engage in predictive and tempo-flexible synchronization, but it might be the case that only some vocal learners have the intrinsic motivation to do so (c.f. Merker's concept of conformal motive) [81]. Takeya and colleagues [30, 80] demonstrate that Japanese monkeys (*Macaca fuscata*), vocal nonlearners, can generate both predictive and tempo-flexible saccades when trained on a visual metronome (with spatialized visual stimuli, rather than flashing). As in the sea lion, this behavior required extensive training—a reward was given after every predictive eye movement—but the monkeys could continue this predictive synchronization even after training when a reward was no longer present [30, 80]. Further, the monkeys could generalize these synchronized saccades to untrained tempi outside of the trained range of tempi and could maintain a tempo when the visual stimuli timing was cued by their own eye movements (akin to the continuation phase of a tapping-continuation task), suggesting maintenance of an internal rhythm. These experiments show that the vocal nonlearners can be trained to exhibit predictive and tempo-flexible synchronization to a visual beat, but do not speak to whether this extends to synchronization to an auditory beat, or the *spontaneous* auditory-motor synchronization exhibited by humans.

Recent formulations of the ASAP hypothesis expand their explanations of beat processing to include the subcortical regions proposed by GAE. ASAP, as originally presented, hypothesized that strong connections between motor and auditory regions within the dorsal auditory pathway enabled activity in motor planning regions to support the prediction of rhythmic auditory beats [17]. The periodic actions "simulated" by motor planning regions under the ASAP hypothesis are not necessarily simulations of physical actions, but rather they reflect the simulation of abstract timing patterns [82, 83]. However, the original description of ASAP lacked an explanation for the specific neurophysiological implementation of those simulated actions. Recently, Cannon and Patel [84] have supplied a potential neurological model for the simulated actions proposed in ASAP. They argue that the simulated "actions" of auditory beat perception are the result of precise timing activity generated by SMA (proto-actions) and the sequencing of that timing activity (or of those proto-actions) into longer repeating patterns by the dorsal striatum. Specifically, during the generation of an actual movement sequence, the SMA operates via the direct pathway to send excitatory signals to dorsal striatum, inhibitory signals to internal globus pallidus, inhibiting the thalamus, which ultimately sends excitatory signals to the SMA to generate the next movement in the sequence. When listening to a predictable rhythmic stimulus, this same cortical-subcortical pathway allows for the disinhibition of motor planning activity in SMA to simulate the next "action"—the next beat—in an auditory sequence. Important in this neurophysiological explanation of the simulated actions that enable auditory beat and rhythm perception is the neurotransmitter, dopamine. Dopamine plays a role in both action generation and in prediction reward. Phasic dopamine release in the substantia nigra is recruited for sub-second timing during continuous movement sequences [85]. This mechanism of "fast-timescale motor-control" in the dorsal striatum may provide the necessary temporal resolution for the fast-timescale of simulated actions used to predict incoming auditory beats.

7.5.2 Updates to GAE: The BPS Components Continuum

The GAE hypothesis—proposed in the same year as ASAP—incorporated at the outset gradual increases in the connection between cortical and subcortical motor planning regions to explain BPS differences between humans and nonhuman primates. Recent updates to GAE, like the updates to VLH, have acknowledged a larger continuum of BPS ability. A new conceptual model, presented as the 4-components hypothesis (BPS components summarized in Box 7.2 [23]), provides an update to GAE that explains how animals that are capable of some components of BPS can be trained or may develop full BPS capacity if the BPS activity is rewarding.

Before expanding on this continuum, we must first visit new research on sensorimotor synchronization and beat perception in primates. Rhesus macaques (*Macaca mulatta*) share similar timing abilities to humans in the perception and production of single interval timing tasks [86–89]. Recent studies have shown that when the reward contingencies emphasize predictive and accurate performance, macaques can flexibly and predictively produce periodic intervals in synchrony with auditory and visual metronomes [14, 15], they exhibit a remarkable capability to synchronize across a wide range of tempos (400–1100 ms), and they can use an error correction mechanism to precisely follow the metronome, especially at their preferred tempo [29]. In addition, after tapping synchronization, Rhesus monkeys can continue tapping at a similar tempo without sensory cues [12, 90]. These collections of studies have dramatically changed the initial notion that monkeys only react to rhythmic stimuli and instead strongly emphasize the flexibility, accuracy, and predictive abilities of monkeys to entrain to isochronous sequences, once they understand that the task reward depends on rhythmic prediction. Overall, this implies that macaques have the brain sensorimotor machinery to: (1) extract a rhythm from a continuous stream of sensory events, (2) generate an internal rhythmic signal that predicts the future beat events, and (3) produce anticipatory motor commands such that movements coincide or slightly anticipate the next rhythm. The current frontier is to demonstrate whether monkeys not only perform SMS but also BPS. Indeed, it has been shown that monkeys show EEG modulations closely linked to the subjective accentuation of 1:2 and 1:3 rhythms from auditory metronomes, similar to what has been recorded in humans with the same stimuli [91]. In addition, monkeys not only synchronize to isochronous metronomes, they can modify their tapping tempo by predicting the beat changes of accelerating and decelerating visual metronomes, as humans do [15]. Notably, monkeys can also consistently tap to the subjective beat of different music excerpts [23]. Therefore, monkeys have an audiomotor system with sophisticated abilities to perceive, extract, and entrain to continuous auditory streams with beat information, but they need a rewarding element in the task to perform BPS, they do not possess an intrinsic rewarding signal like in the case of humans [23]. New fascinating observations by biomusicologists have enriched our knowledge about how complex is BPS in other primates in the wild [92], providing more evidence for the GAE hypothesis. For example, chimpanzees that show rhythmic drumming behaviors [16, 93, 94] can learn to execute human-like sequences [94, 95] and have a bias toward isochrony during tree tapping [16]. In addition, lemurs (*Indri indri*) produce in the wild coordinated songs with isochronous and 1:2 rhythmic categories [96], while gibbons (*Nomascus gabriellae, Nomascus leucogenys,* and *Nomascus siki*) show singing with a strong isochrony that gradually increases in tempo (accelerando) [97].

The fact that monkeys can strongly perform SMS with visual and auditory metronomes opens the crucial possibility of studying the neurophysiological bases of sensorimotor synchronization in the mCBGT circuit [36]. Recent neurophysiological studies in macaques indicate that the internal beat representation during rhythmic tapping to visual and auditory metronomes depends on the neural population dynamics in the medial premotor cortex (MPC). A key property of MPC neurons is their relative representation of beat timing. Cells that encode elapsed or remaining time for a tap show up-down ramping profiles that span the produced interval, with speed scaling as a function of beat [98, 99]. Furthermore, these cells are recruited in fast succession, producing an ongoing neural activation pattern that flexibly fills the duration between beats depending on the tapping tempo, thereby providing a relative representation of the evolution of an interval [100–102]. These neural sequences are reset and reproduced in each produced interval in the rhythmic sequence [103, 104]. The hypothesis is that the neural sequence resetting is precisely the signal that generates a phasic internal pulse in mCBGT, which in turn is spread to the auditory system to coordinate the internal beat prediction with the auditory rhythmic input [36, 105]. The neural cyclic evolution and resetting become more evident when the time-varying activity of hundreds of MPC neurons is projected into a low-dimensional state space [14, 64, 106]. The population neural trajectories show the following properties. First, their circular dynamics form a regenerating loop for every produced interval. Second, they converge in a similar state space at tapping times, resetting the beat-based clock at this point. This internal representation of rhythm could be transmitted as a phasic top-down predictive signal to sensory areas [105]. Third, the periodic trajectories increase in amplitude and decrease in speed as a function of the length of the isochronous beat encoding the tempo of the metronome [14]. Fourth, the scalar property, a hallmark of timing behavior, was accounted for by the variability of the curvilinear radii in neural trajectories [106, 107]. Finally, this population rhythmic clock is bimodal, showing the same dynamic principles for auditory and visual metronomes, with more variability in the circular trajectories in the auditory condition as seen in the tapping behavior [14]. Hence, the neural population trajectories in MPC encode beat, tempo, and variability of rhythmic timing with both modalities.

Box 7.2 Components of beat perception and synchronization
(Summarized from Rajendran et al. [23])

Auditory Component—ability to abstract salient temporal patterns from complex acoustic signals

Predictive Component—ability to predict the timing of future stimulus input

Auditory-Motor Connection—ability to use sensory information and feedback to time motor actions and anticipate future sound events

Reward Component—ability to coordinate auditory-motor activity through *either intrinsic or conditioned* reward

The recent 4-components (4C's) model for BPS integrates this newer research, demonstrating that nonhuman primates may share stronger beat-perception and synchronization abilities with humans than previously identified—including the ability to tap to a subjective beat from complex musical stimuli, observed in Rhesus macaques [23]. The 4C's model places beat-perception on a continuum, identifying the important role that reward has in learning to entrain to external rhythmic stimuli. Rajendran and colleagues [23] observe that macaques rely more strongly on extrinsic reward to synchronize to music, while humans have a strong intrinsic drive to synchronize to music. The strong intrinsic motivation for human audiomotor synchronization in humans is due in part to the stronger connection of auditory and motor regions in humans, enabling rhythmic musical stimuli to elicit prediction and reward mechanisms in the brain [108–110]. This (4-components) continuum finds further empirical support in some recent studies in humans [111], human newborns [112], and macaques [70] using a novel odd-ball paradigm, which can distinguish between beat perception, isochrony perception, and statistical learning (i.e., prediction). While probing beat perception and isochrony perception in animals is still in its infancy [113–115], it appears, at least within the primate lineage, that isochrony perception is shared with most primates [15, 70, 91], but that beat perception has evolved gradually, peaking in humans and present only with limitations in chimpanzees [33], bonobos (*Pan paniscus*) [34], macaques [70], and other nonhuman primates [116] (see also Chap. 12 [117]).

7.6 Conclusions: Converging Ideas from VLH/GAE

In this chapter, we have provided an overview of auditory-motor interactions involved in sensorimotor synchronization (SMS) and the more complex process of beat perception and synchronization (BPS). The vocal learning and rhythmic synchronization hypothesis and its corollary, action simulation for auditory prediction hypothesis, posit that strong auditory-motor connections that underlie BPS arose from the strong auditory-motor connections required for vocal learning [17]. In the same year, the gradual auditory motor evolution hypothesis put forth that, rather than vocal learning, the emergence of BPS abilities arose from a gradual strengthening of subcortical and cortical action pathways with auditory pathways, explaining the existence of a high level of BPS abilities in nonhuman primates [25]. Both of these hypotheses rely on the emergence of strong sensorimotor couplings between auditory and motor regions for the perception and production of rhythmic actions. The revised vocal learning hypothesis and ASAP (rVLH/ASAP) and the gradual audiomotor evolution hypothesis and the 4-Components Model (GAE/4Cs) are converging on the idea of a continuum for beat perception and synchronization. However, these hypotheses still differ in that rVLH does make the claim that the continuum of BPS ability is necessary—if not sufficiently—based on differences in vocal learning capacity. Vocal learning capacity is thought to provide the intrinsic rewards necessary for perceiving and producing complex auditory sequences. In contrast, GAE/4Cs argues that spontaneous BPS has the potential to emerge in species that have some of the components of BPS, so long as the BPS activity is rewarded. The activity does not need to be rewarded specifically through vocal learning, and thus vocal learning is not a necessary precursor to BPS. The dorsal auditory pathways proposed by (r)VLH/ASAP and the mCBGT pathways proposed by GAE together could have formed the basis for hierarchical structure building necessary for BPS (and further for the emergence of human language and music). Asano [118] argues that these neural pathways developed as a result of both vocal learning and increasingly complex tool-use and tool-making in the hominin lineage, which both strengthened these pathways and enhanced working memory capability necessary for perceiving and producing complex sequences of sounds. Dorsal pathways contribute to the ability to form hierarchical internal models, while the mCBGT circuits allow for the control of goal-directed actions. These pathways may have been elaborated on through a process of reward/reinforcement, that is, the ability to imitate both vocal and nonvocal movements, and through living and coordinating in increasingly complex social groups.

Both the rVLH and GAE emphasize the role of dopamine and reward in maintaining complex temporal sequences. While BPS in humans (and parrots) may rely strongly on strong audiomotor connections and intrinsic reward developed as a result of vocal learning, it may be possible to develop BPS through extrinsic reward mechanisms in nonvocal learning species. Vocal learning may increase the likelihood/ease of developing BPS along a relative continuum of vocal learning abilities, but such strong sensorimotor coupling may be developed in different ways—as evidenced through strong rhythmic abilities observed across a wide variety of primates.

In conclusion, while VLH and GAE originated as competing hypotheses to explain the emergence of beat perception in vocal and nonvocal learning species, work over the last two decades has led to meaningful convergence between the revised VLH/ASAP and the GAE/4-components model. Both hypotheses explain aspects of the emergence of BPS, including the importance of audiomotor connections with both cortical and subcortical motor structures of the brain. The key distinction now between vocal and nonvocal learners may be the role that dopamine plays as a reward chemical. In line with rVLH, beat perception and synchronization may arise from pre-adaptations for vocal learning, including intrinsic dopamine reward, resulting in a continuum of beat perception abilities among the continuum of vocal learners. However, in line with GAE, beat perception and synchronization may still arise provided the animal has substantial training and extrinsic reward, suggesting a similar continuum of beat perception abilities among nonvocal learners. In all, the empirical and computational work under the VLH and GAE hypotheses has tremendously energized the research on SMS and BPS in different disciplines and with different animal models. Research relying on a combination of ideas from the revised versions of these hypotheses—with a goal toward explaining the continuum of beat perception abilities—will continue to shed light on the complex rhythmic abilities of humans and nonhuman animals alike.

References

1. Mendoza G, Merchant H (2014) Motor system evolution and the emergence of high cognitive functions. Prog Neurobiol 122:73–93
2. Bufacchi R, Battaglia-Mayer A, Iannetti G, Caminiti R (2023) Cortico-spinal modularity in the parieto-frontal system: a new perspective on action control. Prog Neurobiol 231:102537
3. Hage SR, Nieder A (2016) Dual neural network model for the evolution of speech and language. Trends Neurosci 39:813–829
4. Sauciuc G-A, Schneider JN (2026) Chapter 2: How animals produce rhythm – an overview of ethological mechanisms for rhythmic communication. In: Suiciuc G-A, Lameira AR, Persson T (eds) Nature beats: the what, why and how of animal rhythmic behaviours. Springer Nature Switzerland, Cham
5. Sauciuc G-A, Schneider JN, Burchardt LS et al (2026) Chapter 3: The functional and structural diversity of rhythmic communication across species. In: Sauciuc G-A, Lameira AR, Persson T (eds) Nature beats: the what, why and how of animal rhythmic behaviours. Springer Nature Switzerland, Cham
6. Lameira AR, De Gregorio C, Gamba M et al (2026) Chapter 4: Uncharted rhythmic diversity across the animal world. In: Sauciuc G-A, Lameira AR, Persson T (eds) Nature beats: the what, why and how of animal rhythmic behaviour. Springer Nature Switzerland, Cham
7. Balasubramaniam R, Haegens S, Jazayeri M et al (2021) Neural encoding and representation of time for sensorimotor control and learning. J Neurosci 41:866–872
8. Sauciuc G-A, Schneider JN, Hattori Y (2026) Chapter 9: The neuroethology of rhythmic communication: neural and mechanical building blocks and constraints on rhythm production and perception. In: Sauciuc G-A, Lameira AR, Persson T (eds) Nature beats: the what, why and how of animal rhythmic behaviours. Springer Nature Switzerland, Cham
9. Repp BH (2005) Sensorimotor synchronization: a review of the tapping literature. Psychon Bull Rev 12:969–992. https://doi.org/10.3758/bf03206433

10. Repp BH, Su Y-H (2013) Sensorimotor synchronization: a review of recent research (2006–2012). Psychon Bull Rev 20:403–452. https://doi.org/10.3758/s13423-012-0371-2
11. Merchant H, Grahn J, Trainor L et al (2015) Finding the beat: a neural perspective across humans and non-human primates. Philos Trans R Soc B Biol Sci 370:20140093
12. Zarco W, Merchant H, Prado L, Mendez JC (2009) Subsecond timing in primates: comparison of interval production between human subjects and rhesus monkeys. J Neurophysiol 102:3191–3202. https://doi.org/10.1152/jn.00066.2009
13. Whitton SA, Jiang F (2023) Sensorimotor synchronization with visual, auditory, and tactile modalities. Psychol Res 87:2204–2217
14. Betancourt A, Pérez O, Gámez J et al (2023) Amodal population clock in the primate medial premotor system for rhythmic tapping. Cell Rep 42:113234
15. Gámez J, Yc K, Ayala YA et al (2018) Predictive rhythmic tapping to isochronous and tempo changing metronomes in the nonhuman primate. Ann N Y Acad Sci 1423:396–414. https://doi.org/10.1111/nyas.13671
16. Eleuteri V, Van Der Werff J, Wilhelm W et al (2025) Chimpanzee drumming shows rhythmicity and subspecies variation. Curr Biol 35:2448–2456.e4. https://doi.org/10.1016/j.cub.2025.04.019
17. Patel AD, Iversen JR (2014) The evolutionary neuroscience of musical beat perception: the Action Simulation for Auditory Prediction (ASAP) hypothesis. Front Syst Neurosci 8:57
18. Patel AD, Iversen JR, Bregman MR, Schulz I (2009) Experimental evidence for synchronization to a musical beat in a nonhuman animal. Curr Biol 19:827–830. https://doi.org/10.1016/j.cub.2009.03.038
19. Keehn RJJ, Iversen JR, Schulz I, Patel AD (2019) Spontaneity and diversity of movement to music are not uniquely human. Curr Biol 29:R621–R622
20. Cook P, Rouse A, Wilson M, Reichmuth C (2013) A California sea lion (*Zalophus californianus*) can keep the beat: motor entrainment to rhythmic auditory stimuli in a non vocal mimic. J Comp Psychol 127:412–427. https://doi.org/10.1037/a0032345
21. Rouse AA, Cook PF, Large EW, Reichmuth C (2016) Beat keeping in a sea lion as coupled oscillation: implications for comparative understanding of human rhythm. Front Neurosci 10. https://doi.org/10.3389/fnins.2016.00257
22. Cook PF, Hood C, Rouse A, Reichmuth C (2025) Sensorimotor synchronization to rhythm in an experienced sea lion rivals that of humans. Sci Rep 15. https://doi.org/10.1038/s41598-025-95279-1
23. Rajendran VG, Prado L, Marquez JP, Merchant H (2025) Monkeys have rhythm. Science 390(6776). https://doi.org/10.1126/science.adp5220
24. Patel AD (2014) The evolutionary biology of musical rhythm: was Darwin wrong? PLoS Biol 12:e1001821
25. Merchant H, Honing H (2014) Are non-human primates capable of rhythmic entrainment? Evidence for the gradual audiomotor evolution hypothesis. Front Neurosci 7:274
26. Patel AD (2021) Vocal learning as a preadaptation for the evolution of human beat perception and synchronization. Philos Trans R Soc B 376. https://doi.org/10.1098/rstb.2020.0326
27. Hove MJ, Spivey MJ, Krumhansl CL (2010) Compatibility of motion facilitates visuomotor synchronization. J Exp Psychol Hum Percept Perform 36:1525
28. Pérez O, Delle Monache S, Lacquaniti F et al (2023) Rhythmic tapping to a moving beat motion kinematics overrules natural gravity. Iscience 26:107543
29. Castillo-Almazán A, Pérez O, Prado L et al (2025) Flexible tapping synchronization in macaques: dynamic switching of timing strategies within rhythmic sequences. J Neurophysiol 134:580–590. https://doi.org/10.1152/jn.00158.2025
30. Takeya R, Kameda M, Patel AD, Tanaka M (2017) Predictive and tempo-flexible synchronization to a visual metronome in monkeys. Sci Rep 7:6127. https://doi.org/10.1038/s41598-017-06417-3
31. Patel AD (2006) Musical rhythm, linguistic rhythm, and human evolution. Music Percept 24:99–104. https://doi.org/10.1525/mp.2006.24.1.99

32. Hasegawa A, Okanoya K, Hasegawa T, Seki Y (2011) Rhythmic synchronization tapping to an audio–visual metronome in budgerigars. Sci Rep 1. https://doi.org/10.1038/srep00120
33. Hattori Y, Tomonaga M, Matsuzawa T (2013) Spontaneous synchronized tapping to an auditory rhythm in a chimpanzee. Sci Rep 3:1566. https://doi.org/10.1038/srep01566
34. Large EW, Gray PM (2015) Spontaneous tempo and rhythmic entrainment in a bonobo (*Pan paniscus*). J Comp Psychol 129:317–328. https://doi.org/10.1037/com0000011
35. Honing H (2013) Structure and interpretation of rhythm in music. In: The psychology of music, vol 3, pp 369–404. https://doi.org/10.1016/B978-0-12-381460-9.00009-2
36. Lenc T, Merchant H, Keller PE et al (2021) Mapping between sound, brain and behaviour: four-level framework for understanding rhythm processing in humans and non-human primates. Philos Trans R Soc B 376:20200325
37. Grahn JA, Brett M (2007) Rhythm and beat perception in motor areas of the brain. J Cogn Neurosci 19:893–906
38. Grahn JA, Rowe JB (2009) Feeling the beat: premotor and striatal interactions in musicians and nonmusicians during beat perception. J Neurosci 29:7540–7548. https://doi.org/10.1523/JNEUROSCI.2018-08.2009
39. Chen JL, Zatorre RJ, Penhune VB (2006) Interactions between auditory and dorsal premotor cortex during synchronization to musical rhythms. NeuroImage 32:1771–1781
40. Proksch S, Comstock DC, Médé B et al (2020) Motor and predictive processes in auditory beat and rhythm perception. Front Hum Neurosci 14:578546. https://doi.org/10.3389/fnhum.2020.578546
41. Honing H, Merchant H, Haden GP et al (2012) Rhesus monkeys (*Macaca mulatta*) detect rhythmic groups in music, but not the beat. PLoS One 7:e51369
42. Jarvis ED (2019) Evolution of vocal learning and spoken language. Science 366:50–54
43. Zatorre RJ, Chen JL, Penhune VB (2007) When the brain plays music: auditory–motor interactions in music perception and production. Nat Rev Neurosci 8:547–558
44. Nozaradan S (2014) Exploring how musical rhythm entrains brain activity with electroencephalogram frequency-tagging. Philos Trans R Soc B Biol Sci 369:20130393
45. Nozaradan S, Schwartze M, Obermeier C, Kotz SA (2017) Specific contributions of basal ganglia and cerebellum to the neural tracking of rhythm. Cortex 95:156–168. https://doi.org/10.1016/j.cortex.2017.08.015
46. Araneda R, Renier L, Ebner-Karestinos D et al (2017) Hearing, feeling or seeing a beat recruits a supramodal network in the auditory dorsal stream. Eur J Neurosci 45:1439–1450
47. Teki S, Grube M, Kumar S, Griffiths TD (2011) Distinct neural substrates of duration-based and beat-based auditory timing. J Neurosci 31:3805–3812. https://doi.org/10.1523/jneurosci.5561-10.2011
48. Grahn JA, McAuley JD (2009) Neural bases of individual differences in beat perception. NeuroImage 47:1894–1903
49. Chen JL, Penhune VB, Zatorre RJ (2008) Listening to musical rhythms recruits motor regions of the brain. Cereb Cortex 18:2844–2854
50. Gordon CL, Cobb PR, Balasubramaniam R (2018) Recruitment of the motor system during music listening: an ALE meta-analysis of fMRI data. PLoS One 13:e0207213. https://doi.org/10.1371/journal.pone.0207213
51. Sánchez-Moncada I, Concha L, Merchant H (2024) Pre-supplementary motor cortex mediates learning transfer from perceptual to motor timing. J Neurosci 44:e3191202023
52. Hickok G, Poeppel D (2007) The cortical organization of speech processing. Nat Rev Neurosci 8:393–402
53. Pollok B, Gross J, Müller K et al (2005) The cerebral oscillatory network associated with auditorily paced finger movements. NeuroImage 24:646–655
54. Pollok B, Stephan K, Keitel A et al (2017) The posterior parietal cortex subserves precise motor timing in professional drummers. Front Hum Neurosci 11:183
55. Coull JT, Nobre A (2008) Dissociating explicit timing from temporal expectation with fMRI. Curr Opin Neurobiol 18:137–144

56. Coull JT, Cotti J, Vidal F (2016) Differential roles for parietal and frontal cortices in fixed versus evolving temporal expectations: dissociating prior from posterior temporal probabilities with fMRI. NeuroImage 141:40–51
57. Ross JM, Iversen JR, Balasubramaniam R (2018) The role of posterior parietal cortex in beat-based timing perception: a continuous theta burst stimulation study. J Cogn Neurosci 30:634–643
58. Harrington DL, Zimbelman JL, Hinton SC, Rao SM (2010) Neural modulation of temporal encoding, maintenance, and decision processes. Cereb Cortex 20:1274–1285
59. Kung S-J, Chen JL, Zatorre RJ, Penhune VB (2013) Interacting cortical and basal ganglia networks underlying finding and tapping to the musical beat. J Cogn Neurosci 25:401–420. https://doi.org/10.1162/jocn_a_00325
60. Grahn JA, Rowe JB (2013) Finding and feeling the musical beat: striatal dissociations between detection and prediction of regularity. Cereb Cortex 23:913–921. https://doi.org/10.1093/cercor/bhs083
61. Coull JT, Cheng R-K, Meck WH (2011) Neuroanatomical and neurochemical substrates of timing. Neuropsychopharmacology 36:3–25
62. Bartolo R, Prado L, Merchant H (2014) Information processing in the primate basal ganglia during sensory-guided and internally driven rhythmic tapping. J Neurosci 34:3910–3923
63. Bartolo R, Merchant H (2015) β oscillations are linked to the initiation of sensory-cued movement sequences and the internal guidance of regular tapping in the monkey. J Neurosci 35:4635–4640
64. Merchant H, Bartolo R (2018) Primate beta oscillations and rhythmic behaviors. J Neural Transm 125:461–470
65. Jellinger K (2005) The pathology of Parkinson's disease-recent advances. In: Galvez-Jimenez N (ed) Scientific basis for the treatment of Parkinson's disease. Taylor & Francis, New York, 2nd ed, pp 69–116
66. Harrington DL, Haaland KY, Hermanowitz N (1998) Temporal processing in the basal ganglia. Neuropsychology 12:3
67. Merchant H, Luciana M, Hooper C et al (2008) Interval timing and Parkinson's disease: heterogeneity in temporal performance. Exp Brain Res 184:233–248
68. Grahn JA, Brett M (2009) Impairment of beat-based rhythm discrimination in Parkinson's disease. Cortex 45:54–61
69. Honing H, Merchant H (2014) Differences in auditory timing between human and nonhuman primates [commentary on Ackermann et al.]. Behav Brain Sci 37:6
70. Honing H, Bouwer FL, Prado L, Merchant H (2018) Rhesus monkeys (*Macaca mulatta*) sense isochrony in rhythm, but not the beat: additional support for the gradual audiomotor evolution hypothesis. Front Neurosci 12:475
71. Schachner A (2012) If horses entrain, don't entirely reject vocal learning: an experience-based vocal learning hypothesis. Empir Musicol Rev 7:157–159
72. Reichmuth C, Casey C (2014) Vocal learning in seals, sea lions, and walruses. Curr Opin Neurobiol 28:66–71. https://doi.org/10.1016/j.conb.2014.06.011
73. Rauschecker JP (2011) An expanded role for the dorsal auditory pathway in sensorimotor control and integration. Hear Res 271:16–25
74. Chakraborty M, Jarvis ED (2015) Brain evolution by brain pathway duplication. Philos Trans R Soc B Biol Sci 370:20150056
75. Chakraborty M, Walløe S, Nedergaard S et al (2015) Core and shell song systems unique to the parrot brain. PLoS One 10:e0118496. https://doi.org/10.1371/journal.pone.0118496
76. Nieder A, Mooney R (2020) The neurobiology of innate, volitional and learned vocalizations in mammals and birds. Philos Trans R Soc B 375:20190054
77. Rauschecker JP, Tian B (2000) Mechanisms and streams for processing of "what" and "where" in auditory cortex. Proc Natl Acad Sci 97:11800–11806
78. Petkov CI, Jarvis ED (2012) Birds, primates, and spoken language origins: behavioral phenotypes and neurobiological substrates. Front Evol Neurosci 4:12

79. Arriaga G, Zhou EP, Jarvis ED (2012) Of mice, birds, and men: the mouse ultrasonic song system has some features similar to humans and song-learning birds. PLoS One 7(10):e46610
80. Takeya R, Patel AD, Tanaka M (2018) Temporal generalization of synchronized saccades beyond the trained range in monkeys. Front Psychol 9:2172
81. Merker B (2005) The conformal motive in birdsong, music, and language: an introduction. Ann N Y Acad Sci 1060:17–28. https://doi.org/10.1196/annals.1360.003
82. Ross JM, Iversen JR, Balasubramaniam R (2016) Motor simulation theories of musical beat perception. Neurocase 22:558–565
83. Schubotz RI (2007) Prediction of external events with our motor system: towards a new framework. Trends Cogn Sci 11:211–218
84. Cannon JJ, Patel AD (2021) How beat perception co-opts motor neurophysiology. Trends Cogn Sci 25:137–150. https://doi.org/10.1016/j.tics.2020.11.002
85. Howe MW, Dombeck DA (2016) Rapid signalling in distinct dopaminergic axons during locomotion and reward. Nature 535:505–510
86. Merchant H, Battaglia-Mayer A, Georgopoulos AP (2003) Interception of real and apparent motion targets: psychophysics in humans and monkeys. Exp Brain Res 152:106–112
87. Merchant H, Battaglia-Mayer A, Georgopoulos AP (2004) Neural responses during interception of real and apparent circularly moving stimuli in motor cortex and area 7a. Cereb Cortex 14:314–331
88. Merchant H, Zarco W, Prado L, Pérez O (2009) Behavioral and neurophysiological aspects of target interception. In: Sternad D (ed) Progress in motor control: a multidisciplinary perspective. Springer, New York, pp 201–220
89. Mendez JC, Prado L, Mendoza G, Merchant H (2011) Temporal and spatial categorization in human and non-human primates. Front Integr Neurosci 5:50
90. García-Garibay O, Cadena-Valencia J, Merchant H, De Lafuente V (2016) Monkeys share the human ability to internally maintain a temporal rhythm. Front Psychol 7:1971
91. Criscuolo A, Schwartze M, Prado L et al (2023) Macaque monkeys and humans sample temporal regularities in the acoustic environment. Prog Neurobiol 229:102502
92. De Gregorio C, Maiolini M, Raimondi T et al (2024) Isochrony as ancestral condition to call and song in a primate. Ann N Y Acad Sci 1537:41–50. https://doi.org/10.1111/nyas.15151
93. Ravignani A, Olivera VM, Gingras B et al (2013) Primate drum kit: a system for studying acoustic pattern production by non-human primates using acceleration and strain sensors. Sensors 13:9790–9820
94. Fuhrmann D, Ravignani A, Marshall-Pescini S, Whiten A (2014) Synchrony and motor mimicking in chimpanzee observational learning. Sci Rep 4:5283
95. Dufour V, Poulin N, Curé C, Sterck EHM (2015) Chimpanzee drumming: a spontaneous performance with characteristics of human musical drumming. Sci Rep 5:11320. https://doi.org/10.1038/srep11320
96. De Gregorio C, Valente D, Raimondi T et al (2021) Categorical rhythms in a singing primate. Curr Biol 31:R1379–R1380. https://doi.org/10.1016/j.cub.2021.09.032
97. De Gregorio C, Raimondi T, Bevilacqua V et al (2024) Isochronous singing in 3 crested gibbon species (*Nomascus* spp). Curr Zool 70:291–297. https://doi.org/10.1093/cz/zoad029
98. Merchant H, Zarco W, Pérez O et al (2011) Measuring time with different neural chronometers during a synchronization-continuation task. Proc Natl Acad Sci 108:19784–19789
99. Merchant H, Averbeck BB (2017) The computational and neural basis of rhythmic timing in medial premotor cortex. J Neurosci 37:4552–4564
100. Crowe DA, Zarco W, Bartolo R, Merchant H (2014) Dynamic representation of the temporal and sequential structure of rhythmic movements in the primate medial premotor cortex. J Neurosci 34:11972–11983
101. Mello GB, Soares S, Paton JJ (2015) A scalable population code for time in the striatum. Curr Biol 25:1113–1122
102. Merchant H, Pérez O, Bartolo R et al (2015) Sensorimotor neural dynamics during isochronous tapping in the medial premotor cortex of the macaque. Eur J Neurosci 41:586–602
103. Merchant H, Harrington DL, Meck WH (2013) Neural basis of the perception and estimation of time. Annu Rev Neurosci 36:313–336. https://doi.org/10.1146/annurev-neuro-062012-170349

104. Merchant H, Pérez O, Zarco W, Gámez J (2013) Interval tuning in the primate medial premotor cortex as a general timing mechanism. J Neurosci 33:9082–9096
105. Merchant H, Mendoza G, Pérez O, Betancourt A, García-Saldivar P, Prado L (2024) Diverse time encoding strategies within the medial premotor areas of the primate. In: Merchant H, de Lafuente V (eds) Neurobiology of interval timing. Springer, Switzerland, pp 117–140
106. Gámez J, Mendoza G, Prado L et al (2019) The amplitude in periodic neural state trajectories underlies the tempo of rhythmic tapping. PLoS Biol 17:e3000054
107. Pérez O, Merchant H (2018) The synaptic properties of cells define the hallmarks of interval timing in a recurrent neural network. J Neurosci 38:4186–4199
108. Salimpoor VN, Zald DH, Zatorre RJ et al (2015) Predictions and the brain: how musical sounds become rewarding. Trends Cogn Sci 19:86–91. https://doi.org/10.1016/j.tics.2014.12.001
109. Zatorre RJ (2015) Musical pleasure and reward: mechanisms and dysfunction. Ann N Y Acad Sci 1337:202–211
110. Gold BP, Pearce MT, Mas-Herrero E et al (2019) Predictability and uncertainty in the pleasure of music: a reward for learning? J Neurosci 39:9397–9409
111. Bouwer FL, Werner CM, Knetemann M, Honing H (2016) Disentangling beat perception from sequential learning and examining the influence of attention and musical abilities on ERP responses to rhythm. Neuropsychologia 85:80–90
112. Háden GP, Bouwer FL, Honing H, Winkler I (2024) Beat processing in newborn infants cannot be explained by statistical learning based on transition probabilities. Cognition 243:105670
113. Bouwer FL, Nityananda V, Rouse AA, Ten Cate C (2021) Rhythmic abilities in humans and non-human animals: a review and recommendations from a methodological perspective. Philos Trans R Soc B 376:20200335
114. Henry MJ, Cook PF, De Reus K et al (2021) An ecological approach to measuring synchronization abilities across the animal kingdom. Philos Trans R Soc B 376. https://doi.org/10.1098/rstb.2020.0336
115. Wilson M, Cook PF (2016) Rhythmic entrainment: why humans want to, fireflies can't help it, pet birds try, and sea lions have to be bribed. Psychon Bull Rev 23:1647–1659. https://doi.org/10.3758/s13423-016-1013-x
116. Raimondi T, Di Panfilo G, Pasquali M et al (2023) Isochrony and rhythmic interaction in ape duetting. Proc R Soc B 290:20222244. https://doi.org/10.1098/rspb.2022.2244
117. Sauciuc G-A, Persson T, Hattori Y, Lameira AR (2026) Chapter 12: The significance of the great ape heritage in the evolution of human rhythm in speech, music and dance. In: Sauciuc G-A, Lameira AR, Persson T (eds) Nature beats: the what, why and how of animal rhythmic behaviour. Springer Nature Switzerland, Cham
118. Asano R (2022) The evolution of hierarchical structure building capacity for language and music: a bottom-up perspective. Primates 63:417–428

Neural Resonance and Comparative Rhythm Cognition

8

Edward W. Large

Abstract

Neural resonance theory (NRT) suggests that brains and bodies physically embody music by synchronizing neural oscillations with external sounds and with each other. This implies that rhythm perception and performance are rooted in stable temporal patterns that emerge based on principles of self-organization. NRT encompasses self-organization and learning, describing how brains can adapt to new sound patterns through exposure, enabling cultural transmission. And the fundamental principles of neural resonance and plasticity are widespread, implying a biological basis for certain common musical experiences that may be shared with nonhuman animals. This approach offers an intriguing perspective on the comparative study of rhythm cognition by establishing a direct link between neurobiology and structured communication, eliminating the need for assumptions about representations and computations often associated with human cognition. However, these ideas have not been widely applied in comparative and evolutionary approaches to rhythm cognition, prompting this chapter to address this gap. This chapter will review fundamental concepts relevant to understanding embodied dynamic systems as proposed by NRT. It will then delve into the literature on rhythm cognition in animals, interpreting the results from a dynamical systems perspective. Finally, this chapter will conclude by highlighting some open questions in the literature on animal rhythm cognition.

E. W. Large (✉)
Department of Psychological Sciences, University of Connecticut, Storrs, CT, USA

Department of Physics, University of Connecticut, Storrs, CT, USA
e-mail: edward.large@uconn.edu

G.-A. Sauciuc et al. (eds.), *Nature Beats*, Fascinating Life Sciences,
https://doi.org/10.1007/978-3-032-24250-1_8

8.1 Neural Resonance Theory

Neural oscillations are rhythmic patterns of neural activity that can arise in cortex from the coordinated interactions of neurons, particularly through feedback loops between excitatory and inhibitory neurons (Fig. 8.1a, [1, 2]). These interactions can lead to synchronized firing, where neurons tend to fire together, creating oscillations in local field potentials (Fig. 8.1b) that can be measured using EEG, MEG, or in invasive neural recordings. Over the past several decades, the study of neuronal oscillations has successfully established links between cellular function, network dynamics, perception-cognition, and overt behavior [3, 4]. In particular, neural oscillations may explain many aspects of rhythm cognition and rhythmic behavior in humans, such as the perception of a beat in music [5], synchronization of behavior to sounds [6, 7], and allocation of attention to rhythmic sequences [8, 9].

NRT builds upon concepts from dynamical systems and embodied approaches [10–13] to provide a theoretical framework and a set of computational models for generating novel predictions and testable hypotheses. Nonlinear dynamics provides the theoretical language to integrate information from neural and behavioral levels [3]. Dynamical analysis of neurophysiological processes, such as oscillation and Hebbian

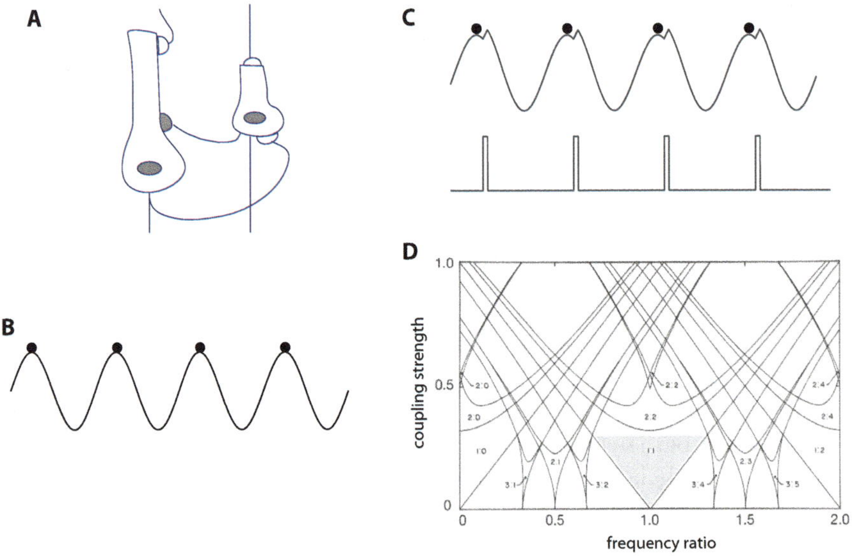

Fig. 8.1 Neural oscillation and mode-locking are phenomena that involve the connectivity between excitatory and inhibitory neurons (**a**). Oscillations can emerge from the coordinated activity of these neurons, resulting in a periodic pattern (**b**). Dots in the figure represent timepoints where motor actions, such as head bobs, might be observed. Phase-coupling in which an oscillation is coupled to a slower periodic signal leads to stable, 1:1 mode-locking; however, motor actions are consistently early because the oscillation is faster than the stimulus (**c**). A bifurcation diagram illustrates the boundaries of stable mode-locking regions. (Adapted from [25]). The gray shaded region in the diagram represents the region where 1:1 mode-locking (phase-locking) is observed (see also Figs. 8.2 and 8.3). More complex mode-locking may be necessary for pulse perception in complex rhythms (**d**)

plasticity, provide a set of general laws and constraints, which enable predictions [1]. Our terminology, derived from physics and applied mathematics, includes concepts such as oscillation, resonance, nonlinearity, stability, attraction, self-organization, and attunement. However, since the vocabulary of NRT sometimes overlaps with biological terminology, we review these terms and concepts first to avoid confusion.

8.1.1 What Counts as Oscillation?

The term "oscillation" is sometimes reserved for biological systems that exhibit endogenous, self-sustained behavior, which occurs in the absence of an external stimulus [14, 15]. Endogenous neural oscillations exist and can be synchronized by external stimuli [9], but this definition is too limiting to explain the variety of rhythmic behaviors exhibited by human and nonhuman animals. For instance, in the auditory system, oscillatory activity can be evoked by a rhythmic stimulus and damps out after the stimulus. By contrast, in a synchronization-continuation task, a motor oscillation begins in response to a rhythmic stimulus but can continue indefinitely when the stimulus stops. All three types of behavior—self-sustained oscillation, damped oscillation, and bistable oscillation—can be observed in neural systems [16, 17]. One key similarity is that they all have an *intrinsic* frequency. This intrinsic frequency can be tuned, and its default value is called the *natural* frequency. The spontaneous tempo of motor movements may be a type of natural frequency [7]. Another similarity is that nonlinear resonance patterns can be found in all three types of neural oscillations [18, 19], which will be described next.

8.1.2 Resonance

Resonance, broadly speaking, refers to the synchronization or entrainment of oscillations with themselves and external stimuli. Resonance causes the oscillator to lock its frequency onto the input, and this is how oscillatory systems communicate with one another. To understand how this concept applies to biological systems, let us consider a radio [10, 20]. In a radio, the receiver must have the appropriate frequency to resonate with the sender, and the signal must be strong enough to make the receiver resonate. If the receiver's frequency is too different from the sender's frequency, or if the signal is too weak, we hear intermittent bursts of signal interrupted by periods of static. This may be similar to the way some animals synchronize intermittently with rhythmic sounds [21, 22].

Dynamic attending [8] suggests that neural oscillations synchronize with rhythmic signals to capture information from the environment. In a radio, the information is conveyed through changes in frequency (FM) or amplitude (AM) of the resonating oscillation. Similarly, in biological organisms that communicate rhythmically, each event within the rhythm can exhibit variations such as being early or late, louder or softer, higher or lower in pitch, or having a different timbre, among others [23].

8.1.3 Nonlinearity

Unlike their linear counterparts, nonlinear oscillators respond to frequencies that are not present in the input but are related to those in the input [24]. One notable type of nonlinear resonance is mode-locking (n:m synchronization; Fig. 8.1c; see, e.g., [25]). Two neural oscillations with distinct natural frequencies can mode-lock when their frequencies are close to a small integer ratio. Stability is enhanced for simpler frequency ratios and stronger coupling. Biologically, nonlinear resonance is observed in critical oscillations of cochlear outer hair cells [26], auditory brainstem neurons [27], nonlinear responses to tones [28], and entrainment of cortical auditory and motor populations to rhythmic sequences [29, 30]. As we will explore further, nonlinear resonance may explain the perception of a pulse and meter in complex rhythms.

8.1.4 Stability

In the realm of dynamical systems, stability emerges as a pivotal concept [31]. It quantifies the resilience of a specific pattern to perturbations in phase or frequency. Dynamical systems naturally gravitate toward and converge toward more stable states. NRT posits that the stability of neural resonances plays a crucial role in shaping our perception of musical structure. Sound sequences that lead to more stable neural patterns are easier to learn, perceive, remember, and reproduce compared to those that form less stable patterns [24]. Examples of stable dynamical states include self-sustained oscillations and synchronized states. Notably, NRT predicts that certain states can be more stable than others. For instance, a rhythmic pattern involving simpler integer ratios is easier to produce and more readily learned than one requiring more complex integer ratios [32].

8.1.5 Self-Organization

Self-organization is a process in which a system's structure and function emerge spontaneously from the interactions of its individual components, as when an oscillation emerges from the interaction between excitatory and inhibitory neurons [33]. Patterns and order arise from the interactions of many individual components rather than being imposed by an external force [13]. No central controller dictates how the system should behave, components interact with their immediate neighbors or within their local environment, leading to global patterns. Self-organizing systems are also highly adaptable, as individual components can respond to local conditions and adapt to changes in their environment.

8.1.6 Attunement

Attunement is a form of self-organization that enables resonating neural circuits to tune themselves, adapting and learning through interactions with the environment [10, 24]. This process involves changes in system parameters that alter the system's stability properties. These processes continuously learn and adapt in an unsupervised manner, forming and adapting neural circuits that embody learned musical patterns. Consequently, patterns that are more frequently experienced become more stable, leading to quicker and more flexible responses. Attunement can unfold over various time scales, including evolutionary, developmental, and behavioral time scales [34]. This chapter will discuss two primary mechanisms of attunement: adaptation of intrinsic oscillator frequency [35] and longer term learning of connections between oscillators through Hebbian synaptic plasticity [36].

8.2 Comparative Approaches

A body of evidence has emerged showing that NRT explains many aspects of human rhythm cognition [24]. Given that its principles are derived from basic neurophysiological mechanisms, such as oscillation and Hebbian plasticity, it makes sense to ask how well the theory explains what is known about rhythm in nonhuman animals. Does it make useful, testable predictions? How much do we know about whether the physiological mechanisms posited by NRT are actually at play?

8.2.1 Synchronization: Frequency and Phase Locking

When a nonlinear oscillator is phase-coupled to a periodic stimulus, such as a metronome, it can synchronize. The stimulus' tempo does not need to precisely match the oscillator's natural frequency; oscillatory synchronization is tempo-flexible, and oscillator models allow us to calculate the conditions under which stable synchronization will occur [31]. Figure 8.2a illustrates regions of stable and unstable behavior, which depend on stimulus tempo and coupling strength. When the coupling is strong enough and the stimulus frequency is close enough, the oscillator locks its frequency to the stimulus. This locked-on frequency, known as the *instantaneous frequency*, matches the frequency of the stimulus, even though the natural frequency may differ. Human adults routinely produce stable frequency-locked synchronization [17]. Stable synchronized behavior has been observed in animals as well. A California sea lion (*Zalophus californianus*; "Ronan") was trained to match head movements to isochronous repeating stimuli, and after training, Ronan showed spontaneous generalization to novel tempos and to the beat of musical rhythms [37]. In all experiments, Ronan displayed stable frequency-locked synchronization (Fig. 8.2b) at tempos ranging from 72 bpm (1.2 Hz) to 137 bpm (2.3 Hz), for entire trials of 20–60 beats, similar to human adults. Ronan was the first nonhuman animal to display stable synchronization for extended periods.

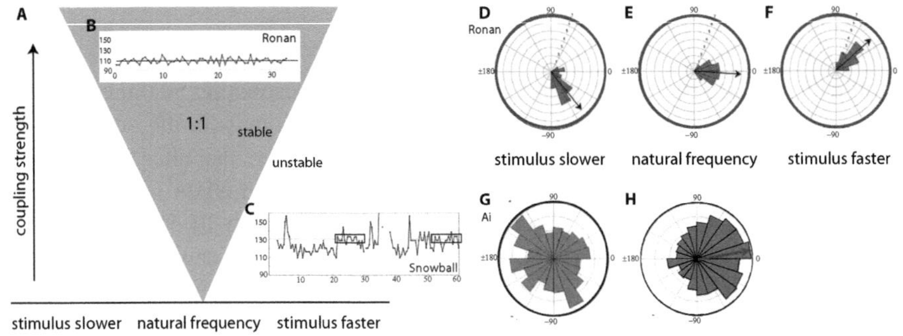

Fig. 8.2 Stable and unstable synchronization in phase-coupled oscillation (Fig. 8.1c). (**a**) Zoom-in on the 1:1 mode-locked region of Fig. 8.1d reveals a stable synchronization region (gray). Within this region, behavior is frequency-locked to the stimulus, similar to Ronan (**b**, data from [37]). In contrast, the unstable region exhibits intermittent frequency-locking, as observed with Snowball (**c**, data from [22]). Within the synchronization region, phase depends on the relationship between the oscillator's natural frequency and the stimulus's tempo, as observed for Ronan (**d–f**, data from [37]). In unstable synchronization, no phase tendency is observed (**g**) until the stimulus is quite close to the natural tempo (**h**, data from [39])

Outside of the synchronization region, when the natural frequency is too far from the stimulus frequency, or the coupling strength is too weak (Fig. 8.2a), one observes unstable behavior. In this region, oscillator frequency may approximately match stimulus tempo for short periods of time, and these are interrupted by periods in which it operates near its natural frequency. This is called intermittent synchronization, and this has been observed in animals as well (Fig. 8.2c). Patel and Iversen [22] reported intermittent synchronization in a Sulfur-crested cockatoo (*Cacatua galerita eleonora*; "Snowball"). Snowball moved to the beat of the song "Everybody" (Backstreet Boys, 1997) by synchronizing his head-bobs, swings, and foot lifts to the music.[1] The experimenters manipulated the tempo of a musical excerpt across a range and showed that the Snowball synchronized with the beat for short amounts of time, called bouts.

Oscillator models also make specific predictions about phase. During stable synchronization, phase clusters tightly around a mean (Fig. 8.2d–f), but the mean phase depends on the relationship between natural frequency (spontaneous tempo) and stimulus tempo. When the stimulus matches the natural tempo precisely, the mean phase is predicted to be zero (e.g., head bobs occur at the same time as the stimulus sounds). When the stimulus is faster than the natural tempo, the mean phase will be positive (head bobs will be late), and when the stimulus is slower, the mean phase will be negative (head bobs will be early). Ronan's behavior seemed to exhibit such behavior [37]. On middle-tempo trials, Ronan's mean phase was zero (Fig. 8.2d), at faster tempi, her mean phase was greater than zero (Fig. 8.2e; she bobbed late), and at slower tempi her

[1] Unlike Ronan, Snowball did not have to be trained to exhibit this behavior; he danced spontaneously, exhibiting creativity and flexibility, with a number of distinct dance moves.

mean phase was less than zero (Fig. 8.2f; she bobbed early). Thus, like humans [38], her behavior matched predictions of an oscillator with a stable natural frequency.

Similar predictions about phase hold in the unstable region, for intermittent synchronization. For Snowball, phase was reported only during synchronized bouts, and spontaneous tempo was not measured, so such predictions could not be evaluated. However, two other animals have been reported to achieve intermittent synchronization. In one experiment, a chimpanzee (*Pan troglodytes*; "Ai") was trained to tap illuminated keys on an electric keyboard and developed a well-defined spontaneous tempo for performing this task [39] (see also Chap. 12 for more details on this study). When an isochronous distractor sound was presented very close to her spontaneous tempo, Ai intermittently aligned her taps with the sounds. Phase was reported for the entire trial, and the mean phase was close to zero, as would be predicted for an oscillator whose natural frequency is close to the stimulus, but has only weak coupling. In another Experiment, Large and Gray [21] directly assessed spontaneous tempo in a bonobo (*Pan paniscus*; "Kuni") during a social drumming interaction. The bonobo intermittently synchronized her drum strikes within a range around her spontaneous motor tempo, again suggesting weak coupling.

8.2.2 Attunement: Adaptation of Natural Frequency

The wide range of tempos for which Ronan displayed synchronization suggested that there may be more to her story. Her fastest tempo was nearly twice that of her slowest tempo, which suggests that her intrinsic frequency may be adapting to the tempo of the stimulus. In one NRT model, intrinsic frequency undergoes short-term attunement, allowing a performer to adapt to the tempo of other performers. The stimulus tempo attracts the intrinsic frequency of the oscillator, and the natural frequency is also an attractor (Fig. 8.3a; Ronan). This model has been shown to closely model human performance in a number of different experiments [35, 40].

A stringent test of the tempo adaptation hypothesis is a perturbation experiment [41–44]. In a perturbation experiment, we wait for the animal to achieve a stable (synchronized) state, and then we perturb either the phase or the frequency of the stimulus. We can then measure the "relaxation" back to a synchronized state. To test the hypothesis, we can fit the parameters of a dynamical model that has both phase entrainment and frequency adaptation.

To test the tempo adaptation hypothesis with Ronan, we presented her with simple rhythmic stimuli at novel tempos. On some trials, we introduced "perturbations," altering either tempo or phase in the middle of a presentation [45]. Ronan adjusted her behavior following all perturbations, recovering her consistent phase and tempo relationships to the stimulus. Ronan's response to phase perturbations was approximately equivalent to human performance, recovering just after one or two events, and consistent with predictions of mathematical models describing coupled oscillation. Model fits also revealed tempo adaptation, but with weaker frequency coupling, making her tempo adaptation slower than most humans [42]. Thus, Ronan was able to adapt her intrinsic tempo, although not as quickly [46] as

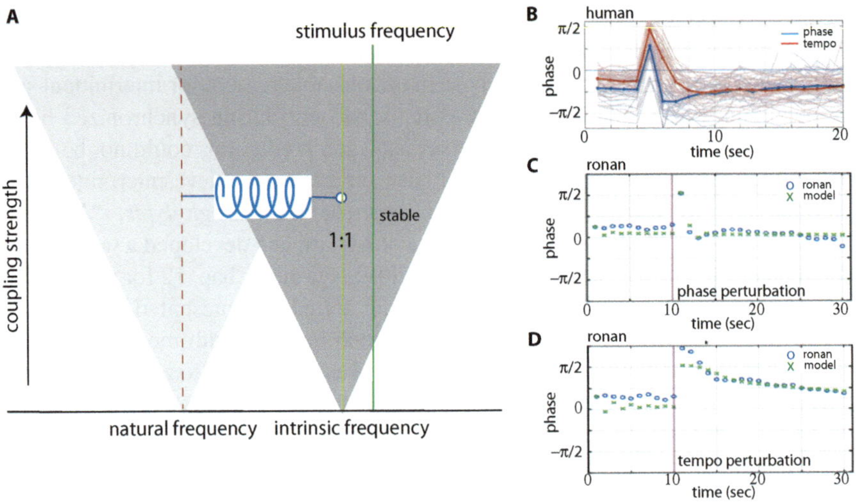

Fig. 8.3 In certain models, the intrinsic frequency of an oscillator undergoes short-term attunement, enabling it to adapt to stimulus tempo. Stimulus tempo naturally attracts the intrinsic frequency of the oscillator, and the natural frequency itself serves as an attractor, facilitating elastic return to the natural tempo (**a**). Data from a perturbation experiment involving a human subject [44] demonstrates both phase synchronization and tempo adaptation (**b**). Ronan's phase synchronization closely aligns with human performance. She exhibits tempo adaptation (**c**), albeit at a slower rate compared to humans (**d**, data from [45]). Ronan's behavior (blue circles) is well described by an oscillator model that incorporates tempo adaptation (green x's)

humans do. A recent head-to-head comparison of Ronan's synchronization to that of humans confirmed a human-like level of performance and suggests that over the past decade or so, Ronan's capacity for tempo adaptation has increased.

8.2.3 Self-Organization of Oscillation

The findings with Ronan were the clearest behavioral evidence yet for parity in human and nonhuman beat keeping and support the view that the human ability to perceive and move in time to rhythm may be rooted in broadly conserved neural mechanisms. However, the neural mechanisms underlying rhythmic synchronization in both humans and nonhuman animals remain elusive. Merchant and colleagues have investigated rhythmic synchronization in macaque monkeys, where detailed neural recordings are feasible [6] (see also Chap. 7 for more details on these studies). In one study, Gamez et al. [47] discovered that during rhythmic tapping, the activity of hundreds of neurons in primate medial premotor cortices, pre-supplementary motor area, and supplementary motor area exhibits a pronounced periodic pattern. This pattern becomes particularly evident when responses are projected into a state space using dimensionality reduction analysis. Distinct tapping tempos were encoded by circular trajectories that traveled at a constant speed but with varying radii. The alterations in the amplitude of the oscillatory dynamics within neuronal state space serve as a signature of duration encoding during rhythmic timing, irrespective of whether it is guided by an external metronome or is internally controlled and not the consequence of repetitive motor commands.

In a recent theoretical model by Zemlianova et al. [48], a recurrent neural network (RNN) consisting of excitatory and inhibitory neurons was trained on the synchronization-continuation task, which required the model to produce an oscillatory output synchronized to a metronome (isochronous) signal and continue the oscillatory output after the signal stopped. Remarkably, the hidden units of the trained RNN showed sequential firing patterns (Fig. 8.4a) that form circular trajectories when projected onto the space spanned by the top principal components (Fig. 8.4b), resembling the neural recordings of macaque monkeys performing the same task [47]. Furthermore, it was shown that the trained RNN consisting of 500 neurons can be reduced to a three-variable oscillator model (Fig. 8.1c). The reduced model revealed an E-I circuit with two distinct inhibitory subpopulations, of which one is tightly synchronized with the excitatory units. A dynamical analysis of the reduced model revealed dynamical mechanisms enabling oscillation.

This combination of detailed neural recording capabilities and cutting-edge dynamical modeling provides the strongest evidence yet that synchronization of periodic motor movements to periodic sound sequences relies on neural oscillation. These combined studies suggest that given the right circumstances, a high-dimensional neural network that reflects realistic neural constraints can self-organize to produce low-dimensional behavior—a nonlinear oscillator—and that this model closely matches the observations of invasive neural recordings in macaque monkeys. Recent evidence of synchronization to a musical beat in macaques suggests that it may be possible to explore the perception of pulse and meter at this level of detail as well.

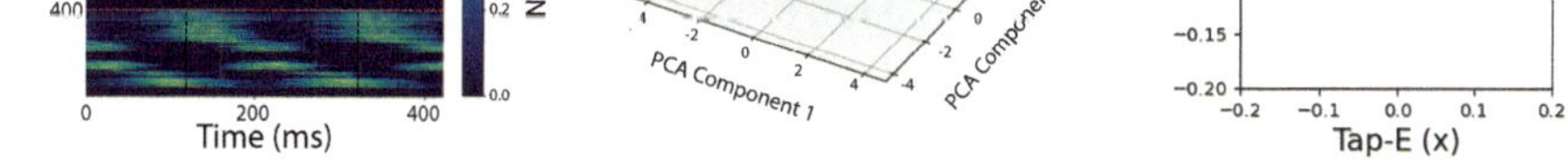

Fig. 8.4 An RNN trained on the synchronization-continuation task. (**a**) Normalized firing rates of excitatory (top 400) and inhibitory (bottom 100) units. (**b**) Firing time courses projected onto the space spanned by the top three principal components (color gradient indicates stimulus frequency; dots indicate model "tap" times). (**c**) Oscillatory trajectories of the reduced model projected to a 2D phase plane (color gradient indicates stimulus frequency). Data replicated independently by Bas E, Roman IR (2024) ZemlianovaRNN. https://github.com/iranroman/ZemlianovaRNN

8.2.4 Self-Organization of Pulse and Meter

While the perception of pulse (the main beat in music) and meter is prevalent in humans, it is not universally observed. While most adults can synchronize periodic movements to isochronous stimuli, some may struggle to discern a beat in the intricate rhythms of music. Moreover, musical rhythm abilities, encompassing the perception of pulse [49], meter [50], and coordinated action to music's rhythmic structure [51], undergo significant transformations throughout human development ([34]; for reviews, see [52], as well as Chap. 10 this volume]). Now, let us consider whether observations of synchronization to the beat in musical recordings provide evidence that animals can perceive the beat in music.

Synchronization experiments with animals have occasionally employed actual musical recordings, but to date, all have utilized musical excerpts with prominent energy at the pulse frequency, as illustrated in Fig. 8.5b, c. This raises the possibility that these animals merely synchronize to the strongest frequency present in the stimulus, rather than perceiving a beat per se. In other words, the observed synchronization essentially mirrors synchronizing to a metronome. However, in highly syncopated rhythms, such as those found in Brazilian samba [53], there may be minimal physical energy at the pulse frequency. These complex rhythms present an opportunity to assess whether our animal participants possess the genuine ability to perceive a beat.

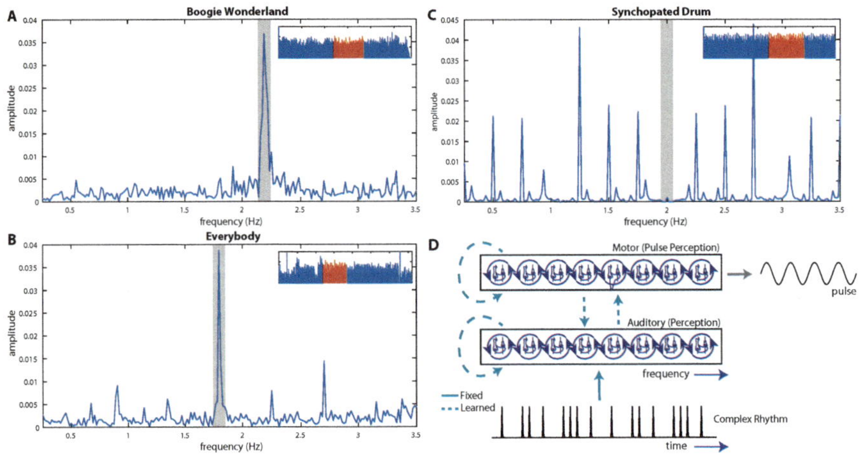

Fig. 8.5 Frequency domain analysis of rhythmic stimuli and an architecture for pulse and meter perception. Frequency domain analyses (FFTs) of the event onsets in Boogie Wonderland (**a**) and Everybody (**b**) reveal energy at the pulse frequency (highlighted in gray) dwarfs the energy at other rhythmic frequencies. In a missing pulse rhythm (**c**), there is virtually no pulse frequency energy. Modeling predicts that mode-locking dynamics in interacting auditory and motor networks could explain pulse perception (**d**; see [5]). Connections within and between networks can be learned via Hebbian plasticity (see [34])

A central hypothesis of neural resonance posits that pulse perception is self-organized, corresponding to a neural oscillation whose frequency may differ from but is related to frequencies in musical rhythm. Interestingly, some syncopated rhythms lack physical energy at the pulse frequency, known as "missing pulse" rhythms [5]. In the absence of a physical pulse in a missing pulse rhythm, the brain might generate the pulse frequency through nonlinear resonance, a form of self-organization. Recent dynamical models demonstrate that oscillator models can indeed explain this phenomenon, and behavioral studies with missing pulse rhythms show that humans perceive the pulse at the frequencies predicted by nonlinear resonance [5]. EEG and MEG SS-EPs revealed the predicted missing frequency, and its amplitude correlates with perception [54, 55]. To our knowledge, no animal has ever been tested for this form of strong, self-organized perception.

If neural oscillation is indeed at play, it is possible that individuals like Ronan might be able to synchronize to missing pulse rhythms. However, for humans, such abilities have a long developmental time course, suggesting that they may need to be learned. The development of appropriate connections between auditory and motor networks and the establishment of strong coupling strength are crucial for nonlinear resonances (e.g., mode-locks) to become stable. Dynamical models, implemented as self-organizing networks of neural oscillations with Hebbian plasticity, predict key features of music development [34]. When formalized in physiologically informed dynamical models, these frameworks offer testable empirical predictions about musical rhythm development. These predictions include the types of native and non-native rhythmic structures infants and children can learn, steady-state evoked potentials to native and non-native musical rhythms, and the effects of short-term [e.g., 36] and long-term [56] learning on music perception-action.

8.2.5 Producing Complex Rhythms

Humans exhibit a preference for rhythms characterized by small integer ratios between inter-onset intervals. While individuals learn the specific rhythms of their native musical cultures, the attraction toward integer ratio categories appears to be a universal aspect of human rhythm learning [57–59]. Interestingly, recent studies have revealed that certain animals, such as songbirds [60] and lemurs [61], also display small integer ratio categories in their vocalizations (for details and more examples, see Chaps. 3 and 10 this volume). However, until recently, explanations of rhythm learning have assumed the ability to learn any arbitrary sequence of inter-onset intervals, which could not account for the preference for integer ratios.

To address this limitation, we recently developed a model that incorporates simple physiological constraints on Hebbian learning in oscillatory networks. This model demonstrates the capacity to account for the learning of complex rhythms and the emergence of integer ratio preferences during the learning process. The mechanism underlying this preference is based on the concept of a central pattern generator responsible for producing complex rhythms. The pattern generator learns to generate novel patterns by establishing connections between oscillators of different frequencies through Hebbian plasticity. Learning results in the formation of connections between an oscillator at the repetition frequency of the rhythm and oscillations at the harmonics activated by listening to the rhythm.

The learning dynamics are constrained to ensure that connections are learned only between excitatory neural populations responsible for generating the oscillations. Consequently, the oscillations can only entrain to each other in-phase or anti-phase, representing the two stable modes of oscillatory entrainment (see HKB, 1985). This constraint facilitates learning while favoring integer ratio relationships between inter-onset intervals.

This model (Fig. 8.6) elucidates several existing behavioral outcomes in terms of well-established neurodynamic mechanisms of oscillation and Hebbian plasticity. It also makes predictions to guide future behavioral and neurophysiological research. The fundamental predictions of this model align with our current understanding. First, the model explains a tendency toward integer ratios, even without enforcing strict integer ratios, due to the intricate dynamics of multi-frequency oscillator interactions. Second, simpler ratios (1:1, 2:1, 3:1) are easier to learn [24], so they would be expected to be more prevalent [60, 61]. Simpler rhythms would also be expected to be observed at faster tempos, as the anti-phase mode of coordination loses stability [60]. Of course, a detailed investigation of this hypothesis would necessitate invasive neural recordings, which would predict oscillations at multiple frequencies combining to generate observed rhythmic patterns.

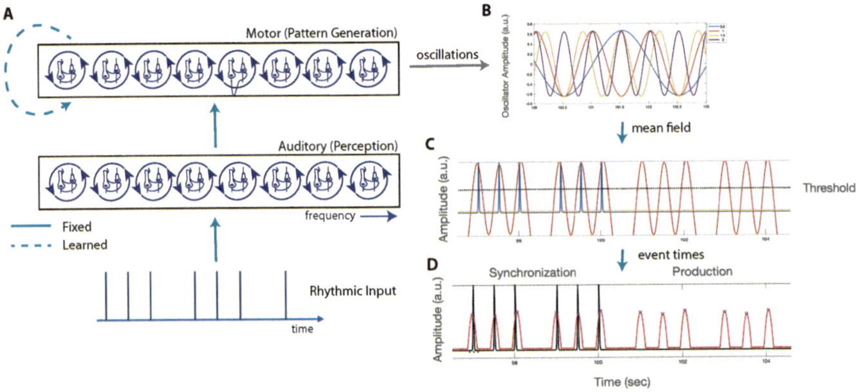

Fig. 8.6 The architecture of the rhythm learning model. A rhythmic input stimulates the perception network, representing the auditory cortex (e.g., A1), and the response of the perception network drives the pattern generation network, representing motor planning cortex (e.g., PMC, SMA). (**a**) The states of the oscillators in the pattern generation network (**b**) are summed to create a mean field response. (**c**) Peak picking is applied to determine the timing of model output (inverted triangles in the bottom plot). (**d**) The model can synchronize to the rhythmic stimulus (Synchronization), as well as produce the rhythm without the stimulus (Production) [62]

8.2.6 Discussion

Comparative research shows that animals such as sea lions, cockatoos, chimpanzees, and monkeys can coordinate movements to rhythmic sounds, sometimes intermittently locking to external rhythms, sometimes exhibiting stable synchronization, and sometimes adapting intrinsic tempos. Such behaviors offer parallels to human beat perception, and the available evidence points to underlying oscillatory neurodynamics. Evidence from neural recordings and computational models provides direct evidence that oscillatory mechanisms may indeed underlie synchronization, at least in macaques. However, dynamical considerations cast doubt on whether pulse and meter perception are present in any of the species tested to date, especially when considering perception of complex, syncopated stimuli such as "missing pulse" rhythms. In humans, such complex forms of perception emerge from long developmental trajectories. Finally, many species produce complex rhythmic vocalizations, and some exhibit preferences for integer-ratio timing patterns. A recent NRT model provides a hypothesized explanation for such integer-ratio preferences, and early characterization of animal vocalizations supports the predictions of this model. However, a great deal of work remains to be done in testing detailed behavioral and neurophysiological predictions.

To more precisely understand the nature of rhythmic coordination with periodic stimuli, significant methodological questions remain to be addressed. Future studies should be explicitly designed to test the detailed quantitative hypotheses that computational models can provide. Experimental paradigms that systematically manipulate stimulus phase and tempo perturbations represent the most stringent behavioral test of oscillator models for animals that exhibit stable synchronization. And if only intermittent synchronization is present, phase dynamics and statistics should be reported for the entire trial. Relatedly, it remains to be determined which nonhuman animals can adjust their intrinsic tempos through short- and long-term attunement in a manner comparable to humans. Only with these empirical tests can the hypotheses proposed by NRT be adequately evaluated. Determining whether such neurodynamic mechanisms are broadly conserved across species, and how coupling strength, flexibility, and adaptive capacity differ between humans and other animals, will be essential for tracing the evolutionary roots of musical rhythm.

From a comparative and evolutionary perspective, another critical question concerns the extent to which nonhuman animals exhibit genuine beat perception, beyond synchronization to surface-level acoustic regularities. Experimental paradigms that systematically manipulate rhythmic complexity and structure, such as syncopation and missing-pulse paradigms, will be crucial for distinguishing between simple entrainment to external energy and genuine beat perception. Clarifying the neural circuits and mechanisms that give rise to nonlinear resonance, particularly in the perception of pulse and meter in missing-pulse rhythms, represents an important challenge.

In developmental and learning contexts, further research is needed to understand how oscillatory circuits become attuned to rhythmic structures over the lifespan. Human infants and children acquire stable resonances through exposure, enculturation, and active movement, and cultural variation has an important role in shaping rhythmic cognition. Further work is needed to establish whether and how exposure and training may influence perception and behavior. Comparing human developmental trajectories with learning processes in nonhuman species may provide unique insights into the adaptive flexibility of neural resonance.

Although evidence suggests that some nonhuman animals exhibit preferences for small integer ratios in rhythmic vocalizations, the underlying mechanisms remain unclear. From a comparative perspective, further work is needed to establish whether integer ratio preferences are widely shared across species that produce complex vocalizations, such as songbirds, primates, and lemurs, and to determine whether such preferences reflect evolutionary homology or convergent adaptation. It also remains unclear whether integer ratio biases are restricted to vocal production or extend broadly to perception and motor synchronization. Finally, methodological and modeling challenges remain. Invasive neural recordings in animal models may help identify oscillatory mechanisms that support integer ratio preferences, while refined computational models are needed to capture both universal biases and cultural variation. Novel experimental paradigms, including perturbation methods, cross-cultural comparisons, and testing with missing-pulse rhythms, will be essential for disentangling innate neurodynamic constraints from learned rhythmic patterns.

Finally, overarching modeling questions remain to be addressed. Invasive neural recordings in animal models may provide direct evidence for the oscillatory mechanisms predicted by NRT, but such work is still in its early stages. Detailed network models can exhibit low-dimensional oscillatory dynamics that can be directly analyzed. However, these models need to inform the development of computational oscillator-level models. And low-dimensional oscillator models, while successful in generating testable hypotheses, require further refinement to more closely approximate real neurophysiological processes. At the network level, it is not yet fully understood how different classes of oscillatory dynamics—endogenous, damped, and bistable—may interact to support rhythm performance, perception, and synchronization. Thus, while significant theoretical work is available to support empirical investigations, we are at the brink of truly understanding the neurodynamic mechanisms that support rhythm cognition in humans and animals.

In summary, despite substantial progress in applying neural resonance theory (NRT) to comparative rhythm cognition, several open questions remain for future research. At the neural level, more evidence is needed to clarify how oscillatory circuits support pulse and meter perception, including nonlinear resonance and integer-ratio preferences. Comparative studies should determine whether nonhuman animals exhibit genuine beat perception and attunement, and to what extent these mechanisms are conserved across species. Developmental research is also required to chart how behavior becomes tuned through exposure and training. Finally, advances in invasive recordings, computational modeling, and experimental paradigms will be crucial for distinguishing simple entrainment from true beat perception. Addressing these open questions will not only advance the theoretical framework of NRT but also shed new light on the neural, developmental, and evolutionary foundations of rhythm cognition.

8.3 Conclusion

Neural resonance theory (NRT) proposes that music perception and performance emerge from neural and bodily oscillations that resonate with external sounds and with each other, forming stable temporal patterns through self-organization and learning. Drawing on dynamical systems theory, NRT explains how oscillations, resonance, nonlinearity, stability, and attunement shape rhythmic cognition, allowing brains to adapt to new sound patterns across evolutionary, developmental, and cultural timescales. By linking neurobiological processes with cultural and behavioral outcomes, NRT provides a unifying framework for studying rhythm cognition in humans and animals, while raising open questions about the neural mechanisms, developmental trajectories, and evolutionary roots of musical rhythm.

References

1. Hoppensteadt FC, Izhikevich EM (1997) Weakly connected neural networks. Applied Mathematical Sciences. Springer, New York. https://doi.org/10.1007/978-1-4612-1828-9
2. Wilson HR, Cowan JD (1973) A mathematical theory of the functional dynamics of cortical and thalamic nervous tissue. Kybernetik 13:55–80
3. Breakspear M (2017) Dynamic models of large-scale brain activity. Nat Neurosci 20:340–352. https://doi.org/10.1038/nn.4497
4. Buzsáki G, Vöröslakos M (2023) Brain rhythms have come of age. Neuron 111:922–926. https://doi.org/10.1016/j.neuron.2023.03.018
5. Large EW, Herrera JA, Velasco MJ (2015) Neural networks for beat perception in musical rhythm. Front Syst Neurosci 9:159. https://doi.org/10.3389/fnsys.2015.00159
6. Bartolo R, Merchant H (2015) Oscillations are linked to the initiation of sensory-cued movement sequences and the internal guidance of regular tapping in the monkey. J Neurosci 35:4635–4640. https://doi.org/10.1523/jneurosci.4570-14.2015
7. Zamm A, Wellman C, Palmer C (2016) Endogenous rhythms influence interpersonal synchrony. J Exp Psychol Hum Percept Perform 42:611–616. https://doi.org/10.1037/xhp0000201
8. Large EW, Jones MR (1999) The dynamics of attending: how people track time-varying events. Psychol Rev 106:119–159. https://doi.org/10.1037/0033-295x.106.1.119
9. Lakatos P, Karmos G, Mehta AD et al (2008) Entrainment of neuronal oscillations as a mechanism of attentional selection. Science 320:110–113. https://doi.org/10.1126/science.1154735
10. Gibson JJ (1966) The senses considered as perceptual systems. Houghton Mifflin, Boston
11. Haken H, Kelso JAS, Bunz H (1985) A theoretical model of phase transitions in human hand movements. Biol Cybern 51:347–356
12. Hoppensteadt F, Izhikevich E (2003) In: Arbib MA (ed) Brain theory and neural networks. 2nd ed (pp 181–187). MIT Press, Cambridge
13. Thompson E, Varela FJ (2001) Radical embodiment: neural dynamics and consciousness. Trends Cogn Sci 5:418–425
14. Lakatos P, Gross J, Thut G (2019) A new unifying account of the roles of neuronal entrainment. Curr Biol 29:R890–R905. https://doi.org/10.1016/j.cub.2019.07.075
15. Pikovsky A, Rosenblum M, Kurths J (2001) Synchronization: a universal concept in nonlinear sciences. Cambridge University Press, Cambridge
16. Izhikevich EM (2004) Which model to use for cortical spiking neurons? IEEE Trans Neural Netw 15:1063–1070. https://doi.org/10.1109/tnn.2004.832719

17. Large EW, Roman I, Kim JC et al (2023) Dynamic models for musical rhythm perception and coordination. Front Comput Neurosci 17:1151895. https://doi.org/10.3389/fncom.2023.1151895
18. Kim JC, Large EW (2015) Signal processing in periodically forced gradient frequency neural networks. Front Comput Neurosci 9:152. https://doi.org/10.3389/fncom.2015.00152
19. Kim JC, Large EW (2019) Mode locking in periodically forced gradient frequency neural networks. Phys Rev E 99:022421. https://doi.org/10.1103/physreve.99.022421
20. Hoppensteadt FC, Izhikevich EM (1998) Thalamo-cortical interactions modeled by weakly connected oscillators: could the brain use FM radio principles? Biosystems 48(1-3):85–94
21. Large EW, Gray PM (2015) Spontaneous tempo and rhythmic entrainment in a bonobo (Pan paniscus). J Comp Psychol. https://doi.org/10.1037/com0000011
22. Patel AD, Iversen JR, Bregman MR, Schulz I (2009) Experimental evidence for synchronization to a musical beat in a nonhuman animal. Curr Biol 19:827–830
23. Barnes R, Jones MR (2000) Expectancy, attention, and time. Cogn Psychol 41:254–311
24. Harding EE, Kim JC, Demos AP et al (2025) Musical neurodynamics. Nat Rev Neurosci 1–15. https://doi.org/10.1038/s41583-025-00915-4
25. Glass L (2001) Synchronization and rhythmic processes in physiology. Nature 410:277–284. https://doi.org/10.1038/35065745
26. Eguiluz VM, Ospeck M, Choe Y et al (2000) Essential nonlinearities in hearing. Arxiv 84:5232–5235. https://doi.org/10.1103/physrevlett.84.5232
27. Laudanski J, Coombes S, Palmer AR, Sumner CJ (2010) Mode-locked spike trains in responses of ventral cochlear nucleus chopper and onset neurons to periodic stimuli. J Neurophysiol 103:1226–1237. https://doi.org/10.1152/jn.00070.2009
28. Meddis R, O'Mard LP (2006) Virtual pitch in a computational physiological model. J Acoust Soc Am 120:3861–3869. https://doi.org/10.1121/1.2372595
29. Fujioka T, Trainor LJ, Large EW, Ross B (2012) Internalized timing of isochronous sounds is represented in neuromagnetic beta oscillations. J Neurosci 32(5):1791–1802
30. Nozaradan S, Peretz I, Keller PE (2016) Individual differences in rhythmic cortical entrainment correlate with predictive behavior in sensorimotor synchronization. Sci Rep 6:20612. https://doi.org/10.1038/srep20612
31. Strogatz SH (2014) Nonlinear dynamics and chaos. Westview Press. Boulder, Colorado
32. Schmidt RC, Beek PJ, Treffner PJ, Turvey MT (1991) Dynamical substructure of coordinated rhythmic movements. J Exp Psychol Hum Percept Perform 17:635–651
33. Kelso JAS (1995) Dynamic patterns: the self-organization of brain and behavior. MIT Press. Boulder, Colorado
34. Tichko P, Kim JC, Large EW (2022) A dynamical, radically embodied, and ecological theory of rhythm development. Front Psychol 13:653696. https://doi.org/10.3389/fpsyg.2022.653696
35. Roman IR, Roman AS, Kim JC, Large EW (2023) Hebbian learning with elasticity explains how the spontaneous motor tempo affects music performance synchronization. PLoS Comput Biol 19:e1011154. https://doi.org/10.1371/journal.pcbi.1011154
36. Tichko P, Kim JC, Large EW (2021) Bouncing the network: a dynamical systems model of auditory–vestibular interactions underlying infants' perception of musical rhythm. Dev Sci. https://doi.org/10.1111/desc.13103
37. Cook P, Rouse A, Wilson M, Reichmuth C (2013) A California sea lion (*Zalophus californianus*) can keep the beat: motor entrainment to rhythmic auditory stimuli in a non vocal mimic. J Comp Psychol. https://doi.org/10.1037/a0032345
38. Zamm A, Wang Y, Palmer C (2018) Musicians' natural frequencies of performance display optimal temporal stability. J Biol Rhythm 33:432–440. https://doi.org/10.1177/0748730418783651
39. Hattori Y, Tomonaga M, Matsuzawa T (2013) Spontaneous synchronized tapping to an auditory rhythm in a chimpanzee. Sci Rep. https://doi.org/10.1038/srep01566
40. Scheurich R, Zamm A, Palmer C (2018) Tapping into rate flexibility: musical training facilitates synchronization around spontaneous production rates. Front Psychol 9:458. https://doi.org/10.3389/fpsyg.2018.00458

41. Large EW, Fink P, Kelso SJ (2002) Tracking simple and complex sequences. Psychol Res 66:3–17. https://doi.org/10.1007/s004260100069
42. Palmer C, Lidji P, Peretz I (2014) Losing the beat: deficits in temporal coordination. Philos Trans R Soc B Biol Sci 369:20130405. https://doi.org/10.1037/a0021435
43. Repp BH (2008) Multiple temporal references in sensorimotor synchronization with metrical auditory sequences. Psychol Res 72:79–98
44. Wei Y, Hancock R, Mozeiko J, Large EW (2022) The relationship between entrainment dynamics and reading fluency assessed by sensorimotor perturbation. Exp Brain Res 240:1775–1790. https://doi.org/10.1007/s00221-022-06369-9
45. Rouse AA, Cook PF, Large EW, Reichmuth C (2016) Beat keeping in a sea lion as coupled oscillation: implications for comparative understanding of human rhythm. Front Neurosci 10:403. https://doi.org/10.1038/nrn2152
46. Cook PF, Hood C, Rouse A, Reichmuth C (2025) Sensorimotor synchronization to rhythm in an experienced sea lion rivals that of humans. Sci Rep 15:12125. https://doi.org/10.1038/s41598-025-95279-1
47. Gámez J, Mendoza G, Prado L et al (2019) The amplitude in periodic neural state trajectories underlies the tempo of rhythmic tapping. PLoS Biol 17:e3000054. https://doi.org/10.1371/journal.pbio.3000054
48. Zemlianova K, Bose A, Rinzel J (2024) A recurrent neural network for rhythmic timing. https://doi.org/10.1101/2024.05.24.595797
49. Hannon EE, Trehub SE (2005) Tuning in to musical rhythms: infants learn more readily than adults. Proc Natl Acad Sci 102:12289–12290
50. Nave-Blodgett JE, Snyder JS, Hannon EE (2021) Hierarchical beat perception develops throughout childhood and adolescence and is enhanced in those with musical training. J Exp Psychol Gen 150:314–339. https://doi.org/10.1037/xge0000903
51. McAuley JD, Jones MR, Holub S et al (2006) The time of our lives: life span development of timing and event tracking. J Exp Psychol Gen 135:348–367. https://doi.org/10.1037/0096-3445.135.3.348
52. Snyder JS, Gordon RL, Hannon EE (2024) Theoretical and empirical advances in understanding musical rhythm, beat and metre. Nat Rev Psychol 3:449–462. https://doi.org/10.1038/s44159-024-00315-y
53. Naveda L, Gouyon F, Guedes C, Leman M (2011) Microtiming patterns and interactions with musical properties in samba music. J New Music Res 40:225–238. https://doi.org/10.1080/09298215.2011.603833
54. Nozaradan S (2012) Selective neuronal synchronization to the beat and meter embedded in a musical rhythm. https://doi.org/10.1523/jneurosci.3203-12.2012
55. Tal I, Large EW, Rabinovitch E et al (2017) Neural entrainment to the beat: the "missing pulse" phenomenon. J Neurosci. https://doi.org/10.1523/jneurosci.2500-16.2017
56. Tichko P, Large EW (2019) Modeling infants' perceptual narrowing to musical rhythms: neural oscillation and Hebbian plasticity. Ann N Y Acad Sci 10:1. https://doi.org/10.1111/j.1467-7687.2009.00912.x
57. Jacoby N, Polak R, Grahn JA et al (2024) Commonality and variation in mental representations of music revealed by a cross-cultural comparison of rhythm priors in 15 countries. Nat Hum Behav 8:846–877. https://doi.org/10.1038/s41562-023-01800-9
58. Jacoby N, McDermott JH (2017) Integer ratio priors on musical rhythm revealed cross-culturally by iterated reproduction. Curr Biol 27:359–370. https://doi.org/10.1016/j.cub.2016.12.031
59. Ravignani A, Delgado T, Kirby S (2016) Musical evolution in the lab exhibits rhythmic universals. Nat Hum Behav 1:0007. https://doi.org/10.1038/s41562-016-0007
60. Roeske TC, Tchernichovski O, Poeppel D, Jacoby N (2020) Categorical rhythms are shared between songbirds and humans. Curr Biol 30:3544–3555.e6. https://doi.org/10.1016/j.cub.2020.06.072
61. De Gregorio C, Valente D, Raimondi T et al (2021) Categorical rhythms in a singing primate. Curr Biol 31:R1379–R1380. https://doi.org/10.1016/j.cub.2021.09.032
62. Brenner HR (2025) Musical rhythm categorization is predicted by neural oscillation and Hebbian learning. Master's thesis, University of Connecticut. ctdigitalarchive.com

9 The Neuroethology of Rhythmic Communication: Neural and Mechanical Building Blocks and Constraints on Rhythm Production and Perception

Gabriela-Alina Sauciuc, Jennifer N. Schneider, and Yuko Hattori

Abstract

Humans and animals communicate with rhythmic behaviours that vary greatly in terms of temporal parameters and patterns. In this chapter, we discuss the neuromuscular architecture underlying rhythmic behaviours and address factors that constrain their rhythmic properties. In Part I of the chapter, we provide background regarding the building blocks of rhythmic communication from the muscles to the brain and then illustrate how these building blocks are coordinated within a single system (e.g., locomotor appendage apparatus, vocal articulatory apparatus, etc.) and across relevant systems (locomotor, respiratory, and vocal). In Part II, we illustrate how variations in anatomical design and physiological function constrain rhythmic production, in particular properties such as temporal rate, stability, versatility, and pattern complexity. We also provide examples of putative mechanisms that influence rhythm perception across species. By reviewing and discussing these aspects of the neuroethology of rhythmic communication, we hope to infuse current mechanistic theorizing on the evolution of rhythmic competence(s) with novel insights from research that, historically, has had less weight in this area of study. We also hope that the design of future experimental studies aimed at testing evolutionary hypotheses will benefit from increased awareness of how variations in aspects of muscular and vocal tract design, along with "low-level" neural mechanisms, shape rhythmic communication.

G.-A. Sauciuc (✉)
Department of Philosophy/Cognitive Science, Lund University, Lund, Sweden
e-mail: Gabriela-Alina.Sauciuc@lucs.lu.se

J. N. Schneider
Department of Psychology, LCC International University, Klaipeda, Lithuania
e-mail: jschneider@lcc.lt

Y. Hattori
Institute for the Evolutionary Origins of Human Behavior (EHUB), Kyoto University, Inuyama, Japan
e-mail: hattori.yuko.8w@kyoto-u.ac.jp

G.-A. Sauciuc et al. (eds.), *Nature Beats*, Fascinating Life Sciences,
https://doi.org/10.1007/978-3-032-24250-1_9

Abbreviations

AMY	Amygdala
ATP	Adenosine triphosphate
BG	Basal ganglia
Cb	Cerebellum
CBGTC	Cortical-basal ganglia-thalamo-cortical
CO	Chordotonal organs
CPG	Central pattern generator
CS	Campaniform sensilla
DA	Dopamine
DLR	Diencephalic locomotor region
DN	Descending neurons
GABA	Gamma amino-butyric acid
GPe	Globus pallidus pars externa
GPi	Globus pallids pars interna
GTO	Golgi tendon organs
latRM	Lateral rostral medulla
LRC	Locomotor-respiratory coupling
LRVC	Locomotor-respiratory-vocal coupling
MCH	Myosin heavy chain
MLR	Mesencephalic locomotor region
MN	Motor neuron
NMDA	*N*-methyl-D-aspartate
PAG	Periaqueductal grey
PBN	Parabrachial nucleus
Pn	Pacemaker nucleus
RVC	Respiratory vocal coupling
SNc	Substantia nigra pars compacta
SNr	Substantia nigra pars reticulata
STN	Subthalamic nucleus
UBG	Unit burst generator
VNC	Ventral nerve cord

9.1 Introduction

As shown in other chapters of this book (Chap. 2 [1], Chap. 3 [2], Chap. 4 [3], Chap. 5 [4], Chap. 6 [5], Chap. 10 [6], Chaps. 12 [7] and 14 [8]), the rhythmic communication of animals varies greatly in terms of temporal parameters and patterns. In this chapter, we discuss the neuromuscular architecture of rhythmic behaviour, focusing on factors that constrain rhythmic properties. No doubt, this is a tremendously complex and extensive topic, whose thorough treatment would require several chapters, if not books. In this chapter, we had to be severely selective, and, therefore, chose to focus on aspects that complement the mechanisms discussed in other chapters (Chap. 5 [4], Chap. 1 [9], Chap. 7 [10], Chap. 8 [11], Chap. 11 [12], and Chap. 14 [8]) and that are less commonly treated in the human literature on rhythm perception and synchronisation. In this endeavour, we organised the chapter into two parts. Part I (Sect. 9.2) is meant to equip the reader with relevant background about the neuromuscular architecture of rhythm production. In this section, we first focus on the building blocks of rhythmic communication from the muscles to the brain (Sect. 9.2.1) and then outline how these building blocks are coordinated within a single system (locomotor appendage apparatus, vocal articulatory apparatus, etc.) and across relevant systems (Sect. 9.2.2). After having provided this background on how rhythm is produced, in Part II (Sect. 9.3), we switch focus to how variations in anatomical design and physiological function constrain rhythmic properties (rate, stability, versatility, and pattern complexity) and also provide examples of putative mechanisms that influence rhythm perception across species. By reviewing and discussing data pertaining to these mechanisms of rhythm production and perception, we hope to infuse current mechanistic theorising on the evolution of rhythmic competence with novel insights from research that, historically, has had less weight in this area of study. We hope that the design of future experimental studies aimed at testing evolutionary hypotheses will benefit from increased awareness of these mechanisms.

9.2 Part I: The Neuroethological Architecture of Rhythmic Communication

9.2.1 The Building Blocks of Rhythm Production

Factors that constrain the rhythmic properties of behaviour can arise at each level of neuroethological organization, from the organelles and proteins that generate mechanical energy in the muscle fibres to the intentional control of motor planning. In this section, we provide a succinct overview of building blocks and levels in the neuroethological architecture of rhythmic communication. This overview is organised bottom-up from peripheral effectors to central commands.

9.2.1.1 Muscular Design: Molecular and Energetic Constraints on Rhythm Production

The mechanical building blocks of behaviour, including rhythmic communication, are the striated muscles. Their contractions generate force on skeletal joints and cartilaginous tissues, which change the shape and position of body parts and produce movement. The principles that govern the structural and functional organisation of the striated muscles are widely conserved across insect and vertebrate species. Within this shared template, however, the physical and dynamic properties of the muscles vary substantially (as detailed further down), affording rhythmic outputs across a great range of rates and patterns. The rhythmic profile of a muscle is determined by several scalable properties—e.g., contraction frequency, shortening velocity, contractile force, fatigability—that constrain one another. These depend on the physical properties (e.g., length, number, density, relative volume) of subcellular components (organelles, proteins) found in the extrafusal fibres (which generate energy through contraction) and the bioenergetic pathways that fuel the activity of these fibres.

Box 9.1 Muscle Mechanisms Beyond MSH Isoform with a Role in Rhythm and Pattern Generation

Muscle tissue consists of long, tubular cells (*myofibres*) grouped in fascicles delimited by highly vascularised and innervated fibrous structures. About 80% of muscular volume is made of *myofibrils*—longitudinally arranged rod-like organelles composed of proteins with contractile and/or structural functions. Each myofibril is surrounded by a net-like structure—the *sarcoplasmic reticulum*—and by conduit-like *T-tubules* that run perpendicular to the fibre's axis, forming the *transverse tubular system*. These two components mediate electrical impulse transmission and the storage and regulation of intracellular calcium levels.

The proteins are organised into myofilaments that form serially arranged *sarcomeres*, which constitute the contractile units of the muscle. The most abundant myofibrillar proteins are the contractile proteins *myosin* and *actin*, which form the *thick* and *thin myofilaments*, respectively. According to a well-accepted theory, muscle contraction is produced by the sliding movement of thin and thick filaments, which causes muscle lengthening or shortening, thereby generating tension. The contractile force is produced at myosin cross-bridge sites, where myosin binds to actin. Myosin cross-bridges are activated by the hydrolysis of adenosine triphosphate (ATP), which is produced by reticular organelles called mitochondria, through the oxidation of nutrients such as carbohydrates and fats. Actin is set into a receptive state by an increase in calcium concentration, which is released from the sarcoplasmic reticulum upon reception of action potentials relayed by T-tubules.

The volume occupied by myofibrils determines the force of contraction, the volume of the sarcoplasmic reticulum influences the contraction frequency, the length of the sarcomere influences shortening velocity and force production, and the volume of mitochondria influences the level of sustained performance. For example, sarcoplasmic reticulum and T-tubules are abundant in muscles that can produce fast (but short-lived) contractions, e.g., the levator pectoralis of gobies that produce rhythmic signals through tremulation [13]. Likewise, the ultrafast muscles that drive swim bladder sound production in fishes tend to have an expanded sarcoplasmic reticulum, and a reduced myofibril and mitochondrial volume [14]. Sarcomere length regulates force-speed trade-offs, whereby longer sarcomeres generate greater force, while shorter ones contract with greater velocity [15, 16]. This variable is more influential in invertebrates, where sarcomere length ranges from 2 to more than 10μm, as opposed to vertebrates, which exhibit a small variation of sarcomere length (2.5–3μm at rest) [15]. Finally, as mitochondria are the biggest consumer of cellular oxygen (up to 98%), high mitochondrial content is associated with high ATP production based on oxidative metabolism, which is the characteristic of type I and IIa fibres and lower contraction velocity.

While numerous mechanical and energetic factors determine the functional profile of a muscle (see Box 9.1), here we focus on the contractile protein myosin, which is a key determinant of contraction speed and force. In over-simplified terms, a contraction occurs when a nervous signal activates the myosin filaments, which transduces chemical energy into mechanical work (for details, see Box 9.1). Like in most cellular processes, chemical energy is shuttled in the form of adenosine triphosphate (ATP) and the specific contractile output will vary depending on the source of ATP, which maps onto molecular variations (i.e., isoforms) in the myosin heavy chain (MHC) component.

The MHC isoform is the main criterion for muscle fibre typing, with the most common striatal subtypes being I, IIa, and IIb/IIx (for reviews, see [17–19]). MHC I is found in type I fibres, which are based on aerobic oxidative pathways and can be active for a long period of time, but produce low-speed, low-power contractions. Hence, these are also termed slow-resistant fibres. MHC IIb and IIx are expressed by type IIb and IIx fibres, which are based on anaerobic glycolytic pathways, and, thus, can produce high-power, high-speed contractions, but only for a short period of time (seconds to a few minutes). Thus, these fibres are also referred to as fast-fatigable fibres. Finally, MHC IIa is expressed by Type IIa fibres (also called fast-resistant fibres), which are an intermediate type with both oxidative and glycolytic capacities and can power relatively fast and prolonged activity.

Notably, muscles are seldom composed of a single fibre type. Even more so, two or more MHC isoforms can occur within the same fibre and the preponderance of "hybrid" fibres within a muscle can range from 0% to 100% [18]. Hybrid fibres abound in the laryngeal muscles and constitute a substrate for rhythmic versatility, enabling a varied range of contraction forces and velocities. This is consistent with the functional diversity of the laryngeal muscles, as they close the glottis to protect the airway, adjust airflow pressure to match respiratory frequency, and control the vocal folds to produce vocalisations (for reviews, see [18, 20]).

The exact fibre-type complement of the laryngeal (and, in avians, syringeal) muscles varies within the species depending on the specific function of the muscle, and across species between homologous muscles. A general mammalian trend is that the fibres of the laryngeal muscles can express the whole spectrum of MHC isoforms that can be found in a species, including ultrafast contracting isoforms that are not found in locomotor muscles [21, 22]. Generally, the vocal fold adductor is composed of faster fibre types than the vocal fold tensor, which expresses the same fibre types as found in the locomotor muscles of the given organism [21, 23]. The differential distribution of fibre-types across specific muscles reflects the differential functional demands placed on each of them. For example, the adductor needs to be faster than the other muscles in the laryngeal group to ensure the protective closure of the glottis. By contrast, slower and less fatigable fibres in the vocal tensor muscle ensure sustained performance (rather than speed), whereas the vocalis muscle has a highly heterogeneous fibre-type complement, which expands its versatility. Finally, as an example of between-species variation, the syringeal muscles of songbirds are much faster than the syringeal muscles of non-singing ring doves (*Streptopelia risoria*), enabling fast modulations with both acoustic and temporal versatility as a behavioural outcome [24].

9.2.1.2 Motor Neuron (MN) Contributions to Rhythm Production

The motor neurons (MNs) relay contraction signals to the muscles. A motor neuron together with the fibres it innervates forms the smallest functional unit of behaviour—the motor unit. The innervation patterns of motor units differ considerably across phyla, while the number and size of motor units also vary within the phylum and among different muscles within a species. These variations have functional significance, constraining the force and speed of muscular contractions, and, thus, the rhythmic capabilities of specific muscles, individuals, or species (as described below and also illustrated in Sect. 9.2.2.1).

In vertebrates, the MNs originate in the brainstem or spinal cord, from where a single axon extends to target effectors. Typically, a single MN innervates multiple fibres of the same type (e.g., type I only), while each fibre receives action potentials from a single MN [22]. When an MN generates an action potential, all of the innervated muscle fibres in the unit contract to produce a twitch. Given this innervation pattern, only fibres of a single type can selectively be activated throughout a muscle. This enables a flexible and efficient recruitment of motor units, such that the speed, force, and endurance of contractions are adequately tuned to changing functional demands.

The number of MNs that innervate a single muscle—and, thus, the number of motor units—ranges from 10–20 to several thousand [25]. This variable constrains the amount of contractile force generated by a muscle, which increases as the number of motor units increases. In turn, the size of motor units varies with the properties of component MNs and muscular fibres. Large motor units are dominated by fast type fibres (IIb/IIx) and are controlled by MNs with large cell bodies and large-diameter axons that can innervate hundreds of muscular fibres. By contrast, small motor units are dominated by slow type (I) fibres, which are innervated by MNs with small-diameter axons. In relative terms, within a given species and individual, the contractions generated by the largest motor units are two to three times stronger and faster compared to the smallest motor units (for a recent review, see [26]). Due to their small cell bodies, small motor units also have low input conductance, meaning that they require less current for reaching firing threshold [27]. They are, thus, the first to be recruited when rhythmic motor output is generated, while larger (and more powerful) motor units are gradually activated as a function of size (for a recent review, see [28]). Overall, the vertebrate design confers consistency across levels of organisation, but also great behavioural flexibility.

The size and number of motor units constrain the versatility and stability of rhythmic output. In the same individual, the motor units of the (smaller) laryngeal muscles comprise 10- to 20-fold fewer fibres compared to the (larger) locomotor muscles [23]. This variation has been related to functional flexibility, where the presence of more—but smaller—motor units allows for more fine-grained control of force and speed. Likewise, the small muscles of human fingers comprise hundreds of small motor units, allowing highly flexible motor tasks such as playing an instrument and typing on a keyboard. Such differences may, thus, account for differences in the precision and stability of rhythmic responses (e.g., eye saccades vs. finger-tapping) used to measure, e.g., rhythmic synchronization and compare it across species.

Contrasting with the complexities illustrated above, the MN organisation is more economic in insects. Insect muscles are typically innervated by only 1–2 MNs, rarely exceeding 10 MNs per muscle [29]. Most fibres have polyneuronal innervation thus contributing to several motor units. Insect MNs are also unipolar, i.e., they only present dendritic arborisation at the tip of the axon. Thus, synaptic information is integrated in dendrites (rather than in the soma) and action potentials are graded rather than all-or-none as in vertebrates [25, 30]. Another important difference is that arthropod MNs can be both excitatory and inhibitory, and, thus, the synaptic information received by muscle fibres is more complex [25]. Overall, this anatomical and functional organisation affords a kind of motor unit recruitment that is graded locally *within* the muscle fibre, resulting from the algebraic summation of the excitatory and inhibitory potentials relayed to the fibre. Thus, differences in the phase, frequency, and type of MN stimulation enable a high-precision graded response. A constraint of this, however, is that the recruitment of fibres with complementary contractile profiles is limited, and, thus, so is the ability of insects to regulate endurance at different levels of motor intensity.

9.2.1.3 The Genesis of Motor Rhythms in CPG Circuits

The activity of motor units is paced and patterned by central pattern generator (CPG) networks. A well-known characteristic of CPG neurons is that they exhibit intrinsic rhythmic activity even when synaptically isolated. The so-called *fictive* rhythms observed in isolated preparations resemble those observed in live behaviour (calls, gaits, etc.), although they tend to lack the smoothness and fine-tuning of their full-bodied counterparts, which in the intact organism is provided by sensory feedback (Sect. 9.2.1.4) and upstream control (Sect. 9.2.1.5).

The CPGs activate variable sets of MNs in particular sequences and at different levels of intensity, generating a diversity of functional patterns of motor output. Notably, different CPG networks pattern different motor functions (ventilation, walking, vocalising, etc.). A widespread cross-species motif in vertebrates is that trunk and appendage muscles are driven by spinal CPGs, which are distributed across cervical segments (forelimb CPGs), and thoracic and lumbar segments (hindlimb CPGs) [31, 33, 38]. In insects, there are analogous CPGs in thoracic and/or abdominal ganglia of the ventral nerve cord (VNC) [42–45]. These CPG mechanisms drive locomotor appendages, including the production of communicative signals, such as body rocking, stridulation, or tremulation (see Chap. 2 [1]). By contrast, the respiratory, laryngeal, and articulatory muscles implicated in vertebrate vocalisation are driven by hindbrain CPGs (for recent reviews, see [46–50]). In all CPG circuits, the molecular, electrophysical, and neuromodulatory properties of component neurons are the crucial determinants of rhythmic genesis, and of the timing and order of neuronal firing (for details, see Boxes 9.2, 9.3, and 9.4).

9.2.1.4 Peripheral Modulation by Mechanoreception

CPG activity and motor unit recruitment are dynamically modulated by proprioceptive feedback from mechanoreceptor organs that monitor muscle status. These receptors can be located within the muscles and tendons (vertebrates) or near the

Box 9.2 CPGs: Neuron Types with a Role in Rhythm and Pattern Generation

CPG circuits are heterogeneous, being composed of molecularly different neuron types that play different roles in the generation, pacing, and patterning of motor rhythms [31]. In vertebrates, most progress in determining the genetic identities of CPG neurons has been made in locomotion research, which has documented several relevant groups of excitatory (glutamatergic) and inhibitory (glycinergic and GABAergic) interneurons that are comparable across species [32]. The selective blockade or deletion of these neurons affects different aspects of rhythmic output, including left–right alternation, amplitude, frequency, and regularity.

The most important excitatory neuronal types that contribute to the rhythmic patterning of trunk or limb muscles are V2, V2a, and neurons that express SHOX2 and HB9 (for recent reviews, see [33, 34]). Some subsets of SHOX2+ neurons exhibit in-phase activity with either extensor or flexor muscles or are implicated in left–right alternation, while V2a populations regulate locomotor speed [34]. Notably, in zebrafish (*Danio rerio*), distinct V2a subsets mediate the gradual recruitment of motor units as a function of speed [35, 36]. As in motor neurons, "slow" V2a neurons have the lowest activity threshold, being the first to be engaged by a descending drive signal. At increased intensity of the signal, "intermediate" neurons are also excited (and recruited), and so forth. V2a interneurons regulate rhythmic frequency, but do not generate rhythmic bursting activity as such.

V0 interneurons are both excitatory (V0D subset) and inhibitory (V0V subset) and, in concert, control left–right alternation in a speed-dependent manner. The selective ablation of certain V0V neurons impairs the left–right pattern at low—but not high—frequencies, while the selective ablation of V0D neurons impairs alternation at high—but not low—speeds [37]. These populations, thus, play a crucial role in the pattern reconfiguration that underlies different gait types and speeds. Relevant inhibitory types are V1 and V2b, which provide inhibition signals to flexors and extensors, respectively [31]. When either population is inactivated, the alternation pattern is lost, being replaced by left–right synchrony [38, 39].

Overall, current data indicates that several neuronal populations contribute to different aspects of rhythm generation and patterning, and that these are differently engaged as a function of amplitude and speed. It is also clear that, across various muscular systems and vertebrate species, the core of CPG circuits consists of glutamatergic excitatory neurons and glycinergic inhibitory interneurons [33, 34, 38, 40, 41]. In vertebrates, ipsilateral excitatory drive is sufficient for rhythm generation and, combined with reciprocal inhibition, underlies basic motor patterns [34].

Box 9.3 CPGs: Electrophysiological Mechanisms That Set the Period and Phase Relationship

The discharge patterns of CPG circuits are governed by cellular events, such as the depolarisation and hyperpolarisation of the membrane potential, whereby the latter becomes more positive or more negative, respectively. These electrophysiological events are constrained by the ionic properties of the component neurons, such as the ion channels expressed in the cellular membrane and the in- or outward movement of ion fluxes between the intra- and extracellular space [51].

Ion currents cause rhythmic membrane fluctuations that can, but do not always, give rise to serial action potentials (bursting). Depolarising events that may result in rhythmic bursting are associated with the activation of inward currents carried by sodium and/or calcium ions or the cessation of outward currents carried by potassium ions [47, 51]. Fundamental to rhythm generation are mechanisms that promote post-inhibitory rebound. In particular, the hyperpolarisation-activated current causes slow depolarisation and the activation of voltage-dependent currents and channels (e.g., low-voltage-activated calcium channels) that are responsible for burst initiation. In concert, these ionic mechanisms can cause looping cycles of de- and hyperpolarisation, resulting in sustained oscillation [47].

Other ionic mechanisms modulate the regularity, frequency, and duration of rhythmic bursting in a time-dependent manner, whereby current injections produce different effects depending on their duration and timing in relation to de- and hyperpolarisation [47, 52]. In spinal neurons, bursting frequency is significantly regulated by the slow afterhyperpolarisation—an outward moving calcium-dependent potassium current—whose amplitude affects the duration of hyperpolarisation [51]. Conversely, the generation of plateau potentials is essential for long-lasting depolarisation, supporting rhythmic stability [34]. Sustained depolarisation is, for example, promoted by the persistent sodium channel, which is activated at subthreshold voltages and amplifies neuronal responses to synaptic input.

The significance of these ionic mechanisms to rhythmic signalling is straightforward, in that the rhythmic patterning of membrane potential oscillations exhibits species-, individual-, and behaviour-specific variations that parallel those observed at behavioural level [50, 53, 54]. At a mechanistic level, these variations have been linked to the differential expression of multiple voltage-gated channels [54].

Box 9.4 Intrinsic Neuromodulators that Influence CPG Activity

Multiple neuromodulatory substances and neurons play a critical role in regulating the cellular processes responsible for rhythm and pattern generation. Some of these are *intrinsic*, i.e., released from the component cells of the circuit, while others are *extrinsic*, i.e., released from sources beyond the CPGs [55, 56]. Neuromodulation contributes to the activation, deactivation, and fine-tuning of CPGs, as a function of behavioural context. It continuously modulates the number and the types of CPG neurons that are active at any given time, and their ionic properties. This way, a single CPG circuit can assume different configurations—and output different motor patterns—which provides a substrate for behavioural flexibility and context-adaptive rhythmic patterns [56].

Several neuromodulatory mechanisms affect rhythmic behaviour by acting on CPG circuits, but much more is yet to be uncovered about their specific cellular targets, release timing, and interactions with other neuromodulators. It is currently known that modulatory projections often feature several co-transmitters, that many neurons and their ion channels exhibit different types of neuromodulator receptors, and that a single neurotransmitter can activate multiple receptor types. The same substance can, thus, elicit different responses in different neuronal populations, which also depends on whether these populations are excitatory or inhibitory.

CPG intrinsic modulators, such as GABA, glutamate, serotonin, and, in some cases, dopamine, can alter the electrophysiological mechanisms of rhythm generation and patterning, such as ion channel properties and the strength of ion fluxes. This way, neuromodulatory mechanisms can modify the frequency, amplitude, and waveform of membrane potential oscillations [57]. For example, GABA-blockade, NMDA current increases, and the activation of glutamate receptors have the effect of speeding up the burst pattern. Burst frequency covaries with the strength of glutamatergic influences, being positively associated with the concentration of glutamate agonists. Glutamatergic synapses can increase calcium entry via NMDA receptors, activating calcium-dependent potassium channels. In turn, this can promote early burst termination and a frequency increase.

Serotonergic mechanisms can have multiple contributions to rhythmic output, including CPG activation, fine-tuning, and stability—a role which is further supported by dopamine. Indeed, in some CPG circuits (e.g., locomotion), the dendrites of some interneuron populations reach varicosities formed by serotonergic neurons that also contain dopamine. The blockade of some serotonin receptors can have the effect of speeding up rhythmic bursting. Conversely, serotonin release in inhibitory populations has stabilising effects on burst amplitude and reduces burst frequency [32]. It is important to note that in different species and different circuits, serotonergic and dopaminergic modulation are also provided by extrinsic mechanisms (see Chap. 11 [12]).

joints or in the exoskeleton (invertebrates) and contain sensory neurons that respond to mechanical stimulation with changes in membrane potential. This initiates action potentials that propagate to MN somata directly or indirectly, via interneurons. Mechanosensory feedback is necessary for tuning motor output to fluctuating functional demands and for coordinating the mechanical roles and contraction timing of interacting muscles (as detailed further down in Sect. 9.2.2.1).

9.2.1.4.1 Insect Mechanoreception and Relevance to Rhythmic Output

Insects exhibit several types of extramuscular tension sensors located near the joints, such as hair plates, campaniform sensilla (CS), and chordotonal organs (COs) (for a recent review, see [58]). These are innervated by a variable number of sensory neurons, ranging from one (e.g., in muscle receptor organs) to several hundred (e.g., the femoral CO of large insects), which are classified as rapidly or slowly adapting. The former respond *phasically* to movements in the monitored structure, i.e., they fire once when activated and once when deactivated, while the latter respond *tonically*, i.e., they fire repeatedly for the entire duration of the stimulation.

In terms of function, the hair plates located at leg joints provide excitatory and inhibitory sensory feedback to MNs that regulate leg muscle extension and flexion, limiting the forward trajectory of the legs. The CS monitors stress and strain generated by load changes in the exoskeleton. They tend to cluster on body appendages, where they contribute to joint coordination by regulating the firing phase of the MNs that control a given wing or leg segment. The COs are the most complex mechanoreceptors of insects. Although they are located inside the body, some respond to mechanical forces that originate from joints and muscles, while others respond to mechanical forces that originate outside of the body, such as substrate-transmitted vibrations and sound [58]. The COs regulate a great diversity of motor functions associated with leg, wing, or mouth muscles. Neuronal populations that are spatially segregated along a dorsal-ventral axis and functionally distinct (i.e., slowly or fast adapting) encode kinetic parameters, such as joint angle, the velocity and acceleration of joint movements, or particular combinations of position, velocity, and acceleration [59, 60]. Experimentally induced CO mutations alter the rhythmic parameters of motor output with respect to speed, regularity, phase, and persistence [61], underscoring the key role of CO feedback to rhythmic output.

9.2.1.4.2 Vertebrate Mechanoreceptors and Relevance to Rhythmic Output Modulation

In vertebrates, motor output is tuned by intramuscular mechanoreceptors, such as the golgi tendon organs (GTOs) and, especially, the muscle spindles. The GTOs contain Ib sensory afferent neurons, whose endings innervate the collagenous tissue of the muscle attachment and tendon. In humans, about 10–20 muscle fibres are connected to a single GTO [62]. Owing to membrane stretch receptors located within their dendritic field, the GTOs can detect the amount of strain developed in the muscle's tendon and monitor muscle force. They can generate both excitatory (positive) and inhibitory (negative) feedback. The former is important to regulating movement phases (e.g., the transition between stance and swing in locomotion [63]), while the latter plays a critical role in limiting the amount of force developed by the muscle, which protects the muscle against damage.

The muscle spindles are small bundles of specialised intrafusal fibres and sensory afferents (Ia and, to a lesser extent, IIa) encapsulated in the extrafusal fibres. The best studied spindle function is the transduction of muscle stretch information into neuronal firing patterns (for a recent review, see [64]). The intrafusal fibres contract along with the extrafusal fibres, and their length changes contribute to conveying information about the mechanical state of the muscle.

The traditional view of muscle spindles as passive sensors of muscle length and velocity has been recently questioned by novel findings on the function of mammalian γ-MNs. These account for about 30% of the mammalian MN pool and provide the intrafusal fibres with their own motor drive. Accumulating evidence suggests that the γ-MNs can also be activated independently of the α-MNs that innervate the extrafusal fibres, being tuned by complex sensorimotor input rather than by mere mechanical stimulation of the muscle (for recent reviews, see [65–67]). This indicates that the mammalian spindles may play a more complex role in the sensorimotor optimisation of behaviour, spanning movement adaptation in ongoing tasks and preparation for future action [65, 67–70]. For example, it has been suggested that subtypes of γ-MNs provide complementary information about the temporal template of the expected movement and departures from it [65]. Since spindle feedback reaches the somatosensory cortex, the evolution of the γ-MNs system has also been linked to enhanced conscious kinaesthesia, voluntary movement control, as well as motor plasticity [67, 71]. Spindle function, however, can also be specialised depending on muscle, e.g., in the upper limbs of humans—and potentially other primates—spindles may play a more predictive role, as forward sensory models, than in the lower limbs [71].

Variations in the properties of muscle spindles can account for variations in rhythmic versatility and precision. A higher number of intrafusal fibres per spindle—as found in humans (8–20) compared to mice (3–5) and fish (1) or reptiles (1)—is thought to contribute to enhanced mechano-sensitivity and, thus, enhanced effector precision. In laryngeal muscles, spindle density is especially high in areas where multiple forces interact [22]. Likewise, human hand muscles exhibit an abundance of spindles [66, 70]. Such constraints are relevant to evolutionary theorising, in particular the conclusions that can be drawn from current cross-species data regarding the evolution of rhythmic synchronisation, given that the few extant studies capitalize on forelimb motor responses, such as finger-tapping [72, 73] or manual drumming [74]. Based on current data, there are substantial differences in terms of tapping stability between humans and chimpanzees (*Pan troglodytes*), with tapping variance being 10 times higher in chimpanzees than in humans [73]. It may be tempting to interpret such performance differences as reflecting cross-species differences in the cognitive abilities that underpin sensorimotor synchronisation. Based on the currently available data, however, these could instead reflect neuromuscular differences at the effector level.

To sum up, research on the role of muscle spindles in motor control has uncovered several functional aspects with implications for the structuring and temporal adaptation of motor rhythms, including inter-muscle and inter-limb coordination, the flexible tuning and voluntary control of rhythmic movement, motor prediction, as well as motor plasticity and learning. Given the heterogeneity of current research, it is premature to formulate general principles regarding spindle constraints on rhythmic behaviour. Future research should strive for more sampling diversification and systematicity, especially considering that each muscle has its own composition of spindles, and each spindle has its own special set of intrafusal fibres, which entails unique morphological and physiological characteristics [75].

9.2.1.5 The Upstream Control of Rhythmic Communication

The activity of CPG circuits is controlled and monitored by an ensemble of upstream regions that play different roles in initiating or halting CPG activity, or modulating output parameters such as frequency, regularity, and amplitude. Several of these structures also function as integration hubs between motor commands and streams from sensory and emotion processing systems. Thus, they contribute to the dynamic adaptation of rhythmic outputs to internal and environmental demands, including sensorimotor synchronisation. In this section, we first discuss upstream controllers in insects. We then switch focus to vertebrates, for which we address mainly components and aspects that have not received attention in other chapters of this volume.

9.2.1.5.1 Descending Neurons as Controllers of Insect Rhythmic Communication

A critical role in the selection, initiation, and fine-tuning of insect rhythmic signals is played by so-called descending neurons (DNs), i.e., neurons that connect specific brain regions with the VNC (reviewed by [76]). The number of DNs is highly conserved across insect species (500–600 pairs), being much lower than the neuronal count of the brain or the VNC, e.g., in fruit flies there is a 10:1 brain-to-DNs ratio [76]. Thus, a few DN pairs—and even a single one—may suffice for exciting specific CPG circuits, as captured by the concept of "command neurons," defined as neurons that are both necessary and sufficient for initiating a certain motor response [77].

Command neurons have been identified for several rhythmic signals, such as the courtship stridulation of grasshoppers [78], the advertisement song of crickets [79], and fruit fly courtship song [80–82]. In field crickets, a single pair of command neurons, whose somata are located in the dorsal brain, reliably elicit advertisement song when activated with depolarised current, while inactivation with hyperpolarising current halts song [79]. Moreover, transient increases in the spiking frequency of these neurons correspond to chirp rate acceleration [79]. In fruit flies, pIP10—a DN that innervates the wings—is both necessary and sufficient for initiating courtship song. However, the primary rhythmic output driven by pIP10 differs across different species—e.g., pulse song in *Drosophila melanogaster*, but clack song in *Drosophila yaguba*—and also depends on activation intensity [82, 83]. For example, in *D. melanogaster*, low activation of pIP10 increases both pulse and sine song production, while high intensity increases pulse song production at the expense of sine song [82].

9.2.1.5.2 Supraspinal and Suprapontine Control of Rhythmic Communication in Vertebrates

The production of vertebrate rhythmic signals is controlled by forebrain-to-hindbrain networks whose overall design and modulatory profile are evolutionarily conserved [84]. Key gates for the regulation of body movements and vocalisation include, for example, brainstem centres for the locomotor (the mesencephalic and diencephalic locomotor regions—MLR and DLR, respectively) and non-locomotor control of limb movements (the lateral rostral medulla, latRM), mid- and forebrain structures with a role in motor output selection (the periaqueductal grey (PAG), the amygdala (AMY), the thalamus and the hypothalamus), and structures implicated in motor command planning and fine-tuning (the basal ganglia (BG), the cerebellum (Cb), the thalamus (T), and forebrain motor regions). In what follows, we briefly discuss how some of these components—MLR, DLR, BG, and Cb—contribute to the control of rhythmic motor output. For a discussion on cortical neural substrates of rhythmic communication, we refer the reader to Chap. 7 [10], while for a discussion of the role of the PAG, AMY, and H, we refer the reader to Chap. 11 [12]. Regarding the roles of BG and Cb, which are also discussed in Chap. 7 [10] and Chap. 11 [12], here we only review complementary aspects to those discussions.

Brainstem Motor Centres for Limb Control Several functionally distinct streams in the brainstem control rhythmic behaviours actuated by the fore- and hindlimbs. The MLR and DLR—which have been identified in vertebrate taxa from the lamprey (*Petromyzon marinus*) to primates—are motor command centres that control locomotor CPGs [33]. The MLR is implicated in both speed and pattern regulation, i.e., the degree of MLR activation corresponds to the degree of locomotor output. This is achieved through an incremental recruitment of MLR neurons or through neuronal specialisation, where certain MLR assemblies activate slower locomotor patterns, while others elicit faster ones [85, 86]. Moreover, it has been suggested that the MLR and the DLR may control locomotor output in distinct behavioural contexts, with DLR being primarily engaged by emotional stimuli [33].

In terms of connectivity, notable MLR projection targets are the thalamus and several brainstem nuclei, with the latter representing a substrate for the well-established association between locomotor activity and the release of noradrenaline and serotonin. The MLR is reciprocally connected with the BG, receiving input related to rhythmic pattern selection and sending back motor information. The MLR projections to BG are specialised, i.e., originate in specific MLR subpopulations and have specific targets in several BG nuclei. These form several independent anatomo-functional circuits that regulate distinct locomotor aspects: fast and slow locomotion, fore- and hindlimb movement, and motor pattern reinforcement [87]. Through these varied inputs, the MLR actively contributes to the control of motor tuning in the BG and beyond. Besides the BG, the MLR also receives input from the hypothalamus, which mediates the initiation of locomotor patterns on the basis of functional demands. DLR connectivity has been mainly studied in the lamprey, where it receives input from several forebrain areas, including the olfactory bulbs, pallium, the striatum, and H, as well as the ventral and dorsal thalami [88].

The MLR and DLR activate CPGs via reticulospinal neurons, which, similar to the insect command neurons, integrate motor commands with sensory feedback and vestibular signals [89]. The reticulospinal neurons are heterogeneous, with specific subpopulations playing roles in initiating, maintaining, and halting locomotion. Distinct assemblies have also been linked to the selective activation of extensors and flexors [90, 91] or the induction of locomotion over a wide range of speeds [92].

Non-locomotor behaviours that preferentially engage the forelimbs are controlled by distinct brainstem mechanisms, such as the lateral rostral medulla (latRM), which is preferentially connected with forelimb MNs [91, 93]. Optogenetic manipulations show that spatially distinct subsets of latRM neurons reliably elicit very specific movements (e.g., reach, retract) or more complex sequences of forelimb behaviour, including tapping movement (reviewed in [94]). This evidence underscores a high specificity, selectivity, and granularity in the brainstem circuits that control a great diversity of fore- and hindlimb behaviours.

The Basal Ganglia and the Cerebellum The basal ganglia (BG) and the cerebellum (Cb) have been implicated in multiple aspects of rhythm competence. Some of these aspects, in particular those hypothesised to play a crucial role in the evolution of human musicality and audiomotor synchronisation—i.e., predictive timing and beat perception—have been discussed in Chap. 7. Here, we discuss complementary roles of the BG and Cb in rhythm production and perception using human and animal data from the broader area of research on motor control, pattern perception, and learning.

The *basal ganglia* consist of six main subcortical nuclei. This structural organisation is highly conserved in vertebrates, which places the evolution of a basic circuit for motor control and time perception as early as 560 Mya [95, 96]. While research on musicality has singled out the BG as a time-keeping mechanism crucial to beat perception and audiomotor synchronisation, the animal literature has examined the BG in the broader framework of voluntary motor control and learning. Of relevance to rhythmic communication, the BG have been shown to control the selection, sequencing, and adaptation of actions as a function of internal and external demands, and to mediate the acquisition of new motor patterns. Two key mechanisms addressed in this context (similar to human research on rhythm competence) are BG connectivity and dopaminergic (DA) signalling (for more details, see Boxes 9.5 and 9.6).

In terms of connectivity, the BG are engaged in the cortico-basal ganglia-thalamic-cortical (CBGTC) loop (see also Chap. 7 [10]) and in several subcortical loops. These loops regulate the execution and learning of sequenced motor output by selecting and concatenating movements into correct behavioural sequences [33, 97, 98]. In songbirds, Area X—the avian counterpart of the striatum, the input nucleus of the BG—mediates the acquisition of song structure. Its surgical or optogenetic manipulation disrupts the regularity and timing of song elements [99, 100]. Once a song is acquired, however, Area X is no longer involved, with song execution being shifted to motor cortical networks.

Box 9.5 Dopaminergic Circuits in the Basal Ganglia

The basal ganglia (BG) comprise six main nuclei: the striatum, which is the main input nucleus that in primates presents multiple subdivisions, three intrinsic nuclei (globus pallidus externa—GPe, the subthalamic nucleus—STN, substantia nigra pars compacta—SNc), and two output nuclei (globus pallidus interna—GPi and the substantia nigra pars reticulata—SNr). Dopaminergic signalling from the SNc to the striatum is critical for motor regulation and exerts its influence along two spatially and functionally segregated pathways to the SNr, known as the *direct* and *indirect* pathways. As a default (i.e., at rest), the SNr and GPi keep thalamic, cortical, and brainstem motor circuits under tonic inhibition. This inhibition, however, can be modulated by dopaminergic activity targeting striatal dopamine receptors (D1 and/or D2) (for more details, see [94, 95, 98, 101]).

Neurons of the *direct* pathway, which express D1 receptors, have direct synaptic connections with SNr and can positively modulate the excitability of downstream motor centres. If sufficiently strong, the stimulation of D1 cells can depolarise neuronal subsets in the SNr and intensify the excitatory signals of glutamatergic projections from the thalamus and cortex. As a result, the default inhibition of motor circuits is selectively lifted, releasing selected motor output.

By contrast, the *indirect* pathway connects to the SNr via the intrinsic BG nuclei (GPe and STN) and expresses D2 receptors. The stimulation of D2 neurons negatively modulates motor output by exciting the SNr and, thus, increasing SNr inhibition on downstream motor centres. Importantly, however, during motor output, both D1 and D2 neurons are active in specific patterns, suggesting that D2 neurons may contribute to enhancing action selectivity by constraining the release of non-planned motor programmes. Consistent with this, the intrinsic nuclei GPe and STN exhibit topographically distinct motor functions [102].

The mammalian CBGTC circuit, however, shows a divergent pattern. Specifically, the striatum and the thalamostriatal projections are critically implicated in both the acquisition and execution of behavioural sequences (e.g., novel stepping sequences, lever press combinations, etc.), whereas the cortico-striatal projections are necessary for learning but not for executing newly learned motor patterns (humans [103], rodents [104–107], macaque monkeys [108]). Indeed, cortical lesions hinder the acquisition of novel motor patterns but not the execution of newly acquired motor skills, whereas disruptions of thalamo-striatal projections impair both [106, 107]. This is consistent with evidence showing that task-related cortical activation decreases over the course of motor learning, i.e., with more automatisation, there is less cortical reliance [109]. Conversely, human patients with BG-related motor dysfunctions do not exhibit an increase in striatal activity when performing automatic action sequences (as healthy individuals do). Instead, they show greater cortical activity, which can be taken as an indication that the cortex never stops playing its tutor role [110].

The acquisition of novel motor patterns is mediated by experience-dependent plasticity of cortico-striatal synapses [104]. This kind of motor cortical guiding also elicits plasticity at thalamostriatal synapses—accounting for the essential role these play in the execution of consolidated skills—and in subcortical circuits—accounting for BG-driven "reprogramming" of circuits that control innate movements [107, 111]. Conversely, the thalamostriatal projections modulate signal flow and plasticity at the corticostriatal synapses, contributing to the formation of associations between state (or context) signals and behavioural responses, which is consistent with the role of the BG in the initiation and execution of action sequences [112].

Different studies provide complementary insights into the more fine-grained details of how the BG regulates motor learning and the execution of consolidated motor sequences, whether innate or learned. For example, some studies demonstrate that the BG encodes and stores action sequences as integrated units [113, 114], thereby enabling behaviour chunking. Other studies highlight the role of the BG in controlling the switching between the elements of a sequence [105] or in specifying the fine-grained and time-varying details of kinematic patterns, including changes of speed and amplitude [106, 107, 115]. These roles also apply to rhythm perception and sensorimotor synchronisation, where the striatum has been found to be involved in maintaining internal representations of isochronous rhythms and in motor response preparation as a function of the perceived rhythm [108] (and references therein). Overall, it is believed that the BG achieves all these outcomes by encoding "policies" that map functional demands (conveyed by cortical and thalamic input) to reward-yielding *kinematic* variants that are conveyed to motor command centres [106].

Box 9.6 Basal Ganglia Connectivity of Relevance to the Production of Rhythmic Communication

The CBGTC loop has been extensively discussed as a prime neural substrate for rhythm perception and sensorimotor synchronisation in humans. This circuit, which has a broader role in motor control, is also found in mammals and has analogous counterparts in non-mammalian taxa (the pallial-basal ganglia-thalamo-pallial loop). In this loop, the motor control functions of the BG rely mainly on cortical and thalamic excitatory input [116]. Cortical projections to the BG have widely distributed origins in motor, sensory, and frontal areas. Some of these projections terminate in the striatum, while others reach the striatum via collaterals of axons that extend further into, e.g., the thalamus, superior colliculus, and brainstem. Most cortical projections to the striatum form specialised cortico-striatal circuits, although some may have overlapping targets. This design enables both sensorimotor integration and specialised segregated streams [117, 118].

Thalamo-striatal inputs originate in several thalamic nuclei, but the main sources are the parafascicular nucleus and the rostral laminar nuclei, which relay signals mostly from subcortical circuits, including the cerebellum (for a review, see [107]). These projections reach somatotopically organised compartments of the dorsolateral striatum comprising forelimb, trunk, hindlimb,

(continued)

Box 9.6 (continued)
and facial areas [119]. Thalamic input to the striatum provides ongoing information about behavioural states and contexts, modulating cortico-striatal inputs through cholinergic interneurons. Overall, these projections provide a substrate for rapid adaptation and behavioural flexibility. Via the output nuclei, the BG projects back to the cortex and the striatum, closing the CBGTC loop. The highly precise spatial organisation of projections noted above is maintained throughout this loop and also characterises projections from the striatum to the intrinsic and the output nuclei [102], from the SNr to the thalamus and their connectivity back to the cortex. In essence, the CBGTC loop consists of multiple, specialised parallel loops.

Of particular relevance for the control of rhythmic behaviour are also the subcortical loops that connect the BG with the superior colliculus [120], the cerebellum [94], the PAG [121], and over 40 brainstem targets [94]. Through these circuits, the BG regulates the motor command centres that control the activation of CPGs, thereby regulating behaviour patterns in the various effector systems [122]. Even here, the BG output has a high degree of granularity and specialisation, allowing the SNr to function as a "keyboard for movement control, where each key can elicit a specific type of movement" [111, p. 1]. Moreover, many SNr neurons that project to motor command centres also send axonal branches to the thalamus and further on to the cortex, conveying efference copies of motor commands, which constitutes a mechanism for monitoring ongoing movement and for learning [123].

The *cerebellum (Cb)* is a small, but highly compact structure that, depending on species, may account for 50–80% of the total neuronal count of the brain [124–126]. The grey matter that forms the cerebellar cortex is highly convoluted around a white matter core that contains the afferent mossy and climbing fibres projecting to the cortex, as well as the efferent Purkinje cell axons projecting to the deep cerebellar nuclei (DCN). The DCN, embedded within the white matter, receive strong inhibitory input from Purkinje cells and excitatory collaterals from mossy and climbing fibres. They also receive additional inputs from brainstem structures including vestibular, reticular, and pontine nuclei. The DCN constitute the primary output of Cb, sending efferent projections to the thalamus, reticular formation and other brainstem targets [127, 128].

The Cb has a highly conserved three-layered cortical organisation, consisting of *granule* cells, *Purkinje* cells, and a *molecular* layer where axons from the granule cells synapse on Purkinje cell dendrites, while basket and stellate cells provide inhibitory modulation of Purkinje cells. However, other characteristics of the cerebellum—size, functional subdivisions, structural elaboration—have undergone substantial evolutionary changes [129–131]. These changes have a high relevance for the evolution of rhythmic competence, insofar as they reflect increasing layers of sensorimotor integration and motor control. The oldest subdivision, the *archicerebellum*, has a vestibular role—a function which is considered the cradle of rhythmic competence in human prenatal development (see Chap. 10 [6]). The more recent *spinocerebellar* subdivision receives proprioceptive input from limb, trunk, and orofacial muscles, along with visual and auditory information, which is integrated at this level and relayed to cortical and brainstem areas that control motor output. Finally, the most recent *cerebrocerebellar* section underpins more advanced forms of motor planning[1] as a function of sensory input and is connected to the premotor and motor cortex via the thalamus.

Added to this phylogenetic gradient of functional subdivisions, variation in relative cerebellar size and foliation reflects variation in functional capability linked to different behavioural specialisations (for a review, see [127]). Of relevance to rhythm production and control, some of these variations regard the complexification of (loco)motor behaviour and sensory acquisition, both of which require increasingly precise temporal prediction. This also applies to the deep nuclei of the Cb. For example, the dentate nucleus is highly expanded and folded in humans and chimpanzees, with pathways that connect the highly expanded cerebellar hemispheres of these species with a highly expanded prefrontal cortex [128]. This nucleus is involved in limb motor control and cognition [132], and connected to brain areas known to be crucial to beat perception and sensorimotor synchronisation (for details, see Chap. 7 [10]), such as the putamen in the BG, as well as the premotor and motor cortex., supporting the fine-grained timing demands of rhythmic coordination.

Taken together, the examples mentioned above position the evolution of the Cb as a key driver of evolutionary changes in rhythmic competence, in particular owing to an enhancement of multisensory and sensorimotor integration, along with capacities for increased fine-tuning and correction of motor output. Indeed, via spinocerebellar tracts, the Cb receives detailed motor and proprioceptive information from the spinal cord about ongoing and planned movements. Conversely, the Cb can directly influence CPGs and MNs to provide motor corrections via specialised pathways as a function of movement—and, thus, rhythmic—phase (for details, see [33] and references therein), supporting the temporal precision required for rhythmic coordination. The role of the Cb in this context has been described as "inserting the timing of sensory inputs into the motor output" in a predictive manner, whereby the Cb can anticipate the sensory consequences of planned motor actions [133] through a short-term storage of sensory stimulus patterns [134].

9.2.2 The Coordination of Building Blocks Within and Across Systems

In Sect. 9.2.1, we discussed neuromechanical building blocks of rhythmic communication and their individual roles in the generation of rhythmic behaviour. In this

[1]The implication of the cerebellum in motor coordination has been known for two centuries. Research conducted during the past few decades has tremendously expanded our understanding of the functional role of the cerebellum to also include a role in learning, sensory integration, emotion, speech, and memory [10, 125] (this volume).

section, we focus on the functional coordination of such building blocks. Like any natural behaviour, the production of a rhythmic signal requires the precise coordination of multiple muscles and, thus, of the neuronal circuits that dictate or modulate individual phases and patterns of muscle contraction. Coordination is achieved at multiple levels of neuromuscular organisation—e.g., between the muscles that power a single joint, across multiple joints within a limb, between limbs, as well as between postural and limb muscles (detailed in Sect. 9.2.2.1). Moreover, functional coordination is also achieved between distinct systems (detailed in Sect. 9.2.2.2), since the episodic rhythms of communication (whether in the form of vocalisation, stridulation, wing beating, foot stomping, etc.) are generally based on neuromuscular substrates that overlap with those that underlie breathing (see Chap. 2 [1]).

9.2.2.1 Within-System Coordination

It is well-known that most striated muscles operate in antagonistic pairs, where an agonist (or prime mover) muscle contracts in antiphase with an antagonist muscle. With few exceptions, however, such as the swim bladder drumming of some fish species [135], this is not sufficient to generate a rhythmic behaviour. Locomotor behaviours, for example, are powered by hundreds of muscles [33]. Even a seemingly simple rhythmic response such as sequential finger tapping implicates 27 hand muscles, along with many others in the forearm, upper arm, and shoulder, all of which are activated in a highly complex and precisely timed sequence [136]. Indeed, in the generation of behaviours, the muscles can play different functional roles, besides the commonly known roles of agonist and antagonist. Agonists can be aided by synergist muscles and constrained by neutralisers, which prevent agonists from executing the full range of possible motions. In addition, joint proximal stabilisers provide further support, while fixators block force production on joints that should not move [137, 138].

Depending on how they change the state of joints or sphincters, muscles also play different mechanical roles, e.g., as extensor or flexor, as adductor or abductor, as protractor or retractor, as rotator or supinator, as sphincter-dilator or sphincter-constrictor. Thus, a single rhythmic movement often implicates multiple muscle groups across several (e.g., limb) segments, and the number of implicated muscles varies greatly depending on the complexity of motor output (for reviews, see [139, 140]).

While muscles do not have fixed functional roles (e.g., flexors are agonists in flexing movements, extensors are agonists in extension movements), they tend to have fixed mechanical roles, which correlate with neuromechanical properties that are relevant for rhythmic output, including synchronisation. In many joints, the flexors are anti-gravitational muscles, being stronger than their extensor counterparts[2] and they may require less cortical processing, especially for simpler movements

[2] It is important to acknowledge that the opposite profile can be recorded in certain joints or activities, e.g., there is evidence that cervical and knee extensors are stronger than their flexor counterparts. Flexor-extensor force ratio can be shaped through training, e.g., stronger extensors facilitate explosive movement in sprinters, while stronger flexors support graceful movement in gymnasts.

such as, for example, finger tapping. Flexion and extension, thus, can have different profiles of energy expenditure, which constrains the preferred kind of motor output. In finger-tapping tasks, for example, individuals generate more stable synchronisation when instructed to synchronise finger flexion (as opposed to finger extension) and, if instructed to synchronise finger extension to an auditory beat, they may even tend to switch to flexion, but not the other way around [141].

Although finger-tapping has generated most of the data on audiomotor synchronisation, it is a relatively simple movement, and thus, its informative value for evolutionary theorising about rhythmic competence may be limited. Humans and animals can produce far more complex rhythmic outputs, which pose far more complex demands in terms of within-system neuromechanical coordination. Thus far, very little research has been conducted on this topic in the context of rhythmic signalling. Important exceptions are the laryngeal/syringeal component of vocal production and research on locomotor patterning. To illustrate the incredible demands for coordination that are posed by rhythmic output, we discuss relevant examples from the locomotor research, which has a greater taxonomic spread compared to research on other rhythmic behaviours.

Locomotor modes vary greatly across taxa, exhibiting many coordination patterns. Undulatory swimming, for example, which is an ancestral model of vertebrate locomotion, is based on a complex rhythmic pattern consisting of a succession of contraction "waves" transmitted along the body, which are generated by the sequential and orderly contraction of trunk muscles from head to tail, with a constant phase lag. From a neural perspective, this pattern has been modelled as a series of segmentally repeated CPGs, where each CPG acts as a "half-centre oscillator." That is, each of these is composed of two hemi-segments that are alternatively activated through mutual inhibition (for recent reviews, see [142, 143]), and abolishing the inhibition also abolishes the rhythmic pattern.

Limbed locomotion (e.g., as found in tetrapods) adds substantial complexity, requiring the intra-limb coordination of multiple limb segments, as well as interlimb coordination across the body midline (between the left and right limbs) and, in quadrupeds, also between the fore- and hindlimbs. Intra-limb coordination has been successfully modelled using the unit burst generator (UBG) model proposed by Sten Grillner [144, 145]. This model captures the fact that locomotor CPGs do not generate simple alternate patterns of flexion-extension but a complex pattern that sequentially starts and terminates activity in synergist muscles in a given order and with a certain timing. Thus, different CPG units interact as coupled oscillators through reciprocal inhibition (e.g., hip flexor UBG—hip extensor UBG) or reciprocal excitation (hip extensor UBG—knee extensor UBG).

Left–right coordination is essentially achieved at each girdle through either alternated (as in walk and trot) or simultaneous activation (as in bound or flying), and the relevant CPGs are linked across the body midline by commissural connections. With added demands for fore- and hindlimb coordination, quadrupeds exhibit complex locomotor patterns, where the interlimb relative timing and sequencing varies in taxon-specific ways, as a function of anatomical design, and as a function of speed (for a review, see [146]). Locomotor patterns can be symmetrical (e.g., walking, trot) or asymmetrical (e.g., bound, gallop), depending on whether fore- and hindlimb footfalls are evenly or unevenly spaced in time. The limbs can be coupled in a lateral sequence, when a hindlimb footfall is followed by an ipsilateral forelimb footfall, or in a diagonal sequence, when a hindlimb footfall is followed by a contralateral forelimb footfall. It is important to note, however, that the limbs implicated in such lateral and diagonal couplets are slightly asynchronous. To exemplify, walking (e.g., equines and felids) has a four-beat symmetrical pattern, where the limbs are activated one at a time in a rotatory way, with two or three limbs providing support in each phase. The high-speed gallop is also a four-beat rotatory gait, but the temporal patterning is asymmetrical. By contrast, the medium-speed trot has a two-beat symmetrical pattern with diagonal couplets. The CPGs controlling for fore- and hindlimb engagement are distributed in the spinal cord with a rostral versus caudal location.

Note that, despite bipedalism, the fore- and hindlimbs continue to co-engage in human locomotion and, even more so, in dance. Rhythmic hand [147] or arm movement can alter the excitability of leg muscle activity [148, 149], and arm swing facilitates the generation of leg locomotor movement (for a recent review, see [47]). These findings are generally explained as neural coupling effects, mainly attributed to the propriospinal neurons located between the cervical and lumbar segments [150], which, ancestrally, ensured fore- and hindlimb coupling in quadrupedal locomotion. These continue to do so in humans during bipedal locomotion [151].

An additional layer of within-system coordination results from the inherent multifunctionality of muscles. The limbs implement multiple locomotor modes and communicative signals and, in some species, such as primates, locomotor limbs can also be used for reaching, stroking, tapping, and more. Examples of within-system multifunctionality abound in arthropods, where locomotor appendages are used for, e.g., percussion, tremulation, or stridulation (see Chap. 2 [1]). For example, in fruit flies and crickets, the muscles that power flight also power the rhythmic sounds of tremulation and stridulation, respectively (reviewed in Chap. 2 [1]). Moreover, flight and song rhythms share MN populations [152] and, in some taxa, even CPG circuits [153]. Other taxa, however, such as crickets, have specialised CPG circuits for song as opposed to locomotion [152] and for different types of song [42]. This specialisation for different kinds of acoustic output presents parallels to vertebrate vocal rhythm, where different song-types result from distinct coupled CPGs that control shared multifunctional muscles (for a recent review, see [50]).

Much continues to be unknown regarding the coordination of competing rhythmic outputs in shared muscular substrates, but in all systems sensory feedback plays a crucial role in rhythm coherence and flexibility within and across locomotor programmes. At peripheral level, in both vertebrates and invertebrates, sensory afferents from tactile organs and/or mechanoreceptors synapse into interneurons or directly onto MNs. This exchange of information helps coordinate motor activity between the segments of a given appendage and across the body midline and also regulates motor recruitment. At central level, motor outputs are modulated by signals from additional sensory modalities, such as vision and olfaction, with audio-motor synchronisation being an example of this process. Overall, this creates additional coordination complexities, since motor signals and different sensory modalities operate not only on different timescales, but within different coordinate frames (e.g., retina-centric versus leg-centric coordinates).

9.2.2.2 Between-System Coordination

Due to overlapping muscular substrates, breathing rhythms compete with locomotor and vocal rhythms, respectively, imposing constraints for between-system coordination. This is often resolved through neuromechanical coupling in the form of locomotor-respiratory coupling (LRC, discussed in Sect. 9.2.2.2.1) and respiratory-vocal coupling (RVC, discussed in Sect. 9.2.2.2.2) at integer ratios, as well as different forms of LRVC (discussed in Sect. 9.2.2.2.3). A general trend is that locomotion constraints respiration as a function of energetic demands, with LRC becoming increasingly necessary as locomotion becomes faster and more intensive [154], and respiration constrains vocalisation, since it provides the air supply required to generate vocal sounds. Between-system coupling provides the benefits of reduced mechanical interference and energy consumption, enabling prolonged motor activity [155].

9.2.2.2.1 Locomotor-Respiratory Coupling

Evolutionarily, the recruitment of locomotor and postural muscles from the thoracic and abdominal regions in breathing [156] generated novel demands for LRC in vertebrate taxa. Converging evidence from multiple species and methodologies shows that in tetrapods breathing entrains to locomotor rhythms [155, 157], while the scarce LRC data from fish species suggests that this directionality could be reversed in some species [158, 159].

Most evidence of LRC comes from tetrapods. In mammals, LRC occurs mainly during high-speed exertion with a 1:1 ratio, where one locomotor cycle is completed per breathing cycle (i.e., between two inspiratory onsets). This has been found in a great range of taxa (insectivores, rodents, lagomorphs, carnivores, ungulates, etc.) and for different locomotor modes, such as quadrupedal running, skipping, hopping, and flight [154, 160]. In birds, LRC occurs during both flight and running with variable ratios, e.g., 1:1, 2:1, 3:1, 4:1, and 5:1 (for a review, see [160]). For flight, there seems to be a tendency for a 3:1 LRC ratio in larger birds (e.g., gees, swans) and for a 1:1 ratio in smaller birds (e.g., pigeons, crows, magpies). Interestingly, there is also some evidence of LRC from fish and insect species. Some perciform species flap their pectoral fin in synchrony with mouth opening at a 1:1 ratio [158] or with operculum opening at ratios of 2:1 or 3:1 [159] and in crickets, ventilation can be coupled with stridulation [161–163].

LRC is also found in humans, where it occurs during high-speed locomotion (i.e., running) [164]—like in mammalian quadrupeds—but also during other kinds of locomotion, such as cycling (reviewed in [165]), wheelchair propulsion (reviewed in [166]), rowing [167], or swimming [154]. LRC is also present during non-locomotor rhythmic movements executed by the upper limbs, including finger-tapping [168], piano playing [169], or co-speech gestures [170]. Unlike other mammalian species, humans exhibit more flexible LRC ratios, although 2:1 seems to dominate across several locomotor modes.

In humans, LRC can be influenced by experience. For example, experienced runners show more stable LRC compared to inexperienced runners. This effect could not be explained by fitness or generic rhythmic practise, given that experienced swimmers only exhibit intermittent LRC during running [154, 171]. According to some data, such expertise effects on LRC stability could be a by-product of an enhanced capability to quickly adapt breathing to changing levels of speed and/or exercise intensity [172].

The strength of LRC is also influenced by the degree of biomechanical constraints that locomotion places on breathing. It, thus, varies depending on anatomical design and movement type. In line with this, the 1:1 LRC ratio of quadrupedal running has been attributed to various forms of mechanical impact on the thoracic complex. In equines, for example, the forelimbs exert an *impact loading* effect on the thoracic complex upon touching the ground, which forces expiration [154, 160]. Data from rabbits suggests that a *visceral piston* can have a similar effect: as the visceral mass shifts position during running, it alters intra-abdominal and intra-thoracic pressure and thus affects the timing of breathing [154, 173–175]. In small- and medium-sized mammals with a limber back (e.g., felids), *spinal flexion and extension* [154, 176, 177] causes the thoracic and abdominal cavities to change size and shape, which influences the breathing rhythm.

Due to bipedal locomotion, such biomechanical constraints are relaxed in humans [154]. Nevertheless, humans can experience high mechanical constraints for LRC during rhythmic movements that more strongly involve the upper limbs, since these movements obligatorily recruit muscles that have dual respiratory and postural functions, i.e., abdominal and pectoral muscles, as well as the diaphragm [178]. Consistent with this, human LRC is very robust during rowing, swimming, or wheel-chair propulsion (for reviews, see [154, 177]).

While biomechanical constraints create demands for different rhythmic behaviours to be carried out in tandem, such coordination outcomes would not be possible without the mediation of neural mechanisms. This is clearly shown by evidence that, during fictive motion, MN activity in the stimulated limb is coordinated with respiratory MN activity [179]. Several neural mechanisms have been shown to contribute to this coordination, including projections from motor to respiratory areas, inputs to motor and respiratory CPGs from a common origin, or signals relayed by proprioceptive sensory afferents. For example, excitatory commands from the MLR (which activate locomotor CPGs) are associated with a rapid acceleration of respiration in fish and mammals [39, 180]. In rodent brainstem-spinal cord preparations, the pharmacological activation of the hindlimb and forelimb CPGs in the lumbar and cervical regions, respectively, modulates the frequency of fictive respiration [157, 181]. The parabrachial/Kölliker-Fuse (PB/KF) nucleus plays a crucial role in LRC, as its integrity seems to be a precondition for the entrainment of respiratory rhythms to locomotor rhythms (see [155] for a discussion).

The role of proprioceptive circuitry in LRC is supported by evidence from multiple sources and experimental manipulations. Sensory afferents in the fore- and hindlimbs become rhythmically activated during rhythmic limb movement and have access not only to the locomotor CPGs, but also to the phrenic (respiratory) nerves, modulating their activity [154, 182, 183]. The rhythmic stimulation of sensory afferents through passive limb activation can reset respiratory rhythm in several species, causing it to entrain 1:1 with the rhythm of the stimulated limb(s) [154, 183–185]. Moreover, neuronal assemblies from the thoracic regions that participate in the propriospinal coordination of the fore- and hindlimb CPGs contribute indirect modulations of respiratory frequency during fictive locomotion [157].

9.2.2.2.2 Respiratory-Vocal Coupling (RVC)

As discussed in Chap. 2 [1], vocal communication is a tetrapod innovation that has largely co-opted neural circuits and peripheral structures (chest wall muscles, larynx, oropharyngeal, and nasal cavities) originally evolved for breathing and feeding. This imposes demands for respiratory-vocal coupling (RVC), where call initiation is coordinated with the activity onset of the expiratory muscle [186, 187]. Consequently, the temporal patterning of vocal communication is constrained by the rhythmic properties of bodily and neuronal oscillators implicated in breathing.[3]

RVC can exhibit different ratios, including 1:1, where a single continuous vocal emission is produced during a breathing cycle, as well as sub-harmonic ratios, where multiple vocal elements are emitted during a single breathing cycle. A relatively basic elaboration is when two clearly audible units—corresponding to inhalation and exhalation, respectively—are produced during a single breathing cycle. This is, for instance, found in pants—a vocalisation type that is widespread within the primate lineage (for details and examples, see Chap. 3 [2]). More elaborate vocalisations, such as avian and mammalian song or human speech, exhibit a great range of vocal patterns, which are produced with characteristic respiratory patterns [190, 191]. Phonation occurs during several successive expiratory pulses, and each of these is, in turn, patterned by several "mini breaths" (vocally silent inspirations) [192, 193]. As a result, sound units are emitted during a single breathing cycle, in species- and song-specific ways. This requires the highly accurate coordination between expiratory and inspiratory muscles and between complementary laryngeal muscles, in order to quickly switch between exhalation and mini breaths.

At a neural level, data from several mammalian species (e.g., cats, rodents, bats) highlight the parabrachial nucleus (PBN) as a key structure underlying RVC (for a

[3] Breathing is in itself a motor function that requires between-system coupling, involving neural oscillators that control ventilatory and orofacial muscles. Nonmammalian vertebrates exhibit both active inspiration and active expiration, which involves the alternative activation of complementary muscles that expand and compress the thoracic-abdominal cavity. A mammalian innovation is triphasic breathing, involving at least three colocalized CPGs: a CPG in the pre-Bötzinger complex in the medulla that drives the inspiratory pump muscles (diaphragm and external intercostals); a CPG in the lateral parafacial nucleus that patterns the contractions of expiratory (abdominal) muscles; a CPG in the post-inspiratory complex that controls a post-inspiratory phase and contributes to vocalization by slowing down the expiratory flow, more specifically by lengthening diaphragm contraction and shortening laryngeal muscle adduction [47, 186]. Notably, mammals do not exhibit active expiration at rest, but this changes in conditions of increased arousal and metabolic demands (e.g., vigorous exercise), as well as when vocalizing [47, 187].

review, see [194]). The PBN is the recipient of a great array of descending motor commands and somatosensory feedback related to lung and larynx status. In turn, the PBN projects to both laryngeal and respiratory premotor neuronal assemblies (reviewed in [194]). This connectivity pattern, along with PBN activity during both breathing and vocalisation, suggests that RVC is mediated by somatosensory feedback that influences the respiratory phase-switching in the PBN [194]. Evidence from a recent virus-tracing study in Alston's singing mice shows that the PBN may have an even more complex role during vocal production, by coordinating CPGs across muscular groups in the larynx, respiratory system, and jaws [195]. In addition, a recent study on pup cries and courtship vocalisations in mice [196] demonstrates how the larynx receives inputs from several interneuron groups, including a reticular oscillator that also innervates the tongue, controls glottal opening, and provides input to the pre-Bötzinger complex, triggering minibreaths. A similar mechanism is thought to be involved in humans during speech production (for a discussion, see [40]). In avian species, a similar role to the mammalian PBN could be played by the nucleus retroambiguus (RAm). This structure receives input from HVC (proper name)—a key upstream component of the song system—and projects to vocal and respiratory CPGs in the hindbrain [197, 198]. In turn, neuronal assemblies in the ventrolateral medulla that are implicated in respiratory control project back to the HVC via a thalamic nucleus.

9.2.2.2.3 Locomotor-Respiratory-Vocal Coupling (LRVC)

Since breathing is constrained by locomotion and, in turn, constrains vocalisation, the demands for between-system coordination can lead to locomotor-respiratory-vocal coupling (LRVC). This is often the case of in-flight calling in bird and bat species: wing flapping mechanically expands and contracts the thoracic cavity, which dictates the phases of the breathing cycles, and, thus, the timing of vocal emissions. LRVC, thus, is present at a 1:1:1 ratio. More flexible ratios can also occur, which, thus far, have been only demonstrated for vocal learning species [199, 200]. In bats, for example, wing stroke, breathing, and biosonar emission are typically synchronised at 1:1:1, but 1:1:2 ratio is also frequent [200] (and references therein). There is also evidence that a species may flexibly switch between different ratios, as it is the case with *Phyllostomus hastatus*, where LRVC ratio can change from a 1:1:1 to 1:1:4, where four sonar pulses are emitted for each wing stroke and breathing cycle [201]. This creates a salient rhythmic pattern where the grouping of vocal units is imposed by the cycle of LRC.

In bats, LRVC is present when the energetic demands for calling are high, which, in turn, is related to the type of foraging behaviour. Accordingly, LRVC occurs during the "search" phase, when bats take generic "snapshots" of the environment by producing long-range sonar, which consists of slow and strong pulses that are energetically costly [202] (and references therein). By contrast, LRVC is absent during active prey pursuit, when the bats use fast sonar pulses (10–15:1:1) which are weaker and energetically less costly [202]. It has been, thus, hypothesised that LRVC constitutes an energy-saving strategy, whereby the energy generated by wing flapping is sufficient to also sustain breathing and vocalisation [203, 204]. Consistent with this, in common pipistrelles (*Pipistrellus pipistrellus*), echolocating during flight does not add any energy expenditure to that necessary for flight alone [203, 205].

Evidence from some vocal learning species shows that locomotor maturation and, consequently, the emergence of LRVC are crucial for acquiring the rhythmic structure of species-specific calls. In the green-rump parrotlet (*Forpus passerinus*), the nestlings begin to exhibit the correctly structured contact calls shortly after they begin to use wing flapping, while the fully fledged adult form of the call is achieved only when the nestlings begin to fly [206]. This is reminiscent of the ontogenetic development of human speech, where the emergence of "canonical babbling"—i.e., rhythmic series of syllables with a well-formed (i.e., speech-like) structure—is preceded by "manual babbling," i.e., a series of rhythmic arm movement (for details, see Chap. 10 [6]).

LRVC has also been described for other locomotor modes than flight. For example, during jumping and foot drumming, Mongolian gerbils (*Meriones unguiculatus*) emit a single nasal ultrasound vocalisation whenever the feet touch the ground [207]. In rats, LRVC has been reported under high-arousal conditions, such as mating, when ultrasonic vocalisations were found to be coordinated with the hind- and forelimb motions produced during copulation [208]. While it is possible that such vocalisations are a biomechanical by-product of locomotion activity, which compresses the thoracic cavity and may cause a forceful air expulsion through the larynx and nostrils [207, 208], this does not preclude that such vocal emissions fulfil a communicative function. For example, in Mongolian gerbils, these vocal emissions occur primarily when the animals move in pairs, meaning that LRVC is modulated by social context and is not an obligatory by-product of mechanical forces.

In other cases, the three systems can be flexibly coupled and decoupled for more efficient resource partitioning. Rats, for example, exhibit LRC at high locomotor speeds, but during exploratory locomotion, the production of ultrasonic vocalisation slows down the breathing rhythm, thereby disrupting LRC and enforcing RVC [209]. It has therefore been proposed that rats exhibit a hierarchy of rhythmic coupling, whereby breathing and vocalisation are intimately bound, while their coupling with locomotor rhythms is secondary and flexible [209].

LRVC is an important aspect of multimodal displays that combine vocal and bodily rhythms, especially considering the high energetic costs and complexity that are typically associated with these. Although multimodal displays are common in birds [210–214], detailed analyses of rhythmic patterning and coupling across the implicated effectors and functional systems are scarce. Among these, a seminal study on the multimodal displays of cowbirds (*Molothrus ater*) shows how locomotor (wing) displays may both enhance and constrain song patterns [215]. In these displays, wing movement, song, and breathing are coordinated during certain phases of the display, but vocalisation is briefly decoupled during the production of certain wing movements that have an adverse biomechanical impact on respiration. This example suggests that the evolution of the temporal patterning of song did not evolve independently of the rhythmic pattern of the wing display, but that it may have been constrained by preexisting LRC during flight (as reviewed in Sect. 9.2.2.1).

9.3 Part II: Neuroethological Constraints and Modulators of Rhythm Production and Perception

9.3.1 Size Effects on the Frequency Rate of Rhythmic Behaviour

As shown in Chap. 3 [2], frequency rate is the rhythmic parameter with the greatest cross-species variation. Evidence from multiple taxa shows that a species—including humans—exhibits a very narrow range of preferred frequency rates, i.e., an *optimal rate.* In this section, we focus on this aspect of rhythmic communication, showing how mechanical and neural variables associated with body size determine its between- and within-species variation. We begin by illustrating the notion of preferential (optimal) rates with examples from humans, nonhuman primates, and other taxa (Sect. 9.3.1.1). We then continue by discussing how size effects on the frequency rate of rhythmic communication are associated with factors related to the molecular, anatomical, and physiological properties of muscles and MNs (Sect. 9.3.1.2), to signal conduction delays at the level of muscles and MNs (Sect. 9.3.1.3), and to the structure of CPGs circuits (Sect. 9.3.1.4).

9.3.1.1 Preferential Rates in Humans, Nonhuman Primates, and Other Taxa

In adult humans, several communicative rhythms produced by locomotor appendages (e.g., finger-tapping, clapping, walking, manual drumming, dancing) [216–218] have a preferential rate of about 2 Hz. Likewise, the frequency rate of phonemic articulation—which is arguably the most common form of vocal communication in our species—centres on 5 Hz [219, 220]. All these rhythmic behaviours can also be produced at faster or slower rates: e.g., 1–3 Hz for drumming (extended to 0.5–5 Hz in experts) and 3–8 Hz for speech [219, 220] (see also Chap. 12 [7]). Notably, the optimal rate of limb-based communicative rhythms matches the preferred rate of human walking [221], whereas mandibular oscillations during speech are much faster than those produced during mastication [219, 220, 222]. This suggests that rhythmic movements hypothesised to be foundational to communicative rhythms (see Chap. 2 [1]) may place stronger or more relaxed constraints on the latter. Thus, conclusions drawn from research on a given communicative rhythm are not automatically valid for other communicative rhythms.

Optimal rates have been examined in nonhuman primates as well, with findings that exhibit both intriguing similarities and dissimilarities. For example, multiple orofacial signals produced by different primate species are underpinned by mandibular oscillations averaging 5 Hz, including orangutans (*Pongo pygmaeus*), gibbons (e.g. *Hylobates lar*), gelada (*Theropithecus gelada*), and macaque (*Macaca* spp.) monkeys [206, 223–226] (see also Chap. 12 [7]). Although this has been hypothesised to be a precursor trait for evolving speech, this frequency rate is not universal for the orofacial communicative rhythms of primates. Chimpanzee lip-smacking, for example, has an average rate of 3 Hz [227] or lower (Gabriela-Alina Sauciuc, *unpublished dataset*). Moreover, the gibbon song exhibits distinct elements with distinct average rates of 2 Hz, 3 Hz, and 5 Hz [225, 228].

Limb-based rhythmic behaviour has been studied to a far lesser extent, but extant data suggest that chimpanzee finger-tapping may exhibit a similar preferential rate to that of adult humans [72]. By contrast, manual drumming has been found to be faster in bonobos (4.5 Hz [74]) and also highly variable in chimpanzees (0.9–4.2 Hz [229]) and gorillas (4–6 Hz [230]). Diverging rates have also been found for chimpanzees' bipedal walking during synchronised tandem walking (0.89 Hz [231]), whereas solo bipedal walking can vary widely between individuals (e.g., 0.3–1.99 Hz [232]). Quadrupedal walking, instead, shows a pattern where chimpanzees walk slightly faster compared to their individual preferential rate when coordinating walking with conspecifics [233].

Preferential rates centred around 2 and 5 Hz can also be found beyond the primate lineage. The 2 Hz rate has been reported for communicative signals in phylogenetically distant species, such as cephalopod chromogenic and bioluminescent displays, insect stridulation, fish vocalisation, and cockatoo drumming, while the 5 Hz rate has been reported in, e.g., rodents (see Chap. 3 [2] and Chap. 4 [3]). Generally, however, the rhythmic signals of animals can have highly diverging optimal rates, which can be both much faster—15–26 Hz in woodpecker drumming, 30–60 Hz in the pulsating song in aquatic frogs, or 10–30 Hz in the sonar pulses emitted by bats (10–30 Hz)—and much slower (e.g., 0.5 Hz in the song of humpback whales) than those encountered in humans (for example, see Chap. 3 [2] and Chap. 4 [3]).

9.3.1.2 Speed-Force and Speed-Resistance Trade-Offs Related to Motor Unit Properties

The examples given above show that the preferential rates of rhythmic communication vary widely across species and significant variations also occur between different kinds of communicative rhythms within a species. Body size plays a major role in determining the preferential rate of species or individuals, through constraints that arise at muscular and MN level. Two well-established cross-taxon motifs in this respect are that smaller species produce fast but short rhythmic signals, while larger species exhibit the opposite pattern. For example, in tanagers—a speciose family that comprises nearly 10% of all songbird species—larger species sing slower songs [234]. Among fishes, drumming in toadfish species [235] and pectoral stridulation in catfish [236, 237] have been shown to have slower rates in larger individuals.[4] In bathyergid mole-rats, the small sized (180 g) Cape mole-rat (*Georychus capensis*) drums at a rate of 26 strikes/second but can sustain drumming only for a couple of minutes. By contrast, the Cape dune mole-rat (*Bathyergus suillus*) foot drums with a slower, non-burst pattern of 2 strikes-pause-2 strikes that can be sustained for a long time [239]. Similar effects have been observed at within-species level, e.g., in humans shorter leg length is associated with faster preferential rates for walking, dancing, and tapping [240–242].

---

[4] In catfish, the slower stridulation of larger individuals has been explained both as an outcome of slower muscular contractions (the mechanism discussed in this section), but also as a consequence of the fact that, as catfish grow, the ridges on their pectoral spines get farther apart [237].

The two trends illustrated above are generally explained in terms of biomechanical efficiency, as reflected in the molecular and energetic properties of effector muscles. In mammals, the laryngeal muscles of small species express faster fibre types than their counterparts in larger species. The vocal fold adductor—the fastest of the laryngeal muscles—expresses increasingly slower fibres as animal size increases: from MHC-EO in rats and rabbits, to MHC-IIb in cats and dogs, to MHC-IIx in baboons, horses, and humans [23]. Similarly, the much slower contracting vocal tensor expresses type IIb in the rat, type IIx in the rabbit, and type I and IIa in baboons and cats, albeit with a higher proportion of IIa in the cat [23]. In the same vein, locomotor muscles express mainly type IIb fibres in small rodents, while type I and IIa are more common in larger species [17]. For example, the hind leg muscles (used for foot drumming) express a significantly higher concentration of type II fibres in the small-sized Cape mole-rat compared to the large-sized Cape dune mole-rat [243].

Differences consistent with these trends are further observed at the motor unit level. Large motor units are generally composed of (up to hundreds of) fast type fibres (IIb/IIx) that are controlled by MNs with large cell bodies and axonal diameter. By contrast, small motor units are dominated by the slow type (I) fibres, which are innervated by small neurons with small-diameter axons. In relative terms, within a given species and individual, the contractions generated by the largest motor units are two to three times stronger and faster compared to the smallest motor units [26]. Motor unit number decreases in ageing mammals, causing muscle function decline [244]. This provides a mechanistic explanation for the function of rhythmic displays as an indicator of the signaller's age and fitness.

MN size influences frequency rate also via differences in persistent inward currents, which shape neuron firing patterns. These currents decay at different rates depending on MN size, and, thus, they determine how fast the MNs stop firing and whether they can sustain tonic firing. Large MNs typically fire short, high-frequency bursts of action potentials, while small MNs discharge tonically for long periods [245]. This mechanism also mediates muscle tissue plasticity, as observed with altered patterns of use, such as intense practice or the cessation of practice. Indeed, experimentally manipulated firing patterns can promote selective changes in gene expression, which result in alterations of fibre characteristics. For example, fast-type fibres can be converted into slow-type fibres following sustained stimulation with low-frequency impulses [246], and conversely, slow-type fibres can shift toward faster phenotypes when stimulated with impulses patterned as high-frequency bursts.

9.3.1.3 Size Effects: Conduction Delays at Muscle and Motor Neuron Level

Size can also impact the frequency rate of rhythmic behaviour in terms of signal conduction delay at the level of muscles (force generation and electromechanical delay) and MNs (nerve conduction delay) [247]. Force generation delay refers to the time needed for the muscle to develop peak twitch force. Relatively to body size, larger animals have shorter muscles compared to smaller ones, and, thus, a greater proportion of the muscle has to shorten before maximum force is balanced by the force in the tendons [247]. As this affects the time interval between successive contractions, it also affects the frequency rate of rhythmic behaviour.

Electromechanical delay—i.e., the time required for an action potential to propagate along the muscle fibre to activate the contractile proteins—and nerve conduction delay—i.e., the time required for transmitting an action potential along a MN axon [247, 248] are both determined by conduction distance. In mammals, this is largely determined by body size simply because, in absolute terms, larger mammals have longer extrafusal fibres and longer MN axons. Electromechanical delay, however, increases less steeply with size, given that, as noted above, larger mammals have relatively shorter muscles than smaller mammals. By contrast, nerve conduction delay has a substantial contribution to the slower movement of larger animals. Long conduction delay has been recently discussed as a putative selective pressure that favored the emergence of feed-forward (predictive) motor responding in large mammals, as a way to ensure rapid motor tuning to environmental changes despite the slow responsiveness of sensorimotor processing [249, 250]. This kind of predictive processing is a crucial feature of rhythm perception and synchronisation (as also discussed in Chap. 5 [4], Chap. 1 [9], Chap. 7 [10], Chap. 8 [11], and Chap. 14 [8]).

9.3.1.4 Size Effects Related to CPG Circuits

Research on electric communication in weakly electric fish (*Apteronotus leptorhynchus*) shows that size effects can arise beyond the muscular and neuro-peripheral level. In this (and other) species of weakly electric fish, rhythmic chirps are generated by an electric discharge organ that is controlled by a CPG circuit known as the pacemaker nucleus (Pn). The Pn network is formed by pacemaker and relay cells, and its overall size varies considerably between individual fish, increasing as fish size increases [251]. In turn, the size of the Pn network and, in particular, the proportion of pacemaker-to-relay cells determines the firing patterns and behavioural outputs. Larger Pn networks also have tremendously enhanced connectivity compared to smaller networks since the number of connections is known to increase with network size. In turn, these structural properties have been linked to higher frequency (i.e., lower inter-spike interval) and increased precision (i.e., low variability) [252].

Interestingly, a similar relationship between size and frequency rate has also been noted in other fish taxa. In gobies (family Gobiidae), which produce acoustic signals through tremulation [13], larger species produce faster signals than smaller ones, and frequency rate can be reliably used for species identification [253, 254]. Similar differences have also been found in individuals of the same species, as a function of size. In some of these cases, large body size appears to be a morphological constraint for achieving ultrafast tonal frequencies [253]. Putative factors explaining such differences could be differences in CPG network size and complexity, as well as/or hormonal factors (as discussed in Sect. 9.3.3).

9.3.1.5 Size Effects: Other Anatomical Differences

Differences in body size can have an effect on the rate of rhythmic signals via additional anatomical structures. Such differences would be particularly important for rhythms created via membrane pulsing (e.g., fish swimbladder size), stridulation, and percussion of body parts other than limbs. For example, the drumming rate of wolf spiders (*Hydrolycosa rubrofasciata*) is limited by metabolic constraints, which is a function of abdominal size [255], which is in turn a function of nutritional quality [256]. The sperm whale (*Physeter macrocephalus*) sound production mechanism is unique in the way that size constraints rhythm production. In odontocetes, sounds are transferred to water via a fatty sac called the melon (see Chap. 2 [1]). However, in sperm whales, vibrations first pass through an additional fatty sac, the spermaceti organ. Sounds reverberate off of air sacs at each end of this organ, creating a rhythmic burst-pulse, rather than a single click like in other odontocetes [257, 258]. Inter-pulse intervals are constrained by spermaceti organ length [259–261].

9.3.2 Air-Supply Constraints and Adaptations that Pattern Phonetic Rhythm Production

Respiratory constraints can have a profound influence on the temporal organisation of acoustic signals such as song. Air needs to be put under sufficient pressure to excite the various types of vibratory tissue—affecting unit duration—and needs to be replenished—affecting inter-unit duration. Therefore, lung and/or air-sac capacity adds an additional constraint to phonetically produced rhythms (see Chap. 2 [1]) for a detailed description of this mechanism). For example, the inter-syllable interval in grosbeak song (*Pheucticus ludovicianus*) is positively correlated with the length of the preceding syllable [262], possibly because longer units use more air, requiring more time to replenish the air supply. Mysticetes differ greatly in song structure: most rorquals sing simple, stereotyped songs, while humpbacks (*Megaptera novaeangliae*) and bowheads (*Balaena mysticetus*) sing complex, variable songs. Whether simple or complex, songs can be divided into repeated phrases based on their acoustic properties and the temporal spacing of the units. The number and duration of each sound unit of a song will depend on the amount of air flow necessary for producing the sound and the capacity of the air source, constraining the content of each phrase of the song [263, 264].

Trill production in songbirds is constrained by the available air volume, leading to two production mechanisms. The fastest trills are produced with a single exhalation and therefore are limited in duration. These pulsative trills are amplitude modulated. They are produced by careful control of the highly specialised muscles of the syrinx. For sustained trills many songbirds take mini-breaths (15–90 ms) between repeated syllables to replenish the air supply [265, 266]. Trill rate is constrained by syllable length as longer syllables require more air to produce and longer mini-breaths to replenish [267, 268]. Hence, mini-breath trills are slower than pulsative trills within species [187, 269], and smaller species produce mini-breath trills at faster rates than larger species [270].

While most phonetic sounds are produced during egression, it is possible for some animals to produce sounds during ingression of air, in some cases, incidentally, as a result of increased arousal [271, 272]. Bidirectional sound production may increase the duration of rhythmic signals, as is well documented in true purrs (as reviewed in [273]). Similarly, nightjars (Caprimulgidae) produce continuous trills with periods of short, low-amplitude pulses hypothesised to be produced during ingression [274]. Zebra finches (*Taeniopygia guttata*) produce some song elements during ingression. These sounds have different acoustic qualities than sound produced by egression, therefore adding to the complexity of vocal rhythms. However, this phenomenon has not been observed in trills [275]. Mysticetes differ in how air flow induces sounds. Testing and modelling of rorqual larynxes only produced sounds during egression, which appears to be an anatomical limitation [276, 277]. On the other hand, there is evidence that bowheads phonate during ingression, but not egression [278]. If sound production is unidirectional, then the silences between song units are a by-product of moving air back into the lungs of rorquals or laryngeal sac in bowheads [276], giving songs their characteristic phase structure.

In humpback whales, the internal recycling of air and the position of the U-fold led to the hypothesis that sound is produced bidirectionally [279–281]. Different alternating sound types in song (rhyme) may be the product of different sounds being produced during egression (lungs to sac) and ingression (sac to lungs [263]). However, the similarity in inter-unit interval between bowhead songs and humpback whale songs [282] indicates that unidirectional sound production likely occurs in both species. The type of sounds humpback whales produce can be explained by changes in configuration, without the necessity of sound production during ingression [281], however, this does not explain the regularity of sounds of certain types. Further research is necessary to understand which features of rhythm (and rhyme) are constrained and which are under the whale's control.

Both songbirds and odontocetes can create sounds independently in the right and left sides of their vocal organ, the syrinx and the phonic lips [283], respectively. The separate sides of the syrinx can be used to add variation to categorical rhythms by producing sound simultaneously or alternately [284]. Canaries (*Serinus canaria*, L.) increase their bandwidth during trills by producing alternating high and low frequency notes with different sides of the syrinx [285]. Odontocetes produce echolocation clicks with the right phonic lips and communication whistles [286] and (in belugas [*Delphinapterus leucas*]) clicks and with their left, such that mixed click/whistle calls can be produced [287]. However, it is not clear if odontocetes use bilateral sound production rhythmically.

9.3.3 Sex-Steroid Effects on the Neuro-Mechanics of Rhythmic Signals

As shown in Chap. 3 [2], the rhythmic signals of many species are, to various degrees, sexually dimorphic. Some display types or elements are either male- or female-exclusive or occur more frequently in one of the sexes. This suggests that sex steroids play a critical role in the regulation of rhythmic communication [288], given that sex-typical behaviours generally arise from differences in how sex steroids promote the emergence of sex-specific features in muscular and neural substrates [289]. The gross distribution of sex steroid receptors across the systems that generate rhythmic communication is a highly conserved vertebrate trait, identified in fish, frogs, birds, and mammals [50, 289–291]. It must be acknowledged, however, that, within this broadly shared organisational motif, there is also tremendous cross-species variability in the specific details of sex steroid organisation, which impacts the timing, quantity, and quality of rhythmic communication [289, 292].

The effects of sex steroids on behaviour can be organisational—i.e., exerted during the ontogenetic development of the individual—or activational, i.e., reflecting the level of sex steroids that circulate at a given moment in a given tissue. Activational effects can, in turn, be genomic, where they develop over days to weeks, or neuromodulatory, where they arise and fluctuate more rapidly, on a moment-to-moment basis. Across these temporal scales, the sex steroids can affect motivational, structural, and perceptual aspects of rhythmic signalling, by acting on central neural circuits and peripheral organs that express sex steroid receptors [50, 291, 293–295].

On the developmental scale, the sex steroids contribute to the organisation of the reproductive phenotype, including the development of genitals, secondary sexual characteristics, and the neural circuits that underlie reproductive behaviour. Consequently, sex steroids contribute to the sexual differentiation of rhythmic communication through effects exerted at the central and/or peripheral level. In frogs, for example, males have a higher number of laryngeal MNs, owing to greater androgen levels that promote greater ontogenetic cell survival [296]. Similarly, cell type, number, and size in the muscular and cartilaginous tissues of the larynx are greater in males compared to females, and the administration of androgen antagonists during development blocks larynx masculinisation [297]. In songbirds, gonadal and brain-synthesised sex steroids affect the anatomical and functional plasticity of the song control system, the growth of the syrinx, and song learning [289, 292, 294]. Generally, during the song learning period, oestrogen promotes vocal plasticity, while testosterone promotes song crystallisation. Suppressing steroidal activity in this sensitive period prevents the acquisition of the correct song template, while androgen hyperactivation induces exaggerated stereotypy (for reviews, see [292, 294]).

In adulthood, many of the rhythmic signals reviewed in Chap. 3 [2] are affected by seasonal fluctuations in gonadal steroids, which regulate the physiology and behaviour of reproduction under the influence of environmental factors. The most significant among the latter is photoperiod, whose increase promotes gonadal recrudescence, sex steroid stimulation, and the activation, intensification, and/or complexification of behavioural displays with reproductive and territorial functions [50, 292]. Laboratory research with various vertebrate taxa provides causal evidence for the activational effects of sex steroids on courtship and territorial displays via both genomic and nongenomic mechanisms. The surgical or pharmacological suppression of sex steroid availability abolishes or significantly diminishes the animals' capabilities to produce rhythmic displays, which has been shown in fish [293], frogs

[53], birds [292, 294], and mammals [298]. Conversely, androgen administration to castrated animals restores displaying functions [50, 294]. Moreover, in species with male-biased rhythmic displays, sex steroid administration to females has masculinising effects on the brain and behaviour, enhancing the neural circuits that underlie the production of rhythmic displays and inducing females to produce male-typical displays [50, 289, 290, 292, 294].

The effects of sex steroid manipulations vary depending on the targeted sites. Across taxa, the hypothalamus and the central nucleus of AMY regulate the drive to display, but not the structural aspects of rhythmic signals [50, 291, 294, 299]. Conversely, manipulating sex steroid activity in motor circuits (CPGs, song control system, etc.) affects the temporal and structural patterning of behavioural displays, both in development and in adulthood [53, 292–294]. In intact fish, global or selective androgen administration at peripheral or central sites elicits advertisement calls that are longer and exhibit more pulses than in nontreated individuals [293]. In birds, disrupting sex steroid function in the HVC nucleus of the birdsong control system compromises song stereotypy, resulting in reduced periodicity and nontypical patterning [294]. In the Alston's singing mouse (*Scotinomys teguina*), androgen depletion results in less stereotypical song with a lengthened period and reduced overall duration [298].

The sex steroids can influence rhythmic displays even outside the breeding season, when steroidal blood levels are basal and the gonads are regressed. Currently, such effects have been recorded in the context of territorial or social regulation and have been linked to non-genomic effects exerted by neurally active oestrogens, as shown in fish [290, 300], birds [292, 301], and mammals [302]. Pharmacological studies in electrically weak fish suggest that oestrogen synthesised from extragonadal testosterone may play a crucial role in mediating the shift to rhythmic discharge patterns that generate nonbreeding aggressive signals [290, 300]. Conversely, in both electric fish and some bird species, substances that inhibit the conversion of testosterone into oestrogen reduce aggressive displays, while estradiol administration restores them [290, 300].

Sex steroid effects are also observed in the peripheral effectors of rhythmic signals, in the form of changes in muscle mass and composition. In canaries, blocking androgen receptor activity at the peripheral level reduces syrinx size and alters the performance features of so-called special trills [303]. Normally, these are produced at extremely fast rates, but androgen depletion in the syrinx lowers the repetition rate of trills and lengthens their inter-note interval. Peripheral androgen mechanisms also influence the rhythmic parameters of nonvocal displays. For example, pectoral fin displays in fish and complex wing displays in birds are influenced by the levels and activity of androgen receptors expressed in these effectors [291].

In humans, the steroidal regulation of rhythmic communication remains understudied. Suggestive data indicate that rhythmic movement (e.g., tango dancing) may enhance testosterone levels in both males and females [304]. Theoretically, it has been argued that sex steroids exert an important role in rhythmic communication, given the abundance of sex steroid receptors in the neural circuitry underlying human speech, co-speech gestures, and motor performance (as reviewed in [291]).

9.3.4 Effects of Body Condition on Rhythm Performance

As we have covered above (e.g., muscle refresh rate), variation in the amount, persistence, and periodicity of a system's action will have side-stream effects on how it may constrain the action of other systems. An animal's physical condition will, thus, be an important factor in its rhythmic performance. Short-term stress can have a negative effect on speed, performance (stability and flexibility), and effort. Individuals that are stressed have limited energy reserves, limiting their ability to sustain prolonged signalling and precise neuro-muscular control. Moreover, energy allocation during development due to nutritional deficits or other forms of stress can result in physiological and neurological differences that affect sensory-motor learning and performance ability throughout an individual's life [305, 306]. Possible costs include insufficient development of the neural circuitry, signal architecture, and behaviour necessary to produce quality rhythmic signals. If the quality and structure of a signal are constrained by the signaller's physical condition, and there are sufficient detectable variations between signallers, then conspecifics may be able to judge the quality of a signaller as a potential mate or competitor.

Sexual and territorial signalling is often energetically costly. For example, the energetic cost of claw waving in fiddler crabs is indicated by increased lactic acid levels during waving, presumably because of anaerobic respiration [307]. The correlation between modes of claw signalling—waving and drumming—in banana fiddler crabs (*Uca mjoebergi*) has energetic costs that affect subsequent sprinting ability after two hours of courtship [308]; signalling vigor correlates with locomotor vigor (i.e., sprint speed). Females prefer males that wave at a higher rate, have larger claws, or drum more vigorosly, depending on species [308, 309]. But note that, in crabs, drumming or waving with a larger claw may entail further increases in energy expenditure. Likewise, faster note production rates have been shown to be metabolically costly in insects [310] and arachnids [311].

A signaller's physical size may be a result of past or present nutritional intake; therefore, large males are often in better condition allowing them to dedicate more effort to signalling, which leads to longer songs/bouts (as shown in, e.g., rock hyrax (*Procavia capensis*) [312] or ovenbirds (Furnariidae) [313]), faster note production rates, and/or higher duty cycle. Vocalisation rate is a common performance cue in many taxa, including tortoises [314], frogs [315], and deer [316], which may seem paradoxical given the review above indicating that larger animals produce slower rates due to anatomical limitations. However, fat reserves and muscle mass contribute significantly to overall body mass/size in many species, such that animals with similarly sized anatomic structures differ greatly in body size.

Older animals often show more consistency in performance, while a lack of consistency may be a signal of young age or poor physical condition. For example, banded wrens (*Thryophilus pleurostictus*) produce trills with more consistency as they age [317]. Lower quality animals are more prone to make mistakes [318]. Recipients may compare the consistent performance of fine scale (spectral, temporal, and/or amplitude characteristics of a specific element [317]) or gross scale (e.g., song length, production rate [319]), and absolute or relative (e.g., note pitch ratios) features [320]. Quality may also be communicated through the production of signal features that are physically difficult to produce. While birdsong may not be metabolically costly in terms of overall calorie use [321], some rhythmic signals, such as trills, require precise neuromuscular control for ultimate performance, possibly leading to motor fatigue [322, 323]. In many species trill rate, consistency, and amplitude decline toward the end of sustained trills [324]. Many species of songbirds achieve greater mating success if they sing fast trills, and/or behave more territorially if they hear songs with fast trills (for review, see [325]), but such responses can be moderated by trill quality [324].

Trade-off constraints occur when the production of one parameter of a signal is constrained by the production of another parameter. Smaller animals may sacrifice one performance parameter to increase the performance of another. For example, leopard seals (*Hydrurga leptonyx*) produce long rhythmic mating calls while underwater, necessitating breath-holding. Large leopard seals maintain consistent vocal output across the breeding season, possibly due to increased stamina, while during peak season, smaller seals produce shorter but more frequent bouts with more breaks [326] (though this may be due to social as well as physiological constraints). During song learning, swamp sparrows (*Melospiza georgiana*) sacrifice copying accuracy (of artificially fast and slow tutors) in favor of producing the best trill-rate performance for their capabilities [325, 327].

Trill's production incurs a rate versus bandwidth trade-off [328]. Presumably, the bigger the frequency change of a unit, the more difficult it is to reproduce the unit quickly, because either the duration of the unit or the gap between units is neuromuscularly constrained. Individuals with better neuromuscular control—and presumably better developmental condition—can minimise this trade-off as has been found in numerous songbirds [329], great Himalayan leaf-nosed bats (*Hipposideros armiger*) [330], Neotropical singing mice (*Scotinomys*) [298], and female Bornean gibbons (*Hylobates muelleri*) [331]. Alternatively, canaries increase their bandwidth by producing alternating high and low frequency notes with different sides of the syrinx; bilateral motor coordination then becomes the fitness indicator [285].

It is important to note that individuals do not typically perform at their maximum potential. Sustaining such a performance would be energetically and physically taxing. Rather, performance should approach maximum when fitness really matters [325], such as in the presence of competition (e.g., Bush crickets (Tettigoniidae) [332], skylarks (*Alauda arvensis*) [333]).

9.3.5 Biological and Cognitive Constraints on Rhythm Perception

As discussed in other chapters of this volume (Chap. 3 [2], Chap. 1 [9], Chap. 7 [10], Chap. 8 [11], and Chap. 14 [8]), rhythm perception is not a given in animals that produce rhythmically structured signals. In such signals, biologically significant cues—i.e., features that trigger adaptive responses from conspecifics—can be a certain tonal frequency, interval length, or stimulus density rather than the perception of a species-specific rhythm. Rhythm perception requires that a stimulus series is perceived as an overall temporal pattern—an ability also discussed as relational (or beat) timing ability [334–336] (and as discussed in Chap. 7 [10] and Chap. 12 [7]). Most of the data acquired thus far suggest that this may be a major neurocognitive bottleneck in the evolution of more advanced rhythmic competencies. For example, when animals learn to discriminate isochronous from non-isochronous series, most species appear to do so by using diagnostic features of single elements (usually the first element in a series) rather than the overall relationship between the elements in the series (for a review, see [336]). More recent studies, however, show that this may be an artifact of the experimental design or training regime. For example, Rhesus macaques (*Macaca mulatta*)—a species initially found incapable of beat perception—can exhibit beat timing abilities if trained with methods that increase the salience of the relationship between units in a series [337, 338] (see also Chap. 7 [10]).

Adequate training, however, cannot abolish the entirety of cross-species differences related to this perceptual ability. For example, the human auditory system captures sound streams in "chunks," which are based on a fixed "time window" of about 200 ms [339]. Accordingly, auditory information acquired within this time window is often processed as belonging to a single auditory event. When listening to music, beat and rhythm perception is aided by grouping effects generated by the human-specific time window. Time windows are an important element of auditory processing, not only in humans but also in other animals, and provide clues about how sounds are processed and integrated in the brain. Recent comparative research shows that humans exhibit significantly longer cortical auditory processing times compared to nonhuman primates like marmosets, rhesus monkeys, and chimpanzees [340, 341]. These studies focused on measuring auditory-evoked potentials, in particular latencies of the P1 and N1 components, which indicate how long it takes for certain brain areas to respond to auditory stimuli. In humans, the auditory evoked potentials (such as the N1 component) diminished as the sound duration shortened, showing that humans rely on longer temporal integration. In contrast, chimpanzees showed a marked response to short sounds, indicating a short integration window. These differences are increasingly pronounced in phylogenetically more distant primates, such as macaque monkeys and marmosets. The elongation of cerebral cortical processing time is viewed as a necessary consequence of the human brain's evolutionary enlargement, contributing to our advanced cognitive abilities, particularly in auditory processing such as music perception.

Besides differences in signal parsing, as enabled by relative timing or species-specific time windows, additional constraints on rhythm perception result from species-specific perceptual thresholds and resolution in relevant sensory modalities. In other words, perceptual systems are optimised for stimulus perception in species-specific temporal (or spectral, etc.) ranges that may differ across different sensory modalities. It is well-known, for instance, that very small and very large animals communicate in ultra- and infra-sound frequency bands that remain imperceptible to humans. Likewise, in the temporal domain, there are species-characteristic preferential frequency rates. Human auditory perception, for example, is optimised around 1.7–2 Hz, although humans can also perceive rhythm in faster or slower auditory streams, with a range that in trained individuals spans rhythms as fast as 6 Hz or as slow as 0.5 Hz [342]. Interestingly, the preferential rate of rhythm perception is, just like rhythm production, influenced by anthropometric variables (body mass, height, leg length, and shoulder width) [241].

Preferential rates and ranges of rhythm perception have been recently uncovered in other species as well. Zebra finches, for example, can form rhythm percepts and perform correct rhythm discriminations as long as the stimuli have a frequency rate between 4 and 10 Hz [343]. Smaller animals, such as insects and anurans, may have even narrower ranges of preferential perceptual rates. For example, in aquatic *Xenopus* frogs, males and females produce behavioural (approaching, counter-calling, etc.) or neural responses only when the sound emission has the species-specific call frequency [50]. Similar findings have been obtained in insects, where neural mechanisms of song perception become active only when the acoustic stimulus exhibits a certain inter-unit interval (for a review, see [344]). Additional examples of this phenomenon come from experiments that investigate the effects of music on nonhuman species (cats, pigs, primates, etc.), where behavioural and physiological responses are only recorded for stimuli with ethologically relevant frequency rates or in species that exhibit preferential rates similar to humans (for a review and examples, see Chap. 11 [12]). These findings can be explained as neural resonance effects, whereby neuronal oscillations in sensory processing brain areas (which have their own intrinsic frequencies) can phase-lock to exogenous rhythms at integer ratios (as detailed in Chap. 8 [11] and reviewed in [345]). It is important to emphasise that in many species these temporal preferences may reflect rate-tuned sensory filtering (rather than relational rhythm perception). They nonetheless illustrate how neuromechanical and sensory constraints shape the temporal ranges in which signals are effectively processed.

Consistent with the mechanisms discussed above, rhythm perception can also be constrained by the broadcasting modality. For example, macaque monkeys seem to be more apt at synchronising motor responses (hand tapping, eye saccades) with visual—compared to auditory—rhythms [337, 338, 346, 347]. By contrast, humans seem to show the opposite pattern (for a review, see [216]), albeit such differences can be significantly attenuated if the visual stimulus is dynamic (for a review, see [348]). Since macaque monkeys show comparable performance in two behavioural modalities (manual tapping, ocular responses), the findings above may stem from perceptual modality biases. One possibility could be that human rhythm perception is optimised in the auditory domain, and macaque rhythm perception is optimised in the visual domain.

An alternative possibility could be that there are differences in the extent to which different sensory modalities modulate motor control during rhythm production, i.e., visual rhythms may have a stronger effect on motor rhythms in monkeys, while auditory rhythms may have a stronger effect on motor rhythms in humans. For example, mounting evidence shows that auditory rhythms can facilitate and enhance the presence of LRC in humans during activities as diverse as running [164, 349, 350], finger tapping [168], or cycling [351, 352]. This effect is most prominent for frequency rates that are close to the preferred rate of either locomotion or respiration, with LRC being weaker for faster or slower auditory tempi [352]. Interestingly, LRC stabilisation through rhythmic sound can have energetic benefits, by decreasing oxygen volume per minute [351]—a finding relevant to functional theories regarding the evolution of audiomotor synchronisation in humans.

Finally, rhythm perception can be constrained by the physiological status of the individual, in particular, by sex steroid effects (for a recent cross-species review see [50]). In anurans, conspecific signals induce changes in sex steroid levels in both males and females, and this affects how species-specific rhythms are perceived and responded to (for reviews, see [50, 296, 299, 353]). Indeed, sex steroids modulate auditory processing in females, enhancing their responsiveness to species-specific reproductive calls. Sex steroids in interaction with other neuromodulators contribute to the fine-tuning of the torus (corresponding to the mammalian inferior colliculus), which is crucial for the processing and recognition of species-specific temporal patterns [296]. In fish, the local increase of estradiol (resulting from the conversion of circulating testosterone) upregulates the auditory system of females, by increasing their sensitivity to the timing pattern of male advertisement calls [290]. In songbirds, rapid estradiol activation regulates various aspects of auditory processing, of which, most relevant here, is the females' capacity to decode aspects of temporal structure in male song [354].

9.4 Conclusion

Across the animal kingdom, rhythmic communication varies considerably in terms of modality of production and/or perception, frequency rate, and patterning. We dedicated this chapter to discussing how rhythmic communication is constrained by the underlying neuromuscular architecture. In Part I (Sect. 9.2), we first provided a quick overview of the mechanical and neural building-blocks that constitute this architecture, and of how these need to coordinate at multiple levels of organisation. In Part II (Sect. 9.3), we then discussed specific constraints that affect specific parameters of rhythmic communication. In this endeavour, we focused primarily on aspects of muscular and vocal tract design, along with "low-level" neural mechanisms. On the one hand, we aimed to provide a complementary perspective to that offered by other chapters of this volume, which mainly focus on motor cortical mechanisms. On the other hand, this focus was motivated by the underrepresentation of the topics addressed here in evolutionary theories that address the rhythmic competence(s) that enable musicality and speech in humans.

The review presented in this chapter exposes several axes of variation in neuromuscular design that account for differences registered at behavioural level in terms of rhythmic stability, temporal rate, patterning, and versatility. Using multiple examples, we demonstrate how such differences can already arise at the effector level, being constrained by features of the molecular design of muscles, and how such differences can then be reinforced or mitigated by properties of the "upper" level components. For example, the evidence reviewed in this chapter demonstrates a remarkably systematic relationship between size, rate, and the molecular, anatomical, and physiological properties of motor substrates, from the muscles to the CPG circuits.

This chapter also shows that the neuromechanical constraints that shape rhythmic communication go beyond factors pertaining to the design and neural control of the specific effector organ that executes a rhythmic behaviour. Breathing, in particular, can shape multiple aspects of rhythmic communication. On the one hand, the rhythms of breathing, locomotion, and vocalisation share muscular substrates. This reflects a deep evolutionary connection resulting from the fact that ancestral locomotor and postural muscles acquired novel ventilatory functions in vertebrates and the emergence of vocalisation co-opted oral and thoracic muscles involved in ventilatory control (see also Chap. 2 [1]). On the other hand, oxygen availability sets the upper limits of the physiological potential with respect to locomotor speed and endurance, as well as the duration and patterning of vocal production. Overall, this underscores the fact that a thorough account of the mechanisms that shape rhythmic signals must also encompass the constraints and affordances that result from breathing anatomy, physiology, and neural control.

By illustrating how the properties of neuromuscular substrates affect the temporal parameters of rhythmic signals, this chapter substantiates functional hypotheses regarding the evolution of rhythmic communication. At the within-species level, rhythmic parameters can vary between individuals as a function of neurophysiological and muscular status, such that rhythmic stability, pattern accuracy and temporal rate can function as honest indicators of fitness. At the between-species level, these rhythmic features likely evolved as an integral part of speciation, currently allowing reliable species recognition. However, as reiterated by several other chapters, an important limitation in the current literature is the absence of perception studies aimed at attesting that animals use rhythm—rather than single intervals—as a diagnostic feature for species recognition and individual assessment. We hope that future research will address this substantial gap, which is critical for formulating valid evolutionary hypotheses regarding the evolution of rhythmic communication.

Overall, the examples discussed in this chapter are aimed at raising awareness of the wide variation of rhythmic patterns and preferential rates that occur naturalistically in animal communication and about the diverse mechanisms that may account for these variations. The point is that animals are optimised to produce and perceive certain rhythms and the mechanisms that constrain the rhythmic parameters of communication may be more or less amenable to change, which may also depend on the modality of production and/or perception.

Two important lessons can be derived from these insights. First, data on a specific rhythmic competence and behaviour in one species should not be used to draw general conclusions about the whole genus and beyond. Generating viable hypotheses on the evolution of human rhythmic competences will require substantial research aimed at mapping rhythmic variables in a wide range of species in a given genus or family and across the potentially multiple behaviours that actuate rhythmic communication in each species, from locomotor modes (quadrupedal, bipedal, and brachiation) to gestural and vocal communication. Second, this species-specific optimisation and the nature of the mechanisms that underpin it ought to be considered when designing cross-species comparative studies and interpreting their results. Neuromechanical constraints that affect the measured behaviours may constitute multiple sources of confounds, biasing the conclusions drawn from the comparison of rhythmic performance across species and behaviours.

Acknowledgments G-A Sauciuc's contribution to this chapter has been funded by the Bank of Sweden Tercentenary foundation (grant nr. P17-0942:1) and the Swedish Research Council (grant 2019-03104).

References

1. Sauciuc G-A, Schneider JN (2026) Chapter 2: How animals produce rhythm – an overview of ethological mechanisms for rhythmic communication. In: Nature beats: the what, why and how of animal rhythmic behaviours. Springer Nature Switzerland, Cham
2. Sauciuc G-A, Schneider JN, Burchardt LS et al (2026) Chapter 3: The functional and structural diversity of rhythmic communication across species. In: Sauciuc G-A, Lameira AR, Persson T (eds) Nature beats: the what, why and how of animal rhythmic behaviours. Springer Nature Switzerland, Cham
3. Lameira AR, De Gregorio C, Gamba M et al (2026) Chapter 4: Uncharted rhythmic diversity across the animal world. In: Sauciuc G-A, Lameira AR, Persson T (eds) Nature beats: the what, why and how of animal rhythmic behaviours. Springer International Publishing, Cham
4. Greenfield M, Merker B (2026) Chapter 5: Endogenous rhythm, isochrony, and the path to perfect synchrony. In: Nature beats: the what, why and how of animal rhythmic behaviours. Springer Nature Switzerland, Cham
5. Schneider JN, Burchardt LS, Sauciuc G-A, Lameira AR (2026) Chapter 6: How environmental and social contexts shape and constrain rhythm structure and production. In: Sauciuc G-A, Lameira AR, Persson T (eds) Nature beats: the what, why and how of animal rhythmic behaviours. Springer Nature Switzerland, Cham
6. Sauciuc G-A, Raimondi T, De Gregorio C (2026) Chapter 10: The developmental trajectory of rhythmic abilities in humans and singing primates. In: Sauciuc G-A, Lameira AR, Persson T (eds) Nature beats: the what, why and how of animal rhythmic behaviours. Springer Nature Switzerland, Cham
7. Sauciuc G-A, Persson T, Hattori Y, Lameira AR (2025) Chapter 12: The significance of the great ape heritage in the evolution of human rhythm in speech, music and dance. In: Sauciuc G-A, Lameira AR, Persson T (eds) Nature beats: the what, why and how of animal rhythmic behaviour. Springer Nature Switzerland
8. Merker B, Cook PF, Reichmuth C et al (2026) Chapter 14: Human group synchrony in light of the comparative method: what we know and would like to know. A peer commentary treatment. In: Sauciuc G-A, Lameira AR, Persson T (eds) Nature beats: the what, why and how of animal rhythmic behaviours. Springer Nature Switzerland, Cham
9. Burchardt LS, Hoeschele M, Honing H et al (2026) Chapter 1: What is rhythm and how can we study it? In: Sauciuc G-A, Lameira AR, Persson T (eds) Nature beats: the what, why and how of animal rhythmic behaviours. Springer Nature Switzerland, Cham
10. Proksch S, Honing H, Ravignani A, Merchant H (2026) Chapter 7: Converging notions on the main hypotheses regarding the origins of beat perception and synchronization: the key role of audiomotor interactions. In: Nature beats: the what, why and how of animal rhythmic behaviours. Springer Nature Switzerland, Cham

11. Large EW (2026) Chapter 8: Neural resonance. In: Sauciuc G-A, Lameira AR, Persson T (eds) Nature beats: the what, why and how of animal rhythmic behaviours. Springer Nature Switzerland, Cham
12. Gustison ML, Sauciuc G-A (2026) Chapter 11: Moving the body and the mind – links between emotion and rhythmic communication and the underlying neurobiological mechanisms. In: Sauciuc G-A, Lameira AR, Persson T (eds) Nature beats: the what, why and how of animal rhythmic behaviours. Springer Nature Switzerland, Cham
13. Parmentier E, Kéver L, Boyle K et al (2013) Sound production mechanism in *Gobius paganellus* (Gobiidae). J Exp Biol 216:3189–3199. https://doi.org/10.1242/jeb.087205
14. Fine ML, Parmentier E (2015) Mechanisms of fish sound production. In: Ladich F (ed) Sound communication in fishes. Springer, Vienna, pp 77–126
15. Medler S, Mykles D (2015) Muscle structure, fiber types, and physiology. In: Chang ES, Thiel M (eds) Natural history of Crustacea. Oxford University Press, pp 103–133
16. Shaffer JF, Kier WM (2016) Tuning of shortening speed in coleoid cephalopod muscle: no evidence for tissue-specific muscle myosin heavy chain isoforms. Invertebr Biol 135:3–12. https://doi.org/10.1111/ivb.12111
17. Schiaffino S, Reggiani C (2011) Fiber types in mammalian skeletal muscles. Physiol Rev 91:1447–1531. https://doi.org/10.1152/physrev.00031.2010
18. Medler S (2019) Mixing it up: the biological significance of hybrid skeletal muscle fibers. J Exp Biol 222:jeb200832. https://doi.org/10.1242/jeb.200832
19. Mukund K, Subramaniam S (2020) Skeletal muscle: a review of molecular structure and function, in health and disease. WIREs Syst Biol Med 12:e1462. https://doi.org/10.1002/wsbm.1462
20. Sandage MJ, Smith AG (2017) Muscle bioenergetic considerations for intrinsic laryngeal skeletal muscle physiology. J Speech Lang Hear Res 60:1254–1263. https://doi.org/10.1044/2016_JSLHR-S-16-0192
21. Hoh JFY (2005) Laryngeal muscle fibre types. Acta Physiol Scand 183:133–149. https://doi.org/10.1111/j.1365-201X.2004.01402.x
22. Tracicaru R-V, Bräuer L, Döllinger M et al (2024) Muscular and neuronal control of voice production – forgotten findings, current concepts, and new developments. Ann Anat Anat Anz 255:152283. https://doi.org/10.1016/j.aanat.2024.152283
23. Hoh JFY (2010) Laryngeal muscles as highly specialized organs in airway protection, respiration and phonation. In: Handbook of behavioral neuroscience. Elsevier, pp 13–21
24. Elemans CPH, Mead AF, Rome LC, Goller F (2008) Superfast vocal muscles control song production in songbirds. PLoS One 3:e2581. https://doi.org/10.1371/journal.pone.0002581
25. Biewener A, Patek S (2018) Animal locomotion. Oxford University Press
26. Bączyk M, Manuel M, Roselli F, Zytnicki D (2022) Diversity of mammalian motoneurons and motor units. In: O'Donovan MJ, Falgairolle M (eds) Vertebrate motoneurons. Springer International Publishing, Cham, pp 131–150
27. Burke RE (1981) Motor units: anatomy, physiology, and functional organization. In: Terjung R (ed) Comprehensive physiology, 1st edn. Wiley, pp 345–422
28. Manuel M, Chardon M, Tysseling V, Heckman CJ (2019) Scaling of motor output, from mouse to humans. Physiology 34:5–13. https://doi.org/10.1152/physiol.00021.2018
29. Hoyle G (1983) Muscles and their neural control. Wiley, New York
30. Smarandache-Wellmann CR (2016) Arthropod neurons and nervous system. Curr Biol 26:R960–R965. https://doi.org/10.1016/j.cub.2016.07.063
31. Dubuc R, Cabelguen J-M, Ryczko D (2023) Locomotor pattern generation and descending control: a historical perspective. J Neurophysiol 130:401–416. https://doi.org/10.1152/jn.00204.2023
32. Grillner S (2006) Biological pattern generation: the cellular and computational logic of networks in motion. Neuron 52:751–766. https://doi.org/10.1016/j.neuron.2006.11.008
33. Grillner S, El Manira A (2020) Current principles of motor control, with special reference to vertebrate locomotion. Physiol Rev 100:271–320. https://doi.org/10.1152/physrev.00015.2019

34. Rancic V, Gosgnach S (2021) Recent insights into the rhythmogenic core of the locomotor CPG. Int J Mol Sci 22:1394. https://doi.org/10.3390/ijms22031394
35. Björnfors ER, Picton LD, Song J, El Manira A (2019) Diversity of neurons and circuits controlling the speed and coordination of locomotion. Curr Opin Physiol 8:170–176. https://doi.org/10.1016/j.cophys.2019.02.006
36. Song J, Pallucchi I, Ausborn J et al (2020) Multiple rhythm-generating circuits act in tandem with pacemaker properties to control the start and speed of locomotion. Neuron 105:1048–1061.e4. https://doi.org/10.1016/j.neuron.2019.12.030
37. Talpalar AE, Bouvier J, Borgius L et al (2013) Dual-mode operation of neuronal networks involved in left–right alternation. Nature 500:85–88. https://doi.org/10.1038/nature12286
38. Grillner S, Kozlov A (2021) The CPGs for limbed locomotion–facts and fiction. Int J Mol Sci 22:5882. https://doi.org/10.3390/ijms22115882
39. Gariépy J-F, Missaghi K, Dubuc R (2010) The interactions between locomotion and respiration. In: Progress in brain research. Elsevier, pp 173–188
40. Barlow SM, Lund JP, Estep M, Kolta A (2010) Central pattern generators for orofacial movements and speech. In: Handbook of behavioral neuroscience. Elsevier, pp 351–369
41. Sweeney LB, Kelley DB (2014) Harnessing vocal patterns for social communication. Curr Opin Neurobiol 28:34–41. https://doi.org/10.1016/j.conb.2014.06.006
42. Jacob PF, Hedwig B (2016) Acoustic signalling for mate attraction in crickets: abdominal ganglia control the timing of the calling song pattern. Behav Brain Res 309:51–66. https://doi.org/10.1016/j.bbr.2016.04.025
43. Mhatre N, Malkin R, Deb R et al (2017) Tree crickets optimize the acoustics of baffles to exaggerate their mate-attraction signal. Elife 6:e32763. https://doi.org/10.7554/eLife.32763
44. Jacob PF, Hedwig B (2020) Modular timer networks: abdominal interneurons controlling the chirp and pulse pattern in a cricket calling song. J Comp Physiol A 206:921–938. https://doi.org/10.1007/s00359-020-01448-0
45. Schöneich S (2020) Neuroethology of acoustic communication in field crickets – from signal generation to song recognition in an insect brain. Prog Neurobiol 194:101882. https://doi.org/10.1016/j.pneurobio.2020.101882
46. Bass AH (2014) Central pattern generator for vocalization: is there a vertebrate morphotype? Curr Opin Neurobiol 28:94–100. https://doi.org/10.1016/j.conb.2014.06.012
47. Steuer I, Guertin PA (2019) Central pattern generators in the brainstem and spinal cord: an overview of basic principles, similarities and differences. Rev Neurosci 30:107–164. https://doi.org/10.1515/revneuro-2017-0102
48. Del Negro CA, Funk GD, Feldman JL (2018) Breathing matters. Nat Rev Neurosci 19:351–367. https://doi.org/10.1038/s41583-018-0003-6
49. Mooney R (2020) The neurobiology of innate and learned vocalizations in rodents and songbirds. Curr Opin Neurobiol 64:24–31. https://doi.org/10.1016/j.conb.2020.01.004
50. Kelley DB (2022) Convergent and divergent neural circuit architectures that support acoustic communication. Front Neural Circuits 16. https://doi.org/10.3389/fncir.2022.976789
51. Grillner S (2003) The motor infrastructure: from ion channels to neuronal networks. Nat Rev Neurosci 4:573–586. https://doi.org/10.1038/nrn1137
52. Harris-Warrick RM (2010) General principles of rhythmogenesis in central pattern generator networks. In: Progress in brain research. Elsevier, pp 213–222
53. Kelley DB, Ballagh IH, Barkan CL et al (2020) Generation, coordination, and evolution of neural circuits for vocal communication. J Neurosci 40:22–36. https://doi.org/10.1523/JNEUROSCI.0736-19.2019
54. Yamaguchi A, Peltier M (2023) Two conserved vocal central pattern generators broadly tuned for fast and slow rates generate species-specific vocalizations in *Xenopus* clawed frogs. Elife 12:e86299. https://doi.org/10.7554/eLife.86299
55. Katz PS, Frost WN (1996) Intrinsic neuromodulation: altering neuronal circuits from within. Trends Neurosci 19:54–61. https://doi.org/10.1016/0166-2236(96)89621-4
56. Marder E, Bucher D, Schulz DJ, Taylor AL (2005) Invertebrate central pattern generation moves along. Curr Biol 15:R685–R699. https://doi.org/10.1016/j.cub.2005.08.022

57. Marder E, Calabrese RL (1996) Principles of rhythmic motor pattern generation. Physiol Rev 76:687–717. https://doi.org/10.1152/physrev.1996.76.3.687
58. Tuthill JC, Wilson RI (2016) Mechanosensation and adaptive motor control in insects. Curr Biol 26:R1022–R1038. https://doi.org/10.1016/j.cub.2016.06.070
59. Field LH, Pflüger H-J (1989) The femoral chordotonal organ: a bifunctional orthopteran (*Locusta migratoria*) sense organ? Comp Biochem Physiol A Physiol 93:729–743. https://doi.org/10.1016/0300-9629(89)90494-5
60. Kittmann R, Schmitz J (1992) Functional specialization of the scoloparia of the femoral chordotonal organ in stick insects. J Exp Biol 173:91–108. https://doi.org/10.1242/jeb.173.1.91
61. Caldwell JC, Miller MM, Wing S et al (2003) Dynamic analysis of larval locomotion in *Drosophila* chordotonal organ mutants. Proc Natl Acad Sci 100:16053–16058. https://doi.org/10.1073/pnas.2535546100
62. Purves D, Augustine GJ, Fitzpatrick D, et al (2001) Neuroscience (2nd ed.). Sunderland, Sinauer Associates
63. Conway BA, Hultborn H, Kiehn O (1987) Proprioceptive input resets central locomotor rhythm in the spinal cat. Exp Brain Res 68. https://doi.org/10.1007/BF00249807
64. Bewick GS, Banks RW (2015) Mechanotransduction in the muscle spindle. Pflüg Arch Eur J Physiol 467:175–190. https://doi.org/10.1007/s00424-014-1536-9
65. Ellaway PH, Taylor A, Durbaba R (2015) Muscle spindle and fusimotor activity in locomotion. J Anat 227:157–166. https://doi.org/10.1111/joa.12299
66. Macefield VG, Knellwolf TP (2018) Functional properties of human muscle spindles. J Neurophysiol 120:452–467. https://doi.org/10.1152/jn.00071.2018
67. Dimitriou M (2022) Human muscle spindles are wired to function as controllable signal-processing devices. Elife 11:e78091. https://doi.org/10.7554/eLife.78091
68. Blum KP, Campbell KS, Horslen BC et al (2020) Diverse and complex muscle spindle afferent firing properties emerge from multiscale muscle mechanics. Elife 9:e55177. https://doi.org/10.7554/eLife.55177
69. Papaioannou S, Dimitriou M (2021) Goal-dependent tuning of muscle spindle receptors during movement preparation. Sci Adv 7:eabe0401. https://doi.org/10.1126/sciadv.abe0401
70. Kissane RWP, Charles JP, Banks RW, Bates KT (2022) Skeletal muscle function underpins muscle spindle abundance. Proc R Soc B Biol Sci 289:20220622. https://doi.org/10.1098/rspb.2022.0622
71. Burke D (2021) Crosstalk opposing view: independent fusimotor control of muscle spindles in humans: there is little to gain. J Physiol 599:2505–2508. https://doi.org/10.1113/JP281337
72. Hattori Y, Tomonaga M, Matsuzawa T (2013) Spontaneous synchronized tapping to an auditory rhythm in a chimpanzee. Sci Rep 3:1566. https://doi.org/10.1038/srep01566
73. Hattori Y, Tomonaga M, Matsuzawa T (2015) Distractor effect of auditory rhythms on self-paced tapping in chimpanzees and humans. PLoS One 10:e0130682. https://doi.org/10.1371/journal.pone.0130682
74. Large EW, Gray PM (2015) Spontaneous tempo and rhythmic entrainment in a bonobo (*Pan paniscus*). J Comp Psychol 129:317–328. https://doi.org/10.1037/com0000011
75. Thornell L, Carlsson L, Eriksson P et al (2015) Fibre typing of intrafusal fibres. J Anat 227:136–156. https://doi.org/10.1111/joa.12338
76. Buschges A, Ache JM (2025) Motor control on the move: from insights in insects to general mechanisms. Physiol Rev 105:975–1031. https://doi.org/10.1152/physrev.00009.2024
77. Kupfermann I, Weiss KR (1978) The command neuron concept. Behav Brain Sci 1:3–10. https://doi.org/10.1017/S0140525X00059057
78. Hedwig B, Heinrich R (1997) Identified descending brain neurons control different stridulatory motor patterns in an acridid grasshopper. J Comp Physiol [A] 180:285–294. https://doi.org/10.1007/s003590050048
79. Hedwig B (2000) Control of cricket stridulation by a command neuron: efficacy depends on the behavioral state. J Neurophysiol 83:712–722. https://doi.org/10.1152/jn.2000.83.2.712
80. Von Philipsborn AC, Liu T, Yu JY et al (2011) Neuronal control of Drosophila courtship song. Neuron 69:509–522. https://doi.org/10.1016/j.neuron.2011.01.011

81. Büschges A, Gorostiza EA (2023) Neurons with names: descending control and sensorimotor processing in insect motor control. Curr Opin Neurobiol 83:102766. https://doi.org/10.1016/j.conb.2023.102766
82. Lillvis JL, Wang K, Shiozaki HM et al (2024) Nested neural circuits generate distinct acoustic signals during Drosophila courtship. Curr Biol 34:808–824.e6. https://doi.org/10.1016/j.cub.2024.01.015
83. Ding Y, Lillvis JL, Cande J et al (2019) Neural evolution of context-dependent fly song. Curr Biol 29:1089–1099.e7. https://doi.org/10.1016/j.cub.2019.02.019
84. Grillner S (2021) Evolution of the vertebrate motor system – from forebrain to spinal cord. Curr Opin Neurobiol 71:11–18. https://doi.org/10.1016/j.conb.2021.07.016
85. Caggiano V, Leiras R, Goñi-Erro H et al (2018) Midbrain circuits that set locomotor speed and gait selection. Nature 553:455–460. https://doi.org/10.1038/nature25448
86. Josset N, Roussel M, Lemieux M et al (2018) Distinct contributions of mesencephalic locomotor region nuclei to locomotor control in the freely behaving mouse. Curr Biol 28:884–901.e3. https://doi.org/10.1016/j.cub.2018.02.007
87. Morgenstern NA, Esposito MS (2024) The basal ganglia and mesencephalic locomotor region connectivity matrix. Curr Neuropharmacol 22:1454–1472. https://doi.org/10.2174/1570159X21666230809112840
88. Ménard A, Grillner S (2008) Diencephalic locomotor region in the lamprey–afferents and efferent control. J Neurophysiol 100:1343–1353. https://doi.org/10.1152/jn.01128.2007
89. Zelenin PV, Orlovsky GN, Deliagina TG (2007) Sensory-motor transformation by individual command neurons. J Neurosci 27:1024–1032. https://doi.org/10.1523/JNEUROSCI.4925-06.2007
90. Arber S (2012) Motor circuits in action: specification, connectivity, and function. Neuron 74:975–989. https://doi.org/10.1016/j.neuron.2012.05.011
91. Esposito MS, Capelli P, Arber S (2014) Brainstem nucleus MdV mediates skilled forelimb motor tasks. Nature 508:351–356. https://doi.org/10.1038/nature13023
92. Capelli P, Pivetta C, Soledad Esposito M, Arber S (2017) Locomotor speed control circuits in the caudal brainstem. Nature 551:373–377. https://doi.org/10.1038/nature24064
93. Ruder L, Schina R, Kanodia H et al (2021) A functional map for diverse forelimb actions within brainstem circuitry. Nature 590:445–450. https://doi.org/10.1038/s41586-020-03080-z
94. Arber S, Costa RM (2022) Networking brainstem and basal ganglia circuits for movement. Nat Rev Neurosci 23:342–360. https://doi.org/10.1038/s41583-022-00581-w
95. Grillner S, Robertson B (2016) The basal ganglia over 500 million years. Curr Biol 26:R1088–R1100. https://doi.org/10.1016/j.cub.2016.06.041
96. Suryanarayana SM, Robertson B, Grillner S (2022) The neural bases of vertebrate motor behaviour through the lens of evolution. Philos Trans R Soc B Biol Sci 377:20200521. https://doi.org/10.1098/rstb.2020.0521
97. Park J, Coddington LT, Dudman JT (2020) Basal ganglia circuits for action specification. Annu Rev Neurosci 43:485–507. https://doi.org/10.1146/annurev-neuro-070918-050452
98. Favila N, Gurney K, Overton PG (2024) Role of the basal ganglia in innate and learned behavioural sequences. Rev Neurosci 35:35–55. https://doi.org/10.1515/revneuro-2023-0038
99. Scharff C, Nottebohm F (1991) A comparative study of the behavioral deficits following lesions of various parts of the zebra finch song system: implications for vocal learning. J Neurosci 11:2896–2913. https://doi.org/10.1523/JNEUROSCI.11-09-02896.1991
100. Xiao L, Chattree G, Oscos FG et al (2018) A basal ganglia circuit sufficient to guide birdsong learning. Neuron 98:208–221.e5. https://doi.org/10.1016/j.neuron.2018.02.020
101. Rocha GS, Freire MAM, Britto AM et al (2023) Basal ganglia for beginners: the basic concepts you need to know and their role in movement control. Front Syst Neurosci 17:1242929. https://doi.org/10.3389/fnsys.2023.1242929
102. Foster NN, Barry J, Korobkova L et al (2021) The mouse cortico–basal ganglia–thalamic network. Nature 598:188–194. https://doi.org/10.1038/s41586-021-03993-3

103. Janacsek K, Shattuck KF, Tagarelli KM et al (2020) Sequence learning in the human brain: a functional neuroanatomical meta-analysis of serial reaction time studies. NeuroImage 207:116387. https://doi.org/10.1016/j.neuroimage.2019.116387
104. Nakamura T, Nagata M, Yagi T et al (2017) Learning new sequential stepping patterns requires striatal plasticity during the earliest phase of acquisition. Eur J Neurosci 45:901–911. https://doi.org/10.1111/ejn.13537
105. Geddes CE, Li H, Jin X (2018) Optogenetic editing reveals the hierarchical organization of learned action sequences. Cell 174:32–43.e15. https://doi.org/10.1016/j.cell.2018.06.012
106. Dhawale AK, Wolff SBE, Ko R, Ölveczky BP (2021) The basal ganglia control the detailed kinematics of learned motor skills. Nat Neurosci 24:1256–1269. https://doi.org/10.1038/s41593-021-00889-3
107. Wolff SBE, Ko R, Ölveczky BP (2022) Distinct roles for motor cortical and thalamic inputs to striatum during motor skill learning and execution. Sci Adv 8:eabk0231. https://doi.org/10.1126/sciadv.abk0231
108. Kameda M, Niikawa K, Uematsu A, Tanaka M (2023) Sensory and motor representations of internalized rhythms in the cerebellum and basal ganglia. Proc Natl Acad Sci 120:e2221641120. https://doi.org/10.1073/pnas.2221641120
109. Hwang EJ, Dahlen JE, Hu YY et al (2019) Disengagement of motor cortex from movement control during long-term learning. Sci Adv 5:eaay0001. https://doi.org/10.1126/sciadv.aay0001
110. Wu T, Long X, Wang L et al (2011) Functional connectivity of cortical motor areas in the resting state in Parkinson's disease. Hum Brain Mapp 32:1443–1457. https://doi.org/10.1002/hbm.21118
111. Grillner S (2025) How circuits for habits are formed within the basal ganglia. Proc Natl Acad Sci 122:e2423068122. https://doi.org/10.1073/pnas.2423068122
112. Díaz-Hernández E, Contreras-López R, Sánchez-Fuentes A et al (2018) The thalamostriatal projections contribute to the initiation and execution of a sequence of movements. Neuron 100:739–752.e5. https://doi.org/10.1016/j.neuron.2018.09.052
113. Jin X, Tecuapetla F, Costa RM (2014) Basal ganglia subcircuits distinctively encode the parsing and concatenation of action sequences. Nat Neurosci 17:423–430. https://doi.org/10.1038/nn.3632
114. Jin X, Costa RM (2015) Shaping action sequences in basal ganglia circuits. Curr Opin Neurobiol 33:188–196. https://doi.org/10.1016/j.conb.2015.06.011
115. Jurado-Parras M-T, Safaie M, Sarno S et al (2020) The dorsal striatum energizes motor routines. Curr Biol 30:4362–4372.e6. https://doi.org/10.1016/j.cub.2020.08.049
116. Guo Q, Wang D, He X et al (2015) Whole-brain mapping of inputs to projection neurons and cholinergic interneurons in the dorsal striatum. PLoS One 10:e0123381. https://doi.org/10.1371/journal.pone.0123381
117. Hintiryan H, Foster NN, Bowman I et al (2016) The mouse cortico-striatal projectome. Nat Neurosci 19:1100–1114. https://doi.org/10.1038/nn.4332
118. Hooks BM, Papale AE, Paletzki RF et al (2018) Topographic precision in sensory and motor corticostriatal projections varies across cell type and cortical area. Nat Commun 9:3549. https://doi.org/10.1038/s41467-018-05780-7
119. Mandelbaum G, Taranda J, Haynes TM et al (2019) Distinct cortical-thalamic-striatal circuits through the parafascicular nucleus. Neuron 102:636–652.e7. https://doi.org/10.1016/j.neuron.2019.02.035
120. Shires J, Joshi S, Basso MA (2010) Shedding new light on the role of the basal ganglia-superior colliculus pathway in eye movements. Curr Opin Neurobiol 20:717–725. https://doi.org/10.1016/j.conb.2010.08.008
121. Mchaffie J, Stanford T, Stein B et al (2005) Subcortical loops through the basal ganglia. Trends Neurosci 28:401–407. https://doi.org/10.1016/j.tins.2005.06.006
122. McElvain LE, Chen Y, Moore JD et al (2021) Specific populations of basal ganglia output neurons target distinct brain stem areas while collateralizing throughout the diencephalon. Neuron 109:1721–1738.e4. https://doi.org/10.1016/j.neuron.2021.03.017

123. Lalive AL, Lien AD, Roseberry TK et al (2018) Motor thalamus supports striatum-driven reinforcement. Elife 7:e34032. https://doi.org/10.7554/eLife.34032
124. Azevedo FAC, Carvalho LRB, Grinberg LT et al (2009) Equal numbers of neuronal and nonneuronal cells make the human brain an isometrically scaled-up primate brain. J Comp Neurol 513:532–541. https://doi.org/10.1002/cne.21974
125. Herculano-Houzel S (2010) Coordinated scaling of cortical and cerebellar numbers of neurons. Front Neuroanat. https://doi.org/10.3389/fnana.2010.00012
126. Von Bartheld CS, Bahney J, Herculano-Houzel S (2016) The search for true numbers of neurons and glial cells in the human brain: a review of 150 years of cell counting. J Comp Neurol 524:3865–3895. https://doi.org/10.1002/cne.24040
127. Yopak KE, Pakan JMP, Wylie D (2020) The cerebellum of nonmammalian vertebrates. In: Evolutionary neuroscience. Elsevier, pp 247–260
128. Kebschull JM, Casoni F, Consalez GG et al (2023) Cerebellum lecture: the cerebellar nuclei–core of the cerebellum. Cerebellum 23:620–677. https://doi.org/10.1007/s12311-022-01506-0
129. Murakami Y, Sugahara F (2021) Evolutionary and developmental perspectives on the origin and diversification of the vertebrate cerebellum. In: Mizusawa H, Kakei S (eds) Cerebellum as a CNS hub. Springer International Publishing, Cham, pp 3–24
130. Heuer K, Traut N, De Sousa AA et al (2023) Diversity and evolution of cerebellar folding in mammals. Elife 12:e85907. https://doi.org/10.7554/eLife.85907
131. Carey MR (2024) The cerebellum. Curr Biol 34:R7–R11. https://doi.org/10.1016/j.cub.2023.11.048
132. Manto M, Bower JM, Conforto AB et al (2012) Consensus paper: roles of the cerebellum in motor control–the diversity of ideas on cerebellar involvement in movement. Cerebellum 11:457–487. https://doi.org/10.1007/s12311-011-0331-9
133. Pisotta I, Molinari M (2014) Cerebellar contribution to feedforward control of locomotion. Front Hum Neurosci 8. https://doi.org/10.3389/fnhum.2014.00475
134. Gao J-H, Parsons LM, Bower JM et al (1996) Cerebellum implicated in sensory acquisition and discrimination rather than motor control. Science 272:545–547. https://doi.org/10.1126/science.272.5261.545
135. Bass AH, Chagnaud BP, Feng NY (2015) Comparative neurobiology of sound production in fishes. In: Ladich F (ed) Sound communication in fishes. Springer Vienna, Vienna, pp 35–75
136. Arunachalam R, Weerasinghe VS, Mills KR (2005) Motor control of rapid sequential finger tapping in humans. J Neurophysiol 94:2162–2170. https://doi.org/10.1152/jn.01173.2004
137. Guillot A, Lebon F, Rouffet D et al (2007) Muscular responses during motor imagery as a function of muscle contraction types. Int J Psychophysiol 66:18–27. https://doi.org/10.1016/j.ijpsycho.2007.05.009
138. Sangwan S, Green RA, Taylor NF (2014) Characteristics of stabilizer muscles: a systematic review. Physiother Can 66:348–358. https://doi.org/10.3138/ptc.2013-51
139. Ting LH (2007) Dimensional reduction in sensorimotor systems: a framework for understanding muscle coordination of posture. In: Progress in brain research. Elsevier, pp 299–321
140. Tresch MC, Jarc A (2009) The case for and against muscle synergies. Curr Opin Neurobiol 19:601–607. https://doi.org/10.1016/j.conb.2009.09.002
141. Scott Kelso JA, Fink PW, DeLaplain CR, Carson RG (2001) Haptic information stabilizes and destabilizes coordination dynamics. Proc R Soc Lond B Biol Sci 268:1207–1213. https://doi.org/10.1098/rspb.2001.1620
142. McCrea DA, Rybak IA (2008) Organization of mammalian locomotor rhythm and pattern generation. Brain Res Rev 57:134–146. https://doi.org/10.1016/j.brainresrev.2007.08.006
143. Bucher D, Haspel G, Golowasch J, Nadim F (2015) Central pattern generators. In: Encyclopedia of life sciences, 1st edn. Wiley, pp 1–12
144. Grillner S (1981) Control of locomotion in bipeds, tetrapods, and fish. In: Terjung R (ed) Comprehensive physiology, 1st edn. Wiley, pp 1179–1236
145. Grillner S (1985) Neurobiological bases of rhythmic motor acts in vertebrates. Science 228:143–149. https://doi.org/10.1126/science.3975635

146. Granatosky MC (2018) A review of locomotor diversity in mammals with analyses exploring the influence of substrate use, body mass and intermembral index in primates. J Zool 306:207–216. https://doi.org/10.1111/jzo.12608
147. Sylos-Labini F, Ivanenko YP, MacLellan MJ et al (2014) Locomotor-like leg movements evoked by rhythmic arm movements in humans. PLoS One 9:e90775. https://doi.org/10.1371/journal.pone.0090775
148. Eke-Okoro S, Gregoric M, Larsson L (1997) Alterations in gait resulting from deliberate changes of arm-swing amplitude and phase. Clin Biomech 12:516–521. https://doi.org/10.1016/S0268-0033(97)00050-8
149. La Scaleia V, Sylos-Labini F, Hoellinger T et al (2014) Control of leg movements driven by EMG activity of shoulder muscles. Front Hum Neurosci 8. https://doi.org/10.3389/fnhum.2014.00838
150. Avaltroni P, Cappellini G, Sylos-Labini F et al (2024) Spinal maps of motoneuron activity during human locomotion: neuromechanical considerations. Front Physiol 15:1389436. https://doi.org/10.3389/fphys.2024.1389436
151. Dietz V, Michel J (2009) Human bipeds use quadrupedal coordination during locomotion. Ann N Y Acad Sci 1164:97–103. https://doi.org/10.1111/j.1749-6632.2008.03710.x
152. Hennig RM (1989) Neuromuscular activity during stridulation in the cricket *Teleogryllus commodus*. J Comp Physiol A 165:837–846. https://doi.org/10.1007/BF00610882
153. Elsner N (1974) Neuroethology of sound production in gomphocerine grasshoppers (Orthoptera: Acrididae): I. Song patterns and stridulatory movements. J Comp Physiol 88:67–102. https://doi.org/10.1007/BF00695923
154. Bramble DM, Carrier DR (1983) Running and breathing in mammals. Science 219:251–256. https://doi.org/10.1126/science.6849136
155. Juvin L, Colnot E, Barrière G et al (2022) Neurogenic mechanisms for locomotor-respiratory coordination in mammals. Front Neuroanat 16:953746. https://doi.org/10.3389/fnana.2022.953746
156. Milsom WK (2008) Evolutionary trends in respiratory mechanisms. In: Poulin MJ, Wilson RJA (eds) Integration in respiratory control. Springer New York, New York, pp 293–298
157. Le Gal J, Colnot E, Cardoit L et al (2020) Modulation of respiratory network activity by forelimb and hindlimb locomotor generators. Eur J Neurosci 52:3181–3195. https://doi.org/10.1111/ejn.14717
158. Webb PW (1975) Synchrony of locomotion and ventilation in *Cymatogaster aggregata*. Can J Zool 53:904–907. https://doi.org/10.1139/z75-104
159. Tytell ED, Alexander JK (2007) Bluegill *Lepomis macrochirus* synchronize pectoral fin motion and opercular pumping. J Fish Biol 70:1268–1279. https://doi.org/10.1111/j.1095-8649.2007.01416.x
160. Boggs DF (2002) Interactions between locomotion and ventilation in tetrapods. Comp Biochem Physiol A Mol Integr Physiol 133:269–288. https://doi.org/10.1016/S1095-6433(02)00160-5
161. Otto D, Hennig RM (1993) Interneurons descending from the cricket subesophageal ganglion control stridulation and ventilation. Naturwissenschaften 80:36–38. https://doi.org/10.1007/BF01139757
162. Paripovic I, Hennig RM, Otto D (1996) Abdominal ventilatory pattern in crickets depends on the stridulatory motor pattern. Physiol Entomol 21:223–230. https://doi.org/10.1111/j.1365-3032.1996.tb00859.x
163. Schöneich S, Hedwig B (2019) Feedforward discharges couple the singing central pattern generator and ventilation central pattern generator in the cricket abdominal central nervous system. J Comp Physiol A 205:881–895. https://doi.org/10.1007/s00359-019-01377-7
164. Bernasconi P, Kohl J (1993) Analysis of co-ordination between breathing and exercise rhythms in man. J Physiol 471:693–706. https://doi.org/10.1113/jphysiol.1993.sp019923
165. Bardy BG, Hoffmann CP, Moens B et al (2015) Sound-induced stabilization of breathing and moving. Ann N Y Acad Sci 1337:94–100. https://doi.org/10.1111/nyas.12650

166. Perret C, Wenger M, Leicht CA, Goosey-Tolfrey VL (2016) Locomotor-respiratory coupling in wheelchair racing athletes: a pilot study. Front Physiol 7. https://doi.org/10.3389/fphys.2016.00011
167. Siegmund GP, Edwards MR, Moore KS et al (1999) Ventilation and locomotion coupling in varsity male rowers. J Appl Physiol 87:233–242. https://doi.org/10.1152/jappl.1999.87.1.233
168. Haas F, Distenfeld S, Axen K (1986) Effects of perceived musical rhythm on respiratory pattern. J Appl Physiol 61:1185–1191. https://doi.org/10.1152/jappl.1986.61.3.1185
169. Ebert D, Hefter H, Binkofski F, Freund H-J (2002) Coordination between breathing and mental grouping of pianistic finger movements. Percept Mot Skills 95:339–353. https://doi.org/10.2466/pms.2002.95.2.339
170. Pouw W, Fuchs S (2022) Origins of vocal-entangled gesture. Neurosci Biobehav Rev 141:104836. https://doi.org/10.1016/j.neubiorev.2022.104836
171. Fulton TJ, Paris HL, Stickford ASL et al (2018) Locomotor-respiratory coupling is maintained in simulated moderate altitude in trained distance runners. J Appl Physiol 125:1–7. https://doi.org/10.1152/japplphysiol.01122.2017
172. McDermott WJ, Van Emmerik REA, Hamill J (2003) Running training and adaptive strategies of locomotor-respiratory coordination. Eur J Appl Physiol 89:435–444. https://doi.org/10.1007/s00421-003-0831-5
173. Young IS, Alexander RMCN, Woakes AJ et al (1992) The synchronization of ventilation and locomotion in horses (*Equus caballus*). J Exp Biol 166:19–31. https://doi.org/10.1242/jeb.166.1.19
174. Bramble DM, Jenkins FA (1993) Mammalian locomotor-respiratory integration: implications for diaphragmatic and pulmonary design. Science 262:235–240. https://doi.org/10.1126/science.8211141
175. Simons RS (1999) Running, breathing and visceral motion in the domestic rabbit (*Oryctolagus cuniculus*): testing visceral displacement hypotheses. J Exp Biol 202:563–577. https://doi.org/10.1242/jeb.202.5.563
176. Bramble DM (1989) Axial-appendicular dynamics and the integration of breathing and gait in mammal. Am Zool 29:171–186. https://doi.org/10.1093/icb/29.1.171
177. Stickford ASL, Stickford JL (2014) Ventilation and locomotion in humans: mechanisms, implications, and perturbations to the coupling of these two rhythms. Springer Sci Rev 2:95–118. https://doi.org/10.1007/s40362-014-0020-4
178. Hodges PW, Gandevia SC (2000) Activation of the human diaphragm during a repetitive postural task. J Physiol 522:165–175. https://doi.org/10.1111/j.1469-7793.2000.t01-1-00165.xm
179. Lee H, Banzett R (1997) Mechanical links between locomotion and breathing: can you breathe with your legs? Physiology 12:273–278. https://doi.org/10.1152/physiologyonline.1997.12.6.273
180. Kawahara K, Kumagai S, Nakazono Y, Miyamoto Y (1989) Coupling between respiratory and stepping rhythms during locomotion in decerebrate cats. J Appl Physiol 67:110–115. https://doi.org/10.1152/jappl.1989.67.1.110
181. Le Gal J-P, Juvin L, Cardoit L et al (2014) Remote control of respiratory neural network by spinal locomotor generators. PLoS One 9:e89670. https://doi.org/10.1371/journal.pone.0089670
182. Palisses R, Persegol L, Viala D, Viala G (1988) Reflex modulation of phrenic activity through hindlimb passive motion in decorticate and spinal rabbit preparation. Neuroscience 24:719–728. https://doi.org/10.1016/0306-4522(88)90364-8
183. Funk GD, Milsom WK, Steeves JD (1992) Coordination of wingbeat and respiration in the Canada goose. I. Passive wing flapping. J Appl Physiol 73:1014–1024. https://doi.org/10.1152/jappl.1992.73.3.1014
184. Morin D, Viala D (2002) Coordinations of locomotor and respiratory rhythms *in vitro* are critically dependent on hindlimb sensory inputs. J Neurosci 22:4756–4765. https://doi.org/10.1523/JNEUROSCI.22-11-04756.2002
185. Potts JT, Rybak IA, Paton JFR (2005) Respiratory rhythm entrainment by somatic afferent stimulation. J Neurosci 25:1965–1978. https://doi.org/10.1523/JNEUROSCI.3881-04.2005

186. Häusler U (2000) Vocalization-correlated respiratory movements in the squirrel monkey. J Acoust Soc Am 108:1443–1450. https://doi.org/10.1121/1.1289209
187. Hartley RS (1990) Expiratory muscle activity during song production in the canary. Respir Physiol 81:177–187. https://doi.org/10.1016/0034-5687(90)90044-Y
188. Anderson TM, Ramirez J-M (2017) Respiratory rhythm generation: triple oscillator hypothesis. F1000Res 6:139. https://doi.org/10.12688/f1000research.10193.1
189. Ashhad S, Kam K, Del Negro CA, Feldman JL (2022) Breathing rhythm and pattern and their influence on emotion. Annu Rev Neurosci 45:223–247. https://doi.org/10.1146/annurev-neuro-090121-014424
190. Allan SE, Suthers RA (1994) Lateralization and motor stereotype of song production in the brown-headed cowbird. J Neurobiol 25:1154–1166. https://doi.org/10.1002/neu.480250910
191. Wild JM, Goller F, Suthers RA (1998) Inspiratory muscle activity during bird song. J Neurobiol 36:441–453. https://doi.org/10.1002/(SICI)1097-4695(19980905)36:3<441::AID-NEU11>3.0.CO;2-E
192. Suthers R, Goller F, Pytte C (1999) The neuromuscular control of birdsong. Philos Trans R Soc Lond Ser B Biol Sci 354:927–939. https://doi.org/10.1098/rstb.1999.0444
193. Hoese WJ, Podos J, Boetticher NC, Nowicki S (2000) Vocal tract function in birdsong production: experimental manipulation of beak movements. J Exp Biol 203:1845–1855. https://doi.org/10.1242/jeb.203.12.1845
194. Smotherman M, Kobayasi K, Ma J et al (2006) A mechanism for vocal-respiratory coupling in the mammalian parabrachial nucleus. J Neurosci 26:4860–4869. https://doi.org/10.1523/JNEUROSCI.4607-05.2006
195. Zheng D, Okobi DE, Shu R et al (2022) Mapping the vocal circuitry of Alston's singing mouse with pseudorabies virus. J Comp Neurol 530:2075–2099. https://doi.org/10.1002/cne.25321
196. Wei XP, Collie M, Dempsey B et al (2022) A novel reticular node in the brainstem synchronizes neonatal mouse crying with breathing. Neuron 110:644–657.e6. https://doi.org/10.1016/j.neuron.2021.12.014
197. Ashmore RC, Wild JM, Schmidt MF (2005) Brainstem and forebrain contributions to the generation of learned motor behaviors for song. J Neurosci 25:8543–8554. https://doi.org/10.1523/JNEUROSCI.1668-05.2005
198. McLean J, Bricault S, Schmidt MF (2013) Characterization of respiratory neurons in the rostral ventrolateral medulla, an area critical for vocal production in songbirds. J Neurophysiol 109:948–957. https://doi.org/10.1152/jn.00595.2012
199. Berg KS, Delgado S, Mata-Betancourt A (2019) Phylogenetic and kinematic constraints on avian flight signals. Proc R Soc B Biol Sci 286:20191083. https://doi.org/10.1098/rspb.2019.1083
200. Burchardt LS, Norton P, Behr O et al (2019) General isochronous rhythm in echolocation calls and social vocalizations of the bat *Saccopteryx bilineata*. R Soc Open Sci 6:181076. https://doi.org/10.1098/rsos.181076
201. Suthers RA, Thomas SP, Suthers BJ (1972) Respiration, wing-beat and ultrasonic pulse emission in an echo-locating bat. J Exp Biol 56:37–48. https://doi.org/10.1242/jeb.56.1.37
202. Stidsholt L, Greif S, Goerlitz HR et al (2021) Hunting bats adjust their echolocation to receive weak prey echoes for clutter reduction. Sci Adv 7:eabf1367. https://doi.org/10.1126/sciadv.abf1367
203. Speakman JR, Racey PA (1991) No cost of echolocation for bats in flight. Nature 350:421–423. https://doi.org/10.1038/350421a0
204. Lancaster WC, Henson OW, Keating AW (1995) Respiratory muscle activity in relation to vocalization in flying bats. J Exp Biol 198:175–191. https://doi.org/10.1242/jeb.198.1.175
205. Speakman JR, Anderson ME, Racey PA (1989) The energy cost of echolocation in pipistrelle bats (*Pipistrettus pipistrellus*). J Comp Physiol A 165:679–685. https://doi.org/10.1007/BF00610999
206. Bergman TJ (2013) Speech-like vocalized lip-smacking in geladas. Curr Biol 23:R268–R269. https://doi.org/10.1016/j.cub.2013.02.038

207. Thiessen DD, Melanie E, Kittrell W, Graham JM (1980) Biomechanics of ultrasound emissions in the Mongolian gerbil, *Meriones unguiculatus*. Behav Neural Biol 29:415–429. https://doi.org/10.1016/S0163-1047(80)92597-2
208. Sales GD (née Sewell) (1972) Ultrasound and mating behaviour in rodents with some observations on other behavioural situations. J Zool 168:149–164. https://doi.org/10.1111/j.1469-7998.1972.tb01345.x
209. Alves JA, Boerner BC, Laplagne DA (2016) Flexible coupling of respiration and vocalizations with locomotion and head movements in the freely behaving rat. Neural Plast 2016:1–16. https://doi.org/10.1155/2016/4065073
210. Dantzker MS, Deane GB, Bradbury JW (1999) Directional acoustic radiation in the strut display of male sage grouse *Centrocercus urophasianus*. J Exp Biol 202:2893–2909. https://doi.org/10.1242/jeb.202.21.2893
211. Fusani L, Hutchison RE, Hutchison JB (1997) Vocal-postural co-ordination of a sexually dimorphic display in a monomorphic species: the barbary dove. Behaviour 134:321–335. https://doi.org/10.1163/156853997X00566
212. Williams H (2001) Choreography of song, dance and beak movements in the zebra finch (*Taeniopygia guttata*). J Exp Biol 204:3497–3506. https://doi.org/10.1242/jeb.204.20.3497
213. Eens M, Pinxten R, Verheyen RF (1992) Song learning in captive European starlings, *Sturnus vulgaris*. Anim Behav 44:1131–1143. https://doi.org/10.1016/S0003-3472(05)80325-2
214. Price JJ, Lanyon SM (2002) Reconstructing the evolution of complex bird song in the oropendolas. Evolution 56:1514–1529. https://doi.org/10.1111/j.0014-3820.2002.tb01462.x
215. Cooper BG, Goller F (2004) Multimodal signals: enhancement and constraint of song motor patterns by visual display. Science 303:544–546. https://doi.org/10.1126/science.1091099
216. Repp BH, Su Y-H (2013) Sensorimotor synchronization: a review of recent research (2006–2012). Psychon Bull Rev 20:403–452. https://doi.org/10.3758/s13423-012-0371-2
217. Engler BH, Zamm A, Møller C (2024) Spontaneous rates exhibit high intra-individual stability across movements involving different biomechanical systems and cognitive demands. Sci Rep 14:14876. https://doi.org/10.1038/s41598-024-65788-6
218. Leman M, Moelants D, Varewyck M et al (2013) Activating and relaxing music entrains the speed of beat synchronized walking. PLoS One 8:e67932. https://doi.org/10.1371/journal.pone.0067932
219. Chandrasekaran C, Trubanova A, Stillittano S et al (2009) The natural statistics of audiovisual speech. PLoS Comput Biol 5:e1000436. https://doi.org/10.1371/journal.pcbi.1000436
220. Ding N, Patel AD, Chen L et al (2017) Temporal modulations in speech and music. Neurosci Biobehav Rev 81:181–187. https://doi.org/10.1016/j.neubiorev.2017.02.011
221. MacDougall HG, Moore ST (2005) Marching to the beat of the same drummer: the spontaneous tempo of human locomotion. J Appl Physiol 99:1164–1173. https://doi.org/10.1152/japplphysiol.00138.2005
222. Farooq M, Sazonov E (2016) Automatic measurement of chew count and chewing rate during food intake. Electronics 5:62. https://doi.org/10.3390/electronics5040062
223. Ghazanfar AA, Takahashi DY, Mathur N, Fitch WT (2012) Cineradiography of monkey lip-smacking reveals putative precursors of speech dynamics. Curr Biol 22:1176–1182. https://doi.org/10.1016/j.cub.2012.04.055
224. Lameira AR, Hardus ME, Bartlett AM et al (2015) Speech-like rhythm in a voiced and voiceless orangutan call. PLoS One 10:e116136. https://doi.org/10.1371/journal.pone.0116136
225. Terleph TA, Malaivijitnond S, Reichard UH (2018) An analysis of white-handed gibbon male song reveals speech-like phrases. Am J Phys Anthropol 166:649–660. https://doi.org/10.1002/ajpa.23451
226. Risueno-Segovia C, Hage SR (2020) Theta synchronization of phonatory and articulatory systems in marmoset monkey vocal production. Curr Biol 30:4276–4283.e3. https://doi.org/10.1016/j.cub.2020.08.019
227. Pereira AS, Kavanagh E, Hobaiter C et al (2020) Chimpanzee lip-smacks confirm primate continuity for speech-rhythm evolution. Biol Lett 16:20200232. https://doi.org/10.1098/rsbl.2020.0232

228. Raimondi T, Di Panfilo G, Pasquali M et al (2023) Isochrony and rhythmic interaction in ape duetting. Proc R Soc B Biol Sci 290:20222244. https://doi.org/10.1098/rspb.2022.2244
229. Lindelöf M, Sauciuc G-A, Persson T, Toumine M (2025) Rhythmic parameters across contexts in the percussive behaviour of chimpanzees (Pan troglodytes). In: Proceedings of the 20th edition of the Swedish Cognitive Science Society conference, Lund
230. Wright E, Grawunder S, Ndayishimiye E et al (2021) Chest beats as an honest signal of body size in male mountain gorillas (Gorilla beringei beringei). Sci Rep 11:6879. https://doi.org/10.1038/s41598-021-86261-8
231. Lameira AR, Eerola T, Ravignani A (2019) Coupled whole-body rhythmic entrainment between two chimpanzees. Sci Rep 9:18914. https://doi.org/10.1038/s41598-019-55360-y
232. Pontzer H, Raichlen DA, Rodman PS (2014) Bipedal and quadrupedal locomotion in chimpanzees. J Hum Evol 66:64–82. https://doi.org/10.1016/j.jhevol.2013.10.002
233. Schweinfurth MK, Baldridge DB, Finnerty K et al (2022) Inter-individual coordination in walking chimpanzees. Curr Biol 32:5138–43.e3. https://doi.org/10.1016/j.cub.2022.09.059
234. Mason NA, Burns KJ (2015) The effect of habitat and body size on the evolution of vocal displays in Thraupidae (tanagers), the largest family of songbirds: habitat, body size, and vocal evolution in tanagers. Biol J Linn Soc 114:538–551. https://doi.org/10.1111/bij.12455
235. Vasconcelos RO, Ladich F (2008) Development of vocalization, auditory sensitivity and acoustic communication in the Lusitanian toadfish *Halobatrachus didactylus*. J Exp Biol 211:502–509. https://doi.org/10.1242/jeb.008474
236. Vance T (2000) Variability in stridulatory sound production in the channel catfish, *Ictalurus punctatus*. Bios 71:79–84
237. Ghahramani ZN, Mohajer Y, Fine ML (2014) Developmental variation in sound production in water and air in the blue catfish *Ictalurus furcatus*. J Exp Biol:jeb.112946. https://doi.org/10.1242/jeb.112946
238. Webb McAdams AL, Smith ME (2023) The relationship between body size and stridulatory sound production in loricariid catfishes. J Acoust Soc Am 154:3672–3683. https://doi.org/10.1121/10.0022575
239. Bennett NC, Faulkes CG, Hart L, Jarvis JUM (2009) *Bathyergus suillus* (Rodentia: Bathyergidae). Mamm Species 828:1–7. https://doi.org/10.1644/828.1
240. Styns F, Van Noorden L, Moelants D, Leman M (2007) Walking on music. Hum Mov Sci 26:769–785. https://doi.org/10.1016/j.humov.2007.07.007
241. Todd NPM, Cousins R, Lee CS (2007) The contribution of anthropometric factors to individual differences in the perception of rhythm. Empir Musicol Rev 2:1–13. https://doi.org/10.18061/1811/24478
242. Dahl S, Huron D, Brod G, Altenmüller E (2014) Preferred dance tempo: does sex or body morphology influence how we groove? J New Music Res 43:214–223. https://doi.org/10.1080/09298215.2014.884144
243. Sahd L, Doubell N, Bennett NC, Kotzé SH (2022) Hind foot drumming: myosin heavy chain muscle fiber distribution in the hind limb muscles of three African mole-rat species (Bathyergidae). Anat Rec 305:170–183. https://doi.org/10.1002/ar.24712
244. Sheth KA, Iyer CC, Wier CG et al (2018) Muscle strength and size are associated with motor unit connectivity in aged mice. Neurobiol Aging 67:128–136. https://doi.org/10.1016/j.neurobiolaging.2018.03.016
245. Manuel M, Meunier C, Donnet M, Zytnicki D (2007) Resonant or not, two amplification modes of proprioceptive inputs by persistent inward currents in spinal motoneurons. J Neurosci 27:12977–12988. https://doi.org/10.1523/JNEUROSCI.3299-07.2007
246. Tothova J, Blaauw B, Pallafacchina G et al (2006) NFATc1 nucleocytoplasmic shuttling is controlled by nerve activity in skeletal muscle. J Cell Sci 119:1604–1611. https://doi.org/10.1242/jcs.02875
247. More HL, Donelan JM (2018) Scaling of sensorimotor delays in terrestrial mammals. Proc R Soc B Biol Sci 285:20180613. https://doi.org/10.1098/rspb.2018.0613
248. More HL, Hutchinson JR, Collins DF et al (2010) Scaling of sensorimotor control in terrestrial mammals. Proc R Soc B Biol Sci 277:3563–3568. https://doi.org/10.1098/rspb.2010.0898

249. Biewener AA, Daley MA (2007) Unsteady locomotion: integrating muscle function with whole body dynamics and neuromuscular control. J Exp Biol 210:2949–2960. https://doi.org/10.1242/jeb.005801
250. Wolpert DM, Ghahramani Z (2000) Computational principles of movement neuroscience. Nat Neurosci 3:1212–1217. https://doi.org/10.1038/81497
251. Dye J, Heiligenberg W (1987) Intracellular recording in the medullary pacemaker nucleus of the weakly electric fish, *Apteronotus,* during modulatory behaviors. J Comp Physiol A 161:187–200. https://doi.org/10.1007/BF00615240
252. Ilieş I, Zupanc GKH (2023) Computational modeling predicts regulation of central pattern generator oscillations by size and density of the underlying heterogenous network. J Comput Neurosci 51:87–105. https://doi.org/10.1007/s10827-022-00835-7
253. Malavasi S, Collatuzzo S, Torricelli P (2008) Interspecific variation of acoustic signals in Mediterranean gobies (Perciformes, Gobiidae): comparative analysis and evolutionary outlook: acoustic signals in Mediterranean gobies. Biol J Linn Soc 93:763–778. https://doi.org/10.1111/j.1095-8312.2008.00947.x
254. Zeyl JN, Malavasi S, Holt DE et al (2016) Convergent aspects of acoustic communication in darters, sculpins, and gobies. In: Sisneros JA (ed) Fish hearing and bioacoustics. Springer International Publishing, Cham, pp 93–120
255. Kotiaho J, Alatalo RV, Mappes J et al (1998) Male mating success and risk of predation in a wolf spider: a balance between sexual and natural selection? J Anim Ecol 67:287–291. https://doi.org/10.1046/j.1365-2656.1998.00192.x
256. Moya-Laraño J, Macías-Ordóñez R, Blanckenhorn WU, Fernández-Montraveta C (2008) Analysing body condition: mass, volume or density? J Anim Ecol 77:1099–1108. https://doi.org/10.1111/j.1365-2656.2008.01433.x
257. Norris KS, Harvey GW (1972) A theory for the function of the spermaceti organ of the sperm whale (physeter Catodon L). NASA Special Publication, p 397
258. Møhl B (2001) Sound transmission in the nose of the sperm whale *Physeter catodon.* A post mortem study. J Comp Physiol [A] 187:335–340. https://doi.org/10.1007/s003590100205
259. Rhinelander MQ, Dawson SM (2004) Measuring sperm whales from their clicks: stability of interpulse intervals and validation that they indicate whale length. J Acoust Soc Am 115:1826–1831. https://doi.org/10.1121/1.1689346
260. Growcott A, Miller B, Sirguey P et al (2011) Measuring body length of male sperm whales from their clicks: the relationship between inter-pulse intervals and photogrammetrically measured lengths. J Acoust Soc Am 130:568–573. https://doi.org/10.1121/1.3578455
261. Dickson T, Rayment W, Dawson S (2021) Drone photogrammetry allows refinement of acoustically derived length estimation for male sperm whales. Mar Mamm Sci 37:1150–1158. https://doi.org/10.1111/mms.12795
262. Lemon RE, Chatfield C (1973) Organization of song of rose-breasted grosbeaks. Anim Behav 21:28–44. https://doi.org/10.1016/S0003-3472(73)80037-5
263. Mercado E, III, Handel S (2012) Understanding the structure of humpback whale songs (L). J Acoust Soc Am 132:2947–2950. https://doi.org/10.1121/1.4757643
264. Adam O, Cazau D, Gandilhon N et al (2013) New acoustic model for humpback whale sound production. Appl Acoust 74:1182–1190. https://doi.org/10.1016/j.apacoust.2013.04.007
265. Calder WA (1970) Respiration during song in the canary (Serinus canaria). Comp Biochem Physiol 32:251–258. https://doi.org/10.1016/0010-406X(70)90938-2
266. Hartley RS, Suthers RA (1989) Airflow and pressure during canary song: direct evidence for mini-breaths. J Comp Physiol A 165:15–26. https://doi.org/10.1007/BF00613795
267. Cardoso GC, Atwell JW, Ketterson ED, Price TD (2007) Inferring performance in the songs of dark-eyed juncos (*Junco hyemalis*). Behav Ecol 18:1051–1057. https://doi.org/10.1093/beheco/arm078
268. Logue DM, Sheppard JA, Walton B et al (2020) An analysis of avian vocal performance at the note and song levels. Bioacoustics 29:709–730. https://doi.org/10.1080/09524622.2019.1674693

269. Goller F (2022) Vocal athletics – from birdsong production mechanisms to sexy songs. Anim Behav 184:173–184. https://doi.org/10.1016/j.anbehav.2021.04.009
270. Zollinger SA, Suthers RA (2004) Motor mechanisms of a vocal mimic: implications for birdsong production. Proc R Soc Lond B Biol Sci 271:483–491. https://doi.org/10.1098/rspb.2003.2598
271. Gaunt AS, Gaunt SLL (1985) Syringeal structure and avian phonation. In: Johnston RF (ed) Current ornithology. Springer US, Boston, pp 213–245
272. Amaya JP, Zufiaurre E, Areta JI, Abba AM (2019) The weeping vocalization of the screaming hairy armadillo (*Chaetophractus vellerosus*), a distress call. J Mammal 100:1427–1435. https://doi.org/10.1093/jmammal/gyz108
273. Peters G (2002) Purring and similar vocalizations in mammals. Mammal Rev 32:245–271. https://doi.org/10.1046/j.1365-2907.2002.00113.x
274. Hunter ML (1980) Vocalization during inhalation in a nightjar. Condor 82:101–103. https://doi.org/10.2307/1366795
275. Goller F, Daley MA (2001) Novel motor gestures for phonation during inspiration enhance the acoustic complexity of birdsong. Proc R Soc Lond B Biol Sci 268:2301–2305. https://doi.org/10.1098/rspb.2001.1805
276. Damien J, Adam O, Cazau D et al (2019) Anatomy and functional morphology of the mysticete rorqual whale larynx: phonation positions of the U-fold. Anat Rec 302:703–717. https://doi.org/10.1002/ar.24034
277. Elemans CPH, Jiang W, Jensen MH et al (2024) Evolutionary novelties underlie sound production in baleen whales. Nature 627:123–129. https://doi.org/10.1038/s41586-024-07080-1
278. Schoenfuss HL, Bragulla HH, Schumacher J et al (2014) The anatomy of the larynx of the bowhead whale, *Balaena mysticetus*, and its sound-producing functions. Anat Rec 297:1316–1330. https://doi.org/10.1002/ar.22907
279. Reidenberg JS, Laitman JT (2007) Discovery of a low frequency sound source in Mysticeti (baleen whales): anatomical establishment of a vocal fold homolog. Anat Rec 290:745–759. https://doi.org/10.1002/ar.20544
280. Mercado E, III, Schneider JN, Pack AA, Herman LM (2010) Sound production by singing humpback whales. J Acoust Soc Am 127:2678–2691. https://doi.org/10.1121/1.3309453
281. Cazau D, Adam O, Laitman JT, Reidenberg JS (2013) Understanding the intentional acoustic behavior of humpback whales: a production-based approach. J Acoust Soc Am 134:2268–2273. https://doi.org/10.1121/1.4816403
282. Bass AH, Clark CW (2003) The physical acoustics of underwater sound communication. In: Simmons AM, Fay RR, Popper AN (eds) Acoustic communication. Springer-Verlag, New York, pp 15–64
283. Cranford TW, Elsberry WR, Van Bonn WG et al (2011) Observation and analysis of sonar signal generation in the bottlenose dolphin (*Tursiops truncatus*): evidence for two sonar sources. J Exp Mar Biol Ecol 407:81–96. https://doi.org/10.1016/j.jembe.2011.07.010
284. Suthers RA (1990) Contributions to birdsong from the left and right sides of the intact syrinx. Nature 347:473–477. https://doi.org/10.1038/347473a0
285. Suthers RA, Vallet E, Kreutzer M (2012) Bilateral coordination and the motor basis of female preference for sexual signals in canary song. J Exp Biol 215:2950–2959. https://doi.org/10.1242/jeb.071944
286. Madsen PT, Lammers M, Wisniewska D, Beedholm K (2013) Nasal sound production in echolocating delphinids (*Tursiops truncatus* and *Pseudorca crassidens*) is dynamic, but unilateral: clicking on the right side and whistling on the left side. J Exp Biol 216:4091–4102. https://doi.org/10.1242/jeb.091306
287. Ames AE, Beedholm K, Madsen PT (2020) Lateralized sound production in the beluga whale (*Delphinapterus leucas*). J Exp Biol:jeb.226316. https://doi.org/10.1242/jeb.226316
288. Adkins-Regan E (2005) Hormones and animal social behavior. Princeton University Press, Princeton

289. Rose EM, Haakenson CM, Ball GF (2022) Sex differences in seasonal brain plasticity and the neuroendocrine regulation of vocal behavior in songbirds. Horm Behav 142:105160. https://doi.org/10.1016/j.yhbeh.2022.105160
290. Dunlap KD, Koukos HM, Chagnaud BP et al (2021) Vocal and electric fish: revisiting a comparison of two teleost models in the neuroethology of social behavior. Front Neural Circuits 15. https://doi.org/10.3389/fncir.2021.713105
291. Tobiansky DJ, Fuxjager MJ (2020) Sex steroids as regulators of gestural communication. Endocrinology 161:bqaa064. https://doi.org/10.1210/endocr/bqaa064
292. Schlinger BA, Brenowitz EA (2002) Neural and hormonal control of birdsong. In: Hormones, brain and behavior. Elsevier, pp 799–839
293. Bass AH (2008) Steroid-dependent plasticity of vocal motor systems: novel insights from teleost fish. Brain Res Rev 57:299–308. https://doi.org/10.1016/j.brainresrev.2007.04.006
294. Alward BA, Cornil CA, Balthazart J, Ball GF (2018) The regulation of birdsong by testosterone: multiple time-scales and multiple sites of action. Horm Behav 104:32–40. https://doi.org/10.1016/j.yhbeh.2018.04.010
295. Ball GF, Madison FN, Balthazart J, Alward BA (2020) How does testosterone act to regulate a multifaceted adaptive response? Lessons from studies of the avian song system. J Neuroendocrinol 32:e12793. https://doi.org/10.1111/jne.12793
296. Hall IC, Kelley DB (2021) Endocrine modulation of acoustic communication: *Xenopus laevis* as a model system. In: Neuroendocrine regulation of animal vocalization. Elsevier, pp 81–100
297. Sassoon D, Kelley DB (1986) The sexually dimorphic larynx of *Xenopus laevis*: development and androgen regulation. Am J Anat 177:457–472. https://doi.org/10.1002/aja.1001770404
298. Pasch B, George AS, Campbell P, Phelps SM (2011) Androgen-dependent male vocal performance influences female preference in Neotropical singing mice. Anim Behav 82:177–183. https://doi.org/10.1016/j.anbehav.2011.04.018
299. Arch VS, Narins PM (2009) Sexual hearing: the influence of sex hormones on acoustic communication in frogs. Hear Res 252:15–20. https://doi.org/10.1016/j.heares.2009.01.001
300. Quintana L, Jalabert C, Fokidis HB et al (2021) Neuroendocrine mechanisms underlying non-breeding aggression: common strategies between birds and fish. Front Neural Circuits 15:716605. https://doi.org/10.3389/fncir.2021.716605
301. Heimovics SA, Merritt JR, Jalabert C et al (2018) Rapid effects of 17β-estradiol on aggressive behavior in songbirds: environmental and genetic influences. Horm Behav 104:41–51. https://doi.org/10.1016/j.yhbeh.2018.03.010
302. Fernández-Vargas M (2017) Rapid effects of estrogens and androgens on temporal and spectral features in ultrasonic vocalizations. Horm Behav 94:69–83. https://doi.org/10.1016/j.yhbeh.2017.06.010
303. Alward BA, Madison FN, Gravley WT, Ball GF (2016) Antagonism of syringeal androgen receptors reduces the quality of female-preferred male song in canaries. Anim Behav 119:201–212. https://doi.org/10.1016/j.anbehav.2016.07.010
304. Quiroga Murcia C, Bongard S, Kreutz G (2009) Emotional and neurohumoral responses to dancing tango argentino: the effects of music and partner. Music Med 1:14–21. https://doi.org/10.1177/1943862109335064
305. Nowicki S, Searcy W, Peters S (2002) Brain development, song learning and mate choice in birds: a review and experimental test of the "nutritional stress hypothesis". J Comp Physiol [A] 188:1003–1014. https://doi.org/10.1007/s00359-002-0361-3
306. Spencer KA, MacDougall-Shackleton SA (2011) Indicators of development as sexually selected traits: the developmental stress hypothesis in context. Behav Ecol 22:1–9. https://doi.org/10.1093/beheco/arq068
307. Matsumasa M, Murai M (2005) Changes in blood glucose and lactate levels of male fiddler crabs: effects of aggression and claw waving. Anim Behav 69:569–577. https://doi.org/10.1016/j.anbehav.2004.06.017
308. Mowles SL, Jennions M, Backwell PRY (2017) Multimodal communication in courting fiddler crabs reveals male performance capacities. R Soc Open Sci 4:161093. https://doi.org/10.1098/rsos.161093

309. Takeshita F, Murai M (2016) The vibrational signals that male fiddler crabs (*Uca lactea*) use to attract females into their burrows. Sci Nat 103. https://doi.org/10.1007/s00114-016-1371-2
310. Mowles SL (2014) The physiological cost of courtship: field cricket song results in anaerobic metabolism. Anim Behav 89:39–43. https://doi.org/10.1016/j.anbehav.2013.12.014
311. Mappes J, Alatalo R, Kotiaho J, Parri S (1996) Viability costs of condition-dependent sexual male display in a drumming wolf spider. Proc R Soc Lond B Biol Sci 263:785–789. https://doi.org/10.1098/rspb.1996.0117
312. Koren L, Geffen E (2009) Complex call in male rock hyrax (*Procavia capensis*): a multi-information distributing channel. Behav Ecol Sociobiol 63:581–590. https://doi.org/10.1007/s00265-008-0693-2
313. Derryberry EP, Seddon N, Derryberry GE et al (2018) Ecological drivers of song evolution in birds: disentangling the effects of habitat and morphology. Ecol Evol 8:1890–1905. https://doi.org/10.1002/ece3.3760
314. Galeotti P, Sacchi R, Rosa DP, Fasola M (2005) Female preference for fast-rate, high-pitched calls in Hermann's tortoises *Testudo hermanni*. Behav Ecol 16:301–308. https://doi.org/10.1093/beheco/arh165
315. Castellano S, Rosso A (2007) Female preferences for multiple attributes in the acoustic signals of the Italian treefrog, *Hyla intermedia*. Behav Ecol Sociobiol 61:1293–1302. https://doi.org/10.1007/s00265-007-0360-z
316. Clutton-Brock TH, Albon SD (1979) The roaring of red deer and the evolution of honest advertisement. Behaviour 69:145–170. https://doi.org/10.1163/156853979x00449
317. De Kort SR, Eldermire ERB, Valderrama S et al (2009) Trill consistency is an age-related assessment signal in banded wrens. Proc R Soc B Biol Sci 276:2315–2321. https://doi.org/10.1098/rspb.2009.0127
318. Cardoso GC (2013) Sexual signals as advertisers of resistance to mistakes. Ethology 119:1035–1043. https://doi.org/10.1111/eth.12165
319. Byers BE (2007) Extrapair paternity in chestnut-sided warblers is correlated with consistent vocal performance. Behav Ecol 18:130–136. https://doi.org/10.1093/beheco/arl058
320. Akre KL, Farris HE, Lea AM et al (2011) Signal perception in frogs and bats and the evolution of mating signals. Science 333:751–752. https://doi.org/10.1126/science.1205623
321. Oberweger K, Goller F (2001) The metabolic cost of birdsong production. J Exp Biol 204:3379–3388. https://doi.org/10.1242/jeb.204.19.3379
322. Lambrechts M, Dhondt AA (1988) The anti-exhaustion hypothesis: a new hypothesis to explain song performance and song switching in the great tit. Anim Behav 36:327–334. https://doi.org/10.1016/s0003-3472(88)80002-2
323. Brumm H, Slater PJB (2006) Ambient noise, motor fatigue, and serial redundancy in chaffinch song. Behav Ecol Sociobiol 60:475–481. https://doi.org/10.1007/s00265-006-0188-y
324. Sierro J, De Kort SR, Hartley IR (2023) Sexual selection for both diversity and repetition in birdsong. Nat Commun 14:3600. https://doi.org/10.1038/s41467-023-39308-5
325. Podos J, Lahti DC, Moseley DL (2009) Chapter 5: Vocal performance and sensorimotor learning in songbirds. In: Advances in the study of behavior. Elsevier, pp 159–195
326. Rogers TL (2017) Calling underwater is a costly signal: size-related differences in the call rates of Antarctic leopard seals. Curr Zool 63:433–443. https://doi.org/10.1093/cz/zox028
327. Lahti DC, Moseley DL, Podos J (2011) A tradeoff between performance and accuracy in bird song learning. Ethology 117:802–811. https://doi.org/10.1111/j.1439-0310.2011.01930.x
328. Podos J (1996) Motor constraints on vocal development in a songbird. Anim Behav 51:1061–1070. https://doi.org/10.1006/anbe.1996.0107
329. Podos J (2010) Acoustic discrimination of sympatric morphs in Darwin's finches: a behavioural mechanism for assortative mating? Philos Trans R Soc B Biol Sci 365:1031–1039. https://doi.org/10.1098/rstb.2009.0289
330. Sun C, Zhang C, Lucas JR et al (2022) Vocal performance reflects individual quality in male great Himalayan leaf-nosed bats (*Hipposideros armiger*). Integr Zool 17:731–740. https://doi.org/10.1111/1749-4877.12545

331. Clink DJ, Charif RA, Crofoot MC, Marshall AJ (2018) Evidence for vocal performance constraints in a female nonhuman primate. Anim Behav 141:85–94. https://doi.org/10.1016/j.anbehav.2018.05.002
332. Hartbauer M, Kratzer S, Römer H (2006) Chirp rate is independent of male condition in a synchronising bushcricket. J Insect Physiol 52:221–230. https://doi.org/10.1016/j.jinsphys.2005.10.006
333. Geberzahn N, Aubin T (2014) Assessing vocal performance in complex birdsong: a novel approach. BMC Biol 12:58. https://doi.org/10.1186/s12915-014-0058-4
334. Merchant H, Honing H (2014) Are non-human primates capable of rhythmic entrainment? Evidence for the gradual audiomotor evolution hypothesis. Front Neurosci 7. https://doi.org/10.3389/fnins.2013.00274
335. Teki S, Grube M, Kumar S, Griffiths TD (2011) Distinct neural substrates of duration-based and beat-based auditory timing. J Neurosci 31:3805–3812. https://doi.org/10.1523/jneurosci.5561-10.2011
336. Ten Cate C, Spierings M (2019) Rules, rhythm and grouping: auditory pattern perception by birds. Anim Behav 151:249–257. https://doi.org/10.1016/j.anbehav.2018.11.010
337. Takeya R, Kameda M, Patel AD, Tanaka M (2017) Predictive and tempo-flexible synchronization to a visual metronome in monkeys. Sci Rep 7:6127. https://doi.org/10.1038/s41598-017-06417-3
338. Gámez J, Yc K, Ayala YA et al (2018) Predictive rhythmic tapping to isochronous and tempo changing metronomes in the nonhuman primate. Ann N Y Acad Sci 1423:396–414. https://doi.org/10.1111/nyas.13671
339. Toida K, Ueno K, Shimada S (2016) Neural basis of the time window for subjective motor-auditory integration. Front Hum Neurosci 9. https://doi.org/10.3389/fnhum.2015.00688
340. Itoh K, Nejime M, Konoike N et al (2019) Evolutionary elongation of the time window of integration in auditory cortex: macaque vs. human comparison of the effects of sound duration on auditory evoked potentials. Front Neurosci 13:630. https://doi.org/10.3389/fnins.2019.00630
341. Itoh K, Konoike N, Nejime M et al (2022) Cerebral cortical processing time is elongated in human brain evolution. Sci Rep 12:1103. https://doi.org/10.1038/s41598-022-05053-w
342. Large EW, Roman I, Kim JC et al (2023) Dynamic models for musical rhythm perception and coordination. Front Comput Neurosci 17:1151895. https://doi.org/10.3389/fncom.2023.1151895
343. Rouse AA, Patel AD, Kao MH (2021) Vocal learning and flexible rhythm pattern perception are linked: evidence from songbirds. Proc Natl Acad Sci 118:e2026130118. https://doi.org/10.1073/pnas.2026130118
344. Hartbauer M, Römer H (2016) Rhythm generation and rhythm perception in insects: the evolution of synchronous choruses. Front Neurosci 10. https://doi.org/10.3389/fnins.2016.00223
345. Harding EE, Kim JC, Demos AP et al (2025) Musical neurodynamics. Nat Rev Neurosci 26:293–307. https://doi.org/10.1038/s41583-025-00915-4
346. Nagasaka Y, Chao ZC, Hasegawa N et al (2013) Spontaneous synchronization of arm motion between Japanese macaques. Sci Rep 3:1151. https://doi.org/10.1038/srep01151
347. Tomatsu S, Isoda M (2023) Tuning in to real-time social interactions in macaques. Proc Natl Acad Sci 120:e2301614120. https://doi.org/10.1073/pnas.2301614120
348. Iversen JR, Balasubramaniam R (2016) Synchronization and temporal processing. Curr Opin Behav Sci 8:175–180. https://doi.org/10.1016/j.cobeha.2016.02.027
349. Byblow WD, Carson RG, Goodman D (1994) Expressions of asymmetries and anchoring in bimanual coordination. Hum Mov Sci 13:3–28. https://doi.org/10.1016/0167-9457(94)90027-2
350. Fink PW, Jirsa VK, Foo P, Kelso JAS (2000) Local and global stabilization of coordination by sensory information. Exp Brain Res 134:9–20. https://doi.org/10.1007/s002210000439
351. Hoffmann CP, Torregrosa G, Bardy BG (2012) Sound stabilizes locomotor-respiratory coupling and reduces energy cost. PLoS One 7:e45206. https://doi.org/10.1371/journal.pone.0045206

352. Hoffmann CP, Bardy BG (2015) Dynamics of the locomotor–respiratory coupling at different frequencies. Exp Brain Res 233:1551–1561. https://doi.org/10.1007/s00221-015-4229-5
353. Muñoz MI, Goutte S, Ellers J, Halfwerk W (2020) Environmental and morphological constraints interact to drive the evolution of communication signals in frogs. J Evol Biol 33:1749–1757. https://doi.org/10.1111/jeb.13713
354. Maney DL, Rodriguez-Saltos CA (2016) Hormones and the incentive salience of bird song. In: Bass AH, Sisneros JA, Popper AN, Fay RR (eds) Hearing and hormones. Springer International Publishing, Cham, pp 101–132

The Developmental Trajectory of Rhythmic Abilities in Humans and Singing Primates

10

Gabriela-Alina Sauciuc, Teresa Raimondi, and Chiara De Gregorio

Abstract

Rhythmicity is a composite competence that spans perceptual, motor, and socio-cognitive abilities. This is highly salient when adopting a developmental perspective, as we do in this chapter, which reviews data from humans and other singing primate taxa with a close (gibbons) or remote (indri lemurs) phylogenetic relatedness to our species. The findings we review show that rhythmicity develops as an interplay between innate maturational processes and socio-ecological influences, and that its underlying mechanisms have distinct developmental trajectories. Generally, however, perceptual skills emerge earlier than motor ones, and rhythmic proficiency improves as (possibly innate) motor rhythms are brought under increased cognitive control. We surmise that developmental primate data are essential for understanding the underlying mechanisms and adaptive benefits of human rhythmic competence and the human-definitory forms of communication and cooperation that it enables. As this kind of data is exceedingly scarce, a substantial task of future research will be to reduce this gap. We also identify substantial gaps in the developmental research with human children, including the need to go beyond auditory rhythms and investigate rhythms broadcasted in other modalities, to increase the cross-cultural representativeness of the samples, and to examine longitudinally the relationships between elements of the rhythmic composite.

G.-A. Sauciuc (✉)
Department of Philosophy/Cognitive Science, Lund University, Lund, Sweden
e-mail: Gabriela-Alina.Sauciuc@lucs.lu.se

T. Raimondi
Department of Life Sciences and Systems Biology, University of Turin, Torino, Italy

Department of Human Neurosciences, Sapienza University of Rome, Rome, Italy
e-mail: teresa.raimondi@uniroma1.it

C. De Gregorio
Department of Psychology, University of Warwick, Coventry, UK
e-mail: Chiara.De-Gregorio@warwick.ac.uk

G.-A. Sauciuc et al. (eds.), *Nature Beats*, Fascinating Life Sciences,
https://doi.org/10.1007/978-3-032-24250-1_10

10.1 Introduction

The previous chapters of this volume have introduced the reader to a staggering diversity of rhythmic signals, emitted by a staggering diversity of taxa in a great range of contexts (Chap. 3 [1]). They have also presented fascinating data about the complex neuroethological machinery that enables the production and reception of these signals (Chap. 7 [2], Chap. 8 [3], Chap. 9 [4], Chap. 11 [5]). Moreover, they have shown that rhythmic competence is not monolithic, but can be decomposed into multiple components, each generated by specific mechanisms Chap. 1 [6]). Some of these components and mechanisms have been shown to be widespread in the animal kingdom, while others are rarer, or even hypothesized to be human-unique. All these chapters are invaluable resources for a holistic understanding of the evolution of rhythmic communication and, in particular, the outstanding rhythmic abilities that, in humans, enable speech and music.

In this chapter, we switch focus to another area of study that contributes key complementary insights to this enormous and complex endeavor—the ontogenetic emergence and development of the abilities implicated in rhythmic communication. Given the centrality of rhythmic abilities to speech and musicality, most research in this area has concentrated on human children and auditory rhythms. Recently, however, there has been a renewed interest in studying the development of nonvocal rhythmic behavior in human children, but this topic remains unstudied in other primates. This status quo is reflected by the content and organization of this chapter, with most space being occupied by findings related to milestones in the development of human rhythm perception and production (Sect. 10.2). In addition, in Sect. 10.3, we review emerging developmental findings in nonhuman primate taxa, such as in gibbons (Hylobatidae) and Indri lemurs (*Indri indri*). These are of particular relevance to the evolution of human rhythmic competence, not only because of their phylogenetic relatedness to our species, but also because they share with humans a trait that, otherwise, is rare among primates ($\approx$16% of the primate species [7]), that is, the song. Even though many definitions of animal "song" exist, most authors agree in identifying the crucial key features of a song as (1) long duration, (2) prominent harmonicity and frequency modulation of elements, (3) nonrandom temporal and structural hierarchy between elements, which form a precise sequence in time, (4) non-referentiality of the elements, (5) long-distance communicative function and social context of emission [8–11].[1]

[1] Historically, singing primates have been identified mostly based on their production of tonal, harmonically rich vocalizations, often in parallel with bird song, and for this reason, species such as chimpanzees and orangutans have been largely excluded. Their vocalizations, like the harsh and noisy pant-hoots of chimpanzees or the highly stereotyped and repetitive long calls of orangutans, did not fit traditional criteria despite their evident temporal complexity. Drawing a clear line between "singing" and "nonsinging" primates remains challenging and is beyond the scope of this chapter. Recent studies, however, have revealed that both chimpanzee and orangutan vocalizations do exhibit rhythmic structure. Chimpanzee pant-hoots show context-dependent rhythmic variation [12], while orangutan long calls include pulse-like segments produced with remarkable isochrony [13], and double meter [14], while alarm calls have a third-order nested isochronous rhythm [15]. These findings suggest a broader and more inclusive view of vocal rhythmicity among primates, underscoring the importance of considering such species within discussions of primate musicality.

We conclude (Sect. 10.4) with a call for expanding developmental research with nonhuman primates by following the conceptual and methodological grid provided by human developmental research on rhythmic competence. With respect to human developmental research, we call for research on more culturally diverse samples in order to determine which laboratory findings maintain their validity in a cross-cultural perspective and can, thus, be considered species-level traits. Finally, we point out that currently there is a need for more systematic developmental research on the production of nonvocal rhythms and the perception of rhythms in other stimuli than speech and music. Bodily movement is an integral part of audiomotor synchronization, the perception and enjoyment of visual rhythms is central to human dance, while the production of acoustic rhythms with the help of tools is a universal aspect of music. Thus, by carefully charting the emergence and developmental trajectories of these less studied rhythmic modalities, we will be in a better position to understand where musicality comes from, both developmentally and evolutionarily.

10.2 The Ontogenetic Trajectory of Rhythmic Abilities in Humans

The rhythmic competence of humans spans a range of perceptual and motor abilities that are crucial to species-definitory forms of communication, such as language, music, turn-taking, and synchronization. Several major developmental disorders that affect these communicative and cooperative abilities (dyslexia, stuttering, autism, ADD, developmental coordination disorder, self-regulation disorder, and more) have been linked to deficits in rhythm perception and motor rhythm control [16–19]. Capitalizing on these insights, rhythm-based interventions have been recently devised to counter such deficits in at-risk children [20–22]. Thus, the significance of developmental research on human rhythmic competence cannot be overstated. In this section, we review crucial findings regarding milestones in the developmental trajectories of several perceptual and motor skills that constitute this competence. In Sect. 10.2.1, we focus on the fetal stage to address the ontogenetic emergence of rhythmic movement and the rhythmic environment of the womb. We then switch focus to infancy and childhood, to review crucial findings regarding the development of rhythm perception (Sect. 10.2.2), rhythmic vocalizations and rhythmic bodily movements (Sect. 10.2.3), and rhythmic responses to rhythmic stimuli (Sect. 10.2.4).

10.2.1 Prenatal Rhythmic Stimulation and Its Significance to Neurotypical Development

The development of rhythmic abilities is an integral part of sensorimotor development that starts during the prenatal months. In humans, neuromotor activity emerges with motoneuron differentiation around 7-weeks of gestational age (GA), initially as transient rhythmic bursts. As these turn into more coherent and widespread activity, "motor babbling" emerges, which enables the fetus to explore movement patterns and the womb environment. Motor exploration has sensory (e.g., tactile, vestibular, proprioceptive) effects that contribute to the growth and fine-tuning of the motor system and to the development of sensory systems and sensorimotor interactions. The sensory systems become functional in a fixed sequence, with peripheral organs developing ahead and independently from the neural circuits to which they are later connected [23, 24]. The first rhythm-relevant sensory systems to become functional are the somaesthetic and vestibular systems at 8–16 weeks GA, while the visual and auditory systems develop at around 24 weeks GA [24–27].

The sensory environment of the womb is essentially rhythmic. The fetus is exposed to the movement, vibrations, and sounds produced by continuous intrauterine rhythms (e.g., maternal heartbeat and breathing) and episodic extrauterine rhythms (e.g., maternal walking, talking). Initially, maternal rhythms are experienced by the fetus in the vestibular and vibrotactile modalities and, subsequently, as the auditory system also becomes functional, as multimodal rhythms [28, 29]. Regarding the temporal properties of this prenatal rhythmic stimulation, maternal heartbeat averages around 91 beats/min (range 68–115, [30]) while breathing averages at 14–19 breaths/min [29]. Both rhythms are patterned by amplitude and durational contrasts, for example, inspiration has a higher amplitude but is twice as short as expiration [28]. The heartbeat, however, dominates the auditory experience of the fetus, as it supersedes background noise by 25 dB [31]. It has therefore been suggested that maternal heartbeat sets the preferences of very young infants for binary patterns of moderate tempo that match the frequency of the maternal heartbeat [28, 32–34].

The two intrauterine rhythms are also coordinated: heart rate increases during inspiration and decreases during expiration, and, moreover, the onsets of heartbeat and inspiration are synchronized [35]. The structural complexity generated by this coupling has been hypothesized to lay the foundation for meter perception. For instance, heartbeats that co-occur with inspiration are louder and faster (i.e., "strong beats") compared to the slower and softer heartbeats produced between inspiratory onsets [28]. In addition, these two rhythms may provide a foundation for inter-individual coordination, as fetal heartbeat entrains to the maternal heartbeat (for recent reviews see [36, 37]). This effect is especially enhanced at faster heart rates that approach the rate of fetal heartbeat [38] and becomes more pronounced as the fetal auditory system becomes functional [36].

Maternal activities (walking, talking, etc.) provide episodic bouts of rhythmic stimulation with variable tempi. With its left-right alternation of the legs and trunk, walking generates a binary rhythm, which in pregnant women averages 50 bpm/min [39]. There is suggestive evidence that fetuses distinguish the vibrotactile patterns generated by different maternal (loco)motor activities [40], as well as the auditory rhythms of nursery rhymes, melodies, syllables, etc. (for a recent review, see [41]).

Summing up, in the third trimester, the womb environment provides the human fetus with abundant and multilayered rhythms that occur at different timescales and levels of intensity. The co-perception of rhythmic patterns in multiple modalities is hypothesized to cultivate intersensory and sensorimotor associations that scaffold the development of rhythmic abilities for speech, musicality, and interaction in the postnatal life [42, 43]). Fetuses also co-experience rhythmic stimulation with interoceptive physiological changes induced by maternal emotional states through chemical transplacental transfer [44, 45], which could provide a foundation for the well-established link between rhythm and arousal (for details, see Chap. 11 [5]).

In preterm infants, this beneficial rhythmic environment is replaced by inadequate kinds and levels of stimulation, posing risks for neurosensory and neuromotor dysfunctions [46–48]. In recent years, increased awareness of such deleterious consequences has prompted researchers and medical practitioners to devise interventions based on rhythmic stimulation in the tactile (rhythmic stroking), vestibular (carrying, water mattresses, rockers), and auditory (singing, maternal heartbeat, music) modalities (for recent reviews, see [49–52]. These are generally found to improve emotional regulation, motor, and respiratory development.

10.2.2 Rhythm Perception in Infancy and Childhood

After birth, human infants continue to be immersed in a rhythm-rich environment that provides significant vestibular and tactile stimulation (from being carried, rocked, patted, and bounced), along with auditory stimulation in the form of speech and song. In this section, we review major findings pertaining to how rhythm perception develops from infancy to childhood, starting with the perception of auditory rhythms (Sect. 10.2.2.1), and continuing with research focused on visual and cross-modal rhythms (Sect. 10.2.2.2).

10.2.2.1 Auditory Rhythm Perception

Most developmental research on rhythm perception has focused on the auditory channel, being informed by questions about the development of verbal and musical skills. Milestones in this development can be tracked already from birth, as neonates are sensitive to various aspects of rhythmic structure. They can discriminate languages (e.g., French vs. Japanese) based on rhythmic parameters such as syllable or stress frequency [53, 54] and are sensitive to tempo differences in musical stimuli [55] and to beat disruptions [56]. At 2–5 months of age, infants can distinguish periodic from nonperiodic stimuli, recognize different kinds of rhythmic grouping based on the relative size and order of inter-onset intervals (IOIs), and detect tempo changes as small as 15% acceleration [57–59]. While the latter skill is mostly confined to a reference tempo of 600 ms IOI, young infants show an attentional preference for salient (faster/slower) tempi rather than those they can process best [60]. By contrast, in human adults and older children, tempo preference and optimal processing ranges overlap, which may indicate divergences in the developmental processes that shape rhythm preferences and processing ability.

Between 5- and 12 months of age, infants' abilities for auditory perception become increasingly sophisticated. At 5 months of age, infants can detect disruptions in rhythmic patterns structured by interval ratios of low (2:1) and moderate (3:2) complexity [61], while 6-month-olds detect metric violations in musical stimuli from their own and other cultures [62]. At 7 months of age, infants show an emergent ability to categorize rhythmic sequences based on duple or triple meter across stimuli with variable frequency and tempo [63]. EEG studies further show that the neural activity of 5-to-8-month-olds tracks beat and/or meter in both naturalistic (e.g., speech, music) and artificially constructed rhythms [64–67]. Already at 6 months of age, metric perception is driven by both bottom-up stimulus cues (e.g., loudness accents) and by top-down endogenous metric representations [66, 67].

Starting at 5 months of age, infants also begin to exhibit adult-like perceptual biases and rhythmic preferences driven by enculturation. For example, as in adults, neural tracking is enhanced for slow-tempo compared to fast-tempo music [68], low pitch facilitates meter processing [67], and regular and/or simple rhythms are preferred to those that are more variable and/or complex [69]. In addition, infants prefer to listen to culturally familiar rhythms [69] or exhibit enhanced meter processing for the metrical structures that dominate their musical environment [64, 66, 70]. At 12 months of age, infants begin to evidence perceptual narrowing in the rhythm domain, as they prevalently discriminate metric violations in enculturated—as opposed to foreign—rhythms [71].

Based on current data, the period from early to middle childhood spans developmental trends related to an increase in the range of discriminated tempi [72]. For example, the rhythmic discrimination ability of 4-year-olds is confined to a tempo (300 ms IOI) that is close to their spontaneous motor tempo (SMT). By contrast, 6-year-olds perform successfully both with slower (600 ms) and faster (100 ms) tempi, while 10-year-olds approach the level of adult-like performance, being able to discriminate rhythmic sequences with a 1500 ms IOI.

10.2.2.2 Visual and Cross-Modal Rhythm Perception

Compared to auditory rhythm, the development of rhythm perception in dynamic visual stimuli has been far less studied. Current data suggest that 4- and 8-month-olds discriminate between different rhythmic patterns (2-2, 3-1, etc.) in movements (waving, pounding), with 4-month-olds showing a preference for rhythms of medium complexity and 8-month-olds preferring complex rhythms [73]. In infants (as in adults), moving stimuli facilitate the perception of visual rhythms [74] and so does intersensory redundancy, where rhythmic patterns are synchronously delivered in two or more modalities. Indeed, 3–5 months who failed in a rhythm discrimination task administered in either the auditory or visual modality, succeeded when this was given bimodally [75–77].

Young infants also evidence an ability for rhythmic cross-modal transfer. For example, 4-month-olds habituated to an auditory rhythm (repeated syllable) show a recovery response (i.e., regained visual interest for the stimulus presentation) when tested with a novel rhythm in the visual modality [78], while 7-month-olds habituated to either visual or auditory rhythmic patterns exhibit a recovery response when presented with the corresponding patterns in the novel modality [79]. Moreover, 7-month-olds who were bounced at either a duple or triple meter, subsequently showed a preference for the auditory rhythm that matched the meter of the vestibular stimulation [80]. Similarly, 10-month-olds execute motor rhythms (drumming) at a faster or slower pace, in congruence with the faster or slower tempo of prior vestibular stimulation [81].

As discussed above, infants can discriminate between rhythmic patterns in audiovisual stimuli, and, moreover, multimodal presentation facilitates rhythm perception as such. Surprisingly, however, infants younger than 10 months fail to detect audiovisual rhythmic asynchrony (range of temporal offsets: 200–600 ms) in behavioral tests [82], although neural evidence points to an incipient sensitivity to rhythmic asynchrony starting with 6 months of age [83, 84]. This burgeoning ability can be enhanced with active experience. For example, 6-month-olds who engaged in a 5-min drumming bout—as opposed to infants who only observed drumming or did not receive any stimulation—were able to detect asynchronies in the 300–600 ms range in an audiovisual drumming stimulus [85]. Finally, at 12 months, infants reliably recognize audiovisual asynchrony in complex stimuli (dancing to music), an ability which at 8 months of age is in an emergent phase [86].

10.2.3 The Development of Rhythmic Movements and Rhythmic Vocalizations

Young infants can spend up to 40% of their awake time performing rhythmic movements (RMs) of the limbs and torso [87, 88]. They also produce a range of orofacial rhythmic movements (RMs) (e.g., sucking, tongue protrusion, mouth gaping) and emit rhythmic vocalizations (RVs), such as crying and babbling. Most research has focused on these RMs and RVs separately, while studies on cross-modal rhythmic interactions remain scarce.

10.2.3.1 The Onset and Peak of Infant Rhythmic Behaviors

The rhythmic behaviors of infancy exhibit divergent developmental trajectories with respect to onset and peak of occurrence. For example, leg and finger RMs are the first to be recorded, but the incidence of finger RMs remains relatively low in the first year of life. By contrast, leg RMs increase considerably at 10 weeks, peaking between 3 and 7 months of age, while torso, arm, and hand RMs peak between 6 and 12 months of age [87, 88]. In absolute quantitative terms, arm and hand RMs predominate by far in this age span. Among orofacial RMs, tongue protrusion is present at birth to decline considerably by 3 months of age and disappear by 6 months of age. By contrast, non-nutritive sucking is maintained into toddlerhood, being a commonly used habituation measure.

Among RVs, crying is present at birth and its incidence increases considerably until 6 weeks of age, to then decline steadily until 3 months [89]. By contrast, laughter emerges at 4–5 months of age, and its incidence continues to increase with age. From birth, infants also produce proto-phones—sounds with features relevant to language development. Initially, they produce mainly vowel-like sounds that may occur in sequences of four to eight units per expiration [90]. Gradually, they start to also produce consonant-like sounds that, occasionally, give rise to rhythmic series (reviewed by [91]). Around 6 months of age (range 4–10 months) children begin to produce canonical (or reduplicative) babbling—rhythmic syllables with a consonant–vowel structure that exhibit spectral and temporal features similar to adult speech [92–95].

Such developments are driven by the maturation of neuromechanical substrates (e.g., larynx descent), which paves the way for the development of lingual, labial, and mandibular motility, and, thus, for new articulatory possibilities [91, 95]. Auditory experience is also an important factor in the development of RV, as deaf infants exhibit deficits [96, 97]. Of equal importance is the emergence and consolidation of neuromotor mechanisms for the coordination of functional modules from across the respiratory, locomotor, axial, laryngeal, and orofacial systems. Laughter, for example, requires the rhythmic coordination of exhalation with the activity of laryngeal adductor muscles [98, 99]. Likewise, canonical babbling requires the coordination of rhythmic contractions across orofacial (mandibular, lingual, labial, etc.) muscles, which enables rapid cycles of mouth opening and closing, and, thus, the articulatory patterning of sound. These are, moreover, temporally coordinated with the respiratory and laryngeal muscles, which determine the temporal properties of expiration and drive rapidly patterned movements of the glottis and vocal folds (for details, see Chap. 9 [4]).

RVs can also engage body parts that do not seem to play a functional role in vocal production, as shown by the co-occurrence of canonical babbling with RMs [87, 93, 94, 100, 101]. Longitudinal and cross-sectional studies have established that the frequency of RMs intensifies a month before the onset of canonical babbling [94, 102–104]. RMs coordinated with RVs continue to occur by 2 years of age [105, 106], with object-associated RMs at 9 months of age predicting proximal deictic gesturing at 12 months [105].

10.2.3.2 The Temporal Characteristics of Infant Rhythmic Behaviors

While the presence of rhythmic behaviors in human infancy has attracted considerable research attention, detailed analyses of their temporal characteristics are scarce. Albeit few, these analyses nevertheless suggest that the pacing and patterning of some communicative rhythms undergo important changes at different transitional points: from early to late infancy, from infancy to childhood, from childhood to adulthood. For instance, the rhythm of crying is affected by respiratory maturation, with the duration of its expiratory—but not inspiratory—subunits increasing until 8 months of age (reviewed in [91]).

Regarding canonical babbling, most research has focused on positioning the phenomenon as a language precursor, thereby primarily measuring syllable duration. Based on this information, the frequency rate (i.e., cycles/s) of babbling in early to late infancy averages around 3 Hz [107–110]—to be contrasted with the average 5 Hz of adult speech. Noteworthy, however, these average rates mask substantial inter- and within-individual variability that ranges from 2 to nearly 6 Hz [110]. Some studies also suggest a developmental trend toward rhythm acceleration in the same period, accomplished by syllable shortening [109]. Moreover, syllable duration is significantly shorter when RVs co-occur with RMs [94]. Deaf infants, however, exhibit a significantly slower (2.01–2.77 Hz) babbling rhythm, a phenomenon also found in infants with Down Syndrome (1.82 Hz, [108]).

Besides syllable duration, other reported variables include rhythmic variability and the temporal relationship between RVs and RMs. Dolata and colleagues [110], for example, show that the variability of babbling IOIs is about 3,5 higher than that measured in adult speech, while [101] show that the phase relationship between RMs and RVs resembles that found in adult speech, where RMs are either synchronous or slightly preceding RVs [101]. The temporal precision of this synchronization increases steadily from 9 to 24 months of age [106].

With respect to RMs, the main reported variable is spontaneous motor tempo (SMT). For example, non-nutritive sucking becomes slightly faster from birth (550 ms IOI) to 2 months of age (500 ms IOI), a trend which is paralleled by an increase in variability [111]. For manual RMs, the developmental trajectory may vary with task. Hand tapping shows a gradual acceleration from early (660 ms IOI) to late infancy (530 ms IOI) and gradual deceleration in childhood [72, 112], from about 450 ms IOI in 1½-year-olds to 500 ms IOI in 5-year-olds [112]. By contrast, the SMT of hand clapping has been found to remain constant across childhood (3- to 7-year-old) at an average of 2.8 Hz [113].

With age—and neuromechanical maturation—rhythmic consistency increases. For instance, 3–4-year-olds exhibit 50% less variable tapping IOIs compared to 1½-year-olds ([81, 112]. At the same time, the range of inter-individual variability expands (e.g., 400–476 ms at 1½ years vs. 386–577 ms at 3½ years, [112]) and so does the range of accessible tempi, which nearly doubles at 6 years of age compared to 4–5 years [72, 114]. The range of slow tempi increases steeply in this period and continues to do so even after 10 years of age, while the range of fast tempi exhibits a more gradual incrementation [72].

10.2.3.3 The Functional Significance of Early Rhythmic Behaviors

The RMs and RVs of early infancy are considered to be rigid motor stereotypies driven by low-level neuromotor mechanisms that function as fixed release mechanisms with an arousal regulatory function during transition states such as fussiness, postural change, and prior and after feeding, [87, 88]. It is hypothesized that, similar to prenatal motility, these behaviors promote neuromuscular calibration through kinaesthetic, tactile, and auditory feedback [87, 93–95, 115, 116]. Moreover, it has been argued that earlier rhythmic behaviors (e.g., manual RMs) train motor timing skills in later emerging ones (e.g., canonical babbling) via entrainment [101].

Several lines of evidence support such proposals. The onset of individual RMs correlates with the neuromuscular development of effectors, and the initial rigidity of individual RMs is gradually supplanted by more flexible and complex forms, indicative of higher degrees of volitional motor control [87, 88]. Moreover, early RMs are correlated with later skilled behavior ([105] and references therein), while vocal articulation disorders are linked to developmental deficits in the temporal coordination of articulatory modules. Interestingly, the incidence and characteristics of RM are dependent on the vestibular stimulation (i.e., rocking, bouncing, carrying) received by infants, being higher in infants who receive lower vestibular stimulation [88].

The rhythmic behaviors of early infancy are also central to the infants' development of social competence. As the timing precision of RMs and RVs matures, caregivers increasingly treat these behaviors as communicative and respond to them with contingent utterances and imitative gestures (reviewed in [117]). Contingent responses socially reinforce infant RMs and RVs and embed them in rudimentary interaction sequences. This way, the functional specialization of certain rhythmic behaviors gradually increases, and, by 5–6 months of age, the incidence of socially elicited rhythmic behaviors compares to that of rhythmic behaviors elicited by fussiness [88]. With increased motor fluency of arm and hand movements, there is also a sharp increase in rhythmic behaviors elicited by objects [88], which scaffolds manual deixis [105].

10.2.4 Rhythmic Interactions with Exogenous Rhythms

As for the other aspects of rhythmic competence surveyed in this section, research on children's abilities to interact and, indeed, synchronize with exogenous rhythms has mainly focused on auditory rhythms and, in particular, music. The developmental trajectory of these interactions begins with *rhythmic engagement*, whereby children produce RMs in response to auditory rhythms. This response precedes *tempo adjustment*, that is, the ability to flexibly adapt the tempo of one's motor responses toward the tempo of the stimulus. In turn, tempo adjustment scaffolds rhythmic audiomotor synchronization, whereby children also become able to time their motor response to match the phase of auditory stimuli.

Rhythmic engagement has been documented from at least 3 months of age [118–120]. This response becomes more frequent and complex with age [121, 122]. In a recent survey study, a quarter of American responders reported that their 3-month-olds showed rhythmic engagement, a proportion that doubled for 6-month-olds, to reach 90% for 12-month-olds [122]. The range of distinct RMs and body parts involved in this response was also reported to increase [122].

The incidence of rhythmic engagement varies across studies depending on various kinds of familiarity: with the environment, auditory stimulus, or the measured response. First, the incidence of rhythmic engagement is higher in the home environment [121–123] compared to the lab [118, 119, 124]. Second, infants and young children engage in longer bouts of RMs when presented with familiar tunes, which also elicit faster RMs than those produced in response to unfamiliar music [121, 123, 125]. Third, having experienced a familiar tune in conjunction with RMs—either seen in a caregiver or produced jointly by the caregiver and infant—may increase the likelihood of RM beyond mere familiarity effects [121]. This is consistent with data suggesting that cultural settings that expose infants to more RM-accompanied music facilitate rhythmic engagement [120]. This effect is likely mediated by imitation, given that, starting with 12 months of age, children often respond to music with "dance moves" they have seen in adults [122].

Rhythmic engagement is also moderated by the acoustic characteristics of the stimulus. It is more robust when infants are exposed to music compared to rhythmic speech [118], to soundtracks with pronounced beat salience (drumming vs. classical music [118, 120]) or moderate-to-fast tempo (play song vs. lullaby, [68]). Such effects could be mediated by positive arousal, given that smiling frequently accompanies music-elicited RMs in children [118, 121, 123, 125].

Tempo adjustment begins to be consistently documented from about 1 year of age, as infants show an emergent ability to accelerate their motor response in response to auditory rhythms that are faster than their SMT [112, 126, 127]. Interestingly, neonates and 2-month-olds also appear capable of a similar ability in an RM they master well, that is, non-nutritive sucking [111]. It is, thus, possible that tempo acceleration emerges earlier, and its manifestation depends on the degree of maturation of the specific behavior. By contrast, RM deceleration develops considerably between 24 and 30 months of age [126, 127] in a social context and between 3 and 4 years of age in a nonsocial context [112, 128].

At 30 months of age, toddlers show an emergent ability to also time the onset of RMs with the phase of auditory rhythms, evidencing sensorimotor synchronisation (SMS) [126, 128, 129]. This has been observed both in the social context of joint drumming [126, 129] and in a nonsocial task in which children's anticipatory taps to rhythmic animal vocalizations were "rewarded" with the appearance of an animal picture on a screen [128].

SMS develops substantially between 2½ and 10 years of age [112, 126]. Before 6 years of age, children exhibit more accurate SMS with moderate tempi and exhibit mainly direct SMS, that is, they synchronize their taps with every sound in a sequence, in a strict bottom-up fashion [72, 113, 114]. With age, the use of SMS at higher levels of metrical organization increases, and so does the accuracy of SMS with very slow or very fast tempi. In preschoolers, SMS is also moderated by the perceptual salience of the metrical structure. For example, 4-year-olds show ceiling performance when synchronizing to a musical sequence in which the hierarchical structure is richly marked by different kinds of accents, but their performance is poorer if this is only marked by an intensity accent, or altogether unsuccessful if the meter is encoded by durational values [72].

10.3 The Developmental Trajectory of Rhythmic Abilities in Singing Nonhuman Primates

As shown in Sect. 10.2, different rhythmic abilities may present specific developmental trajectories. In a similar vein, they may also have distinct evolutionary histories, but for the time being, this is mostly unknown. For instance, we know very little about which building blocks of human rhythmic competence are shared with other primates. In this section, we examine these issues by reviewing findings on the development of rhythmic abilities in nonhuman primates, with a particular focus on the development of abilities for periodic pattern generation and the emergence of song.

Periodic pattern generation, that is, the ability to produce patterns that are repeated over time, is a definitory component of rhythmic competence. In a selected few primate taxa, periodic patterns can coalesce into highly complex signals, commonly referred to as *song* (for details, see Chaps. 1 [6], 2 [130], and 3 [1], as well as [11]). Such songs have been proposed to represent the closest behavioral output to human musicality, in terms of behavioral context, melodic and temporal features [10]. As in human music, primate songs can be emitted in duets and choruses, demand synchrony, show long duration, frequency modulation and harmonicity, and display a temporal and structural hierarchization of elements [10]. They serve functions related to mate guarding and defence, territory advertisement and territory and resource defence [131].

The species we will focus on in this section are gibbons (Sect. 10.3.1) and indri lemurs (Sect. 10.3.2). While gibbons are estimated to have shared a common ancestor with humans as late as 17 million years ago [132], the indri belong to a primate suborder (Strepsirrhini) that is hypothesized to have been the first to diverge from the common ancestor of modern primates, sharing a last common ancestor with humans around 74 million years ago [133]. As the targeted taxa span nearly the whole phylogenetic arc of primates, it is plausible that in this clade the generation of flexible and complex periodic patterns is a homologous trait with archaic origins.

10.3.1 The Ontogeny of Song in Gibbons

Gibbons (family Hylobatidae) are small apes inhabiting Asian tropical forests, where their incredible songs resonate in the humid, early mornings. Gibbon songs are composed of a series of different notes uttered in a precise sequence, often taking the form of duets performed by the reproductive pair, but solo songs are common too [11]. Studies of vocal rhythms in gibbons have evidenced that spontaneous isochrony is not a human prerogative, but strongly characterizes the songs of all the gibbon species investigated so far (*Hylobates lar* [134]; *Nomascus gabriellae*, *N. leucogenys*, *N. siki* [135]; *N. nasutus*, *N. hainanus*, *N. concolor* [136]). Additionally, the singing behavior of the Hainan black crested gibbon (*N. haianus*) shows an additional level of complexity, being characterized by an additional layer of rhythmic organization [136].

Despite the parallel between the duetting behavior of gibbons and that of humans, and the importance of carrying out comparative work on nonhuman primate species to understand how coordinated vocal display evolved in the primate clade, not much is known about the developmental trajectory of rhythmic abilities in these animals. However, different pieces of evidence suggest that both genetics and vocal plasticity play an important role in the process. On the one hand, seminal work by [137] has shown that the acoustic pattern of songs in hybrid individuals seems to be under strong genetic control and is not influenced by the social setting. Specifically, hybrid gibbons perform songs with a rate of note emission halfway between those of the parental species [137].

On the other hand, fine-tuned analyses of co-singing dynamics between mothers and daughters suggest that this vocal interaction is critical for song development in daughters, and in particular for rhythmic vocal control, which is necessary to attain accurate temporal synchronization during duets [138]. Intriguingly, both male and female offspring co-vocalize with their mother. Initially, immature males sing female-specific song types, likely under the influence of low androgen levels [139]. Such vocal interactions even take place in gibbon species that normally do not duet as adults [140]. It has been suggested that there is a general maternal influence on song acquisition [139] and that there is a certain degree of plasticity in gibbons' vocal behavior, as juvenile males improve with time the structure of their calls [141]. Mother-daughter co-singing influences the acquisition of the correct sequence of notes over time [142]. In males, the transition from female-specific calls to male-specific ones occurs gradually with time, and this process can take up to 5 years [143].

These works provide indications that, besides the overall song template, which appears to be under genetic and physiological influence, there is an important developmental process guided by practice and vocal interactions with mothers. It is likely that the rhythmic properties of gibbon song continue to be shaped by vocal developmental plasticity in adulthood [144]. The strong isochronous component of gibbons' song has been only partially explained by physiological constraints (e.g., call rate), while social factors influence how isochronous animals are [134]. The hypothesis that the rhythm of gibbons' songs is not fixed at birth but develops plastically across time is in line with evidence that immature individuals gradually acquire the ability to adjust the timing of notes, which, in turn, facilitates song coordination and synchronization between pair-mates, among group members, and even members from other groups [134, 136, 145]. Such temporal plasticity of acoustic output has also been documented in other singing primates, such as coppery titi monkeys (*Plecturocebus cupreus*) [146], as well as indri lemurs (as will be discussed below).

10.3.2 The Ontogeny of Periodic Pattern Generation in Indri Lemurs

The indris are the only lemur species traditionally included among the singing primates [11, 147]. Their advertisement song (henceforth, song) is used for inter- and intra-group communication, playing a role in broadcasting territorial occupancy and defence during group encounters without escalation to physical conflict [148, 149]. Indris live in small family groups of two to six individuals, usually composed of a reproductive pair and their offspring [150]. The species is critically endangered and does not survive ex situ but lives solely in the hard-to-access primary forests of Madagascar. The challenge of obtaining developmental, longitudinal data—and, indeed, any kind of systematic data—on such a species is significant.

Over the past decade, long-term data collection conducted by the Ethology Lab at the University of Torino in the primary forests of Maromizaha (Madagascar) has generated a database comprising songs from more than 60 individuals of both sexes. This has enabled systematic and detailed quantifications of song features, and analyses of developmental trends [151].

Currently, it has been established that the song of adult indris is composed of units, which in adulthood are clustered into phrases, separated by silent gaps [148, 152]. Recent work has further revealed that the indri displays a rhythmic trait that is universally found in the temporal structure of human music, that is, categorical rhythms. The relationship between adjacent IOIs in indri song can be described by small-integer ratios, such as 1:1 (isochrony), 1:2, and 2:1 [151, 153], which overlaps largely with those that are most prevalent in human music [154]. Based on current data, however, it is possible that this trait has a more local nature (i.e., spanning adjacent phrases) in nonhuman primates and other vertebrates [155], compared to humans where rhythmic categories also characterize the global organization of a musical piece (for a more detailed discussion, see Chap. 1 [6] and Chap. 5 [156]).

Young indris start engaging in song duets and choruses with their parents at around 1 year of age (females at 11.88 months and males at 14.76) [151]. The spectro-temporal parameters of the song, however, undergo radical changes during growth. This finding aligns with the view that vocal production is partially innate and genetically determined in nonhuman primates [157, 158], while factors such as practice and social influences may also affect the features of song [151, 159–161].

Longitudinal research has also investigated the ontogeny of small-integer ratios in indri song [162]. To this end, the individual contributions to song choruses of 62 individuals of both sexes were analyzed at three developmental stages: 0–2 years, 3–4 years, and adults. These analyses focused on quantifying the IOIs of note series in order to compute rhythmic ratios between two adjacent IOIs and on evaluating the rhythmic stability of these patterns. This work revealed that the 1:1 ratio—isochrony—is produced from the earliest developmental stage in both sexes. Basically, the indris start singing on an isochronous temporal grid, emitting units whose onsets are separated by intervals of equal duration, like the beat of a metronome. Isochronous patterns are plausibly the result of neural and physiological oscillations that are shared among a broad range of animal species and longitudinally across developmental stages [163]. However, this feature is not universal in the vocalizations of mammalian immatures, as harbor seal pup calls, for instance, do not contain any rhythmic categories coinciding with small-integer ratios in their call [164].

In addition, the production of isochronous patterns in indri is flexible, that is, it occurs at different *tempi* within- and across individuals rather than being fixed [153]. Tempo flexibility is of paramount importance in choral displays, such as the indri song, as it facilitates mutual synchronization between co-singing individuals [11, 134, 152]. In these cases, periodicity is also particularly relevant, as it makes the contribution to the song predictable, and thus synchrony possible [165–167]. Given its presence in a taxon that diverged so early in the evolution of primates, it has been proposed that periodicity, and its simplest form consisting in isochrony, may be an ancestral rhythmic trait that has been conserved in the evolution of primate communication to facilitate the acquisition of species-specific vocal behaviors, as well as coordination in the earliest life stages [151, 168].

Crucially, the rhythmic patterns produced by immature indri undergo ontogenetic transformation along sexually dimorphic trajectories. The rhythmic structure of the song develops with progress toward sexual maturity, in agreement with evidence from other vertebrate species [169–171]. In particular, both sexes reach the complete rhythmic configuration (1:2, 1:1, and 2:1) only in adulthood. The emergence of other rhythmic categories besides isochrony is likely the result of the omission of one element in the isochronous temporal grid deriving from endogenous neural oscillations. As the 1:2 and 2:1 rhythmic categories derive from the alternation, respectively, of short/long and long/short intervals [151, 153], any omitted element in the temporal grid produces 1:2 and 2:1 rhythmic ratios. The presence of isochrony from the earliest song stages in both sexes is in line with previous evidence showing that young indris emit more single notes—units that are not embedded in phrases—compared to adults [151]. In this respect, the close connection between the organization of song units and the resulting rhythmic categories is paramount.

As shown in Sect. 10.2, another relevant parameter in the ontogenetic development of rhythmic abilities is rhythmic stability, which can be evaluated at different levels of structural organization. Analyses focused on quantifying the stability of rhythmic ratios around small-integer values revealed the same degree of stability at the earliest stages of their production as in adulthood, in both sexes. In other words, there is no gradual convergence from simple isochrony toward typical adult intervals, but a sudden shift in the song organization: at the early stages of song development, units are unorganized and lack grouping into phrases and become distinctly organized into structured phrases as development progresses. This finding contrasts with evidence from Cape fur seals (*Arctocephalus pusillus*)—another species that produces isochronous patterns—which shows higher period stability in adults than in pups [172]. However, when it comes to other forms of rhythmic stability—the stability of inter-phrase intervals—young indris too exhibit greater variability compared to adults [151].

Such an abrupt appearance of combinatoriality in song is not limited to indri but has been observed in other singing species, such as zebra finches [173]. This suggests a potentially innate ability for precise temporal tuning of local song isochrony [174]. More broadly, in many species, rhythmic stability is an important parameter in adult communication, where it conveys the quality of the emitter and is shaped by sexual selection [175, 176] (Chap. 9). Indeed, there is direct evidence linking impaired rhythmic stability to genetic deficits that affect fitness [176]. The rhythmic stability of indri songs is likely to be likewise influenced by sexual selection, but further research is needed to determine if it predicts reproductive success in this species.

To sum up, data on the development of indri song shows that the juveniles are capable of producing periodic patterns based on small-integer ratios from the earliest stages in the developmental emergence of song. However, as in humans and other mammals, the production of complex rhythms develops with age. In particular, the rhythmic patterning of song undergoes a maturational process in which the hierarchical organization of the temporal structure is enhanced [173]. Overall, the current developmental data, along with analyses of adult song, suggest that quantized rhythm patterning is an adaptive trait by virtue of facilitating flexible turn-taking and coordination dynamics in duets and choruses. The temporal song organization is indeed known to follow sex-, age-, and status-related rules [177, 178].

10.4 Conclusions and Future Directions

As reiterated throughout this volume, rhythmicity is a composite competence spanning multiple perceptual, motor, cognitive, and socio-cognitive traits that enable the production and processing of periodic patterns. The developmental perspective underscores this compositionality—probably more clearly than any other perspective. Adopting a developmental angle, in this chapter, we reviewed relevant data from humans and singing primate taxa with a close (gibbons) or remote (indri lemurs) phylogenetic relatedness to our species. Besides being consistent with the compositionality hypothesis, the reviewed data show that rhythm-based communication develops as an interplay between innate maturational processes and socio-ecological influences. In concert, these shape a range of neuromechanical substrates and socio-cognitive mechanisms that enable the acquisition of species-specific rhythmic structures and ways of interacting through rhythm.

The evidence reviewed in this chapter also shows that the underlying mechanisms of rhythmicity have distinct developmental trajectories. Some emerge earlier than others, develop faster than others, and progress gradually or in steep leaps. Two general trends, however, can be distinguished. First, perceptual skills tend to offset motor ones and, in humans at least, from very early on, rhythm perception involves an interplay between bottom-up processing of sensory input and top-down filtering through endogenous rhythms driven by neurocognitive constraints, experience, and enculturation. Second, rhythmic proficiency seems to be intimately tied to the ability to bring simple—possibly innate—motor rhythms under increased cognitive control, which is hypothesized to permit the production of coordinated rhythmic behaviors with complex temporal sequences. In humans and other animals, the ability to synchronize and adapt turn-taking during social interactions is crucial for effective communication, for avoiding or managing overlap, for conveying emotions, and for fostering social connections [179].

With data from nonhuman primates, our review can also bridge these developmental trends with evolutionary ones. Of particular relevance is the capacity to flexibly omit (or not omit) elements in an isochronous sequence, as a mechanism of complexification observed in both rhythm perception and production. Interestingly, developmental changes of rhythmic structure have been documented in multiple species, where they are closely linked to the combinatorial features of species-specific signals [173]. For most species, the development of an adult-like rhythmic structure entails enhanced information encoding capabilities through the sequential arrangement of elements. This enhancement is both quantitative and qualitative and provides a significant adaptive advantage [180]. The gradual incorporation of increasingly complex rhythms and combinatorial possibilities into communication mediates the range of socio-dynamic affordances of the individual—a relationship that can also be evaluated comparatively at the species level [173].

As suggested above, data from nonhuman primate species are essential for understanding the adaptive benefits and the evolutionary roots of the mechanisms that enable human speech and musicality. However, this kind of data is currently so scarce that the field can rightfully be characterized as being in its infancy. For the ambitious goals of reconstructing the evolution of rhythm in the primate clade, there is a tremendously long way to go. Despite the prestige associated with experimental work, the more urgent needs in this area of research are those of covering the grounds through bottom-up approaches aimed at quantifying the emergence, pace, frequency, forms, and contextual distribution of periodical patterns in the ontogenetic development of multiple nonhuman primate species. The tenacious and detailed work carried out in the indri and gibbon species (reviewed in Sect. 10.3) provides valuable guidelines in this respect. Such data will help shed light on the evolutionary history, putative shared selective pressures, and the proximate causes related to periodic pattern generation in the primate clade [181]. The comparative study of rhythm ontogeny represents a pivotal step toward uncovering the biological foundations of rhythmicity. While field access is difficult and the conservation status of many primates is precarious [182], they are invaluable models worthy of protection and focused investment in the near future.

We hope that this review will facilitate progress and future research in this area. In this endeavor, the human developmental data reviewed in Sect. 10.2 could provide a suitable conceptual and methodological grid for tailoring research with other primate species. That being said, this review reveals important gaps in human research as well, where the great majority of studies have focused on the perception of—and rhythmic motor responses to—auditory rhythms, given their importance to speech and music. Thus, there is a considerable need for research that targets rhythms broadcasted in other modalities (visual, vestibular, tactile), as well as the cross-modal interplay of these. The field also suffers from a bias that is widespread in human psychology and cognitive science, that is, the overwhelming reliance on Western populations. Of course, there are also exceptions, which highlight how, in the absence of more robust cross-cultural representation, the findings reviewed in this chapter cannot (yet) provide valid conclusions about species-universal developmental trends. More research is also needed to address whether—and if so how—the rhythmic environment of the womb scaffolds the rhythmic preferences of early infancy, along with species-specific forms of social interactions that rely on accurate predictive timing. As the developmental trajectories of certain rhythmic abilities begin to be clarified, it may be tempting to regard abilities that emerge early as foundation for those that emerge later. Such conclusions, however, await suitable longitudinal data to assess how the elements of the rhythmic composite relate to one another, and how both maturational processes and socio-ecological factors contribute to their emergence, enhancement, and, indeed, their potential coupling.

Acknowledgments GA Sauciuc's contribution to this chapter has been funded by the Bank of Sweden Tercentenary foundation (grant nr. P17-0942:1) and the Swedish Research Council (grant 2019-03104). T Raimondi's contribution to this chapter has been funded by the European Union (ERC, TOHR, 101041885). The authors thank the Malagasy field guides, the field assistants help during the data collection, and the GERP (Groupe d'Étude et des Recherche sur les Primates de Madagascar) for their support during the field activities in the Maromizaha forests.

Ethics Approval The current review is based primarily on secondary sources, which does not entail requirements for ethical permits. Some of the reviewed data, however, are novel and come from studies that are currently under review. Permits for this research have been obtained for each year of data collection from "Direction des Eaux et Forêts" and "Madagascar National Parks" (formerly ANGAP) [2011 (N274/11/MEF/ SG/DGF/DCB.SAP/SCB), 2012 (N245/12/MEF/SG/DGF/DCB.SAP/SCB), 2014 (N066/14/MEF/SG/DGF/DCB.SAP/SCB), 2015 (N180/15/MEEMF/SG/DGF/DAPT/SCBT), 2016 (N98/16/MEEMF/SG/DGF/ DAPT/SCB.Re and N217/16/MEEMF/SG/DGF/DSAP/SCB.Re)], 2017 (73/17/MEEF/SG /DGF/ DSAP/SCB.RE). 2018: 91/18/MEEF/SG/DGF/DSAP/SCB.Re; 2019: 118/19/MEDD/SG/DGEF/DSAP/DGRNE and 284/19/MEDD/SG/DGEF/DSAP/DGRNE; 2019/2020: 338/19/MEDD/SG/DGEF/DSAP/DGRNE]. Data collection did not require a permit for 2013 because it was performed by Malagasy citizens only.

References

1. Sauciuc G-A, Schneider JJN, Burchardt LS et al (2025) Chapter 3: The functional and structural diversity of rhythmic communication across species. In: Sauciuc G-A, Lameira AR, Persson T (eds) Nature beats: the what, why and how of animal rhythmic behaviours. Springer Nature Switzerland, Cham
2. Proksch S, Honing H, Ravignani A, Merchant H (2025) Chapter 7: Converging notions on the main hypotheses regarding the origins of beat perception and synchronization: the key role of audiomotor interactions. In: Nature beats: the what, why and how of animal rhythmic behaviour. Springer Nature Switzerland, Cham
3. Large EW (2025) Chapter 8: Neural resonance and comparative rhythm cognition. In: Sauciuc G-A, Lameira AR, Persson T (eds) Nature beats: the what, why and how of animal rhythmic behaviours. Springer Nature Switzerland, Cham
4. Sauciuc G-A, Schneider JJN, Hattori Y (2025) Chapter 9: The neuroethology of rhythmic communication: neural and mechanical building blocks and constraints on rhythm production and perception. In: Sauciuc G-A, Lameira AR, Persson T (eds) Nature beats: the what, why and how of animal rhythmic behaviours. Springer Nature Switzerland, Cham
5. Gustison ML, Sauciuc G-A (2025) Chapter 11: Moving the body and the mind—links between emotion and rhythmic communication and the underlying neurobiological mechanisms. In: Sauciuc G-A, Lameira AR, Persson T (eds) Nature beats: the what, why and how of animal rhythmic behaviours. Springer Nature Switzerland, Cham
6. Burchardt LS, Hoeschele M, Honing H et al (2025) Chapter 1: What is rhythm and how can we study it? Center for Open Science. https://doi.org/10.31219/osf.io/69v5a_v1
7. Haimoff EH (1983) Gibbon songs: an acoustical, organizational, and behavioural analysis. Doctoral thesis, University of Cambridge
8. Spector DA (1994) Definition in biology: the case of "bird song". J Theor Biol 168:373–381. https://doi.org/10.1006/jtbi.1994.1117
9. Thorpe CW (1961) Bird-song: the biology of vocal communication and expression in birds. Cambridge monographs in experimental biology no. 12. Cambridge University Press, Cambridge
10. Geissmann T (1999) Gibbon songs and human music from an evolutionary perspective. In: Wallin NL, Merker B, Brown S (eds) The origins of music. The MIT Press, pp 103–124
11. De Gregorio C, Carugati F, Valente D et al (2022) Notes on a tree: reframing the relevance of primate choruses, duets, and solo songs. Ethol Ecol Evol 34:205–219. https://doi.org/10.1080/03949370.2021.2015451
12. Van Der Vleuten BJR, Hovenkamp VA, Varkevisser JM, Spierings MJ (2024) Context-dependent rhythmicity in chimpanzee displays. Proc R Soc B 291:20242200. https://doi.org/10.1098/rspb.2024.2200

13. Lameira AR, Hardus ME, Ravignani A et al (2024) Recursive self-embedded vocal motifs in wild orangutans. eLife 12:RP88348. https://doi.org/10.7554/eLife.88348
14. De Gregorio C, Lameira AR (2026) Twice times two: dual mechanism for double rhythmic meter in orangutans and the evolution of human song. iScience 29:114273. https://doi.org/10.1016/j.isci.2025.114273
15. De Gregorio C, Gamba M, Lameira AR (2025) Third-order self-embedded vocal motifs in wild orangutans, and the selective evolution of recursion. Ann N Y Acad Sci nyas.15373. https://doi.org/10.1111/nyas.15373
16. Falk S, Müller T, Dalla Bella S (2015) Non-verbal sensorimotor timing deficits in children and adolescents who stutter. Front Psychol 6:847. https://doi.org/10.3389/fpsyg.2015.00847
17. Cumming R, Wilson A, Leong V et al (2015) Awareness of rhythm patterns in speech and music in children with specific language impairments. Front Hum Neurosci 9:672. https://doi.org/10.3389/fnhum.2015.00672
18. Trainor LJ, Chang A, Cairney J, Li Y (2018) Is auditory perceptual timing a core deficit of developmental coordination disorder? Ann N Y Acad Sci 1423:30–39. https://doi.org/10.1111/nyas.13701
19. Ladányi E, Persici V, Fiveash A et al (2020) Is atypical rhythm a risk factor for developmental speech and language disorders? WIRES Cogn Sci 11:e1528. https://doi.org/10.1002/wcs.1528
20. Hidalgo C, Falk S, Schön D (2017) Speak on time! Effects of a musical rhythmic training on children with hearing loss. Hear Res 351:11–18. https://doi.org/10.1016/j.heares.2017.05.006
21. Morris PO, Hope E, Foulsham T, Mills JP (2021) Dance, rhythm, and autism spectrum disorder: an explorative study. Arts Psychother 73:101755. https://doi.org/10.1016/j.aip.2020.101755
22. Williams KE, Berthelsen D (2019) Implementation of a rhythm and movement intervention to support self-regulation skills of preschool-aged children in disadvantaged communities. Psychol Music 47:800–820. https://doi.org/10.1177/0305735619861433
23. Valman HB, Pearson JF (1980) What the fetus feels. BMJ 280:233–234. https://doi.org/10.1136/bmj.280.6209.233
24. Kisilevsky BS, Low JA (1998) Human fetal behavior: 100 years of study. Dev Rev 18:1–29. https://doi.org/10.1006/drev.1998.0452
25. Hepper PG, Shahidullah BS (1994) Development of fetal hearing. Arch Dis Child Fetal Neonatal Ed 71:F81–F87. https://doi.org/10.1136/fn.71.2.F81
26. Lecanuet J-P, Schaal B (2002) Sensory performances in the human foetus: a brief summary of research. Intellectica 34:29–56. https://doi.org/10.3406/intel.2002.1072
27. Zimmerman E, Barlow SM (2012) The effects of vestibular stimulation rate and magnitude of acceleration on central pattern generation for chest wall kinematics in preterm infants. J Perinatol 32:614–620. https://doi.org/10.1038/jp.2011.177
28. Teie D (2016) A comparative analysis of the universal elements of music and the fetal environment. Front Psychol 7:1158. https://doi.org/10.3389/fpsyg.2016.01158
29. Bossung V, Singer A, Ratz T et al (2023) Changes in heart rate, heart rate variability, breathing rate, and skin temperature throughout pregnancy and the impact of emotions—a longitudinal evaluation using a sensor bracelet. Sensors 23:6620. https://doi.org/10.3390/s23146620
30. Coad F, Frise C (2021) Tachycardia in pregnancy: when to worry? Clin Med 21:e434–e437. https://doi.org/10.7861/clinmed.2021-0495
31. Querleu D, Renard X, Versyp F et al (1988) Fetal hearing. Eur J Obstet Gynecol Reprod Biol 28:191–212. https://doi.org/10.1016/0028-2243(88)90030-5
32. Parncutt R (2006) Prenatal development. In: McPherson G (ed) The child as musician. Oxford University Press, pp 1–32
33. Ullal-Gupta S, Vanden Bosch Der Nederlanden CM, Tichko P et al (2013) Linking prenatal experience to the emerging musical mind. Front Syst Neurosci 7:48. https://doi.org/10.3389/fnsys.2013.00048
34. Pino O (2016) Fetal memory: the effects of prenatal auditory experience on human development. BAOJMN 2:2. https://doi.org/10.24947/baojmn/2/2/120

35. Elstad M, O'Callaghan EL, Smith AJ et al (2018) Cardiorespiratory interactions in humans and animals: rhythms for life. Am J Phys Heart Circ Phys 315:H6–H17. https://doi.org/10.1152/ajpheart.00701.2017
36. Marzbanrad F, Kimura Y, Palaniswami M, Khandoker AH (2015) Quantifying the interactions between maternal and fetal heart rates by transfer entropy. PLoS One 10:e0145672. https://doi.org/10.1371/journal.pone.0145672
37. Nichting TJ, Bester M, Joshi R et al (2023) Evidence and clinical relevance of maternal-fetal cardiac coupling: a scoping review. PLoS One 18:e0287245. https://doi.org/10.1371/journal.pone.0287245
38. Van Leeuwen P, Geue D, Thiel M et al (2009) Influence of paced maternal breathing on fetal–maternal heart rate coordination. Proc Natl Acad Sci USA 106:13661–13666. https://doi.org/10.1073/pnas.0901049106
39. Marshall MR, Montoye AHK, George AJ (2018) Pregnancy walking cadence does not vary by trimester. Gait Posture 65:81–85. https://doi.org/10.1016/j.gaitpost.2018.06.175
40. Lecanuet J, Jacquet A (2002) Fetal responsiveness to maternal passive swinging in low heart rate variability state: effects of stimulation direction and duration. Dev Psychobiol 40:57–67. https://doi.org/10.1002/dev.10013
41. Movalled K, Sani A, Nikniaz L, Ghojazadeh M (2023) The impact of sound stimulations during pregnancy on fetal learning: a systematic review. BMC Pediatr 23:183. https://doi.org/10.1186/s12887-023-03990-7
42. Larsson M, Richter J, Ravignani A (2019) Bipedal steps in the development of rhythmic behavior in humans. Music Sci 2:2059204319892617. https://doi.org/10.1177/2059204319892617
43. Parncutt R, Chuckrow R (2019) Chuckrow's theory of the prenatal origin of music. Music Sci 23:403–425. https://doi.org/10.1177/1029864917738130
44. van den Bergh B (1992) Maternal emotions during pregnancy and fetal and neonatal behavior. In: Fetal behaviour: developmental and perinatal aspects. Oxford University Press, Oxford, pp 157–178
45. Parncutt R (2015) Prenatal development. In: McPherson GE (ed) The child as musician. Oxford University Press, pp 3–30
46. Ryckman J, Hilton C, Rogers C, Pineda R (2017) Sensory processing disorder in preterm infants during early childhood and relationships to early neurobehavior. Early Hum Dev 113:18–22. https://doi.org/10.1016/j.earlhumdev.2017.07.012
47. Chorna OD, Slaughter JC, Wang L et al (2014) A pacifier-activated music player with mother's voice improves oral feeding in preterm infants. Pediatrics 133:462–468. https://doi.org/10.1542/peds.2013-2547
48. Olsen A, Dennis EL, Evensen KAI et al (2018) Preterm birth leads to hyper-reactive cognitive control processing and poor white matter organization in adulthood. NeuroImage 167:419–428. https://doi.org/10.1016/j.neuroimage.2017.11.055
49. Cramer SJE, Dekker J, Dankelman J et al (2018) Effect of tactile stimulation on termination and prevention of Apnea of prematurity: a systematic review. Front Pediatr 6:45. https://doi.org/10.3389/fped.2018.00045
50. Provasi J, Blanc L, Carchon I (2021) The importance of rhythmic stimulation for preterm infants in the NICU. Children 8:660. https://doi.org/10.3390/children8080660
51. Haslbeck FB (2012) Music therapy for premature infants and their parents: an integrative review. Nord J Music Ther 21:203–226. https://doi.org/10.1080/08098131.2011.648653
52. Bieleninik Ł, Ghetti C, Gold C (2016) Music therapy for preterm infants and their parents: a meta-analysis. Pediatrics 138:e20160971. https://doi.org/10.1542/peds.2016-0971
53. Mehler J, Jusczyk P, Lambertz G et al (1988) A precursor of language acquisition in young infants. Cognition 29:143–178. https://doi.org/10.1016/0010-0277(88)90035-2
54. Nazzi T, Bertoncini J, Mehler J (1998) Language discrimination by newborns: toward an understanding of the role of rhythm. J Exp Psychol Hum Percept Perform 24:756–766. https://doi.org/10.1037/0096-1523.24.3.756

55. Háden GP, Honing H, Török M, Winkler I (2015) Detecting the temporal structure of sound sequences in newborn infants. Int J Psychophysiol 96:23–28. https://doi.org/10.1016/j.ijpsycho.2015.02.024
56. Winkler I, Háden GP, Ladinig O et al (2009) Newborn infants detect the beat in music. Proc Natl Acad Sci USA 106:2468–2471. https://doi.org/10.1073/pnas.0809035106
57. Demany L, Mckenzie B, Vurpillot E (1977) Rhythm perception in early infancy. Nature 266:718–719. https://doi.org/10.1038/266718a0
58. Chang H-W, Trehub SE (1977) Infants' perception of temporal grouping in auditory patterns. Child Dev 48:1666. https://doi.org/10.2307/1128532
59. Baruch C, Drake C (1997) Tempo discrimination in infants. Infant Behav Dev 20:573–577. https://doi.org/10.1016/S0163-6383(97)90049-7
60. Baruch C, Panissal-Vieu N, Drake C (2004) Preferred perceptual tempo for sound sequences: comparison of adults, children, and infants. Percept Mot Skills 98:325–339. https://doi.org/10.2466/pms.98.1.325-339
61. Hannon EE, Soley G, Levine RS (2011) Constraints on infants' musical rhythm perception: effects of interval ratio complexity and enculturation. Dev Sci 14:865–872. https://doi.org/10.1111/j.1467-7687.2011.01036.x
62. Hannon EE, Trehub SE (2005) Metrical categories in infancy and adulthood. Psychol Sci 16:48–55. https://doi.org/10.1111/j.0956-7976.2005.00779.x
63. Hannon EE, Johnson SP (2005) Infants use meter to categorize rhythms and melodies: implications for musical structure learning. Cogn Psychol 50:354–377. https://doi.org/10.1016/j.cogpsych.2004.09.003
64. Cirelli LK, Spinelli C, Nozaradan S, Trainor LJ (2016) Measuring neural entrainment to beat and meter in infants: effects of music background. Front Neurosci 10:229. https://doi.org/10.3389/fnins.2016.00229
65. Cantiani C, Dondena C, Molteni M et al (2022) Synchronizing with the rhythm: infant neural entrainment to complex musical and speech stimuli. Front Psychol 13:944670. https://doi.org/10.3389/fpsyg.2022.944670
66. Flaten E, Marshall SA, Dittrich A, Trainor LJ (2022) Evidence for top-down metre perception in infancy as shown by primed neural responses to an ambiguous rhythm. Eur J Neurosci 55:2003–2023. https://doi.org/10.1111/ejn.15671
67. Lenc T, Peter V, Hooper C et al (2023) Infants show enhanced neural responses to musical meter frequencies beyond low-level features. Dev Sci 26:e13353. https://doi.org/10.1111/desc.13353
68. Nguyen T, Reisner S, Lueger A et al (2023) Sing to me, baby: infants show neural tracking and rhythmic movements to live and dynamic maternal singing. Dev Cogn Neurosci 64:101313. https://doi.org/10.1016/j.dcn.2023.101313
69. Soley G, Hannon EE (2010) Infants prefer the musical meter of their own culture: a cross-cultural comparison. Dev Psychol 46:286–292. https://doi.org/10.1037/a0017555
70. Bergeson TR, Trehub SE (2006) Infants perception of rhythmic patterns. Music Percept 23:345–360. https://doi.org/10.1525/mp.2006.23.4.345
71. Hannon EE, Trehub SE (2005) Tuning in to musical rhythms: infants learn more readily than adults. Proc Natl Acad Sci USA 102:12639–12643. https://doi.org/10.1073/pnas.0504254102
72. Drake C, Jones MR, Baruch C (2000) The development of rhythmic attending in auditory sequences: attunement, referent period, focal attending. Cognition 77:251–288. https://doi.org/10.1016/S0010-0277(00)00106-2
73. Mendelson MJ (1986) Perception of the temporal pattern of motion in infancy. Infant Behav Dev 9:231–243. https://doi.org/10.1016/0163-6383(86)90031-7
74. Brandon M, Saffran JR (2011) Apparent motion enhances visual rhythm discrimination in infancy. Atten Percept Psychophys 73:1016–1020. https://doi.org/10.3758/s13414-011-0106-x
75. Bahrick LE, Lickliter R (2000) Intersensory redundancy guides attentional selectivity and perceptual learning in infancy. Dev Psychol 36:190–201. https://doi.org/10.1037/0012-1649.36.2.190

76. Lickliter R, Bahrick LE (2001) The salience of multimodal sensory stimulation in early development: implications for the issue of ecological validity. Infancy 2:451–467. https://doi.org/10.1207/S15327078IN0204_04
77. Bahrick LE, Lickliter R (2003) Intersensory redundancy guides early perceptual and cognitive development. In: Advances in child development and behavior. Elsevier, pp 153–187
78. Mendelson MJ, Ferland MB (1982) Auditory-visual transfer in four-month-old infants. Child Dev 53:1022. https://doi.org/10.2307/1129143
79. Allen TW, Walker K, Symonds L, Marcell M (1977) Intrasensory and intersensory perception of temporal sequences during infancy. Dev Psychol 13:225–229. https://doi.org/10.1037/0012-1649.13.3.225
80. Phillips-Silver J, Trainor LJ (2005) Feeling the beat: movement influences infant rhythm perception. Science 308:1430–1430. https://doi.org/10.1126/science.1110922
81. Rocha S, Southgate V, Mareschal D (2021) Rate of infant carrying impacts infant spontaneous motor tempo. R Soc Open Sci 8:210608. https://doi.org/10.1098/rsos.210608
82. Lewkowicz DJ (2003) Learning and discrimination of audiovisual events in human infants: the hierarchical relation between intersensory temporal synchrony and rhythmic pattern cues. Dev Psychol 39:795–804. https://doi.org/10.1037/0012-1649.39.5.795
83. Kopp F, Dietrich C (2013) Neural dynamics of Audiovisual synchrony and asynchrony perception in 6-month-old infants. Front Psychol 4:2. https://doi.org/10.3389/fpsyg.2013.00002
84. Kopp F (2014) Audiovisual temporal fusion in 6-month-old infants. Dev Cogn Neurosci 9:56–67. https://doi.org/10.1016/j.dcn.2014.01.001
85. Gerson SA, Schiavio A, Timmers R, Hunnius S (2015) Active drumming experience increases infants' sensitivity to audiovisual synchrony during observed drumming actions. PLoS One 10:e0130960. https://doi.org/10.1371/journal.pone.0130960
86. Hannon EE, Schachner A, Nave-Blodgett JE (2017) Babies know bad dancing when they see it: older but not younger infants discriminate between synchronous and asynchronous audiovisual musical displays. J Exp Child Psychol 159:159–174. https://doi.org/10.1016/j.jecp.2017.01.006
87. Thelen E (1979) Rhythmical stereotypies in normal human infants. Anim Behav 27:699–715. https://doi.org/10.1016/0003-3472(79)90006-X
88. Thelen E (1981) Kicking, rocking, and waving: contextual analysis of rhythmical stereotypies in normal human infants. Anim Behav 29:3–11. https://doi.org/10.1016/S0003-3472(81)80146-7
89. Barr RG (1990) The early crying paradox: a modest proposal. Hum Nat 1:355–389. https://doi.org/10.1007/BF02734051
90. Papousek M, Papousek H (1981) Musical elements in the infant's vocalization: their significance for communication, cognition, and creativity. Adv Infancy Res 1:163–224
91. Kent RD, Mitchell PR, Sancier M (1991) Evidence and role of rhythmic organization in early vocal development in human infants. In: Advances in psychology. Elsevier, pp 135–149
92. Oiler DK (1986) Metaphonology and infant vocalizations. In: Lindblom B, Zetterström R (eds) Precursors of early speech. Palgrave Macmillan, London, pp 21–35
93. Iverson JM, Hall AJ, Nickel L, Wozniak RH (2007) The relationship between reduplicated babble onset and laterality biases in infant rhythmic arm movements. Brain Lang 101:198–207. https://doi.org/10.1016/j.bandl.2006.11.004
94. Ejiri K, Masataka N (2001) Co-occurences of preverbal vocal behavior and motor action in early infancy. Dev Sci 4:40–48. https://doi.org/10.1111/1467-7687.00147
95. Kent RD (2022) The maturational gradient of infant vocalizations: developmental stages and functional modules. Infant Behav Dev 66:101682. https://doi.org/10.1016/j.infbeh.2021.101682
96. Oller DK, Eilers RE (1988) The role of audition in infant babbling. Child Dev 59:441. https://doi.org/10.2307/1130323
97. van den Dikkenberg-Plot I, Koopmans-van Beinum F (1988) Influence of lack of auditory feedback: vocalizations of deaf and hearing infants compared. In: Institute of Phonetic

Sciences, University of Amsterdam, Proceedings 22. Institute of Phonetic Sciences, Amsterdam, pp 47–60
98. Luschei ES, Ramig LO, Finnegan EM et al (2006) Patterns of laryngeal electromyography and the activity of the respiratory system during spontaneous laughter. J Neurophysiol 96:442–450. https://doi.org/10.1152/jn.00102.2006
99. Provine RR (2017) Laughter as an approach to vocal evolution: the bipedal theory. Psychon Bull Rev 24:238–244. https://doi.org/10.3758/s13423-016-1089-3
100. Eilers RE, Oller DK, Levine S et al (1993) The role of prematurity and socioeconomic status in the onset of canonical babbling in infants. Infant Behav Dev 16:297–315. https://doi.org/10.1016/0163-6383(93)80037-9
101. Iverson JM, Fagan MK (2004) Infant vocal–motor coordination: precursor to the gesture–speech system? Child Dev 75:1053–1066. https://doi.org/10.1111/j.1467-8624.2004.00725.x
102. Locke JL, Bekken KE, Mcminnlarson L, Wein D (1995) Emergent control of manual and vocal-motor activity in relation to the development of speech. Brain Lang 51:498–508. https://doi.org/10.1006/brln.1995.1073
103. Cobo-Lewis AB, Oller DK, Lynch MP, Levine SL (1996) Relations of motor and vocal milestones in typically developing infants and infants with Down syndrome. Am J Ment Retard AJMR 100:456–467
104. Ejiri K (1998) Relationship between rhythmic behavior and canonical babbling in infant vocal development. Phonetica 55:226–237. https://doi.org/10.1159/000028434
105. Murillo E, Montero I, Casla M (2021) On the multimodal path to language: the relationship between rhythmic movements and deictic gestures at the end of the first year. Front Psychol 12:616812. https://doi.org/10.3389/fpsyg.2021.616812
106. Borjon JI, Abney DH, Yu C, Smith LB (2024) Infant vocal productions coincide with body movements. Dev Sci 27:e13491. https://doi.org/10.1111/desc.13491
107. Levitt AG, Wang Q (1991) Evidence for language-specific rhythmic influences in the reduplicative babbling of French-and English-learning infants. Lang Speech 34:235–249. https://doi.org/10.1177/002383099103400302
108. Lynch MP, Oller DK, Steffens ML, Buder EH (1995) Phrasing in prelinguistic vocalizations. Dev Psychobiol 28:3–25. https://doi.org/10.1002/dev.420280103
109. Nathani S, Kimbrough Oller D, Cobo-Lewis AB (2003) Final syllable lengthening (FSL) in infant vocalizations. J Child Lang 30:3–25. https://doi.org/10.1017/S0305000902005433
110. Dolata JK, Davis BL, MacNeilage PF (2008) Characteristics of the rhythmic organization of vocal babbling: implications for an amodal linguistic rhythm. Infant Behav Dev 31:422–431. https://doi.org/10.1016/j.infbeh.2007.12.014
111. Bobin-Bègue A, Provasi J, Marks A, Pouthas V (2006) Influence of auditory tempo on the endogenous rhythm of non-nutritive sucking. Eur Rev Appl Psychol 56:239–245. https://doi.org/10.1016/j.erap.2005.09.006
112. Bobin-Bègue A, Provasi J (2008) Régulation rythmique avant 4 ans: effet d'un tempo auditif sur le tempo moteur. Annee Psychol 108:631. https://doi.org/10.4074/S000350330800403X
113. Fitzpatrick P, Schmidt RC, Lockman JJ (1996) Dynamical patterns in the development of clapping. Child Dev 67:2691. https://doi.org/10.2307/1131747
114. McAuley JD, Jones MR, Holub S et al (2006) The time of our lives: life span development of timing and event tracking. J Exp Psychol Gen 135:348–367. https://doi.org/10.1037/0096-3445.135.3.348
115. Iverson JM (2010) Developing language in a developing body: the relationship between motor development and language development. J Child Lang 37:229–261. https://doi.org/10.1017/S0305000909990432
116. Lanuza MA, Tomàs J, Garcia N et al (2018) Axonal competition and synapse elimination during neuromuscular junction development. Curr Opin Physio 4:25–31. https://doi.org/10.1016/j.cophys.2018.04.001
117. Sauciuc G-A, Zlakowska J, Persson T et al (2020) Imitation recognition and its prosocial effects in 6-month old infants. PLoS One 15:e0232717. https://doi.org/10.1371/journal.pone.0232717

118. Zentner M, Eerola T (2010) Rhythmic engagement with music in infancy. Proc Natl Acad Sci USA 107:5768–5773. https://doi.org/10.1073/pnas.1000121107
119. Fujii S, Watanabe H, Oohashi H et al (2014) Precursors of dancing and singing to music in three- to four-months-old infants. PLoS One 9:e97680. https://doi.org/10.1371/journal.pone.0097680
120. Ilari B (2015) Rhythmic engagement with music in early childhood: a replication and extension. J Res Music Educ 62:332–343. https://doi.org/10.1177/0022429414555984
121. Cirelli LK, Trehub SE (2019) Dancing to Metallica and Dora: case study of a 19-month-old. Front Psychol 10:1073. https://doi.org/10.3389/fpsyg.2019.01073
122. Kim M, Schachner A (2023) The origins of dance: characterizing the development of infants' earliest dance behavior. Dev Psychol 59:691–706. https://doi.org/10.1037/dev0001436
123. Kragness HE, Ullah F, Chan E et al (2022) Tiny dancers: effects of musical familiarity and tempo on children's free dancing. Dev Psychol 58:1277–1285. https://doi.org/10.1037/dev0001363
124. De l'Etoile SK, Bennett C, Zopluoglu C (2020) Infant movement response to auditory rhythm. Percept Mot Skills 127:651–670. https://doi.org/10.1177/0031512520922642
125. Kragness HE, Johnson EK, Cirelli LK (2022) The song, not the singer: infants prefer to listen to familiar songs, regardless of singer identity. Dev Sci 25:e13149. https://doi.org/10.1111/desc.13149
126. Yu L, Myowa M (2021) The early development of tempo adjustment and synchronization during joint drumming: a study of 18- to 42-month-old children. Infancy 26:635–646. https://doi.org/10.1111/infa.12403
127. Yu L, Todoriki K, Myowa M (2022) From spontaneous rhythmic engagement to joint drumming: a gradual development of flexible coordination at approximately 24 months of age. Front Psychol 13:907834. https://doi.org/10.3389/fpsyg.2022.907834
128. Provasi J, Bobin-Bègue A (2003) Spontaneous motor tempo and rhythmical synchronisation in 2½- and 4-year-old children. Int J Behav Dev 27:220–231. https://doi.org/10.1080/01650250244000290
129. Kirschner S, Tomasello M (2009) Joint drumming: social context facilitates synchronization in preschool children. J Exp Child Psychol 102:299–314. https://doi.org/10.1016/j.jecp.2008.07.005
130. Sauciuc G-A, Schneider JJN (2025) Chapter 2: How animals produce rhythm—an overview of ethological mechanisms for rhythmic communication. In: Nature beats: the what, why and how of animal rhythmic behaviours. Springer Nature Switzerland, Cham
131. De Gregorio C, Valente D, Ferrario V et al (2024) Who you live with and what you duet for: a review of the function of primate duets in relation to their social organization. J Comp Physiol A 210:281–294. https://doi.org/10.1007/s00359-023-01689-9
132. Carbone L, Alan Harris R, Gnerre S et al (2014) Gibbon genome and the fast karyotype evolution of small apes. Nature 513:195–201. https://doi.org/10.1038/nature13679
133. Pozzi L, Hodgson JA, Burrell AS et al (2014) Primate phylogenetic relationships and divergence dates inferred from complete mitochondrial genomes. Mol Phylogenet Evol 75:165–183. https://doi.org/10.1016/j.ympev.2014.02.023
134. Raimondi T, Di Panfilo G, Pasquali M et al (2023) Isochrony and rhythmic interaction in ape duetting. Proc R Soc B 290:20222244. https://doi.org/10.1098/rspb.2022.2244
135. De Gregorio C, Raimondi T, Bevilacqua V et al (2024) Isochronous singing in 3 crested gibbon species (*Nomascus* spp.). Curr Zool 70:291–297. https://doi.org/10.1093/cz/zoad029
136. Ma H, Wang Z, Han P et al (2024) Small apes adjust rhythms to facilitate song coordination. Curr Biol 34:935.e3–945.e3. https://doi.org/10.1016/j.cub.2023.12.071
137. Brockelman WY, Schilling D (1984) Inheritance of stereotyped gibbon calls. Nature 312:634–636. https://doi.org/10.1038/312634a0
138. Koda H, Lemasson A, Oyakawa C et al (2013) Possible role of mother-daughter vocal interactions on the development of species-specific song in gibbons. PLoS One 8:e71432. https://doi.org/10.1371/journal.pone.0071432

139. Koda H, Oyakawa C, Kato A et al (2014) Immature male gibbons produce female-specific songs. Primates 55:13–17. https://doi.org/10.1007/s10329-013-0390-2
140. Yi Y, Choi A, Lee S et al (2022) Transient co-singing of offspring and mothers in non-duetting Javan gibbons (*Hylobates moloch*). Front Ecol Evol 10:910260. https://doi.org/10.3389/fevo.2022.910260
141. Hradec M, Bolechová P, Svobodová I (2016) Production of a female-specific great call in an immature male gibbon, the N*omascus* genus. Primates 57:445–448. https://doi.org/10.1007/s10329-016-0569-4
142. Merker B, Cox C (1999) Development of the female great call in *Hylobates gabriellae*: a case study. Int J Fluid Power 70:97–106. https://doi.org/10.1159/000021680
143. Hradec M, Illmann G, Bartoš L, Bolechová P (2021) The transition from the female-like great calls to male calls during ontogeny in southern yellow-cheeked gibbon males (*Nomascus gabriellae*). Sci Rep 11:22040. https://doi.org/10.1038/s41598-021-01648-x
144. Adret P (2022) Developmental plasticity in primate coordinated song: parallels and divergences with duetting songbirds. Front Ecol Evol 10:862196. https://doi.org/10.3389/fevo.2022.862196
145. Terleph TA, Malaivijitnond S, Reichard UH (2018) Male white-handed gibbons flexibly time duet contributions. Behav Ecol Sociobiol 72:16. https://doi.org/10.1007/s00265-017-2432-z
146. De Gregorio C, Atonini P, Heymann E, Gamba M (2025) Isochrony in titi monkeys duets: social context as a proximate cause of duets' rhythm and regularity. Proc R Soc B Biol Sci 292:20242805. https://doi.org/10.1098/rspb.2024.2805
147. Geissmann T (2000) Duet songs of the siamang (*Hylobates syndactylus*): I. Structure and organisation. Primate Rep 56:33–60
148. Torti V, Gamba M, Rabemananjara ZH, Giacoma C (2013) The songs of the indris (Mammalia: primates: Indridae): contextual variation in the long-distance calls of a lemur. Ital J Zool 80:596–607. https://doi.org/10.1080/11250003.2013.845261
149. Bonadonna G, Zaccagno M, Torti V et al (2020) Intra- and intergroup spatial dynamics of a pair-living singing primate, *Indri indri*: a multiannual study of three indri groups in Maromizaha Forest, Madagascar. Int J Primatol 41:224–245. https://doi.org/10.1007/s10764-019-00127-5
150. Torti V, Bonadonna G, De Gregorio C et al (2017) An intra-population analysis of the indris' song dissimilarity in the light of genetic distance. Sci Rep 7:10140. https://doi.org/10.1038/s41598-017-10656-9
151. De Gregorio C, Carugati F, Estienne V et al (2021) Born to sing! Song development in a singing primate. Curr Zool 67:597–608. https://doi.org/10.1093/cz/zoab018
152. Gamba M, Torti V, Estienne V et al (2016) The indris have got rhythm! Timing and pitch variation of a primate song examined between sexes and age classes. Front Neurosci 10:249. https://doi.org/10.3389/fnins.2016.00249
153. De Gregorio C, Maiolini M, Raimondi T et al (2024) Isochrony as ancestral condition to call and song in a primate. Ann N Y Acad Sci 1537:41–50. https://doi.org/10.1111/nyas.15151
154. Savage PE, Brown S, Sakai E, Currie TE (2015) Statistical universals reveal the structures and functions of human music. Proc Natl Acad Sci USA 112:8987–8992. https://doi.org/10.1073/pnas.1414495112
155. Roeske T, Larrouy-Maestri P, Sakamoto Y, Poeppel D (2020) Listening to birdsong reveals basic features of rate perception and aesthetic judgements. Proc R Soc B 287:20193010. https://doi.org/10.1098/rspb.2019.3010
156. Greenfield M, Merker B (2025) Chapter 5: Endogenous rhythm, isochrony and the path to perfect synchrony. In: Nature beats: the what, why and how of animal rhythmic behaviour. Springer Nature Switzerland, Cham
157. Hammerschmidt K, Fischer J (2008) Constraints in primate vocal production. In: Oller DK, Griebel U (eds) Evolution of communicative flexibility. The MIT Press, pp 92–119
158. Seyfarth RM, Cheney DL (1997) Behavioral mechanisms underlying vocal communication in nonhuman primates. Anim Learn Behav 25:249–267. https://doi.org/10.3758/BF03199083

159. Hammerschmidt K, Newman JD, Champoux M, Suomi SJ (2000) Changes in rhesus macaque 'coo' vocalizations during early development. Ethology 106:873–886. https://doi.org/10.1046/j.1439-0310.2000.00611.x
160. Lameira AR, Santamaría-Bonfil G, Galeone D et al (2022) Sociality predicts orangutan vocal phenotype. Nat Ecol Evol 6:644–652. https://doi.org/10.1038/s41559-022-01689-z
161. Takahashi DY, Fenley AR, Teramoto Y et al (2015) The developmental dynamics of marmoset monkey vocal production. Science 349:734–738. https://doi.org/10.1126/science.aab1058
162. Laffi L, Raimondi T, De Gregorio C et al (2026) The ontogeny of vocal rhythms in a nonhuman primate. Dev Sci 29:e70189. https://doi.org/10.1111/desc.70189
163. Kotz SA, Ravignani A, Fitch WT (2018) The evolution of rhythm processing. Trends Cogn Sci 22:896–910. https://doi.org/10.1016/j.tics.2018.08.002
164. Anichini M, De Reus K, Hersh TA et al (2023) Measuring rhythms of vocal interactions: a proof of principle in harbour seal pups. Philos Trans R Soc B 378:20210477. https://doi.org/10.1098/rstb.2021.0477
165. Merker BH, Madison GS, Eckerdal P (2009) On the role and origin of isochrony in human rhythmic entrainment. Cortex 45:4–17. https://doi.org/10.1016/j.cortex.2008.06.011
166. Ravignani A, De Reus K (2019) Modelling animal interactive rhythms in communication. Evol Bioinformatics Online 15:1176934318823558. https://doi.org/10.1177/1176934318823558
167. Ravignani A, Madison G (2017) The paradox of isochrony in the evolution of human rhythm. Front Psychol 8:1820. https://doi.org/10.3389/fpsyg.2017.01820
168. Ravignani A (2021) Isochrony, vocal learning, and the acquisition of rhythm and melody. Behav Brain Sci 44:e88. https://doi.org/10.1017/S0140525X20001478
169. Brenowitz EA, Beecher MD (2005) Song learning in birds: diversity and plasticity, opportunities and challenges. Trends Neurosci 28:127–132. https://doi.org/10.1016/j.tins.2005.01.004
170. Doupe AJ, Kuhl PK (1999) Birdsong and human speech: common themes and mechanisms. Annu Rev Neurosci 22:567–631. https://doi.org/10.1146/annurev.neuro.22.1.567
171. Fischer J, Kitchen DM, Seyfarth RM, Cheney DL (2004) Baboon loud calls advertise male quality: acoustic features and their relation to rank, age, and exhaustion. Behav Ecol Sociobiol 56:140–148. https://doi.org/10.1007/s00265-003-0739-4
172. Osiecka AN, Fearey J, Ravignani A, Burchardt LS (2024) Isochrony in barks of cape fur seal (*Arctocephalus pusillus pusillus*) pups and adults. Ecol Evol 14:e11085. https://doi.org/10.1002/ece3.11085
173. Lipkind D, Marcus GF, Bemis DK et al (2013) Stepwise acquisition of vocal combinatorial capacity in songbirds and human infants. Nature 498:104–108. https://doi.org/10.1038/nature12173
174. Geissmann T (1984) Inheritance of song parameters in the gibbon song, analysed in 2 hybrid gibbons (*Hylobates pileatus* × *H. Lar*). Int J Fluid Power 42:216–235. https://doi.org/10.1159/000156165
175. Sebastianelli M, Lukhele SM, Secomandi S et al (2024) A genomic basis of vocal rhythm in birds. Nat Commun 15:3095. https://doi.org/10.1038/s41467-024-47305-5
176. Van Den Broek E, Todd PM (2003) Piep Piep Piep—ich Hab' dich Lieb: rhythm as an indicator of mate quality. In: Banzhaf W, Ziegler J, Christaller T et al (eds) Advances in artificial life. Springer, Berlin/Heidelberg, pp 425–433
177. De Gregorio C, Zanoli A, Carugati F et al (2022) Parent-offspring turn-taking dynamics influence parents' song structure and elaboration in a singing primate. Front Ecol Evol 10:906322. https://doi.org/10.3389/fevo.2022.906322
178. Zanoli A, Raimondi T, De Gregorio C et al (2023) "The song remains the same": not really! Vocal flexibility in the song of the indris. Anim Cogn 26:2009–2021. https://doi.org/10.1007/s10071-023-01826-6
179. De Reus K, Soma M, Anichini M et al (2021) Rhythm in dyadic interactions. Philos Trans R Soc B Biol Sci 376:20200337
180. Engesser S, Townsend SW (2019) Combinatoriality in the vocal systems of nonhuman animals. WIRES Cogn Sci 10:e1493. https://doi.org/10.1002/wcs.1493

181. Fitch WT (2015) Four principles of bio-musicology. Philos Trans R Soc B 370:20140091. https://doi.org/10.1098/rstb.2014.0091
182. Fernández D, Kerhoas D, Dempsey A et al (2022) The current status of the world's primates: mapping threats to understand priorities for primate conservation. Int J Primatol 43:15–39. https://doi.org/10.1007/s10764-021-00242-2

Moving the Body and the Mind: Links Between Emotion and Rhythmic Communication and Underlying Neurobiological Mechanisms

11

Morgan L. Gustison and Gabriela-Alina Sauciuc

Abstract

Rhythmic patterning is a widespread structural trait of animal (including human) communication. Concomitantly, a key role of communication functions to convey and induce emotion. The expression and modulation of emotion is also hypothesised to be a key mechanism underlying the evolution of rhythmic synchronisation in humans. Despite the centrality of both rhythm and emotion to human and nonhuman communication, a systematic treatment of the link between them is currently lacking. In this chapter, we bridge these two areas of research by reviewing findings from human and animal studies. These studies how rhythmic properties encode and communicate emotional states and how variations in these properties influence the emotional states of receivers. Taken together, this review of literature demonstrates that the main rhythmic parameter currently evaluated in emotion research is temporal rate (often referred to as frequency rate), which is reliably associated with levels of emotional arousal in all surveyed taxa. We discuss a range of neurobiological mechanisms that could link rhythm and emotion. We also identify several avenues for future research that address current limitations, such as the overwhelming focus on the acoustic modality and frequency rate, while other communication modalities and rhythmic feature remain understudied.

M. L. Gustison (✉)
Department of Psychology, Western University, London, ON, Canada
e-mail: mgustiso@uwo.ca

G.-A. Sauciuc (✉)
Department of Philosophy/Cognitive Science, Lund University, Lund, Sweden
e-mail: gabriela-alina.sauciuc@lucs.lu.se

G.-A. Sauciuc et al. (eds.), *Nature Beats*, Fascinating Life Sciences,
https://doi.org/10.1007/978-3-032-24250-1_11

Abbreviations

ERC-link	The Emotion-Rhythmic Communication link
CNS	Central Nervous System
ANS	Autonomic Nervous System
SNS	Sympathetic branch of the ANS
PSNS	Parasympathetic branch of the ANS

Brain Areas

ACC	Anterior cingulate cortex
AMY	Amygdala
BG	The basal ganglia
BLA	Basolateral amygdala
BrS	Brainstem
CeA	Central nucleus of the amygdala
H	Hypothalamus
Hb	Habenula
HPC	Hippocampus
HPA	Hypothalamic–pituitary-adrenal axis
IC	Insular cortex
LC	Locus coeruleus
NAc	Nucleus accumbens
OFC	Orbitofrontal cortex
P	Pituitary gland
PFC	Prefrontal cortex
PGA	Periaqueductal grey
preBötC	Pre-Bötzinger complex
RAS	The reticular activating system
RN	Raphe nuclei
STN	Subthalamic nucleus
T	Thalamus
VTA	Ventral tegmental area
CPG	Central pattern generator circuit
d	Dorsal
l	Lateral
m	Medial
v	ve

Neuromodulators

5-HT	Serotonin
ACh	Acetylcholine
ACTH	Adrenocorticotropin
ß-END	Beta-endorphin
C	Cortisol
CRH	Corticotropin-releasing hormone
DA	Dopamine
E	Oestrogen
GABA	Gamma aminobutyric acid
MOR	μ-opioid receptors
NE	Norepinephrine
OT	Oxytocin
T	Testosterone
VP	Arginine-vasopressin

11.1 Introduction

As shown in Chap. 3 [1], rhythmic communication is widespread across animal taxa. Decades of research also highlight the centrality of emotion as a widespread aspect of both human and nonhuman communication. In this regard, it has long been conjectured that the *acoustic features* that convey emotion in animal vocalisations constitute the evolutionary foundation for those that convey emotion in human speech and music [2–8]. Research inspired by these ideas, however, has focused on spectral rather than rhythmic qualities. Arguments of remote ancestry have also been made about the capacity of rhythmic *movements* to express emotions in relation to human musical behaviours [9–11]. Moreover, emotion is an important element in several evolutionary theories of human rhythmic synchronisation, which emphasise, for example, its affiliative effects and benefits to emotion regulation, or the formidability of communal rhythmic displays (reviewed in [12]).

Despite the centrality of both rhythm and emotion to human and nonhuman communication, a systematic treatment of the link between them is currently lacking. In this chapter, we address this gap by reviewing relevant findings from these two literatures. To provide the reader with a suitable background for interpreting these findings, we start by discussing how emotion is defined in contemporary emotion research and review relevant neurobiological mechanisms of emotion (Sect. 11.2). We then map the Emotion and Rhythmic Communication link (henceforth ERC-link) at the behavioural level, by focusing on the encoding and decoding of emotion in rhythmic parameters (Sect. 11.3) and on the emotional effects of these rhythmic parameters (Sect. 11.4). Finally, we address the ERC-link from the perspective of underlying neurobiological mechanisms (Sect. 11.5) and end with a brief overview of current limitations in ERC research and suggestions for future work (Sect. 11.6). In line with the broader literature, in this chapter we use rhythmic features to refer to descriptors computed from the data (e.g., onset density, IOI, Hz, etc.) and rhythmic *parameters* to refer to variables that define or manipulate rhythm (e.g., tempo, meter, etc.)

11.2 Emotions: Brief Background on Research Perspectives and Neurobiological Mechanisms

11.2.1 Main Perspectives on Emotion

Emotion has fascinated human thinkers for millennia, yet attempting to capture its essence in a commonly accepted definition has proven to be elusive [13]. Traditionally, emotions have been contrasted to cognition, with most definitional efforts being devoted to capturing the essential features of subjective experiences that are associated with individual emotional states (joy, anger, etc.). Given this focus, the prospect of emotional experiences in nonhuman species has long been controversial, and animal research has concentrated much on emotional behaviours (startle, approach, withdrawal, etc.), eliciting conditions and neurobiological substrates.

For more than a century, emotion research has been dominated by two broad perspectives, known as the *categorical* and the *constructionist* (or *dimensional*) perspectives. The central assumption of the *categorical perspective* is that discrete emotion categories (fear, love, etc.) constitute natural types that are necessarily and sufficiently defined by mutually exclusive patterns of behavioural, physiological and neural responses. Best known within this perspective is the Basic Emotions Theory, which posits that a small number of emotions are biologically and psychologically foundational. These basic emotions are described in three main ways: as tightly linked to survival drives, as the building blocks of 'secondary' emotions, and as cross-culturally universal and phylogenetically continuous (e.g. [14, 15]). Basic emotion theorists, however, disagree on criteria for 'basicness' and the number of basic emotions, with inventories ranging from as few as three (fear, love, rage: [16]) to as many as ten (interest, joy, surprise, sadness, anger, disgust, contempt, fear, shame and guilt: [17]) discrete categories (for a review and comparison of other inventories, see [18]).

By contrast, the *constructionist perspective* - of which *dimensional* models are a central component - operates under the general assumption that a common neurophysiological system underlies a constantly shifting spectrum of emotional states. Accordingly, emotion categorisation is viewed as an emergent and flexible process of psychological construction rather than as a inflexible process imposed by biologically pre-defined modules. A central tenet of this perspective is that the 'core affect' of emotional states are characterised along several *dimensions*, with valence and arousal being the most fundamental (e.g. [19]). Valence captures the variability of emotional states across the pain-pleasure continuum, while arousal captures the variability of emotion in terms of activation levels, the latter of which generally (but not always) overlaps with felt intensity and the degree of behavioural manifestation.

Within this broader debate, there is a perplexing variety of theories regarding the borders between emotion and other kinds of mental states, the fundamental constituents of emotional states, and the primacy of certain constituents over others. Across this vast diversity, however, it is generally accepted that emotions are multifaceted, scalable and valenced states that implicate the organism on a global level and take processing priority over other ongoing processes. Besides valence and arousal, commonly accepted (and evaluated) aspects include subjective feeling, physiological activity at brain and peripheral level, behavioural responses, action tendencies (the urge to carry out a behaviour), eliciting conditions and appraisal (assessing a situation with respect to goals, expectations, controllability, causing agent, etc.). Many of the objectively quantifiable aspects of emotion can be addressed comparatively, in both humans and nonhuman species. Owing to methodological hindrances, however, some aspects can only be addressed in humans (e.g. subjective feeling and appraisal), whereas others (e.g. neural microcircuits) can only be examined in nonhuman species.

Overall, there is a general agreement concerning the *function* of emotions, toward a view that mitigates the traditional contrast between emotion and cognition and acknowledges the adaptive role of emotions in the service of *homeostasis* or *allostasis* (e.g [20, 21].). Both homeostasis and allostasis concern organismic equilibrium in terms of physical integrity, osmotic, chemical and metabolic balance, which fluctuate with changes in the internal and external environment. Homeostasis, however, is conceptualised as a more static and reactive process of maintaining organismic systems within a set point, while allostasis is conceptualised as the dynamic and predictive process of tuning organismic systems *among* multiple viable operating ranges, as appropriate to environmental conditions and life-history demands ([22] and references therein). From this perspective, emotional states are signals to self and others of relevant changes that have happened or are predicted to happen, and these signals mobilise the appropriate physiological and behavioural responses. This means that emotions are adaptive both at the individual and the social levels, where the perceived expressions of emotional behaviour in others (e.g. vocalisations and bodily displays) elicit adaptive emotional responses, which can coordinate group behaviour.

Owing to such regulatory functions, emotional states inherently involve activity changes within the central, somatic and autonomic nervous systems which, in turn, changes orchestrate responses in the neuroendocrine, cardiovascular, respiratory, metabolic and musculoskeletal systems. Many mechanisms underlying these processes are thought to have evolutionarily ancient origins, but they are also dynamic. With the increased complexity of nervous systems, ancestral and widely conserved mechanisms become intimately entwined with evolutionarily novel ones. As emotions implicate homeostatic and allostatic adjustments, prolonged exposure to environments that continuously solicit such adjustments presents the risk of emotional dysregulation. This emotional dysregulation, in turn, is at the core of many major psychiatric pathologies and is often accompanied by somatic, physiological and immune dysregulation.

11.2.2 Brief Overview of the Neurobiological Architecture of Emotion

Recent advances have vastly improved our understanding of the underlying neural mechanisms of emotional processing, by permitting a more precise identification of neuronal populations based on their function, afferent pathways, projection targets, molecular structure and genetic markers. Overall, current research shows that emotions are underpinned by an array of functions (e.g. stimulus valuation, learning and motor drive, etc.). These functions are subserved by a range of processes implemented by multi-layered neural networks that exhibit widely distributed patterns of activation and neuromodulation. Given the comparative focus of this book and the current state of research, which trends toward decreased support for the categorical perspective, in this subsection we give a quick neurobiological overview of valence and arousal and neuromodulation.

The current view in the emotion literature is that arousal and valence are dissociable dimensions that follow a specific order of operations (for reviews, see, e.g. [23–25]). Arousal modulation can be viewed as a process by which sensory and interoceptive input is 'tagged' according to the magnitude of its potential impact on the organism, so that more survival-impactful input is prioritised for additional neural processing and the preparation of context-adaptive responses. Valence further classifies sensory input in terms of its aversive/harmful as opposed to appetitive/beneficial consequences for the organism.

Arousal At the neural level, arousal arises from the interaction of phylogenetically conserved parts of the brain, most prominently the reticular activating system (RAS) and the hypothalamic–pituitary (HP) neuroendocrine system. In concert, these systems regulate the autonomic nervous system (ANS), which is constituted by spinal neuronal hubs (ganglions), preganglionic neurons that project from the central nervous system (CNS) to the ganglions, and postganglionic neurons that project from ganglions to target organs throughout the body.

The RAS spans a system of interconnected brainstem (BrS) nuclei and their projection fibres to sites in the basal forebrain, thalamus and hypothalamus (H). Through an extensive network of afferent and efferent connections, the RAS reaches throughout the brain and body, to activate or deactivate a range of cognitive, neuroendocrine, autonomic and somatic processes [23]. RAS activity is modulated by sensory stimuli that convey threat, novelty, danger or reproductive relevance and by inputs from other emotion processing areas.

The HP complex contributes to physiological and behavioural regulation by synthesising neuromodulators that are transmitted within the brain and by regulating hormone release in the bloodstream. HP activity is modulated by afferent inputs that arise from cortical emotion 'processors' (amygdala—AMY, hippocampus—HPC, etc.) and subcortical sensory and/or motor structures, such as the thalamus, the basal ganglia (BG), the septum and the BrS. Among efferent pathways, the dorsal longitudinal fasciculus connects the HP to various BrS nuclei, thereby playing a crucial role in the integration of ANS and neuromodulator activity.

The ANS regulates physiological responses through the activity of its sympathetic (SNS) and parasympathetic (PSNS) branches. The SNS, which is associated with the 'flight-or-fight' response, is activated by central signals (e.g. from the HP) which stimulate epinephrine release from the adrenal medulla and norepinephrine

(NE) from sympathetic nerve terminals. Released into the bloodstream, these hormones promote a mobilisation response characterised by increased blood pressure (vasoconstriction), heartbeat and breathing rate which, in turn, increases blood and oxygen flow to the muscular systems. By contrast, PSNS is associated with the 'rest-and-digest' response that lowers oxygen use and prepares the organism for rest.

It is important to expand on the above overgeneralisation with three details. First, SNS and PSNS components are not activated in an 'all-or-none' fashion. Instead, pools of SNS or PSNS neurons can be selectively engaged and, thus, their target organs or systems can be independently activated [26]. Second, ANS activity is orchestrated as an allostatic response. This means that the ANS has multiple points of stability, which are achieved through dynamic resource recruitment as a function of situational needs [27]. Third, besides its flexible and adaptive mobilisation to unpredictable environmental factors, the activity of ANS is also organised by multiple intrinsic rhythms on both longer timescales (e.g. the circadian rhythm, the monthly or annual rhythms of reproduction) and on the (sub-)second range, including blood pressure (0.1–0.15 Hz in humans), breathing (0.2 Hz in humans), heartbeat (1.6 Hz at rest) and electroencephalographic oscillations (1–100 Hz). According to the Neurovisceral Integration model [26, 27], the ANS is tonically regulated by the brain, and the scalable and selective suppression of this inhibition allows for context-adaptive mobilisation of ANS functions.

Valence Valence processing encompasses a range of widely distributed processes, with critical hubs in the mesocorticolimbic dopaminergic system (e.g. nucleus accumbens (NAc) and ventral tegmental area (VTA)), nuclei of the thalamus (Th) and amygdala (AMY), the hippocampus (HPC), the insular cortex (IC), the orbitofrontal cortex (OFC), the anterior cingulate cortex (ACC) and the medial prefrontal cortex (mPFC) (for recent reviews, see [23–25, 28–33]). A range of valence-related functions is flexibly implemented within this network. These functions include, for example, the assessment of pain, threat, pleasure and incentive salience, associative valence learning and the formation of affective memories and the selection of adaptive behavioural output and context-dependent stimulus valuation [24, 25, 32].

Processes that promote valence-related functions take place at various levels of brain organisation, from small neuronal assemblies to global inter-region connectivity that spans multiple processing hubs. For example, within the basolateral nuclei of the AMY (BLA), distinct neuronal populations signal valence and subserve associative valence learning, while the PFC integrates multiple information streams to provide context-tuned stimulus valuation [30]. At a more local level, neuronal populations within the BLA are spatially segregated along a posterior-to-anterior axis and respond to appetitive and aversive stimuli, respectively [31]. Similarly, small shifts along various spatial axes in the NAc correspond to shifts in positive or negative valence signalling. Such assemblies, however, exhibit dynamic affective *modes* and, under certain conditions, may also respond to the opposite valence [24, 25]. At both local and global levels, valence processing is underpinned by several neural mechanisms including 'labelled lines' circuits, divergent and/or convergent

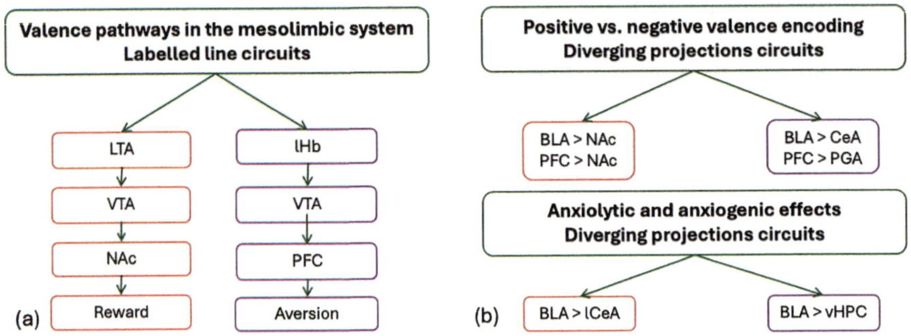

Fig. 11.1 Examples of valence pathways (**a**) Labelled line circuits consisting of relatively separate processing streams and projections to distinct motor output systems (approach or avoidance) are typical for the rapid and energetically effective processing of innately valenced stimuli. (**b**) Diverging projections circuits often arise from emotion processing hubs and can mediate opposite outcomes through associative learning, which increases the flexibility of valence processing. See Box 11.1 for an abbreviation list

pathways, and neuromodulators [24] (for schematic examples of 'labelled lines' and divergent circuits, see Fig. 11.1; for more details on neuromodulators, see below and Sect. 11.5.4).

Neuromodulation A wide range of neuromodulators affect both valence and arousal processing by enhancing or inhibiting neural activity in the circuits described above. In the CNS, the neuromodulators circulate through both synaptic transmission and extrasynaptic diffusion (via the extracellular fluid) and exert their effects through tonic or phasic release, by acting either directly on target tissues where they bind to specialised receptors or by indirectly influencing the synthesis and activity of other (neuro)hormones. They can affect neural activity and communication locally or globally by modulating intrinsic firing patterns and synaptic connectivity. In turn, this contributes to changes in the signal-to-noise ratio—which results in signal boosting, signal attenuation and neuronal (de)synchronisation, the latter of which enables activity coordination within and across widely distributed neural networks.

Providing details on each neuromodulator system that acts on all thinkable aspects of emotional processing is beyond the scope of this brief review. In this chapter, we focus on a select group of well-known systems that, in the light of current knowledge, appear to be of highest relevance to the ERC-link, including monoamines such as norepinephrine (NE), acetylcholine (ACh), serotonin (5-HT) and dopamine (DA); neuropeptides, such as endogenous opioids, oxytocin (OT) and vasopressin (VP); and steroids, such as cortisol (C), testosterone (T) and estrogen (E). In the interest of space and readability, brief overviews of these neuromodulators are provided in Boxes 11.1, 11.2, and 11.3. In Sect. 11.5.4, we describe how these neuromodulator systems participate in the ERC-link, with monoamines being discussed in Sect. 11.5.4.1, neuropeptides in Sect. 11.5.4.2 and lipid-derived steroids in Sect. 11.5.4.3.

11.3 Expression of Emotion in Rhythm

The emotional expression potential of rhythmic activities has been acknowledged for centuries in human modes for expression, ranging from music to voice. In the past several decades, scientific research has begun to systematically explore how this happens, with a focus on how emotional expression varies within and between cultures. A complementary line of research has emerged on how emotional states are expressed in nonhuman animals, with a focus on the evolutionary origins of emotional states and how they are communicated. For both humans and nonhuman species, research oriented exclusively on rhythm as a means to express emotions is relatively sparse, and rhythmic elements are typically included as part of a larger list of measures. There are considerable differences between human and nonhuman research in terms of the research goals, methods and results, and as such, we review these two lines of work separately.

11.3.1 Research with Humans

This section explores the complex ways that humans express their emotions through rhythmic behaviour. Most findings relevant to this section are supplied by interrelated research on voice and music. Much of this work can be broken down by how humans communicate emotions to others and how they interpret emotions in others' behaviour. These encoding and decoding perspectives are important both for voice (reviewed in Sect. 11.3.1.1) and for music (reviewed in Sect. 11.3.1.2). Although the vast majority of work on emotional expression in rhythm comes from acoustic modalities like voice and music, another relevant modality is movement (reviewed in Sect. 11.3.1.3). These strands of research largely explore emotional expression in full-grown adults. Finally, it is critical to take a complementary developmental perspective, which provides new insights into when and how emotional expression through rhythmic behaviour is learned (reviewed in Sect. 11.3.1.4).

11.3.1.1 Encoding and Decoding Emotional States in Speech Rhythm

The voice is a preponderant stimulus in humans' everyday acoustic environment. Classically referred to as 'the mirror to our soul' [34], it is well-known that the human voice can communicate emotional states [5, 35]. The vast majority of research focusing on emotional expression in voice examines temporal properties of speech like speech rate and pauses [5]. Speech rate can be measured as units per duration or overall duration, including voiced segments or full utterances. Pauses include the number or duration of silences within a vocal utterance. In this section, we explore early and current research on the production and perception of rhythmic cues in speech and voice, as well as some theoretical models used to explore emotion communication.

Early evidence that the velocity of speech encodes emotional expression comes from two main lines of research focused on the encoding and decoding of emotion

[3, 5, 36]. The first line of research, focused on emotion encoding, has examined whether discrete emotions can be communicated in vocal cues. These studies involve natural and simulated vocal expression and also vocal expression during induced emotions or 'form-function' mappings [36]. In a foundational study by Scherer and colleagues [37], professional actors were told to speak in ways that reflected an expanded set of 14 emotional states (hot anger, cold anger, panic, anxiety, despair, sadness, elation, happiness, interest, boredom, shame, pride, disgust and contempt). Distinct patterns emerged. Decreased speech rate was associated with sadness and to a lesser extent despair. On the other hand, an increase in speech rate was associated with intense emotions like anger, panic, elation and happiness. Overall, these studies suggest that increased speech rate occurs during high-arousal emotions like anger and happiness, while speech rate is decreased during low-arousal emotions like sadness and tenderness [5].

Another temporal property that appears to correspond to emotional states is the regularity of duration in speech utterances. Negative emotions like anger, fear and sadness occur during more irregular speech than do positive emotions like happiness and tenderness [5]. Some temporal properties seem to differ across negative emotions (reviewed in [5]). For example, anger tends to be associated with a small proportion of pauses, while sadness is associated with a larger proportion of pauses compared to other emotions. Together, these early studies on emotion encoding revealed that discrete emotions can be communicated by vocal cues related to rhythm.

The second line of early research focused on the decoding of emotional states in vocal cues. Much of this work was built on basic research that examined perceptions of emotion in synthetic auditory stimuli [38]. In one key study by Scherer and Oshinsky [39], the tempo of electronically synthesised tone sequences was the most powerful predictor for emotion ratings, accounting for around 30% of the variance across all rating scales. Slower tempos were associated with listener ratings of sadness, boredom and disgust, and faster tempos were associated with ratings of surprise, happiness, fear and anger.

Studies on emotion detection in human speech support these findings from artificial stimuli. Listeners interpret emotional states from voice cues at better than chance levels, with accuracy rates higher in within-cultural than cross-cultural contexts [3, 5, 35, 40]. In Banse and Scherer [37], for example, professional actors encoded 14 different emotions into speech utterances. Listeners inferred these emotions accurately with a recognition rate of around 50%, with shame and disgust being the most difficult to identify. In general, this study and others (see [41] for an example with vocalisations from naturalistic interactions) show that anger and sadness are easiest to identify from voice cues, followed by fear and happiness [5].

Although few studies on emotion decoding examine specific acoustic features, separate lines of inferential work show the importance of temporal features like speech rate [3, 36]. Studies that manipulate speech stimuli have helped to parse out the relative importance of different acoustic properties [3]. Such studies frequently use low-pass filtering techniques to mask verbal information, and the results show that the temporal structure is critical for emotion decoding (reviewed in [3]). In sum, these studies provide further support for the notion that discrete emotions can be communicated in vocal cues that are related to rhythm.

In recent years, research on emotional expression in voice has expanded in several ways. One key theme has been to advance the classification techniques used to detect emotions from specific acoustic features [42]. This work is derived from speech databases compiled from natural and simulated emotional speech, in addition to speech recorded during induced emotions. This body of work uses both linear classifiers (e.g. logistic regression, Bayes classifiers and linear support vector machines) and non-linear classifiers (e.g. hidden Markov models, *k*-nearest neighbour and non-linear support vector machines) [42, 43]. Current approaches are moving towards machine learning and neural network tools [44].

Another approach has been to generate synthesised speech based on existing corpora, which shows that listeners use duration features (e.g. duration of vocal units) to differentiate between basic emotions like anger (shortest duration), joy and sadness (longest duration) [45]. Overall, comprehensive reviews of emotion classification studies show that temporal features like speaking rate, sound unit duration and articulation rate contain critical information to differentiate between emotions [42, 43]. However, it is still important to keep in mind that the precise ways in which temporal features differentiate between emotions may depend on language and cultural 'display rules' for emotion. Research on the recognition rates of basic emotions across languages has shown that speaking rate can help differentiate some emotions like sadness and fear across languages but not other emotions like anger [46].

Another key theme in recent years has been to explore how unique emotional dimensions (e.g. activation, valence, potency and intensity) are encoded and perceived in speech. Activation and valence are the two most common dimensions studied, referring to the subjective energy and pleasure of emotions, respectively (for details, see Sect. 11.2). Potency corresponds to the cognitive appraisal of control, and intensity corresponds to the quantity of an emotion. Research shows that these emotion dimensions are associated with a wide array of acoustic features [47]. Activation, potency and intensity are negatively associated with speaking rate, while valence is positively correlated with speaking rate. When combined with other acoustic features, temporal features like speech rate help listeners differentiate between these dimensions. Temporal features can also enable differentiation between emotions with a similar valence. For example, several studies show that speech rate is increased during positive emotions like pride, relief and joy and decreased during pleasure, contentment and admiration [48]. Such findings suggest that within a similar valence, speech rate encodes different levels of arousal.

The research summarised above has inspired several theoretical models to predict how emotional states are encoded in speech and other auditory stimuli. The Component Process model of vocal affect expression, proposed in the 1980s [35, 40], puts forth predictions for how patterns in voice cues map onto different emotions. The reasoning behind this theory is that emotional states are a response to cognitive appraisals of a situation that impact the ANS. These physiological changes, in turn, influence the vocal anatomy that controls voice production.

Emotional states are thought to be encoded within three main voice types: lax-tense, narrow-wide and full-thin. The best studied of these, lax-tense, involves the rhythmic parameter of tempo, or speech rate. From this model, speech rate is hypothesised to increase in a tense voice during high-arousal emotional states like joy/elation, fear/terror, irritation/cold anger, rage/hot anger. Conversely, speech rate is hypothesised to decrease in a lax voice that occurs during low-arousal emotional states like sadness/dejection. Another model, the Brunswikian Lens Model [36], is more comprehensive in that it includes both encoding and decoding of emotion communication in voice. This model provides a framework to explore how information is inferred by the listener through distal voice cues and the perception of these cues. Application and testing of these models will be an important goal in future work.

11.3.1.2 Encoding and Decoding Emotional States in Musical Rhythm

Aside from speech, research on emotional states in human behaviour has focused extensively on singing and instrumental music performance [49]. The majority of research focusing on emotional expression in music rhythm examines temporal properties like tempo, articulation and timing [5, 50]. Tempo can be measured by the number of beats per min, or as event density measured by the number of note onsets per second. Articulation refers to the proportion of sound to silence in adjacent notes, as well as the clarity and stability of the beat (i.e. pulse clarity). Timing refers to the deviation, or variation, in tempo and rhythm and includes measures like attack time (i.e. duration from initial note onset to its peak time) and rhythm novelty (i.e. degree of temporal repetition in rhythm). In this section, we explore early and current research on the production and perception of rhythmic cues in music performance, including instrumental performance, singing and combinations of the two. We also explore theoretical frameworks that compare emotion communication in music and speech.

Similar to vocal expression, early evidence that temporal properties of music performance correspond to emotional expression came from two main lines of research on the encoding and decoding of emotion [3, 5]. The first line of research, on emotion encoding, focused on demonstrations of whether discrete (e.g. joy, sadness and anger) or dimensional (e.g. valence, energy/arousal and tension) aspects of emotions could be communicated in musical elements. Overall, these studies suggest that increased music tempo occurs during anger and happiness, while music tempo is decreased during sadness and tenderness [5].

Another temporal property that appears to correspond to emotional states is the timing variation. Negative emotions like anger, fear and sadness are encoded in more irregular tempos than positive emotions like happiness and tenderness [5]. These patterns mirror what has been found with speech rate. However, there are also temporal properties that are unique to musical performance (reviewed in [5]). For example, staccato articulation has been associated with anger, fear and happiness, while legato articulation has been associated with sadness and tenderness. Happiness is further associated with variability in articulation. Together, these studies revealed that discrete emotions can be communicated within musical properties that are related to rhythm.

The second line of early research focused on the timing and accuracy of decoding emotional states in music. As with speech research, it is thought that temporal properties are important in the perception of emotion in other types of auditory stimuli like music [39]. Early studies of cross-cultural music showed that listeners are sensitive to both tempo and rhythmic complexity in unfamiliar music, which helps them to perceive basic emotions like joy, sadness and anger [51]. For example, tempo ratings are positively associated with ratings of joy and negatively associated with ratings of sadness. Rhythmic complexity ratings are found to be negatively associated with ratings of joy and positively associated with ratings of sadness. In sum, these studies provide further support for the notion that discrete emotions can be communicated in musical properties related to rhythm.

In recent years, research on emotional expression in music has expanded in several ways. One key theme has been to advance understanding of universal elements of music that communicate similar emotions across cultures and music genres [52, 53]. This work shows that people readily recognise intended emotions in both familiar and unfamiliar music. For example, Japanese people listening to Japanese, Western and Hindustani music recognise joy and sadness based on features like tempo [54]. Across all music sets, music pieces perceived as slow were rated high in sadness, while music pieces perceived as fast were rated high in joy. In another study, it was found that Mafas people, an indigenous population in Cameroon, judged Western music with faster tempos as happy and music with slower tempos as fearful [55]. In opera singing, tempo appears to reflect emotions associated with different arousal levels [56]. For example, sadness and tenderness were associated with slower tempos, while anger was associated with faster tempos. Together, these studies show that many basic emotions have similar patterns of musical tempo across cultures.

Another key theme has been to explore how a variety of temporal music elements mediate both encoding and decoding of emotions. These associations, between the performance and perception of emotional expression in music, have been studied in depth on professional bowed-string musicians and listeners across cultures [52]. In this cross-cultural research, musicians were asked to convey 11 basic (e.g. happiness, anger and fear) and non-basic (e.g. humour) emotions in Swedish folk music, Hindustani classical music, Japanese traditional music and Western classical music. Then, Swedish, Indian and Japanese participants judged the emotional content of these music pieces. This research shows that slower attack times are associated with affection, neutrality and peacefulness in music production and perception. On the other hand, faster attack times are associated with anger and fear. In addition, pulse clarity is positively associated with anger, happiness and humour but negatively associated with longing, peacefulness and spirituality. Low event densities occur with production and perception of peacefulness and sadness, while high event densities occur with anger and fear. Finally, rhythm complexity (i.e. novelty in tempo repetitions) tends to be higher with emotional states like affection, longing, neutrality and peacefulness but lower with anger, happiness and humour. In sum, several temporal properties related to rhythm may help facilitate both the production and perception of emotion in music.

The research summarised above has inspired several comparisons between music and other auditory stimuli like speech. There is a great degree of overlap in how emotions are communicated in speech and music. For example, professional singers can use tempo in similar ways to speech to convey basic emotions like fear (fast tempo) and sorrow (slow tempo) [3]. Moreover, this ability is not limited to professional musicians. Even nonmusicians use tempo to express basic emotions (e.g. faster tempo for joy and anger and slower tempo for peace and sadness) when producing music (e.g. [57].). This line of work has led to the hypothesis that speech and music represent different channels for emotion that use similar cues [5]. The listener's ability to produce and perceive emotion in music may have come from conserved mechanisms for auditory processing that evolved to be sensitive to a wide range of auditory cues [51, 58]. This acoustic sensitivity may be especially true for more 'basic' emotions, where there is a high degree of universal expression in auditory cues across cultures [50]. This line of thinking also aligns with the Lens-model approach, which puts forth that auditory cues for emotion are neither universal nor culture-specific [50, 59]. According to the Lens-model approach, the universality of emotion expression depends on the emotion. Basic emotions may be easier to communicate across modalities because they are expressed through generalisable acoustic cues like tempo, while other types of emotions may rely on culturally specific cues. Thus, music and speech can be used in similar ways to communicate emotions, but this communication may be more straightforward for some emotions than others.

Although there is overlap in how emotions are expressed in music and speech, there are also notable differences that alter how effective these channels are in different contexts. Listeners are better at identifying emotions in familiar within-cultural speech than in musical performances, and yet, they are also better at identifying emotions in musical performances than in unfamiliar cross-cultural speech [5]. Thus, it may be the case that music provides a cross-cultural way to express emotion, which may be facilitated by unique cues not found in speech. Professional singers, for example, can use music-specific acoustic cues like inter-syllable interval length and vibrato to communicate emotion (e.g. longer inter-syllable intervals for fear) [3, 56]. This aligns with the Cue Redundancy model, which puts forth that musicians use a wide range of conventional and unique acoustic cues to express emotional meaning [51, 54]. Many of these acoustic cues (e.g. tempo, complexity, loudness and timbre) are useful because they do not require musical knowledge and a combination of cues can be used to interpret emotional meaning [54]. Taken together, the Lens Model approach and Cue Redundancy Model provide frameworks for how emotions are expressed in music and how this expression compares with other channels like speech. Although there is a large degree of overlap in emotional expression, this overlap is not entirely universal and varies with emotion and culture.

11.3.1.3 Beyond Speech and Music: Emotion Expression Through Movement

Voice and music have received a great deal of attention regarding emotional expression through rhythm, but there is also a growing interest in other modalities. Movement of different parts of the body provides an alternative way to express emotion through rhythm. Such movements can occur as a complement to music or on their own. Research shows that movement is an important part of emotional expression for performing musicians. When asked to perform musical excerpts with emotional expression, as compared to without expression, ensemble musicians engaged in more synchronised body sway [60]. This body synchrony also tended to increase with happy music rather than sad music. Synchrony of movement can also happen without music. For example, groups induced with positive-valenced sounds produced faster and more synchronised finger oscillation movements than when induced with negative-valenced sounds [61].

Research shows that movement shares properties with music for emotional expression, even when they occur separately. In a cross-cultural study, US and Cambodian participants created artificial stimuli of bouncing balls and piano melodies by manipulating a range of dynamic features, some of which were related to rhythm (e.g. rate and jitter) [10]. Movement and music had similar features for each of the five emotions studied (i.e. sad, peaceful, happy, angry and scared). For example, 'sad' had a low rate (beats per minute) and low jitter (beat regularity) in both modalities. These studies suggest that rhythmic body movement is another modality, in addition to speech and music, for human emotional expression.

Another type of form of movement shown to be important for emotional communication is facial expression. Historically, facial expression has been studied in static forms, but there is increasing interest in facial expression as a dynamic movement involving temporal cues, with research showing that the speed of facial expression influences perceptions of emotion [62]. For example, people are more likely to perceive anger or happiness in faster facial movements and sadness in slower movements [63]. This also occurs when facial movements are artificially manipulated. People are better at recognising anger and happiness when movies of facial expressions are sped up, and they are better at recognising sadness when facial expressions are slowed down [63]. When taken together, these lines of research suggest that temporal cues are important for emotion communication in a range of behavioural modalities, including voice, music performance, body movement and facial expression.

11.3.1.4 Development of Emotional Expression in Rhythmic Behaviours

The ability to encode and decode emotions in rhythmic behaviours begins to develop in childhood. Early research from the 1980s and 90s showed that young children (around 4 years old) can identify basic emotions in vocal expression and music, and this ability improves with age (reviewed in [5]). Even infants differentially respond to emotions in vocal, as well as musical stimuli. This work has suggested that rhythmic acoustic cues may help children identify emotions (e.g. [64]). These findings are supported by research on adults, who exaggerate acoustic cues like tempo when communicating specific emotions (e.g. love-comfort) in infant-directed speech [65]. Together, this early work suggests that children can decode emotions in vocal and musical expression. There were few studies, however, on the specific rhythmic cues that children use in emotion encoding and decoding or how this usage changes across development.

More recent studies on music production and perception in children show that rhythmic cues like tempo are used to differentiate emotions, and the use of these cues varies with age. In one study by Kragness and colleagues [66], three age groups (3-, 5- and 7-year-olds) of children produced musical sequences to communicate emotions that differed in valence (e.g. anger and joy) and arousal (e.g. sadness and anger). A key result was that there were developmental differences in the use of acoustic cues. Of the three acoustic parameters that could be altered—tempo, loudness and articulation—tempo and loudness were used by all age groups to differentiate emotions, while only older children used articulation. Interestingly, however, older children were better at diversifying the three acoustic cues to differentiate emotions. For example, the differences in tempo between negative (slower) and positive (faster) valence emotions became more extreme in older children. In another study, children aged 4–6 years old were able to identify intended emotions in happy (fast tempo) and sad (slow tempo) tracks of child-directed vocal and instrumental music [67]. In the same study, the ability to identify emotions varied with age. Recognition accuracy tended to increase with age (from 3 to 6 years old), and 3-year-olds reliably identified happy music but not sad music.

Together, early and recent developmental studies suggest that tempo is one of the earliest rhythmic cues used by children for emotion expression and recognition. Children use tempo even as toddlers, and they become more skilled in using auditory cues like tempo as they get older. Such findings have been used as developmental evidence to support the hypothesis that temporal properties in music and non-music sounds emerged from preexisting adaptations for auditory scene analysis [58]. However, it is still unclear why and how emotional expression in different behaviours could benefit survival in social groups. Future work will continue to illuminate why an ability to use temporal (and other cues) to communicate emotion exists in early development and why this ability becomes increasingly sophisticated with age.

11.3.2 Research with Nonhuman Animals

Nonhuman animals are thought to experience emotions, or at least basic emotions, but the subjective experience of these emotions is debated [7]. In other words, animals show neurophysiological, behavioural and cognitive signs that they experience different emotions in different contexts, but it is unclear whether their experiences can or even should map onto discrete emotions explored in human studies [7]. To circumvent this issue, animal studies tend to apply a dimensional approach (reviewed in Sects. 11.2 and 11.3.1), whereby arousal and valence have the capacity to capture all discrete emotions without making assumptions based on the human experience [7, 68]. In this dimensional approach, also known as the Bipolar Valence-Arousal Model' negative emotions make up an unpleasant-defensive motivational system and positive emotions make up a pleasant-appetitive motivational system [104]. Emotional states within these two valence systems then vary from low to high arousal levels.

This section reviews how nonhuman animals express emotions in rhythmic behaviour from a dimensional perspective. Most research relevant to this section involves the expression of arousal in vocalisations (reviewed in Sect. 11.3.2.1). However, a considerable amount of work has also been carried out on the vocal expression of valence (reviewed in Sect. 11.3.2.2). As in humans, nonhuman animals engage in rhythmic behaviour in modalities other than the acoustic domain, and so a final area of review explores what is known about emotional expression through movement (reviewed in Sect. 11.3.2.3).

11.3.2.1 Expression of Arousal in Vocalisations

Variation in arousal levels occurs across a wide range of contexts in nonhuman animals. Situations thought to cause distress are one of the most common. In naturalistic environments, distressing contexts include environmental disturbances like predator threat, intruder threat and unusual events, during which arousal and alertness levels are thought to increase with urgency (e.g. predator distance) and risk (e.g. predator type) [69–73]. For many animals, interactions with humans are also thought to cause high arousal, including human presence [74], animal capture [75], handling [76, 77] and physical restraint [78–80].

Other common situations thought to affect arousal involve social interaction. Many species become stressed when isolated from, and excited when reunited with, their social group [79, 81–84]. Agonistic encounters are another common type of social context thought to increase arousal, especially when the intensity increases (e.g. threat vs contact aggression) [85–88]. Other high-arousal social contexts include group instability [89], group movement [90], infant handling [90] and mating events [91]. Bringing all these situations together, arousal-inducing contexts are typically those that elicit stress or alertness.

Although it is not possible to ask nonhuman animals about their feelings, researchers have relied on other strategies to infer affective dimensions like arousal. The simplest strategy is form-function mapping, where inferences from human experience translate to nonhuman experience [7]. For example, it can be assumed that stressful experiences cause elevated arousal levels across species. Other strategies involve the use of behavioural or physiological variables to validate these form-function mappings. Typical behavioural indicators of elevated arousal include increased activity levels that involve head, limb or body movements [74, 76, 80, 84, 92, 93]. Other behavioural indicators include postural displays like elevated head, forward ears, upright tail and bristled fur [73, 81, 84, 94].

A handful of studies have included physiological indicators of arousal such as heart rate, respiration and heart rate variability. Elevated arousal is associated with increased heart rate and respiration, and decreased heart rate variability [84, 95, 96].

Other physiological indicators include circulating corticosterone and adrenocorticotropin levels, which are thought to be positively associated with arousal [78, 83]. Researchers have even manipulated corticosterone levels to induce higher arousal in animals (e.g. zebra finches (*Taeniopygia guttata*) [83]). These behavioural and physiological indicators have helped to validate that several of the distressing and social situations mentioned above induce increased arousal in nonhuman animals.

There are many potential behavioural outlets for nonhuman animals to express arousal, and the best studied of these is vocalisation (reviewed in [7]). Researchers have identified a range of rhythmic components in vocalisation that appear to reflect arousal. One of the most basic of these components is the vocalisation quantity, often measured by call duration. However, the way in which these measures map onto arousal is not universal, and instead, these associations vary by species and even context (reviewed in [7]). One line of evidence shows that longer duration calls, bouts and songs are associated with elevated arousal, which is found in common marmoset (*Callithrix jacchus*) phee calls [82], common raven (*Corvus corax*) defensive calls [88], domestic kitten (*Felis silvestris catus*) isolation calls [77], hamadryas baboon (*Papio ursinus*) grunts [90], piglet (*Sus scrofa*) distress calls [79], silver fox (*Vulpes vulpes*) human-directed calls [74], giant panda (*Ailuropoda melanoleuca*) infant calls [76], mongoose (*Suricatta suricatta*) alarm call bouts [71], rock hyrax (*Procavia capensis*) calls [73], tree shrew (*Tupaia belangeri*) chitters [94] and zebra finch songs [83]. And yet, there are examples where call duration is not associated with arousal levels, for example, in horse (*Equus caballus*) whinnies [95] and piglet grunts and screams [80]. Therefore, longer call durations are typically associated with increased arousal, but this pattern is not ubiquitous.

Another common rhythmic component studied alongside arousal is the number of calls, sometimes referred to as 'call rates', within a call bout or song. There is good evidence showing that higher numbers of calls per bout or bouts per song occur during elevated arousal. Notable examples include bottlenose dolphin (*Tursiops truncatus*) signature whistle loops [75], common marmoset phee calls [82], common raven defensive calls [88], fox human-directed calls [74], giant panda infant calls [76], goat (*Capra hircus*) food and isolation calls [84], greater false vampire bat (*Megaderma lyra*) agonistic call syllables and sequences [86], hamadryas baboon grunts [90], piglet distress calls, screams and grunts [79, 80], rock hyraxes songs [73], tree shrew chatter call syllables [94]. Like call duration, however, patterns that involve call or bout numbers and arousal have some mixed evidence. For example, zebra finches produce fewer calls during isolation versus control contexts [83], and domestic pigs produce fewer grunts during a separation and restraint context as compared to separation only [78]. Combining these results with call duration suggests that vocalisation quantity increases with arousal levels but that these patterns are species-specific.

Similar to human speech, tempo and its association with emotional state are also studied in nonhuman animal vocal communication. Measures of tempo in animal vocalisations include intercall interval lengths and rate of calls per unit of time. In contrast to duration-based measures, studies show that increased tempo is a universal indicator of increased arousal across a wide range of contexts and species (reviewed in [7, 68]). Shorter intercall interval lengths and faster call rates are more likely to occur during distressing contexts such as capture-release in bottlenose dolphins [75], human handling in cats [77], human intruders in domestic dogs (*Canis lupus familiaris*) [72], and new environments in tree shrews [94], as well as high predator risk in marmots [70], mongoose [71] and rock hyraxes [73].

In addition, faster calling tempos occur during a variety of 'high-arousal' social contexts such as conflict. Species that demonstrate this form-function mapping include banded wrens (*Thryophilus pleurostictus*) [87], Cape fur seals (*Arctocephalus pusillus pusillus*) [91], chimpanzees (*Pan troglodytes*) [85], greater false vampire bats [86], rock hyraxes [73], and spotted hyenas [81]. For example, greater false vampire bats produce faster call rates during high compared to low intensity agonistic displays [86]. Other social situations associated with faster calling tempos include isolation (common marmosets [82]), group movement (e.g. chacma baboons (*Papio ursinus*) [97]; hamadryas baboons [90]), infant handling (chacma baboons [97]; hamadryas baboons [90]) and play (domestic dogs [72]). These studies suggest that calling tempo consistently reflects arousal level.

More research will need to be done to understand why calling tempo appears to be a universal acoustic feature that reflects nonhuman animal arousal. Some promising explanations involve physiological mechanisms. For instance, increased respiration rate occurs during heightened arousal, which would have an immediate influence on behaviours that involve breathing like vocalisation [7, 68]. Whether these physiological changes also translate to more complex forms of rhythm in vocalisation will be a promising question for future research.

11.3.2.2 Expression of Valence in Vocalisations

Valence is the other primary emotion dimension that is studied in nonhuman animals. Although there is less research on the expression of valence in animal vocalisations than on arousal expression, valence has been receiving more attention in recent years (reviewed in [7]). Valence is similar to arousal in that it is studied across a wide range of contexts and species. Many studies focus on contexts thought to elicit negative valence, such as physical discomfort [98, 99], social isolation [68, 99] and predator risk [98]. Other studies focus on contexts thought to elicit positive valence, such as social contact [99], food reward [92], mating [98] and social reunion [98, 99]. A limitation of this approach, however, is that interpretations are restricted to one end of the valence spectrum, either negative or positive.

A common research approach has been to compare vocalisations produced during contexts on different ends of the valence spectrum. Comparisons of affiliative interactions (positive valence) with agonistic or dominance interactions (negative valence) have been conducted in several species, including African elephants (*Loxodonta africana*) [100], Cape fur seals [91], domestic dogs [101, 102], lab rats (*Rattus norvegicus*) [103] and wild boars [92]. Other social comparisons involve isolation (negative valence) and affiliative interactions (positive valence) like reunion in horses [95] and social foraging in pigs [96]. Less commonly, comparisons are made between different feeding situations, like food types in spotted pacas (*Cuniculus paca*) [93] and food reward versus frustration in goats [84]. Or, a comparison is made between individuals that have contrasting responses to the same situation. Aggressive strains of silver foxes have negative responses to human interaction, while tame strains have positive responses [74]. Bridging these studies, there is much potential to explore the vocal expression of valence by investigating a range of social contexts.

Similar to the arousal dimension, researchers have used inferences from form-function mappings, as well as behavioural and physiological metrics, to place different contexts on the positive to negative valence spectrum. Behavioural indicators of negative valence include defensive body postures (e.g. crouched body, lowered head and freezing) and vocalisations [96, 98]. Behavioural indicators of positive valence include chewing, forward ears and lifted tails [84, 96]. In some laboratory environments, researchers can even measure how much an individual seeks or avoids neural stimulation in specific brain areas [104]. Researchers can also use physiological manipulations to induce positive or negative valence states. One strategy is through neural stimulation of pleasurable or aversive hedonic brain regions [104]. Another strategy is to activate the dopaminergic (positive) or cholinergic (negative) systems [98]. Similar to arousal, these behavioural and physiological indicators help to validate the valence of different social and environmental contexts.

Vocalisation quantity, often measured by call duration, is a common rhythmic feature studied in relation to valence expression. In reviews of several species, there is strong evidence to suggest that shorter vocalisations reflect positive valence while longer vocalisations reflect negative valence [7, 68]. These trends have been shown in dog growls [100, 101], human-directed fox calls [74], horse whinnies [95], pig grunts [96], piglet calls [99], rat ultrasonic vocalisations [98] and squirrel monkey (*Saimiri* spp.) shrieks and bawls [104]. There is even some support for this trend when controlling for arousal level. For example, wild boars produce longer call types (screams and squeals) in negative contexts and shorter call types (grunts) in positive contexts, even while controlling for body activity as a behavioural indicator of arousal level [92].

On the other hand, there are some species for which the call duration does not fit these trends. In African elephants, rumble calls have a longer duration in both negative and positive valence contexts as compared to rumble calls given in neutral contexts [100]. One interpretation for this is that negative valence contexts had higher affect intensity than the positive valence contexts, which could mean that call duration was influenced by arousal levels. In the spotted paca, valence expression through call duration depended on the call type [93]. Roar calls and snort phrases were longer in negative and ambiguous contexts than in positive contexts, but the opposite was true for bark phrases. When combined, these examples present strong evidence that call duration is a reliable acoustic cue for valence expression. However, it is important to consider instances where valence expression depends on the species [68]. In future studies, it will be important to include multiple call types and to control for arousal level.

Aside from call duration, other rhythmic feature have received less attention in valence expression studies. There are a handful of studies on 'call rates', and these studies yield mixed findings. In domestic dogs, positive play contexts are associated with higher numbers of growls than negative aggression contexts, but the opposite is true for barking bouts [101]. Spotted pacas produce more snort phrases in a negative feeding condition than other call types, but they also produce more bark phrases in a positive feeding condition [93]. Moreover, in laboratory rats, higher rates of frequency-modulated 50-kHz calls occurred during positive affective states, while higher rates of flat 22-kHz calls occurred during negative affective states [103].

Mixed evidence is also found for tempo. In domestic dogs, inter-growl intervals are shorter during positive play contexts than during negative aggressive contexts, which suggests that positive valence is associated with a faster tempo [101]. However, in spotted pacas, bark phrases are faster in negative feeding contexts than in positive feeding contexts, but there was no evidence that roar or snort phrase tempos varied by the valence of the context [93]. From these species, it appears that valence expression through call rate or tempo is dependent on the call type and species rather than showing a general pattern across taxa.

Valence and arousal expression are typically studied separately. However, some recent research directions aim to compare and integrate these two dimensions. Two main hypotheses have been proposed to explain how both dimensions can be expressed in vocalisation. The 'segregation of information' hypothesis puts forth that arousal and valence can be communicated simultaneously in vocalisations through the use of independent acoustic cues [95]. This hypothesis is supported in at least a few different species. In domestic horses, arousal seems to be expressed primarily in F0 fundamental frequency and the energy spectrum, while valence is expressed in G0 fundamental frequency and call duration [95]. In foxes, calling rates and proportion of time calling are linked to arousal, while amplitude frequency differs mostly by valence [74]. And, in wild boars, call duration varies with valence even while controlling for arousal levels [92].

Conversely, the 'trade-off' hypothesis puts forth that acoustic cues in vocalisations either communicate one dimension or the other, and the acoustic cues are not independent [95]. Potential support for this hypothesis comes from African elephants. Elephant rumble calls, and their durations, seem to primarily communicate affect intensity rather than valence [100]. Although there appears to be more preliminary evidence for the segregation of information hypothesis, future work is needed to disentangle the two hypotheses. A key challenge for this future work will be to infer negative and positive valence for similar arousal levels, since negative valence contexts tend to be more intense than positive contexts [7].

Compared to arousal, much less is known about how vocalisations communicate valence. Based on current findings, it appears that the acoustic cues for valence are species-specific, while arousal cues are more universal across species [68]. However, these interpretations are based mostly upon non-rhythmic acoustic features, as few rhythmic features are studied alongside valence. Promising directions for future research will include both valence and arousal in studies of animal communication, as well as include a variety of temporal acoustic features.

11.3.2.3 Emotion Expression Through Movement

Although nonhuman animal vocalisation has received a great deal of attention regarding emotional expression, there is also a growing interest in other modalities where rhythm can occur. Similar to humans, movement of different body parts provides nonhuman animals with an alternative way to express emotion through rhythm. Animals engage in a variety of rhythmic movements, including walking, flying and chewing, which are controlled by the CNS and are therefore thought to be sensitive to internal states [105]. On the other hand, it is uncommon to study animal movement from the perspective of emotional expression. One of the exceptions to this rule is dancing behaviour in birds and specifically cordon-bleus (*Uraeginthus* spp.).

Cordon-bleus are a group of small birds from sub-Saharan Africa known for tap-dancing displays [106]. Their tap-dancing displays involve multimodal rhythmic elements within the stepping, bobbing and singing parts of the display. Although cordon-bleu tap-dancing has not been linked directly to arousal or valence, tap-dancing varies across contexts in ways that suggest a link with emotional expression. Captive blue-capped cordon-bleus (*U. cyanocephalus*) will intensify their tap-dancing display when they are on the same perch as their mate [106]. The intensified displays involve more steps per bob and a faster bobbing tempo. Step amplitude and stepping tempo also increase during courtship contexts [107]. Indeed, in the wild, male blue-capped cordon-bleus and red-cheeked cordon-bleus (*U. bengalus*) are more likely to give multimodal tap-dancing displays when a female is on the same branch [108]. It is thought that this shift in display during courtship may relate to increases in motivation [106], which could also be linked to a higher arousal level or positive valence.

Interestingly, the finding that movement quantity and tempo increase during motivating contexts parallels research on vocal expression, where increases in arousal also occur with increases in tempo (reviewed in Sect. 11.3.3.1). It may be the case that nonhuman animal movement and vocalisation rhythms provide cues about internal emotional states like arousal and, thus, could involve a signalling function (as described in Sect. 11.3.2 regarding human communication). A promising future direction will be to directly explore the role of rhythmic parameters in the expression of arousal and valence in animal movements.

11.4 The Influence of Rhythm on Emotion

The emotion-induction potential of rhythmic activities has been acknowledged for centuries and abundantly addressed in treatises of poetry, rhetorics, musical composition, artistic performance or philosophy. While some of these ideas continue to enjoy popularity in modern conventional wisdom, scientific research focusing on whether and how rhythmic stimulation influences the receiver's emotions remains scarce, as most existing studies embed rhythm within complex musical stimuli rather than testing it independently. For both humans and nonhuman species, findings relevant to this question come mainly from studies that incidentally quantify the temporal properties of eliciting stimuli while addressing other research questions. Due to considerable divergences between the human and nonhuman research with respect to research questions, approaches and measured responses, we will review human and nonhuman literature separately.

11.4.1 Research with Humans

Most findings of relevance to this section are supplied by research on music and emotion - an interdisciplinary field of research with origins in music psychology. A considerable portion of this research uses psychophysiological measures to examine the kinds of emotions or emotion phenomena that are elicited when participants are exposed to musical pieces (reviewed in Sect. 11.4.1.1). Another relevant strand in this field addresses music and emotion from a predictive coding perspective, combining behavioural and neural measures with computational modelling and a cognitive science perspective (reviewed in Sect. 11.4.1.2). Finally, a third relevant strand brings into the mix the fields of social psychology and social neuroscience and focuses on the role of rhythmic synchronisation in the elicitation of positive affect and prosocial emotions (reviewed in Sect. 11.4.1.3).

11.4.1.1 Emotion-Elicitation Studies Using Rhythmic Stimuli

A common method for examining the emotional effects of various forms of stimulation is to present individuals with purpose-designed stimuli (sounds, images, short films, etc.) and assess the emotion-elicitation potential of the stimuli through physiological (heart rate, skin conductance, etc.) or subjective (e.g. ratings of experienced arousal and valence or discrete emotions) measures. In a small number of studies, this approach has been used to determine how variations in the rhythmic parameters of auditory stimuli influence human emotional states, primarily using musical excerpts with variable tempos. Overall, this research suggests that (musical) rhythm influences experienced valence and arousal. On the valence dimension, faster tempo tends to be associated with higher ratings of experienced happiness, whereas slower tempo tends to be associated with higher ratings of experienced sadness. This relationship, however, is modulated by other acoustic (e.g. mode and pitch) and exogenous (e.g. familiarity) factors, as shown by a study with American college students, in which the positive association between fast tempo and happiness was reversed when musical pieces were rendered in minor mode [109].

On the arousal dimension, a consistent finding relayed by both self-report and physiological variables is that a faster tempo is associated with higher levels of arousal, as compared to a slow tempo and vice versa [110–114]. Fast rhythms are consistently associated with higher ratings of emotional intensity, as well as increases in skin conductance, breathing and heart rate, while slow rhythms tend to have the opposite effect [111–113, 115–117]. Interestingly, these effects extend across cross-cultural samples [116] and to non-musical rhythmic stimuli like environmental and metronome sounds [111, 118]. Moreover, rhythmic tempo is shown to drive the physiological synchronisation of emotional states within musical concert audiences [114] (see also Sect. 11.5.1.3). The emotional effects of other rhythmic parameters than tempo remain unexplored, except for a handful of studies showing that more pronounced rhythmic accentuation and articulation are associated with higher arousal levels [111, 119].

Overall, owing to research scarcity, the conclusions of this body of research are limited. While the extant literature suggests that changes of both emotional valence and emotional arousal are induced by variations of rhythmic tempo, the effects of the former fluctuate considerably and appear to be influenced by cultural and life-history dependent variables (e.g. enculturation and familiarity). By contrast, tempo effects are reliable, suggesting that tempo-modulated arousal could be an ancestral mechanism that emerges independent of learning and cultural influences. It must be emphasised that much research remains to be conducted before drawing firm conclusions about how rhythmic parameters affect human emotion. A critical limitation to be surmounted by future research is the lack of cross-cultural representation in sampling procedures and administered stimuli. Current data fail to represent the great majority of the human population and cannot generalise to the species level. This inability to generalise jeopardises the accuracy of cross-species comparisons, thereby hampering evolutionary theory.

11.4.1.2 Chills and Groove: Emotions of Rhythmic 'Sweet Anticipation'

In his book titled *Sweet Anticipation*, David Huron elaborated on the theory that much of the emotional power of musical rhythm is derived from anticipatory processes that rhythmicity shares with a fundamental principle of brain function known as 'predictive coding' ([120], but see also [49, 121, 122]). In line with this principle, the brain works as a 'hypothesis-tester' that continuously generates predictions about incoming sensory input based on prior experiences, and the brain updates such predictions upon error feedback, that is, when top-down predictions do not match bottom-up input. Importantly, there are 'first-order' predictions about sensory content and 'second-order' predictions about the precision of first-order predictions. In this process, both correct and incorrect predictions engage the reward system in distinct ways. Correct predictions are rewarded as a way to reinforce a valid predictive model, whereas prediction errors (at moderate levels) are rewarded to promote learning [120, 123].

Within the predictive coding framework, musical rhythm is regarded as an inherently rewarding stimulus, given that it continuously feeds the brain with prediction material. Rhythm unfolds as a sequence of patterned acoustic events that occur with high statistical regularity, thereby generating a need to predict *which* events will occur *when* [124, 125]. ERC-relevant phenomena examined from a predictive coding perspective are chills (Sect. 11.4.1.2.1), groove (Sect. 11.4.1.2.2) and entrainment (Sect. 11.4.1.3), with chills and groove seen as emerging from expectancy violations, and entrainment-induced pleasure as emerging from fulfilled predictions [123, 125, 126]. According to this framework, rhythm processing spans the full reward cycle across phases of incentive salience and expectancy ('wanting'), peak pleasure ('liking') and learning [125, 127], thereby having the potential to engage the brain in a continuous reward cycle. This capacity may, in turn, explain why individuals get absorbed by musical behaviours for prolonged periods of time [128] and may hypothetically explain the preservation of musical behaviour in human evolution.

11.4.1.2.1 Musical Chills

'Chills' (also known as 'frissons' or 'thrills') are (typically) intense emotional experiences that occur in response to music, as well as other acoustic events (e.g. loud sounds and compact timbre [129]) and aesthetic objects (e.g. poetry, novels and paintings, reviewed in [130]). The 'chills' experience is typically accompanied by piloerection and a shivering or tingling sensation in various locations on the body. This physiological response is produced via SNS mobilisation, which results in increased skin conductance and rapid contraction of hair erector muscles. There is, however, considerable variation in the extent to which chill self-reports co-occur with a physiologically measurable response. The debate stirred by such inconsistencies has led to various attempts to identify stable variations within this, in appearance, heterogeneous cluster of experiences. Some have proposed dividing this experiential domain into entirely distinct phenomena deserving of their own alternative labels ('thrills', 'frissons', etc.), while others have proposed sub-classes of 'chills' (for details, see [131]). For example, Sumpf et al. [132] differentiate between 'subjective' and 'objective' chills based on whether piloerection can be measured, while Bannister and Eerola [133]—building on earlier work by Panksepp [134] and Huron [120]—differentiate between 'vigilance' and 'social' chills based on piloerection function.

According to the latter proposal, 'vigilance' chills would be grounded in the threat-signalling function of piloerection, being triggered by acoustic events that violate expectations, such as loudness, sudden rhythm changes and the entry of new instruments. Huron [120] proposed that musical stimuli that possess such qualities engage a dual processing track. Specifically, an initial fast-track response with negative valence would be processed subcortically, within 150 ms of event onset, triggering a resource mobilisation reaction in the body, which materialises as piloerection and a shiver-like sensation. Subsequently, a slower-track response would be processed cortically, underlying a reflective appraisal process that mitigates the initially negative response, resulting in an overall experience of positive emotion [120]. By contrast, 'social chills' would be grounded in the thermoregulatory function of piloerection, being an artefact of crossover effects between neural and behavioural systems that underlie thermal and socioemotional regulation [133]. Consequently, 'social chills' have been theorised—and empirically demonstrated—to be associated with a 'being moved' feeling in response to musical stimuli, which engages socioemotional mechanisms that promote shared states, such as empathy, belonging and entrainment [133].

Whereas the theoretical accounts reviewed above suggest at least two major rhythmic sources of chills—expectancy violation and entrainment—the empirical research on this topic lags behind. According to one of the few studies on the topic, chills are more likely to be associated with slow tempos [135]. More recent findings suggest that the chill response is accompanied by decreased neural oscillatory activity in the theta range (4–8 Hz) in the right central region, which may reflect the processing of rhythmic expectancies in the supplementary motor area [136]. Corroborating this hypothesis, rhythmic uncertainty and the violation of rhythmic expectancies have been recently highlighted as primary sources of 'chills' [137]. Finally, the role of rhythmic expectancies in chill induction is suggested by data showing that the DA-reward system is associated, on the one hand, with the experience of chills and, on the other hand, with rhythm processing (for more details, see Sect. 11.5.4.1).

11.4.1.2.2 Groove

'Groove' is a music-specific emotional experience that captures the pleasurable urge to move in response to music [122, 138, 139]. Groove may be covertly experienced or overtly expressed as foot tapping, head bobbing, dance, etc. (for a recent review, see [140]). The subjective experience of groove correlates highly with measures of motor engagement and emotion, including the occurrence of spontaneous head movement and its energy [141, 142], pleasure ratings [143, 144] and increased neural activity in motor and reward processing areas [145].

In the predictive coding framework, the experience of groove is regarded as an adaptive mechanism implemented by the motor system, whose function is to support prediction. This proposal is grounded in findings that motor brain areas are implicated in auditory rhythm processing. Corroborating this view, audiomotor and interpersonal synchronisation across diverse motor responses (tapping, head movement, walking) are enhanced by high—compared to low—groove rhythms, that is, by rhythms that elicit a stronger urge to move [138, 141, 142, 146].

The experience of groove is not an obligatory response to rhythmic stimulation but depends on both stimulus features and listener psychometrics. Accordingly, groove is facilitated by rhythmic features such as highly repetitive patterns, moderate complexity, beat salience (for a recent review see [140]) or tempo in the 107–126 bpm range [138, 147]. Acoustic features such as high energy in the low frequency bands also enhance the groove response [148, 149]. Finally, psychometric traits that promote groove include high levels of empathy [150], groove experience acquired through formal musical education or informal embodied practice (e.g. having the habit of dancing or tapping to music) and familiarity with a particular musical piece or style [151, 152].

Of all these groove-moderating variables, the most investigated to date is rhythmic complexity, in particular, syncopation, that is, unexpected shifts of rhythmic emphasis that violate the listener's metric expectations [140, 143]. To a lesser extent, researchers also examine other groove-inducing sources of rhythmic complexity such as *polyrhythm*, which results from the overlap of multiple rhythmic patterns, and *microtiming*, which is created through minor asynchronies between beat-keeping instruments [153]. From a phenomenological perspective, it has been argued that rhythmic complexity (i.e. shifts in the expected rhythmic structure) generates perceptual gaps to be filled in by the listener, thereby creating a participatory frame of engagement ([153] and references therein).

The relationship between groove and rhythmic complexity conforms to a U-shaped curve, whereby groove measures (e.g. pleasure ratings and urge to move) peak in response to rhythmic patterns of medium complexity [139, 143, 152]. This pattern is similar to the one that, more broadly, describes the relationship between arousal and complexity in other artistic domains [154, 155]. This relationship is, moreover, consistent with a predictive coding explanation, in which medium levels of rhythmic complexity maximise the weighted predicted error and cause the listener to embody the rhythm through movement to reduce error [122, 128, 139]. By contrast, low rhythmic complexity entails small prediction error, which reduces the need to recruit movement in the service of prediction. Conversely, high rhythmic complexity entails high entropy and, thus, high prediction uncertainty. This uncertainty reduces the brain's ability to predict rhythmic regularity at timescales that permit sensorimotor synchronisation [122, 139]. Intermediate degrees of complexity are thought to produce a balanced tension between sensory input and predictive model (expected rhythmic structure), thereby generating the pleasurable feeling of groove.

11.4.1.2.3 Rhythmic Entrainment as a Source of Positive Affect and Prosocial Emotions

Rhythmic entrainment is often discussed as an important source of emotion, which is supported by evidence that a sense of pleasure arises when neural or motor responses align with music or between individuals [156–159]. Besides generating positive affect, inter-individual synchronisation of vocal (e.g. singing and chanting) or bodily actions (e.g. stepping, finger-tapping and rowing) also evokes a plethora of prosocial emotions (e.g. affiliation, liking and trust) and enhances prosocial behaviours such as helping, generosity and cooperation (for recent reviews see [12, 160]). Several non-exclusive theories have been advanced to account for the emotional effects of rhythmic synchronisation, which has been proposed as an emotional substrate for social connectedness and prosocial responses.

Prominent mechanisms championed by these theories include predictive coding and proprioceptive feedback, as well as a range of socio-cognitive processes (emotional contagion, shared intentionality, etc.) theorised to potentiate the emotional effects of rhythmic synchronisation through various forms of 'state sharing'. From a predictive coding perspective, the rewardness of entraining to (musical) rhythms emerges from predictive processes (described further above), in particular the fulfilment of predictions related to rhythmic expectancies. Proprioceptive feedback theories argue that the activation of a single emotional component activates additional associated components. For example, holding a pencil in the mouth—which causes a mouth configuration reminiscent of smiling—biases subjective ratings of emotion towards positive valence (for a recent multi-lab study and review, see [127]). In line with this, it has been hypothesised that rhythmic synchronisation derives its rewardness from neural and somatic responses associated with pleasure that are triggered by rhythmic movement [156, 161, 162].

Theories centred on social processes emphasise 'shared states' as mediating mechanisms for the emotional effects of rhythmic synchronisation. For example, it has been suggested that rhythmic synchronisation may elicit emotions via emotional contagion, which arises from the perception of emotional cues in the kinematic profiles of bodily movement or musical rhythm [161]. Alternatively, during inter-individual rhythmic synchronisation, shared emotions could arise from shared intentions to engage in rhythmic synchronisation [163] or from the mere recognition that individuals are moving in synchrony [164]. Overall, the idea of group-level 'shared states' is supported by evidence of emotional synchronisation between individuals who are collectively engaged in an activity, as measured through subjective rating and (neuro)physiological measures. Emotional synchronisation has been reported in the context of rituals [165, 166], music performance (e.g. choir singing [167] and drumming [168]) and within audiences of sport competitions [169], theatre performance [170], film shows [171] and musical concerts [114, 172–175]. Interestingly, the phenomenon even occurs in 6-month-old infants attending a live musical concert [176].

Although multiple emotion-induction mechanisms may simultaneously contribute to the positive emotional effects of rhythmic synchronisation, several studies have reported that rhythmic synchronisation may enhance prosocial behaviour without the mediation of emotion [177, 178]. An important factor that may explain such inconsistencies is group size [160, 178]. According to a recent meta-analysis [160], rhythmic synchronisation is consistently associated with positive emotion in large-group studies, but less so in dyadic studies. Large group size has, moreover, been found to enhance the effects of rhythmic synchronisation on prosocial behaviour. From an evolutionary perspective, this pattern is consistent with theories proposing that the adaptive significance of rhythmic synchronisation resides in group-level benefits that are derived from socioemotional harmonisation, which in turn promotes social cohesion, cooperation and efficient decision-making (for a review, see [12]).

Intriguingly, however, group size does not seem to mediate the effects of rhythmic synchronisation on social bonding [160]. This runs against the popular theory that human rhythmic synchronisation has emerged as a mechanism of social bonding under selection pressures exerted by large group-size (e.g [178]. and references therein). To account for this contradiction, it has been suggested that distinct mechanisms could moderate the effects of rhythmic synchronisation on social bonding at the dyadic and group levels [160, 178]. Specifically, dyadic bonding may result from non-emotional processes, such as increased attention to the social partner or the blurring of the self-other distinction (e.g. via motor mimicry), which are effective only when the number of interactants is small. By contrast, social bonding in large groups is more efficiently mediated by shared emotional states arising from energising group activity that instils a sense of 'collective effervescence' [165, 179, 180]. Overall, the large group setting may represent a more naturalistic—and therefore more effective—context for engaging ancestral emotional mechanisms thought to underlie the survival benefits of rhythmic synchronisation.

Besides group size in itself, a range of methodological choices can also influence whether the participants experience enhanced positive emotion in the context of rhythmic synchronisation. For example, studies with larger groups are often conducted in naturalistic settings (e.g. concert hall) and focus more on behaviours that naturally emerge in such contexts [114, 142, 181]. By contrast, dyadic studies are carried out in laboratory settings, requiring the performance of actions (e.g. cup moving) imposed by the experimental protocol, which may be experienced as tedious, thereby affecting emotion ratings [177]. Moreover, since emotions are transient and subjected to psychological construction, continuous emotional tracking across the task using both self-report and physiological measures is likely to be a more accurate reflection of emotional experience than self-reports collected after task completion [177]. Finally, emotional engagement is influenced by stimulus properties, being more pronounced in response to groovier rhythms [142, 182], faster tempo centred around 120 bmp [114] or rhythms with simpler structure [173].

11.4.2 Research with Animals

In nonhuman species, the emotional effects of rhythmic communication can be gauged from research that integrates rhythmic stimulation and assessments of behavioural responses that are commonly used to measure valence or arousal. Emotional valence is typically operationalised by the approach-avoidance dimension of behaviour, where approach indicates positive valence elicited by rewarding or appetitive stimuli, while avoidance indicates negative valence elicited by threatening or unpleasant stimuli. Levels of emotional arousal are inferred from the levels and type of activity in which animals engage after stimulus exposure. For example, a calming effect and a low level of arousal are inferred when animals lie down and relax. By contrast, arousing effects are inferred when levels of overall or specific activities (e.g. locomotion and social interaction) increase in response to given stimuli.

The main bodies of research that fulfil the elicitation and assessment criteria mentioned above are, on the one hand, research on the rhythm perception abilities of animals and, on the other hand, research on animals' reactions to human music. Since the rhythm perception literature is covered more extensively in other chapters of this volume, here we only briefly outline how this body of research applies to the effects of rhythm on the emotional states of nonhuman animals. As shown in Chap. 3 [1], in a wide range of taxa, features (pattern, tempo, etc.) structure communicative signals that are produced in contexts regulated by affective processes, including reproduction, caregiving, aggression and affiliation. Discriminant analyses and playback studies suggest that certain rhythmic qualities not only convey the emotional states of signallers (as detailed in Sect. 11.3), but also influence the emotional states of conspecifics, bringing about context-adaptive behaviour. This is commonly observed in contexts such as alarm calling, male dominance display and group territorial display, where specific variations of rhythmic pattern or temporal rate reliably elicit distinct behaviours in conspecifics, including vigilance, avoidance, flight, counter-display or aggression. In addition, a recent analysis of rhythmic trajectories in thrush nightingale song (*Luscinia luscinia*) suggests that birdsong exhibits features of rhythm dynamics that are also found in human music and are known to induce rhythmic expectancies, tension and pleasure [183].

The research on how animals respond to music has been primarily conducted in the field of applied ethology, with the aim of assessing if music is a welfare-compliant stimulus for species that have close contact with humans, including company, farm and zoo animals. The main approach in this research is to expose animals to different musical and control stimuli, such as white noise, nature sounds and silence. An overarching hypothesis driving stimulus design in this field is that the same acoustic features of music that induce and express emotional states in humans are also effective in other species, provided that they are adapted to species-specific perceptual, physiological and motor ranges [3, 4, 184, 185]. Specific rhythm-focused predictions within this hypothesis are that fast tempo and staccato articulation are arousing, whereas slow tempo and legato notes are calming, and that rhythmic fluency is associated with positive emotion, while rhythmic irregularity is associated with negative emotion [184].

As long as species-specific perceptual constraints are considered in the experimental design, the empirical data accumulated to date are generally supportive of these predictions. For example, cotton-top tamarins (*Sanguis oedipus*) exhibit systematic behavioural changes when exposed to musical stimuli designed with an awareness of species-specific physiological rhythms (resting state heartbeat) and sensory ranges [186]. Accordingly, irrespective of spectral features, musical stimuli with a tempo that exceeds the heart rate of resting tamarins enhance arousal, while stimuli with a tempo that matches heart rate at rest induce relaxation. In the same vein, domestic cats show higher levels of affiliative responses (approach, rubbing) and lower stress levels during medical examination when exposed to musical stimuli designed to match species-relevant rhythmic tempos (purring: 1380 bpm; suckling: 250 bpm) as opposed to music played at 60 bpm—a tempo that is considered as relaxing by humans [187, 188]. Likewise, newly hatched chicks prefer acoustic stimuli that are pulsating, have a faster tempo and are more similar to hen calls [189].

Supporting evidence also comes from studies in which unaltered musical stimuli are presented to species whose acoustic perception range or physiological rhythms are relatively similar to those exhibited by humans. As such, fast tempo music has been reported to enhance positive arousal in chimpanzees, gorillas and pigs [190–193], as indicated by increased levels of activity or species-specific communicative behaviours (e.g. piglet tail wagging). By contrast, slow tempo music has been reported to decrease levels of arousal, being associated with lying down, stress and agonism reduction, and low levels of overall activity [190–195]. Similarly, decreased levels of arousal induced by exposure to slow-tempo music (compared to silence) have been reported in dogs [196].

On the valence dimension, two approach-avoidance studies with chimpanzees suggest that chimpanzees prefer slower rhythmic tempos, as well as rhythmic stimuli without a salient beat. In one of these studies, the chimpanzees were more likely to enter a music booth when it broadcasted classical music (tempo range 50–90 bpm) and were more likely to exit it when the booth broadcasted music with a tempo that exceeded 90 bpm [197]. In the other study, chimpanzees were found to spend more time in the broadcasting area during bouts of West African Akan music and North Indian raga as compared to bouts of Japanese Taiko or silence [198]. This pattern of behaviour has been interpreted as reflecting a preference for rhythmic stimuli that lack a salient beat, the latter being highly prominent in Taiko, but masked or infrequent in Akan music and Indian raga, respectively.

At present, relatively little is known about whether and how rhythmic parameters influence emotions in nonhuman species. Extant data are generally consistent with theoretical proposals regarding the phylogenetic continuity of an ERC-link, in particular when it comes to the tempo-arousal association, whereby fast tempos increase arousal level, and slow tempos diminish it. It is, however, important to note that contradicting findings are also reported for several species (e.g. dogs [199], pigs [200] and macaques [201]).

A range of methodological factors could account for the inconsistencies noted above. First, most studies focus on the effects of music genres, with rhythmic variables being incidentally mentioned rather than specifically controlled for. Even when rhythmic parameters (e.g. tempo) are purposely manipulated, other acoustic features vary widely, which may constitute a confound. For example, Zapata Cardona et al. [202] found that a higher tempo generated higher arousal and a higher positive valence response in piglets only in a condition where the music was played by two to four instruments, while a slower tempo generated lower arousal and a positive valence response only in a condition where the music was played by four to six instruments [202]. Second, the individuals' history of acoustic stimulation is likely to affect behavioural responses, introducing enculturation-like effects [203]. Third, characteristics of the broadcasting environment may impact sound propagation or affect the performance of the animals. Echo and reverberations, for example, may hinder sound source location, which is crucial in the approach-avoidance paradigm. Open or closed spaces may inhibit or stimulate the animals' movements in species- and individual-specific ways, independent of the broadcasted stimulus. Fourth, inconsistencies can result from failure to control for the baseline emotional states of subjects prior to exposure to the experimental conditions. Fifth, differential responses can arise from differences in the administration protocols. For example, exposing the animals to a variety of stimuli featuring a variety of tempos in a single experimental session may yield mixed responses due to interference effects [199]. Finally, effects may be masked by sexually dimorphic responses to rhythmic stimuli. For instance, Hoeschele and Bowling [204] found that female budgerigars (*Melopsittacus undulatus*) exhibited a preference for rhythmic—compared to arhythmic—stimuli, while the males showed the opposite response pattern.

11.5 Neurobiological Mechanisms Underlying the ERC-Link

As shown in Sects. 11.3 and 11.4, at the behavioural level, the ERC-link is bidirectional: rhythmic tempo, duration and articulation convey the emotional state of the signaller and, conversely, the rhythmic parameters of signals can affect the emotional state of the receivers. Consistent with this, early research in the neuroscience of music has shown that music listening implicates neuroanatomical regions commonly associated with *arousal* and *valence* processing, including NAc, H, AMY,

HPC, IC, vmPFC, OFC and ACC (for reviews, see [122, 157]). While the exact role of rhythm in these associations is not well understood, recent findings suggest several potential neurobiological substrates that could underpin the ERC-link: autonomic mediation (Sect. 11.5.1), breathing (Sect. 11.5.2), sensorimotor and emotion integration hubs (Sect. 11.5.3) and neuromodulation (Sect. 11.5.4). Because these mechanisms are still being actively investigated, the pathways outlined below should be viewed as plausible candidates rather than definitive explanations.

11.5.1 Autonomic Mediation of the ERC-Link

As shown in Sects. 11.3 and 11.4, the most stable component of the ERC-link is the bidirectional connection between *temporal frequency rate* and *arousal*, whereby fast and slow rhythms are associated with physiological markers of SNS and PSNS dominance, respectively. This connection is likely established already at the level of the RAS, which orchestrates ANS activity (detailed in Sect. 11.2.2) and features convergence sites for afferent streams of interoceptive and exteroceptive information (relevant to emotion processing and rhythm perception) and efferent streams that regulate autonomic and somatic functions, relevant to the production of emotional expressions and rhythmic behaviours [21]. Relevant afferent pathways include the cochlear and vestibular nerves, which relay auditory and motion information, while efferent pathways include the motor nerves that innervate the skeletal muscles, which generate rhythmic behaviours (for details, see Chap. 9 [205]). Moreover, neuronal assemblies in the RAS are particularly responsive to novelty and expectancy violation, threat cues and sexual information, which is relevant not only to arousal processing, but also to aspects of RC. Indeed, as discussed in Sect. 11.4.1.2, rhythmic expectancy violation is an important source of music-induced emotion and, as shown in Sects. 11.3.2 and 11.4.2 (see also Chap. 3 [1]), courtship and aggressive-defensive interactions are major contexts of rhythmic communication.

Although SNS and PSNS activity are traditionally depicted in antagonistic terms, the two branches of the ANS can also be co-activated. This may occur in musical behaviours [23] and, more broadly, in rhythmic practices aimed at emotional regulation, such as rhythmic meditation that induces slow breathing [206]. On a related note, rhythmic exertion mobilises SNS and C response with positive emotional effects. In humans, such effects have been established for both physical training activities (e.g. running), as well as RC behaviours (dancing, drumming, singing, etc.), which can be carried out for extended periods of time in both leisurely and ritualistic contexts (for details, see Chap. 13 [207]). Although the cardiorespiratory profile of such metabolically demanding activities suggests SNS hyperactivation, these can selectively activate certain SNS components to result in beneficial effects. Rhythmic stimulation can, for example, selectively enhance an epinephrine response without concomitantly enhancing NE and C, which has positive health effects [208]. Thus, it is plausible that the widely documented beneficial effects of music and other rhythmic practices (for reviews see [23, 157, 206]) are not merely due to

PSNS dominance, as has been suggested in the past. Instead, through repeated practice, such activities seem to promote a more efficient autonomic reactivity to stressors and overall ANS balance.

11.5.2 Breathing as an Interface for the ERC-Link

Breathing constitutes a key intermediary in the ERC-link. On the one hand, it is neuro-mechanically coupled with the locomotor and vocal systems that actuate rhythmic communication (for details, see Chap. 2 [209] and Chap. 9 [205]) by virtue of sharing muscular substrates. In many vertebrate taxa, breathing parameters (rate, the relative duration of inspiration and expiration, etc.) constrain the features of vocal output, whereas breathing rhythms are strongly influenced by the speed, intensity and type of locomotor work. While at the neural level these various rhythms are generated by distinct central pattern generators (CPGs), they nevertheless interact to ensure the smooth coordination of diverging behavioural outputs in shared effectors (for more details, see Chap. 9 [205]). It has been suggested that the inspiratory CPG provides a clock signal for other CPGs localised at brainstem level, which contributes to the integration of breathing and other motor rhythms [210].

On the other hand, the rhythmic features of breathing are bidirectionally coupled with arousal and, to a lesser degree, valence. Fast and slow breathing is reliably associated with high and low arousal, respectively, while combinations of breathing velocities and depth can reliably convey emotional valence. For instance, deep and slow breathing is associated with pleasant relaxation, while shallow and fast breathing is associated with negative states [211].

Anatomical and functional connectivity data suggest that multiple mechanisms contribute to the link between emotion and breathing. Most structures and circuits that are implicated in arousal regulation (Sect. 11.2.2) also participate in the extended network that contributes to generating or modulating the features of breathing (for a comprehensive review, see [212]). The rhythmogenic mechanisms of breathing are bidirectionally connected to areas that supply neuromodulators (LC, dNR, H) and are implicated in arousal and valence regulation. In addition, there are bidirectional projections between areas that generate breathing rhythms and areas involved in valence processing (PAG, AMY subdivisions and IC). Neuronal assemblies that control features of breathing interact with motor circuits that innervate the major muscular effectors of rhythmic communication at the level of the larynx, trunk and limbs [212]. Finally, it is believed that the motor cortex transmits efference copies (i.e. a duplicate of the motor output) to the inspiratory CPG (the pre-Bötzinger complex, preBötC), which fine-tunes the coordination of breathing to other motor outputs in a context-adaptive fashion [210].

The specific causal mechanisms operating within these pathways remain to be clarified. A recent discovery is that a small assembly of glutamatergic neurons located within the preBötC directly modulates NE activity in the LC (a major regulator of arousal), while neurons within the LC influence inspiratory rhythms [210,

213, 214]. This pathway has been suggested as a potential underlying mechanism for the arousal-dampening effects of slow breathing in meditative and therapeutic settings. In mice, inactivating this neuronal subset induces slow respiration that, neurophysiologically, correlates with an overall decrease in oscillatory brain activity from the theta (6–10 Hz) to the delta band (2–4 Hz) and, behaviourally, correlates with an exaggerated state of calm [213].

Another potential mechanism that contributes to the effect described above is localised at AMY level, whose activation increases and decreases as a function of breathing rate [211]. It has been therefore proposed that the voluntary reduction of breathing rate reduces AMY activation, thereby reducing levels of arousal and promoting a state of calm. The AMY is bidirectionally connected with respiratory mechanisms in the medulla and the pons, which generate and modulate the rate of inspiratory and expiratory rhythms. Thus, it has been further proposed that, when experiencing emotions, brain signals associated with emotional processing in the AMY have the capacity to override the rhythms of autonomic breathing [211]. In turn, this mechanism could contribute to the ERC-link by exploiting the neuromechanical couplings that exist between breathing rhythms and communicative rhythms.

Finally, breathing rhythms may facilitate the integration of neural processes that underpin emotion and rhythmic communication through effects exerted on neural synchronisation. Mounting evidence from rodent and human studies shows that a widespread network of cortical and subcortical structures tracks the slow breathing cycle (0.2 Hz in humans, at rest), promoting breathing-locked oscillations that are synchronised across large distances across the brain [210, 215–218]. On a local level, breathing rhythms entrain gamma-band oscillations and the fast transitions between neuronal synchronisation and desynchronisation, such that gamma power increases occur preferentially during certain breathing phases. This mechanism could underpin the integration of locally generated gamma assemblies at global level, that is, across distant brain regions [216, 218]. Thus, the incessant rhythm of breathing can act as an organising hierarchical principle for neuronal oscillations throughout the brain and across sensory, motor, emotional and cognitive functions [216, 217]. Interestingly, this local and global scale entrainment is not driven by the brainstem CPGs that regulate the activity of breathing muscles, but by re-afferent inputs originating in the sensory epithelia of the nasal cavity. The phenomenon is suppressed by manipulations that block nasal breathing, while air puffs delivered to the nasal cavity restore it (reviewed in [218]). Moreover, breathing route (nasal vs oral) and phase (inspiration vs expiration) modulate emotional recognition, such that emotional recognition is selectively enhanced by nasal inspiration [215].

11.5.3 Where Motion Meets Emotion: Putative Hubs for the ERC-Link

Both emotion and rhythmic communication are complex phenomena, underpinned by distinct processing streams that span cortical and subcortical brain regions. Their neural paths, however, meet in certain brain areas—for example, the PAG, the basal ganglia (BG) and the cerebellum—shown to play a direct role in both emotional processes and aspects of rhythmic communication.

11.5.3.1 The ERC-Link in the PAG

The PAG is a convergence site for streams implicated in emotion processing, breathing regulation and rhythmic communication. As noted in Sect. 11.2.2, the PAG contributes to arousal and valence processing by integrating afferent streams from the H, AMY and the PFC and efferent output that regulates defensive, socioemotional (altruism, maternal care, etc.) and sexual behaviour. Moreover, PAG exerts a gating function over vocal behaviour via projections to vocal CPGs and the caudal ventral respiratory group, which has direct projections to all motor areas recruited for vocalisation (for reviews see [212, 219]). The PAG is also an important hub for ANS integration: it receives input from autonomic nerves and centres for autonomic regulation (LC, caudal and dorsal RN) and provides output to areas that participate in the regulation of breathing rate, cardiovascular response and motor coordination (e.g. inferior olive and cerebellum). The participation of PAG in the above functions is flexibly coordinated by a range of emotion-relevant neuromodulators, including NE, 5-HT, endorphins, OT and VP. The PAG is, thus, a major interface between various aspects of emotional processing, neuromodulation, vocal and breathing CPGs, as well as non-vocal motor output.

11.5.3.2 The Basal Ganglia and the Cerebellum

Traditionally, the functions of the BG and the cerebellum were thought to be limited to motor processing. Consistent with this, both play a prominent role in rhythm processing, even when this does not involve an overt motor component, such as during music listening [220]. Functionally connected with regions of the auditory and (pre)motor cortex, the BG and the cerebellum are implicated in the analysis of temporal sequences, and in particular, in extracting features of rhythmic patterning and hierarchical organisation (for details, see Chap. 7 [221] and Chap. 9 [205]). The BG have been reported to play an important role in beat extraction and the formation of a pulse percept [222], as well as in beat prediction [223]. The cerebellum is thought to play a major role in fine-tuning motor output. Notably, both beat perception and motor fine-tuning are viewed as crucial for rhythmic synchronisation and, historically, have been considered to be rare in the animal kingdom, if not unique to humans (for details, see Chap. 1 [224], Chap. 5 [225] and Chap. 14 [226]).

Recently, the neurofunctional repertoire of the BG and the cerebellum has been augmented with several roles in emotion processing. Methodological advances have permitted the identification of novel bidirectional projections that link the BG and the cerebellum to each other and with nearly all other anatomical brain structures (for a recent review, see [227]). These discoveries position the BG and the cerebellum as convergence hubs for sensorimotor, learning and emotional processing, including the modulation of breathing and other autonomic functions [210, 212, 227]. Besides the NAc (mentioned in Sect. 11.2.2), several nuclei of the striatum—the putamen, the caudate nucleus and the substantia nigra—contribute to emotion functions such as approach-avoidance behaviour, attachment and reinforcement learning. Overall, the striatum plays a key role in reward processing and reward-based learning mediated by DA activation in the VTA and substantia nigra. Since this process involves positive or negative feedback to exteroceptive and interoceptive cues, it has been argued to contribute to valence modulation via predictive coding [227]. The striatum, moreover, contributes to the selection of context-appropriate behaviours through DA-sensitive paths that differentially potentiate or inhibit certain motor outputs (for details, see Chap. 9 [205]). Consistent with this, the BG underlies the automatic production of motor responses to emotional cues, including laughter, facial and bodily expressions [228], which, theoretically, can also include bodily movement in response to music.

Beyond the striatum, the subthalamic nucleus (STN) —another component of the BG—has been highlighted as a key integration site for input from sensorimotor, emotional and associative networks. This is crucial for the emergence of cross-functional influences across the brain, as the STN, through back-projections, exerts a concerted modulation of neuronal oscillations across multiple areas that input to the BG [227] (and references therein; see also [229] for examples related to music processing). This is consistent with the 'switch hub' role attributed to the BG, as a vertebrate brain structure that monitors input from different functional networks to resolve the allocation of limited motor, cognitive and, ultimately, energetic resources in a context-adaptive fashion [230].

Regarding the contributions of the cerebellum to emotion processing, it is relevant to note its role as a convergence site for BrS streams implicated in autonomic regulation, the AMY and PFC areas implicated in the contextual tuning of emotion. Research suggests that the cerebellum is involved in threat assessment and the regulation of negative emotion via connections with the PFC and AMY. Moreover, it is involved in modulating the salience of emotional cues and in minimising prediction error in reward processing, owing to connections with DA neurons in the VTA and the BG. This function is believed to rely on broad monitoring of current versus expected or desired states and rapid feedback to cortical input areas, which fine-tunes emotion recognition, motor output, emotion encoding and decision-making [231]. The role of the cerebellum in emotion processing is, thus, largely complementary to the role of the BG, and together, they contribute to the selection and precise execution of motor responses in a context-adaptive fashion.

11.5.4 The Neurochemical Mediation of the ERC-Link

To map the neurochemical mediation of the ERC-link, we rely on the brief reviews regarding the role of neuromodulators in emotion processing (provided in Boxes 11.1, 11.2, and 11.3), complemented with theoretical considerations and empirical findings relevant to tracing the role of neuromodulators in rhythmic communication.

11.5.4.1 The Role of Monoamines in Mediating the ERC-Link

The LC, which sources NE projections throughout the brain, is richly connected with both emotion (reviewed in Box 11.1) and rhythm processing areas. Particularly relevant are excitatory projections to the cerebellum, which is implicated in motor planning and coordination, and to the cochlear nuclei, which are implicated in auditory processing. Theoretically, given this connectivity pattern and role in arousal upregulation, the NE system could be broadly involved in mediating the relationship between arousal and rhythmic communication (as shown in Sects. 11.3 and 11.4). Indeed, pupils dilate upon detecting violations of rhythmic expectancies, which may reflect NE-driven emotional effects that arise from rhythm processing [232–235]. However, since emotion-evoked pupillary dilation can also be

Box 11.1 Monoaminergic Modulators of Emotion

The monoamines are synthesised by small neuronal pools located in subcortical nuclei that also function as convergence sites for sensory and motor information. These nuclei have widely distributed projections throughout the brain. Monoamine molecules (or their precursors) can also be synthesised by peripheral organs and released into the bloodstream. Recent phylogenomic evidence positions the evolutionary emergence of the monoaminergic system at the *Cryogenian / Ediacaran* boundary (650–600 Mya), with most of the genes involved in monoamine production, modulation and reception being present in stem bilaterians [236]. From a functional perspective, the monoamines are involved to some degree in novelty detection and uncertainty computation and, thus, in the elicitation of somatic responses that, in oversimplified terms, are associated with vigilance (largely driven by NE and ACh), approach (significantly promoted by DA) and aversion (flexibly modulated by 5-HT). Ascending monoaminergic pathways mediate the activating and deactivating effects that RAS exerts on a global brain level [23].

Norepinephrine and acetylcholine. The NE system (originating mainly in the locus coeruleus, LC, within the BrS) and the ACh system (originating from the basal forebrain and the mesopontine tegmentum) have overlapping emotional effects. Both NE and ACh are released in response to salient (e.g. unexpected and novel) stimuli, which upregulates arousal and supports selective attention, learning and the reinforcement of behaviour associated with rewarding or aversive cues [237–240]. These effects are achieved through influences exerted on RAS, the HPA axis and an array of valence processing hubs (e.g. mPFF, OFC, ACC, AMY, HPC and VTA). NE (and its precursor, epinephrine) and ACh are also the chief neurotransmitters of the ANS. Whereas NE is released within most postganglionic segments of the SNS, ACh is released within the preganglionic segments of the SNS, as well as both pre- and postganglionic segments of the PSNS. Overall, for optimal emotion modulation, NE and ACh levels follow an inverted-U model. Overly high or low

(continued)

Box 11.1 (continued)

release, impaired breakdown and the dysregulation of NE release relative to ACh and vice-versa have effects (e.g. interference with fear extinction and stress over-reactivity) that are central to the aetiology and symptomatology of anxiety disorders [240].

Serotonin. 5-HT exhibits largely conserved functions across animal phyla, despite a great diversity of neural organisation [241]. In vertebrates, 5-HT is synthesised within the raphe nuclei (RN) of the BrS, reaching brain-wide targets, where it binds to an array of 5-HT receptor types to flexibly mediate a range of emotional effects. Notable 5-HT targets are areas implicated in harm aversion and in risk and reward assessment (OFC, AMY and VTA), ANS regulation (the HPA axis) and the selection of context-adaptive motor output (e.g. PAG) within the axis of defensive-aggressive-cooperative behaviour [242–244]. Low 5-HT synthesis promotes risk-taking and aggressive behaviour by impairing the processing of harmful stimuli and learning from negative outcomes [243]. By contrast, enhanced 5-HT in the BLA promotes aversion processing, fear learning, vigilance and evasion. Owing to these roles, 5-HT plays a major role in mood pathologies (depression, anxiety, PTSD, etc.), certain psychiatric disorders (e.g. schizophrenia) and in the emotional effects of certain recreational and ritualistic drugs, such as stimulants (MDMA) and psychedelics (psylocibin, LSD and MDMA).

Dopamine. DA neurons originate primarily from the VTA (mesocorticolimbic pathway) and substantia nigra pars compacta (nigrostriatal pathway). Within the mesocorticolimbic pathway, DA neurons chiefly ascend to emotional valence hubs, such as the NAc, AMY and PFC. This circuit is a key modulator of the 'wanting' component of the reward cycle, with DA release up- and down-regulating appetitive responses for a broad spectrum of rewards, spanning food, sex, gambling, laughter, music, etc. (reviewed in [123, 245]). Modulatory outcomes are dependent on the output path from NAc and the DA receptor types expressed on these paths. The activation of D1 receptors found on the 'direct' output path to VTA is sufficient for the upregulation of appetitive responses. By contrast, aversive responses require the activation of both D1 and D2 receptor types, where the latter mostly belong to the 'indirect' output path that projects to the ventral pallidum in the BrS and H [245]. In addition, DA modulation is linked to the *learning* component of the reward cycle, with DA release encoding reward prediction error, that is, the discrepancy between reward predictions and actual reward experience [246]. Finally, recent data from humans suggest that DA modulates the *liking* component of the reward cycle, specifically for abstract—as opposed to sensory—rewards [247]. Given the significant roles the mesocorticolimbic DA pathway plays in the reward cycle for a wide range of sensory rewards, its dysregulation is widely implicated in pathologies of addiction.

associated with the release of ACh and, less commonly, 5-HT, it may be premature to attribute the arousing effects of rhythmic expectancy violations solely to NE. While animal research has found that music selectively enhances NE—but not 5-HT—turnover in birds [4], ACh has not been measured, nor has it been examined how the rhythmic qualities of the musical stimuli may contribute to such effects.

Similar challenges arise when addressing the potential role of DA as a mediator of the ERC-link. Evidence linking DA with the pleasurable experience of music has accumulated in the past two decades, but the contribution of rhythm remains unclear. Theoretically, several kinds of DA-dependent computations could mediate the ERC-link, given that interactions between NE and DA underlie rhythmic expectancies induced by rhythmic sound and movement, and are believed to constitute an ancestral pre-adaptation for the emotional processing of music [4]. First, as shown in Box 11.1, DA mediates the *wanting* and *learning* components of the reward cycle for sensory rewards and, in humans, also the *liking* component for abstract rewards, with music being a prime example of the latter. Second, the DA system has been implicated in coding rhythmic predictions and prediction error, which is intimately linked with the reward cycle [123, 125, 248, 249]. Third, DA modulates temporal processing in the second- and millisecond-range—which is the main temporal scale exploited by rhythmic communication [250, 251] —and regulates the fine-tuning of movement control, which is crucial to rhythmic synchronisation.

Beyond such theoretical considerations, very little research has examined the role of DA modulation on rhythmic communication from an emotional perspective. For example, Koshimori and colleagues [252] found that rhythmic cueing contributes to reduced DA response in the ventral striatum during a taping task, a finding which may suggest reduced motivational (and attentional) effort for maintaining motor control during rhythm production. In addition, [253] found that listening to highly rhythmic music induces altered activity and connectivity in the NAc, which is primarily implicated in the *liking* component of the pleasure cycle and is rich in DA receptors.

The potential implication of ACh and 5-HT in the ERC-link is far less discussed, but the roles they play in arousal regulation (see Box 11.1) and their interactions with DA-driven reward, movement and temporal processing [254–257] are relevant. Recent data show that ACh modulates the precision or weighting of prediction error in the striatum [257] and auditory cortex [258], while 5-HT has been implicated in signalling aversive outcomes and aspects of punishment-related prediction errors [259]. Of relevance to rhythmic tempo and patterning, both ACh and 5-HT modulate motoneuron firing patterns and, thus, the patterning of muscular activity [260, 261]. In the vertebrate brain, 5-HT neuronal projections link sites implicated in arousal and valence processing with motor areas that control behavioural output, in particular vocal CPGs (see [262] and Box 11.1). For example, the application of 5-HT to the frog vocal CPG evokes laryngeal patterned activity that mirrors advertisement calls, suggesting that 5-HT plays a role in the initiation of rhythmic signals [263]. Findings from avian research further suggest the implication of 5-HT in the initiation, patterning and learning of birdsong [264], while research with electric fish shows that 5-HT modulates pulse amplitude, shape and patterning, as well as the electrosensory perception of electric signals [265]. 5-HT has also been linked to dance in at least two ways. First, the polymorphism of SLC6A5 (a 5-HT transporter gene) is associated with creative and emotional aspects of the human dance phenotype—a complex trait spanning physical, cognitive and socioemotional factors that define an individual's dancing ability [266]. Second, dance-based therapy has been shown to optimise the 5-HT–DA balance in teenagers with mild depression [267]. Finally, 5-HT plays a crucial role in the emotional regulation of behavioural responses in sexual, agonistic and maternal care contexts associated with rhythmic communication (as mentioned in Box 11.1).

11.5.4.2 Neuropeptidergic Mediation of the ERC-Link

Two main avenues have been considered for the potential peptidergic mediation of the ERC-link: HPA axis regulation and socioemotional regulation. As reviewed in Box 11.2, β-END is credited with anxiolytic, antinociceptive and pleasure-boosting effects, as well as the facilitation of approach behaviour, all of which have also been linked to rhythmic stimulation in humans and other species. At the most general level, music exerts a strong influence on AMY clusters that contribute to emotion regulation through the release of endorphins [157]. β-END, in particular, has been discussed as the mechanism that underlies the anxiolytic and mood uplifting effects of rhythmic exertion across a broad range of behaviours (dance, stereotypic behaviour, workout, etc.). It has also been invoked by evolutionary theories of human musicality that link proto-musical rhythmic movement to behavioural stereotypies [268, 269], which are considered a stress-buffering mechanism (see [270] and chapters therein). Interestingly, this is one of the few contexts where spontaneous rhythmic synchronisation has, thus far, been documented in chimpanzees ([269], as well as Chaps. 12 and 13 this volume).

The analgesic effects of β-END have been invoked to explain why certain rhythmic displays—from human ritualic practices to animal social play and courtship displays—are persistently deployed for a prolonged amount of time despite being metabolically demanding and known to cause respiratory and muscular discomfort. In ritualic contexts (e.g. meditation and trance), β-END has also been linked to the experience of euphoric feelings associated with practices such as chanting and dancing (for details, see Chap. 13 [207]). Arguments along these lines, however, ought to be taken with caution, as the exact roles of β-END and rhythmic stimulation in such contexts, as well as the link between the two, are yet to be charted.

Empirically, the most compelling evidence for a putative opioidergic mediation of the ERC-link comes from research that examines the effects of rhythmic behaviour and synchronisation on prosocial emotions and pain tolerance as a proxy of β-END-release [271, 272]. This research shows that rhythmic synchronisation across a range of behaviours (singing, chanting, dancing, drumming, rowing, etc.) is consistently associated with higher pain thresholds and prosocial emotions, as compared to non-synchronised rhythmic exertion, solitary rhythmic behaviour or passive exposure to rhythmic stimuli. Interestingly, the administration of μ-opioid receptor antagonists has been found to block the analgesic effects of rhythmic synchronisation, but not the enhanced perception of social bonding [273]. This may either suggest that the β-END mediates only the antinociceptive effects of rhythmic synchronisation, or alternatively that β-END exerts analgesic and socioemotional effects through distinct opioidergic mechanisms or through synergistic effects with other neuromodulators.

As reviewed in Box 11.2, the nonapeptides are implicated in socioemotional regulation in a great range of contexts, including courtship and mating, territorial advertising and defence, maternal care and mother-offspring interaction, dominance and affiliative interactions with- or beyond kin. Intriguingly, this functional spread mirrors the full range of functional contexts in which rhythmic communication has been studied thus far (as reviewed in Chap. 3 [1]), as well as the plethora of emotional—both therapeutic and social—benefits ascribed to human musicality (for a recent review, see [12]). In theory, thus, OT and VP (and their non-mammalian counterparts) could modulate rhythmic communication in all these contexts, including musical rhythmic behaviours in humans. Indeed, a potential connection between OT and music has been repeatedly contemplated in the past two decades, given their overlapping anxiolytic and social bonding effects [274–277] and the overlapping neural substrates associated with these benefits.

Currently, however, it remains unclear if OT mediates the positive emotional effects of music. Supporting evidence comes from research that reports increased OT levels in response to music listening [278], especially slow tempo music [279], or after active music making, including solo or group singing, drumming and jazz improvisation [280–285]. Yet, some studies fail to find evidence of both OT stimulation and HPA axis downregulation, which is the mechanism typically assumed to underpin the positive effects of music through OT mediation. For example, listening to fast-tempo music, which increases arousal, has a dampening effect on C, but no apparent influence on OT [279]. Likewise, singing may result in decreased C levels, without a concomitant increase in OT [283] and OT levels may even decrease after group singing [283, 286] or other kind of musical performance [282] compared to baseline levels.

The variable results reviewed above could be explained by methodological variability with respect to OT-assay methods, baseline stress levels, as well as what is performed, how and for what purpose. For example, following a score and rehearsal for a competition entails performance stress and more individualistic goals, which appear to inhibit OT release, while leisurely music performance and improvisation have a more relaxed and participatory character that promotes OT [282, 283, 285]. Consistent with this pattern of findings, OT-administration improves rhythmic synchronisation in finger-tapping [287] and dancing [288] tasks by inducing cooperative and altruistic predispositions.

By contrast to OT, research on a potential link between VP and rhythmic communication remains scarce. At the molecular level, polymorphisms of the VP receptor *AVPR1a* have been linked to musical aptitude, as well as music and dance creativity [266, 289] (and references therein). Moreover, a case study suggests that male sexual arousal (mediated by VP) enhances rhythmic synchronisation [290].

Although widely known as regulators of socioemotional life, the role of VP and OT in nonhuman rhythmic communication has only recently begun to be examined.

In fish species, VP- and OT-homologs inhibit acoustic rhythmic communication with a reproductive function in males and females, respectively [265]. In electric fish, however, VT-homologs only inhibit aggressive displays, while enhancing reproductive signals. In birds, the avian homolog of OT is widely distributed in brain areas that underlie the motivation to sing and, in some species, also in motor areas that control song patterning and perception [291]. This distribution suggests that OT may play a role in the ERC-link by modulating both the emotional and structural aspects of birdsong. Consistent with this, another recent study has shown that OT contributes to song learning by enhancing approach and attention to tutors [292]. While direct empirical evidence remains limited, it nevertheless supports the view that the nonapeptides provide an ancestral interface for the ERC-link,

Box 11.2 Opioidergic and Nonapeptidergic Modulation of Emotional Functions

The endogenous opioidergic system comprises widely scattered neurons that produce a family of molecules (e.g. endorphins and enkephalins), which are grouped together due to sharing a common amino-terminal sequence known as the opioid motif [294]. Of most relevance to the present chapter are the endorphins, in particular beta-endorphin (β-END). Endogenous opioids are produced within the pituitary and bind preferentially to μ-opioid receptors (MOR), which are abundantly expressed in emotion processing areas such as LC, H, NAc, PAG, OFC, AMY, HPC and ACC. Notably, β-END and MOR stimulation produce analgesic effects in response to pain, euphoria in response to physical exertion and primary rewards, and they regulate arousal physiology (temperature, breathing, vasoconstriction) and valenced behaviour (approach vs avoidance). Generally, elevated opioid levels enhance the motivation to approach rewarding or threat stimuli, diminish the aversiveness of negative experience (e.g. pain and loss), and boost the experience of rewarding stimuli. By contrast, opioid depletion diminishes the attractivity and pleasantness of rewarding stimuli and increases nociception (for recent reviews, see [271, 295]). In humans and other social species, the opioidergic system modulates the sense of pleasure triggered by social contact, promoting approach and socioemotional bonding [272, 296, 297]. Notably, however, MOR agonists decrease both sexual desire and pleasure, which can be due to suppressive effects exerted by opioids on OT and T [298].

Oxytocin and vasopressin. OT and VP are closely related mammalian nonapeptides with analogous counterparts in other vertebrate and invertebrate classes (reviewed in [299, 300]). Synthesised in H, they can bind to a broad range of chemical entities throughout the brain and body and are implicated in the regulation of HPA, ANS and socioemotional control [301]. The function and mechanisms of OT/VT synthesis have undergone an evolutionary process of diversification and complexification that is paralleled by increased encephalisation [302]. The most ancestral functions of the nonapeptides are related to osmoregulation (through antidiuretic effects) and reproduction,

(continued)

Box 11.2 (continued)

where egg laying is foundational and the regulation of physiological and behavioural processes associated with mating and courtship are derived functions [293, 299, 300, 303, 304].

In mammals, VP and OT regulate an array of socioemotional functions. Specifically, VP regulates defensive behaviour (offspring protection, as well as mate and territorial guarding), which capitalises on VP effects on HPA upregulation and SNS mobilisation. OT has mammal-specific reproductive functions in preparing the body and brain for parturition and nursing, which facilitate mother-offspring attachment. This attachment has been theorised to be the forerunner of many complex social phenomena, such as empathy, trust and cooperation [305, 306]. The functional diversification of OT corresponds to an expansion of modulated targets from subcortical (e.g. forebrain and AMY) to cortical regions implicated in processing social reward, emotion recognition and empathy [307, 308]. Such effects are further mediated by OT influences on HPA downregulation and PSNS activation. Notably, however, VT and OT effects vary across species depending on the number of releasing neurons, targets, receptor abundance and interactions with other neuromodulators [299]. For example, depending on conditions, DA and NE may stimulate or dampen OT. Moreover, VP and OT effects can synergise, owing to the co-activation of SNS and PSNS, as reported for sexual arousal and orgasm, nurturing and protective maternal behaviours, and some responses to music [23].

considering that, from insects to primates, rhythmic communication is present in contexts where behaviour and emotion are regulated by VT and OT. Moreover, across evolutionary time, the diversification of the behavioural contexts regulated by VT and OT parallels the diversification of behavioural contexts regulated by rhythmic communication ([293] and Chap. 3 this volume).

11.5.4.3 Role of Sex Steroids in the ERC-Link

As shown in Box 11.3, the effects of sex steroids on emotion and behaviour can be organisational, owing to developmental effects, or activational, reflecting the circulating levels of sex steroids at a given moment, in a given tissue. Activational effects can, in turn, be genomic, developing over days to weeks, or can arise rapidly on a moment-to-moment basis. Across these temporal scales, sex steroids can affect motivational, structural and perceptual aspects of rhythmic communication, through effects exerted on central neural circuits and peripheral organs, where steroidal receptors are widely distributed and richly expressed [262, 309–312]. Here, we focus on the motivational aspect of steroidal regulation, given that the aim of the chapter is to map the ERC-link and that the other two aspects are addressed in Chap. 9 [205]. Research trends, foci and traditions vary greatly depending on the targeted taxon, which means that knowledge about the mechanistic details underlying steroidal contributions to the ERC-link is unevenly spread across taxa. Nevertheless, each finding on this topic provides important pieces of the larger puzzle.

First, the sex steroids mediate the ERC-link by regulating the motivation to display. Consistent with this view, the range of social contexts in which emotional states are modulated by sex steroids (reviewed in Box 11.3) largely overlaps with contexts where rhythmic communication is used (reviewed in Chap. 3 [1]).

Box 11.3 The Role of Steroid Hormones in the Neuromodulation of Emotion

The steroids of relevance to the ERC-link—Cortisol (C), testosterone (T) and estrogen (E)—are synthesised at peripheral sites that are under HP regulation. Corticotropin-releasing hormone released from H promotes adrenocorticotropin secretion in P, which stimulates C production in the adrenal glands. Although commonly known as the 'stress' hormone, C is not the trigger of the stress response, rather it sustains metabolic mobilisation by promoting glucose production and availability within the blood and in the brain [313]. This can occur in response to stressors, but also during intensive physical exertion for other purposes (e.g. mating display and play). The regulation of C through the up- and down-stimulation of the HPA axis by an array of neuromodulators constitutes a core aspect of emotional processing. Chronic activation of the HPA axis and C release has widespread clinical implications for the aetiology of mood disorders, as well as metabolic, addictive and (auto)immune pathologies [314].

The sex steroids T and E are produced by the gonads through cholesterol conversion, under the stimulation of hormones released within the HP complex. Circulating steroids modulate behaviour by binding to intranuclear steroid receptors in brain areas that activate and pattern behaviour, as well as on motoneurons and muscle fibres that execute behaviours. Smaller amounts of sex steroids are synthesised within the brain, and the significance of these steroid supplies to behaviour modulation has been increasingly recognised in recent years.

The effects of sex steroids can be indirect and organisational, that is, exerted on brain sise and organisation during periods of tissue formation in embryogenesis or other key developmental periods, such as puberty, or more direct and activational, that is, related to circulating levels of sex steroid hormones at a given time [315, 316]. Sex steroid levels fluctuate periodically across the reproductive cycle, being influenced by environmental factors that regulate the seasonal resurgence of reproductive behaviours. Traditionally, the activational effects of sex steroids were considered to be based on genomic mechanisms, whereby steroids bind to androgen or estrogen intranuclear receptors and change neuronal gene expression, with behavioural effects arising in the course of hours to weeks. Currently, however, it is recognised that sex steroids can also bind to membrane receptors, which rapidly activate other cellular signalling pathways and modify behaviour within minutes or seconds.

(continued)

Box 11.3 (continued)

The sex steroids modulate emotional states and expression in many species (including humans) through effects exerted on socioemotional perception, emotional reactivity and behaviour in a range of contexts like mating, parenting, social bonding and social conflict (for reviews see [317–319]). Sex steroid levels co-vary with the degree of activation and connectivity in emotion-relevant circuits (e.g. AMY, LC, NAcc, HC, OFC and PFC), and they modulate the release or expression of other neuromodulators (OT, VP and DA) implicated in emotional experience and perception (reviewed in [316–318, 320]). Beyond the brain, the sex steroids affect ANS activation through OT and VT mediation patterns that depend on the socio-environmental challenge at hand [318]. Generally, elevated T levels activate VP, thereby leading to HPA upregulation and SNS activation. This process is consistent with the association of T with aggression, trust reduction, energetic mobilisation and increased motivation. By contrast, E stimulates OT release, thereby facilitating PSNS dominance and HPA downregulation, which promotes a shift toward affiliative behaviour. Both T and E (indirectly) engage DA-based reward-seeking behaviour, but with distinct mediation patterns (VP vs OT) that have contrasting behavioural outputs (aggression vs affiliation).

Moreover, several emotional processing hubs (the CeA and the preoptic area of the hypothalamus) are sex steroid-sensitive areas that regulate the production and perception of rhythmic signals. For example, the CeA is implicated by research with frogs, birds and mammals [262, 312]. Additionally, suppressing sex steroid activity in the preoptic area of the hypothalamus diminishes courtship behaviour like singing [310].

Second, the sex steroids modulate mechanisms that underpin the encoding and decoding of emotion in rhythmic signals. In several taxa, rhythmic features that encode arousal (such as those reviewed in Sects. 11.3 and 11.4), can be affected by spontaneous or experimentally induced changes in sex steroid levels (for details, see Chap. 9 [205]). This is consistent with data from brain connectivity research in several vertebrate taxa, spanning fish, frog, bird and mammalian species. Specifically, neuronal assemblies implicated in valence and arousal computations project to putative ERC hubs (e.g. PAG and BG, for details see Sect. 11.5.3.3) that regulate the activation or inhibition of rhythmic communication behaviour. These hubs, in turn, project to neuronal assemblies that regulate the temporal patterning of signals [265, 309, 310, 312, 321, 322]. Sex steroids may therefore contribute to the ERC-link by influencing both the timing of rhythmic signals (i.e. when they are produced) and their temporal characteristics as a function of the signaller's motivational state.

While certain sex-steroidal effects on the motivational and structural aspects of birdsong can be temporally and spatially segregated [323], this does not entail that the sex-steroidal modulation of neuromotor tuning is entirely independent of the sex-steroidal regulation of emotion. For example, T-administration to the preoptic area of H—which regulates the motivation to sing—augments brain volume both in the treated region and in motor areas that control song stereotypy [323]. This outcome, however, may be partly due to an indirect effect of motivation, whereby increased sex steroid levels enhance the motivation to sing and singing activity induces neuroplasticity in the song control system.

11.5.4.4 Neuromodulator Interactions in the ERC-Link

An important caveat to interpreting the findings reviewed above—and, thus, to hypotheses regarding the neuromodulator mediation of the ERC-link—is that, traditionally, the effects of neuromodulators have been investigated in isolation. Despite early awareness that certain neuromodulators act as precursors or catalysts for the release of other neuromodulators, research on their concerted activity patterns is still in its infancy and is propelled by new technical and computational advances. As an illustrative example, NE released from sympathetic nerves that innervate the pineal gland activates an enzyme that converts 5-HT into melatonin (MT), a process that can be inhibited by DA activity [324]. In turn, MT can have suppressive effects on T and E, which modulate the OT and VT systems that up- and downregulate C. In addition, T and E are known to enhance activity in other neuromodulator systems in various ways (e.g. by potentiating release, turnover or receptors), including 5-HT, ACh, NE and DA [325]. Moreover, cells that synthesise NE, 5-HT and DA are sensitive to estradiol [326].

The occurrence of such interactions blurs the attribution of specific effects to specific neurochemical mechanisms. Currently, this complex picture of widely distributed and cascading interactions is assembled from limited findings in a handful of model species. It is the task of future integrative research to investigate which specific elements of the complex picture co-occur, how they set each other in motion, and under what conditions. Thus, despite all the findings reviewed in Sect. 11.5.4, there is still much to uncover about how neuromodulators contribute to the ERC-link—and also separately to emotion processing and rhythmic communication.

11.6 Conclusions and Future Directions

The evolution, development and mechanisms of emotion have fascinated scholars for centuries, resulting in opposing perspectives on how to study it. Traditionally, emotions have been described from a categorical perspective. And yet, the need for objective studies of emotion across species has resulted in an emphasis on a dimensional perspective. From this perspective, emotional states lie along a continuum across multiple features, such as arousal and valence. A broad range of evolutionarily conserved neurobiological mechanisms is involved in the generation and processing of emotional responses across these dimensions. These mechanisms span multiple biological levels, from neuromodulators in the brain to endocrine systems throughout the body.

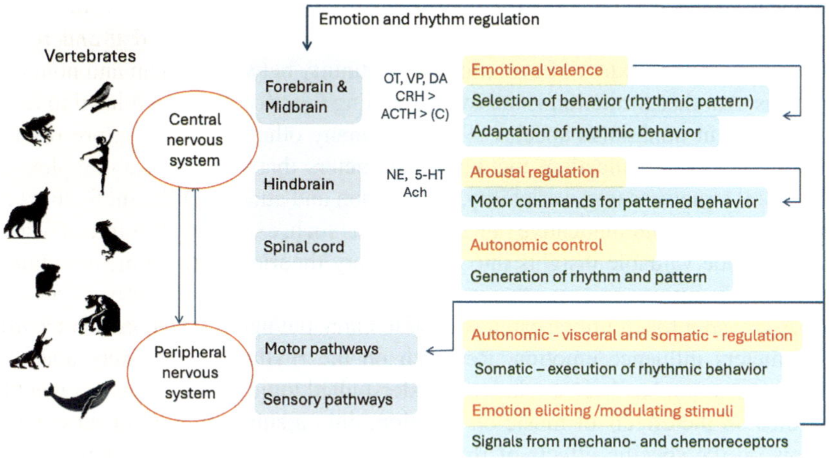

Fig. 11.2 Schematic representation of the neurobiological architecture of the ERC-link, showing levels of organisation and interplay, including autonomic control, neuromodulation, arousal regulation, and valence processing. *Disclaimer regarding AI-tool use: The black silhouette figures used in the composition of* Fig. 11.2 *were generated with Microsoft Copilot and do not reproduce or derive from any copyrighted source material*

The intricate ways in which individuals communicate emotional states remain a puzzle that scientists continue to solve. The growing body of research reviewed in this chapter demonstrates that there are profound links between emotion and rhythmic communication (ERC-link). In human research, these links often arise through musical stimuli and other acoustic stimuli like speech. And yet, rhythmic features of communication channels, such as tempo, also occur in nonhuman communication. Such similarities suggest ancestral origins and an evolutionary continuity of the ERC-link. Supporting this continuity, rhythmic features enable emotional expression, and these features can induce changes in emotional states, in both human and nonhuman animals. Moreover, recent advances in affective neuroscience have vastly increased our understanding of proximate mechanisms that underlie emotional processing through rhythmic communication. Similar to emotional states, the mechanisms underpinning the ERC-link span several biological levels from the autonomic nervous system to putative hubs in the brain and a variety of neuromodulator systems—a complexity that we briefly summarise in Fig. 11.2.

Despite the wealth of information currently available on the ERC-link, there remain several limitations that set up promising avenues for future research. One of the most obvious of these limitations is the overwhelming focus on rhythmic communication in the acoustic domain, which includes music and voice in humans and vocalisations in nonhuman animals. Rhythm is not exclusive to acoustic signals but is also an important component of visual displays that involve body movements.

Expanding research into the visual domain, particularly with studies of rhythmicity-induced emotion, will be critical to assess universal mechanisms that underpin the ERC-link and the extent of evolutionary continuity between human and nonhuman species. Another limitation is the overwhelming focus on features related to tempo, especially in nonhuman species. There are many other ways to explore rhythmic features of communication, including parameters that characterise complexity in temporal structures (e.g. variability, articulation and attack). Different features may serve different communicative functions, and as such, expanding this line of research will provide valuable insights into evolutionary theories on rhythmic communication and cognitive processes that involve emotion.

A key area for future research will be to carry out new studies on how rhythmic parameters influence emotion. Research on the rhythmic parameters that affect human emotion has been a relatively understudied topic. While there is a wealth of studies on the effects of music on emotion, only a small fraction of these studies focus on the specific effects of rhythm. Studies on rhythmicity-induced emotion beyond the realm of music are scarse. Such studies will have important implications in applied settings like music therapy and psychotherapy. It is unclear which musical features—and corresponding mechanisms—underlie the emotional benefits of musicality, resulting in a reduction of symptoms of chronic stress, depression and loneliness.

Future research on rhythmically-induced emotions will also have extensive benefits for nonhuman research and evolutionary frameworks. The vast majority of studies on how rhythmicity affects emotion have been conducted with human-like stimuli (i.e. music). Future studies will need to adapt stimuli to suit species-specific perceptual ranges and sensitivity levels. The use of species-specific stimuli can be enriched by paradigms used in human studies to explore pleasure, groove, chills and prosociality. These phenomena are grounded in ancestral neurobiological mechanisms that are likely to be widespread in the animal kingdom. Such ancestral mechanisms include entrainment of endogenous periodicities to exogenous ones, emotional contagion and even more advanced forms of empathy like empathic concern. By capitalising on such premises in research design and by measuring emotional responses, future research will be in a better position to identify mechanistic continuities and discontinuities across taxa and to assess the evolutionary variability or consistency of the biological significance of rhythmicity.

A final promising area for future research will be to increase research that incorporates a multi-dimensional approach to emotion. Due to a predominant focus on the arousal dimension, especially in nonhuman animals, there remains a critical need for research that examines arousal and valence simultaneously. Furthermore, within the valence dimension, most research focuses on negative emotional states. However, there exists many opportunities to understand how rhythm communicates and influences positive emotional states. Research is needed, for example, to elucidate how and when rhythmic synchronisation produces positive emotion, and which other emotional responses can arise or be enhanced by this synchronisation. Integration of these valence and arousal dimensions will expand understanding of the ERC-link within species, as well as cross-species comparisons to build evolutionary theory on the ECR-link.

Acknowledgements G-A Sauciuc's contribution to this chapter has been funded by the Bank of Sweden Tercentenary foundation (grant nr. P17-0942:1) and the Swedish Research Council (grant 2019-03104).

References

1. Sauciuc G-A, Schneider JN, Burchardt LS et al (2026) Chapter 3: the functional and structural diversity of rhythmic communication across species. In: Sauciuc G-A, Lameira AR, Persson T (eds) Nature beats: the what, why and how of animal rhythmic behaviours. Springer Nature Switzerland
2. Darwin CR (1872) The expression of the emotions in man and animals. John Murray, London
3. Scherer KR (1995) Expression of emotion in voice and music. J Voice 9:235–248. https://doi.org/10.1016/S0892-1997(05)80231-0
4. Panksepp J, Bernatzky G (2002) Emotional sounds and the brain: the neuro-affective foundations of musical appreciation. Behav Process 60:133–155. https://doi.org/10.1016/s0376-6357(02)00080-3
5. Juslin PN, Laukka P (2003) Communication of emotions in vocal expression and music performance: different channels, same code? Psychol Bull 129:770–814. https://doi.org/10.1037/0033-2909.129.5.770
6. Bachorowski J-A, Owren MJ (2008) Vocal expressions of emotion. In: Lewis M, Haviland-Jones JM, Barrett LF (eds) Handbook of emotions, 3rd edn. The Guilford Press, pp 196–2010
7. Briefer EF (2012) Vocal expression of emotions in mammals: mechanisms of production and evidence. J Zool 288:1–20. https://doi.org/10.1111/j.1469-7998.2012.00920.x
8. Filippi P, Congdon JV, Hoang J et al (2017) Humans recognize emotional arousal in vocalizations across all classes of terrestrial vertebrates: evidence for acoustic universals. Proc R Soc B Biol Sci 284:20170990. https://doi.org/10.1098/rspb.2017.0990
9. Clynes M (1977) Sentics: the touch of emotions. Doubleday Anchor, New York
10. Sievers B, Polansky L, Casey M, Wheatley T (2013) Music and movement share a dynamic structure that supports universal expressions of emotion. Proc Natl Acad Sci USA 110:70–75. https://doi.org/10.1073/pnas.1209023110
11. Fink B, Bläsing B, Ravignani A, Shackelford TK (2021) Evolution and functions of human dance. Evol Hum Behav 42:351–360. https://doi.org/10.1016/j.evolhumbehav.2021.01.003
12. Sauciuc G-A, Persson T, Madsen E (2022) The natural history of musical rhythm: functional and mechanistic theories on the evolution of human rhythm cognition and the relevance of rhythmic animal behaviors. Comp Cogn Behav Rev 17:33–87. https://doi.org/10.3819/CCBR.2022.170003
13. Haviland-Jones JM, Lewis M, Barrett LF (2016) Handbook of emotions, 4th edn. Guilford Press, New York
14. Panksepp J (1982) Toward a general psychobiological theory of emotions. Behav Brain Sci 5:407–422. https://doi.org/10.1017/S0140525X00012759
15. Ekman P (1992) An argument for basic emotions. Cogn Emot 6:169–200. https://doi.org/10.1080/02699939208411068
16. Watson JB (1930) Behaviorism, Rev. WW Norton & Co
17. Izard C (1977) Human emotions. Plenum Press, New York
18. Ortony A, Turner TJ (1990) What's basic about basic emotions? Psychol Rev 97:315–331. https://doi.org/10.1037/0033-295X.97.3.315
19. Russell JA (1980) A circumplex model of affect. J Pers Soc Psychol 39:1161–1178. https://doi.org/10.1037/h0077714

20. Man K, Damasio A (2019) Homeostasis and soft robotics in the design of feeling machines. Nat Mach Intell 1:446–452. https://doi.org/10.1038/s42256-019-0103-7
21. Barrett LF, Westlin C (2021) Chapter 2 – navigating the science of emotion. In: Meiselman HL (ed) Emotion measurement, 2nd edn. Woodhead Publishing, pp 39–84
22. Schulkin J, Sterling P (2019) Allostasis: a brain-centered, predictive mode of physiological regulation. Trends Neurosci 42:740–752. https://doi.org/10.1016/j.tins.2019.07.010
23. Koelsch S, Jacobs AM, Menninghaus W et al (2015) The quartet theory of human emotions: an integrative and neurofunctional model. Phys Life Rev 13:1–27. https://doi.org/10.1016/j.plrev.2015.03.001
24. Tye KM (2018) Neural circuit motifs in valence processing. Neuron 100:436–452. https://doi.org/10.1016/j.neuron.2018.10.001
25. Malezieux M, Klein AS, Gogolla N (2023) Neural circuits for emotion. Annu Rev Neurosci 46:211–231. https://doi.org/10.1146/annurev-neuro-111020-103314
26. Kreibig SD (2010) Autonomic nervous system activity in emotion: a review. Biol Psychol 84:394–421. https://doi.org/10.1016/j.biopsycho.2010.03.010
27. Ellis RJ, Thayer JF (2010) Music and autonomic nervous system (dys)function. Music Percept 27:317–326
28. Thayer JF, Lane RD (2000) A model of neurovisceral integration in emotion regulation and dysregulation. J Affect Disord 61:201–216. https://doi.org/10.1016/S0165-0327(00)00338-4
29. Condy EE, Friedman BH, Gandjbakhche A (2020) Probing neurovisceral integration via functional near-infrared spectroscopy and heart rate variability. Front Neurosci 14. https://doi.org/10.3389/fnins.2020.575589
30. Coley AA, Padilla-Coreano N, Patel R, Tye KM (2021) Valence processing in the PFC: reconciling circuit-level and systems-level views. Int Rev Neurobiol 158:171–212. https://doi.org/10.1016/bs.irn.2020.12.002
31. Smith DM, Torregrossa MM (2021) Valence encoding in the amygdala influences motivated behavior. Behav Brain Res 411:113370. https://doi.org/10.1016/j.bbr.2021.113370
32. Flavell SW, Gogolla N, Lovett-Barron M, Zelikowsky M (2022) The emergence and influence of internal states. Neuron 110:2545–2570. https://doi.org/10.1016/j.neuron.2022.04.030
33. Wassum KM (2022) Amygdala-cortical collaboration in reward learning and decision making. eLife 11:e80926. https://doi.org/10.7554/eLife.80926
34. Sundberg J (1998) Expressivity in singing. A review of some recent investigations. Logoped Phoniatr Vocol 23:121–127. https://doi.org/10.1080/140154398434130
35. Scherer KR (1986) Vocal affect expression: a review and a model for future research. Psychol Bull 99:143–165. https://doi.org/10.1037/0033-2909.99.2.143
36. Scherer KR (2003) Vocal communication of emotion: a review of research paradigms. Speech Comm 40:227–256. https://doi.org/10.1016/S0167-6393(02)00084-5
37. Banse R, Scherer KR (1996) Acoustic profiles in vocal emotion expression. J Pers Soc Psychol 70:614–636. https://doi.org/10.1037/0022-3514.70.3.614
38. Murray IR, Arnott JL (1993) Toward the simulation of emotion in synthetic speech: a review of the literature on human vocal emotion. J Acoust Soc Am 93:1097–1108. https://doi.org/10.1121/1.405558
39. Scherer KR, Oshinsky JS (1977) Cue utilization in emotion attribution from auditory stimuli. Motiv Emot 1:331–346. https://doi.org/10.1007/BF00992539
40. Scherer KR (1985) Vocal affect signaling: a comparative approach. In: Rosenblatt JS, Beer C, Busnel M-C, Slater PJB (eds) Advances in the study of behavior. Academic Press, pp 189–244
41. Anikin A, Persson T (2017) Nonlinguistic vocalizations from online amateur videos for emotion research: a validated corpus. Behav Res Methods 49:758–771. https://doi.org/10.3758/s13428-016-0736-y
42. Koolagudi SG, Rao KS (2012) Emotion recognition from speech: a review. Int J Speech Technol 15:99–117. https://doi.org/10.1007/s10772-011-9125-1
43. Ververidis D, Kotropoulos C (2006) Emotional speech recognition: resources, features, and methods. Speech Comm 48:1162–1181. https://doi.org/10.1016/j.specom.2006.04.003

44. de Lope J, Graña M (2023) An ongoing review of speech emotion recognition. Neurocomputing 528:1–11. https://doi.org/10.1016/j.neucom.2023.01.002
45. Iida A, Campbell N, Higuchi F, Yasumura M (2003) A corpus-based speech synthesis system with emotion. Speech Comm 40:161–187. https://doi.org/10.1016/S0167-6393(02)00081-X
46. Pell MD, Paulmann S, Dara C et al (2009) Factors in the recognition of vocally expressed emotions: a comparison of four languages. J Phon 37:417–435. https://doi.org/10.1016/j.wocn.2009.07.005
47. Laukka P, Juslin PN, Bresin R (2005) A dimensional approach to vocal expression of emotion. Cogn Emot 19:633–653. https://doi.org/10.1080/02699930441000445
48. Kamiloğlu RG, Fischer AH, Sauter DA (2020) Good vibrations: a review of vocal expressions of positive emotions. Psychon Bull Rev 27:237–265. https://doi.org/10.3758/s13423-019-01701-x
49. Meyer LB (1956) Emotion and meaning in music. University of Chicago Press, Chicago
50. Laukka P, Eerola T, Thingujam NS et al (2013) Universal and culture-specific factors in the recognition and performance of musical affect expressions. Emotion 13:434–449. https://doi.org/10.1037/a0031388
51. Balkwill L-L, Thompson WF (1999) A cross-cultural investigation of the perception of emotion in music: psychophysical and cultural cues. Music Percept 17:43–64. https://doi.org/10.2307/40285811
52. Swaminathan S, Schellenberg EG (2015) Current emotion research in music psychology. Emot Rev 7:189–197. https://doi.org/10.1177/1754073914558282
53. Mehr SA, Singh M, Knox D et al (2019) Universality and diversity in human song. Science 366:eaax0868. https://doi.org/10.1126/science.aax0868
54. Balkwill L-L, Thompson WF, Matsunaga R (2004) Recognition of emotion in Japanese, Western, and Hindustani music by Japanese listeners. Jpn Psychol Res 46:337–349. https://doi.org/10.1111/j.1468-5584.2004.00265.x
55. Fritz T, Jentschke S, Gosselin N et al (2009) Universal recognition of three basic emotions in music. Curr Biol 19:573–576. https://doi.org/10.1016/j.cub.2009.02.058
56. Scherer KR, Sundberg J, Fantini B et al (2017) The expression of emotion in the singing voice: acoustic patterns in vocal performance. J Acoust Soc Am 142:1805. https://doi.org/10.1121/1.5002886
57. Kragness HE, Trainor LJ (2019) Nonmusicians express emotions in musical productions using conventional cues. Music Sci 2:1–13. https://doi.org/10.1177/2059204319834943
58. Trainor LJ (2015) The origins of music in auditory scene analysis and the roles of evolution and culture in musical creation. Philos Trans R Soc B Biol Sci 370:20140089. https://doi.org/10.1098/rstb.2014.0089
59. Brunswik E (1956) Perception and the representative design of psychological experiments. University of California Press, Berkley, CA
60. Chang A, Kragness HE, Livingstone SR et al (2019) Body sway reflects joint emotional expression in music ensemble performance. Sci Rep 9:205. https://doi.org/10.1038/s41598-018-36358-4
61. Bieńkiewicz MMN, Janaqi S, Jean P, Bardy BG (2023) Impact of emotion-laden acoustic stimuli on group synchronisation performance. Sci Rep 13:7094. https://doi.org/10.1038/s41598-023-34406-2
62. Krumhuber EG, Skora LI, Hill HCH, Lander K (2023) The role of facial movements in emotion recognition. Nat Rev Psychol 2:283–296
63. Sowden S, Schuster BA, Keating CT et al (2021) The role of movement kinematics in facial emotion expression production and recognition. Emotion 21:1041–1061. https://doi.org/10.1037/emo0000835
64. Dalla Bella S, Peretz I, Rousseau L, Gosselin N (2001) A developmental study of the affective value of tempo and mode in music. Cognition 80:B1–B10. https://doi.org/10.1016/s0010-0277(00)00136-0

65. Trainor LJ, Austin CM, Desjardins RN (2000) Is infant-directed speech prosody a result of the vocal expression of emotion? Psychol Sci 11:188–195. https://doi.org/10.1111/1467-9280.00240
66. Kragness HE, Eitel MJ, Baksh AM, Trainor LJ (2021) Evidence for early arousal-based differentiation of emotions in children's musical production. Dev Sci 24:e12982. https://doi.org/10.1111/desc.12982
67. Franco F, Chew M, Swaine JS (2017) Preschoolers' attribution of affect to music: a comparison between vocal and instrumental performance. Psychol Music 45:131–149. https://doi.org/10.1177/0305735616652954
68. Briefer EF (2020) Coding for 'dynamic' information: vocal expression of emotional arousal and valence in non-human animals. In: Aubin T, Mathevon N (eds) Coding strategies in vertebrate acoustic communication. Springer International Publishing, Cham, pp 137–162
69. Fischer J, Hammerschmidt K, Todt D (1995) Factors affecting acoustic variation in Barbary-macaque (Macaca sylvanus) disturbance calls. Ethology 101:51–66. https://doi.org/10.1111/j.1439-0310.1995.tb00345.x
70. Blumstein DT, Armitage KB (1997) Alarm calling in yellow-bellied marmots: I. The meaning of situationally variable alarm calls. Anim Behav 53:143–171. https://doi.org/10.1006/anbe.1996.0285
71. Manser MB (2001) The acoustic structure of suricates' alarm calls varies with predator type and the level of response urgency. Proc R Soc Lond B Biol Sci 268:2315–2324. https://doi.org/10.1098/rspb.2001.1773
72. Yin S, McCowan B (2004) Barking in domestic dogs: context specificity and individual identification. Anim Behav 68:343–355. https://doi.org/10.1016/j.anbehav.2003.07.016
73. Demartsev V, Kershenbaum A, Ilany A et al (2014) Male hyraxes increase song complexity and duration in the presence of alert individuals. Behav Ecol 25:1451–1458. https://doi.org/10.1093/beheco/aru155
74. Gogoleva SS, Volodin IA, Volodina EV et al (2010) Sign and strength of emotional arousal: vocal correlates of positive and negative attitudes to humans in silver foxes (*Vulpes vulpes*). Behaviour 147:1713–1736. https://doi.org/10.1163/000579510X528242
75. Esch HC, Sayigh LS, Blum JE, Wells RS (2009) Whistles as potential indicators of stress in bottlenose dolphins (*Tursiops truncatus*). J Mammal 90:638–650. https://doi.org/10.1644/08-MAMM-A-069R.1
76. Stoeger AS, Baotic A, Li D, Charlton BD (2012) Acoustic features indicate arousal in infant giant panda vocalisations. Ethology 118:896–905. https://doi.org/10.1111/j.1439-0310.2012.02080.x
77. Scheumann M, Roser A-E, Konerding W et al (2012) Vocal correlates of sender-identity and arousal in the isolation calls of domestic kitten (*Felis silvestris catus*). Front Zool 9:36. https://doi.org/10.1186/1742-9994-9-36
78. Schrader L, Todt D (1998) Vocal quality is correlated with levels of stress hormones in domestic pigs. Ethology 104:859–876. https://doi.org/10.1111/j.1439-0310.1998.tb00036.x
79. Illmann G, Hammerschmidt K, Špinka M, Tallet C (2013) Calling by domestic piglets during simulated crushing and isolation: a signal of need? PLoS One 8:e83529. https://doi.org/10.1371/journal.pone.0083529
80. Linhart P, Ratcliffe VF, Reby D, Špinka M (2015) Expression of emotional arousal in two different piglet call types. PLoS One 10:e0135414. https://doi.org/10.1371/journal.pone.0135414
81. Theis KR, Greene KM, Benson-Amram SR, Holekamp KE (2007) Sources of variation in the long-distance vocalizations of spotted hyenas. Behaviour 144:557–584
82. Yamaguchi C, Izumi A, Nakamura K (2010) Time course of vocal modulation during isolation in common marmosets (*Callithrix jacchus*). Am J Primatol 72:681–688. https://doi.org/10.1002/ajp.20824
83. Perez EC, Elie JE, Soulage CO et al (2012) The acoustic expression of stress in a songbird: does corticosterone drive isolation-induced modifications of zebra finch calls? Horm Behav 61:573–581. https://doi.org/10.1016/j.yhbeh.2012.02.004

84. Briefer EF, Tettamanti F, McElligott AG (2015) Emotions in goats: mapping physiological, behavioural and vocal profiles. Anim Behav 99:131–143. https://doi.org/10.1016/j.anbehav.2014.11.002
85. Siebert ER, Parr LA (2003) A structural and contextual analysis of chimpanzee screams. Ann N Y Acad Sci 1000:104–109. https://doi.org/10.1196/annals.1280.022
86. Bastian A, Schmidt S (2008) Affect cues in vocalizations of the bat, *Megaderma lyra*, during agonistic interactions. J Acoust Soc Am 124:598–608. https://doi.org/10.1121/1.2924123
87. Vehrencamp SL, Yantachka J, Hall ML, De Kort SR (2013) Trill performance components vary with age, season, and motivation in the banded wren. Behav Ecol Sociobiol 67:409–419. https://doi.org/10.1007/s00265-012-1461-x
88. Szipl G, Ringler E, Spreafico M, Bugnyar T (2017) Calls during agonistic interactions vary with arousal and raise audience attention in ravens. Front Zool 14:57. https://doi.org/10.1186/s12983-017-0244-7
89. Dunlop RA (2017) Potential motivational information encoded within humpback whale non-song vocal sounds. J Acoust Soc Am 141:2204. https://doi.org/10.1121/1.4978615
90. Rendall D (2003) Acoustic correlates of caller identity and affect intensity in the vowel-like grunt vocalizations of baboons. J Acoust Soc Am 113:3390–3402. https://doi.org/10.1121/1.1568942
91. Martin M, Gridley T, Elwen SH, Charrier I (2022) Feel the beat: cape fur seal males encode their arousal state in their bark rate. Sci Nat 109:5. https://doi.org/10.1007/s00114-021-01778-2
92. Maigrot A-L, Hillmann E, Briefer EF (2018) Encoding of emotional valence in wild boar (*Sus scrofa*) calls. Animals 8:85. https://doi.org/10.3390/ani8060085
93. Lima AF, Lima SGC, Nogueira-Filho SLG et al (2022) Vocal expression of emotions in farmed spotted paca (*Cuniculus paca*). Appl Anim Behav Sci 256:105753. https://doi.org/10.1016/j.applanim.2022.105753
94. Schehka S, Zimmermann E (2009) Acoustic features to arousal and identity in disturbance calls of tree shrews (*Tupaia belangeri*). Behav Brain Res 203:223–231. https://doi.org/10.1016/j.bbr.2009.05.007
95. Briefer EF, Maigrot A-L, Mandel R et al (2015) Segregation of information about emotional arousal and valence in horse whinnies. Sci Rep 5:9989. https://doi.org/10.1038/srep09989
96. Briefer EF, Vizier E, Gygax L, Hillmann E (2019) Expression of emotional valence in pig closed-mouth grunts: involvement of both source- and filter-related parameters. J Acoust Soc Am 145:2895. https://doi.org/10.1121/1.5100612
97. Meise K, Keller C, Cowlishaw G, Fischer J (2011) Sources of acoustic variation: implications for production specificity and call categorization in chacma baboon (*Papio ursinus*) grunts. J Acoust Soc Am 129:1631–1641. https://doi.org/10.1121/1.3531944
98. Brudzynski SM (2007) Ultrasonic calls of rats as indicator variables of negative or positive states: acetylcholine–dopamine interaction and acoustic coding. Behav Brain Res 182:261–273. https://doi.org/10.1016/j.bbr.2007.03.004
99. Tallet C, Linhart P, Policht R et al (2013) Encoding of situations in the vocal repertoire of piglets (*Sus scrofa*): a comparison of discrete and graded classifications. PLoS One 8:e71841. https://doi.org/10.1371/journal.pone.0071841
100. Soltis J, Blowers TE, Savage A (2011) Measuring positive and negative affect in the voiced sounds of African elephants (*Loxodonta africana*). J Acoust Soc Am 129:1059–1066. https://doi.org/10.1121/1.3531798
101. Taylor AM, Reby D, McComb K (2009) Context-related variation in the vocal growling behaviour of the domestic dog (*Canis familiaris*). Ethology 115:905–915. https://doi.org/10.1111/j.1439-0310.2009.01681.x
102. Faragó T, Pongrácz P, Range F et al (2010) 'The bone is mine': affective and referential aspects of dog growls. Anim Behav 79:917–925. https://doi.org/10.1016/j.anbehav.2010.01.005
103. Burgdorf J, Brudzynski S, Moskal J (2020) Using rat ultrasonic vocalization to study the neurobiology of emotion: from basic science to the development of novel therapeutics for affective disorders. Curr Opin Neurobiol 60:192–200. https://doi.org/10.1016/j.conb.2019.12.008

104. Fichtel C, Hammerschmidt K, Jürgens U (2001) On the vocal expression of emotion. A multi-parametric analysis of different states of aversion in the squirrel monkey. Behaviour 138:97–116. https://doi.org/10.1163/15685390151067094
105. Delcomyn F (1980) Neural basis of rhythmic behavior in animals. Science 210:492–498. https://doi.org/10.1126/science.7423199
106. Ota N, Gahr M, Soma M (2015) Tap dancing birds: the multimodal mutual courtship display of males and females in a socially monogamous songbird. Sci Rep 5:16614. https://doi.org/10.1038/srep16614
107. Ota N, Gahr M, Soma M (2017) Songbird tap dancing produces non-vocal sounds. Bioacoustics 26:161–168. https://doi.org/10.1080/09524622.2016.1231080
108. Ota N (2020) Tap dancers in the wild: field observations of multimodal courtship displays in socially monogamous songbirds. Sci Nat 107:30. https://doi.org/10.1007/s00114-020-01686-x
109. Webster GD, Weir CG (2005) Emotional responses to music: interactive effects of mode, texture, and tempo. Motiv Emot 29:19–39. https://doi.org/10.1007/s11031-005-4414-0
110. Husain G, Thompson WF, Schellenberg EG (2002) Effects of musical tempo and mode on arousal, mood, and spatial abilities. Music Percept 20:151–171. https://doi.org/10.1525/mp.2002.20.2.151
111. Gomez P, Danuser B (2007) Relationships between musical structure and psychophysiological measures of emotion. Emotion 7:377–387. https://doi.org/10.1037/1528-3542.7.2.377
112. van der Zwaag MD, Westerink JHDM, van den Broek EL (2011) Emotional and psychophysiological responses to tempo, mode, and percussiveness. Music Sci 15:250–269. https://doi.org/10.1177/1029864911403364
113. Suguna S, Deepika K (2017) The effects of music on pulse rate and blood pressure in healthy young adults. Int J Res Med Sci 5:5268–5272. https://doi.org/10.18203/2320-6012.ijrms20175438
114. Czepiel A, Fink LK, Fink LT et al (2021) Synchrony in the periphery: inter-subject correlation of physiological responses during live music concerts. Sci Rep 11:22457. https://doi.org/10.1038/s41598-021-00492-3
115. Dillman Carpentier FR, Potter RF (2007) Effects of music on physiological arousal: explorations into tempo and genre. Media Psychol 10:339–363. https://doi.org/10.1080/15213260701533045
116. Egermann H, Fernando N, Chuen L, McAdams S (2015) Music induces universal emotion-related psychophysiological responses: comparing Canadian listeners to Congolese pygmies. Front Psychol 5:1341. https://doi.org/10.3389/fpsyg.2014.01341
117. Kim S-G, Mueller, Lepsien J et al (2019) Brain networks underlying aesthetic appreciation as modulated by interaction of the spectral and temporal organisations of music. Sci Rep 9:19446. https://doi.org/10.1038/s41598-019-55781-9
118. Haas F, Distenfeld S, Axen K (1986) Effects of perceived musical rhythm on respiratory pattern. J Appl Physiol 61:1185–1191. https://doi.org/10.1152/jappl.1986.61.3.1185
119. Luck G, Toiviainen P, Erkkilä J et al (2008) Modelling the relationships between emotional responses to, and musical content of, music therapy improvisations. Psychol Music 36:25–45. https://doi.org/10.1177/0305735607079714
120. Huron D (2006) Sweet anticipation: music and the psychology of expectation. The MIT Press, Cambridge, MA
121. Vuust P, Ostergaard L, Pallesen KJ et al (2009) Predictive coding of music—brain responses to rhythmic incongruity. Cortex 45:80–92. https://doi.org/10.1016/j.cortex.2008.05.014
122. Vuust P, Heggli OA, Friston KJ, Kringelbach ML (2022) Music in the brain. Nat Rev Neurosci 23:287–305. https://doi.org/10.1038/s41583-022-00578-5
123. Gebauer L, Kringelbach ML, Vuust P (2012) Ever-changing cycles of musical pleasure: the role of dopamine and anticipation. Psychomusicol Music Mind Brain 22:152–167. https://doi.org/10.1037/a0031126
124. Large EW, Kolen JF (1994) Resonance and the perception of musical meter. Connect Sci 6:177–208. https://doi.org/10.1080/09540099408915723

125. Vuust P, Kringelbach ML (2010) The pleasure of music. In: Kringelbach ML, Berridge KC (eds) Pleasures of the brain. Oxford University Press, New York, pp 255–269
126. Vuust P, Frith CD (2008) Anticipation is the key to understanding music and the effects of music on emotion. Behav Brain Sci 31:599–600. https://doi.org/10.1017/S0140525X08005542
127. Coles NA, March DS, Marmolejo-Ramos F et al (2022) A multi-lab test of the facial feedback hypothesis by the many smiles collaboration. Nat Hum Behav 6:1731–1742. https://doi.org/10.1038/s41562-022-01458-9
128. Elst OFV, Vuust P, Kringelbach ML (2021) Sweet anticipation and positive emotions in music, groove, and dance. Curr Opin Behav Sci 39:79–84. https://doi.org/10.1016/j.cobeha.2021.02.016
129. Koumura T, Nakatani M, Liao H-I, Kondo HM (2021) Dark, loud, and compact sounds induce frisson. Q J Exp Psychol 74:1140–1152. https://doi.org/10.1177/1747021820977174
130. Bignardi G, Chamberlain R, Kevenaar ST et al (2022) On the etiology of aesthetic chills: a behavioral genetic study. Sci Rep 12:3247. https://doi.org/10.1038/s41598-022-07161-z
131. Harrison L, Loui P (2014) Thrills, chills, frissons, and skin orgasms: toward an integrative model of transcendent psychophysiological experiences in music. Front Psychol 5:790. https://doi.org/10.3389/fpsyg.2014.00790
132. Sumpf M, Jentschke S, Koelsch S (2015) Effects of aesthetic chills on a cardiac signature of emotionality. PLoS One 10:e0130117. https://doi.org/10.1371/journal.pone.0130117
133. Bannister S, Eerola T (2023) Vigilance and social chills with music: evidence for two types of musical chills. Psychol Aesthet Creat Arts 17:242–258. https://doi.org/10.1037/aca0000421
134. Panksepp J (1998) Affective neuroscience: the foundations of human and animal emotions. Oxford University Press, New York, NY
135. Guhn M, Hamm A, Zentner M (2007) Physiological and musico-acoustic correlates of the chill response. Music Percept 24:473–484. https://doi.org/10.1525/mp.2007.24.5.473
136. Chabin T, Gabriel D, Chansophonkul T et al (2020) Cortical patterns of pleasurable musical chills revealed by high-density EEG. Front Neurosci 14. https://doi.org/10.3389/fnins.2020.565815
137. Mori K (2022) Decoding peak emotional responses to music from computational acoustic and lyrical features. Cognition 222:105010. https://doi.org/10.1016/j.cognition.2021.105010
138. Janata P, Tomic ST, Haberman JM (2012) Sensorimotor coupling in music and the psychology of the groove. J Exp Psychol Gen 141:54–75. https://doi.org/10.1037/a0024208
139. Vuust P, Witek MAG (2014) Rhythmic complexity and predictive coding: a novel approach to modeling rhythm and meter perception in music. Front Psychol 5. https://doi.org/10.3389/fpsyg.2014.01111
140. Etani T, Miura A, Kawase S et al (2024) A review of psychological and neuroscientific research on musical groove. Neurosci Biobehav Rev 158:105522. https://doi.org/10.1016/j.neubiorev.2023.105522
141. Hurley BK, Martens PA, Janata P (2014) Spontaneous sensorimotor coupling with multipart music. J Exp Psychol Hum Percept Perform 40:1679–1696. https://doi.org/10.1037/a0037154
142. Dotov D, Bosnyak D, Trainor LJ (2021) Collective music listening: movement energy is enhanced by groove and visual social cues. Q J Exp Psychol 74:1037–1053. https://doi.org/10.1177/1747021821991793
143. Witek MAG, Clarke EF, Wallentin M et al (2014) Syncopation, body-movement and pleasure in groove music. PLoS One 9:e94446. https://doi.org/10.1371/journal.pone.0094446
144. Senn O, Bechtold T, Rose D et al (2020) Experience of groove questionnaire: instrument development and initial validation. Music Percept Interdiscip J 38:46–65
145. Matthews TE, Witek MAG, Lund T et al (2020) The sensation of groove engages motor and reward networks. NeuroImage 214:116768. https://doi.org/10.1016/j.neuroimage.2020.116768
146. Leow L-A, Parrott T, Grahn JA (2014) Individual differences in beat perception affect gait responses to low- and high-groove music. Front Hum Neurosci 8. https://doi.org/10.3389/fnhum.2014.00811

147. Etani T, Marui A, Kawase S, Keller PE (2018) Optimal tempo for groove: its relation to directions of body movement and Japanese nori. Front Psychol 9. https://doi.org/10.3389/fpsyg.2018.00462
148. Stupacher J, Hove MJ, Janata P (2016) Audio features underlying perceived groove and sensorimotor synchronization in music. Music Percept 33:571–589. https://doi.org/10.1525/mp.2016.33.5.571
149. Lustig E, Tan I (2020) All about that bass: audio filters on basslines determine groove and liking in electronic dance music. Psychol Music 48:861–875. https://doi.org/10.1177/0305735619836275
150. Zelechowska A, Gonzalez Sanchez VE, Laeng B et al (2020) Who moves to music? Empathic concern predicts spontaneous movement responses to rhythm and music. Music Sci 3:1–16. https://doi.org/10.1177/2059204320974216
151. Witek MAG, Clarke EF, Kringelbach ML, Vuust P (2014) Effects of polyphonic context, instrumentation, and metrical location on syncopation in music. Music Percept 32:201–217. https://doi.org/10.1525/mp.2014.32.2.201
152. Witek MAG, Liu J, Kuubertzie J et al (2020) A critical cross-cultural study of sensorimotor and groove responses to syncopation among Ghanaian and American university students and staff. Music Percept 37:278–297. https://doi.org/10.1525/mp.2020.37.4.278
153. Witek MAG (2017) Filling in: syncopation, pleasure and distributed embodiment in groove. Music Anal 36:138–160. https://doi.org/10.1111/musa.12082
154. Berlyne DE (1971) Aesthetics and psychobiology. Appleton-Century-Crofts, New York, NY
155. Orr MG, Ohlsson S (2005) Relationship between complexity and liking as a function of expertise. Music Percept 22:583–611. https://doi.org/10.1525/mp.2005.22.4.583
156. Juslin PN, Liljeström S, Västfjäll D, Lundqvist L-O (2010) How does music evoke emotions? Exploring the underlying mechanisms. In: Juslin PN, Sloboda JA (eds) Handbook of music and emotion: theory, research, applications. Oxford University Press, pp 605–642
157. Koelsch S (2014) Brain correlates of music-evoked emotions. Nat Rev Neurosci 15:170–180. https://doi.org/10.1038/nrn3666
158. Thaut MH, McIntosh GC, Hoemberg V (2015) Neurobiological foundations of neurologic music therapy: rhythmic entrainment and the motor system. Front Psychol 5. https://doi.org/10.3389/fpsyg.2014.01185
159. J. Trost W, Labbé C, Grandjean D (2017) Rhythmic entrainment as a musical affect induction mechanism. Neuropsychologia 96:96–110. https://doi.org/10.1016/j.neuropsychologia.2017.01.004
160. Mogan R, Fischer R, Bulbulia JA (2017) To be in synchrony or not? A meta-analysis of synchrony's effects on behavior, perception, cognition and affect. J Exp Soc Psychol 72:13–20. https://doi.org/10.1016/j.jesp.2017.03.009
161. Scherer KR, Zentner MR (2001) Emotional effects of music: production rules. In: Juslin PN, Sloboda JA (eds) Music and emotion: Theory and research. Oxford University Press, pp 361–392
162. Scherer KR, Coutinho E (2013) How music creates emotion: a multifactorial process approach. In: Cochrane T, Fantini B, Scherer KR (eds) The emotional power of music: multidisciplinary perspectives on musical arousal, expression, and social control. Oxford University Press, pp 121–145
163. Reddish P, Fischer R, Bulbulia J (2013) Let's dance together: synchrony, shared intentionality and cooperation. PLoS One 8:e71182. https://doi.org/10.1371/journal.pone.0071182
164. Bharucha JJ, Curtis M (2008) Affective spectra, synchronization, and motion: aspects of the emotional response to music. Behav Brain Sci 31:579–579. https://doi.org/10.1017/S0140525X08005335
165. Konvalinka I, Xygalatas D, Bulbulia J et al (2011) Synchronized arousal between performers and related spectators in a fire-walking ritual. Proc Natl Acad Sci 108:8514–8519. https://doi.org/10.1073/pnas.1016955108

166. Fischer R, Xygalatas D, Mitkidis P et al (2014) The fire-Walker's high: affect and physiological responses in an extreme collective ritual. PLoS One 9:e88355. https://doi.org/10.1371/journal.pone.0088355
167. Müller V, Lindenberger U (2011) Cardiac and respiratory patterns synchronize between persons during choir singing. PLoS One 6:e24893. https://doi.org/10.1371/journal.pone.0024893
168. Gordon I, Gilboa A, Cohen S et al (2020) Physiological and behavioral synchrony predict group cohesion and performance. Sci Rep 10:8484. https://doi.org/10.1038/s41598-020-65670-1
169. Baranowski-Pinto G, Profeta VLS, Newson M et al (2022) Being in a crowd bonds people via physiological synchrony. Sci Rep 12:613. https://doi.org/10.1038/s41598-021-04548-2
170. Ardizzi M, Calbi M, Tavaglione S et al (2020) Audience spontaneous entrainment during the collective enjoyment of live performances: physiological and behavioral measurements. Sci Rep 10:3813. https://doi.org/10.1038/s41598-020-60832-7
171. Golland Y, Arzouan Y, Levit-Binnun N (2015) The mere co-presence: synchronization of autonomic signals and emotional responses across co-present individuals not engaged in direct interaction. PLoS One 10:e0125804. https://doi.org/10.1371/journal.pone.0125804
172. Shoda H, Adachi M, Umeda T (2016) How live performance moves the human heart. PLoS One 11:e0154322. https://doi.org/10.1371/journal.pone.0154322
173. Bernardi NF, Codrons E, di Leo R et al (2017) Increase in synchronization of autonomic rhythms between individuals when listening to music. Front Physiol 8. https://doi.org/10.3389/fphys.2017.00785
174. Chabin T, Gabriel D, Comte A, Pazart L (2022) Audience interbrain synchrony during live music is shaped by both the number of people sharing pleasure and the strength of this pleasure. Front Hum Neurosci 16. https://doi.org/10.3389/fnhum.2022.855778
175. Chabin T, Gabriel D, Comte A et al (2022) Interbrain emotional connection during music performances is driven by physical proximity and individual traits. Ann N Y Acad Sci 1508:178–195. https://doi.org/10.1111/nyas.14711
176. Kragness HE, Eitel MJ, Anantharajan F et al (2023) An itsy bitsy audience: live performance facilitates infants' attention and heart rate synchronization. Psychol Aesthet Creat Arts. https://doi.org/10.1037/aca0000597
177. Wiltermuth SS, Heath C (2009) Synchrony and cooperation. Psychol Sci 20:1–5. https://doi.org/10.1111/j.1467-9280.2008.02253.x
178. Tarr B, Launay J, Dunbar RIM (2016) Silent disco: dancing in synchrony leads to elevated pain thresholds and social closeness. Evol Hum Behav 37:343–349. https://doi.org/10.1016/j.evolhumbehav.2016.02.004
179. Durkheim E (2008) The elementary forms of religious life (C. Cosman, Trans.). Oxford University Press
180. Jackson JC, Jong J, Bilkey D et al (2018) Synchrony and physiological arousal increase cohesion and cooperation in large naturalistic groups. Sci Rep 8:127. https://doi.org/10.1038/s41598-017-18023-4
181. Swarbrick D, Bosnyak D, Livingstone SR et al (2019) How live music moves us: head movement differences in audiences to live versus recorded music. Front Psychol 9. https://doi.org/10.3389/fpsyg.2018.02682
182. Bernardi NF, Bellemare-Pepin A, Peretz I (2017) Enhancement of pleasure during spontaneous dance. Front Hum Neurosci 11. https://doi.org/10.3389/fnhum.2017.00572
183. Rothenberg D, Roeske TC, Voss HU et al (2014) Investigation of musicality in birdsong. Hear Res 308:71–83. https://doi.org/10.1016/j.heares.2013.08.016
184. Snowdon CT, Teie D (2013) Emotional communication in monkeys: music to their ears? In: Altenmüller E, Schmidt S, Zimmermann E (eds) Evolution of emotional communication. Oxford University Press, pp 133–151
185. Snowdon CT (2021) Animal signals, music and emotional well-being. Animals 11:2670. https://doi.org/10.3390/ani11092670
186. Snowdon CT, Teie D (2010) Affective responses in tamarins elicited by species-specific music. Biol Lett 6:30–32. https://doi.org/10.1098/rsbl.2009.0593

187. Snowdon CT, Teie D, Savage M (2015) Cats prefer species-appropriate music. Appl Anim Behav Sci 166:106–111. https://doi.org/10.1016/j.applanim.2015.02.012
188. Hampton A, Ford A, Cox RE et al (2020) Effects of music on behavior and physiological stress response of domestic cats in a veterinary clinic. J Feline Med Surg 22:122–128. https://doi.org/10.1177/1098612X19828131
189. De Tommaso M, Kaplan G, Chiandetti C, Vallortigara G (2019) Naïve 3-day-old domestic chicks (Gallus gallus) are attracted to discrete acoustic patterns characterizing natural vocalizations. J Comp Psychol 133:118–131. https://doi.org/10.1037/com0000132
190. Harvey H, Rice R, Kayhart K, Torres C (2000) The effects of specific types of music on the activity levels of singly housed chimpanzees (Pan troglodytes). Am J Primatol 51(Suppl. 1):60
191. Brooker JS (2016) An investigation of the auditory perception of western lowland gorillas in an enrichment study. Zoo Biol 35:398–408. https://doi.org/10.1002/zoo.21312
192. Li X, Zhao JN, Zhao P et al (2019) Behavioural responses of piglets to different types of music. Animal 13:2319–2326. https://doi.org/10.1017/S1751731119000260
193. Zapata Cardona J, Ceballos MC, Tarazona Morales AM et al (2023) Spectro-temporal acoustic elements of music interact in an integrated way to modulate emotional responses in pigs. Sci Rep 13:2994. https://doi.org/10.1038/s41598-023-30057-5
194. Wells DL, Coleman D, Challis MG (2006) A note on the effect of auditory stimulation on the behaviour and welfare of zoo-housed gorillas. Appl Anim Behav Sci 100:327–332. https://doi.org/10.1016/j.applanim.2005.12.003
195. Videan EN, Fritz J, Howell S, Murphy J (2007) Effects of two types and two genre of music on social behavior in captive chimpanzees (Pan troglodytes). J Am Assoc Lab Anim Sci JAALAS 46:66–70
196. Bowman A, Scottish SPCA, Dowell FJ, Evans NP (2015) Four seasons' in an animal rescue centre; classical music reduces environmental stress in kennelled dogs. Physiol Behav 143:70–82. https://doi.org/10.1016/j.physbeh.2015.02.035
197. Wallace EK, Altschul D, Körfer K et al (2017) Is music enriching for group-housed captive chimpanzees (pan troglodytes)? PLoS One 12:e0172672. https://doi.org/10.1371/journal.pone.0172672
198. Mingle ME, Eppley TM, Campbell MW et al (2014) Chimpanzees prefer African and Indian music over silence. J Exp Psychol Anim Learn Cogn 40:502–505. https://doi.org/10.1037/xan0000032
199. Amaya V, Descovich K, Paterson MBA, Phillips CJC (2021) Effects of music pitch and tempo on the behaviour of kennelled dogs. Animals 11:10. https://doi.org/10.3390/ani11010010
200. Zhao P, Zhao J, Liu H et al (2020) Effects of long-term exposure to music on behaviour, immunity and performance of piglets. Anim Prod Sci 61:532–539. https://doi.org/10.1071/AN20407
201. Kunimatsu J, Akiyama Y, Toyoshima O, Matsumoto M (2022) A noninvasive method for monitoring breathing patterns in nonhuman primates using a nasal thermosensor. eNeuro 9. https://doi.org/10.1523/ENEURO.0352-22.2022
202. Zapata Cardona J, Ceballos MC, Tarazona Morales AM et al (2022) Music modulates emotional responses in growing pigs. Sci Rep 12:3382. https://doi.org/10.1038/s41598-022-07300-6
203. Castelhano-Carlos MJ, Baumans V (2009) The impact of light, noise, cage cleaning and in-house transport on welfare and stress of laboratory rats. Lab Anim 43:311–327. https://doi.org/10.1258/la.2009.0080098
204. Hoeschele M, Bowling DL (2016) Sex differences in rhythmic preferences in the budgerigar (Melopsittacus undulatus): a comparative study with humans. Front Psychol 7. https://doi.org/10.3389/fpsyg.2016.01543
205. Sauciuc G-A, Schneider JN, Hattori Y (2025) Chapter 9: the neuroethology of rhythmic communication: neural and mechanical building blocks and constraints on rhythm production and perception. In: Sauciuc G-A, Lameira AR, Persson T (eds) Nature beats: the what, why and how of animal rhythmic behaviours. Springer Nature Switzerland
206. Russo MA, Santarelli DM, O'Rourke D (2017) The physiological effects of slow breathing in the healthy human. Breathe 13:298–309. https://doi.org/10.1183/20734735.009817

207. Lameira AR, Perlman M, Sauciuc G-A, Persson T (forthcoming) Chapter 13: rhythm-induced altered states and the evolution of human dance and music. In: Sauciuc G-A, Lameira AR, Persson T (eds) Nature beats: the what, why and how of animal rhythmic behaviour. Springer Nature Switzerland
208. Kox M, van Eijk LT, Zwaag J et al (2014) Voluntary activation of the sympathetic nervous system and attenuation of the innate immune response in humans. Proc Natl Acad Sci 111:7379–7384. https://doi.org/10.1073/pnas.1322174111
209. Sauciuc G-A, Schneider JN (2025) Chapter 2: how animals produce rhythm—an overview of ethological mechanisms for rhythmic communication. In: Nature beats: the what, why and how of animal rhythmic behaviours. Springer Nature Switzerland
210. Ashhad S, Kam K, Del Negro CA, Feldman JL (2022) Breathing rhythm and pattern and their influence on emotion. Annu Rev Neurosci 45:223–247. https://doi.org/10.1146/annurev-neuro-090121-014424
211. Masaoka Y, Izumizaki M, Homma I (2014) Where is the rhythm generator for emotional breathing? In: Holstege G, Beers CM, Subramanian HH (eds) Progress in brain research. Elsevier, pp 367–377
212. Krohn F, Novello M, van der Giessen RS et al (2023) The integrated brain network that controls respiration. eLife 12:e83654. https://doi.org/10.7554/eLife.83654
213. Yackle K, Schwarz LA, Kam K et al (2017) Breathing control center neurons that promote arousal in mice. Science 355:1411–1415. https://doi.org/10.1126/science.aai7984
214. Del Negro CA, Funk GD, Feldman JL (2018) Breathing matters. Nat Rev Neurosci 19:351–367. https://doi.org/10.1038/s41583-018-0003-6
215. Zelano C, Jiang H, Zhou G et al (2016) Nasal respiration entrains human limbic oscillations and modulates cognitive function. J Neurosci 36:12448–12467. https://doi.org/10.1523/JNEUROSCI.2586-16.2016
216. Heck DH, McAfee SS, Liu Y et al (2017) Breathing as a fundamental rhythm of brain function. Front Neural Circ 10. https://doi.org/10.3389/fncir.2016.00115
217. Herrero JL, Khuvis S, Yeagle E et al (2018) Breathing above the brain stem: volitional control and attentional modulation in humans. J Neurophysiol 119:145–159. https://doi.org/10.1152/jn.00551.2017
218. Tort ABL, Brankačk J, Draguhn A (2018) Respiration-entrained brain rhythms are global but often overlooked. Trends Neurosci 41:186–197. https://doi.org/10.1016/j.tins.2018.01.007
219. Gruber-Dujardin E (2010) Role of the periaqueductal gray in expressing vocalization. In: Brudzynski SM (ed) Handbook of behavioral neuroscience. Elsevier, pp 313–327
220. Grahn JA (2012) Neural mechanisms of rhythm perception: current findings and future perspectives. Top Cogn Sci 4:585–606. https://doi.org/10.1111/j.1756-8765.2012.01213.x
221. Proksch S, Honing H, Ravignani A, Merchant H (2025) Chapter 7: converging notions on the main hypotheses regarding the origins of beat perception and synchronization: the key role of audiomotor interactions. In: Nature beats: the what, why and how of animal rhythmic behaviour. Springer Nature Switzerland
222. Kung S-J, Chen JL, Zatorre RJ, Penhune VB (2013) Interacting cortical and basal ganglia networks underlying finding and tapping to the musical beat. J Cogn Neurosci 25:401–420. https://doi.org/10.1162/jocn_a_00325
223. Grahn JA, Rowe JB (2009) Feeling the beat: premotor and striatal interactions in musicians and nonmusicians during beat perception. J Neurosci 29:7540–7548. https://doi.org/10.1523/JNEUROSCI.2018-08.2009
224. Burchardt LS, Hoeschele M, Honing H, et al (2025) Chapter 1: what is rhythm and how can we study it?
225. Greenfield M, Merker B (2025) Chapter 5: endogenous rhythm, isochrony and the path to perfect synchrony. In: Nature beats: the what, why and how of animal rhythmic behaviour. Springer Nature Switzerland
226. Merker B, Cook PF, Reichmuth C et al (2025) Chapter 14: human group synchrony in light of the comparative method: what we know and would like to know. A peer commentary treat-

ment. In: Sauciuc G-A, Lameira AR, Persson T (eds) Nature beats: the what, why and how of animal rhythmic behaviour. Springer

227. Pierce JE, Péron J (2020) The basal ganglia and the cerebellum in human emotion. Soc Cogn Affect Neurosci 15:599–613. https://doi.org/10.1093/scan/nsaa076
228. Tamietto M, de Gelder B (2010) Neural bases of the non-conscious perception of emotional signals. Nat Rev Neurosci 11:697–709. https://doi.org/10.1038/nrn2889
229. Koelsch S, Skouras S (2014) Functional centrality of amygdala, striatum and hypothalamus in a "small-world" network underlying joy: an fMRI study with music. Hum Brain Mapp 35:3485–3498. https://doi.org/10.1002/hbm.22416
230. Grillner S, Robertson B, Stephenson-Jones M (2013) The evolutionary origin of the vertebrate basal ganglia and its role in action selection. J Physiol 591:5425–5431. https://doi.org/10.1113/jphysiol.2012.246660
231. Schmahmann JD (2019) The cerebellum and cognition. Neurosci Lett 688:62–75. https://doi.org/10.1016/j.neulet.2018.07.005
232. Damsma A, van Rijn H (2017) Pupillary response indexes the metrical hierarchy of unattended rhythmic violations. Brain Cogn 111:95–103. https://doi.org/10.1016/j.bandc.2016.10.004
233. Bowling DL, Graf Ancochea P, Hove MJ, Fitch WT (2019) Pupillometry of groove: evidence for noradrenergic arousal in the link between music and movement. Front Neurosci 12. https://doi.org/10.3389/fnins.2018.01039
234. Spiech C, Sioros G, Endestad T et al (2022) Pupil drift rate indexes groove ratings. Sci Rep 12:11620. https://doi.org/10.1038/s41598-022-15763-w
235. Skaansar JF, Laeng B, Danielsen A (2024) Microtiming and mental effort: onset asynchronies in musical rhythm modulate pupil size. Music Percept 37:111–133
236. Goulty M, Botton-Amiot G, Rosato E et al (2023) The monoaminergic system is a bilaterian innovation. Nat Commun 14:3284. https://doi.org/10.1038/s41467-023-39030-2
237. Avery MC, Nitz DA, Chiba AA, Krichmar JL (2012) Simulation of cholinergic and noradrenergic modulation of behavior in uncertain environments. Front Comput Neurosci 6. https://doi.org/10.3389/fncom.2012.00005
238. Ruivo LMT-G, Baker KL, Conway MW et al (2017) Coordinated acetylcholine release in prefrontal cortex and hippocampus is associated with arousal and reward on distinct timescales. Cell Rep 18:905. https://doi.org/10.1016/j.celrep.2016.12.085
239. Mineur YS, Cahuzac EL, Mose TN et al (2018) Interaction between noradrenergic and cholinergic signaling in amygdala regulates anxiety- and depression-related behaviors in mice. Neuropsychopharmacology 43:2118–2125. https://doi.org/10.1038/s41386-018-0024-x
240. Mineur YS, Picciotto MR (2021) The role of acetylcholine in negative encoding bias: too much of a good thing? Eur J Neurosci 53:114–125. https://doi.org/10.1111/ejn.14641
241. Bacqué-Cazenave J, Bharatiya R, Barrière G et al (2020) Serotonin in animal cognition and behavior. Int J Mol Sci 21:1649. https://doi.org/10.3390/ijms21051649
242. Crockett MJ, Siegel JZ, Kurth-Nelson Z et al (2015) Dissociable effects of serotonin and dopamine on the valuation of harm in moral decision making. Curr Biol 25:1852. https://doi.org/10.1016/j.cub.2015.05.021
243. Avery MC, Krichmar JL (2017) Neuromodulatory systems and their interactions: a review of models, theories, and experiments. Front Neural Circ 11. https://doi.org/10.3389/fncir.2017.00108
244. Vázquez-León P, Miranda-Páez A, Valencia-Flores K, Sánchez-Castillo H (2023) Defensive and emotional behavior modulation by serotonin in the periaqueductal gray. Cell Mol Neurobiol 43:1453–1468. https://doi.org/10.1007/s10571-022-01262-z
245. Berridge KC, Kringelbach ML (2015) Pleasure systems in the brain. Neuron 86:646. https://doi.org/10.1016/j.neuron.2015.02.018
246. Keiflin R, Janak PH (2015) Dopamine prediction errors in reward learning and addiction: from theory to neural circuitry. Neuron 88:247–263. https://doi.org/10.1016/j.neuron.2015.08.037

247. Ferreri L, Mas-Herrero E, Zatorre RJ et al (2019) Dopamine modulates the reward experiences elicited by music. Proc Natl Acad Sci 116:3793–3798. https://doi.org/10.1073/pnas.1811878116
248. Zatorre RJ, Salimpoor VN (2013) From perception to pleasure: music and its neural substrates. Proc Natl Acad Sci 110:10430–10437. https://doi.org/10.1073/pnas.1301228110
249. Salimpoor VN, Zald DH, Zatorre RJ et al (2015) Predictions and the brain: how musical sounds become rewarding. Trends Cogn Sci 19:86–91. https://doi.org/10.1016/j.tics.2014.12.001
250. Merchant H, Honing H (2014) Are non-human primates capable of rhythmic entrainment? Evidence for the gradual audiomotor evolution hypothesis. Front Neurosci 7. https://doi.org/10.3389/fnins.2013.00274
251. Marinho V, Oliveira T, Rocha K et al (2018) The dopaminergic system dynamic in the time perception: a review of the evidence. Int J Neurosci 128:262–282. https://doi.org/10.1080/00207454.2017.1385614
252. Koshimori Y, Strafella AP, Valli M et al (2019) Motor synchronization to rhythmic auditory stimulation (RAS) attenuates dopaminergic responses in ventral striatum in young healthy adults: [11C]-(+)-PHNO PET study. Front Neurosci 13. https://doi.org/10.3389/fnins.2019.00106
253. Brodal HP, Osnes B, Specht K (2017) Listening to rhythmic music reduces connectivity within the basal ganglia and the reward system. Front Neurosci 11. https://doi.org/10.3389/fnins.2017.00153
254. Howe M, Ridouh I, Allegra Mascaro AL et al (2019) Coordination of rapid cholinergic and dopaminergic signaling in striatum during spontaneous movement. eLife 8:e44903. https://doi.org/10.7554/eLife.44903
255. Bang D, Kishida KT, Lohrenz T et al (2020) Sub-second dopamine and serotonin signaling in human striatum during perceptual decision-making. Neuron 108:999–1010.e6. https://doi.org/10.1016/j.neuron.2020.09.015
256. Hamid AA, Frank MJ, Moore CI (2021) Wave-like dopamine dynamics as a mechanism for spatiotemporal credit assignment. Cell 184:2733–2749.e16. https://doi.org/10.1016/j.cell.2021.03.046
257. Matityahu L, Gilin N, Sarpong GA et al (2023) Acetylcholine waves and dopamine release in the striatum. Nat Commun 14:6852. https://doi.org/10.1038/s41467-023-42311-5
258. Pérez-González D, Lao-Rodríguez AB, Aedo-Sánchez C, Malmierca MS (2024) Acetylcholine modulates the precision of prediction error in the auditory cortex. eLife 12:RP91475. https://doi.org/10.7554/eLife.91475
259. Frey A-L, McCabe C (2020) Effects of serotonin and dopamine depletion on neural prediction computations during social learning. Neuropsychopharmacology 45:1431–1437. https://doi.org/10.1038/s41386-020-0678-z
260. Kuo IY, Ehrlich BE (2015) Signaling in muscle contraction. Cold Spring Harb Perspect Biol 7:a006023. https://doi.org/10.1101/cshperspect.a006023
261. Kavanagh JJ, Taylor JL (2022) Voluntary activation of muscle in humans: does serotonergic neuromodulation matter? J Physiol 600:3657–3670. https://doi.org/10.1113/JP282565
262. Kelley DB (2022) Convergent and divergent neural circuit architectures that support acoustic communication. Front Neural Circ 16. https://doi.org/10.3389/fncir.2022.976789
263. Kelley DB, Ballagh IH, Barkan CL et al (2020) Generation, coordination, and evolution of neural circuits for vocal communication. J Neurosci 40:22–36. https://doi.org/10.1523/JNEUROSCI.0736-19.2019
264. Wood WE, Lovell PV, Mello CV, Perkel DJ (2011) Serotonin, via HTR2 receptors, excites neurons in a cortical-like premotor nucleus necessary for song learning and production. J Neurosci 31:13808–13815. https://doi.org/10.1523/JNEUROSCI.2281-11.2011
265. Dunlap KD, Koukos HM, Chagnaud BP et al (2021) Vocal and electric fish: revisiting a comparison of two teleost models in the neuroethology of social behavior. Front Neural Circ 15. https://doi.org/10.3389/fncir.2021.713105

266. Bachner-Melman R, Dina C, Zohar AH et al (2005) AVPR1a and SLC6A4 gene polymorphisms are associated with creative dance performance. PLoS Genet 1:e42. https://doi.org/10.1371/journal.pgen.0010042
267. Jeong Y-J, Hong S-C, Lee MS et al (2005) Dance movement therapy improves emotional responses and modulates neurohormones in adolescents with mild depression. Int J Neurosci 115:1711–1720. https://doi.org/10.1080/00207450590958574
268. Dissanayake E (2009) Root, leaf, blossom, or bole: concerning the origin and adaptive function of music. In: Communicative musicality: exploring the basis of human companionship. Oxford University Press, New York, pp 17–30
269. Lameira AR, Eerola T, Ravignani A (2019) Coupled whole-body rhythmic entrainment between two chimpanzees. Sci Rep 9:18914. https://doi.org/10.1038/s41598-019-55360-y
270. Mason G, Rushen J (2008) Stereotypic animal behaviour: fundamentals and applications to welfare. CABI
271. Nummenmaa L, Tuominen L (2018) Opioid system and human emotions. Br J Pharmacol 175:2737–2749. https://doi.org/10.1111/bph.13812
272. Panksepp J, Siviy SM, Normansell LA (1985) Brain opioids and social emotions. In: Reite M, Field T (eds) The psychobiology of attachment and separation. Academic Press, pp 3–49
273. Tarr B, Launay J, Benson C, Dunbar RIM (2017) Naltrexone blocks endorphins released when dancing in synchrony. Adapt Hum Behav Physiol 3:241–254. https://doi.org/10.1007/s40750-017-0067-y
274. Freeman W (2000) A neurobiological role of music in social bonding. In: The origins of music. The MIT Press, Cambridge, MA, pp 411–424
275. Harvey AR (2020) Links between the neurobiology of oxytocin and human musicality. Front Hum Neurosci 14. https://doi.org/10.3389/fnhum.2020.00350
276. Greenberg DM, Decety J, Gordon I (2021) The social neuroscience of music: understanding the social brain through human song. Am Psychol 76:1172–1185. https://doi.org/10.1037/amp0000819
277. Fukui H, Toyoshima K (2023) Testosterone, oxytocin and co-operation: a hypothesis for the origin and function of music. Front Psychol 14. https://doi.org/10.3389/fpsyg.2023.1055827
278. Nilsson U (2009) Soothing music can increase oxytocin levels during bed rest after open-heart surgery: a randomised control trial. J Clin Nurs 18:2153–2161. https://doi.org/10.1111/j.1365-2702.2008.02718.x
279. Ooishi Y, Mukai H, Watanabe K et al (2017) Increase in salivary oxytocin and decrease in salivary cortisol after listening to relaxing slow-tempo and exciting fast-tempo music. PLoS One 12:e0189075. https://doi.org/10.1371/journal.pone.0189075
280. Grape C, Sandgren M, Hansson L-O et al (2002) Does singing promote Well-being?: an empirical study of professional and amateur singers during a singing lesson. Integr Physiol Behav Sci 38:65–74. https://doi.org/10.1007/BF02734261
281. Kreutz G (2014) Does singing facilitate social bonding? Music Med 6:51–60. https://doi.org/10.47513/mmd.v6i2.180
282. Keeler JR, Roth EA, Neuser BL et al (2015) The neurochemistry and social flow of singing: bonding and oxytocin. Front Hum Neurosci 9. https://doi.org/10.3389/fnhum.2015.00518
283. Schladt TM, Nordmann GC, Emilius R et al (2017) Choir versus solo singing: effects on mood, and salivary oxytocin and cortisol concentrations. Front Hum Neurosci 11. https://doi.org/10.3389/fnhum.2017.00430
284. Yuhi T, Kyuta H, Mori H et al (2017) Salivary oxytocin concentration changes during a group drumming intervention for maltreated school children. Brain Sci 7:152. https://doi.org/10.3390/brainsci7110152
285. Good A, Russo FA (2022) Changes in mood, oxytocin, and cortisol following group and individual singing: a pilot study. Psychol Music 50:1340–1347. https://doi.org/10.1177/03057356211042668
286. Bowling DL, Gahr J, Ancochea PG et al (2022) Endogenous oxytocin, cortisol, and testosterone in response to group singing. Horm Behav 139:105105. https://doi.org/10.1016/j.yhbeh.2021.105105

287. Gebauer L, Witek MAG, Hansen NC et al (2016) Oxytocin improves synchronisation in leader-follower interaction. Sci Rep 6:38416. https://doi.org/10.1038/srep38416
288. Josef L, Goldstein P, Mayseless N et al (2019) The oxytocinergic system mediates synchronized interpersonal movement during dance. Sci Rep 9:1894. https://doi.org/10.1038/s41598-018-37141-1
289. Oikkonen J, Kuusi T, Peltonen P et al (2016) Creative activities in music—a genome-wide linkage analysis. PLoS One 11:e0148679. https://doi.org/10.1371/journal.pone.0148679
290. Miani A (2016) Sexual arousal and rhythmic synchronization: a possible effect of vasopressin. Med Hypotheses 93:122–125. https://doi.org/10.1016/j.mehy.2016.05.030
291. Haakenson CM, Balthazart J, Madison FN, Ball GF (2022) The neural distribution of the avian homologue of oxytocin, mesotocin, in two songbird species, the zebra finch and the canary: a potential role in song perception and production. J Comp Neurol 530:2402–2414. https://doi.org/10.1002/cne.25338
292. Pilgeram NR, Baran NM, Bhise A et al (2023) Oxytocin receptor antagonism during early vocal learning reduces song preference and imitation in zebra finches. Sci Rep 13:6627. https://doi.org/10.1038/s41598-023-33340-7
293. Godwin J, Thompson R (2012) Nonapeptides and social behavior in fishes. Horm Behav 61:230–238. https://doi.org/10.1016/j.yhbeh.2011.12.016
294. Shenoy SS, Lui F (2024) Biochemistry, endogenous opioids. In: StatPearls. StatPearls Publishing, Treasure Island (FL)
295. Meier IM, van Honk J, Bos PA, Terburg D (2021) A mu-opioid feedback model of human social behavior. Neurosci Biobehav Rev 121:250–258. https://doi.org/10.1016/j.neubiorev.2020.12.013
296. Achterberg EJM, van Swieten MMH, Houwing DJ et al (2019) Opioid modulation of social play reward in juvenile rats. Neuropharmacology 159:107332. https://doi.org/10.1016/j.neuropharm.2018.09.007
297. Inagaki TK, Hazlett LI, Andreescu C (2020) Opioids and social bonding: effect of naltrexone on feelings of social connection and ventral striatum activity to close others. J Exp Psychol Gen 149:732–745. https://doi.org/10.1037/xge0000674
298. Argiolas A, Melis MR (2013) Neuropeptides and central control of sexual behaviour from the past to the present: a review. Prog Neurobiol 108:80–107. https://doi.org/10.1016/j.pneurobio.2013.06.006
299. Carter CS (2017) The role of oxytocin and vasopressin in attachment. Psychodyn Psychiatry 45:499–517. https://doi.org/10.1521/pdps.2017.45.4.499
300. Banerjee P, Joy KP, Chaube R (2017) Structural and functional diversity of nonapeptide hormones from an evolutionary perspective: a review. Gen Comp Endocrinol 241:4–23. https://doi.org/10.1016/j.ygcen.2016.04.025
301. Stoop R (2012) Neuromodulation by oxytocin and vasopressin. Neuron 76:142–159. https://doi.org/10.1016/j.neuron.2012.09.025
302. Albers HE (2015) Species, sex and individual differences in the vasotocin/vasopressin system: relationship to neurochemical signaling in the social behavior neural network. Front Neuroendocrinol 36:49–71. https://doi.org/10.1016/j.yfrne.2014.07.001
303. Goodson JL, Kelly AM, Kingsbury MA (2012) Evolving nonapeptide mechanisms of gregariousness and social diversity in birds. Horm Behav 61:239–250. https://doi.org/10.1016/j.yhbeh.2012.01.005
304. Grinevich V, Knobloch-Bollmann HS, Eliava M et al (2016) Assembling the puzzle: pathways of oxytocin signaling in the brain. Biol Psychiatry 79:155–164. https://doi.org/10.1016/j.biopsych.2015.04.013
305. Hostinar CE, Sullivan RM, Gunnar MR (2014) Psychobiological mechanisms underlying the social buffering of the hypothalamic–pituitary–adrenocortical axis: a review of animal models and human studies across development. Psychol Bull 140:256–282. https://doi.org/10.1037/a0032671

306. Ziegler J, Kretzschmar H, Stachniss C et al (2011) Accurate human motion capture in large areas by combining IMU- and laser-based people tracking. In: 2011 IEEE/RSJ international conference on intelligent robots and systems. IEEE, San Francisco, pp 86–91
307. Gunnar MR, Hostinar CE (2015) The social buffering of the hypothalamic–pituitary–adrenocortical axis in humans: developmental and experiential determinants. Soc Neurosci 10:479–488. https://doi.org/10.1080/17470919.2015.1070747
308. Rogers CN, Ross AP, Sahu SP et al (2018) Oxytocin- and arginine vasopressin-containing fibers in the cortex of humans, chimpanzees, and rhesus macaques. Am J Primatol 80:e22875. https://doi.org/10.1002/ajp.22875
309. Bass AH (2008) Steroid-dependent plasticity of vocal motor systems: novel insights from teleost fish. Brain Res Rev 57:299–308. https://doi.org/10.1016/j.brainresrev.2007.04.006
310. Alward BA, Cornil CA, Balthazart J, Ball GF (2018) The regulation of birdsong by testosterone: multiple time-scales and multiple sites of action. Horm Behav 104:32–40. https://doi.org/10.1016/j.yhbeh.2018.04.010
311. Ball GF, Madison FN, Balthazart J, Alward BA (2020) How does testosterone act to regulate a multifaceted adaptive response? Lessons from studies of the avian song system. J Neuroendocrinol 32:e12793. https://doi.org/10.1111/jne.12793
312. Tobiansky DJ, Fuxjager MJ (2020) Sex steroids as regulators of gestural communication. Endocrinology 161:bqaa064. https://doi.org/10.1210/endocr/bqaa064
313. Haase CG, Long AK, Gillooly JF (2016) Energetics of stress: linking plasma cortisol levels to metabolic rate in mammals. Biol Lett 12:20150867. https://doi.org/10.1098/rsbl.2015.0867
314. Karin O, Raz M, Tendler A et al (2020) A new model for the HPA axis explains dysregulation of stress hormones on the timescale of weeks. Mol Syst Biol 16:e9510. https://doi.org/10.15252/msb.20209510
315. Valentino RJ, Reyes B, Van Bockstaele E, Bangasser D (2012) Molecular and cellular sex differences at the intersection of stress and arousal. Neuropharmacology 62:13–20. https://doi.org/10.1016/j.neuropharm.2011.06.004
316. Kret ME, De Gelder B (2012) A review on sex differences in processing emotional signals. Neuropsychologia 50:1211–1221. https://doi.org/10.1016/j.neuropsychologia.2011.12.022
317. van Wingen GA, Ossewaarde L, Bäckström T et al (2011) Gonadal hormone regulation of the emotion circuitry in humans. Neuroscience 191:38–45. https://doi.org/10.1016/j.neuroscience.2011.04.042
318. Bos PA, Panksepp J, Bluthé R-M, van Honk J (2012) Acute effects of steroid hormones and neuropeptides on human social-emotional behavior: a review of single administration studies. Front Neuroendocrinol 33:17–35. https://doi.org/10.1016/j.yfrne.2011.01.002
319. Zilioli S, Bird BM (2017) Functional significance of men's testosterone reactivity to social stimuli. Front Neuroendocrinol 47:1–18. https://doi.org/10.1016/j.yfrne.2017.06.002
320. Little AC (2013) The influence of steroid sex hormones on the cognitive and emotional processing of visual stimuli in humans. Front Neuroendocrinol 34:315–328. https://doi.org/10.1016/j.yfrne.2013.07.009
321. Brudzynski SM (2013) Ethotransmission: communication of emotional states through ultrasonic vocalization in rats. Curr Opin Neurobiol 23:310–317. https://doi.org/10.1016/j.conb.2013.01.014
322. Zheng D-J, Singh A, Phelps SM (2021) Conservation and dimorphism in androgen receptor distribution in Alston's singing mouse (*Scotinomys teguina*). J Comp Neurol 529:2539–2557. https://doi.org/10.1002/cne.25108
323. Alward BA, Balthazart J, Ball GF (2013) Differential effects of global versus local testosterone on singing behavior and its underlying neural substrate. Proc Natl Acad Sci 110:19573–19578. https://doi.org/10.1073/pnas.1311371110
324. González S, Moreno-Delgado D, Moreno E et al (2012) Circadian-related heteromerization of adrenergic and dopamine D_4 receptors modulates melatonin synthesis and release in the pineal gland. PLoS Biol 10:e1001347. https://doi.org/10.1371/journal.pbio.1001347
325. Nguyen T-V, Ducharme S, Karama S (2017) Effects of sex steroids in the human brain. Mol Neurobiol 54:7507–7519. https://doi.org/10.1007/s12035-016-0198-3
326. Maney DL, Rodriguez-Saltos CA (2016) Hormones and the incentive salience of Bird song. In: Bass AH, Sisneros JA, Popper AN, Fay RR (eds) Hearing and hormones. Springer International Publishing, Cham, pp 101–132

The Significance of the Ape Heritage in the Evolution of Human Rhythm in Speech, Music and Dance

12

Gabriela-Alina Sauciuc, Tomas Persson, Yuko Hattori, and Adriano R. Lameira

Abstract

The significance of the great ape heritage to the evolution of human rhythmic competence—as manifested in speech, music and dance—has been greatly underestimated in both past and contemporary research. In this chapter, we deconstruct received misconceptions about the rhythmic competence of the apes and defend the position that, contrary to the received view, the great apes likely constitute the most important model species for research on the evolution of human rhythmicity as manifested in speech, dance and music. To this end, we review data on a range of relevant aspects of great ape behaviour and cognition, including evidence of tool-assisted sound production, the production of rhythmic body movements (including rhythmic engagement and rhythmic synchronisation) and rhythmic vocalisations. We also discuss two cognitive mechanisms—relational processing and shared intentionality—that have been theorised to underlie rhythm perception and sensorimotor synchronisation in humans. For each of these behavioural and cognitive aspects, we discuss their relevance to the evolution of human speech, music and dance and highlight new avenues for future research. Understanding the evolution of these human communicative rhythms and disentangling classical evolutionary conundrums regarding the speech–music and the dance–music relationship hinge on understanding their genesis in the precursor traits of the Hominid family.

G.-A. Sauciuc (✉) · T. Persson
Department of Philosophy/Cognitive Science, Lund University, Lund, Sweden
e-mail: Gabriela-Alina.Sauciuc@lucs.lu.se; Tomas.Persson@lucs.lu.se

Y. Hattori
Institute for the Evolutionary Origins of Human Behavior (EHUB), Kyoto University, Inuyama, Japan
e-mail: hattori.yuko.8w@kyoto-u.ac.jp

A. R. Lameira
Department of Psychology, University of Warwick, Coventry, UK
e-mail: Adriano.Lameira@warwick.ac.uk

G.-A. Sauciuc et al. (eds.), *Nature Beats*, Fascinating Life Sciences,
https://doi.org/10.1007/978-3-032-24250-1_12

12.1 Introduction

The significance of the great ape heritage to the evolution of human rhythmic competence—as manifested in speech, music and dance—has been greatly underestimated in both past and contemporary research. Arguments that the nonhuman great apes (henceforth 'apes') lack pertinent forms of vocal behaviour or rhythmic abilities have deflected research attention from our closest living relatives, showcasing them as irrelevant comparative models for research on the topic. In this chapter, we argue that this traditional view neglects currently available evidence, while also hampering new research on a plethora of relevant ape behaviours that continue to remain understudied. We, moreover, refute this view as being radical, given that it implies that the rhythmic features of speech and musicality would have evolved without a role of descent with modification and shared ancestry. This has left the evolution of human rhythm with nebulous scenarios of origin and evolution that themselves challenge the mechanics and principles of evolution by natural selection. Since rhythmic behaviours (e.g. sound patterns, bodily movements) do not fossilise, studying rhythmic behaviour and cognition in our closest living relatives will allow us to partly recover the extinct roots of musical and speech rhythms within our lineage. Understanding the evolution of these human communicative rhythms hinges on understanding their genesis in the precursor traits of the Hominid family.

As shown in previous chapters, speech and musical competence spans multiple abilities for rhythm production and perception, which, in turn, are underpinned by a great array of neural, physiological, cognitive and socio-cognitive mechanisms (Chaps. 1 [1], 2 [2], 3 [3], 7 [4], 8 [5], 10 [6], and 11 [7]). Some of these abilities and mechanisms are widely shared with other species (e.g. neural resonance, percussive behaviour and isochrony), while others are perceived as either uniquely human or marginally found in some selected few non-ape species (e.g. beat perception, audiomotor synchronisation). Capitalising on the updated view of rhythmic competence provided in previous chapters, in the present chapter, we focus on relevant aspects of great ape behaviour and cognition. In particular, we examine tool-assisted sound production (Sect. 12.2), rhythmic body movements, including rhythmic engagement and rhythmic synchronisation (Sect. 12.3), as well as vocal rhythm production and its relation to speech and music evolution (Sect. 12.4). In Sect. 12.5, we briefly discuss two cognitive mechanisms—relational processing and shared intentionality—that have been theorised to underlie rhythm perception and sensorimotor synchronisation in humans. Traditionally, these have been posited as uniquely human, but currently, there is growing evidence that these may also be present in apes. Additional ape behaviours of relevance, which are related to the induction of altered states of proprioception and consciousness, are examined in Chap. 13 [8]. Our aim here is to deconstruct received misconceptions about the rhythmic competence of the apes, in order to renew our views of human rhythm evolution in light of our ape heritage and appreciate once more its significance for evolutionary theories of human behaviour.

12.2 Tool-Assisted and Non-vocal Sound Production: Isochrony, Variability and Contexts

The use of instruments for sound production is a universal feature across musical cultures of the world. This includes cases in which the body itself is used as a tool for sound production, as in hand clapping or leg slapping. Besides the body, in so-called traditional societies, the most prevalent and simple instruments are the drums, as well as various forms of idiophones, such as rattles, maracas, castanets and wooden blocks [9]. They are usually constructed from perishable materials encountered in the natural environment, such as wood and bone, and may require little to no processing. Since such materials rarely fossilise, it is difficult to get an idea when the use of instruments became a widespread feature in the lives of human ancestors.

Percussive behaviours are documented in chimpanzees (*Pan troglodytes*) and gorillas (*Gorilla* spp.), which has prompted the suggestion that *manual* drumming could be a preserved trait underlying human musicality [10–12]. Ground-slapping and branch shaking are also encountered in other primate species, e.g. bonnet macaques (*Macaca radiata*) [13] and orangutans (*Pongo* spp.) [14], suggesting that human forms of manipulating the physical environment for the purpose of rhythmic sound production have ancestral origins deep within the primate order.

The ape percussive behaviour that has been most studied and most frequently featured in evolutionary theories on human musicality is the chimpanzee *buttress-drumming* [15–18]. In the wild, this behaviour consists of rapid kicks or hand slaps on a resonant tree trunk or buttress. A buttress-drumming bout is brief, typically lasting 1–2 s, and includes a small number of strikes, which, however, varies across chimpanzee communities (from ≈3 strikes/bout [17] to ≈5–7 strikes/bout [19]). These strikes, which exhibit various degrees of isochrony [19], are generally produced as a dominance display. Some analyses have revealed that the limb sequence in a 'drumming' bout closely resembles the limb sequence of a gallop, suggesting that this type of percussive behaviour may be an exaptation of gait patterns [20]. Other studies have suggested that each chimpanzee produces its own drumming sequence as an individual sound signature [16]. Since other forms of chimpanzee rhythmicity are rarely mentioned in evolutionary scenarios of musicality, and since buttress-drumming is brief and, until recently, thought to lack isochrony, it is commonly concluded that chimpanzees—and by extension the other apes too—are a-rhythmical.

Chimpanzee drumming is also common in captivity, where it can last longer (up to minutes) and be isochronous [21, 22]. The frequency rate of drumming is also more variable (0.9–4.2 Hz [22]) and the patterning of foot and hand strikes is clearly dissociated from the gait pattern. A potential explanation may be that larger resonant surfaces (e.g. walls, cage mesh and platforms) are available in captivity, disposed both vertically (as in the wild) but also horizontally, affording longer and more stable drumming. The absence, in captivity, of predators or other potentially dangerous neighbouring chimpanzee groups may also facilitate more attention-drawing behaviours, such as loud drumming, compared to in the wild. Moreover, differences in activity budgets and reinforcement contexts between wild and captive apes may set different stages for the development and performance of percussive behaviours, e.g. promoting the occurrence of drumming in contexts in which it is less documented in the wild.

Indeed, in captivity, all ape genera engage in tool-assisted sound-making as a means for attention-getting, primarily from humans. Apes may repeatedly strike the mesh, walls or windows of the enclosure with their hands or fists or may swipe the mesh with their fingers, resulting in sustained rhythms. Additional tools, such as sticks, containers or other loose enrichment objects, may be used, and the free availability of these may more easily prompt percussive behaviours in captivity. Hand clapping is another attention-getter that apes often use for requesting desired items, which has also been observed in the wild in the form of play invitations [23].

Chimpanzees have also been shown to be able to engage in percussive behaviour that exhibits features reminiscent of human music, such as intentionality, decontextualisation and formality [21]. For example, Dufour and colleagues report a drumming bout by a chimpanzee that lasted several minutes, exhibited isochrony and occurred in a calm, leisurely context. While this chimpanzee's performance was described as exceptional [24], other observations indicate that this behaviour – and individual – is not a unique or odd case. For example, we have documented chimpanzee rhythmic sound produced by sliding back and forth a plastic box, which the chimpanzee had modified to attach to his toe [25]. This episode lasted nearly 20 minutes and included several isochronous bouts. It was also accompanied by a relaxed open-mouth face, indicating a play context and suggesting enjoyment [26]. Such sustained, constant behaviour is distinct from those typically produced hastily during arousal and has been interpreted as an indicator of intentional behaviour [21].

Playful drumming by chimpanzees has also been recorded in the wild. Matsusaka [27] reports two cases in which a juvenile (in the first case) and an adolescent (in the second case) were observed to drum on large clay pots left behind by humans. Like in captivity, these were lengthy episodes (up to 6 min) that involved strikes made by using a variety of effectors (e.g. hands, fists, heels and soles). Moreover, the relaxed open-mouth expression was visible in one case, while bodily or facial expressions of arousal (e.g. piloerection, compressed lips) were absent and the chimpanzees appeared absorbed in the activity. The behaviours were, thus, described as object or sound play [27]. The observation that sound production in these examples is a case of solitary activity suggests that rhythmic sounds are enjoyed individually, other than being necessarily part of a social affair. As such, rhythmic sound production and perception may be intrinsically rewarding for apes. It is unclear whether such a putative reward might stem from the exploration of unusual sounds, the production of rhythm per se, or both.

Another famous percussive ape behaviour often mentioned in evolutionary studies of musical rhythm is gorilla bimanual chest beating—short bursts of hand beats against the chest at a rapid pace (mean of 8 beats at $\approx$14 beats/s [28]). This behaviour is produced by both males and females in a multitude of contexts, such as dominance displays, play and long-distance communication. In adult males, the behaviour seems an honest signal of body size, which, in turn, correlates with male dominance [28]. Gorillas also repeatedly slap other body parts, other individuals, objects and surfaces [23, 29], but no studies seem to exist on the rhythmicity of these behaviours.

The ability of apes to produce a diversity of percussive behaviours has been exploited in experimental studies aimed at evaluating if apes are capable of sensorimotor entrainment, i.e. of aligning a motor response with the phase and period of an exogenous rhythm [30–33]. These studies show that both chimpanzees and bonobos (*Pan paniscus*) can entrain percussive behaviour (finger tapping on a keyboard and unimanual drumming, respectively) with auditory rhythms. However, in these studies, phase-locking occurred when the apes heard rhythms that were close to their own spontaneous motor tempi for the respective percussive behaviour [30, 31, 33]. For chimpanzee finger tapping, this tempo was found to be similar to human finger tapping, i.e. in the range of 400–600 ms IOI [30, 31]. For manual drumming, however, the spontaneous motor tempo was faster in both the bonobos (216 ms) and chimpanzees ($\approx$250 ms) [24, 33].

Besides drumming, there is a broad array of other object-assisted and/or sound-producing behaviours in apes. For example, in the wild, chimpanzees throw rocks at particularly resonant trees as the culminating part of displays, similarly to buttress drumming, but is a non-rhythmic manner given that a thrown stone only produces one strike [34]. Rock throwing is practiced, however, only by some populations. Given that trees and stones are widely available across chimpanzee territory, this suggests the behaviour may be cultural. Chimpanzees can also tear apart a leaf repeatedly and audibly with the mouth without ingesting it as a prelude to pant hoots in some populations [35], but also as part of play invites or courtship displays in others [36]. The rhythmicity of these behaviours is yet to be studied. Rhythmic sound-making in chimpanzees can, however, also occur as a by-product of tool-use in a foraging context, such as manual fruit cracking by hammering [37]. These behaviours, which tend to be exhibited more equally by males and females, typically consist of a short succession of strong strikes or sequences where a strong strike is followed by one or two soft strikes.

Thus far, there are no published studies of percussive rhythms in wild orangutans. However, orangutans exhibit what might represent the only tool-assisted modification of species-specific vocalisations found in apes. Orangutans strip off the leaves of a branch and position them in a bouquet against the mouth while producing an alarm call, which lowers the pitch of the call [38]. This behaviour occurs more often in situations perceived to be more dangerous, and because in nature large body size is associated with lower-pitched calls, orangutans seem to be attempting to acoustically deceive potential predators by sounding larger than they are. Leaf-use differs, however, between populations, suggesting the behaviour is cultural [39].

To sum up, both in the wild and in captivity, tool-assisted sound production is relatively common in apes, and it may result in acoustic rhythms. These behaviours occur across several contexts, but are most commonly associated with play, attention-getting and display towards conspecifics or other species—which is reminiscent of the contexts of human rhythmic behaviours. The prevalence of drumming in display contexts may also help explain the traditionally observed male bias in great ape percussive behaviour. However, it is important to note that this perceived male bias may have been exaggerated by methodological choices, as more recent and exploratory studies provide mounting evidence of percussive behaviours in females. The tool-assisted production of sounds and rhythms in apes often involves inventive ways of using the body and the physical environment, sometimes comprising part of a community's local culture, suggesting a certain degree of control and intentionality, even when produced in high arousal contexts.

12.3 Rhythms in Bodily Movements: Forms, Contexts and Coordination

Besides drumming and other tool-assisted forms of sound production, the apes also exhibit various forms of rhythmic body movements, such as swaying, rocking, head bobbing, arm swinging, [25, 40–42], stereotypic rocking [43], tandem walking in the zoos [44] and synchronised walking in sanctuaries [45]. Moreover, in the wild, they exhibit serial bodily movements during rainstorms and close to waterfalls, which may potentially be rhythmic ('rain dance' [46–49]). Beyond great apes, small apes (i.e. gibbons) show highly structured rhythmicity across modalities. Recent work demonstrates isochrony and rhythmic interaction in duetting [50], robust isochronous singing across three crested gibbon species singing [51], along with female 'dance' displays that constitute structured, intentional rhythmic body movements [52]. These various examples of rhythmic body movements in the great apes have rarely been discussed in relation to the evolutionary roots of human musicality. Nevertheless, there are several aspects that make them relevant to this discussion, including the presence of *rhythmic engagement*, i.e. rhythmic movement elicited by rhythmic sounds or other rhythmic stimuli that is not synchronised with the eliciting stimulus, as well as entrainment.

A landmark paper in this respect is the experimental study of Hattori and colleagues on chimpanzees [40], showing that swaying of the whole body, as well as other rhythmic movements (rocking, foot tapping, etc.) can be induced by auditory stimuli. This effect was found in both sexes but was more pronounced in males. Further experiments with the male chimpanzee with the most pronounced responses revealed a positive correlation between the tempi of rhythmic sound stimuli and the speed of swaying movements [40]. This suggests that chimpanzees not only exhibit rhythmic engagement with auditory rhythms, but to some extent, they can also entrain the period of rhythmic movement to external sound rhythms. In human development, this kind of period alignment emerges later than the more generic rhythmic engagement shown in infancy, being the closest precursor of accurate audiomotor synchronisation (Chap. 10 [6]). In chimpanzees, however, rhythmic body movements were elicited not only by strongly regular beats but also by irregular rhythmic stimuli. This differs from the pattern of rhythmic engagement observed in human development, whereby human infants produce more rhythmic movement when presented with highly regular rhythmic stimuli (music, drumming patterns) compared to the less regular rhythm of speech [53]. This indicates some limitations that have to be acknowledged when drawing parallels between the evolutionary and developmental trajectories of certain rhythmic abilities.

Swaying and rocking in response to sound stimuli have also been reported as stereotypic responses in solitarily reared chimpanzees [43, 54], but this behaviour is likely based on different mechanisms than the rhythmic engagement reported by Hattori and Tomonaga [40]. Indeed, the chimpanzees in Ref. [40] voluntarily came to the experimental booth and sought spatial proximity to the sound source, besides showing a flexible response to the sound. That the chimpanzees voluntarily entered the experimental booth and moved closer to a louder sound, is entirely different from the response animals would have to aversive stimuli. It is therefore likely that the rhythmic response was not due to being distressed by an aversive stimulus.

Outside of the laboratory setting, in more naturalistic and social situations, there have been several investigations of rhythmic body movement and coordination. For example, Sauciuc et al. [41] documented several cases of rhythmic engagement, in which the rhythmic behaviour initiated by a chimpanzee (e.g. ground slapping and rocking) subsequently overlapped with or was followed by the same or another rhythmic behaviour in another individual. Such rhythmic engagements occurred predominantly at a dyadic level, between same-sex or different-sex pairs, and did not display any obvious functional purpose or specialisation. Rhythmic engagements beyond dyads are commonly documented in display contexts, with both sexes participating in them and involving behaviours such as pant hooting, rocking and drumming. The same study [41] also found that the chimpanzees routinely exhibited a broad range of rhythmic behaviours in a diversity of social contexts (initiation of social interaction, play, display, grooming, feeding or courtship). Interestingly, while both males and females produced rhythmic behaviours, males produced the great majority. Thus far, the available evidence, which comes exclusively from African apes, seems to hint at a greater predisposition in males to engage in rhythmic behaviour, in both social and solitary contexts.

However, a most compelling case of rhythmic entrainment between social peers has been recently described in two captive chimpanzee females [44], a behaviour known as tandem walking. The two females were described to routinely engage in synchronised bouts of bipedal co-locomotion, where one female closely followed another in a way that resembled a human conga-line. In these conga-lines, continuous bodily contact was maintained between the chimpanzees, whereby the second in line held the first. This behaviour has been described as a potential form of proto-dance, as the chimpanzee's rhythmic movement exhibited isochrony and mutual synchronisation across tempi, with the females switching roles as 'pacemakers' of the behaviour. Interestingly, a similar behaviour referred to as the 'snake dance' [29] has been documented in young gorillas in the wild, where it has been described as a form of play involving more than two individuals. Unfortunately, most of the information related to this report boils down to a sketch and details about its rhythmic properties are lacking.

Besides tandem walking, chimpanzees may exhibit additional forms of inter-individual rhythmic phase-matching. For example, in a face-to-face interaction documented in Ref. [41], the completion of a binary rhythmic cycle associated with the rocking movement of a male chimpanzee aligned with the completion of a ternary rhythmic cycle generated by the prancing movement of a female. Moreover, in the semi-natural environment of a sanctuary, some of the chimpanzees occasionally coordinated their walking patterns [45]. Specifically, the chimpanzees were found to adapt their individual steps to temporally coordinate the timing with others, which might provide a basis for engaging in other more complex forms of joint action. Again, the gibbon literature provides convergent evidence at the vocal level, where duet partners mutually adjust timing and maintain consistent synchrony [50], and at the species-comparative level, where isochrony is shared in Nomascus singing [51].

Accordingly, even though there is a general sense that male great apes may individually engage more often in rhythmic behaviour, cases of rhythmic entrainment or coordination seem to involve both males and females. As such, it is unclear whether there would have been determinant sex differences in proto-music or proto-dance, as currently proposed by some evolutionary theories of musicality (e.g [55–57]). Findings from gibbons also underscore this point by documenting female participation and leadership in rhythmic behaviours across modalities [50, 52].

To sum up, rhythmic behaviours have been frequently documented in chimpanzees, as well as other ape species. Such behaviours can take on a variety of forms and occur in a variety of contexts. Moreover, contrary to the received view, the great apes—in particular chimpanzees—spontaneously exhibit rhythmic coordination. Although further research is needed to elucidate the cognitive neural mechanisms underlying these behaviours and their functions, the reports described so far indicate that the great apes are a sufficiently relevant and important species when discussing the evolutionary origins of human musicality and the question of which came first, dance or music.

12.4 Vocal and Orofacial Rhythms

Body movement has proven crucial to understanding ape's latent capacities for rhythmic production and rhythmic coordination; however, speech and song production in humans rest on a different type of rhythm expression—vocal rhythms. Any evolutionary account of human rhythm will be required to enlighten whether or how similar vocal rhythms occur in apes, and if present, what may have been the earliest forms and functions of song-like and speech-like behaviours within the human lineage. In this section, we cover available evidence on great ape vocal rhythms.

Around the world, despite some variation, articulatory (i.e. syllable) rate in spoken languages gravitates around 4–5 Hz (range 2–7 Hz or 3–8 Hz) rhythm [58, 59]. That is, across regions and language families, speakers tend to open and close their mouths for an average of 4-to-5 times per second during speech production. This rhythm has now been identified in two different ape orofacial behaviours—chimpanzee lip-smacking [60] and orangutan faux-speech [61]. Chimpanzee lip-smacking (sometimes also termed 'teeth-clacking') is a multimodal orofacial expression that occurs in a grooming context, exhibited specifically by the groomer.

The signal consists of a rapid succession of mouth opening and closing cycles, with either the lips or the teeth producing a smacking or clacking sound. Lip-smacking is selective in that it primarily occurs between valuable partners, either affiliates or higher-ranking individuals [62]. Grooming bouts that are initiated or accompanied by lip-smacking also typically last longer, confirming their affiliative function [62]. The production of grooming sounds varies, however, per population [63], suggesting that there may be ecological or cultural factors affecting whether and how these sounds are produced.

In the remainder of the primate order, speech-like rhythm has been identified in macaque lip-smacks [64] and geladas [65]. Despite the same name as chimpanzee grooming sounds, these behaviours are mechanically distinct between genera (e.g. produced with or without vocal cord involvement) and occur in different contexts. Teeth-clacking in other macaque species [66] and some gibbon song elements also show a speech-like rhythm [67].

Together, these primate orofacial behaviours (including vocalisations) indicate that speech-like rhythm has old roots within the primate orofacial repertoire. They are population- or species-wide behaviours, and their use covers various contexts, suggesting that primate vocal rhythms may be underpinned by specific biological mechanisms (e.g. neurological, mechanical) that have facilitated the repeated emergence of this rhythm at $\approx$5 Hz across independent primate branches.

In addition, speech-like rhythm has been described in a vocally learned sound emitted by a captive orangutan [61], thus acquired through the same means as humans acquire the rhythm of their own language. New instances of similar babbling vocal behaviours at speech-like rhythm exist in other orangutans (Lameira, personal observation), suggesting again that there may be inherent, built-in production and/or perception benefits for primates to adopt vocal behaviours around this rhythmic band, whether acquired through genetic or cultural inheritance.

Most primate behaviours characterised thus far by a speech-like rhythm seem to be affiliative signals. Accordingly, the consensus is that rapid sonorous orofacial movements between 3 and 8 Hz produced between peers in an ancestral species are likely to have offered a precursor to the rhythm of human speech.

In spite of both speech and song being rhythmic, an important feature differentiates the two: isochrony. Speech rhythm is irregular, whereas song rhythm tends to be constant, and thus it often exhibits a tempo and a 'beat'. This differentiation is a direct consequence of the duration of the sounds and their spacing when combined to make up speech or song. Songs may have lexical lyrics; however, to preserve isochrony, the duration of the speech sounds is altered, typically that of the vowel sounds (something well noticeable, for example, in opera).

Within the hominid family, vocal isochrony is not, however, exclusive to human song and has been identified in the long-distance calls of male orangutans [68], in the alarm calls of female orangutans [69], as well as the pants, hoots and grunts of chimpanzees [42]. The vocal behaviour of other ape vocalisations and species (bonobos, gorillas) awaits rhythmic analysis; however, isochrony has now been extensively described across the primate order, to an extent that isochrony is now proposed to represent the putative ancestral state of long-distance calls and/or song in primates [70]. Human vocal isochrony is, thus, well aligned with the behaviour of its extended primate tree.

Rhythmic analyses have also revealed other deeper homologies between ape and human vocal isochrony. For example, one of the few universal features of human song (as well as instrumental music) is the 'duple meter', where a sound sequence is structured by a binary pattern with a strong and a weak beat. Among primates, this feature was first found in lemur song [71], followed by its discovery in gibbon song [72]. Similarly to humans, lemurs and gibbons are singing species; thus, the presence of duple meter in these species, as well as humans, could suggest that the evolution of duple meter among primates depends on the presence of singing behaviour, even if not all primate songs evolve to exhibit duple meter. Though logical, this proposition raises immediately a conundrum for the evolution of song and duple meter in the human lineage—all great apes are non-singing species, and thus, the last ape common ancestor is also predicted to have been non-singing. To address this problem, a re-assessment of orangutan vocal rhythm was conducted, given that duple meter depends on isochrony and given that vocal isochrony has been previously confirmed in orangutans. Analysis of a new set of orangutan male long calls has revealed that ape non-song vocal production can exhibit duple meter [69]. Indeed, orangutans were shown to be able to produce duple meter by two different mechanisms: through tempo variations, as achieved by other singing primates, but also by voiced in-exhaled alternations, as still achieved by some human song traditions [69]. Even though ancient hominids were not singing species, their vocal behaviour can be confidently presumed to have included isochronous displays and duple meter, providing a rich basis for the evolution of what would one day become the amazing diversity of human song displayed across the world's music traditions.

Rhythmic analyses of ape vocal isochrony have also revealed for the first time a feature that is central for song, as well as speech and language—the presence of vocal hierarchies [68]. Analyses of orangutan long calls have shown that isochrony can occur on various timescales. Notably, a call that comprises an isochronous vocal string can itself decompose into an isochronous vocal string. Isochrony can, hence, be nested within isochrony, indicating that, even if all call units of a long call are produced linearly, they are in fact structurally organised into layers or levels. This finding has now been followed by the discovery of third-order self-embedded isochrony in orangutan female alarm calls [69]: isochronous calls compose isochronous combinations that compose isochronous bouts. This evidence raises new interesting questions about song and speech evolution, given that both depend on hierarchies that embed signal strings within other signal strings (e.g. clauses within clauses and phrases within phrases). Orangutan vocal hierarchies exhibit, and have been identified through, isochrony but these vocal hierarchies do not represent songs, and speech is not isochronous. These observations may finally provide empirically support to Darwin's suggestion that music and language derived from the same undifferentiated behaviour forms.

12.5 Great Ape Cognitive Heritage for Human Rhythm Perception and Synchronisation

Before concluding this chapter, we highlight two cognitive mechanisms that have been theorised to support rhythm perception and sensorimotor synchronisation in humans: relational processing and shared intentionality. These have been proclaimed to be human-unique features not found in the other apes (or other nonhuman species), yet the extant empirical evidence points to the contrary. Like most other mechanisms underlying rhythm perception and synchronisation [73] (and references therein), relational processing and shared intentionality are domain-general cognitive processes that are not specialised for musicality or speech. This clarification is important to justify their inclusion in this chapter, given that research that directly links these cognitive abilities to rhythm processing in apes is currently lacking, but there is evidence that apes evince these abilities in other domains. Thus, great ape studies that evaluate relational processing and shared intentionality in rhythmic tasks constitute an important topic for future research regarding the great ape heritage of speech and musicality. In what follows, we will briefly introduce each of the two mechanisms, the role they play in human rhythm perception and synchronisation, and the (currently only non-rhythm related) evidence that the great apes exhibit these relevant cognitive abilities.

12.5.1 Relational Processing: A Cognitive Substrate for Isochrony, Rhythmic Grouping and Recursion

Relational processing, conceived broadly as the ability to perceive relations between items or events, has been studied in various domains (space, time, social relations, etc.). A core distinction in the field is that between first-order relational processing (e.g. recognising that a sound lasts just as long as another sound) and higher-order processing of relations between relations, i.e. recognising a drum pattern as similar to a bass pattern based on the shared temporal relationships [74]. Higher-order relational processing has been showcased as being uniquely human and linked to cognitive capacity, in particular, working memory and fluid intelligence (for a recent overview, see Ref. [75]).

Relational processing has been discussed as underlying several cognitive processes that enable rhythm perception, including relative timing, hierarchical grouping and recursion (see also Chap. 1 [1]). *Relative timing* captures the ability to encode temporal patterns by abstracting the relations between temporal intervals, as opposed to *absolute* (or *duration-based*) timing which captures the encoding of temporal sequences as series of absolute durations [76–78]. Relative timing subserves the perception of isochrony, i.e. the encoding of the relative similarity of successive inter-stimulus intervals [78, 79] (see also Chap. 1 [1]), as well as rhythmic grouping, i.e. the cognitive process of organising temporal units into chains of 'chunks', which is often based on relational similarity [79, 80]. In turn, rhythmic grouping affords metre structuring, i.e. the organisation of beats and 'chunks' into rhythmic hierarchies [80]. The perception of metrical structure, however, may require even more complex forms of relational processing based on recursion, whereby rhythmic patterns are embedded into other rhythmic patterns and so forth [80–83] (for relevant discussions and examples, see Chaps. 1 [1] and 3 [3], respectively).

As already noted above, relational processing strictly related to rhythm processing has been hardly studied in nonhuman great apes. Suggestive evidence comes from the handful of reports of audiomotor and coupled-body synchronisation in chimpanzees and a bonobo (reviewed in Sects. 12.2 and 12.3), given that isochrony perception is considered a prerequisite for audio-motor and inter-individual synchronisation (as detailed in Chap. 1 [1] and above), and isochrony perception is based on relative timing. Relevant is also the recent finding of recursive rhythmic structure in the orangutan long call [68]. An important follow-up to the latter study would be to establish to what extent orangutans perceive such complex relational structures and, if so, what is the biological significance of it.

Besides the suggestive data mentioned above, there is additional evidence from non-rhythmic tasks that apes possess at least some of the cognitive prerequisite for recursion. For example, they can recognise distant dependencies, i.e. they can identify recurrent relations between spatially and/or temporally distant items. Of high relevance to the topics of this volume, is also the finding that chimpanzees can map cross-modal relations between relations, i.e. they can match auditory patterns (e.g. high tone—low tone—high tone) with relationally similar visual patterns (e.g. triangle—square—triangle [84]). Similar higher-order relational processing capabilities extend even to other domains that are of less relevance to musical abilities, e.g. visual and spatial domain [85] (and references therein).

Overall, the findings reviewed above suggest that the relational processing capabilities of apes span several domains and, thus, may also extend to rhythm processing. Notably, recursive abilities, which are regarded by some as the pinnacle of the rhythmic organisation of speech and music, appear to be an ancestral component of the relational processing kit of the apes, possibly being present in the last common ancestor of great apes (ca 20 mya). To more closely address the evolution of relational processing as a mechanism of rhythm cognition, future studies with nonhuman apes should implicate relative timing in distant dependency detection and recursion tasks. There is also a need to conduct direct tests of isochrony perception and rhythmic grouping with great apes.

12.5.2 Shared Intentionality as a Mechanism of Rhythmic Engagement and Synchronisation

Shared intentionality is typically defined as the motivation to share—and co-represent—mental states with others, which, in turn, is viewed as a prerequisite for joint coordinated action (for a recent review see [86]). Therefore, it has been proposed that shared intentionality constitutes a socio-cognitive mechanism of human rhythmic synchronisation, including dancing to music [87, 88]. For example, situations that implicate shared intentionality (i.e. drumming along with a social partner as opposed to a soundtrack) promote the precocious emergence of accurate rhythmic synchronisation in human development [87]. This catalysing effect is likely mediated by the prosocial mindset induced by interindividual synchronisation [87, 88]. This phenomenon is present before the emergence of the ability to synchronise with exogenous rhythms, i.e. young infants—who are not yet capable of synchronisation—exhibit prosocial behaviour towards individuals who synchronised with them [89, 90].

Shared intentionality has been long postulated as a uniquely human set of sociocognitive abilities [91]. This long-held view, however, is currently questioned by a growing body of empirical findings which shows that the nonhuman great apes exhibit nearly all the behavioural markers that, in the developmental literature, have been associated with shared intentionality in children (for an extensive review, see Ref. [86]). Thus far, the putative link between shared intentionality and rhythmic synchronisation has not been directly studied in nonhuman apes. Future research with these species, however, will be crucial to establishing the role that shared intentionality may have played in the evolution of human rhythmic synchronisation and, thus, musicality. Speculating on the currently available data, shared intentionality could underlie the propensity for rhythmic engagement shown by apes (Sect. 12.3) and young children (Chap. 10 [6]). Moreover, the relatively rudimentary forms of shared intentionality documented in apes and young children could partly explain the pattern of success or failure across the rhythmic synchronisation tasks and contexts thus far examined in these populations.

12.6 Conclusions

In this chapter, we set out to defend the position that, contrary to the received view, the great apes likely constitute the most important model species for research on the evolution of human rhythmicity as manifested in speech, dance and music. Based on earlier evolutionary theories, the relevance of the ape heritage to the evolution of human rhythmicity was limited to homologies such as the presence of manual drumming, illustrated by chimpanzee buttress drumming and gorilla chest beating [10–12] or the motivation to display as a group [55]. Capitalising on recent findings and our many years of experience working closely with apes, we greatly expand the confines of these earlier proposals by reviewing data on rhythm in tool-assisted sound making, bodily movements and vocalisations. All these behavioural modalities have been traditionally characterised as being essentially arrhythmic in great apes.

For tool-assisted sound making and bodily movements, we show that there is considerable variety in behavioural forms and contexts of production. We also review evidence that apes exhibit different forms (and degrees) of rhythmic coordination—rhythmic engagement, tempo alignment and sensorimotor synchronisation—that in humans are associated with different developmental stages (Chap. 10 [6]). Notably, this evidence spans behaviour synchronisation with rhythmic stimuli presented in various perceptual modalities (auditory, audiovisual, tactile), which match those used in the human literature. Finally, it is also noteworthy that some aspects of the circumstances surrounding rhythm production in these two behavioural modalities present features reminiscent of musicality: attentional focus (in some cases to the point of absorption), exploration, inventiveness, enjoyment and cultural variability.

For vocal rhythms we highlight music- and speech-like features identified by recent research in the temporal and structural patterning of vocal and orofacial rhythms produced by apes. In particular, we discuss vocal—and more broadly orofacial—communicative rhythms whose frequency rates match the frequency rate of speech, as well as structural patterns encountered in music (isochrony, integer ratios) or in both speech and music (recursion).

Finally, the chapter also discusses data suggesting that cognitive mechanisms theorised to uniquely underpin human rhythm cognition and audiomotor synchronisation are shared with the apes. However, a significant limitation of the current ape research on these topics is that it focuses on other domains than rhythm cognition. This opens up new avenues for future research with apes, aimed, for example, at examining if ape relational processing abilities extend to the temporal domain and if the forms of shared intentionality documented in apes mediate inter-individual synchronisation.

As hinted by some of the examples discussed in this chapter, future research on the rhythmic capabilities of apes will contribute to creating solid empirical grounds for valid theorising on classical evolutionary conundrums regarding the relationship between music and dance and between speech and music. By studying the rhythmic properties of vocal-bodily displays (e.g. the rocked pant-hoots of chimpanzees) or situations in which auditory and motor rhythms are co-perceived by the apes (as in the case of incidental sounds produced by bodily movements), we can gain insight into whether dance and music have likely emerged as an integrated, organic process, or as distinct systems that more recently have coalesced. Likewise, by evaluating rhythmic parameters and rhythmic patterning in ape vocalisations and by comparing these with those that prevail cross-culturally in speech and music, we can gain insight into whether speech and music have arisen from the differentiation of an originally common communication system.

Also important in terms of future research is the need to balance the empirical record with respect to species representation. Currently, most studies on rhythms in tool-assisted sound production and bodily movement have been conducted in the African ape genera. By contrast, most data on the rhythmic properties and patterning of vocalisations have been generated by studies with orangutans. Due to this sampling discrepancy, it is currently difficult to theorise about the evolutionary trajectories of rhythmic competences regarding distinct behavioural modalities.

Acknowledgments The contributions to this chapter by G-A Sauciuc and Tomas Persson have been funded by the Bank of Sweden Tercentenary foundation (grant nr. P17-0942:1) and the Swedish Research Council (grant 2019-03104). The contributions of Adriano R. Lameira have been funded by UK Research & Innovation, Future Leaders Fellowship (grant nr MR/T04229X/1).

References

1. Burchardt LS, Hoeschele M, Honing H et al (2026) Chapter 1: What is rhythm and how can we study it? In: Nature beats: the what, why and how of animal rhythmic behaviours. Springer Nature Switzerland
2. Sauciuc G-A, Schneider JN (2026) Chapter 2: How animals produce rhythm – an overview of ethological mechanisms for rhythmic communication. In: Nature beats: the what, why and how of animal rhythmic behaviours. Springer Nature Switzerland
3. Sauciuc G-A, Schneider JN, Burchardt LS et al (2026) Chapter 3: The functional and structural diversity of rhythmic communication across species. In: Sauciuc G-A, Lameira AR, Persson T (eds) Nature beats: the what, why and how of animal rhythmic behaviours. Springer Nature Switzerland
4. Proksch S, Honing H, Ravignani A, Merchant H (2026) Chapter 7: Converging notions on the main hypotheses regarding the origins of beat perception and synchronization: the key role of audiomotor interactions. In: Nature beats: the what, why and how of animal rhythmic behaviour. Springer Nature Switzerland
5. Large EW (2026) Chapter 8: Neural resonance and comparative rhythm cognition. In: Sauciuc G-A, Lameira AR, Persson T (eds) Nature beats: the what, why and how of animal rhythmic behaviours. Springer Nature Switzerland
6. Sauciuc G-A, Raimondi T, De Gregorio C (2026) Chapter 10: The developmental trajectory of rhythmic abilities in humans and singing primates. In: Sauciuc G-A, Lameira AR, Persson T (eds) Nature beats: the what, why and how of animal rhythmic behaviour. Springer Nature Switzerland
7. Gustison ML, Sauciuc G-A (2026) Chapter 11: Moving the body and the mind - links between emotion and rhythmic communication and the underlying neurobiological mechanisms. In: Sauciuc G-A, Lameira AR, Persson T (eds) Nature beats: the what, why and how of animal rhythmic behaviours. Springer Nature Switzerland
8. Lameira AR, Perlman M, Sauciuc G-A, Persson T (2026) Chapter 13: Rhythm-induced altered states and the evolution of human dance and music. In: Sauciuc G-A, Lameira AR, Persson T (eds) Nature beats: the what, why and how of animal rhythmic behaviour. Springer Nature Switzerland
9. Morley I (2013) The prehistory of music: human evolution, archaeology, and the origins of musicality, 1st edn. Oxford University Press, Oxford
10. Fitch WT (2011) The biology and evolution of rhythm: unravelling a paradox. In: Rebuschat P, Rohmeier M, Hawkins JA, Cross I (eds) Language and music as cognitive systems. Oxford University Press, pp 73–95
11. Fritz T, Koelsch S (2013) Acoustically mediated emotional contagion as an across-species homology underlying music processing. In: Altenmüller E, Schmidt S, Zimmermann E (eds) Evolution of emotional communication, 1st edn. Oxford University Press, Oxford, pp 300–312
12. Ravignani A, Fitch WT, Hanke FD et al (2016) What pinnipeds have to say about human speech, music, and the evolution of rhythm. Front Neurosci 10. https://doi.org/10.3389/fnins.2016.00274
13. Gupta S, Sinha A (2024) Structure and functions of gesture sequences in wild bonnet macaques (*Macaca radiata*). BioRxiv
14. Mackinnon J (1974) The behaviour and ecology of wild orang-utans (*Pongo pygmaeus*). Anim Behav 22:3–74. https://doi.org/10.1016/S0003-3472(74)80054-0
15. Goodall J (1986) Grooming. In: The chimpanzees of Gombe: patterns of behavior. The Belkna, Cambridge, MA, pp 387–408
16. Arcadi AC, Robert D, Boesch C (1998) Buttress drumming by wild chimpanzees: temporal patterning, phrase integration into loud calls, and preliminary evidence for individual distinctiveness. Primates 39:505–518. https://doi.org/10.1007/BF02557572
17. Arcadi A, Robert D, Mugurusi F (2004) A comparison of buttress drumming by male chimpanzees from two populations. Primates 45:135–139. https://doi.org/10.1007/s10329-003-0070-8

18. Eleuteri V, Henderson M, Soldati A et al (2022) The form and function of chimpanzee buttress drumming. Anim Behav 192:189–205. https://doi.org/10.1016/j.anbehav.2022.07.013
19. Eleuteri V, Van Der Werff J, Wilhelm W et al (2025) Chimpanzee drumming shows rhythmicity and subspecies variation. Curr Biol 35:2448–2456.e4. https://doi.org/10.1016/j.cub.2025.04.019
20. Arcadi AC, Wallauer W (2013) They wallop like they gallop: audiovisual analysis reveals the influence of gait on buttress drumming by wild chimpanzees (*Pan troglodytes*). Int J Primatol 34:194–215. https://doi.org/10.1007/s10764-013-9656-1
21. Dufour V, Poulin N, Curé C, Sterck EHM (2015) Chimpanzee drumming: a spontaneous performance with characteristics of human musical drumming. Sci Rep 5:11320. https://doi.org/10.1038/srep11320
22. Lindelöf M, Sauciuc G-A, Persson T, Toumine M (2025) Rhythmic parameters across contexts in the percussive behaviour of chimpanzees (*Pan troglodytes*). In: Proceedings of the 20th edition of the Swedish Cognitive Science Society Conference. Lund
23. Salmi R, Muñoz M (2020) The context of chest beating and hand clapping in wild western gorillas (*Gorilla gorilla gorilla*). Primates 61:225–235. https://doi.org/10.1007/s10329-019-00782-5
24. Dufour V, Pasquaretta C, Gayet P, Sterck EHM (2017) The extraordinary nature of Barney's drumming: a complementary study of ordinary noise making in chimpanzees. Front Neurosci 11. https://doi.org/10.3389/fnins.2017.00002
25. Sauciuc G-A, Zlakowska J, Klang K-J et al (2019) Social antecedents and consequences of rhythmic behaviours in chimpanzees (*Pan troglodytes*). University of Skövde, Umeå, p 8
26. Ross KM, Bard KA, Matsuzawa T (2014) Playful expressions of one-year-old chimpanzee infants in social and solitary play contexts. Front Psychol 5. https://doi.org/10.3389/fpsyg.2014.00741
27. Matsusaka T (2012) Playful drumming by immature wild chimpanzees at Mahale: do they enjoy making sounds? Pan Afr News 19:23–25
28. Wright E, Grawunder S, Ndayishimiye E et al (2021) Chest beats as an honest signal of body size in male mountain gorillas (*Gorilla beringei beringei*). Sci Rep 11:6879. https://doi.org/10.1038/s41598-021-86261-8
29. Schaller GB (1963) The mountain gorilla, 5. impr. University of Chicago Press, Chicago
30. Hattori Y, Tomonaga M, Matsuzawa T (2013) Spontaneous synchronized tapping to an auditory rhythm in a chimpanzee. Sci Rep 3:1566. https://doi.org/10.1038/srep01566
31. Hattori Y, Tomonaga M, Matsuzawa T (2015) Distractor effect of auditory rhythms on self-paced tapping in chimpanzees and humans. PLoS One 10:e0130682. https://doi.org/10.1371/journal.pone.0130682
32. Yu L, Tomonaga M (2015) Interactional synchrony in chimpanzees: examination through a finger-tapping experiment. Sci Rep 5:10218. https://doi.org/10.1038/srep10218
33. Large EW, Gray PM (2015) Spontaneous tempo and rhythmic entrainment in a bonobo (*Pan paniscus*). J Comp Psychol 129:317–328. https://doi.org/10.1037/com0000011
34. Kalan AK, Carmignani E, Kronland-Martinet R et al (2019) Chimpanzees use tree species with a resonant timbre for accumulative stone throwing. Biol Lett 15:20190747. https://doi.org/10.1098/rsbl.2019.0747
35. Kalan AK, Boesch C (2018) Re-emergence of the leaf clip gesture during an alpha takeover affects variation in male chimpanzee loud calls. PeerJ 6:e5079. https://doi.org/10.7717/peerj.5079
36. Matsumoto-Oda A, Tomonaga M (2005) "Intentional" control of sound production found in leaf-clipping display of Mahale chimpanzees. J Ethol 23:109–112. https://doi.org/10.1007/s10164-004-0133-3
37. Meguerditchian A, Vuillemin A, Pruetz JD (2018) Identifying the ape beat in the wild: rhythmic individual signatures from the sounds of manual fruit cracking in Fongoli chimpanzees. In: Proceedings of the 12th international conference on the evolution of language (Evolang12). Wydawnictwo Naukowe Uniwersytetu Mikołaja Kopernika

38. Hardus ME, Lameira AR, Van Schaik CP, Wich SA (2009) Tool use in wild orang-utans modifies sound production: a functionally deceptive innovation? Proc R Soc B Biol Sci 276:3689–3694. https://doi.org/10.1098/rspb.2009.1027
39. Lameira AR, Hardus ME, Nouwen KJJM et al (2013) Population-specific use of the same tool-assisted alarm call between two wild orangutan populations (*Pongo pygmaeus wurmbii*) indicates functional arbitrariness. PLoS One 8:e69749. https://doi.org/10.1371/journal.pone.0069749
40. Hattori Y, Tomonaga M (2020) Rhythmic swaying induced by sound in chimpanzees (*Pan troglodytes*). Proc Natl Acad Sci 117:936–942. https://doi.org/10.1073/pnas.1910318116
41. Sauciuc G-A, Zlakowska J, Klang K-J et al (2021) Rhythmic behaviors in chimpanzees: range, functional contexts, sex differences and emotional correlates. In: Proceedings of the annual meeting of the cognitive science society, vol 43. Retrieved from https://escholarship.org/uc/item/3pv0q6kw
42. Van Der Vleuten BJR, Hovenkamp VA, Varkevisser JM, Spierings MJ (2024) Context-dependent rhythmicity in chimpanzee displays. Proc R Soc B Biol Sci 291:20242200. https://doi.org/10.1098/rspb.2024.2200
43. Berkson G, Mason WA (1964) Stereotyped behaviors of chimpanzees: relation to general arousal and alternative activities. Percept Mot Skills 19:635–652. https://doi.org/10.2466/pms.1964.19.2.635
44. Lameira AR, Eerola T, Ravignani A (2019) Coupled whole-body rhythmic entrainment between two chimpanzees. Sci Rep 9:18914. https://doi.org/10.1038/s41598-019-55360-y
45. Schweinfurth MK, Baldridge DB, Finnerty K et al (2022) Inter-individual coordination in walking chimpanzees. Curr Biol 32:5138–5143.e3. https://doi.org/10.1016/j.cub.2022.09.059
46. Van Lawick-Goodall J (1968) The behaviour of free-living Chimpanzees in the Gombe Stream Reserve. Anim Behav Monogr 1:161–IN12. https://doi.org/10.1016/S0066-1856(68)80003-2
47. Hashimoto C (1998) Chimpanzees of the Kalinzu Forest, Uganda. Pan Afr News 5:6–8. https://doi.org/10.5134/143364
48. Nishida T, Zamma K, Matsusaka T et al (2010) Chimpanzee behavior in the wild: an audio-visual encyclopedia. Springer Japan, Tokyo
49. Whiten, Goodall, McGrew WC et al (2001) Charting cultural variation in chimpanzees. Behaviour 138:1481–1516. https://doi.org/10.1163/156853901317367717
50. Raimondi T, Di Panfilo G, Pasquali M et al (2023) Isochrony and rhythmic interaction in ape duetting. Proc R Soc B Biol Sci 290:20222244. https://doi.org/10.1098/rspb.2022.2244
51. De Gregorio C, Raimondi T, Bevilacqua V et al (2024) Isochronous singing in 3 crested gibbon species (*Nomascus* spp.). Curr Zool 70:291–297. https://doi.org/10.1093/cz/zoad029
52. Coye C, Caspar KR, Patel-Grosz P (2025) Dance displays in gibbons: biological and linguistic perspectives on structured, intentional, and rhythmic body movement. Primates 66:61–73. https://doi.org/10.1007/s10329-024-01154-4
53. Zentner M, Eerola T (2010) Rhythmic engagement with music in infancy. Proc Natl Acad Sci 107:5768–5773. https://doi.org/10.1073/pnas.1000121107
54. Bertolo M, Singh M, Mehr SA (2021) Sound-induced motion in chimpanzees does not imply shared ancestry for music or dance. Proc Natl Acad Sci 118:e2015664118. https://doi.org/10.1073/pnas.2015664118
55. Merker B (1999) Synchronous chorusing and the origins of music. Music Sci 3:59–73. https://doi.org/10.1177/10298649000030S105
56. Merker B (2025) Human group synchrony in light of the comparative method: what we know and would like to know. A peer commentary treatment. In: Nature beats: the what, why and how of animal rhythmic behaviour. Springer Nature Switzerland
57. Miller G (1999) Evolution of human music through sexual selection. In: Wallin NL, Merker B, Brown S (eds) The origins of music. The MIT Press, pp 329–360
58. Chandrasekaran C, Trubanova A, Stillittano S et al (2009) The natural statistics of audiovisual speech. PLoS Comput Biol 5:e1000436. https://doi.org/10.1371/journal.pcbi.1000436
59. Ding N, Patel AD, Chen L et al (2017) Temporal modulations in speech and music. Neurosci Biobehav Rev 81:181–187. https://doi.org/10.1016/j.neubiorev.2017.02.011

60. Pereira AS, Kavanagh E, Hobaiter C et al (2020) Chimpanzee lip-smacks confirm primate continuity for speech-rhythm evolution. Biol Lett 16:20200232. https://doi.org/10.1098/rsbl.2020.0232
61. Lameira AR, Hardus ME, Bartlett AM et al (2015) Speech-like rhythm in a voiced and voiceless orangutan call. PLoS One 10:e116136. https://doi.org/10.1371/journal.pone.0116136
62. Fedurek P, Slocombe KE, Hartel JA, Zuberbühler K (2015) Chimpanzee lip-smacking facilitates cooperative behaviour. Sci Rep 5:13460. https://doi.org/10.1038/srep13460
63. Watts DP (2016) Production of grooming-associated sounds by chimpanzees (*Pan troglodytes*) at Ngogo: variation, social learning, and possible functions. Primates 57:61–72. https://doi.org/10.1007/s10329-015-0497-8
64. Ghazanfar AA, Takahashi DY, Mathur N, Fitch WT (2012) Cineradiography of monkey lip-smacking reveals putative precursors of speech dynamics. Curr Biol 22:1176–1182. https://doi.org/10.1016/j.cub.2012.04.055
65. Bergman TJ (2013) Speech-like vocalized lip-smacking in geladas. Curr Biol 23:R268–R269. https://doi.org/10.1016/j.cub.2013.02.038
66. Toyoda A, Maruhashi T, Malaivijitnond S, Koda H (2017) Speech-like orofacial oscillations in stump-tailed macaque (*Macaca arctoides*) facial and vocal signals. Am J Phys Anthropol 164:435–439. https://doi.org/10.1002/ajpa.23276
67. Terleph TA, Malaivijitnond S, Reichard UH (2018) An analysis of white-handed gibbon male song reveals speech-like phrases. Am J Phys Anthropol 166:649–660. https://doi.org/10.1002/ajpa.23451
68. Lameira AR, Hardus ME, Ravignani A et al (2023) Recursive self-embedded vocal motifs in wild orangutans. Elife 12:RP88348
69. De Gregorio C, Gamba M, Lameira AR (2025) Third-order self-embedded vocal motifs in wild orangutans, and the selective evolution of recursion. Ann N Y Acad Sci 1549:219–229. https://doi.org/10.1111/nyas.15373
70. De Gregorio C, Maiolini M, Raimondi T et al (2024) Isochrony as ancestral condition to call and song in a primate. Ann N Y Acad Sci 1537:41–50. https://doi.org/10.1111/nyas.15151
71. De Gregorio C, Valente D, Raimondi T et al (2021) Categorical rhythms in a singing primate. Curr Biol 31:R1379–R1380. https://doi.org/10.1016/j.cub.2021.09.032
72. Ma H, Wang Z, Han P et al (2024) Small apes adjust rhythms to facilitate song coordination. Curr Biol 34:935–945.e3. https://doi.org/10.1016/j.cub.2023.12.071
73. Trainor LJ (2015) The origins of music in auditory scene analysis and the roles of evolution and culture in musical creation. Philos Trans R Soc B Biol Sci 370:20140089. https://doi.org/10.1098/rstb.2014.0089
74. Flemming TM, Beran MJ, Thompson RKR et al (2008) What meaning means for same and different: analogical reasoning in humans (*Homo sapiens*), chimpanzees (*Pan troglodytes*), and rhesus monkeys (*Macaca mulatta*). J Comp Psychol 122:176–185. https://doi.org/10.1037/0735-7036.122.2.176
75. Gray ME, Holyoak KJ (2020) Individual differences in relational reasoning. Mem Cogn 48:96–110. https://doi.org/10.3758/s13421-019-00964-y
76. Teki S, Grube M, Griffiths TD (2012) A unified model of time perception accounts for duration-based and beat-based timing mechanisms. Front Integr Neurosci 5. https://doi.org/10.3389/fnint.2011.00090
77. Leow L-A, Grahn JA (2014) Neural mechanisms of rhythm perception: present findings and future directions. In: Merchant H, De Lafuente V (eds) Neurobiology of interval timing. Springer New York, New York, NY, pp 325–338
78. Merchant H, Honing H (2014) Are non-human primates capable of rhythmic entrainment? Evidence for the gradual audiomotor evolution hypothesis. Front Neurosci 7. https://doi.org/10.3389/fnins.2013.00274
79. Ten Cate C, Spierings M (2019) Rules, rhythm and grouping: auditory pattern perception by birds. Anim Behav 151:249–257. https://doi.org/10.1016/j.anbehav.2018.11.010
80. Fitch WT (2016) Dance, music, meter and groove: a forgotten partnership. Front Hum Neurosci 10. https://doi.org/10.3389/fnhum.2016.00064

81. Fitch WT (2013) Rhythmic cognition in humans and animals: distinguishing meter and pulse perception. Front Syst Neurosci 7. https://doi.org/10.3389/fnsys.2013.00068
82. Patel AD (2006) Musical rhythm, linguistic rhythm, and human evolution. Music Percept 24:99–104. https://doi.org/10.1525/mp.2006.24.1.99
83. Iversen JR, Balasubramaniam R (2016) Synchronization and temporal processing. Curr Opin Behav Sci 8:175–180. https://doi.org/10.1016/j.cobeha.2016.02.027
84. Ravignani A, Sonnweber R (2017) Chimpanzees process structural isomorphisms across sensory modalities. Cognition 161:74–79. https://doi.org/10.1016/j.cognition.2017.01.005
85. Hribar A (2011) Comparing Great Apes' and children's ablitiy to reason by analogy. Doctoral thesis, Ruprecht-Karls-Universität Heidelberg
86. Sauciuc G-A, Persson T (2023) Empirical challenges from the comparative and developmental literature to the shared intentionality theory – a review of alternative data on recursive mind reading, prosociality, imitation and cumulative culture. Front Psychol 14:1157137. https://doi.org/10.3389/fpsyg.2023.1157137
87. Kirschner S, Tomasello M (2009) Joint drumming: social context facilitates synchronization in preschool children. J Exp Child Psychol 102:299–314. https://doi.org/10.1016/j.jecp.2008.07.005
88. Reddish P, Fischer R, Bulbulia J (2013) Let's dance together: synchrony, shared intentionality and cooperation. PLoS One 8:e71182. https://doi.org/10.1371/journal.pone.0071182
89. Cirelli LK, Einarson KM, Trainor LJ (2014) Interpersonal synchrony increases prosocial behavior in infants. Dev Sci 17:1003–1011. https://doi.org/10.1111/desc.12193
90. Tunçgenç B, Cohen E, Fawcett C (2015) Rock with me: the role of movement synchrony in infants' social and nonsocial choices. Child Dev 86:976–984. https://doi.org/10.1111/cdev.12354
91. Tomasello M (2018) A natural history of human thinking, Paperback edn. Harvard University Press, Cambridge, Masschusetts London

Rhythm-Induced Altered States and the Evolution of Human Dance and Music

13

Adriano R. Lameira, Marcus Perlman, Gabriela-Alina Sauciuc, and Tomas Persson

Abstract

Across human cultures, rhythmic movement appears to be consistently associated with the induction of altered states of consciousness (ASCs), yet research on the topic remains scarce. In this chapter, we examine the potential role of rhythmic movement in ASC induction and its significance for understanding the evolution of dance and music. We discuss the cross-cultural spread of the association between rhythmic movement (e.g. dancing, chanting, breathwork, etc.) and various categories of ASCs and outline four major classes of neurobiological mechanisms that may underlie this association. First, rhythmic movement can accelerate or decelerate breathing and, thus, modulate autonomic activity towards sympathetic or parasympathetic dominance. Second, these can be psychologically amplified through expectation and fear priming. Third, rhythmic practices can induce vestibular perturbations that distort proprioception and multisensory integration. Fourth, through neural entrainment and neural resonance, rhythmic movement can affect brain oscillatory dynamics, as well as the strength and nature of functional connectivity. Building on these mechanistic considerations, we also review emerging comparative evidence that rhythmic movement may induce altered proprioception, arousal and social attunement in the nonhuman great apes, paralleling some of the ASC features reported in humans. We discuss rhythmic stereotypies (e.g. rocking, pacing) as potential analogues of rhythm-

A. R. Lameira (✉)
Department of Psychology, University of Warwick, Coventry, UK
e-mail: Adriano.Lameira@warwick.ac.uk

M. Perlman
Department of Linguistics and Communication, University of Birmingham, Birmingham, UK
e-mail: m.perlman@bham.ac.uk

G.-A. Sauciuc · T. Persson
Department of Philosophy/Cognitive Science, Lund University, Lund, Sweden
e-mail: Gabriela-Alina.Sauciuc@lucs.lu.se; Tomas.Persson@lucs.lu.se

G.-A. Sauciuc et al. (eds.), *Nature Beats*, Fascinating Life Sciences,
https://doi.org/10.1007/978-3-032-24250-1_13

based ASC inductors, with implications for understanding the neural circuits involved. Socially synchronised rhythmic behaviours in apes further suggest evolutionary roots for collective ASC practices. These findings underscore potentially deep phylogenetic origins of rhythm-induced state changes and call for integrative research bridging neurobiology, comparative psychology and evolutionary theory to elucidate the role of rhythm in shaping minds across species.

13.1 Introduction

So far in this volume, many rhythmic patterns, mechanisms and capacities have been shown to be found in the natural world and shared across species. These examples from the animal kingdom offer solid grounds for researchers to gradually make headway towards explaining how rhythmic behaviour evolved in our own species and why it ultimately took on the intricate, remarkable and stunning forms found today across the world's dance and music traditions. Drawing from a wide array of taxa allows understanding biological constraints and selective pressures that apply to any animal's brain, body and behaviour, including humans and their ancestors.

In this chapter, we explore a facet of human rhythm that, thus far, has eluded comparative research and remained overlooked by major theories of dance and music evolution—the induction of altered states of mind by the means of rhythm. Here, we explore the role that rhythmic stimulation may play in precipitating such states. We provide a short overview of what is known about the potential link between rhythm and altered states of consciousness (ASCs) in humans, theorise about plausible biological mechanisms underlying this link and review new evidence for potentially equivalent data in nonhuman animals. We conclude with suggestions for future research and opportunities to better understand the role of rhythm for ASC induction in human evolution.

13.2 Human Rhythm as a Vehicle to Altered States of Consciousness

Practices for inducing ASCs have a high prevalence across cultures of the world and have been found in at least 90% of a cross-culturally representative sample of societies [1, 2], but the prevalence of *rhythmic techniques* for ASC induction is currently unknown. The association between rhythmic stimulation and ASCs, however, has not escaped the attention of scholars [3–6]. In this review, we will focus on recent literature on the topic, spanning disciplines such as ethnography, neurobiology and (comparative) psychology. We will start with a short conceptual and terminological discussion of ASCs to then examine more closely the association between rhythmic stimulations and ASCs.

ASC is an umbrella term for states that individuals experience as qualitatively distinct from a baseline of "normal" consciousness with respect to patterns and contents of experience and the relationship between self, body and the surrounding world. In its broadest sense, the term spans voluntarily and involuntarily induced states. Voluntarily induced ASCs include those that are mediated by exogenous methods, e.g. through the consumption of pharmacologically psychoactive substances, hypnotic procedures or behavioural practices such as dance, music, meditation and fasting. Involuntarily induced ASCs include those that arise naturally during an individual's daily life (e.g. transition between wakefulness and sleep and vice versa), as well as a consequence of pathological syndromes that (partially) suspend reflective consciousness and/or distort perception (e.g. psychosis, epilepsy [7–9]).

ASCs can vary greatly in kind, intensity and duration due to individual and practice differences, from dizziness after spinning to drug-induced trance involving out-of-body experiences (for a recent phenomenological taxonomy of ASCs, see [10]). At the same time, ASCs do not represent full-or-null responses but are instead graded along multiple axes. In this chapter we focus primarily on ASCs *voluntarily induced by different rhythmic practices* around the world, defining ASCs along the lines of one of the very first definitions given, as "any mental state(s), induced by various physiological [...] manoeuvres or agents, which can be recognized subjectively by the individual himself (or by an objective observer of the individual) as representing a sufficient deviation in subjective experience or psychological functioning from certain general norms for that individual during alert, waking consciousness" [11 p. 225].

Voluntary ASC induction through rhythmic stimulation spans a great many experiences across human populations. How different rhythms and movements directly correlate (or not) with different states or experiences of ASCs is currently poorly understood; however, cultural practices and techniques around the world show that rhythm may be causally related to ASCs. For example, in athletes, intense rhythmic activity can produce "runner's high" or "rower's high", a state in which individuals experience time dilation, narrow attentional focus, analgesia and/or effortless peak performance [12, 13]. Less intense, but potentially equally transformative, experiences are attributed to long-range walking (e.g. pilgrimage) and hiking [14]. This suggests that, despite individual predispositions, it is fairly common that once one "steps" into a rhythmic movement, some sort and degree of ASCs follow.

The connection between bodily rhythm and ASCs is best attested in spiritual and religious contexts, where the purposeful engagement in repetitive and rhythmic behaviours co-occurs with feelings of ecstasy, transcendence, mysticism, nirvana and enlightenment [15]. Rhythmic practices around the world's major religions include the repetitive reciting of prayers, sacred passages, chants and mantras declaimed with prescribed tempo, bead counting (e.g. the Rosary in Catholicism, the Tasbih in Islamism, the Japa Mala in Hinduism and Buddhism) and repetitive bowing and prostration. In many of these cases, rhythmic musical instrumentation (e.g. bells, drums, rattles) can accompany the rites. A remarkable example is the whirling dervishes of the Mevlevi order, who use sustained spinning to achieve a state of spiritual ecstasy. Rhythms, body parts and movements vary across different world traditions, but repetition is often *explicitly* stated by the initiated and practitioners to be an aid to spiritual awakening, helping elevate the soul or enhance a deeper connection with the divine [16, 17].

The association between rhythm and heightened spiritual experiences extends well beyond the major Abrahamic and Dharmic religions. Rhythm is prevalent across indigenous, folk and pagan religions, cults and esoteric sects. Such practices take the form of circle dances, body rocking, chants, drumming, clapping, stamping and stereotypical movements. Ethnography attests that in many cases, engaging in rhythmic behaviour is done with the explicit objective of evading the human world. The archetype of the shaman, sorcerer or witch is, for instance, typically associated with rhythmic movement and performance. Dancing and rhythmic stimulation is, of course, also purposely sought to enter states of relaxation, elation and ecstasy in modern culture, in clubs, discos, concerts and festivals. Some rhythmic practices have also been repurposed in psychotherapy to alleviate depression, anxiety, post-traumatic stress disorder and other psychological conditions [18].

Sitting in quiet meditation is probably the most notable voluntary practice designed to induce ASCs that seemingly does not involve rhythmic movement. However, even in this case, dismissing the role of rhythm may be premature. In one way or another, it is common that meditation techniques instruct the meditators to become aware of their breathing, which is in itself an inherently rhythmic behaviour. Voluntary control over breathing requires primarily motor control of the diaphragm, which is a large muscle with a large degree of connectivity with other body systems (described in Chap. 2 [19] and Chap. 9), something that meditators are instructed to perfect through various breathing techniques (e.g. "belly breathing"). Thus, even when sitting in quiet meditation, a closer look reveals a potential connection between rhythmic movement and the "manipulation" of mental states.

13.3 Putative Mechanisms Underlying the Role of Rhythm in the Induction of Altered States of Consciousness

There are at least four major categories of mechanisms through which rhythm may trigger ASCs. These mechanisms can work in conjunction with each other, and different rhythmic rituals seem to exploit different mechanisms for inducing different types of ASC experiences.

13.3.1 Mechanisms Connected to Breathing

A basic mechanism for rhythm to induce ASCs is through the alteration of nominal breathing parameters. Anatomically, breathing and body rhythm are mechanically coupled by virtue of sharing substantial muscular substrates (for details, see Chap. 2 [19, 20] and Chap. 9 [20]). The muscles used for breathing are tightly connected and centrally positioned in the body in such a way that any substantial movement tends to affect breathing dynamics. At the same time, vigorous muscle engagement for limb or body movement alters the position, shape and/or rigidity of the thoracic box; thus, any non-trivial rhythmic movement will predictably and inexorably impact breathing depth and speed (Chap. 9 [20]). When rhythmic movement is sustained, energetic demands also increase, triggering further physiological responses throughout the body designed to ensure adequate blood flow, oxygen and nutrient supply to muscles, heart and brain.

Human rhythmic rituals and practices around the world can have both decelerating and accelerating effects on breathing rate, causing it to exceed the lower or upper limits of an individual's normal range. Both techniques can have a range of different experiences, from causing short light-headedness to sustained hallucinations, as well as, in extreme cases, fainting and unconsciousness.

13.3.1.1 Increasing Breathing Rate

Many rhythmic rituals (e.g. dance traditions) engage large parts of the body through sustained periods of time, which results in physical exertion, thus, increasing the amplitude and rate of breathing. These breathing changes often have an activating effect on the sympathetic nervous system. The upregulation of this system elicits the "fight-or-flight" response—an evolutionarily conserved mechanism that engenders hyperarousal, mobilising the brain and the body to deal with imminent threats. A key neurophysiological mediator of this response is norepinephrine (also known as noradrenaline), which functions both as a neurotransmitter and a hormone (for details, see Chap. 11 [21]). As a neurotransmitter, norepinephrine is mainly produced in the brainstem (particularly in the locus coeruleus), from where it can diffuse throughout the brain, regulating arousal, attention, vigilance, alertness and overall focus. This may result in so-called tunnel vision, whereby sensory and cognitive focus is narrowly channelled on specific tasks to avoid distractions, but with the natural trade-off of causing impaired situational awareness. As a hormone, norepinephrine is produced by the adrenal glands and released into the bloodstream, reaching multiple targets throughout the body. Its release increases vasoconstriction, increasing blood pressure, and increases cardiac contractility to help maintain blood flow to, e.g., muscular tissues.

Other neuromodulators upregulated by hyperarousal include beta endorphins and endocannabinoids that help the body respond to stress and possible damage (for details, see Chap. 11 [21], as well as [22]). These are natural analgesics that increase pain thresholds and tolerance for discomfort that could otherwise be physically or psychologically challenging. In the context of the rhythmic practices of high motoric intensity and/or prolonged duration, this mechanism helps, for example, practitioners better withstand the painful symptoms of muscular fatigue and exhaustion. There is also direct evidence that analgesic effects are enhanced when individuals synchronise their movement to one another (reviewed in Chap. 11 [21]). As part of their pain-coping function, beta endorphins and endocannabinoids also trigger intense positive emotions, such as euphoria and exaltation, often reported during rhythmic ceremonies.

Modern practices, such as holotropic breathwork—a technique developed by Stanislav and Christina Grof in the 1970s [23] and popularised by Wim Hof, "the Iceman"—trigger some of these effects, not through body movement, but by mimicking the breathing effects that vigorous physical movement would otherwise have. In holotropic breath, the pace and amplitude of the breathing rhythm are consciously manipulated in order to achieve hyperventilation, which practitioners report to trigger states of heightened lightheadedness, giddiness, numbness, emotional release, introspective experiences and even temporary loss of consciousness.

13.3.1.2 Decreasing Breathing Rate

In contrast to the breathing acceleration mechanisms, many rhythmic practices seek to slow down the breathing rate, decreasing energy consumption and enhancing the activity of the parasympathetic nervous system responsible for the so-called rest and digest response. Besides slow breathing, these responses are also characterised by increased breathing amplitude, slow heart rate, muscular relaxation and overall feelings of well-being and calm. For example, in various traditions, breathing rate is voluntarily slowed down from the average normal rate of 12–18 breaths/min to about six [24]. Slow breathing practices can also bring about temporal coherence between respiratory, cardiac and blood pressure phases (for a review, see [25]. Other traditions, such as mantra chanting in monasteries or davening at the Wailing Wall in Jerusalem, carried out for tens of minutes, hours or days, may lead practitioners to a state of mental torpor and meditative trance through lowering and lengthening breathing rates. Indeed, these ASCs may be one of the main reasons why practitioners are able to sustain and deliver rituals for such lengths of time.

13.3.2 Psychological Mechanisms

In humans, hyperarousal can be triggered by affective forecasting, that is, predictive simulations of emotions that could be experienced in certain situations. This suggests that, in certain conditions, psychological and physical stress could potentiate each other to elicit extreme hyperarousal. Such compounding effects may occur in certain ceremonial traditions that feature intense and prolonged rhythmic exertion and culminate in ASCs. Indeed, in this context, there is a known culture of anticipation, fear and terror around many rhythm rituals and practices performed with the explicit purpose of inducing ASCs, such as rites of passage. For example, devotees who participate in spirit possession ceremonies experience extreme fear upon entering trance, which they attribute to the expectation of losing body control [26]. Extreme fear is also reported by highly experienced trancers, as indicated by ethnographers of the Great Dance of the San people [27, 28], where fear (whether of losing control or of the entities they expect to encounter in trance) is seen as a necessary springboard into ASCs [29]. Some world traditions have, thus, built a culture around these feelings of unease, which could enhance the ASC experience itself via psychological mechanisms that compound on the effects of the rhythmic components of the practice or ritual.

13.3.3 Vestibular Mechanisms

The vestibular system comprises peripheral organs located in the inner ear and central components, such as, the vestibular nuclei in the brainstem, the thalamus, the cerebellum and the spinal cord [30, 31]. The vestibular system encodes linear and rotational acceleration of the head, contributing to the sense of balance, equilibrium and spatial orientation, ultimately ensuring body and movement precision, stability and coordination. Besides equilibrioception, it is also involved in rhythm perception [31] and the perception of the bodily self, which is considered to be the foundation of self-consciousness [32].

Regarding rhythm perception, it has been shown that the central vestibular system largely overlaps with the system that mediates rhythm cognition [31]. Both natural and artificial stimulation of the vestibular system (e.g. moving up and down or electric current injections to vestibular afferents) influence the kind of rhythmic structure (e.g. binary, ternary) that is perceived in a metrically ambiguous auditory stimulus [33, 34]. Conversely, rhythmic stimulation activates the vestibular system, eliciting responses that are more intense and more sensitive to temporal patterning than those elicited in other sensory modalities, including the auditory modality, which has traditionally been considered central to music perception [35].

With respect to the bodily self, the vestibular system contributes to aspects such as the body schema, body ownership and agency, first-person perspective and perspective taking, as well as the relationship between self, environment and others [30, 32, 36]. Such self-percepts, however, do not rely solely on vestibular input, but depend on the multisensory integration of vestibular, visual, auditory, somatosensory and proprioceptive signals. These signals provide complementary, modality-specific information, along with partially redundant information about the orientation and location of the body in space. The vestibular system, however, is the only modality that provides spatial and motion information based on a self-centred (i.e. head-centred) frame of reference, as opposed to the other modalities that detect location and motion cues relative to an external frame of reference [37].

"Normal" states of consciousness are dependent on unperturbed vestibular functioning and the congruence of signals from across sensory modalities. Distorted vestibular input—as found in patients with vestibular dysfunction or generated by different forms of vestibular stimulation (e.g. centrifugation)—interferes with this seamless process and can induce alterations in proprioception and conscious experience. Notably, the perceptual distortions documented in these contexts are very similar to those described in the ASC context, spanning out-of-body experience, time dilation, body part deformation, increased pain threshold, etc. (for reviews, see [32, 38, 39]).

Given its intimate association with rhythm perception and consciousness, the vestibular system emerges as a prime candidate for mediating the putative role of rhythmic stimulation in ASC induction. This could be achieved through several mechanisms, and we first consider how distorted vestibular perception caused by unusual characteristics of rhythmic movement can lead to *altered proprioception*. Any practice that involves the rhythmic and repetitive movement of the head activates the vestibular organs in the inner ear and, in certain cases, can result in alterations of proprioception and even consciousness via *structural changes* in the peripheral vestibular organs.

The peripheral vestibular organs comprise three semi-circular canals that sense angular acceleration (i.e. head rotation) and two otolith organs (the utricle and the saccule) that detect linear acceleration (i.e. the effect of gravity or inertia). The utricles and saccules contain a sensory epithelium composed of hair cell receptors, whose tips are embedded in a gelatinous layer—or otolithic membrane—that is weighted with calcium carbonate deposits called otoliths ("ear stones"). The ducts of the semicircular canals contain a fluid (the endolymph), and each canal presents an enlargement end with microscopic hair cells embedded in a gelatinous substance (the cupula). The hair cells are extremely sensitive mechanoreceptors that can detect mechanical stimulation of less than 1 nm in amplitude [40].

Changes in head position cause structural changes in the peripheral vestibular organs. In the saccule and utricle, both the otolithic membrane and the sensory epithelium are displaced. Due to the mass difference between them, this results in a shearing motion that alters the orientation of the hair cells. In the semicircular canals, head movements displace the endolymph, which in turn bends the cupula and the hair cell receptors within it. Due to inertia, however, the endolymph and the otoliths are initially displaced in the opposite direction of the head movement. The bending of the hair cells and their return to the initial position trigger a neural signal that informs the brain about the head movement.

The minuscule lag that arises due to endolymph inertia provides an entry point to "bluff" the vestibular system and misinform the brain. Spinning, used for example by whirling dervishes for spiritual elation and by children around the world for momentary vertigo, illustrates this point well. When an individual begins to spin on its own axis, the circular motion displaces the fluid inside the vestibular canals. This movement bends the hair cells, which signal to the brain that the body is rotating. However, if an individual keeps on spinning, within less than 10 s of motion, the endolymph catches up with the motion due to inertia, and the hair cells stop bending. As a result, the sensation of acceleration stops, even though the body is still in motion [41].

The opposite occurs when an individual stops spinning. While the body halts quickly, the endolymph continues to maintain acceleration, bending the hair cells in the opposite direction. This creates the illusion that the body is still spinning. Meanwhile, visual and proprioceptive inputs—from the eyes, muscles, tendons and joints—indicate that the body is stationary. This sensory mismatch causes individuals to experience dizziness, nausea, or "motion sickness". Thus, spinning is a particularly effective means to trigger altered states of proprioception because the body can reach very high accelerations and velocities within short time spans.

Experimental evidence shows that the stimulation of either the semicircular canals or the utriculi (e.g. centrifugation) may, in addition, generate a range of experiences also reported in ASC (reviewed in [39]). In these vestibular stimulation experiments, participants commonly report sensations such as altered bodily awareness, slowed time perception, heightened focus, elevated anxiety and difficulty imagining an object from a different viewpoint. Such symptoms are also characteristic of patients with peripheral vestibular dysfunctions [38, 39].

Similar vestibular mechanisms may underlie altered proprioception and consciousness even for other rhythmic movements that exceed the evolutionary optimisation of the vestibular system, such as prolonged head shaking, head banging and prostration. In contrast, rhythmic movements derived from locomotor rhythms (e.g. rocking, swaying, etc.), which are widespread in an ASC context, are less likely to impair *proprioception,* since the vestibular system is evolutionarily tuned to handle the perturbations associated with such movements (for a recent review and discussion, see e.g [31].).

Even when rhythmic movements do not impair proprioception, the vestibular system may still mediate *consciousness-altering* effects through *mechanisms involving the central vestibular pathways.* On the one hand, the vestibular network largely overlaps with the network that underlies rhythm perception, and the vestibular modality is more sensitive to rhythmic stimulation than other sensory modalities [35]. On the other hand, the vestibular system projects to brain regions involved in reward processing and autonomic regulation [42–44] whose putative role in ASC precipitation has been discussed in Sect. 13.3.1. When vestibular input is (near) normal, the vestibular-reward pathways contribute to the intrinsic sense of pleasure evoked by dance-like movement. This could explain why rhythmic behaviours such as swaying, rocking or pacing are commonly sought as a source of self-stimulation [31]. In contrast, distorted vestibular input can trigger abnormal autonomic responses, expressed as feelings of panic and anxiety. This is consistent with clinical findings that link otolith dysfunction to abnormal cardiovascular and respiratory responses [45] and anxiety [39].

The connection between vestibular stimulation and reward is evolutionarily ancient, being inherited from anamniote ancestors whose vestibular system responded to both self-motion and acoustic stimulation [40]. This duality is retained in mammals, with the saccule being the vestibular organ that is the most sensitive to sound (for a review, see [46]). Such vestibular responsiveness to sound is well documented in humans [31] and is evoked by loud stimuli above 90 dB (sound pressure level), with frequencies between 100 and 300 Hz. These values correspond to those commonly found in dance clubs and rock concerts (reviewed by Todd and Cody [46]) and plausibly also in ritual contexts for ASC. Thus, in ASC settings, the effects of rhythmic stimulation on the vestibular system are likely to be multi-modal—induced by both rhythmic head motion and auditory rhythms that exceed the vestibular threshold. Moreover, ASC ceremonies often feature multiple, overlapping auditory rhythms from various sources, which may overwhelm the vestibular system with divergent frequencies and rhythmic inputs. This can plausibly lead to sensory overload and/or sensory mismatches, offering another potential route to ASC-like effects. Such effects may be further amplified in collective ceremonies through visuomotor synchronisation among participants, as both rhythmic synchronisation and vestibular stimulation are known to contribute to self-other blurring effects.

13.3.4 Neural Mechanisms

An enduring idea in both popular belief and research is that rhythmic stimulation induces ASCs by virtue of affecting neuronal activity. Neurons naturally fire in rhythmic patterns, and when groups of neurons synchronise, they produce oscillatory activity—commonly referred to as brain waves—across distinct frequency bands (for details, see Chap. 8 [47]). These oscillations are not isolated. As shown in other chapters of this volume (Chap. 7 [48], Chap. 8 [47] and Chap. 9 [20]), neural oscillations interact locally and across distributed networks and regulate the rhythmicity of behaviours, sensory sampling and visceral processes. In this way, they shape how we perceive, feel and behave. Specific patterns of oscillatory activity across frequency bands support distinct modes of functional brain connectivity, which are fundamental to both rhythm perception and production, as well as to the experience of normal and altered states of consciousness. Given that conscious perception may be inherently cyclic—i.e. it is based on the cyclical updating of neural states [49, 50]—the widespread association between ASCs and rhythmic stimulation may reflect the influence of external rhythmic perturbations on consciousness. In the following sub-sections, we review research on the neural mechanisms underlying ASCs, tracing the chronological development of major frameworks that have shaped the field, including the sensory driving hypothesis, the transient hypofrontality hypothesis, the entropic brain hypothesis, the thalamic gating hypothesis and the predictive coding hypothesis.

13.3.4.1 The Sensory Driving Hypothesis

Historically, Neher [51] is considered the first to propose that rhythmic auditory stimulation induces ASCs by altering the properties of neuronal activity. In a seminal study conducted in 1961, Neher exposed 10 participants to fast drum sounds at 3, 4, 6 or 8 beats/s (3–8 Hz) and used electroencephalography (EEG) to measure neural activity across the right frontoparietal area, as well as bilaterally across the parieto-occipital and parieto-temporal areas [51]. Neher found that neuronal oscillations in the alpha band (8–12 Hz) aligned (i.e. were phase-coupled) with the drumbeats in the auditory cortex and, to a lesser extent, in occipital areas. In addition, involuntary eye blinking synchronised with the beats in half of the participants, especially those who also exhibited a pronounced EEG response. The participants also reported various emotional and perceptual experiences—e.g. fear, astonishment, amusement, pulsations in the back, muscle tightening, visual imagery and auditory imagery. These, however, were not related to the intensity of the EEG responses.

Neher interpreted these findings as evidence of “auditory driving”, given their similarity to those reported in studies of photic driving—a phenomenon where neuronal oscillations in the occipital lobe become time-locked with a rhythmic light stimulus, especially at high stimulus frequencies (8–30 Hz) [52, 53]. In a subsequent theoretical piece published in 1962, Neher capitalised on the (highly limited) evidence produced by his EEG study to propose “auditory driving” as a general neural mechanism of ASC induction in all traditional ceremonies that involve drums [3].

Neher's proposal was picked up by the anthropologist Michael Harner, who, starting in the 1980s, popularised the idea that fast drumbeats at a 4–8 Hz rate induce trance [4]. Unfortunately, these popularisation efforts were not accompanied by continued empirical efforts to understand the neurobiological mechanisms of rhythm-associated ASCs. Several decades passed until scientific interest in the topic was rekindled. In the meantime, however, our knowledge about the neurobiological effects of rhythmic stimulation has grown substantially owing to neurobiological and psychological research on music, locomotion and physical exercise. Now, we know that the time-locking of neural oscillations to rhythmic stimuli is a default response of the brain and that rhythmic stimulation has remarkable effects on cognitive, emotional and physiological states (for details, see Chap. 7 [48], Chap. 8 [47], Chap. 9 [20] and Chap. 11). Thus, it has become clear that "auditory driving"—a phenomenon currently captured by notions such as "neural entrainment" and "neural resonance"—cannot precipitate ASCs *on its own*. The question we ought to ask now is not if rhythmic stimulation affects brain rhythms, but whether—and if so, in which circumstances—it can affect brain activity towards patterns that produce ASCs.

13.3.4.2 The Transient Hypofrontality Hypothesis (THH)

In the early 2000s, Arne Dietrich formulated the influential "transient hypofrontality" hypothesis (THH) of ASC [54]. This hypothesis posits that ASCs result from reduced activity in the prefrontal cortex (PFC)—a brain region that is central to executive control, self-awareness and temporal orientation, and is known for its remarkably high energetic demands (for a recent review, see [55]). According to the THH, intense physical or psychological engagement (e.g. endurance running, ritual dancing, meditation, etc.) can temporarily redirect metabolic resources away from the PFC towards bottom-up functions.[1]

This shift is hypothesised to produce a corresponding change in consciousness, characterised by diminished executive oversight and enhanced sensorimotor and autonomic dominance. As a result, individuals may experience hallmark features of ASCs, such as reduced self-awareness, impaired cognitive control, a sense of timelessness and ego dissolution and unusual perceptual experiences. The exact phenomenological profile of the ASC would then depend on the nature of metabolic redistribution, which in turn is determined by the sensorimotor demands and autonomic impact of the involved activity.

The THH is relatively well supported by neurobiological data. Brain imaging studies show that flow states induced by movement correlate with enhanced activity in the motor cortex, cerebellum and the basal ganglia and downregulation in the

---

[1] The THH premise builds on the fundamental principle that brain processes compete for access to consciousness and for finite metabolic resources [56]. Indeed, independent of effort level, the energetic resources of the brain remain constant at the global level and, thus, need to be re-allocated depending on fluctuating functional demands [57].

PFC [57–60]. Similar activation patterns have been reported by research on meditation [61, 62], so-called shamanic and possession trance [63, 64] and music-induced flow [65–67]. Generally, the PFC—and the executive control network—is dampened and its connectivity with other areas is weakened. By contrast, sensorimotor and, in some cases, also autonomic areas are upregulated, and their dominance is hypothesised to account for an ASC-like phenomenology, characterised by imagery and certain emotional experiences.

The functional connectivity profile described above is consistent with anatomical findings and EEG research. For example, Whirling Dervishes—as compared with a control group—exhibit lower cortical thickness in relevant PFC areas [68]. In electrophysical research, a recurring pattern is that neural oscillations in the frontal region are enhanced for the theta (4–8 Hz) and alpha (8–12 Hz) bands during deep meditation [61, 69–71], trance [72–74] and music- or movement-induced flow [58, 75–77]. Concomitantly, oscillatory activity in the low beta band (12–20 Hz) decreases and is desynchronised or suppressed. An exception is the case of movement-induced flow, during which maintained or enhanced beta activity is required for sustaining (goal-directed) motor performance in the ongoing activity, e.g. running, playing an instrument, etc.

As increased alpha power is an indication of cortical inhibition, the enhanced frontal alpha described above suggests a temporal dampening—or even suppression—of top-down executive functions. Conversely, enhancement of frontal midline theta activity has been linked to sustained—yet effortless—attentional focus, internal monitoring and a disengagement from external sensory input. These theta oscillations are also associated with relaxation and autonomic balance, often observed during meditative states [61]. Elevated theta frequently coincides with a reduction in high-frequency (beta) power, a pattern interpreted as the brain operating in a low-effort, high-efficiency mode. This shift is also associated with increased suggestibility and reduced critical evaluation, as seen in hypnosis and other ASCs.

In summary, there is convergent evidence suggesting that an alteration in the connectivity between intention and action plays a central role in the emergence of ASCs. Conventional executive oversight may be bypassed to various degrees, with motor control being increasingly guided by bottom-up processes. This shift may contribute to the unusual phenomenology of the ASCs, including experiences of automaticity, altered agency or effortless performance, as for example reported during meditative and flow states, possession and shamanic trances [8, 54].

Although the evidence discussed above is all drawn from studies where ASCs were elicited through rhythmic stimulation, hypofrontality can also occur in ASCs elicited with non-rhythmic techniques [54, 61]. Importantly, however, hypofrontality appears to be more pronounced and more consistently observed in rhythm-induced ASCs, especially when it comes to meditative and flow states [8]. This indicates that rhythmic stimulation—whether in the form of running, dancing, chanting, singing or breathing—is an effective way to induce hypofrontality and "quiet" executive oversight. This effectiveness is likely driven by neural entrainment—which, as discussed in Sect. 13.3.4.1, is a default brain response—in combination with the fact that this response is stronger and more coherent for the lower frequency bands (delta to alpha) compared to the higher ones (beta, gamma) [78–80]. In turn, this kind of neural oscillatory profile generates patterns of functional connectivity that correspond to various forms of hypofrontality, as well as various degrees of sensorimotor and autonomic dominance.

Hypofrontality is putatively reinforced when the stimulation is not merely perceived (e.g. being exposed to music) but also embodied, such as when chanting or dancing. By involving a prolonged motor output, rhythmic embodiment generates a context that favours the allocation of metabolic sources towards cortical-subcortical motor loops (as detailed in Chap. 7 [48] and Chap. 9 [20]). Rhythm also creates a predictable temporal scaffold which, in line with the predictive coding framework (see Sect. 13.3.4.6), reduces the need for top-down control and makes attention tracking more fluid. Hypofrontality as a response to rhythmic stimulation could also be enhanced—or even induced—via vagal pathways consistent with data showing that rhythmic stimulation (e.g. breathing, chanting, etc.) often enhances the parasympathetic tone [81, 82].

Finally, it is important to clarify that rhythmic stimulation by itself—i.e. in the absence of a meaningful context—does not reliably induce transient hypofrontality [80, 83, 84]. Thus, rhythmic stimulation—and the ensuing neural entrainment—can be a potent trigger of hypofrontality, provided that it is embedded in a meaningful, emotionally salient context (ritual, music, sports).

13.3.4.3 The Entropic Brain Hypothesis (EBH)

In 2014, in a study that examined the neural and phenomenological effects of psilocybin, Carhart-Harris and colleagues [85] proposed the "entropic brain hypothesis" (EBH) as a novel neuro-mechanistic account of ASC. The EBH posits that the quality of conscious experience is determined by the degree of neural entropy—that is, the *richness* and *variability* of spontaneous brain activity. In this framework, "normal" states of consciousness correspond to brain activity that is neither too orderly nor too chaotic, i.e. maintained within a *critical* zone of moderate entropy [85, 86]. By contrast, during ASCs brain entropy increases or decreases approaching the edge of the criticality zone or even surpassing it. On the one hand, low entropy, defined by reduced variability of functional connectivity and oscillatory activity in the lowest frequency bands, would underlie highly constrained states of consciousness, e.g. deep sleep, hypnosis, anaesthesia. On the other hand, high brain entropy—defined by increased EEG signal diversity, heightened inter-regional communication and a broader repertoire of functional connectivity patterns—would correspond to "expansive states", such as those found in the psychedelic states, and certain meditative and trance practices [85, 86].

As compared to the THH, the evidence supporting the EBH is less extensive, which can be explained by the novelty of the hypothesis and, thus, the scarcity of relevant research. Nevertheless, recent studies provide suggestive evidence that states of consciousness covary with shifts across the brain entropy continuum. Empirical work (EEG/MEG, fMRI, computational modelling) shows that repetitive rhythms—such as those found in breathing and chanting meditation or drumming rituals—increase entropy, while also producing ASC-characteristic phenomenological experiences [87–90].

These findings indicate that rhythmic stimulation can effectively tune the brain dynamics, biasing it towards particular spectral regimes, which in turn modulate functional connectivity and signal complexity. The direction of these entropy changes and the exact profile of ensuing phenomenological experiences potentially covary with arousal and depend on tempo, regularity and complexity. Data from [83, 88] suggest, however, that meaningful intentionality may be a critical moderator of the EBH effect in the sense that pure rhythmic entrainment in the absence of the cultural-meaning component may not produce a sufficiently robust entropy change, nor ASCs.

13.3.4.4 The Thalamic Gating Hypothesis (TGH) of ASC

Situated at the intersection of sensory pathways and cortical processing, the thalamus plays a pivotal role in regulating the flow of sensory information to the cortex. The thalamus is an important node in the cortical–basal ganglia–thalamic–cortical loop (the CBGTC loop), which also underlies rhythmic production and perception (for details, see Chap. 7 [48] and Chap. 9 [20]). As part of the CBGTC loop, the thalamus is subjected to inhibitory modulation by the basal ganglia [91, 92]. This inhibitory mechanism is hypothesised to serve a gating function, selectively filtering out task-irrelevant stimuli and allowing task-pertinent input to reach conscious awareness. Such gating is essential for maintaining focused attention, increasing perceptual salience and supporting executive function and behavioural regulation.

The concept of thalamic gating was introduced to the field of ASCs by Franz Vollenweider through a series of psychedelic induction studies conducted during the late 1990s and early 2000s (reviewed in [93]). In a model that integrated the CBGTC loop with major neuromodulatory systems (in particular serotonergic, dopaminergic and glutamatergic ones), Vollenweider proposed that psychedelic drugs cause disruptions in the thalamus' filter function, leading to sensory overflooding of the cortex and, thus, altered perception. The hypothesis has since been supported by multiple studies demonstrating altered thalamo-cortical communication under psychedelics (for a recent review, see [94]).

Framed as a generalised theoretical framework of ASCs [95, 96], the TGH posits that an overly permissive thalamic gate enhances thalamo-cortical transmission and overfloods the cortical network with sensory information. This, in turn, could produce ASC-like phenomenology characterised by vivid imagery, hallucinations and a sense of expanded awareness. Conversely, an overly restrictive thalamic gate dampens thalamo-cortical transmission, effectively cutting off external input. Conscious experience then turns inward, generating phenomenological experiences similar to those associated with meditative and shamanic practices, such as attentional narrowing, dream-like experiences and a loss of self-boundary [96].

Although the TGH has not been directly tested in the context of rhythm-associated ASCs, converging evidence from studies on rhythmic ASC practices offers indirect support. For example, it has been reported that long-term practitioners of Vedic chanting exhibit increased grey-matter volume and white-matter integrity within the thalamus [97], and thalamic activity is enhanced during mantra meditation [98]. A comprehensive review [99] notes that slow, controlled breathing consistently modulates thalamic activity across multiple neuroimaging studies. Likewise, the finding that shamanic trance induced by drumming downregulates the auditory cortex activity is consistent with the idea of a restrictive thalamic gate that suppresses exogenous auditory input [63]. In addition, a recent synthesis of EEG and fMRI research on shamanic and possession trance practices identifies activation patterns indicative of sensory input modulation [100].

How changes in thalamic activity induced by rhythmic stimulation may alter consciousness remains an open question for future research. There is broad consensus that the CBGTC loop is crucially involved in rhythm production and perception, which indicates that its components—including the thalamus—are sensitive to rhythmic information (see Chap. 7 [48] and Chap. 9 [20]). Rhythm could influence conscious experience by acting as a timing scaffold for the thalamic gating function, through at least two mechanisms: phase-resetting of thalamo-cortical oscillations and adaptive synchrony. Through phase-resetting, rhythmic stimulation can alter the timing of thalamic neuron firing, thereby increasing or decreasing the excitability of downstream cortical populations [101]. In contrast, adaptive synchrony refers to how rhythmic stimulation modulates the degree of oscillatory synchrony within the thalamus and influences the frequency of neural oscillations, thereby also shaping the gating function [102]. Animal research has, for example, demonstrated that auditory rhythmic stimulation can entrain the non-lemniscal neurons in the auditory thalamus, providing direct evidence of thalamic sensitivity to rhythmic input [103, 104]. These findings support the view that rhythmic stimulation may modulate consciousness by influencing thalamic gating dynamics.

Currently, the accepted view is that the TGH can be considered as a *conceptually* distinct hypothesis that, however, is *mechanistically intertwined* with other models of ASCs. It is conceptually distinct because it emphasises a subcortical brain structure, in contrast to THH—which focuses on the PFC—and to the EBH, which focuses on statistical properties of whole-brain activity. It is mechanistically intertwined because the thalamus operates within the CBGTC loop and, thus, hypofrontality may arise as a downstream effect of thalamic gating [105]. Furthermore, modulations in thalamic gating likely contribute to shifts in brain entropy [106]. While it is increasingly accepted that the thalamus plays a critical role in consciousness modulation [96, 107], it remains unclear whether thalamic gate alterations induced by rhythmic stimulation are primary drivers of ASCs or ancillary mechanisms. An important caveat is raised by methodological challenges posed by the deep location and small size of the thalamic nuclei, which complicates efforts to isolate pure thalamic effects from those of adjacent structures, such as the basal ganglia [108].

13.3.4.5 Other Subcortical Gating Mechanisms

Besides the thalamus, two additional subcortical structures considered to play an important role in shaping conscious experience are the ascending reticular activating system (ARAS) and the claustrum. Neither of these has yet been discussed in the context of rhythm-induced ASC, but they could be relevant. Like the thalamus, the ARAS and the claustrum contribute to the processing streams underlying consciousness, owing to their roles as signal gate-keepers, and as integration and relay hubs.

Located in the brainstem, the ARAS serves as the brain's arousal system (for further details, see Chap. 11 [21]), regulating alertness, attention and the transitions between wakefulness and other states of consciousness [109]. The ARAS functions as a filter for sensory information. It sends (de)activating signals to the thalamus and cortex, modulating the brain's response to internal and external stimuli. Through its noradrenergic and cholinergic pathways that reach broadly across the brain, the ARAS regulates that the brain remains responsive to sensory input and maintains a baseline level of alertness. Disruption of this system can lead to profound unconsciousness, such as in a coma.

Several models of ASC regard arousal as a foundational dimension, whereby different ASC categories correspond to distinct arousal profiles (for examples and a critical discussion, see [10]). Consequently, the ARAS is arguably a critical structure in ASCs. The ARAS—and arousal—are strongly modulated by rhythmic stimuli (as reviewed in Chap. 11 [21]). On these theoretical and empirical bases—i.e. arousal-based models of ASC and data linking rhythm and arousal—slow, steady rhythmic stimulation may have the capacity to induce relaxation and meditative states, while fast, intense rhythms would heighten arousal and lead to ecstatic states. Additionally, rhythmic stimuli may also affect the salience modulation and sensory gating functions of ARAS, facilitating shifts towards states of either present-centred awareness or immersive inner journeys.

The claustrum is a thin, sheet-like structure nestled between the insular cortex and the putamen that is bidirectionally connected with nearly all regions of the cerebral cortex. Given its anatomical connectivity, the claustrum has been hypothesised to be the key mechanism underlying the binding of sensory inputs into a unified perceptual experience [110]. Currently, however, data from lesion and functional imaging studies instead point to a more modest role for the claustrum in regulating cortical excitability, modulating internal awareness, arousal, perceptual salience and attention [111, 112].

As a structure that is nevertheless implicated in shaping conscious experience, the claustrum has also been discussed for its role in ASCs [113, 114]. Although this theory does not address alterations in consciousness in the context of rhythmic practices, the claustrum may be relevant. If rhythmic stimuli entrain claustral neurons, this could modulate oscillatory coherence and synchronisation across the many disparate brain areas that are connected with the claustrum, facilitating shifts in consciousness. Additionally, the claustrum's involvement in salience detection and attentional gating may help prioritise rhythmic input over other competing stimuli. In ritual settings, this could mean that the rhythmic elements become the dominant focus of awareness, allowing participants to disengage from external distractions and enter narrowly focused states as commonly reported for certain categories of ASCs.

13.3.4.6 The Predictive Processing Hypothesis (PP)

At the core of the predictive processing hypothesis (PP) is the popular idea that the brain continuously generates predictions about sensory input and updates them upon error feedback, i.e. when top-down predictions based on prior experiences do not match bottom-up input. In 2010, Friston[2] published an influential PP version (known as the *free energy principle*-based PP or the *active inference framework*) that was readily adopted in the ASC field [115]. According to this version, it is proposed that conscious experience arises from the brain's continual effort to minimise prediction error across hierarchical generative models, which comprise cortical layers for prediction generation and cortical-subcortical layers for precision weighing. Friston explicitly linked PP to the *contents* and *levels* of consciousness, suggesting that the predictive operations that underlie perception determine *what and how much sensory input reaches awareness.*

Shortly after, Seth and colleagues [120] applied the PP to explain consciousness alterations that resulted from a distorted sense of presence, i.e. distortions of the feeling of being in the world and in one's body. This work linked the PP with interoception, proprioception and affective states, laying a foundation for understanding how disruptions in predictive coding of bodily signals can lead to dissociative experiences—a key "symptom" in the phenomenology of many ASCs. Later on, Carhart-Harris suggested that imbalances in the predictive coding machinery can induce brain entropy shifts that are conducive to ASCs [86]. In particular, he theorised that psychedelic drugs diminish the precision of top-down priors, tipping the predictive coding balance towards bottom-up sensory activity, thereby increasing neural entropy. This idea was further elaborated into the REBUS model of psychedelic states [121], according to which serotonergic psychedelics relax the precision weighting of high-level priors. This "loosening" lets otherwise suppressed prediction errors propagate upward, increasing the overall entropy of brain activity and producing the characteristic psychedelic phenomenology.

Capitalising on this work, the PP can be further linked to rhythm-induced ASCs, based on at least three premises. First, rhythm-induced ASCs share core phenomenological features with the psychedelic states, e.g. reduced sense of self, expanded perceptual bandwidth, increased emotional fluidity, etc. Second, through entrainment, rhythmic stimulation has effects on brain activity patterns (as discussed in the previous sections), and this has the potential to modulate—and even disrupt—the

[2] While Friston [115] is often referred to as foundational to this theoretical framework, the predictive coding hypothesis (PP) was already articulated in the early 1990s [116], being first applied to vision modelling [117]. The seminal paper of Friston, however, contributed to popularising the PP by integrating it with Bayesian statistics and the *free-energy* principle, whereby the brain's overarching "goal" is to minimise variational free energy—a variable that bounds prediction error. By providing a rigorous mathematical backbone, Friston [115] spurred the expansion of PP beyond vision and towards becoming a generalised explanatory model for attention, perception, action, language, psychiatric disorders—and, more broadly, consciousness. It is, however, important to note that this popular version of PP has been criticised for being too broad, abstract or difficult to falsify (among other things) and it currently coexists with several other versions and interpretations of the PP principle [118, 119].

balance between top-down predictions and precision weighing. Third, rhythmic stimulation engages the predictive machinery of the brain in ways that modulate affective states, having the potential to induce persistent reward loops that implicate the full reward cycle—across phases of incentive salience ("wanting"), peak pleasure ("liking") and learning (as detailed in Chap. 11 [21] this volume).

In line with the above, a highly repetitive rhythm (as in e.g. running, breathing meditation, mantra, shamanic drumming, etc.) would be so predictable that the brain may "deem" incoming sensory information as uninformative, registering very low prediction error at the level of timing. This could generate predictive imbalance by reducing the precision assigned to—and thus the brain's confidence in—higher-order priors and, therefore, top-down models (e.g. self, semantic explanations) can run unchecked [122]. In line with the REBUS model, precision relaxation opens a "computational window" where bottom-up sensory fluctuations (including spontaneous neural "noise") could become excessively influential [121].

Very few studies thus far have directly tested PP predictions in relation to rhythm-induced ASCs. For example, a modelling study of EEG data collected during binaural beat presentation found that the best-fitting model required lowered precision on top-down predictions [123]. It is, however, frequent in the recent literature on rhythm and ASCs that authors interpret findings through a predictive coding lens, especially within the framework set by the REBUS model [84, 88, 124]. Just like the TGH (discussed in the previous section), the PP can be best described as complementary to the other hypotheses discussed in this chapter and mechanistically intertwined with them. Predictive coding mechanisms, such as those suggested by the REBUS model, can indirectly underlie the emergence of ASCs by modulating the thalamic gate, contributing to shifts in brain entropy or network stability and may result in hypofrontality.

Despite this recent trend of accounting for ASC in terms of PP, the explanatory power of PP may be limited. Recent work by Westerberg and Roelfsema [125] shows that the sensory cortices often exhibit feedback dynamics that contradict the inhibitory suppression assumed in PP. Instead, they propose the BELIEF framework ("backward enhancement of limited information and inhibition of expected information in feedforward pathways"), reconceptualising feedback as a constructive and excitatory mechanism that enhances and integrates sensory signals rather than merely minimising prediction error. Applied to rhythm-induced ASCs, this alternative view would predict that during repetitive stimulation, feedback loops sustain and enhance activity in sensory and motor cortices, reinforcing rhythm processing, attentional entrainment and perceptual stability. This may offer a richer account of how rhythmic stimulation may promote immersive, trance-like experiences.

13.3.5 Limitations in Neuropsychological Research on Rhythm and Altered States in Humans

Current research on rhythm-associated ASCs—and on ASCs more generally—is highly heterogeneous. Across studies, there is a great variety in participant demographics (e.g. sex, expertise, spiritual background, etc.), experimental setup (e.g. session duration, control conditions), measured variables and methods (e.g., cardiovascular parameters, number and placement of electrodes, oscillatory band) and data analysis techniques. Altogether, this hinders comparability, limiting possible mechanistic conclusions.

Most studies conducted thus far also have low ecological validity. Rhythmic stimuli used in experimental settings do not typically reflect those used traditionally in "the wild". In a traditional ritualistic context, rhythmic stimulation associated with ASC induction may last for hours, sometimes days, following prescribed phases. By contrast, the stimuli presented in the lab tend to be short and stripped of multi-modality and multi-layering. In [83], for instance, rhythmic stimulation consisted of 60-s bouts of an unspecified instrument or sound that alternated with 30-s intervals of silence. Thus far, there has been little attempt to re-create traditional rituals in experimental contexts.

To date, there has also been a significant sampling bias, where the majority of studies on rhythm-associated ASCs has been carried out with "experts", individuals who have practised certain types of meditation or trancing extensively. No doubt, such sampling decisions maximise the likelihood and readiness of ASC induction. However, this methodological choice creates a confound, whereby the putative effects of rhythmic stimulation in ASC induction are conflated with those of enculturation and expertise.

Enculturation and expertise result in neuroplasticity, habituation to induction practices and the formation of associations between cues and induced states. In this way, in the rare cases when control populations are used, "shamanic practitioners" differ from controls with respect to the effects of drumbeats on EEG frequencies [88]. As noted by ethnographers [26], experience has priming effects on practitioners. That is, the individual becomes predisposed to experience ASC as soon as they hear an ASC-associated rhythm.

This culturally-shaped predisposition may also be compounded by individual variables (e.g. general differences in emotional sensitivity to music). For example, as an additional control condition, Huels and colleagues exposed both "shamans" and control participants to classical music and specifically instructed the subjects *not* to enter a trance [88]. Nevertheless, even in this condition the "shamans" had distinct EEG profiles compared to control subjects, and some of their responses were similar in the drumming and classical music conditions. This speaks to the need for proper experimental control, both by including subjects who do not self-identify as shamanic practitioners or "psychonauts", as well as other (non-ASC-associated) musical stimuli in order to delineate the contribution of rhythmic sounds on brain responses and phenomenal experiences and begin to disentangle learning effects.

Another significant limitation in the experimental study of rhythm-associated ASCs is the reliance on case studies and small sample sizes. This obscures the effects of individual variability and could lead to erroneous interpretations or conclusions. For example, in an EEG study of ASC induction through sound and tactile-transmitted vibrations, the outcomes for the two participants (one male and one female) were highly divergent [126]. Both participants showed a decrease in theta band activity and an increase in the beta band (attributed to sound processing), but the alpha band was significantly engaged only in the female participant, while the male participant showed a significant increase only in the beta band. In the study, these divergent effects were interpreted as representing individual differences in susceptibility to hypnosis [126], an effect also observed in other studies [127]. A sample of two, however, cannot help extricating an "individual difference" from sex, age, religious identity, ethnic background, history of illness, the effect of diet or time of the day during which the experiments took place.

There is, however, a general notion that there may exist personality traits with increased susceptibility to rhythmically induced ASCs. For example, people who score medium or high on hypnoidal scores, that is, with high "hypnotisability", tend to be affected more strongly by the effect of drumming [127]. Interestingly, in [127], hypnotisability across participants seemed to increase when hypnosis sessions were preceded by a drumming condition. This suggests that, despite different innate levels of hypnotisability, these can be modified through priming effects. This supports the view that specific rhythmic media can facilitate, trigger or lower the threshold of certain neural phenomena in a specific fashion.

Ideally, authentic rhythmic practices and rituals from around the world could be brought into the lab to investigate the effects of different components in a controlled environment. Many of the relevant ingredients, however, could be a fruit of specific belief systems and cultural contexts that are impossible to replicate in a lab. The social and physical setting of a ritual is known to have a determinant impact on how individuals experience ASCs.

To conclude, ASC induction practices are widespread across the world's cultures [1, 2], but little is known about the presence and properties of rhythmic stimulation in this context. Are there technical commonalities across cultures to expect similar mental effects? Or should differences in techniques predict that the altered experiences also will be different? Cross-cultural analyses are required to find out how and why different communities and societies have developed different rhythmic practices in association with different ASCs.

13.4 Retracing the Form and Function of Rhythm-Induced Altered States in Human Evolutionary Past

If rhythmic motion and sensory stimulation indeed prove to be effective means for inducing divergent mental states in humans, a new evolutionary model is required to explain why humans evolved as a hominid species that dances and makes music—a model that accounts for the effects of rhythm on the mind. Such a model will need to focus on the axes linking kinetics, neurophysiology, perception, mood and pleasure in hominids more than any other hypothesis has done so far. Moving back along the human evolutionary timeline, the first fundamental question becomes, "Assuming that rhythm is exploited by modern humans as a vehicle to alter mental states, can we retrace similar behaviours back to an ancestral hominid?"

Rhythmic behaviour does not fossilise; therefore, nonhuman great apes—our closest living relatives and last remaining surviving members of the Hominid family—provide the most realistic living model of rhythmic behaviour in an ancient and ancestral hominid. Rhythm in great ape behaviour can, thus, help bring new insight and clarity on the role of repetitive, cyclic movement on altered states at the basis of the human phylogenetic branch.

Recent observations show that there are, indeed, *natural* ways (i.e. in the absence of human intervention, training or priming) in which great apes spontaneously engage in rhythmic behaviour. These behaviours offer a high chance to better understand why and how extant nonhuman hominids engage in movements that potentially affect mental states, and thus, by extension, why and how extinct ancestral hominids may have started to do so in ancient times. Such studies hold the potential of clarifying whether rhythmic behaviours had enough impact to shift the evolution of the human mind in ways still under-appreciated.

13.4.1 Higher Apes

The first observations that great apes engage in rhythmic movement for the purpose of altering or shifting their view and subjective experience of the world have come from a behaviour originally observed in captivity—spinning. The behaviour of rapidly rotating around one's vertical bodily axis is readily and directly recognised by any human observer. In form, it is indistinguishable from spinning during dance or children's play. Just like in humans, it serves no obvious function, purpose or direct physical goal. Individuals use it mostly as an "aimless" form of play. Lameira and Perlman [128] analysed dozens of videos uploaded to YouTube of great apes spinning on ropes, including chimpanzees, bonobos, gorillas and orangutans. The results showed that great apes typically spun for five or more rotations per bout, and several bouts in a row, reaching vertiginous rotational velocities of 2–3 rotations per second. Any human without dedicated or long-term training would be knocked to the ground when undergoing similar speeds for a similar number of revolutions. It should be noted, however, that among humans and great apes, there may be individuals with atypical vestibular processing profiles that can be completely immune to dizziness and nausea from fast spinning [129].

A further comparison by the authors with various human world dance and performance traditions that involve spinning revealed that great apes can attain speeds comparable to skilled human professionals. This did not come without consequences. After spinning, great apes often showed evidence of dizziness. After releasing the rope following spinning, animals usually sat or laid down, sometimes after stumbling clumsily. In many instances, after taking a few moments to recover, they got back up for another spin. YouTube videos mainly captured this behaviour among captive animals in zoos and sanctuaries, but there were also cases of young mountain gorillas in the wild playfully spinning themselves dizzy on vines hanging from trees. Further field observations confirm that other great apes, such as chimpanzees, also spin in the wild (C. Hobaiter, personal communication).

Even though spinning differs structurally from more typical rhythmic behaviours and stereotypies found in animals, such as rocking, swaying, bouncing and pacing, spinning is rhythmic because each pirouette or revolution sets the tempo of the behaviour, which repeats cyclically. As we have covered, in some world traditions, the effects of spinning are used to directly induce a meditative and spiritual trance, such as the spinning of Sufi Whirling Dervishes, as well as some Yoga traditions. Based on a homologous and overall similar inner ear physiology and anatomy, the sustained rotational movement taps into the vestibular system in fundamentally similar ways in great apes as it does in humans to induce dizziness. It seems, thus, that our closest great ape relatives are hacking the same mechanisms as typically developing children and dervishes to tap into heightened sensations and experiences.

YouTube videos also suggest that other primates, including gibbons and monkeys, are similarly inclined towards rope spinning, although it remains to be determined whether they experience or seek dizziness. Primate ear and eye anatomy would suggest that spinning could also potentially produce disorienting effects in gibbons and monkeys. However, being arboreal, smaller and much more agile than great apes could also mean that their biology has adapted to prevent such effects. In point, if monkeys and small apes got dizzy up in the canopy, it could prove fatal by a lethal fall. One should thus expect a strong degree of accommodation or adaptation against dizziness among arboreal primates. Any tendency or pre-aptitude for it would have been strongly selected against over evolutionary time in these taxa.

However, primates are not alone in their propensity for rapid spinning. A search through YouTube reveals that taxonomically diverse species spin, from pandas to ostriches. Dogs sometimes spin in repeated circles during excited play, at times with impressive torque and speed. Whether dogs and other animals find part of the "fun" of spinning to be its perceptual stimulation and dizziness is a question for future research. In the case of humans and great apes, there is now suggestive evidence that, at least through spinning, rhythm can stimulate the vestibular system and alter individuals' state of mind. And that great apes are sufficiently aware of this effect to purposely seek it.

The observation of spinning across phylogenetically distant species begs the question of why animals might engage in the behaviour, and which common socio-ecological settings may drive or predispose animals to do so. Spinning is conspicuously similar even between animals with markedly distinct body designs, representing a behaviour patently shared between animal behaviour and human dance. Besides [128], animal spinning has received virtually no comparative or evolutionary research interest or effort. There are, thus, promising new horizons for future research (see also Chap. 4 [130]).

There are (at least) two other categories of rhythmic behaviours in animals that suggest an evolutionarily ancient link between rhythm and ASCs. A link that extends well beyond the Hominid family and that is empirically accessible but, until now, virtually unexplored—animal stereotypical and hedonic behaviours.

13.4.2 Animal Vices and Stereotypies

Stereotypical behaviours include rocking, swaying, bouncing, bobbing, pacing and other repetitive body movements. They are known for a great variety of animals living in poor captive and husbandry conditions, such as barren enclosures, confined housing, or environments overall deprived of physical and/or social stimulation. For this reason, stereotypies are classically seen as pathological and/or maladaptive behaviours that need to be reduced (for recent reviews on primates, see [131–133]). Enrichment provided to animals to avert or reduce the incidence of stereotypies is desirable for their improved welfare and is warranted for the ethical and responsible holding of animals in captivity. Yet, suppressing stereotypic behaviours obscures the underlying mechanisms at work. Why sensorily deprived animals develop ritualistic rhythmic behaviours, such as rocking and pacing, instead of any other kind of behaviour(s), requires further research.

A connection between these behaviours and animals' mental states is, however, evident and plain to see by any passer-by; when animals are engaged in a stereotypy, they often look as if they are somewhere else, lost in a self-immersive state where attention and interest in the surrounding environment are diminished. As exemplified by some of the grimmest cases of animal stereotypies in dire housing conditions, the engagement in cyclic, cadenced, clockwork behaviours has the power to potentially withdraw individuals from their proximate reality and transport them to an experience detached from the here and now. This speaks to a part of vertebrate physiology that remains largely unknown and little understood, yet widely distributed taxonomically. Stereotypies can occur in species ranging from fish and snakes to birds and primates, demonstrating that an effect of rhythm on an individual's self and world perception is not necessarily dependent on large brains or human-like cognition.

How exactly body movement and the senses interplay and relay information to the brain to alter attention and (proprio)perception remains a tantalising question, but in Sect. 13.3, we discuss several possibilities. Stereotypies have been associated with lesions or dysfunctions of the cerebellum, basal ganglia and other components of the CBGTC loop [134, 135]. Of relevance to the topic of this chapter, these structures have been linked to reward and rhythm processing (for details, see Chap. 7 [48], Chap. 9 [20] and Chap. 11 [21]), as well as ASC (see Sect. 13.3.4). Their study could provide an entry point to gain new knowledge from relatively simple model species that would nonetheless apply to other vertebrates, including primates and great apes, ultimately providing potential new insights into what drove human ancestors to engage in rhythmic and mind-altering rituals.

13.4.3 Apes in Sync

A new piece in this evolutionary puzzle has come in the form of the discovery of a "social stereotypy" in captive chimpanzees. Lameira and colleagues [136] described two chimpanzees moving bipedally one after the other in close coordination.

Rhythmic analysis showed that the two females were indeed moving in isochrony and in strict synchrony with each other across various tempos. Causality analyses revealed that sometimes one of the individuals took the lead in setting the tempo, effectively functioning as the pace-maker of the behaviour, whereas sometimes the other individual took on this role. Yet on some occasions, they both shared the role of setting the tempo of the behaviour, showing that the two individuals were dynamically locking into each other's rhythm. The couple performed the behaviour various times per day over the course of years. In itself, this finding was revealing and seminal by disproving a prevalent notion that great apes lacked the neural-wiring necessary to "keep a beat" and to synchronise to external isochronous stimuli (i.e. "the vocal learning and rhythmic synchronisation hypothesis" [137]; for further discussions and examples, see also Chap. 7 [48] and Chap. 12 [138]).

The life history of these two individuals suggests that this behaviour was rooted in early-life trauma due to separation from their mother—as it has been more commonly observed in chimpanzees [139]. This shared trauma seems to have provided a common denominator in both female physiological and psychological conditions, priming them to seek each other as "foster twin-sisters" and create a behaviour that was of some simultaneous consolation to both of them. This behaviour introduces a new facet to our valenced view of animal rhythms in the context of stereotypies. While pacing is typically a non-desirable stereotypy at the level of one individual, being considered an indicator of poor well-being conditions, here we see that, among hominids, engaging in a stereotypy and synchrony with a social partner could become the grounds for (what is most likely) a positive experience and possibly a coping strategy. If engaging in a stereotypy means to momentarily enter an ASC, then sharing a stereotypy means to share an ASC. This fascinating possibility invites future study of other instances of social rhythmic synchrony in great apes.

13.4.4 Comparative Considerations

Until now, there has been a major discrepancy between the conceptualisation of the mental lives of humans and nonhuman animals. On the one hand, each human can personally attest that he or she experiences various and fluctuating states of mind and mood daily. How we feel and perform is affected by the foods and drinks we consume in our diets, by the quality and quantity of our sleep, by the type, level and frequency of our physical activity, by the weather across seasons, whether we are in love or have lost a loved one. Conversely, animals are largely treated as though such variations and shades in awareness quality are inconsequential. With the possible exception of occasions when animals are chronically ill or in oestrus, we generally assume that animals' reactiveness and readiness, temperament and tendencies are fundamentally uniform. This stance makes it theoretically challenging to start pondering whether or how animals may have different *modus operandi* or mental states, and whether or how they may undergo different subjective experiences, as humans do. This means that consciousness is relegated to a human-exclusive feature [140]. If animals are not conscious, it follows that there is no "self" to experience different states, vetoing any attempt to start approaching the form and function of state-altering rhythms in a comparative and evolutionary manner.

Historical ideas that consciousness and sentience are uniquely human and represent the pinnacle of metacognition have implied that these traits were necessarily seated in, or associated with, areas of the brain that showed similarly human-unique size, proportion or profile—typically the prefrontal cortex. This view has gradually changed. Today, consciousness can be linked "down" to fundamental physiological processes of bodies and brain areas with older evolutionary histories, associated, for example, with emotions and feeling, rather than intelligence or rationality [141]. This view is progressing as the conception of brain function is becoming less localised and more closely linked with body function. To gain new knowledge about dance and music evolution, this view seems to be inherently favourable and promising as it directly connects what happens with the body—cyclic, repetitive movement and stimulation—with what happens with the mind-altered states.

Acknowledgments The contributions of GA Sauciuc and Tomas Persson to this chapter have been funded by the Bank of Sweden Tercentenary foundation (grant nr. P17-0942:1) and the Swedish Research Council (grant 2019-03104). The contributions of AR Lameira to this chapter have been funded by UK Research & Innovation, Future Leaders Fellowship (grant nr. MR/T04229X/1).

References

1. Bourguignon E (1973) Religion, altered states of consciousness, and social change. Ohio State Univ. Pr, Columbus
2. Bourguignon E, Evascu TL (1977) Altered states of consciousness within a general evolutionary perspective: a Holocultural analysis. Behav Sci Res 12:197–216. https://doi.org/10.1177/106939717701200303
3. Neher A (1962) A physiological explanation of unusual behavior in ceremonies involving drums. Hum Biol 34:151–160
4. Harner M (1980) The way of the shaman: a guide to power and healing. Harper & Row, San Francisco
5. Rouget G (2005) Music and trance: a theory of the relations between music and possession. Univ. of Chicago Press, Chicago
6. Becker J (2004) Deep listeners: music, emotion, and trancing. Indiana University Press, Bloomington
7. Revonsuo A, Kallio S, Sikka P (2009) What is an altered state of consciousness? Philos Psychol 22:187–204. https://doi.org/10.1080/09515080902802850
8. Dietrich A, Al-Shawaf L (2018) The transient hypofrontality theory of altered states of consciousness. J Conscious Stud 25:226–247
9. Cofré R, Herzog R, Mediano PAM et al (2020) Whole-brain models to explore altered states of consciousness from the bottom up. Brain Sci 10:626. https://doi.org/10.3390/brainsci10090626
10. Cardeña E, Berkovich-Ohana A, Valli K et al (2025) A consensus taxonomy of altered (non-ordinary) states of consciousness: bringing order to disarray. Psychol Conscious Theory Res Pract. https://doi.org/10.1037/cns0000431
11. Ludwig AM (1966) Altered states of consciousness. Arch Gen Psychiatry 15:225. https://doi.org/10.1001/archpsyc.1966.01730150001001

12. Nakamura J, Csikszentmihalyi M (2014) The concept of flow. In: Flow and the foundations of positive psychology. Springer Netherlands, Dordrecht, pp 239–263
13. Sachs M (1991) Running—a psychosocial phenomenon. In: Diamant L (ed) Psychology of sports, exercise, and fitness: social and personal issues. Hemisphere Publishing Corp, pp 237–247
14. Solnit R (2014) Wanderlust: a history of walking. Granta, London
15. Fell J, Axmacher N, Haupt S (2010) From alpha to gamma: electrophysiological correlates of meditation-related states of consciousness. Med Hypotheses 75:218–224. https://doi.org/10.1016/j.mehy.2010.02.025
16. Oman D, Bormann JE, Kane JJ (2022) Mantram repetition as a portable mindfulness practice: applications during the COVID-19 pandemic. Mindfulness 13:1418–1429. https://doi.org/10.1007/s12671-020-01545-w
17. Burke A, Lam CN, Stussman B, Yang H (2017) Prevalence and patterns of use of mantra, mindfulness and spiritual meditation among adults in the United States. BMC Complement Altern Med 17:316. https://doi.org/10.1186/s12906-017-1827-8
18. Brandmeyer T, Delorme A, Wahbeh H (2019) The neuroscience of meditation: classification, phenomenology, correlates, and mechanisms. In: Progress in brain research. Elsevier, pp 1–29
19. Sauciuc G-A, Schneider JJN (2025) Chapter 2: How animals produce rhythm—an overview of ethological mechanisms for rhythmic communication. In: Nature beats: the what, why and how of animal rhythmic behaviours. Springer Nature Switzerland, Cham
20. Sauciuc G-A, Schneider JJN, Hattori Y (2025) Chapter 9: The neuroethology of rhythmic communication: neural and mechanical building blocks and constraints on rhythm production and perception. In: Sauciuc G-A, Lameira AR, Persson T (eds) Nature beats: the what, why and how of animal rhythmic behaviours. Springer Nature Switzerland, Cham
21. Gustison ML, Sauciuc G-A (2025) Chapter 11: Moving the body and the mind - links between emotion and rhythmic communication and the underlying neurobiological mechanisms. In: Sauciuc G-A, Lameira AR, Persson T (eds) Nature beats: the what, why and how of animal rhythmic behaviours. Springer Nature Switzerland, Cham
22. Steardo L, Carbone EA, Menculini G et al (2021) Endocannabinoid system as therapeutic target of PTSD: a systematic review. Life 11:214. https://doi.org/10.3390/life11030214
23. Grof C, Grof S (2017) Spiritual emergency: the understanding and treatment of transpersonal crises. Int J Transpers Stud 36:30–43. https://doi.org/10.24972/ijts.2017.36.2.30
24. Russo MA, Santarelli DM, O'Rourke D (2017) The physiological effects of slow breathing in the healthy human. Breathe 13:298–309. https://doi.org/10.1183/20734735.009817
25. Sevoz-Couche C, Laborde S (2022) Heart rate variability and slow-paced breathing: when coherence meets resonance. Neurosci Biobehav Rev 135:104576. https://doi.org/10.1016/j.neubiorev.2022.104576
26. Fischer R, Tasananukorn S (2018) Altered states of consciousness, spirit mediums, and predictive processing—a cultural cognition model of Spirit possession. J Conscious Stud 25:179–203
27. Katz R (2003) Boiling energy: community healing among the Kalahari Kung, 11th edn. Harvard Univ. Press, Cambridge, MA
28. Campbell B (2023) San trance dance: embodied experience and neurological mechanisms. Relig, Brain Behav 13:18–34. https://doi.org/10.1080/2153599X.2022.2118360
29. Low C (2015) The role of the body in Kalahari San healing dances. Hunt Gather Res 1:29–60. https://doi.org/10.3828/hgr.2015.3
30. Pfeiffer C, Serino A, Blanke O (2014) The vestibular system: a spatial reference for bodily self-consciousness. Front Integr Neurosci 8:31. https://doi.org/10.3389/fnint.2014.00031
31. Todd NPM, Lee CS (2015) The sensory-motor theory of rhythm and beat induction 20 years on: a new synthesis and future perspectives. Front Hum Neurosci 9:105736. https://doi.org/10.3389/fnhum.2015.00444
32. Lenggenhager B, Lopez C (2015) Vestibular contributions to the sense of body, self, and others. In: Open MIND. MIND Group. https://doi.org/10.15502/9783958570023

33. Phillips-Silver J, Trainor LJ (2007) Hearing what the body feels: auditory encoding of rhythmic movement. Cognition 105:533–546. https://doi.org/10.1016/j.cognition.2006.11.006
34. Trainor LJ, Gao X, Lei J et al (2009) The primal role of the vestibular system in determining musical rhythm. Cortex 45:35–43. https://doi.org/10.1016/j.cortex.2007.10.014
35. Todd NPM, Keller PE, Govender S, Colebatch JG (2024) Electrophysiological activity associated with a cross-modal anapaest rhythm: evidence for the vestibular syncopation hypothesis. Timing Time Percept 13:173–226. https://doi.org/10.1163/22134468-bja10108
36. Lopez C, Schreyer H-M, Preuss N, Mast FW (2012) Vestibular stimulation modifies the body schema. Neuropsychologia 50:1830–1837. https://doi.org/10.1016/j.neuropsychologia.2012.04.008
37. Goldberg JM, Wilson VJ, Cullen KE et al (2012) The vestibular system: a sixth sense. Oxford University Press
38. Jáuregui Renaud K (2015) Vestibular function and depersonalization/derealization symptoms. Multisens Res 28:637–651. https://doi.org/10.1163/22134808-00002480
39. Elyoseph Z, Geisinger D, Zaltzman R et al (2023) How vestibular dysfunction transforms into symptoms of depersonalization and derealization? J Neurol Sci 444:120530. https://doi.org/10.1016/j.jns.2022.120530
40. Krey JF, Gillespie PG (2012) Molecular biology of hearing and balance. In: Basic neurochemistry. Elsevier, pp 916–927
41. Saladin KS (2012) Anatomy & physiology: the unity of form and function, 6th edn. McGraw-Hill, New York
42. Balaban C (2002) Neural substrates linking balance control and anxiety. Physiol Behav 77:469–475. https://doi.org/10.1016/S0031-9384(02)00935-6
43. Balaban CD, Yates BJ (2004) Vestibuloautonomic interactions: a teleologic perspective. In: Highstein SM, Fay RR, Popper AN (eds) The vestibular system. Springer New York, New York, pp 286–342
44. Hitier M, Besnard S, Smith PF (2014) Vestibular pathways involved in cognition. Front Integr Neurosci 8:59. https://doi.org/10.3389/fnint.2014.00059
45. Yates BJ, Bronstein AM (2005) The effects of vestibular system lesions on autonomic regulation: observations, mechanisms, and clinical implications. J Vestib Res 15:119–129. https://doi.org/10.3233/VES-2005-15301
46. Todd NPM, Cody FW (2000) Vestibular responses to loud dance music: a physiological basis of the "rock and roll threshold"? J Acoust Soc Am 107:496–500. https://doi.org/10.1121/1.428317
47. Large EW (2025) Chapter 8: Neural resonance and comparative rhythm cognition. In: Sauciuc G-A, Lameira AR, Persson T (eds) Nature beats: the what, why and how of animal rhythmic behaviours. Springer Nature Switzerland, Cham
48. Proksch S, Honing H, Ravignani A, Merchant H (2025) Chapter 7: Converging notions on the main hypotheses regarding the origins of beat perception and synchronization: the key role of audiomotor interactions. In: Nature beats: the what, why and how of animal rhythmic behaviour. Springer Nature Switzerland, Cham
49. Baumgarten TJ, Schnitzler A, Lange J (2015) Beta oscillations define discrete perceptual cycles in the somatosensory domain. Proc Natl Acad Sci USA 112:12187–12192. https://doi.org/10.1073/pnas.1501438112
50. Chakravarthi R, Van Rullen R (2012) Conscious updating is a rhythmic process. Proc Natl Acad Sci USA 109:10599–10604. https://doi.org/10.1073/pnas.1121622109
51. Neher A (1961) Auditory driving observed with scalp electrodes in normal subjects. Electroencephalogr Clin Neurophysiol 13:449–451. https://doi.org/10.1016/0013-4694(61)90014-1
52. Adrian ED, Matthews BHC (1934) The interpretation of potential waves in the cortex. J Physiol 81:440–471. https://doi.org/10.1113/jphysiol.1934.sp003147
53. Walter VJ, Walter WG (1949) The central effects of rhythmic sensory stimulation. Electroencephalogr Clin Neurophysiol 1:57–86. https://doi.org/10.1016/0013-4694(49)90164-9

54. Dietrich A (2003) Functional neuroanatomy of altered states of consciousness: the transient hypofrontality hypothesis. Conscious Cogn 12:231–256. https://doi.org/10.1016/S1053-8100(02)00046-6
55. Jamadar SD, Behler A, Deery H, Breakspear M (2025) The metabolic costs of cognition. Trends Cogn Sci 29:541–555. https://doi.org/10.1016/j.tics.2024.11.010
56. Miller EK, Cohen JD (2001) An integrative theory of prefrontal cortex function. Annu Rev Neurosci 24:167–202. https://doi.org/10.1146/annurev.neuro.24.1.167
57. Dietrich A (2006) Transient hypofrontality as a mechanism for the psychological effects of exercise. Psychiatry Res 145:79–83. https://doi.org/10.1016/j.psychres.2005.07.033
58. Filho E, Dobersek U, Husselman T-A (2021) The role of neural efficiency, transient hypofrontality and neural proficiency in optimal performance in self-paced sports: a meta-analytic review. Exp Brain Res 239:1381–1393. https://doi.org/10.1007/s00221-021-06078-9
59. Jung M, Ryu S, Kang M et al (2022) Evaluation of the transient hypofrontality theory in the context of exercise: a systematic review with meta-analysis. Q J Exp Psychol 75:1193–1214. https://doi.org/10.1177/17470218211048807
60. Jung M (2025) Does prefrontal cortex oxygenation mediate executive function during high-intensity exercise? A systematic review of fNIRS studies. Behav Brain Res 495:115765. https://doi.org/10.1016/j.bbr.2025.115765
61. Lee DJ, Kulubya E, Goldin P et al (2018) Review of the neural oscillations underlying meditation. Front Neurosci 12:178. https://doi.org/10.3389/fnins.2018.00178
62. Schlechta Portella CF, Ghelman R, Abdala V et al (2021) Meditation: evidence map of systematic reviews. Front Public Health 9:742715. https://doi.org/10.3389/fpubh.2021.742715
63. Hove MJ, Stelzer J, Nierhaus T et al (2016) Brain network reconfiguration and perceptual decoupling during an absorptive state of consciousness. Cereb Cortex 26:3116–3124. https://doi.org/10.1093/cercor/bhv137
64. Vazquez JLM, Gonzalez CJC, Wiederhold B K et al (2018) Resting stated-tractography-fMRI in initial phase of spiritual possession—a case report. Trends Med 18:1–6. https://doi.org/10.15761/TiM.1000137
65. Limb CJ, Braun AR (2008) Neural substrates of spontaneous musical performance: an fMRI study of jazz improvisation. PLoS One 3:e1679. https://doi.org/10.1371/journal.pone.0001679
66. Liu S, Chow HM, Xu Y et al (2012) Neural correlates of lyrical improvisation: an fMRI study of freestyle rap. Sci Rep 2:834. https://doi.org/10.1038/srep00834
67. Vergara VM, Norgaard M, Miller R et al (2021) Functional network connectivity during jazz improvisation. Sci Rep 11:19036. https://doi.org/10.1038/s41598-021-98332-x
68. Cakmak YO, Ekinci G, Heinecke A, Çavdar S (2017) A possible role of prolonged whirling episodes on structural plasticity of the cortical networks and altered vertigo perception: the cortex of Sufi whirling dervishes. Front Hum Neurosci 11:3. https://doi.org/10.3389/fnhum.2017.00003
69. Harne BP, Hiwale AS (2018) EEG spectral analysis on OM mantra meditation: a pilot study. Appl Psychophysiol Biofeedback 43:123–129. https://doi.org/10.1007/s10484-018-9391-7
70. Dobrakowski P, Blaszkiewicz M, Skalski S (2020) Changes in the electrical activity of the brain in the alpha and theta bands during prayer and meditation. Int J Environ Res Public Health 17:9567. https://doi.org/10.3390/ijerph17249567
71. Katyal S, Goldin P (2021) Alpha and theta oscillations are inversely related to progressive levels of meditation depth. Neurosci Conscious 2021:niab042. https://doi.org/10.1093/nc/niab042
72. Konopacki M, Madison G (2018) EEG responses to shamanic drumming. Does the suggestion of trance state moderate the strength of frequency components? J Sleep Sleep Disord Res 1:16–25. https://doi.org/10.14302/issn.2574-4518.jsdr-17-1794
73. Flor-Henry P, Shapiro Y, Sombrun C (2017) Brain changes during a shamanic trance: altered modes of consciousness, hemispheric laterality, and systemic psychobiology. Cogent Psychol 4:1313522. https://doi.org/10.1080/23311908.2017.1313522

74. Kawai N, Honda M, Nishina E et al (2017) Electroencephalogram characteristics during possession trances in healthy individuals. Neuroreport 28:949–955. https://doi.org/10.1097/WNR.0000000000000857
75. De Smedt T, Menschaert L, Heremans P et al (2016) An EEG study of creativity in expert classical musicians arXiv:1612.06719
76. Gande N (2022) Neural phenomenon in musicality: the interpretation of dual-processing modes in melodic perception. Front Hum Neurosci 16:823325. https://doi.org/10.3389/fnhum.2022.823325
77. Tan J, Di Bernardi LC, Bhattacharya J (2024) The after-glow of flow: neural correlates of flow in musicians. Creat Res J 36:469–490. https://doi.org/10.1080/10400419.2023.2277042
78. Nozaradan S, Peretz I, Mouraux A (2012) Selective neuronal entrainment to the beat and meter embedded in a musical rhythm. J Neurosci 32:17572–17581. https://doi.org/10.1523/JNEUROSCI.3203-12.2012
79. Zhou H, Melloni L, Poeppel D, Ding N (2016) Interpretations of frequency domain analyses of neural entrainment: periodicity, fundamental frequency, and harmonics. Front Hum Neurosci 10:274. https://doi.org/10.3389/fnhum.2016.00274
80. Duecker K, Doelling KB, Breska A et al (2024) Challenges and approaches in the study of neural entrainment. J Neurosci 44:e1234242024. https://doi.org/10.1523/JNEUROSCI.1234-24.2024
81. Yang X, Nah FF-H, Lin F (2023) A review on the effects of chanting and solfeggio frequencies on Well-being. In: Gao Q, Zhou J, Duffy VG et al (eds) HCI international 2023—late breaking papers. Springer Nature Switzerland, Cham, pp 628–639
82. Sezer I, Sacchet MD (2025) Advanced and long-term meditation and the autonomic nervous system: a review and synthesis. Neurosci Biobehav Rev 173:106141. https://doi.org/10.1016/j.neubiorev.2025.106141
83. Rogerson RG, Barnstaple RE, DeSouza JF (2021) Neural correlates of a trance process and alternative states of consciousness in a traditional healer. Brain Sci 11:497. https://doi.org/10.3390/brainsci11040497
84. Aparicio-Terrés R, López-Mochales S, Díaz-Andreu M, Escera C (2025) The strength of neural entrainment to electronic music correlates with proxies of altered states of consciousness. Front Hum Neurosci 19:1574836. https://doi.org/10.3389/fnhum.2025.1574836
85. Carhart-Harris RL, Leech R, Hellyer PJ et al (2014) The entropic brain: a theory of conscious states informed by neuroimaging research with psychedelic drugs. Front Hum Neurosci 8:55875. https://doi.org/10.3389/fnhum.2014.00020
86. Carhart-Harris RL (2018) The entropic brain—revisited. Neuropharmacology 142:167–178. https://doi.org/10.1016/j.neuropharm.2018.03.010
87. Martínez Vivot R, Pallavicini C, Zamberlan F et al (2020) Meditation increases the entropy of brain oscillatory activity. Neuroscience 431:40–51. https://doi.org/10.1016/j.neuroscience.2020.01.033
88. Huels ER, Kim H, Lee U et al (2021) Neural correlates of the shamanic state of consciousness. Front Hum Neurosci 15:610466. https://doi.org/10.3389/fnhum.2021.610466
89. Young JH, Arterberry ME, Martin JP (2021) Contrasting electroencephalography-derived entropy and neural oscillations with highly skilled meditators. Front Hum Neurosci 15:628417. https://doi.org/10.3389/fnhum.2021.628417
90. Atad DA, Mediano PAM, Rosas FE, Berkovich-Ohana A (2025) Meditation and complexity: a review and synthesis of evidence. Neurosci Conscious 2025:niaf013. https://doi.org/10.1093/nc/niaf013
91. Beiser DG, Houk JC (1998) Model of cortical-basal ganglionic processing: encoding the serial order of sensory events. J Neurophysiol 79:3168–3188. https://doi.org/10.1152/jn.1998.79.6.3168
92. Frank MJ, Loughry B, O'Reilly RC (2001) Interactions between frontal cortex and basal ganglia in working memory: a computational model. Cogn Affect Behav Neurosci 1:137–160. https://doi.org/10.3758/CABN.1.2.137

93. Vollenweider FX (2001) Brain mechanisms of hallucinogens and entactogens. Dialogues Clin Neurosci 3:265–279. https://doi.org/10.31887/DCNS.2001.3.4/fxvollenweider
94. Vollenweider FX, Preller KH (2020) Psychedelic drugs: neurobiology and potential for treatment of psychiatric disorders. Nat Rev Neurosci 21:611–624. https://doi.org/10.1038/s41583-020-0367-2
95. Jerath R, Barnes VA, Dillard-Wright D et al (2012) Dynamic change of awareness during meditation techniques: neural and physiological correlates. Front Hum Neurosci 6:131. https://doi.org/10.3389/fnhum.2012.00131
96. Aparicio-Terrés R, López-Mochales S, Díaz-Andreu M, Escera C (2025) The neurobiology of altered states of consciousness induced by drumming and other rhythmic sound patterns. Ann N Y Acad Sci 1550:55–70. https://doi.org/10.1111/nyas.15403
97. Kumar U, Singh A, Paddakanya P (2021) Extensive long-term verbal memory training is associated with brain plasticity. Sci Rep 11:9712. https://doi.org/10.1038/s41598-021-89248-7
98. Kalyani B, Venkatasubramanian G, Arasappa R et al (2011) Neurohemodynamic correlates of 'OM' chanting: a pilot functional magnetic resonance imaging study. Int J Yoga 4:3. https://doi.org/10.4103/0973-6131.78171
99. Zaccaro A, Piarulli A, Laurino M et al (2018) How breath-control can change your life: a systematic review on psycho-physiological correlates of slow breathing. Front Hum Neurosci 12:353. https://doi.org/10.3389/fnhum.2018.00353
100. Gosseries O, Marie N, Lafon Y et al (2024) Exploration of trance states: phenomenology, brain correlates, and clinical applications. Curr Opin Behav Sci 58:101400. https://doi.org/10.1016/j.cobeha.2024.101400
101. Voloh B, Womelsdorf T (2016) A role of phase-resetting in coordinating large scale neural networks during attention and goal-directed behavior. Front Syst Neurosci 10:18. https://doi.org/10.3389/fnsys.2016.00018
102. Wang Q, Webber RM, Stanley GB (2010) Thalamic synchrony and the adaptive gating of information flow to cortex. Nat Neurosci 13:1534–1541. https://doi.org/10.1038/nn.2670
103. Gao L, Meng X, Ye C et al (2009) Entrainment of slow oscillations of auditory thalamic neurons by repetitive sound stimuli. J Neurosci 29:6013–6021. https://doi.org/10.1523/JNEUROSCI.5733-08.2009
104. Pranjić M, Braun Janzen T, Vukšić N, Thaut M (2024) From sound to movement: mapping the neural mechanisms of auditory–motor entrainment and synchronization. Brain Sci 14:1063. https://doi.org/10.3390/brainsci14111063
105. Preller KH, Razi A, Zeidman P et al (2019) Effective connectivity changes in LSD-induced altered states of consciousness in humans. Proc Natl Acad Sci USA 116:2743–2748. https://doi.org/10.1073/pnas.1815129116
106. Xue S-W, Wang D, Tan Z et al (2019) Disrupted brain entropy and functional connectivity patterns of thalamic subregions in major depressive disorder. Neuropsychiatr Dis Treat 15:2629–2638. https://doi.org/10.2147/NDT.S220743
107. Cacciatore M, Magnani FG, Barbadoro F et al (2025) Thalamus and consciousness: a systematic review on thalamic nuclei associated with consciousness. Front Neurol 16:1509668. https://doi.org/10.3389/fneur.2025.1509668
108. Segobin S, Haast RAM, Kumar VJ et al (2024) A roadmap towards standardized neuroimaging approaches for human thalamic nuclei. Nat Rev Neurosci 25:792–808. https://doi.org/10.1038/s41583-024-00867-1
109. Wijdicks EFM (2019) The ascending reticular activating system. Neurocrit Care 31:419–422. https://doi.org/10.1007/s12028-019-00687-7
110. Crick FC, Koch C (2005) What is the function of the claustrum? Philos Trans R Soc B 360:1271–1279. https://doi.org/10.1098/rstb.2005.1661
111. Atilgan H, Doody M, Oliver DK et al (2022) Human lesions and animal studies link the claustrum to perception, salience, sleep and pain. Brain 145:1610–1623. https://doi.org/10.1093/brain/awac114
112. Smith JB, Watson GDR, Liang Z et al (2019) A role for the claustrum in salience processing? Front Neuroanat 13:64. https://doi.org/10.3389/fnana.2019.00064

113. Yaden DB, Johnson MW, Griffiths RR et al (2021) Psychedelics and consciousness: distinctions, demarcations, and opportunities. Int J Neuropsychopharmacol 24:615–623. https://doi.org/10.1093/ijnp/pyab026
114. Liaw YS, Augustine GJ (2023) The claustrum and consciousness: an update. Int J Clin Health Psychol 23:100405. https://doi.org/10.1016/j.ijchp.2023.100405
115. Friston K (2010) The free-energy principle: a unified brain theory? Nat Rev Neurosci 11:127–138. https://doi.org/10.1038/nrn2787
116. Mumford D (1992) On the computational architecture of the neocortex: II. The role of cortico-cortical loops. Biol Cybern 66:241–251. https://doi.org/10.1007/BF00198477
117. Rao RPN, Ballard DH (1999) Predictive coding in the visual cortex: a functional interpretation of some extra-classical receptive-field effects. Nat Neurosci 2:79–87. https://doi.org/10.1038/4580
118. Hohwy J (2018) The predictive processing hypothesis. In: Newen A, De Bruin L, Gallagher S (eds) The Oxford handbook of 4E cognition, 1st edn. Oxford University Press, pp 129–146
119. Piekarski M (2021) Understanding predictive processing. A review. Avant 12:1–48. https://doi.org/10.26913/avant.2021.01.04
120. Seth AK, Suzuki K, Critchley HD (2012) An interoceptive predictive coding model of conscious presence. Front Psychol 2:395. https://doi.org/10.3389/fpsyg.2011.00395
121. Carhart-Harris RL, Friston KJ (2019) REBUS and the anarchic brain: toward a unified model of the brain action of psychedelics. Pharmacol Rev 71:316–344. https://doi.org/10.1124/pr.118.017160
122. Parr T, Friston KJ (2019) Generalised free energy and active inference. Biol Cybern 113:495–513. https://doi.org/10.1007/s00422-019-00805-w
123. Kim J, Kim H-W, Kovar J, Lee YS (2024) Neural consequences of binaural beat stimulation on auditory sentence comprehension: an EEG study. Cereb Cortex 34:bhad459. https://doi.org/10.1093/cercor/bhad459
124. Pezzulo G, Parr T, Friston K (2022) The evolution of brain architectures for predictive coding and active inference. Philos Trans R Soc B 377:20200531. https://doi.org/10.1098/rstb.2020.0531
125. Westerberg JA, Roelfsema PR (2025) Hierarchical interactions between sensory cortices defy predictive coding. Trends Cogn Sci 30:S1364661325002712. https://doi.org/10.1016/j.tics.2025.09.018
126. Fachner J, Rittner S (2011) Ethno therapy, music and trance: an EEG investigation into a sound-trance induction. In: Cvetkovic D, Cosic I (eds) States of consciousness. Springer, Berlin/Heidelberg, pp 235–256
127. Maurer RL, Kumar VK, Woodside L, Pekala RJ (1997) Phenomenological experience in response to monotonous drumming and hypnotizability. Am J Clin Hypn 40:130–145. https://doi.org/10.1080/00029157.1997.10403417
128. Lameira AR, Perlman M (2023) Great apes reach momentary altered mental states by spinning. Primates 64:319–323. https://doi.org/10.1007/s10329-023-01056-x
129. Ottenbacher K (1982) Vestibular processing dysfunction in children with severe emotional and behavior disorders. Phys Occup Ther Pediatr 2:3–12. https://doi.org/10.1300/J006v02n01_02
130. Lameira AR, De Gregorio C, Gamba M et al (2025) Chapter 4: Uncharted rhythmic diversity across the animal world. In: Sauciuc G-A, Lameira AR, Persson T (eds) Nature beats: the what, why and how of animal rhythmic behaviour. Springer International Publishing
131. Lutz CK (2014) Stereotypic behavior in nonhuman primates as a model for the human condition. ILAR J 55:284–296. https://doi.org/10.1093/ilar/ilu016
132. Poirier C, Bateson M (2017) Pacing stereotypies in laboratory rhesus macaques: implications for animal welfare and the validity of neuroscientific findings. Neurosci Biobehav Rev 83:508–515. https://doi.org/10.1016/j.neubiorev.2017.09.010
133. Lutz CK, Coleman K, Hopper LM et al (2022) Nonhuman primate abnormal behavior: etiology, assessment, and treatment. Am J Primatol 84:e23380. https://doi.org/10.1002/ajp.23380
134. Kitchenham L, Mason GJ (2021) The neurobiology of environmentally induced stereotypic behaviours in captive animals: assessing the basal ganglia pathways and cortico-striatal-

thalamo-cortical circuitry hypotheses. Behaviour 158:1251–1302. https://doi.org/10.1163/1568539X-bja10084
135. Dean SL, Tochen L, Augustine F et al (2022) The role of the cerebellum in repetitive behavior across species: childhood stereotypies and deer mice. Cerebellum 21:440–451. https://doi.org/10.1007/s12311-021-01301-3
136. Lameira AR, Eerola T, Ravignani A (2019) Coupled whole-body rhythmic entrainment between two chimpanzees. Sci Rep 9:18914. https://doi.org/10.1038/s41598-019-55360-y
137. Patel AD, Iversen JR, Bregman MR, Schulz I (2009) Experimental evidence for synchronization to a musical beat in a nonhuman animal. Curr Biol 19:827–830. https://doi.org/10.1016/j.cub.2009.03.038
138. Sauciuc G-A, Persson T, Hattori Y, Lameira AR (2025) Chapter 12: The significance of the great ape heritage in the evolution of human rhythm in speech, music and dance. In: Sauciuc G-A, Lameira AR, Persson T (eds) Nature beats: the what, why and how of animal rhythmic behaviour. Springer Nature Switzerland
139. Spijkerman RP, Dienske H, Van Hooff JARAM, Jens W (1994) Causes of body rocking in chimpanzees (*Pan troglodytes*). Anim Welf 3:193–211. https://doi.org/10.1017/S0962728600016845
140. Andrews K, Birch J, Sebo J (2025) Evaluating animal consciousness. Science 387:822–824. https://doi.org/10.1126/science.adp4990
141. Damasio A, Damasio H (2024) Sensing, feeling and consciousness. Philos Trans R Soc B 379:20230243. https://doi.org/10.1098/rstb.2023.0243

Human Group Synchrony in Light of the Comparative Method: What We Know and Would Like to Know. A Peer Commentary Treatment

14

Bjorn Merker, Peter F. Cook, Colleen Reichmuth, Denise Herzing, Adriano R. Lameira, Kazuo Okanoya, Aniruddh Patel, Tudor Popescu, and Tecumseh W. Fitch

Abstract

The issue addressed in this peer commentary chapter is a rather narrow one. Across the great diversity of animal rhythmic behaviours, which are the ones that might be related by either homology (i.e. by common descent) or analogy (i.e. by convergent evolution) to our capacity for precise and sustained group synchrony such as it occurs in human music, dance, and drill? Candidate species to fill such a role, whether by homology or analogy, would help us understand biological aspects, if any, of the nature and origin of the human capacity. Recent decades have seen an upsurge in proposals regarding species that may fill this role, but so

B. Merker (✉)
Independent Scholar, Kristianstad, Sweden
e-mail: bjornmerker@gmail.com

P. F. Cook
Marine Mammal Science/Neuroscience, New College of Florida, Sarasota, FL, USA

Institute for Marine Sciences, University of California Santa Cruz, Santa Cruz, CA, USA
e-mail: pcook@ncf.edu

C. Reichmuth
Institute for Marine Sciences, University of California Santa Cruz, Santa Cruz, CA, USA
e-mail: coll@ucsc.edu

D. Herzing
Department of Biological Science, Florida Atlantic University, Boca Raton, FL, USA

A. R. Lameira
Department of Psychology, University of Warwick, Coventry, UK
e-mail: Adriano.Lameira@warwick.ac.uk

K. Okanoya
Advanced Comprehensive Research Organization (ACRO), Teikyo University, Tokyo, Japan

G.-A. Sauciuc et al. (eds.), *Nature Beats*, Fascinating Life Sciences,
https://doi.org/10.1007/978-3-032-24250-1_14

far none of them have achieved consensus status regarding their relevance to our understanding of the human capacity. As a step toward resolving some of the outstanding issues, a number of colleagues were invited to comment on the treatment of the comparative study of musical rhythmicity I present in this chapter. Their commentaries and my response follow that treatment. Though one peer commentary exercise of this kind can hardly be expected to settle the many technical and programmatic issues involved, in the setting of the broader treatment of rhythmicity afforded by this volume, it can nevertheless help delineate conceptual and empirical matters that deserve further close scrutiny.

14.1 Target Proposal by Bjorn Merker

14.1.1 Introduction

The world of animal rhythmicity presents us with a bewildering array of heterogeneous phenomena, even if we confine ourselves to rhythms that are expressed in overt behaviour with period lengths from sub-seconds to seconds. They span from the truly ubiquitous periodicities of locomotion, breathing, and chewing, generated by low-level central pattern generators controlled by often simple higher-level signals (for recent reviews, see [1–4], and also, Chap. 9 [5]), to rather rare instances of precise entrainment of individual rhythms in group synchrony, as found in a few species of arthropods and anurans, as well as in human rhythmic ensemble music [6–9].

The question to be addressed in this proposal is a rather narrow one. Across the diversity of animal rhythmic behaviours, which ones might be related by either homology (i.e. by common descent) or analogy (i.e. by convergent evolution) to our capacity for precise group synchrony such as it occurs in human music, dance, and drill (henceforth: human group synchrony)? Credible candidates for such a role, whether by homology or analogy, would help us understand biological aspects, if any, of the nature and origin of the human capacity. To that end, the better we are able to characterize the essential attributes of the temporal coordination that issues in human group synchrony, the more readily will the comparative search proceed

A. Patel
Department of Psychology, Tufts University, Medford, MA, USA

CIFAR Program in Brain Mind and Consciousness, Toronto, ON, Canada
e-mail: a.patel@tufts.edu

T. Popescu
Department of Cognition, Emotion, and Methods in Psychology, University of Vienna, Vienna, Austria
e-mail: tudor.popescu@oxon.org

T. W. Fitch
Department of Behavioural and Cognitive Biology, University of Vienna, Vienna, Austria
e-mail: tecumseh.fitch@univie.ac.at

and the more stringently can candidates be evaluated for relevance. I begin, therefore, with a brief characterization of our capacity for group synchrony by interindividual yoking in time—entrainment—of rhythmic behaviour such as it occurs in human music, dance, and drill (see also Chap. 5 [10]).

14.1.2 The Entrainment Capacity Underlying Human Group Synchrony

Not only does the animal kingdom abound in rhythmic phenomena of diverse kinds but even the small subset of these behaviours that involve temporally coordinated interactive signalling between two or more individuals exhibits diversity across species [11–13]. These behaviours do not all have the same adaptive function, and accordingly differ in their structural characteristics, underlying mechanisms, ontogeny, and distribution on the tree of life. Here we are concerned with only one particular kind of rhythmic temporal coordination among individuals, namely, entrainment of the behaviour of members of a group *to a beat* such as it occurs in human group synchrony, epitomized by human music, dance, and drill.

In its most basic and accessible form, this behaviour is exemplified by a human group clapping in unison to a steady beat. In all its simplicity, the structural device of the beat that enables such elementary group synchrony is, nevertheless, the foundation for all the additional complex timing relations exhibited by human rhythmic music. The *regular, evenly paced, sustained periodicity*, or *isochrony*, of the elementary "one, one, one, one, one..." of the pulse, beat or *tactus* ([14] pp. 179 ff) allows performers in a group to synchronize their behaviour in time with a precision unachievable in any other way.[1] The shared evenly paced tactus makes the time of occurrence of the next beat in the sequence *perfectly predictable.* This temporal predictability allows the otherwise inevitable lag in time—asynchrony—introduced

[1] The isochronous tactus defines an intermediate level between the grouping of beats into rhythmic cycles or other metric arrangements of multiple beats, and subdivisions of the beat constitutive of rhythmic figures and patterns at temporal scales finer than the beat itself (for details, see [15]). Some studies [9, 16] unfortunately obscure the central role of the isochronous tactus for rhythmic ensemble precision synchrony by referring to musical elements at the level of beat *subdivisions* as occurring in "non-isochronous positions," whereas isochrony is a necessary attribute only of the beat-defining tactus itself. All four beats of the four-beat cycles of the four different compositions analysed in the two cited papers are equidistant in time with a precision that not only falls well within the limits of *pulse attribution* reported in [17] but whose minute deviations from isochrony (inevitable for a biological system) in all likelihood are small enough to fall below the threshold for human detection of anisochrony, which requires deviations of 3–6% of the inter-beat interval. In keeping with all this, each of the many recorded drum parts supplied by these authors is eminently "clappable," and thus tactus-bound (see text). For the kind of factors that must be taken into account in describing or explaining the temporal position of musical elements at the level of beat subdivisions, see [15]. Mali ensemble drumming, it may be concluded, furnishes no exception to the principle that rhythmic ensemble music relies on an isochronous tactus as a principal means of ensuring precise group synchrony.

by the reaction time needed to respond to the current stimulus to be eliminated by *predictive timing* to achieve lagless, on-the-beat synchrony [18].

In all its simplicity, this shared beat or *tactus* is the enabling structural device for all precise human group synchrony, from simple unison group clapping to the most complex multipart ensemble polyrhythms composed of differently patterned individual contributions [8, 9, 14–16] (see also footnote 1). Arom [14] (pp. 179 ff), provides a historical and cross-cultural review of the structural role of this beat or *tactus* in rhythmic music, and Table 1 in [9] provides estimates of the precision of interindividual synchrony achieved in different genres of music (see third paragraph below). The tactus must not be confused with the technical musical term *meter*, which refers to ways in which the beats of the tactus may be grouped into cycles of beats, an entirely separate issue.

We shall call the entire genre of tactus-based human group synchrony, whether in music, dance, or drill, beat-bound or *tactus-bound synchrony*, and the lagless coincidence in time of the behaviour of two or more individuals made possible by the predictive timing based upon tactus-bound rhythmicity *lagless synchrony* or—in its simplest manifestation—*on-the-beat synchrony*. As the *primary distinctive feature* of human group synchrony, tactus-bound lagless synchrony is, in other words, the basic behavioural phenomenon for which non-human analogs or homologs are sought. One does not, of course, expect to find the more complex polyrhythmic combinatorics featured in human ensemble music instantiated among non-human animals, but we already know that on-the-beat tactus-bound synchrony exists among them (for details, see Sect. 14.1.4). What we would like to know is the more particular distribution of the phenomenon in the animal kingdom, and what light that distribution might shed on possible biological aspects of our own capacity and propensity for tactus-bound synchrony.

In the human case, the basic sensorimotor capacity on which tactus-bound group synchrony is based has been well characterized empirically and is beginning to be understood in terms of its mechanism. Our ability to make our own behavioural marking of the beat, say by tapping, coincide with that of an external time-giver *on the beat* has been documented in great detail in more than a century's worth of sensorimotor synchronization studies (for reviews, see [18–21]). The inherently predictive nature of the entrained party's behaviour means that it is based on an *endogenously produced isochronous periodicity* aligned with that of the time giver with zero, or even slightly negative (anticipatory), phase lag.[2] When this capacity for *lagless entrainment* is sufficiently flexible to span a range of tempos, as it is in the human case, it is known as *perfect synchrony* [22].

"Perfect synchrony" is a technical term, in which "perfect" does not refer to the mathematical precision with which synchrony is achieved, but only to the fact that the synchronized periodic signals *on average* occur in phase rather than out of phase relative to one another. For group clapping, it means that claps cluster unimodally on the beat in the rhythmic cycle. That is, claps occur on the beat rather than after it,

[2] Negative asynchrony may under certain circumstances be adaptive by compensating for a conduction delay "relativity effect" between participants, for which see Chap. 5 [10].

albeit with some inevitable temporal scatter. When, in the laboratory, stimulus manipulations introduce an unexpected phase offset, the deviation is typically corrected within a few beats without skipping beats [23]. Assuming one-to-one matching at the level of the actual beat, *deviations exceeding half the period length are diagnostic of a system failing to maintain perfect synchrony in the technical sense* [24]. The imprecision of on-the-beat matching, measured in standard deviations, typically amounts to a few (2–5) percent of the period length [19], i.e. some 15–30 ms for inter-beat intervals in the 500–600 ms range. These laboratory results are in good agreement with the most precise synchrony documented for live percussion ensemble music (see Table 1 in [16]).

Such performance cannot be achieved without the conjoint operation of two mechanisms, one capable of adjusting the phase, the other the period length of the entrainee's behavioural output [22, 25–27]. An example of the joint operation of these two mechanisms in cooperative synchrony between two human drummers is illustrated in Figure 1 in [8]. This duality in the timing mechanism, though operationalized differently and given different names by different authors, runs like a ground bass through studies of human sensorimotor timing performance [20, 28–34].

The capacity for period adjustment is mandated by the requirement for tempo flexibility, even when it is doing no more than compensating for the fact that entraining sequences produced by biological creatures will exhibit tempo drift. Hence the need to adjust period length on a running basis to the local average of a few periods in the entraining sequence. Even a perfectly duplicated tempo is indeterminate as to phase, however, which introduces the need for a phase control mechanism to make output coincide with the beat. Some of the control processes ensuring lagless synchrony are automatic and subliminal, i.e. performers tend to duplicate a synchronizing partner's just performed interval (i.e. to "copy" it with "lag one") even for deviations as small as a few milliseconds, which are decidedly below the threshold for conscious detection [27, 35–37].

To sum up: In human rhythmic group synchrony *we align—or latch—our own endogenously produced rhythmic patterns, whether simple or complex, to one another without phase lag by means of an underlying simple isochronous beat or tactus enabling predictive timing.* That tactus can always be made explicit by the performers themselves by clapping. The patterns aligned by means of the tactus stay locked, through both phase correction and period adjustment, to the shared isochrony despite biologically inevitable noise in the form of local deviations from perfect isochrony and tempo drift. Thus equipped, behavioural timing stays well within asynchronies of less than half a period length, that is, it does not "lose its place" in the isochrony, as long as the underlying tactus stays reasonably steady. This, then, is the type of synchronizing capacity an animal must possess, even in as simple a form as the equivalent of unison group clapping, if it is to provide analogs or homologs of human group synchrony.

In asking about a possible biological background to such a capacity, it is worth keeping in mind that we ourselves exercise it almost exclusively in music, dance, and drill. These displays consume time and energy resources that could easily be turned to account for other, more directly survival-related, purposes [38]. The narrow compass within which we exercise our capacity for group synchrony, and the unresolved questions surrounding its biological function (see e.g. relevant portions of [39, 40] and associated peer commentaries), give us scant reason to expect such a capacity to be widely distributed in the animal kingdom. Possibly quite special circumstances and selection pressures are required to equip a species with the capacity for tactus-bound group synchrony, assuming it is in fact a biological adaptation. Such questions invite resort to the comparative method for guidance and evidence.

14.1.3 Human Group Synchrony in Comparative Perspective, I: Heuristics for Narrowing the Search

The comparative search space is vast and highly differentiated. Closely related species may differ in their behavioural adaptations, and traits may be shared by convergent evolution across vast evolutionary distances. There are many tens of thousands of vertebrate species (see Figure 2 in [41]) and several million species of arthropods [42]. Both of these taxonomic groups are known to contain species that engage in interindividual rhythmic synchrony of some sort (reviewed in Chap. 3 [43]), so the task of using the comparative method to shed light on the biology of our own group synchrony is a daunting one. Lightly applied as preliminary screening devices, the following heuristics may make that task more tractable:

The principal source of evidence regarding the comparative biology of human group synchrony is the *natural behaviour of animals in the wild* as character states on a phylogenetic tree. This is after all what evolution, whether by natural or sexual selection, has worked on to shape behavioural phenotypes, rendering behaviour not observed in the wild supplementary or marginal sources of evidence for the task at hand.

Outside the laboratory, the entrainment of two or more individuals requires the individuals themselves to be the source of the evenly paced entraining sequence. That is, they must be capable of *endogenously producing isochronous behavioural output without prior input*.[3] They must, in other words, have an expressive capacity,

[3] When the isochronous *tactus* is hidden in stimulus complexity, the requisite capacity for perceptual beat extraction is needed as well (for reviews of beat extraction models see [44, 45]). This capacity is not, however, a defining aspect of entrainment with perfect synchrony as such, nor is it tapped by most studies of human sensorimotor synchronization [20]. A species that lacks the isochrony production propensity essential for mutual entrainment may nevertheless possess a perceptual beat extraction capacity as a matter of auditory system sophistication alone. That by itself does not make it a synchronizing species. Nor are all cyclical production phenomena based on equal period lengths (isochrony) suitable for interindividual entrainment of behaviour. Those employed by entraining species (from human to firefly) typically have a leisurely tempo of 2 Hz or slower. The exceedingly rapid (10–60 Hz, i.e. 600–3600 beats per minute) pulse grids extractable from sequence timing in the singing of zebra finches (not known to entrain behaviourally) documented in [46] presumably reflect the temporal granularity of their song system motor control rather than anything related to the isochrony of an entrainment beat. As noted in footnote 1, the tactus lies at an intermediate level of time discretization.

even propensity, to spontaneously produce isochronous behavioural signalling of some kind, say chirping, calling, flashing, waving, croaking, hooting, clapping, stomping, or drumming, at a pace within the tempo range of their entrainment mechanism, if present. The occurrence of *spontaneous isochronous signalling behaviour in the wild* can therefore serve as a preliminary screening device for group synchrony candidates.

Precise synchrony is facilitated by signal characteristics that allow the signalling behaviour to be approximated by a point-process, for example, when the elements of an evenly paced signal sequence have well-defined onsets or offsets. This provides a discrete time point for each signal element to which the entrainment mechanism can latch its timing. Signal elements produced for purposes of entrainment, and for lagless (on-the-beat) synchrony in particular, accordingly tend to feature some abrupt, discrete, signal attribute, as in chirps, flashes, barks, claps, or stomps, and the time dimension of human rhythmic music is ubiquitously discretized [15, 47]. Far from all such signals serve entrainment, of course, but where discrete signal elements are lacking, as in fluid, smoothly graded, vocalizations, "entrainment candidacy" is correspondingly weakened.

Should the necessary production capacity be present—discrete repetitive signalling approximating isochrony is hardly rare (see Chap. 3 [43]) —it remains to be determined whether additional individuals ever synchronize their own signalling to the evenly paced time-giver, and if so to characterize the nature of that synchrony parametrically. This allows us to eliminate, for example, the interactive behaviours of the turn-taking and duetting mammals and birds alluded to in Sect. 14.1.5 from consideration, because the coordination of their signalling is not tactus-bound. These turn-taking cases, rather, are analogous to human conversational turn-taking, which also proceeds without being bound to a tactus.[4]

Not every instance of periodic temporal yoking of behaviour across individuals qualifies as group synchrony. The temporal synchrony elicited in members of a group by their equal reaction time to initiate flight from a suddenly appearing threat may induce an initial sequence of synchronized wing, leg, or swimming movements without for that reason exemplifying the phenomenon we are exploring. Even more obviously, joint locomotor or other movements of animals *physically yoked*, such as when in close clasping embrace [53], do not qualify. That does not, however, eliminate, say, human dancers in physical contact from consideration, because in their case, each participant aligns their movements to the shared time-giver of the beat of the music to which they dance and would accordingly be synchronized even when not in physical contact.

To illustrate how the first four points above may guide our search, consider the following. One cannot spend much time in the forests of South-East Asia (point 1) without noticing the conspicuously steady isochrony (points 2 and 3) with which various species of barbets deliver their species specific (innate) call patterns

[4] This is not the place to revisit the "speech isochrony" controversy. Suffice it to say that the purported regular periodicity is often not found when looked for in the speech signal [48–52] and that we typically are not tempted to tap our feet to conversational speech.

(personal observation). There are some 80 different species of barbets world-wide, inhabiting the forests and scrublands of Asia, Africa, and Central and South America. However, in none of the several isochronously calling barbet species I personally witnessed in the forests of South and South-East Asia, was there ever even a hint of interindividual coordination in the timing of calls. Instead, the chance-wise juxtaposition of the sequences of many calling individuals creates endlessly varying combinatorial phase relations which are pleasant to listen to, but never really repeat.

Might there, among the 80 plus species of barbets gracing this planet, be some species that use their capacity for steady and sustained isochronous timing of calls to actually synchronize those calls with one another in joint performances (point 4)? While writing this proposal, I conducted a cursory YouTube search for barbet species other than the ones with which I am familiar, and it took me less than 20 min to locate an intriguing and charming example, featuring a pair of Red-and-yellow barbets recorded in Kenya [54]. I say "intriguing" because of the incipient synchrony, to a humanly recognizable beat, that develops between the two individuals at several points in their duet. Needless to say, plenty of study would have to be devoted to this species to know whether it in fact is a synchronizing one or not. The duets of Black-collared and Toucan barbets (Africa and New World, respectively) as well as the chorusing of Prong-billed barbets (New World) [55] also deserve scrutiny from this point of view.

I tell this anecdote only to illustrate how early in the employment of the comparative method to the biology of rhythmic group synchrony we happen to be, and how simple heuristics such as the above ones may help to shorten and ease the labours of canvassing the behavioural riches of the animal world for relevant evidence.

14.1.4 Human Group Synchrony in Comparative Perspective, II: Animal Parametric Parallels

As already mentioned, there are a number of non-human species whose natural behaviour in the wild features rhythmic synchrony between two or more individuals. Spottily distributed across chordates and arthropods, these behaviours must have evolved independently in those large taxa, and repeatedly within each of them. The interindividual behavioural synchrony of these different species is not all of a kind, however (for recent reviews and discussions, see [12] and Chap. 5 [10]). There are differences in the synchronous behaviours themselves, in the mechanisms underlying them, and in their functional role in the behavioural ecology of the different species, and thus, also in the selection pressures behind their evolution in any given case.

The most striking aspect of the distribution, within these varied forms of interindividual synchrony, of species that exhibit tactus-bound group synchrony is that those who in parametric terms are closest to human levels of entrainment type and precision are also located at a great distance from us on the tree of life. So far, the most precise on-the-beat non-human group synchronizers are three species of tropical lampyrid beetles (fireflies), namely, *Pteroptyx malaccae*, *Pteroptyx tener*, and *Luciola pupilla* [22, 56–58]. Their entrainment is underwritten by a steady isochrony of flash production. Its period length exhibits a standard deviation of less than 3% [22], placing it at the more skilled end of the human range.

Moreover, the lagless entrainment of these beetles, along with a few species of chorusing crickets, is not tied to a fixed tempo, but is maintained without phase lag over a range of tempos [12, 59–61]. In *P. malaccae* this range amounts to some 15% in either direction centred on roughly 2 Hz.[5] The physiological mechanism of this tempo flexibility remains to be unlocked, but the lagless synchrony of these tactus-bound synchronizers across a range of tempos make them members, along with humans, of the so far small club of perfect synchronizers as defined by [22], and detailed in Sect. 14.1.2 for the human case. The behaviour that gives them their claim to membership is equivalent to human unison group clapping, without any further addition of the kind of rhythmic patterning typical of human music.

Besides these insects that engage in entrainment with perfect synchrony, other arthropod species have developed group synchrony without the period adjustment mechanism needed to maintain zero phase lag at different tempos (reviewed in [12]; see also [8, 66], as well as Chaps. 3 [43] and 5 [10]). In the *Mecopoda* bush cricket species studied by [67] (a model of clarity for the presentation of entrainment data), this results in steady locking to the entraining signal over a range of tempos, but without being on-the-beat at any other than the animal's spontaneous tempo. Other entraining tempos introduce a phase offset whose size varies with the difference between spontaneous and entraining tempos. When the latter is faster or slower than what can be accommodated with a phase offset, individuals may nevertheless entrain by means of whole integer beat combination ratios familiar from metrical practice in human music (ibid., figures 2 and 3). Thus, for a fast stimulus, individuals may signal only on every other beat (2:1 ratio), or for a slow one, they may complete five of their own beat cycles for every three stimulus cycles (3:5 ratio), etc.

These examples by no means exhaust the variety of group synchrony phenomena on exhibit in nature. Interindividual entrainment aimed at signal alternation is a major such category covered in Chap. 5 [10], with particular attention to anurans, while some examples drawn from mammals and birds are briefly touched upon in Sect. 14.1.5 below. It is the attempt to understand the tactus-bound, on-the-beat (lagless) synchrony that underlies human group synchrony in music, dance, and drill that motivates a focus on the beetle and cricket "perfect synchronizers" considered here. As insects, these group synchronizers are not even deuterostomes, but protostomes, forcing us to go more than 560 million years back, to *before the start of the Cambrian era*, to locate the last common ancestor between them and our own

[5] This tempo range is narrower than the broad tempo range of human sensorimotor synchrony. It is worth noting, however, that the human range reflects the cognitive strategy of mentally subdividing long periods (so-called subitizing) to maintain precision at low tempos [62, 63]. Moreover, humans have a preferred tempo centred on 2 Hz where precision is highest and around which the tempo of rhythmic music concentrates. In fact, the firefly tempo range of 30% covers the central 58% of the area of the tempo histogram for 74,000 pieces of Western popular music compiled by [64] Dirk Moelants (see also [65]).

species. To find the closest parametric parallels to human perfect synchrony at such evolutionary distances means that we are faced with a striking case of convergent evolution ("analogy" in comparative method terms). The emergence of similar behavioural phenotypes through convergent evolution in distantly related species takes place through adaptation to shared facets of behavioural ecology, introducing shared selection pressures for the function in question.

Profound interest therefore attaches to the behavioural ecology that may have selected for perfect synchrony in these few insect species as a window on the selection pressures that may lie behind our own analogous capacity. In the parallelism between our own rhythmic group synchrony and that of the insect perfect synchronizers there may lie hints, in other words, about the selection pressures behind our own propensity to engage in rhythmic group synchrony. Those pressures are not always easy to discern, particularly in cases such as our own, for which the utility of the behaviour in plain survival terms is moot, as noted at the end of Sect. 14.1.2.

There is currently no settled account of the selection pressures behind the evolution of perfect synchrony in the insect species in question. One theme in their behavioural ecology is, however, that males, for reasons such as patchy resource distribution among others, are aggregated in clumps during the breeding season, aggregates to which ambulatory females are drawn and guided by the male group displays, whether by vision or by hearing. As summarized by [12], under such circumstances, simultaneous male signalling may serve to keep a species-specific male signal from being smeared in time beyond a female's recognition threshold, it may impair the ability of predators to localize individuals in the group, and it may render the group signal a "super-normal stimulus" that penetrates farther in the terrain through amplitude summation of individual contributions or adds visibility through size (a flashing tree, the "beacon effect"). All these selection pressures may of course operate together as well as singly.

It is therefore noteworthy that the very socio-sexual ecology that provides the setting for insect perfect synchrony, namely, aggregated males and wandering females, has a parallel in the behavioural ecology presumed to have framed our own departure from the ancestral state we share with chimpanzees. In Sect. 6, below, the implications of this coincidence of analogous entrainment behaviour as well as socio-sexual ecology in humans and group synchronizing insects are detailed and explored, along with a conjectural scenario for the origin of our own capacity for perfect group synchrony in its light. Before doing so, however, a potential objection must be addressed. It runs as follows: In examples such as Snowball, the "dancing" cockatoo [68, 69], or Ronan, the synchronously head-bobbing sea lion [70], do we not have synchronizers that are far closer to us in evolutionary terms than are fireflies and crickets, and therefore, more germane to our own synchronizing capacity than exceedingly distant insects?

Neither parrots nor otariid seals lie on our ancestral line, of course, so the relevance of their behavioural capacities to our own would, just as in the case of the insects, be by analogy in comparative terms. The question of their relevance therefore devolves upon the nature and quality of their synchrony in comparison to that of the insects and humans, viewed in light of the heuristics sketched in Sect. 14.1.3.

We turn therefore to examine findings and suggestions regarding synchronizing behaviour in a variety of species examined in the wake of [68, 69].

14.1.5 Human Group Synchrony in Comparative Perspective, III: Recent Proposals for Additional Non-human Parallels

In [68, 69], the spontaneous behaviour of individual parrots, living with humans, was examined in the presence of rhythmic human dance music. Relative to the steady beat of the music, the birds exhibited intermittent on-the-beat synchrony in the course of major phase drifts with repeated loss of entrainment. The birds accordingly must have produced repeated asynchronies exceeding half a period length which, as noted in Sect. 14.1.2, is a diagnostic of failing to maintain synchrony [24]. In view of what was covered in Sects. 14.1.2 and 14.1.4, this overall pattern is not what is expected of a species equipped with an entrainment mechanism. As we have seen, entraining fireflies, crickets, and humans, unlike the tested individuals, stay tightly locked to a tactus that stays reasonably steady, and this is true even for the wider range of synchronizing arthropods that lack a period adjustment mechanism [67].

The parrots do, however, show spontaneous responsiveness to and engagement with the rhythmic music to which they are exposed. They are highly encephalized and vocal animals and presumably have the auditory capacity to extract the beat from the human music, to judge by their demeanour.[6] The videos show them visibly captured by the music and expressing their engagement with it through dance-like bodily movements, albeit without the consistent latching to the beat that is the unique contribution of a predictive entrainment mechanism. Both sets of authors proposed that the vocal learning ability of these bird species is the critical factor helping to account for their engagement with rhythmic music, in keeping with the "vocal learning and rhythmic synchronization hypothesis" previously introduced by Patel [71].

Patel proposed his hypothesis on the basis of the following theoretical rationale: "Vocal learning requires a tight coupling between auditory input and motor output in order to match vocal production to a desired model. This online integration of the auditory and motor system places special demands on the nervous system..." and "The resulting tight coupling between auditory input and motor output may be a necessary foundation for beat perception and synchronization" [71 p. 102]. As we have seen, tight online (real time) coupling between sensory (whether visual or auditory) input and motor output is indeed a critical, in fact defining, attribute of interindividual rhythmic synchrony. The latter cannot take place without that coupling, the tightness of which is typically measured in the tens of milliseconds range.

Nothing of the kind takes place, however, in vocal learning, which does not feature on-line (real time) coupling between auditory input and motor output. Instead, vocal learning is a protracted process conspicuous for its radical temporal

[6] See footnote 3, above, and [45].

decoupling of auditory input and motor output. In vocal learning, the perceptual and mnemonic stage of acquisition antedates the motor practice and production stage by days, weeks, and even months. It typically proceeds in three principal stages, the first of which is devoted to committing the to-be-produced vocal pattern to *memory*, a purely perceptual-mnemonic process devoid of motor output or any necessary behavioural engagement with the tutor or other model (as demonstrated by the common practice of tutoring isolated birds via loudspeakers in their cages). This is followed by a protracted period of varied solitary motor practice, again requiring no interaction with a tutor or other external model. It finally, on what is called song crystallization, issues in competent vocal production, typically in a *solo performance*, that duplicates the sound pattern received by ear and committed to memory weeks earlier. As such, vocal learning is a spectacular instance of what has been called *deferred imitation* in a human context (for the concept, see [72–74]; for vocal learning as deferred imitation, see [75–77]).

This is not to say that birds, whether they are vocal learners or not, lack the capacity to couple their vocal output to heard stimuli in real time, but only that that is not a feature of vocal learning. Birds, like many other vocal species, engage socially with conspecifics by calling or singing, and these engagements involve various forms of interindividual temporal coordination of vocal behaviour, as in antiphonal calling, turn-taking, and duetting [11, 13, 78–81] (as well as Chaps. 3 [43] and 5 [10]). As already mentioned in Sect. 14.1.4, a major category of such interindividual coordination features control of the timing of communicative signals so as to *avoid signal overlap* with one or more other signallers, typically by turn-taking [11–13].

The timing of such alternation can be quite precise as well as flexible [82], and some duetting or chorusing bird species produce joint vocal patterns of considerable complexity by interlocking individual contributions in time [83, 84]. Melissa Coleman and colleagues have performed the feat of simultaneously monitoring neural activity in each partner of mated, freely moving, and duetting pairs of plain-tailed wrens, finding a pattern of real time see-saw mutual inhibition in a vocal control nucleus contingent on hearing the partner's voice [85]. Such interindividual coordination is often dyadic, and in some cases may consist of no more than an obligatory pause between the end of a received signal and the initiation of a response. The length of the obligatory pause easily accommodates responding on the basis of ordinary reaction-time, even when the pause is as short as in human conversational turn-taking, i.e. a couple of hundred milliseconds [86].

Turn-taking interaction between individuals accordingly does not require signalling to be bound to a regularly paced periodicity (isochrony), as it must be when synchrony aims at signal coincidence rather than avoidance (see Sects. 14.1.2 and 14.1.4). Turn-taking mammals and birds that signal repetitively, such as bamboo rats [13], meerkats [87], marmosets [88], and starlings [81], therefore have no need to do so with regular periodicity. The same is true of Indri family choruses [89–91], and those of a number of gibbon species [92–94]. The paced repetitive signalling of both Indri and gibbon choruses' features overlapping as well as sequential elements, but the interactive behaviour of these and the other species just cited proceeds without being bound to the isochrony that defines the rhythmic group synchrony with which we are concerned, as is evident from the sound spectrograms of the respective reports.

When repetitive signalling occurs in groups, even turn-taking may nevertheless make the signals of two or more individuals coincide in time by default (i.e. passively), such as when two individuals both alternate with the same third party [12, 95]. In larger groups, such "passive signal coincidence" may lead aggregate group signalling to segregate into two clusters with simultaneous signalling *within* a cluster, and alternating (antiphase) signalling *between* clusters, an outcome that is sensitive, however, to boundary conditions such as the spatial distribution of signallers and even whether the group has an odd or even number of members [96]. This kind of overlap avoidance, even when isochronously paced, will therefore not generate even the simplest case of the kind of rhythmic group synchrony at issue here, namely, that of an entire group signalling in unison on the beat, as in human unison clapping.

Apart from the fact that the interactive modes of real-time temporal coordination between individuals just considered are typically not bound to the isochronous tactus that defines human group synchrony, their species distribution is independent of vocal learning. Untold numbers of species without vocal learning engage in such exchanges. Apparently both vocal learning and the real-time temporal coordination of interactive signalling are behavioural dimensions without any necessary connection to the tactus-bound rhythmicity that underlies human group synchrony. It is hardly surprising, then, that the joint presence of vocal learning and rhythmic engagement in the two parrot species tested in 2009 did not generalize to many other non-parrot vocal learners among the birds [45, 97], quite apart from the counter-evidence to the "vocal learning and rhythmic synchronization hypothesis" represented by entraining insects with perfect synchrony cited in Sect. 14.1.4, species whose behaviour is unrelated to vocal learning.

As a possible explanation for the counterevidence of a lack of rhythmic engagement on the part of oscine songbird vocal learners, Patel, in his "revised vocal learning and rhythmic synchronization hypothesis" has noted that the neural organization of the song system of parrots differs from these other vocal learners, and that this may help explain the behavioural difference [97]. Chakraborty and colleagues have discovered that parrots, over and above the "core" system of song nuclei common to all vocal learners, have an expansion of this system in the form of additional neural territories designated a "shell" system, because it surrounds the core system, and is tightly interconnected with it [98].

The extent of shell-system development in the parrot species examined correlates with the extent of their capacity for mimicry [98]. This presumably also holds true for the many other bird species that are true vocal mimics, such as myna birds, mockingbirds, lyrebirds, pied butcher birds, marsh warblers, and others (reviewed by [99]). The feat of memorizing and perfecting performance of the calls and songs of 179 other species from two continents, which the marsh warbler keeps "on tap" for its singing bouts [100], requires information storage capacity beyond that of ordinary (same-species) vocal learning. That extra capacity presumably is supplied by a "shell" system in these other mimics, no less than in the parrots Chakraborty and colleagues examined. Pending further studies, this sizeable addition to the avian vocal learning system accordingly is most plausibly interpreted as a neural expansion of vocal learning capacity to encompass vocal mimicry of sounds beyond the song patterns of conspecifics.

The likelihood that the vocal learning "shell" system of parrots serves their capacity for vocal mimicry opens the door to an alternative to the Patel hypothesis in trying to account for the engagement of the "parrots of 2009" with human music. The lengthy practice by which vocal learners master the production—by voice and from memory—of stored vocal patterns typically takes place without external reinforcement. Instead, an endogenous motivational mechanism, a "conformal motive," must sustain practice until duplication is achieved [75, 77]. In true mimics, this duplicative propensity and capacity extend beyond the song patterns of conspecifics to incorporate calls and songs of other species, and even arbitrary environmental sounds such as the sound of fluttering wings (imitated *vocally*), snapping branches, chain saw noise, computer game jingles, camera shutter sounds, and of course human speech, all mimicked with a degree of fidelity sufficient to fool the human auditory system.

As I have conjectured regarding humans: "Around the core of learned control of the voice, the same mimetic ('conformal') motive that ensures copying fidelity in song … [c]ould have been linked to other aspects of … behavioural display. … Such a linkage … extending to bodily (postural and gestural) expressive learning would have turned us into a species of 'imitative generalists'" [76 p. 53]. Some mimics among the parrots extend mimicry beyond vocal mimicry to a more general imitative capacity for bodily postural and gestural patterns. For example, monogamous species in the genus *Trichoglossus* engage in joint cooperative pair displays, some of which they learn by imitating one another [101].

Cockatoos appear to possess this generalized imitative capacity. They are accomplished actors, clowns, and hams, employing a range of human postural attitudes and prosodic gestures in interaction with their keepers.[7] They are also highly susceptible to imprinting on their human hosts [102], so much so that commercial bird breeders resort to special isolation containers in order to allow their hatchlings to breed with their own kind on sexual maturity. The suggestion, then, is that the "parrots of 2009," acculturated to humans, have learned the human ways with rhythmic music and are doing their enthusiastic best to dance, but without the benefit of an endowment for entrainment synchrony that would allow them to keep the beat consistently and with precision [45, 66]. Vocal learning would, on this interpretation, supply the key to the imitative propensity these birds do possess, and not to the synchronizing ability in which they are deficient.

The behavioural difference between a cockatoo and oscine vocal learners with respect to rhythmic engagement would thus be accounted for by the generalized imitative capacity that the former possesses, but the latter lack. On adoption by

[7] Examples abound on YouTube, of which the following video provides an example ("Cockatoo Finding Out He Is Going to the Vet," https://www.youtube.com/watch?v=5UUjJysUMTw)

humans an imitative generalist can use its broad imitative capacity to produce striking facsimiles of human leg and head-banging movements to rhythmic music.[8] The fact that in so doing they drift in and out of the beat without apparent discomfiture, as is also the case for human infants, tells us that they lack something we have, at least by the time we reach puberty [8, 66], namely, a propensity and mechanism for tactus-bound on-the-beat entrainment synchrony. We tend to take both the imitative and the synchronizing capacity for granted because we possess them both, but they are not commonplaces in the animal kingdom, as we learn when attempting to induce an animal in which neither capacity is well developed to join a human in rhythmic activity.

An example is provided by the synchronization study with the bonobo Kuni, in which an experimenter tried to get the animal, equipped with its own drum, to join in human drumming [106]. In sharp contrast to the spontaneously performing parrots, the bonobo required various measures of habituation, coaxing, and reinforcement in order to perform. Unfortunately, the hectic human drumming employed in this study, and in particular in Experiment 2, designed to assess synchrony, has a pace typical of ape and human noise making and lacks the very characteristics that induce humans (and parrots) to move to a beat [8, 15, 65, 107, 108]. The design of Experiment 1, in which human drumming stopped as soon as Kuni started drumming (presumably to resemble the "continuation" mode in the standard human sensorimotor "tapping" paradigm), may also have discouraged or confused the animal.

To what extent these circumstances explain Kuni's erratic performance, featuring unmotivated tempo changes, frequent pauses, and phase drifts not expected of a species equipped with an entrainment mechanism, is unknown. They do, however, urge caution in treating her performance as indicative of bonobo rhythmic limitations. For one thing, a common chimpanzee in captivity has spontaneously produced a more than 5 min long solo of untaught drumming [109]. It is not performed to an external beat, but its characteristics are far closer to human musical drumming than anything heard in the samples provided of Kuni's performance. The drumming is bimanual, rather evenly paced and, as indicated by the most evenly paced passages and stretches of duple meter, has a basic tempo centred on 128 beats per minute (bpm), close to the human preferred tempo centred on 120 bpm.

Meanwhile, De Waal's [110] tantalizing report of rhythmic vocal synchrony in captive bonobos still has no follow-up and remains unexplored after close to three decades [8, 66, 111, 112]. According to De Waal, in situations of high excitement or tension, two or more bonobos occasionally stage a loud chorus of evenly paced staccato hooting. In doing so, their hoots are reported to occur either in synchrony or in alternation at a steady rhythm of approximately 2 Hz (120 bpm). This call type has not been reported for common chimpanzees but has been reported under a different

[8] For the intricacies and controversies surrounding the comparative study of human and animal imitation, see, e.g. [103–105]. Without entering details, it is worth noting that the available evidence, both human and animal, falls into place only by drawing a distinction between ritual culture and instrumental culture [38, 75–77].

name for bonobos in the wild, though in this case without any information regarding group synchrony [113].

Recalling the importance of naturalistic behaviour in the wild for comparative analysis noted in point 1 of Sect. 14.1.3, I have concluded that "few issues would seem to provide more leverage for the comparative study of the biology of human musical rhythmicity than a thorough characterization of bonobo staccato hooting in the wild. Should it occur, and serve interindividual entrainment of voices, the genus *Homo* would not be alone among the apes in having evolved a capacity for rhythmic entrainment of voices" [8 p. 7].

The importance of naturalistic behaviour in the wild notwithstanding, observational evidence on captive animals as well as laboratory experimentation can provide information about animal capacities and mechanisms not easily accessible in the wild. Entrainment in non-human animals can, for example, be explored with a variety of conditioning procedures. Such procedures do not, of course, play any role in the ontogeny of insect entrainment, whose stereotypy relies on innate mechanisms, while the human capacity appears to be largely a matter of maturation in which the role of various forms of learning remains to be determined [8]. In non-human animals, operant procedures might nevertheless reveal both limitations and latent capacities with regard to their entrainment capacity, as the following recent examples illustrate.

Zarco and colleagues adapted the human sensorimotor "tapping" task to macaques with the help of operant procedures requiring extensive training [114, 115], as well as Chap. 7 [116]. It took the macaques an average of 16 months of training to master their task. They thus had ample time and opportunity to hit on the strategy of targeting the upcoming stimulus itself, perfectly predictable as far as its time of arrival is concerned. Instead of doing so, as humans do with ease in the same situation, the animals consistently responded *after* the stimulus with response latencies that are compatible with reaction time responding to the current stimulus.[9] Subsequent studies, employing optimized conditioning procedures involving tens of thousands of trials spread over many months of training, have produced results more closely aligned with human performance, while still differing from it in a number of respects [118, 119].

Reaction time considerations may help explain the peculiar results of another study that, like the macaque study, also used operant conditioning procedures to test for entrainment synchrony to short isochronous sequences, this time in budgerigars [120]. A wide tempo range spanning interstimulus intervals from 450 to 1800 ms

[9] See [117] for auditory reaction time. Macaques, with shorter conduction paths, are likely to be faster than humans. The fact that in [114] stimulus presentations with randomized onset intervals produced even greater asynchronies is not evidence of predictive timing in the sense of entrainment synchrony. Reaction times to predictable stimuli are shorter than those to unpredictable ones (cued versus uncued reaction time). A sprinter's dash is initiated by a reaction time limited response to the sound of the start pistol, shortened by the preparatory cues of "ready, steady." The regular stimulus train of the monkeys' task can accordingly be construed as a serial cued reaction time signal. Stimulus presentation with random onset intervals removes this cueing function, and so over-estimates macaque cued reaction time.

(133–33 bpm) was tested. At the faster (apparently preferred) tempos, the birds' inter-response intervals matched those of the stimuli well: they were regular, but asynchronies were large and positive, approaching the birds' reaction time as estimated by responses to randomized intervals. At slow tempos, responses matched inter-stimulus intervals poorly; they were irregular, yet averaged asynchronies were smaller, though still positive.

Assuming that the reaction time estimate, for the same reasons as in the macaque study, is an overestimate (see footnote 9), the birds' fast tempo responses were presumably reaction-time-based and thus not evidence for predictive entrainment synchrony. For the slow tempos, the paradoxical result that more irregular performance resulted in reduced asynchronies requires an explanation. For humans, synchronizing precision declines substantially above 900 ms interval length [62] unless the strategy of mentally subdividing the interval (subitizing) is adopted [63]. The birds cannot be expected to manage long intervals better than humans, and without subitizing, irregular budgerigar performance would be expected, as was indeed the case.

The anomaly of reduced asynchronies with more irregular performance may have been caused as follows. An average phase approximating stimulus onset rather than the point of estimated reaction time (yielding low average asynchrony despite irregular pecking) occurred at 900 ms or longer intervals in 6 of the 46 bird × tempo combinations. Premature (anticipatory, "jumping the gun") pecking during the long "acceptable" period before stimulus onset, as in the trace of female D in figure 3 of [120], is more likely to occur at long intervals (slow tempos). Averaged into irregular performance, these early pecks might explain the anomalous result. If so, little remains of a case for entrainment by predictive synchrony in budgerigars as assessed in this study.

Hattori and colleagues [121] used food reinforcement to train three chimpanzees to alternately tap two illuminating and sounding keyboard keys and then assessed the influence of a separate isochronous sound sequence (a distractor sequence) on their tapping performance. Only one of the three chimpanzees (named Ai) showed any systematic relationship between her tapping and the isochronous distractor, and she did so only at one of the three distractor tempos, namely, the tempo that coincided almost perfectly with the self-paced tempo Ai exhibited during training. At that distractor tempo (100 bpm), Ai showed a very broadly tuned assimilation of her tapping to the distractor stimulus, centred on close to zero asynchrony.

Ai showed no influence of the distractor isochrony at any other tempo, and the other two chimpanzees showed no influence at any distractor tempo. If one of the distractor tempos had not happened to coincide with Ai's self-paced tempo, the study might not have yielded any evidence of synchrony in these animals. In the single instance of synchrony observed, the tempo coincidence made period adjustment unnecessary. Ai's performance is not, however, explicable in terms of the mechanism that allow the *Mecapoda* crickets to synchronize without period correction [67]. As already mentioned, at other than the self-paced tempo, such a mechanism still generates good entrainment, but with a phase offset whose size reflects the extent of departure from that tempo (cf. the sea lion study considered next). In Ai's case, this did not occur.

No known entrainment mechanism works in the way Ai performed, and certainly not the human one. Perhaps the tempo coincidence allowed the distractor sequence to exert a subtle subliminal influence on Ai's tapping. As already noted in Sect. 14.1.2, subliminal effects do make a partial contribution to human tapping synchrony [27, 36, 37], and may help account for Ai's performance.

We come then to the female sea lion Ronan [70]. An operant training regime based on food reinforcement was used to shape her head bobbing to isochronous auditory stimulus trains. Initially this was done on what amounts to a bob-by-bob basis, gradually extended to bobbing sequences. After lengthy training, Ronan bobbed her head reliably to the stimulus train at two tempos. She then readily generalized her performance to five novel tempos but did not generalize to a piece of rhythmic popular music. Further training on this piece brought her to criterion performance. She then readily generalized to different tempos of that piece and, without further training, to two different pop tunes. After further training on one of these, she generalized to five novel tempo versions of that tune as well.

A peculiarity of Ronan's performance across this variety of entrainment stimuli, noted by the authors, is an overall tendency to produce positive asynchronies at faster tempos (i.e. to lag the stimulus, and the more so the faster the tempo) and negative asynchronies at lower tempos (analogously, but leading the stimulus). The tendency is not strictly quantitative, and some reversals between adjacent tempos are in evidence (see figures 2B and 2C in [70]), yet the tendency is unmistakable. The same tendency is evident in a subsequent study in which the animal's response to phase and tempo perturbations was studied [122]; see the panels for the 15% and 25% tempo deviations in either direction in figure 1 of that report for striking illustrations of the effect).

The pattern approximates, in other words, the *Mecapoda* cricket performance documented by [67] (figure 2A), but without the consistency displayed by the insects (see also figure 3 in [12, 123]). Perhaps the lengthy and detailed operant shaping procedure to which Ronan was subjected induced in her brain, equipped as it is with a capacity for learning skills [124], a learned, acquired, and partial version of the circuitry with which evolution has equipped *Mecapoda* crickets on an innate basis. If so, it was the intrepid efforts of the experimenters that gave Ronan her "entrainment capacity." In any case, the characteristics of Ronan's performance underscore what she is doing on the basis of intensive operant training is, not what humans do when they synchronize to a beat, quite apart from the fact that they do not require operant shaping to do so.

It is to be regretted that Ronan was not observed in the presence of highly compelling rhythmic music prior to training and also that no male sea lion has been tested for entrainment. The latter desideratum relates to points 2 and 4 in Sect. 14.1.3 concerning the centrality of a capacity for spontaneous isochronous *production* in any naturally synchronizing species. Male sea lions use a repetitive, evenly paced bark as part of their territorial, herding, and agonistic display behaviours. The tempo of this bark, which in California sea lions has a default around 2 Hz, varies across a considerable range with contextual and motivational factors, with a faster pace signalling higher excitement or stakes [125, 126] (see also Chaps. 3 [43], 6

[127], and 11 [128]). At each tempo, the barking exhibits a steady isochrony, save for the first few barks. These animals come equipped, in other words, with a crucial component of natural synchronous chorusing, but are not on record for exhibiting such group behaviour. The steady pacing of the male bark at different tempos might nevertheless play a role in sea lion entrainment under appropriate testing conditions.

I conclude this brief review of recent studies of non-human entrainment by noting that the study reporting that macaques facing each other in a button press task spontaneously synchronize their behaviour with a precision exceeding that of skilled humans [129] appears to be an artifact of data analysis. The set theory formula specifying how button press delays between the two participants were calculated (by plain subtraction) lacks the device of taking absolute values of the time differences obtained as a basis for calculating asynchronies [24]. Without that device, deviations in opposite directions (before and after button presses by the reference monkey) will tend to cancel one another as data accumulate, generating artifactually low asynchronies even in the absence of synchrony.

For none of the species featured in the studies reviewed in this section are there indications that they engage in tactus-based interindividual entrainment in their natural habitats. The failure of any of these recent candidate synchronizers to match the synchronizing performance of even the less endowed among the insects is therefore hardly surprising, particularly when the heuristics suggested in Sect. 14.1.3 are kept in mind. There is thus no compelling reason to revise our initial supposition that, as far as current evidence is concerned, behaviour analogous to the well documented human capacity for perfect synchrony has been found and characterized only for a few species of insects.[10]

That does not mean, of course, that there might not be "perfect synchronizers" far closer to us than insects. The need to settle the question of possible interindividual entrainment in bonobo staccato hooting has already been mentioned. Moreover, few animals have as sophisticated a control over the timing of their vocal output as species that navigate and hunt with the help of echolocation. As emphasized by Peter Tyack, echolocation skills may also be turned to communicative uses [130], raising the possibility of finding echolocating cetaceans that employ their timing skills to synchronize communicative calls between individuals. There are in fact indications that dolphins may synchronize their voices [131]. They have been claimed to do so in a rhythmic fashion [132], but unfortunately, the latter study's many measurements do not include the one measure needed to assess synchrony, namely, voice onset asynchronies between individuals. It may also be that dolphins time the execution of joint, simultaneous, acrobatic manoeuvres by a vocal signal from one of the performers [133].

Many cetaceans besides dolphins are equipped with echolocation and should be canvassed for interindividual synchrony of voices and behaviour. There are also

[10] Beyond the matter of the nature and precision of the tactus-based entrainment on which human group synchrony is based, the complexity and sophistication of human rhythmic ensemble music have no analog in nature. See Chap. 5 [10] for some of the functional distinctions that come with our advanced capacity for group synchrony.

more than 800 species of echolocating bats to explore from this point of view. Echolocation apart, pinnipeds have been suggested for comparative analysis of rhythmic abilities [134]. As intimated in Sect. 14.1.3, the vastness of the comparative search space may yet turn up some surprises.

Meanwhile, the instance of convergent evolution represented by the group synchrony of insects and humans holds implications for the nature and origin of our curious propensity to entrain with others in the narrow contexts of music, dance, and drill. As discussed in Sect. 14.1.4, evolutionary analogy, however distant, is informative about selection pressures behind a convergently evolved behavioural trait, even when, as is often the case, it is implemented by different mechanisms. The socio-sexual ecology that frames perfect synchrony in the insect examples may in fact give us a hint that voice synchrony, specifically, may have played a part in our own evolutionary parting of ways with what would one day be chimpanzees and bonobos [111].

14.1.6 Human Group Synchrony in Comparative Perspective, IV: A Selection Pressure for Perfect Synchrony at the Human–Chimpanzee Split

Humans are the only primate known to feature beat-based behavioural group synchrony as part of its ethogram, principally in the form of a cross-culturally ubiquitous propensity to gather from time to time for rhythmic singing and dancing in a group. We do so independent of mode of subsistence and other societal characteristics, often in a ritual setting. As already noted in Sect. 14.1.2, the survival utility of the behaviour is moot in the sense that the time and effort it consumes might well be spent on activities more directly related to survival. Its character of aesthetic display places it squarely, it would seem, in the class of traits that forced Darwin to add sexual selection to his theory [135–137]. Given that no other primate has found its way to this peculiar group behaviour, its origin in our own case presumably involved a unique combination of circumstances rarely found together in other primates.

There are nevertheless hints in the behavioural biology of our hominoid relations that may help us trace the origin of our peculiar group displays. The first of these is the association between long-distance calls and physical display among apes. As noted by [94], the long-distance calls used by apes in distance signalling (e.g. orangutan [*Pongo* spp.] long call, chimpanzee pant-hoot, gibbon pair duet, etc.) tend to be accompanied by vigorous physical displays such as locomotor excitement, branch shaking, chest beating, or canopy acrobatics. Though these displays do not feature tactus-based intern-individual synchronization of behaviour, they provide a precedent for the *linkage between voice and bodily movement* featured in human group singing and dancing [94].

A pant-hooting chimpanzee may be joined by others, typically a few near-by individuals, in a pant-hooting "chorus," thought to reflect affiliative relations [138–140]. Individuals accommodate their call characteristics to facilitate such joint vocal displays. They may, for example, prolong the introductory phase of the pant-hoot to leave time for another to join [139]. None of this, however, including the buttress drumming that may accompany pant-hooting [141], features interindividual synchrony timed by a shared tactus. A hint concerning how joint vocal displays of this kind might have been regularized ("rhythmicized") by aligning the timing of the call elements and display behaviour of different individuals to a shared evenly paced time-giver, as in human group singing and dancing, is hidden in the behavioural ecology of the last common ancestor we share with chimpanzees.

A selection pressure for such a development is potentially available in the particular combination of social and mating system that is thought to have characterized the last common ancestor we share with chimpanzees [8, 38, 66, 111, 112], namely, group territoriality with male philopatry and female intergroup migration in the form of female exogamy [142–148]. Chimpanzee male territoriality means that a male is born, lives, and dies on the same patch of rainforest territory he helps his natal group defend against neighbouring groups. Not so for female chimpanzees. About the time of reaching reproductive age, roughly one in every two female chimpanzees disappears from her natal territory into the rainforest, eventually to settle, mate, and rear her young on a different territory [149, 150]. As the larger setting for ancestral long-distance display, this combination of male territoriality and female exogamy is reminiscent of the pattern of clumped males and migrating females that we have encountered as the setting for group synchrony in some synchronously chorusing or flashing insects.

In our own case, a common ancestor subpopulation in which display participants—unlike present-day chimpanzees—managed to synchronize their voices to a common, isochronous pulse (most readily provided by in-place locomotor activity, i.e. stomping [151]), would achieve superposition in time of their voice amplitudes. The resulting joint signal would reach farther afield in the terrain than that of a group with unsynchronized voices. This "supervoice" would carry farther across neighbouring territories to reach a greater number of migrating female ears, attracting them to the synchronizers *at the reproductive expense of non-synchronizing groups*. The more complete the amplitude summation, i.e. the more precise the synchrony, the greater the spatial reach of the joint signal, hence the need for *on-the-beat precise group synchrony*. Thus equipped, the synchronizing sub-population would gain a reproductive advantage over sub-populations that persisted in the non-synchronous ancestral display, setting the stage for a speciation event.[11]

That speciation event, on this account, founded a line of hominin group-synchronizers who hooted in time to a stomping beat, and eventually became

[11] This "inter-communal sexual selection" pressure for synchrony would, once migrating females are "on territory," be joined by an "intra-communal sexual selection" pressure for elaboration of the structural content of the display in the direction of added complexity and sophistication, in both pitch and rhythm domains, in keeping with the aesthetic logic and handicap principle of sexual selection. That, however, is a matter of the subsequent evolution of music proper [38, 66, 77], which lies beyond the scope of the present proposal.

singing and dancing humans. Left behind in the equatorial rain forests of Africa were the non-synchronizers, ancestors of today's chimpanzees and bonobos.[12]

14.1.7 Conclusion

It is worth noting, in conclusion, that in this account of the origin of our capacity to keep on-the-beat time together in a group, it was not the homologies that we share with chimpanzees—male territoriality and female exogamy—that supplied the key to our proposed path to group synchrony. Those adaptations have been well known for decades without ever having been connected to the topic of group synchrony. Instead, it was the analogy with the far distant insect examples of group synchrony that pointed the way to the potential significance of those adaptations as a joint selection pressure for precise group synchrony.

Should the conjectural account of the origin of our synchronizing capacity proposed here turn out to have merit, it holds a programmatic lesson. There is a common, and often tacit, assumption underpinning much comparative work on human origins. It is the assumption that "comparative analysis of the behaviour of modern primates, in conjunction with an accurate phylogenetic tree of relatedness, has the power to chart the early history of human cognitive evolution" [152 p. 543]. I let this quote stand for a pervasive tendency only because of its conciseness. By gratuitously limiting the search space for evolutionary insight into human traits to other primates, such an approach accords a privileged status to homology as compared to analogy in evolutionary analysis, as if we could count on finding every human characteristic of interest at least incipiently represented among other primates.

Not every trait is, however, handed down like an heirloom in phylogeny [153]. Ultimately, any adaptation must at some point be founded de novo in a speciation event. For the analysis of such uniquely derived traits (autapomorphies), other instances of the evolution of the trait, however distantly located in phylogeny, and however independently originated, are invaluable sources of potential insight into the selection pressures that may underlie that trait. The comparative method accordingly gives us analogy as the companion to homology for analysing the evolution of life forms and shows no preference for one over the other. In a sense, it is in the analysis of analogy that the comparative method truly comes into its own, and primates are no exception in that regard [154, 155]. We would be truly indebted to that method if, via the analogy of insect perfect group synchrony as proposed here, it has in fact shown us the origin of that part of our nature which inclines us to gather now and then to sing and dance together to a common beat.

[12] Unless bonobos should turn out to engage in synchronous staccato hooting in the wild, in which case the selection pressure I have proposed would have figured in a second speciation event, namely, that which split chimpanzees from bonobos. Hence, my emphasis on the importance of pursuing [110] regarding staccato hooting.

14.2 Peer Commentaries

14.2.1 Peer Commentary by P. Cook and C. Reichmuth: The Case for Comparative Studies of Mammalian Synchrony

Corresponding address: pcook@ncf.edu

In his provocative proposal (Sect. 14.1), Merker judges laboratory-based comparative attempts to replicate human-like sensorimotor synchrony in birds and mammals and finds them wanting. In this brief response, we make the case that his judgment is premature, both in light of new comparative data for a California sea lion (*Zalophus californianus*) [156] and because of a long-standing tendency to over-essentialize rhythmic behaviour in adult humans and thus ignore its extended and flexible developmental trajectory.

14.2.1.1 The Comparative Project

If rhythmic synchrony is an evolved trait (or relies on evolved traits), we may gain insight into its evolutionary underpinnings through the phylogenetic comparison of species that demonstrate some degree of this capability. According to Merker, only humans and certain insects meet the criteria for "perfect" synchrony [22], that is, lagless one-to-one beat matching across a range of tempos. The absence of perfect synchrony in other species suggests convergent evolution of rhythmic synchrony in humans and select insects.

14.2.1.2 Synchrony in a Non-human Mammal

Although the ecological interest of many researchers, Merker among them, is on naturally occurring group synchrony, the ability to produce behaviour in time to a steady beat has been taken as an experimentally accessible precondition. Among the non-human mammals and birds evaluated thus far, Merker acknowledges that one laboratory subject, California sea lion *Ronan*, has shown notably consistent beat matching performance across a range of tempos, with metronomic and musical stimuli. Initially tested in 2012, *Ronan* came closer to human-style beat keeping than other birds and mammals tested. Although she could entrain the movement of her head to novel rhythmic stimuli, she initially showed a fairly strong (albeit somewhat inconsistent) tempo-phase relationship, such that she was farther ahead of each beat with slower stimuli and farther behind with faster [70]. Merker claims this is not typical of human performance data and likens it to the natural synchrony pattern of the *Mecapoda* cricket [67]. Perhaps, he suggests, through extensive reward training with a few tempos, Ronan's brain developed a "learned, acquired, and partial version of the circuitry with which evolution has equipped *Mecapoda* crickets on an innate basis." He suggests that such a case might be interesting, but not directly relevant to human synchrony.

14.2.1.3 The Rhythmic Apprenticeship

Our primary point of contention with this interpretation is that it assumes typical beat-matching behaviour in adult humans to be fundamental and, in the Fodorian sense [157], modular (though to be fair, early in his proposal (Sect. 14.1) Merker does raise the possibility that human beat keeping might not be an evolved trait). While it is true that the vast majority of adult humans across most human cultures show lagless "perfect" synchrony with tactus-based stimuli, this does not entail perfect synchrony as a specific human evolutionary endowment. Indeed, the developmental literature on human beat keeping emphasizes the long course of learning that leads to adult-style synchrony. Very young children are poor timekeepers, frequently losing the beat. Slightly older children become more consistent but sometimes show phase-tempo relationships [158–160] (see also Chap. 10 [161] and the peer-commentary of Patel in Sect. 14.2.5).

One might argue that this has more to do with slowly developing motor capabilities than a lack of some internal cognitive mechanism for synchrony, but not until at least 10 years or older do typical Western humans reliably demonstrate the capability for lagless synchrony across a wide range of tempos with little tempo-phase dependence. Of course, unlike *Ronan* the sea lion, most humans do not receive specific operant instruction in beat keeping. We believe this point has been overemphasized, both by Merker and other recent reviews of the comparative literature [97, 162]. Most human cultures do extensively enculturate very young children (even babies) into rhythm, with repetitive songs, nursery rhymes, hand clapping games, group dance, and more. This could be considered a "rhythmic apprenticeship" [163]. Further evidence for its importance comes from a range of rhythmic specializations in adults from cultures with different dominant rhythmic characteristics in their music [9, 164, 165].

Consider then: After 6–8 years of extensive rhythmic enculturation, most human children are no better at synchronizing than *Ronan* the sea lion was following her primary training, and they often exhibit similar tempo-phase relationships. While humans have years of rhythmic exposure in socially salient situations, Ronan the sea lion had only 40–60 training sessions over a few months. With this perspective, one might argue *à la* Merker that humans, through extensive social rhythmic training over the first 6 years of life, develop a "learned, acquired, and partial version of the circuitry…" with which evolution has endowed the *Mecapoda*. Certainly, there is no clear categorical difference between *Ronan's* early performance with unfamiliar tempos and that of human children. Given another 6 years of training (some of which may be explicit, in music classes or with instrument training), human teenagers may further develop a "learned, acquired…version of the circuitry…" for perfect synchronization with which evolution has endowed certain fireflies. Given this logic, could an animal like *Ronan* do likewise?

14.2.1.4 Perfect Synchrony in a Non-human Mammal

Ronan's rhythmic performance to a range of tempos was first formally evaluated in 2012 after a few months of regular operant training with simple stimuli presented at rates of 80 or 120 bpm. She has been occasionally presented with similar stimuli over the following 12 years [122]. Thus, her own rhythmic apprenticeship has continued. We recently tested Ronan with three unfamiliar metronomic stimuli head-to-head against ten college aged adult human subjects performing an arm movement task with the same response amplitude as *Ronan's* head bobbing [156]. Ronan showed the same near-perfect entrainment as in 2012, with her body moving at a rate within 1.3% of the presented auditory tempo. This time, however, she showed no clear phase-tempo relationship—the extent to which her bobs fell in front or behind the beat was not dependent on stimulus rate and was indistinguishable from the humans' phase-tempo alignments. Her performance was highly consistent; the standard deviation of her bob period was ≈3%, lower than any of the humans. In this recent directly comparable task, Ronan demonstrated "perfect" synchrony and matched or outperformed all ten typical adult humans on all related measures.

14.2.1.5 A Case for Homology

It takes humans a decade or more to become "perfect synchronizers" with a mix of formal and informal learning and everyday exposure to rhythmic stimuli in the form of music. If, with explicit training but far less exposure and experience, a sea lion can also meet this performance criterion, this supports an argument for homology of an emergent, developmental mechanism for perfect synchrony that may operate across multiple mammalian species. As we have before [122], we tentatively credit loosely coupled oscillation, fundamental to the animal nervous system [166] (see also Chap. 8 [167]) with a top-down capability for rapid alignment and realignment. Brains communicate via rhythm, and this has been clearly demonstrated in the vertebrate sensorimotor system, which uses rhythmic volleys to coordinate temporally dependent activity [168]. Rhythm and synchrony seem to be built into the way that brains flexibly organize movement and perception in time. While animals produce a wide range of rhythms in nature [169] (see also Chaps. 3 [43] and 4 [170]), most animals do not natively synchronize with just any isochronous stimulus—there is clearly more to the picture, and comparative studies will continue to provide insight into underlying mechanisms. Motivation and attention are likely relevant factors [163]. If, as suggested by many [171, 172], synchronization requires covertly simulating a motor sequence at the stimulus period, then prolonged learning may be required for many species to enact this simulation and achieve perfect synchrony. Careful experimental paradigms with motor behaviours and auditory stimuli appropriately selected on a by-species basis are required (for similar arguments, see also Chaps. 9 [5] and 11 [128]). Interspecies comparisons should use similar types of motor output. More attention should also be paid to the developmental and learned course of potentially emergent rhythmic capabilities in humans and comparator species. In any case, it is far too early to close the book on the possibility of perfect synchrony in non-human mammals.

14.2.2 Peer Commentary by D. Herzing: We Need Better Technology and Data to Measure Underwater Mammal Synchrony

Corresponding address: dlherzing@wilddolphinproject.org

In his proposal (Sect. 14.1), Merker rightly noted the importance, and lack, of naturalistic behaviour in the wild for comparative analysis of on-the-beat precise group synchrony. Given the difficulty in obtaining data in aquatic habitats, specifically from dolphins (family Delphinidae) who use sophisticated controls over the timing of their echolocation clicks for communicative purposes, recording voice onset asynchronies between individuals has not been possible. Only recently has technology even been developed that would localize individual vocalizing dolphins with correlated underwater behaviour, potentially providing such onset accuracies along with group behaviour [173]. And with two sound generators belugas (*Delphinapterus leucas*), another toothed whale (suborder Odontoceti) with echolocation, have been reported to synchronize their two independently-produced clicks, merging into one click, showing extreme skill in their internal vocal synchrony [174].

Dolphins have been observed underwater to synchronize physical behaviour and sounds [132, 175], although the resolution required for on-beat synchrony (described here as within less than half a period length, so to not "lose its place" in the beat) has not been reported using existing data collection tools. These "rough" vocal synchronies, described in the literature as joint signals, have already been proposed to be an amplified signal that would carry farther across aquatic distances. If meant to reach females ears, this might have attracted them to the synchronizers at the reproductive expense of non-synchronizing groups. Alternatively, since dolphin synchrony has been observed primarily during intergroup aggressive behaviour of males, it is more likely that synchrony is functioned to repel competing male groups thereby reducing the competition for reproductively active female dolphins.

Another aquatic mammal, specifically male sea lions (family Otariidae), also come equipped with a crucial component of natural synchronous chorusing; they use repetitive evenly paced barks for various individual displays. Male sea lions use these repetitive, evenly paced barks as part of their territorial, herding, and agonistic display behaviours. Such displays are not known to attract females but instead fend off males in competitive situations [126] (see also Chaps. 3 [43] and 6 [127]). Group behaviour of the synchrony of sea lion barks has not been observed, perhaps because this species does not form male alliances for mating purposes as do dolphins.

Selection pressures might be expected to be different between the insect and mammal world, given the large separation in evolutionary time as well as behavioural ecology drivers. Although perfect insect synchrony functions in the framework of aggregated males and wandering females, dolphins have fission/fusion societies with a framework of wandering males and females that show natal philopatry. This suggests a different selection pressure for precise group synchrony within social mammal species and is probably not a homology but instead represents an example of convergent evolution. But until new technology allows us precise localization of a vocalizing dolphin (and therefore precise vocal onset times), coupled with specific underwater behaviour, we cannot answer whether biological entrainment and synchrony as described in these exact terms exists.

14.2.3 Peer Commentary by A. Lameira: The Narrow Search for "Perfect Synchronizers By-What-Ever-Means" Across Animals Is Vacuous About the Nature and Evolution of Human Rhythm and Sync

Corresponding address: adriano.lameira@warwick.ac.uk

In Sect. 14.1, Merker lays a solid technical groundwork for research of rhythmic synchrony in animals, setting a clear, necessarily high benchmark for students and scientists of rhythm and synchrony, handing over to the field a valuable, precise toolbox for the study of biological interaction. To show off the legitimate power of this toolbox, Merker dissects Patel's "vocal learning and rhythmic synchronization hypothesis," dismantling its alleged evolutionary implications and bringing much needed clarity and corrections to the field of rhythm evolution (see also Chap. 5 [10]).

However, in the process, Merker seems to get caught in the headlights of "perfect synchronizers," his heuristics for narrowing the search of this social behaviour becoming all too narrow to gain insight and generate predictions about the evolution of human music, dance, and drill. Eventually, Merker takes the same wrong turn as Patel—allowing distant analogy in species (very, very) far related to humans (*Homo sapiens*) to displace homology within the Hominid family, throwing shared ancestry and descent with modification to the back seat of natural selection and evolution. Here I clarify how Merker's proposition that perfect synchrony arises in "philopatric males on patchy resources" does not survive the level of scrutiny that he invites the field to follow.

Perfect synchrony is required for human group synchrony. Merker proposes that finding other animal cases is the key to understand the evolution of human rhythm and sync. This seems a logical premise. However, of all animals, only three species of fireflies fulfil the criteria of perfect synchronizers. In fact, they are perfect*er* synchronizers than humans (see Sect. 14.1). This infinitesimal sample across 600 million years of evolution drastically reduces what can be comparatively extrapolated. Had multiple lineages, with different divergence dates from humans, been identified, recurring socio-ecological themes across species could hint at why ancient hominids started to synchronizing perfectly. But fireflies weigh less than 20 mg (the same as a few grains of sand), have an exoskeleton, six legs, two pairs of wings, a few hundred thousand neurons arranged in interconnected ganglia [176] (the human brain has ≈86 billion neurons, *five* orders of magnitude larger [177]), and synchronize through light signals that they produce in their abdomens [178]. Perfect synchrony in fireflies and humans is clearly implemented in radically distinct means and mechanisms by radically distinct bodies and brains. The proposition that it evolved for the same reason in both groups should be made with caution since apparent similarity could be purely coincidental. As Merker indicates, the firefly human common ancestor takes us back to the Ediacaran period before the Cambrian era. For perspective, at that time, most organisms were microscopic, most complex multicellular ones had no active mobility, and bilaterians showed little resemblance to modern lifeforms [179].

Nevertheless, Merker risks a trendline connecting “perfect synchrony by-whatever-means” between three insect data points with one hominid data point. According to Merker, fireflies loosely overlap with the last chimpanzee-human common ancestor by exhibiting male-philopatry and patchy resources. This, it is argued, must have thus been the reason why human ancestors become perfect synchronizers.

If philopatric males on patchy resources can allegedly drive perfect synchrony in lineages as far apart as fireflies and humans, then any other philopatric males living on patchy resources should be perfect synchronisers. Merker seems to imply that this socioecological combination is as rare as perfect synchrony, but this is not the case. Many species exhibit male-philopatry and rely on patchy resources [180], *including mammal and primate species* [181], none of which have evolved perfect synchrony according to Merker. Neither chimpanzees nor the putative chimpanzee-human last common ancestors are strictly male-philopatric [181], as suggested by Merker. In chimpanzees (*Pan troglodytes*), highly complex social affiliation and demographic factors determine whether a female or male will disperse from its natal group or not. For example, female chimpanzees are more likely to remain philopatric when they have a high-ranking mother as ally in their natal group [181]. Similar reasons can lead male chimpanzees to disperse, as occurring in other primates [181]. Such dynamics are abysmally different from anything known in insects [182]. Arguably, the firefly’s social behaviour would be dramatically disrupted if some females choose to stay and males choose to disperse, as chimpanzee do. Male philopatry and female dispersal in fireflies and chimpanzees are not alike. This equally applies to humans’ putative ancestral state; both male- and female-philopatry dynamically co-occurred within and across groups [181].

From a feeding ecology perspective, adult fireflies feed on nectar or pollen, they do not eat at all, or females cannibalize males [183]. Exactly how this “patchy resource-ness” converts to the context of chimpanzees and humans is unformulated, if a dependable parallel can be established at all. The case raised by Merker for a hominid ancestor with firefly-like sex dispersal and feeding ecology finds no empirical support.

Ultimately, the search for “perfect synchrony by-whatever-means” proves uncapable of illuminating the states of individual and/or social rhythm that preluded or led to perfect synchrony. For instance, Merker explicitly dismisses the possibility that interactive social behaviours in mammals and primates, such as turn-taking and duetting, could potentially be tactus-bound, with no explicit genuine interest toward quantitatively characterizing these rhythms nor toward whether or how such interactions could one day potentially become perfectly synchronised. Merker’s heuristics and hypothesis cannot make predictions about possible forerunning states of perfect synchrony in fireflies, nor whether such precursors would have been similar or distinct in insects versus apes. Any favourable and advantageous rhythm behaviours, solitary or social, for the evolution of perfect synchrony in animals will remain unexplored, undetected, and unforecasted if one is to adopt Merker’s heuristics (see Sect. 14.1).

Surprisingly, Merker disregards one of the most exciting recent discoveries on chimpanzee rhythmic behaviour, the spontaneous capacity for coupled social entrainment to a degree that matches orchestra players across tempos [53, 184]. It is unclear if Merker overlooks this case because it originated from captivity (even if in the absence of human training or priming). Naturalistic behaviour in captivity allows a glimpse into animal behavioural, cognitive and social capacities that are part of a species' phenotypical potential but not typically expressed in the wild, in part due to the lifting of constrains faced in the wild (e.g. free nutrition) and the imposition of new ones (e.g. manipulated social organization and territorial range). Then again, humans do not live "in the wild," so behavioural parallels between humans and captive great apes are in fact comparatively and evolutionarily pertinent. It is difficult to imagine why one would trade insects for apes to understand rhythm and sync evolution in apes. Students and scientists of rhythm and synchrony should use Merker's technical toolbox but be guided and guarded by much more explorative, analytical, and inquisitive heuristics.

14.2.4 Peer Commentary by K. Okanoya: The "Waiting" Origin of Rhythmicity in Animals

Corresponding address: kazuookanoya@gmail.com

14.2.4.1 Introduction

I am privileged to contribute this commentary, which is a testament to my fruitful collaboration with Bjorn Merker. Our partnership, which began with a visit to Stockholm in 2004, has resulted in two co-authored papers [185, 186]. These papers delve into the possibility that human speech may have evolved from wholistic, non-linguistic singing. We suggested that these songs, sung in various contexts, were mutually segmented into common sub-phrases and sub-contexts, creating a partial phrase with a more specified meaning [185]. We proposed that the neural circuitry supporting this process of mutual segmentation includes the prefrontal-basal ganglia loop for phrasal segmentation and the prefrontal-hippocampus loop for contextual segmentation [186].

Since then, I have been working on the issue of phrasal segmentation in humans and animals. For example, both human adults and newborn babies can segment randomized triplet sound sequences as indicated by mismatch negativity [187, 188]. Non-human animals, like Bengalese finches showed mismatch-like local field potential when listening to the triplet song syllables [189]. Interestingly though, just mixing triplet and quadlet stimuli made this segmentation disappear (Abla and colleagues, unpublished observation). This suggests not only sequential information, but also rhythmic information is crucial for statistical segmentation. This experience made me think about the importance of rhythmic structure in the segmentation of auditory sequences. Thus, we also started to work on rhythm production and perception in animals [120, 190]. In the following section, I present a response to Merker's critique of our paper [120].

14.2.4.2 Reply to Merker's Criticism on Hasegawa et al. (2011)

Bjorn Merker's critiques have questioned whether the rhythmic synchronization behaviours observed in budgerigars (*Melopsittacus undulatus*) reflect genuine predictive entrainment or merely reaction-time-based responses. In this response, I address these concerns based on the experimental findings of [120], which provide multiple lines of evidence supporting the claim that budgerigars exhibit predictive synchronization with rhythmic stimuli. Merker has argued that synchronization behaviour in budgerigars under fast tempo conditions is driven by reaction times, not predictive entrainment. Additionally, it has been claimed that the reduced asynchronies observed under slow tempo conditions are artifacts of anticipatory (premature) pecking. These claims are addressed by revisiting the data and analyses reported by [120].

A critical feature of predictive entrainment is phase matching to the beat, not merely reacting after a stimulus [120]. In 6 out of 46 experimental conditions, budgerigars exhibited phase-matched tapping to the stimulus onset but not to the estimated reaction time (ERT). Moreover, in 32 of 46 conditions, the centre of the significant pecking phase distribution was located prior to the ERT, suggesting anticipation rather than delayed response. Negative mean asynchrony (NMA), a hallmark of predictive synchronization, was observed in all constant inter-onset interval (IOI) conditions except for the fastest (450 ms). This tendency for pecks to precede the stimulus onset strongly supports the interpretation that budgerigars were anticipating rhythmic events rather than simply reacting to them.

Hasegawa et al. conducted extensive simulations (n = 10,000 per condition) across four models: Monte Carlo (random pecking), constant interval reproduction, memorized IOI reproduction, and simple reaction time models. None of these models could reproduce the observed frequency of phase-matching or NMA in the real birds. Particularly, the simple reaction (SR) model failed to account for the higher failure rates and anticipatory timing seen in real subjects.

Contrary to the claim that early pecks in slow tempo conditions explain reduced asynchronies, [120] observed that birds initially produced inter-response intervals (IRIs) shorter than the target IOI and then corrected them over successive pecks. This behaviour indicates ongoing error correction and supports the hypothesis of active tempo tracking rather than passive reaction. Notably, predictive entrainment was also demonstrated under auditory-only conditions. Birds exhibited significant phase matching to auditory stimuli alone, further reinforcing the interpretation that entrainment was not dependent on visual cues or stimulus-response training artifacts.

Collectively, the evidence presented by [120] strongly supports the interpretation that budgerigars are capable of predictive rhythmic entrainment. The presence of NMA, anticipatory phase alignment, failure of alternative models to explain the data, and entrainment under auditory-only conditions, all provide converging support for this conclusion.

14.2.4.3 The "Waiting" Origin of Rhythmicity in Animals

Antiphonal calling occurs in some species of animals when individuals call each other without overlapping. For example, we showed that naked mole rats (*Heterocephalus glaber*) call with an inter-call interval of about 0.8 s, while in the presence of another individual, they call with an intercall interval of 0.3 s [191]. Thus, while isolated naked mole rats call at a slower pace when they get a response, the sender and the responder start antiphonal calling or turn-taking of calls. Considering the average length of the signature call of this species being 0.25 s, the intercall interval of 0.3 s is very short, effectively avoiding overlapping of the signals while maximizing the efficacy of communication while avoiding overlap in signals, as correctly pointed out by Merker in Sect. 14.1).

The interval of 0.3 0.8 s is worth consideration because it involves two types of "waiting," First, the difference between antiphonal calling and spontaneous calling is 0.5 s (= 0.8–0.3 s). If the animal spontaneously called in shorter intervals, the antiphonal calling of the other animal would overlap with it. Thus, there should be a mechanism to suppress spontaneous calling after the animal calls for at least 0.5 s, and if the other calls, the animal should call back immediately. Second, if the response call is reactive, this is too short, considering the average auditory reaction time of humans is 0.23 s [192]. If the response call is predictive, it could be possible to make it sound antiphonally. But how could the response call be predictive and simultaneously avoid overlapping? The two "waiting times" are the times to suppress spontaneous calls and the time to suppress predictive calling.

How is this timing controlled? A recent study on zebra finch (*Taeniopygia guttata*) antiphonal calling is insightful [193]. Electrophysiological studies of the putative pre-motor cortex in zebra finches revealed the following. Gaba normally inhibits premotor neurons. Under these conditions, the response time of the antiphonal response was approximately 0.25 s. However, the response time decreased to approximately 0.17 s when this inhibition was pharmacologically removed. This finding indicates that inhibition circuitry was implemented in the zebra finch's brain to avoid responding too quickly and overlapping with the other's voice. This study found the circuitry for "waiting" in zebra finches' vocalizations. Notably, zebra finches dare to construct a 0.08 s waiting period at the neural circuit level to prevent overlapping of voices, even though they are able to call back in 0.17 s after the other's voice is produced.

"Waiting" facilitates antiphonal responding and makes turn-taking possible. The phase angle of these behaviours is around 180 degrees. By manipulating this phase angle to 360 degrees, the antiphonal response becomes rhythm-taking. The "Waiting" origin of rhythmicity is much simpler than the vocal learning origin of rhythmicity [97]. I hope to advance this idea further in behavioural and neurophysiological domains with Bjorn Merker.

14.2.5 Peer Commentary by A. Patel: The Vocal Learning and Rhythmic Synchronization Hypothesis—Reply to B. Merker

Corresponding address: a.patel@tufts.edu

In his interesting proposal, Merker (Sect. 14.1) critiques the vocal learning and rhythmic synchronization hypothesis [71]. This hypothesis proposes that ability to synchronize movements predictively and flexibility across tempi to auditory beats builds on the brain circuitry for complex vocal learning. Merker also argues parrots lack the capacity to predictively entrain to musical beats because their rhythmic movements to music often drift off the beat. I thank Merker for engaging with the vocal learning hypothesis (henceforth, VLH) and associated research. In this commentary, I explain why I disagree with his critique of the VLH and his interpretation of related parrot studies. I conclude by arguing that the VLH remains viable and by pointing to the revised VLH [97], which makes testable predictions for neural, cross-species, and genetic research on beat-based rhythmic processing.

The VLH was motivated by a synthesis of behavioural and neural research (reviewed in [97], section 2b). As noted in several papers, the hypothesis is concerned with predictive and tempo-flexible synchronization of movement to beats perceived in complex auditory stimuli (such as music), not with synchronization to relatively simple pulse trains as seen in some crickets or other insects [68, 71, 194, 195] (cf. [196]). Since 2005, I have referred to the former type of synchrony as "beat perception and synchronization" or BPS. BPS involves tight online coupling between complex auditory processing and rhythmic motor behaviour and is central to dancing and making music in synchrony with others.

The paper which introduced the VLH [71] stated that vocal learning "requires tight coupling between auditory input and motor output in order to match vocal production to a desired model," and argued that the resulting strong two-way communication between forebrain auditory and motor regions laid the foundation for BPS. Merker objects that vocal learning "Does not feature on-line (real time) coupling between auditory input and motor output" and is instead "a protracted process conspicuous for its radical decoupling of auditory input and motor output." A similar objection was recently voiced by Brown [197], who argues that vocal learning and BPS "occur in strikingly different time domains."

It is true that numerous vocal learning species (including humans and many songbirds) go through an early developmental phase of auditory learning where they listen to sound patterns that they will only later produce with accuracy, after considerable practice [198]. However, as pointed out in a recent review [199], vocal learning is not just about passively forming auditory templates to guide later sound production. Vocal production learning also requires "control over anatomical structures to produce an approximation of the desired sound, and auditory feedback to perceive and adjust vocal output as needed" [200]. Crucially, producing a desired sound (and feedback-based vocal adjustments) can involve dynamic auditory-motor interactions at fast timescales, as seen in research on speech production. For example, bite-block experiments show that when jaw movement is restricted a speaker rapidly and dramatically alters articulation to produce formant patterns similar to those made when their jaw is free, and when speakers are given altered auditory feedback of their own production, they immediately make compensatory gestures to try and maintain the expected formant patterns in light of their altered auditory input, as discussed in [201].

Real-time contributions of auditory feedback to vocal motor control also occur in songbirds [202]. I suspect that such online precise auditory–motor interactions are particularly strong in "high vocal learners" such as parrots, who can learn to mimic novel sounds throughout life and who have a more elaborate neural system for vocal learning than complex vocal learners such as zebra finches (*Taeniopygia guttata*) [97, 98]. In other words, I disagree with the contention that vocal learning does not involve sophisticated real-time coupling between auditory input and motor output.

Turning to Merker's discussion of parrot rhythmic movement to music, he argues that parrots lack a mechanism for synchronizing predictively to a beat because their rhythmic movements to music drift from on-beat synchrony to movements not synchronized to the beat. This type of sporadic synchrony was quantitatively documented in experiments with a Sulphur-crested cockatoo (*Cacattua galerita*) named Snowball, who synchronized his head bobs predictively to musical beats in episodic "bouts" (median = 16 head bobs/bout) and who gravitated to a head bob tempo near 126 BPM during unsynchronized rhythmic movement to music [68, 194]. When the original song Snowball danced to was sped up or slowed down, Snowball exhibited this sporadic synchrony at nine different tempi ranging from 10% slower to 20% faster than the original song. This type of sporadic synchrony differs from the abilities of human adults, who can sustain predictive synchrony for much longer periods, e.g. over an entire song lasting for minutes. Critically, however, statistical methods showed that the amount of predictive synchrony shown by Snowball was highly unlikely to be due to chance alignment of rhythmic head bobs to beats. Schachner and colleagues also applied statistical methods to show significant predictive synchronization of rhythmic movement to musical beats in several other species of parrots [69, 203]. Thus, I disagree that sporadic synchrony to a musical beat is decisive evidence for the lack of a mechanism for predictive synchronization. Rather, such a mechanism could be present but not operating at the level seen in human adults.

Indeed, as noted previously [68] it is possible that children go through a developmental phase where their beat synchronization is sporadic rather than sustained. Research on infants and young children (ages 1.5–6) moving rhythmically to music suggests that they do not show reliable beat synchronization, even if they show tempo flexibility (moving faster to music with faster tempi) [204–207] (see also Chap. 10 [161]). Yet infants and young children grow into adults capable of sustained predictive beat synchronization. The key point is that human development is unlikely to involve a binary transition from no synchronization ability to fully mature abilities. There may be a childhood stage in which sporadic predictive synchronization to musical beats is interspersed with periods of unsynchronized rhythmic movement, though further work is needed to determine if this is the case.

In summary, I believe the VLH remains viable, as argued in a recent paper reviewing the challenges to the hypothesis to date [97]. That paper also proposed a revision of the VLH addressing why BPS develops spontaneously in parrots but not in other vocal learning species raised around humans and their music, such as pet songbirds. Importantly, neither the VLH nor the revised VLH claim that vocal learning is *sufficient* as a neurological foundation for BPS. As noted previously [195] and as emphasized by Merker, other abilities may also be necessary, such as the capacity and proclivity to imitate nonvocal movements (cf. [208]). This might help explain why BPS develops spontaneously in parrots but not in other talented vocal mimics kept as pets, such as minah birds. This is one of the issues I discuss in a recent paper focusing on how and why the evolution of strong forebrain auditory-motor integration in the service of learned *vocal* control led to a capacity and proclivity to synchronize *nonvocal* movements to the beat [209, 210].

14.2.6 Peer Commentary by T. Popescu and T. Fitch: On the Evolution of Human Synchronization Abilities

Corresponding address: tudor.popescu@oxon.org

14.2.6.1 Introduction

Bjorn Merker's proposal (Sect. 14.1) represents the culmination of a stellar career, not just as a neuroscientist, but as one of the deepest thinkers about a problem fundamental to the cognitive evolution of our species: why do humans, quite exceptionally among primates, display an ability and proclivity to precisely synchronize our voices and movements?

In previous work, Merker has both deepened our understanding of what musical factors incline us to move to the beat [8, 15], and what evolutionary forces could have led to the ability to do so [111, 112]. Merker has adopted a wide-ranging comparative perspective to argue that precise entrainment in other species boosts the signal strength of group displays. Synchronization allows groups of males to increase the reach of their calls, whether visual displays in fireflies (family Lampyridae) or acoustic displays in frogs (order Anura) and crickets (family Gryllidae). Amplitude summation of individual displays creates a "super-stimulus" perceptible over greater distances than any single call, attracting mates otherwise beyond reach. Of course, the clustering of males also increases inter-male competition, but clearly in these invertebrate and amphibian cases, the price has been worth paying. The proposal in Sect. 14.1 deepens and convincingly extends this line of reasoning.

We start by commending Merker's use of a broad comparative approach, that considers all relevant species and not just a narrow subset (e.g. non-human primates or mammals). Convergent evolution provides one of the strongest ways to test adaptive hypotheses, and Merker puts this principle to good work here, while carefully considering factors specific to our evolutionary branch. For instance, male chimpanzees (*Pan troglodytes*) are philopatric as adults, staying in the group they are born in, while many females emigrate to other groups upon reaching sexual maturity. In dense rainforest, acoustic cues presumably play an important role in helping emigrating females to find (and perhaps assess) neighbouring groups. The loud pant-hoot displays of male chimpanzees, especially when combined with drumming displays, may provide a key acoustic "beacon" toward which these females can move.

In the key conceptual move, Merker then argues that if we combine the amplitude summation argument with this ape-specific female emigration pattern, the stage was set in the last common ancestor (LCA) of humans and chimpanzees for selection on displaying males to combine their displays synchronously, thus reaching a wider pool of emigrating (or ready-to-emigrate) females. The subset of the LCA population leading to *Homo* were the ones who adopted this strategy.

We note that this argument may have had particular force for those early ancestors who moved out into the drier, mixed woodlands that seem to be a more typical habitat for early australopithecines [211] than the wetter and denser forest typifying most modern populations of African apes. This is because, in these drier more open environments, sound might have a wider reach, while the predation hazards for females moving on open ground would be greater, increasing the incentive to rapidly and efficiently move toward a group of signalling conspecifics.

Overall, we find Merker's arguments compelling, his use of the comparative method rigorous and commendable, and the overall integration of different strands of knowledge admirable. Nonetheless, in a spirit of constructive criticism and open scientific debate, we will now turn to some counterarguments and advance a related but distinct hypothesis for the evolution of human synchronization capabilities.

14.2.6.2 Counter-Arguments to Merker's Beacon Hypothesis

A key problem with Merker's hypothesis, as with other sexual-selection-based accounts [135, 212, 213], is that human females and males have equivalent rhythmic (and generally musical) skills. While in frogs, crickets or fireflies, only males display, musical synchronization in modern humans is thoroughly egalitarian. In Merker's scenario, it seems unlikely that resident females would gain an adaptive advantage by synchronously displaying with other group members to attract new females; indeed, the opposite seems more likely. Of course, in its early evolutionary stages such synchronizing ability could have been limited to males, and only later in our evolutionary history extended to females, but then we require an argument for why this extension to another sex would have been adaptive. Alternative hypotheses, such as the "music for social bonding" hypothesis [40] do not suffer from this counterargument, because the need to create and maintain positive social bonds with other group members applies equally to both sexes in our species.

A second issue is the very early maturation of synchronization behaviour. Hints of neural synchronization to a beat are already visible at birth in human babies (in both sexes) [214], and actual audiomotor synchronization is well documented during relatively early childhood, around age 6 [215] (see also Chap. 10 [161]). These abilities obviously mature well before puberty, which is when we would expect sexual displays to come online during development. Male songbirds (family

Passeridae) learn songs before their first mating season, and if competent adult displays require childhood practice, early maturation seems necessary—but, again, it should be more typical of boys. Although neither of these two arguments is insurmountable, together the dual facts that human entrainment ability develops early and equally in both sexes do not seem consistent with a primary function of this ability as a male sexually selected display. We will next explore several alternative hypotheses that do not in our opinion suffer from this difficulty: The evolution of synchronization in duetting, and/or in group anti-predator displays.

14.2.6.3 Duetting: An Alternative Hypothesis

Another widespread source of comparative data relevant to synchronous behaviour comes from duetting: Tightly synchronized (but not typically simultaneous) displays that are prototypically performed by male-female pairs to advertise both their pair-bond and their occupancy of a territory [216–220]. Duetting, observed in over half of songbird species, is hypothesized to be the ancestral condition for song in songbirds [221]. Duetting is also seen in multiple parrot species, e.g. *Agapornis roseicollis, Brotogeris jugularis* [222, 223] and, while not as common, is also documented across primate male-female pairs, including indri (*Indri indri*), tarsiers (genus *Tarsius*), langurs (genus *Presbytis*), titi monkeys (subfamily Callicebinae), and gibbons (family Hylobatidae) [216, 218, 224] (see also Chaps. 3 [43] and 10 [161]).

Animal duets typically involve antiphonal calling: Synchronized calling at the same period, but with different phases, typically leading to an interlocked F-M-F-M… pattern (where F denotes female syllables and M male syllables). This lack of 0° phase and tactus-based synchronization leads Merker to exclude duets as a model for human rhythmic entrainment. However, there are a few tantalizing exceptions to antiphony, such as the plain-tailed wren (*Pheugopedius euphrys*). These birds live in multi-male, multi-female groups, and the entire group participates in "super duets," where the female syllables are sung in precise simultaneous synchrony, and likewise for the male syllables [225, 226]. This level of rhythmic complexity and precision, combining both simultaneous and interlocking songs, rivals human musical performances. We also note that music and speech each support both types of entrainments, with simultaneity more typical of music and other "choric" performances (e.g. chanting), and conversational or "dialogic" turn-taking more typical of speech, though also observed in some musical contexts [227]. The choric-dialogic continuum illustrates how, for humans, either phase alignment or phase offset are equally possible depending on communicative needs.

We agree with Merker that beat-based synchronization is the key feature of human rhythmic entrainment. However, we know of little evidence that speaks to this form of alternating entrainment in duetting birds or primates. It would certainly be useful to create experimental duet setups, in which one partner's vocal output is sped up or slowed down, to determine whether the other partner is able to achieve tight synchronization regardless of tempo. But as far as we know such experiments have not been performed.

Furthermore, 1:1 simultaneity of sounds (the form envisioned as central in the beacon hypothesis) is by no means typical of human rhythm and would lead to a rather boring musical world if it were. Instead, human rhythmic cognition features a metrical level above the beat, which facilitates grouping and synchronization of non-simultaneous acoustic events at different phases (different slots in Merker's "groove matrix" [15]). This hierarchical structuring of rhythms allows an unbounded set of potential rhythmic patterns, all obeying one over-arching grammar of rhythm [228, 229]. This provides myriad exciting possibilities for dancers to entrain their movements to a musical accompaniment [230], and creates a framework within which different individual contributions to a group rhythm can remain perceptible, while still together forming a cohesive and pleasing overall pattern [40]. From this perspective, duetting seems like an ideal comparative arena for further exploration of human musical rhythm, one that is virtually unexplored.

We thus suggest that the capacity for entrainment may have originated in duetting—whereby individual utterances align in time and possibly also in pitch—in our ancestors, as previously hinted at by many researchers in the context of a "musical protolanguage" [135, 231–235]. This entails precise synchronization of (mostly) interlocking vocalizations and/or noise-producing body movements (claps, drumming, etc), well-attested in the animal world. Being able to create such patterns slower or faster, following a beat, may potentially require a further evolutionary step (though the comparative data on which to base such a conclusion may not yet be present), as would achieving the level of tight 1:1 simultaneity observed in the plain-tailed wren and modern humans.

Increasingly discussed is also the potential role of *spectral* domain synchronicity in social bonding, e.g. singing in tune. As with time-domain entrainment, harmony provides conspecifics' voices a framework to be perceived both independently and fused into a greater whole. In humans, the ability to entrain harmonically may be rooted in the nearly perfect 2:1 adult female:male singing voice range, itself the basis for octave equivalence in humans [236, 237], which allows close and pleasant pitch-matching between males and females, and also between adults and juveniles [238]. Just like—as Merker reminds us—in certain bird species, vocal learning can, under particular circumstances, effectively bootstrap a pseudo-entrainment ability by forming superstimuli of greater amplitude, vocal learning might also have helped buttress *tonal* synchronization: singing in harmony. Indeed, [239] proposed that vocal learning, alongside harmonic clarity, differing vocal ranges, and simultaneous/turn-taking vocalization, represents one of the "pillar" spectral traits of musicality. Together, these traits constrain octave equivalence, consonance preference, and timbre saliency, all traits that contribute to the possibility of group singing creating a harmonic superstimulus. By adopting small integer ratios of fundamentals (1:1, 2:1, 3:2, etc), harmonic singing creates *aesthetic* superstimuli, devoid of the roughness of a chorus of untuned singers.

Early humans might have cultivated octave singing due to the power of chorusing to foster social bonding [40, 240, 241]. Once evolved, this might later have produced a *proclivity* to inter-align in pitch, broadened to include non-octave (but still small-ratio) harmonic intervals like the perfect fifth (3:2). By choosing different sung pitches, syllables that are simultaneously sung now becoming pleasingly separated spectrally. This proclivity for harmony, presumably driven by cultural evolution, is still apparent in modern humans as the social-bonding value of group singing [40, 241], where consonances are still referencing elements, if even only conspicuous by absence. Although speculative, such an account for the basis of harmonically-aligned voices might explain the culturally contingent nature of consonance preference [242–245]. The bonding value of specific horizontal tonal alignments, expanded through complex harmony and voice leading into vast landscapes of tonal tension, lacks however sufficient empirical study for firm conclusions.

Returning to our duetting hypothesis, insofar as duets constitute sexual and/or territorial displays, the early maturation of human entrainment abilities remains inadequately explained by this hypothesis. We therefore now turn to a second alternative hypothesis, which we suggest can encompass both the lack of sex differences, and the early maturation, of human rhythmic entrainment.

14.2.6.4 A Beacon-Oriented Alternative Hypothesis

Our other alternative hypothesis is that synchronization serves to make anti-predator displays more effective. This employs Merker's amplitude summation argument and also considers the evolutionary history of the hominin line, but should favour equal entrainment abilities in both sexes, and from an early age. The evolutionary setting is one that applies to all primates and many other vertebrates as well: The use of group mobbing displays. Small birds commonly vocally harass larger predators, alerting them they have been detected and attracting conspecifics to the display [246]. In chimpanzees, both males and females additionally throw sticks and brandish them as clubs, to more effectively repel big cat predators [247].

We suggest, in exact parallel with Merker's hypothesis, that amplitude summation of vocalizations, along with synchronous motor displays, would render such group displays even more effective than they are already known to be in modern chimpanzees. This capacity, again, might have been particularly relevant to our hominin ancestors that invaded mixed woodlands, known from paleontological data to have harboured a terrifying diversity of large predators [248]. Thus, when the branch of the LCA population that led to *Australopithecus* moved into these more open habitats, the need for ever-more effective anti-predator displays incorporating all group members, including females and juveniles, would have been ubiquitous.

14.2.6.5 Conclusions

In closing, we note that the various hypotheses discussed above need not necessarily be mutually exclusive. For example, it is easy to imagine how Merker's scenario, positing an early male-only display capability, might later have been coopted into anti-predator displays by group members of both sexes. The same is true for an initial duetting stage leading to the type of simultaneous calling required by both "beacon" hypotheses. Indeed, positive feelings resulting from such synchronous group displays may have provided the behavioural "raw material" for social bonding through musical displays based on later-evolved rhythmic and harmonic entrainment. Possibly even rarer than the biological capacity for synchronization and entrainment is the *motivation* to employ this ability specifically to increase social bonding and cooperation. Doing so effectively ratchets up social cognition in humans, and if we are right that the same happens with spectral entrainment, this creates an additional ratchet effect, setting the stage for the cultural invention of tonal music, where time- and pitch-alignment become prediction currencies to be fulfilled or violated. Lacking foresight, evolution rarely proceeds along straight or parsimonious paths. Our main purpose here was to outline some hypotheses that, while borrowing certain elements of Merker's "beacon" hypothesis, perhaps avoid some of its problems.

We end by commending Bjorn Merker on his masterful combination of deep thinking about musical rhythm with the comparative method, which has inspired the entire field of bio-musicology. May his groove continue!

14.3 Author's Response

A few background reflections will allow me to be more concise in responding to individual commentaries, all of whose authors I thank for the food for thought and challenges their efforts have provided me in preparing this response. I proposed a peer commentary format for my contribution to the *Nature Beats* volume as a way to address various unanswered questions, controversies, and outright errors that encumber current efforts to explore the biological background to human musical rhythmicity. That our task is to canvas the animal kingdom for phenomena that bear specifically on musical rhythmicity, and not on rhythm in general, is indicated by the term "beats" featured in the title of our volume. That term is obviously not to be taken in the sense of the "beat frequencies" of physics, but as the informal designation "beat" for the basic structural device of human rhythmic music.

By the same token, the rest of our title gives us the entire animal kingdom as the search space to be mined for illuminating analogies and homologies. Given the many millions of animal species on earth, and the ubiquity and variety of rhythmic behaviours they exhibit (Chaps. 2 [249], 3 [43], and 4 [170]), the scope for potential spurious analogies and false homologies is uncomfortably wide. This imposes *stringent requirements on clarity regarding the nature of the rhythmic behaviour to be subjected to comparative scrutiny.*

An invaluable aid to such clarity would be the presence, in the behaviour of interest, of some kind of "distinctive feature," i.e. an empirical marker that draws a qualitative or categorical distinction between that behaviour and others with which it might be confused. For the case of human rhythmic music, we are fortunate to have such a distinctive marker, intimately bound up with the interindividual synchrony featured in human music, dance, and drill. It is *the human capacity to endogenously produce evenly paced signalling behaviour, along with a propensity to longitudinally track such sequences produced by others by aligning the phase of one's own signalling behaviour to theirs over a range of tempos.* We call this "keeping time" or "keeping the beat," a capacity and behaviour that has few uses outside of human music, dance, and drill.

More than a century of rigorous empirical work on this human capacity for sensorimotor synchronization (for reviews, see [18–21, 250, 251]), probed under a wide variety of experimental conditions and perturbations, has thoroughly characterized and documented this human synchronizing capacity. Section 14.1.2 reviewed the nature and attributes of this capacity at some length, with additional detail available in the Preamble of Chap. 5 [10]. The hallmark of this human synchronizing capacity is the following. *Once synchrony has been achieved, and as long as the external signal remains reasonably steady, the beat-for-beat matching performance is continuous in the sense that it does not—at any point along its tracking trajectory—make gradual, drifting transitions across the 180-degree phase deviation (period half-length) that is the maximum deviation possible under maintained beat matching performance* [24].[13] Whole missing beats (a "pause") do not violate this criterion, as long as performance is still in phase when beat marking resumes. Phase alignment is, in other words, *continuous* throughout beat-tracking performance.

The sustained isochronous pacing that enables this behaviour is a *global* characteristic of the performance in question (for which see [14] pp. 179 ff.), and cannot be assessed by local measures of even pacing or phase matching, nor on solo performance alone, however steady its pacing. Any local measure of "isochrony," such as the one recently introduced into the comparative study of musical rhythmicity in [253] and rapidly adopted by others, cannot therefore capture the distinctive feature of human musical rhythmicity. Moreover, synchronizing behaviour that drifts in and out of phase across the 180-degree phase angle is prima facie *evidence of non-possession of the basic entrainment capacity that enables humans to engage in the lagless and sustained rhythmic interindividual synchrony we exercise in music, dance, and drill.*

This specification of the distinctive feature of human musical entrainment disqualifies the rhythmic behaviour exhibited by parrot species featured in reports published back-to-back in a 2009 issue of *Current Biology* from instantiating human-like entrainment to a musical beat [68, 69] (see also [254]). Not only do the animals repeatedly drift across the half-period "Rubicon of music-relevant synchrony," but they do so with enthusiasm and without apparent discomfiture, indicating a lack of cognizance on their part of what they are failing to accomplish.

To date the only documented instance in a warm-blooded animal other than humans of entrainment to an external rhythmic signal in the musically relevant way, i.e. without crossing the synchronization Rubicon in routine performance, is the head-bobbing performed to metronomic stimuli as well as human rhythmic music by the sea lion Ronan after extensive laboratory training [70, 122]. The case is important, because it demonstrates with didactic clarity what a qualifying behaviour looks like when it actually occurs. Ronan could in fact be incorporated into a human

[13] For the technically minded, the basics of human synchronizing performance can be approximated by the simple engineering device of a phase-locked loop set to in-phase operation [252]. The behaviour of such a device, not to be confused with that of simple coupled oscillators, furnishes a convenient operational definition of the basic entrainment capacity that underlies any and all *additional* rhythmic complexity featured in human music and dance.

ensemble playing rhythmic music without occasioning human embarrassment, while the 2009 parrots placed in the same situation would generate a musical disaster. Such is the difference between the presence or absence of the distinctive feature of human rhythmic entrainment: it makes all the difference in the world as far as the comparative task we have set ourselves. Clarity and progress regarding the evolutionary background to human musicality hinges on taking the distinctive feature of specifically musical synchrony seriously.

This is all the more so in that the distinction delivers far more of comparative interest than a single individual trained to exhibit a behaviour that never occurs among her conspecifics, save for the persistent efforts of human experimenters, in this case working in the rich tradition of animal skill training bequeathed by Ronald Schusterman [255]. Among arthropods and anurans there is a scattering of species that engage in spontaneous interindividual synchrony of a kind that brings them within range of the distinctive feature of human rhythmic synchrony. Reviewed in [162] and Chap. 5 [10], they are epitomized by the precisely synchronized bioluminescent flash displays of the tropical lampyrid beetle *Pteroptyx malaccae*. Their group synchrony, once achieved, stays tightly locked at approximately zero phase difference, accommodates a range of tempos around their free-running periodicity, while staying clear of the synchronization Rubicon of half a period length, unlike the parrots of 2009.

With these prefatory remarks on animal rhythmic synchrony in light of the distinctive feature that sets human interindividual synchrony in music, dance, and drill apart from other kinds of temporal coordination, I turn to specifics in the invited peer commentaries, starting with the already introduced sea lion Ronan.

In their commentary, Peter Cook and Colleen Reichmuth (Sect. 14.2.1) open their critique of my interpretation of their Ronan findings by an apparent equation of "evolved trait" or "specific human evolutionary endowment" with its being present at birth (Sect. 14.2.1.3). There is of course no such equation, as illustrated by our secondary sex characteristics, nor does the presence of learning in the ontogeny of a trait contradict its evolved status. The authors then make a concerted attempt to equate Ronan's ontogenetic path to music-relevant synchrony with that of humans. To do so they are forced to ignore the conspicuous differences between those paths. Ronan's performance was forged entirely by behavioural conditioning, requiring reinforced shaping of behaviour from elementary, single movements that happened to coincide in phase with any one of the beats in a sequence, to its gradual extension to two, three, and more consecutive coincident movements.

This mode of shaping Ronan's behaviour "from the bottom up" was adopted only after Ronan, who first had been trained to bob her head, had failed to show spontaneous entrainment of that head bobbing to rhythmic stimulus sequences, and also had failed to progress toward entrainment to them in a course of 30 multi-trial sessions in which she was reinforced when the rate of her bobbing approached the stimulus rate [70].

It was only after these failures that the experimenters resorted to a training protocol of movement by individual movement reinforcement. It is not possible to do a more detailed and direct shaping of the desired behaviour by experimenter control than that. It means, in effect, that *Ronan herself brought nothing to the exercise besides her trainability*. She came to her long course of training and testing without a propensity to track rhythmic sequences with movements of any kind, let alone to match them in phase and over a range of tempos. Her capacity to eventually do so was acquired as the direct result of the rigorous training she underwent, analogous in many ways to horse dressage.

Needless to say, no such interventions are required for humans to acquire the basic ability to keep behavioural time to a rhythmic stimulus, nor does any formal training have to intervene for us to keep time to a new tune at the same tempo or the same tune at a different tempo, as did Ronan in some, but not all such cases [70]. In our case, formal training is devoted to the acquisition of culture-specific patterns of dance steps or musical parts that are elaborations beyond the entrainment capacity itself and not, as in Ronan's case, formal shaping through extrinsic reinforcement of the very rudiments of beat-based synchrony.

Learning does of course take place along the protracted human path to our full capacity to keep time (see [8] p. 13 and [162] p. 13 and references therein), but that learning is of a *different kind* than the movement-by-movement operant shaping through which Ronan acquired her skill. It is therefore futile to compare, as Cook and Reichmuth attempt to do, the developmental time course over which the capacity emerges in the two cases. We know little about the experiential prerequisites in the human case, because our capacity emerges gradually without explicit instruction in the course of maturation, with a terminus at puberty. This leisurely human path to full synchrony, rather than providing evidence for a long human "learning curve," as Cook and Reichmuth (Sect. 14.2.1) suggest, may simply reflect, as I have proposed, that our synchronizing capacity is an evolved predisposition forged through sexual selection that needs to be fully functional only by the time of puberty [8, 111, 112, 162].

The suggestion that the brain's ubiquitous oscillatory activity is the causal key to interindividual entrainment of behaviour hides massive interpretive difficulties that can be accessed in two publications of mine [256, 257]. What I say there about proposed *causal* roles of neural oscillatory activity in cognition extends effortlessly to its suggested role in behavioural rhythmic entrainment. Finally, it would be interesting to know, in view of what is noted in Sect. 14.1 about vocal sexual dimorphism in sea lions, how male sea lions would behave in the Cook and Reichmuth training situation.

In sum, as far as beat-based rhythmic synchrony is concerned, Ronan appears to have come to her trainers a blank slate, as one would expect of a member of a species whose natural behaviour is devoid of any Ronan-like entrainment. Humans, on the other hand, tend to find ways to engage in such behaviour through some form of rhythmic music, dance, or drill across the vast range of cultures and environments that host their lives. On this interpretation it was the patient state of the art conditioning labours of her trainers that inscribed a workable mechanism of entrainment in Ronan's pinniped mind, after a notable initial conditioning failure. The synchronizing capacity of that mechanism today compares favourably in parametric terms to that of humans engaged in a comparable synchronizing task [156].

While there are no indications that interindividual rhythmic synchrony is part of the natural behaviour of sea lions, there are other aquatic mammals whose untrained behaviour make them promising prospective candidates in this regard. In her commentary, Denise Herzing (Sect. 14.2.2) addresses some of these indications from the world of toothed whales, and dolphins in particular. In doing so she raises the crucial issue of the kind of selection pressures that may operate in different taxa and ecological circumstances to promote precise interindividual synchrony.

That topic did not receive the attention it deserves in Sect. 14.1 even with regard to the species whose interindividual synchrony has forced itself on the attention of biologists by being conspicuous, precise, and sustained, namely, synchronously chorusing arthropods and anurans. Among them it serves mate attraction during the breeding season. In that context, selection pressures for synchrony have been repeatedly discussed in the literature [6, 56, 60, 61, 162, 258–262].

That does not mean, however, that mate attraction is the only behavioural context that can select for synchrony. For the human case, coalition signalling may be enhanced by synchrony [39, 263], and many other routes to synchrony are conceivable. Temporal coordination of interindividual behaviour without sustained phase-locked synchrony is not uncommon among cetaceans, as in dolphin surfacing and the herding of females (described in [264] and Sect. 14.2.2) or baleen whale bubble-net foraging [265]. In circumstances that put a premium on the temporal precision of the joint behaviour, the predictive timing made possible through phase-locked isochrony offers itself as a potential solution. And yes, display behaviour that serves to impress a potential mate may also deter rivals, i.e. have agonistic uses, as we learn from the biology of birdsong [266]. In fact, it commonly has this dual function (see, e.g [267, 268].). Cetacean behavioural biology offers a rich field for exploring such possibilities, and technological advances promise to ease the labours of those who, like Herzing, pursue them.

I appreciate that Adriano Lameira (Sect. 14.2.3) found something of value in the programmatic exercise of my proposal (Sect. 14.1), though he finds fault with the way I apply it in my conjectural account of the origin of the human capacity for rhythmic group synchrony. Perhaps because of the proximity to ourselves of the species they study, primatologists tend to be partial to homology as opposed to the other, coeval, leg of the comparative method, namely, analogy (convergent evolution). As one student of primate cognition quoted in my concluding remarks (Sect. 14.1.7) put it "comparative analysis of the behaviour of modern primates, in conjunction with an accurate phylogenetic tree of relatedness, has the power to chart the early history of human cognitive evolution" [152 p. 543]. As a programmatic prescription this hope fails for uniquely derived traits (autapomorphies) [153, 155, 269]. With millions of years of evolution separating ourselves from the last common ancestor we share with our closest primate relatives, we cannot count on finding every human characteristic of interest prefigured among them. One does not, for example, go looking among the great apes for precursors of our relative human hairlessness, however closely we are related to them.

Needless to say, even novel traits evolve from *some* pre-existing ancestral conditions. These may or may not be present in our closest living relatives, because they too have millions of years of independent evolution behind them. The greater the difference between such candidate ancestral traits and the trait they are presumed to prefigure, the less illuminating they are with regard to the *actual genesis* of the derived trait of interest, because of the many characteristics of the latter that are not present in the former. This is the reason why, in Sect. 14.1, I went to considerable lengths to define the human trait of interest, namely, the kind of interindividual rhythmic synchrony we exercise in music, dance, and drill, and why I introduced my author's response by underlining the distinctive feature of that synchrony. Its sustained, phase-locked, interindividual alignment defines the behaviour for which we are looking for evolutionary homologies or analogies.

If, then, one surveys the evolutionary landscape for species whose members display a "capacity to endogenously produce evenly paced signalling behaviour, along with a propensity to longitudinally track such sequences produced by others by aligning the phase of [their] own signalling behaviour to theirs over a range of tempos" an inescapable fact stares one in the face. There are species who do so in a manner that duplicates the human behaviour in its defining characteristics and do so at a level of parametric precision that rivals our own. Those species are not confined to "three species of fireflies," as Lameira claims, but include some species of bush crickets as well [59, 60] (in Sect. 14.1 I mistakenly included the tree cricket studied by Walker [61] among these).

In citing species differences other than the focal trait of behavioural synchrony as arguments against the relevance of insect synchrony, Lameira betrays a lack of understanding of the analogy leg of the comparative method. *Every true comparative analogy bristles with differences on points other than the analogous trait itself.* It is by finding duplication of a trait *in the midst of all those differences*, and across vast phylogenetic distances, that analogy renders its service to evolutionary understanding, because in so doing *it sets us on the trail of actual selection pressures*. The comparative method thus forces us to take seriously the striking parallelism between insect and human group synchrony, a parallelism that in some species includes both phase and period correction (technically known as "perfect synchrony"), the central output variables of human sensorimotor synchrony.

Nothing of the sort occurs among our primate relatives, with one possible exception, and I am not referring to what Lameira considers to be "one of the most exciting recent discoveries on chimpanzee rhythmic behaviour" [53]. Regarding the latter, Sect. 14.1 cites the very paper Lameira claims I ignore, and it does so to note the irrelevance of that behaviour to the topic at hand (for which see Point 5 of Sect. 3 in Sect. 14.1.3).[14] No, I am referring to a tantalizing possibility I have been at pains to call to the attention of my colleagues for a full quarter century [8, 66, 111, 112] (see also Sect. 14.1.5), namely, the possibility that bonobos might have the

[14] Regarding the report of coordinated walking in chimpanzees also referred to by Lameira in this connection, I note that the phase relations of the coordinated walking reported by [184] are even more broadly distributed than in the case of the budgerigars considered in my response to Okanoya (Sect. 14.2.4), below. Such unintentional walking synchrony may involve the kind of subliminal effects I invoke in addressing the report on tapping synchrony in chimpanzees [121] in Sect. 14.1.5.

capacity to synchronize their voices to a self-generated steady beat, as reported by De Waal for captive bonobos [110]. Should that trail go cold on thorough investigation in the wild, our closest living relatives would seem to have little to offer the comparative biology of rhythmic group synchrony in behavioural terms compared to our fellow group synchronizers among arthropods and anurans.

It is a misapplication of evolutionary reasoning to claim, as Lameira does, that if a circumstance, such as male philopatry, is claimed to play a role in a postulated speciation event, all species with male philopatry should exhibit the trait evolved in dependence on that circumstance. The claim ignores the critical role of reproductive isolating mechanisms in speciation events, which typically leave behind a surviving population subject to the circumstance, but without the trait evolved by the other part of the ancestral population. More generally, the claim ignores the role of contingency in evolution. It is, moreover, female exogamy in the setting of a group territoriality and not "firefly-like sex dispersal and feeding ecology" that plays the key operative role in my conjectural scenario for our path to group synchrony.

The evolutionary analogy is between the synchronizing behaviours, not the behavioural ecology details, though finding female migration as the common denominator in this case makes me confident that I am on the right track. On my scenario, the simple fact of female migration offers synchronizing groups an opportunity to incorporate those migrators—who, as females, are the limiting resource in reproductive terms—into their own group *at the reproductive expense of non-synchronizing groups.* In my conjectural ancestral scenario, this "reproductive supplement" may help the synchronizers break out of the demographic trap of long birth spacing in a highly K-selected taxon [77, 270], leaving the non-synchronizers in their rain forest habitat to pursue their path to modern chimpanzee-hood. Ann Pusey and colleagues estimate that close to half of all females on a territory end up migrating [150], a figure that dwarfs any marginal migration on the part of males, should they survive solitary ventures onto neighbouring territories. In sum, then, Lameira employs flawed evolutionary reasoning to critique a conjecture his arguments show he has not fully understood, and I hope this response helps clarify what I am in fact proposing.

I come then, alphabetically, to Kazuo Okanoya (Sect. 14.2.4), whom I thank for reviving fond memories! In light of his commentary, I add the following reflections on the data reported in [120]. Unlike the human-acculturated parrots of 2009, the 2011 budgerigars have not, as far as is known, displayed dance-like behaviour to human rhythmic music. Like Ronan they were laboratory trained "from the bottom up" by operant conditioning procedures. Thus, as in Ronan's case, they appear to have brought "nothing of their own" to the task set them by the experimenters, except their trainability. The length of the training they underwent to reach criterion at 50 consecutive 6-peck sequences for each of six tempos is not reported in the paper, but details of the stepwise progression of conditioning stages by which they did so is provided.

In the criterion runs, which supplied the data analysed in the report, the eight budgerigars, unlike Ronan, did not produce anything like the phase-locked synchrony characteristic of human sensorimotor synchronization. Their pecks show a wide distribution across the periodic stimulus cycle, covering a range that approaches half the cycle length, as shown in schematic diagrams for the eight birds at one tempo in Figure 4 (even then, there were pecks outside the range marked by red in that figure, a range defined by "a significant number of the pecks" at a $p < 0.01$).

Actual distributions of pecks across the cycle are plotted for only two of the eight birds at one of the 6 tempos in Figure 3. One of these distributions, that of Male B, looks strikingly reminiscent of the simulated distribution for "simple reaction" responding (i.e. reaction time-based responses to stimulus onset) plotted on the same graph, except that the peak of the real bird's distribution is displaced some 60 ms earlier in time.

This may simply mean that the estimated reaction time (ERT) used in the simulations is an overestimate (longer than the birds' actual reaction time). This is a plausible interpretation, given that the reaction time estimate used in the simulations was obtained by presenting birds with sequences with intervals of unequal period length after a long series with fixed period length. Such a change is likely to violate bird expectations and thus to delay their responding, resulting in an inflated reaction time estimate.

The peak of Male B's responding at 120 ms after stimulus onset presumably lies within budgerigar reaction time, given the short conduction paths of their nervous systems, which compared to ours are tiny. Note, in this connection, as shown in the "phase matching" column of Table 1, that over-all, the birds' performance tended to be better aligned, or almost as well aligned, with ERT compared to stimulus onset (it can of course be statistically aligned with both when falling in between them, as would happen in reactive responding when ERT is an over-estimate).

The other individual plotted in Figure 3, Female D, illustrates the broad scatter of pecks across the cycle noted above for the schematic depictions in Figure 4. In the plot of Female D they cover some 600 ms, i.e. exactly half, of the 1200 ms cycle, with a skewed distribution peaking at 120 ms before stimulus onset. This is too early to qualify as the negative mean asynchronies (NMA) of a few tens of milliseconds frequently (but not invariably) observed in human sensorimotor synchronization, and may represent "premature" pecks, as proposed in Sect. 14.1, or timing "carry-over" from the previous sequence, as noted by Okanoya (Sect. 14.2.4), and touched upon below.

In his commentary, Okanoya (Sect. 14.2.4) emphasizes early responding (of any magnitude) as evidence for predictive timing in synchronization, repeating the statement of [120] that NMA "was observed in all constant inter-onset interval (IOI) conditions except for the fastest (450 ms)." This formulation obscures the fact that, as Table 1 shows, though NMA did occur in each of five remaining tempo conditions, only a minority of sequences (always six pecks long) exhibited NMA. On average, across the eight (or seven) birds, it occurred in 105 out of 400 slow tempo sequences, and 4 of the 350 600 ms IOI sequences. Okanoya even includes responding *after* stimulus onset but *earlier than the estimated reaction time* as evidence of anticipatory responding when he states that "in 32 of 46 conditions, the centre of the significant pecking phase distribution was located prior to the ERT, suggesting anticipation rather than delayed response." I suggest that an *estimated* reaction time longer than the actual one, as per above, provides a more straightforward account of the same facts.

The well-tuned maxima centred on zero in Figure 6 of the paper would be evidence for phase-locked synchrony in budgerigars if they represented bird response asynchronies, but they do not. They represent instead the difference, in milliseconds, between the length of stimulus inter-onset intervals (IOI) and the length of bird inter-peck intervals (IRI). Apparently, the birds were good at replicating stimulus pacing in their pecking at fast tempos and less so for slow tempos, the slowest of which covers over 10 s of time for a full six cycle sequence.

The contrast between the birds' relatively precise interval duplication, at least at fast tempos, and their poor synchrony may indicate that their performance was based on attempted retention of the entire stimulus sequence in memory as a template for pecking, a retention that would be easier for the shorter total sequence lengths (i.e. fast tempos). They would then have to decide only "when to start," i.e. align their "pecking package" with stimulus phase at sequence onset, something the broad distribution of their pecking over the cycle period indicates they had trouble doing. This interpretation may shed light on a number of aspects of the budgerigar data, including some of the "carry-over" effects between sequences with different IOIs [120 pp. 5–6], but I leave this conjecture as a mere hint of directions in which the data might be explored to determine what exactly the budgerigars were shaped to do by the conditioning procedure they underwent.

Aniruddh Patel's commentary (Sect. 14.2.5) addresses questions raised in Sect. 14.1 regarding his vocal learning and rhythmic synchronization hypothesis (VLH, revised: rVLH). That hypothesis brings together three general topics in cognition and behaviour, namely:

1. Beat extraction from complex stimuli such as human music.
2. Behavioural entrainment to a rhythmic external time-giver.
3. Vocal learning as a proposed resource for point 2.

These three points are conceptually and empirically independent of one another, as demonstrated by species that excel in precisely timed behavioral group synchrony, such as the lampyrid beetle *Pteroptyx malaccae*, but lack both the capacity to extract the isochronous beat from human music as well as any capacity for vocal learning. Patel nevertheless attempts to link the three by the following measures: (1) restricting the evolutionary search space for phenomena under point 2 to "synchronization of movement to beats perceived in complex auditory stimuli (such as music), not with synchronization to relatively simple pulse trains as seen in some crickets or other insects" (Patel commentary in Sect. 14.2.5, and references therein), that is to species with the capacity listed under point 1 above; (2) treating the dancing behaviour of the parrots of 2009 [68, 69] as if it in fact exhibits the distinctive feature of human rhythmic synchrony in music, dance, and drill; (3) construing the nature of vocal learning to the effect that its "tight coupling between auditory input and motor output" (Patel, 2006, p. 102, and Patel commentary in Sect. 14.2.5) involves online coupling to an external time-giver, which it does not. These commitments are problematic for the following reasons.

There are no principled grounds on which to restrict the comparative search for species whose interindividual behavioural synchrony might shed light on interindividual synchrony in human music to those that in addition to their synchronizing capacity possess the capacity to extract the beat from complex human music. That would constitute a principled restriction for a comparative hypothesis regarding beat extraction, but not for interindividual synchrony. Patel's introduction of such a restriction in the latter case by simple stipulation eliminates not only striking counterexamples to his hypothesis ("some crickets or other insects") by fiat, but it also rules out perfectly good examples of human musical synchrony that do not involve beat extraction from complex stimuli. Thus, in some of the traditional group dances of indigenous Australians, dance steps and unison chanting are synchronized to simple clapstick isochrony. That is, *even for humans, beat perception in complex auditory patterns is by no means "central to dancing and making music in synchrony with others" if "central" implies any kind of obligatory dependence.*

The structural invariant that unites all forms of precise interindividual rhythmic synchrony, whether arthropod, pinniped, or human, is consistent phase-locked latching of behaviour to sustained isochrony as defined in my introductory remarks. That is not, however, what is documented in the primary exhibit of Patel's hypothesis, the "parrots of 2009" [68, 69]. As seen in the sample presented in Figure 3 of [68], the greater part of the parrot's *active dancing time* is spent without such latching. Moreover, the circular plot of behavioural phase relative to the beat in Figure 2 of [68] is based *not on total dancing time*, as it should be in order to reflect the precision of the animal's time-keeping while dancing, but only on the short stretches *within* the dancing time during which an approximation to synchrony in fact occurred. Needless to say, that plot would look very different if based instead on the behaviour at issue, namely, the parrot's actual dancing movements as a whole. It is not the case, in other words, that the animal performs short stretches of synchronized behaviour and then pauses to resume synchrony after the pause. Instead, performance varies in both tempo and phase, causing it to repeatedly drift across the "synchronization Rubicon" of half a period length defined in my introductory remarks.

None of this is what is expected from an animal actually in possession of a mechanism for interindividual synchrony. One plausible interpretation for the failure of the birds to produce this kind of behaviour that is diagnostic for possession of such a mechanism is that, as a consequence of acculturation to humans, they are performing credible imitations of human dancing (for which see [66, 75, 162, 271]). That they show no apparent discomfiture on drifting in and out of synchrony while doing so may mean they have not "gotten" (speaking informally) that a key point of the human behaviour they are enthusiastically imitating is "to keep the beat." As long as evidence is limited to behaviour, we cannot even know that the birds are extracting the actual beat from the complex human music to which they dance. Their rough behavioural adjustment to different tempos at which the same tune was presented to them indicates that they are not without sensitivity to the temporal structure of the music. Human toddlers likewise show an influence of stimulus tempo while being deficient in synchronizing ability [272] (and see also Chap. 10

[161]). Unlike the children the parrots are motorically mature and may, of course, have an accurate perceptual/cognitive representation of the beat—a matter of afferent operations alone—but fail to lock their behaviour to it for lack of an entrainment mechanisms that implements online motoric phase-locking to an external time-giver (as previously suggested [66, 162, 271]; and see also Chap. 5 [10] and Sect. 14.1).

The fact that nothing resembling the dance-mimicry of parrots acculturated to humans occurs in the natural behaviour of these species [68] prevents us from treating their imperfect dance-mimicry as at least a possible stage on an evolutionary path to the "real thing." There is nothing in the biology of parrots that predisposes them to perform a semblance of human dancing *except their capacity for mimicry*. My alternative to Patel's rVLH is that these parrots possess a broad-based capacity for imitation/mimicry [66, 75, 271], plausibly generalized from the mimicry that is intrinsic to vocal learning [75], and that *it is this generalized behaviour-matching capacity that explains the behaviour of these parrots when acculturated to humans* [66, 271] (see also Sect. 14.1).[15] This brings us to the last of the issues enumerated above, namely, the vocal production learning that gives name to Patel's hypothesis.

Patel's VLH in both its original and revised versions proposes that vocal production learning ("high vocal learning" in the revised version[16]), a capacity shared by humans and parrots, provides a resource for implementing entrainment to an external time-giver on account of online "tight coupling between auditory input and motor output in order to match vocal production to a desired model." Patel answers my objection that vocal production learning features extreme temporal *decoupling* between auditory input from an external sound source and motor output by citing a set of diverse examples, in birds as well as humans, of real-time audiomotor interactions involved in or drawing on vocal learning. Those are all examples, however, of the audiomotor feed-back interactions that ever since Thorpe's deafening experiments on songbirds in the 1950s [275] have been integral to the very definition of vocal production learning, namely, *feedback from one's own voice*. The latter serves the crucial function by which vocal learning allows one to sculpt one's own vocal

[15] The "shell system" of high vocal learners appears to implement expansion of vocal production learning from conspecific imitation to allo-mimicry of other species and even ambient sounds, or from shared song repertoire to shared repertoire plus individual vocal signatures [273]. That the shell system may promote extension of imitation in vocal learning to a generalized propensity to imitate [75] is all the more plausible given that the less developed shell system of songbirds is activated not by singing but bodily movement such as trained hopping [273, 274]. This conjecture provides a more direct explanatory framework for interpreting the dancing behavior of the parrots of 2009 (and other imitative expressive gestures on the part of some of their conspecifics also kept as pets) than the rather indirect arguments invoking fortuitous circumstances by which Patel attempts to buttress his vocal learning hypothesis [209].

[16] It should be noted, in addition to what I say about the vocal learning "shell system" in Sect. 14.1, that only some of the parrots equipped with a shell system are known to be "dancers." If, as seems likely, all bird species that engage in allo-mimicry ("high vocal learners")—a group of species that by no means is confined to parrots—are equipped with a shell system, there would be ample comparative material on which to assess competing accounts of the "dancing" displays of the parrots of 2009.

output to conform to a model of tutor song or other heard auditory pattern *stored in memory*, typically long before any vocal production is attempted [275–277].

The sound of one's own voice is not to be equated with the *external time-giver* with which an entrainment mechanism must synchronize. One's own voice is timed by the same individual who receives the resulting auditory feedback. Such a functional loop, by definition operative in all vocal production learners to sculpt their vocal output, does indeed instantiate audio-motor feedback in real time, but without any temporal coupling to an independently timed on-going signal emanating from an *external source* (typically another individual or ambient sound source), the very crux of sensorimotor synchronization.

Thus, for purposes of interindividual entrainment and synchrony, the tight coupling between hearing and voice that is intrinsic to vocal production learning [71 p. 102] is *to the wrong voice*, namely, your own and not that of another, and thus not germane to the topic of our comparative quest.[17] Vocal learning simply does not feature tight coupling between an external time giver and motor output, and the assumption that it does is the central error of Patel's vocal learning and rhythmic synchronization hypothesis.

To the stark contrast between the online matching of an external time giver in music-relevant synchrony and the temporal decoupling from its external sound source that virtually defines vocal production learning we can add their contrasting ontogenies. In the human case, vocal production learning is fully operative in infancy, where it accomplishes the feat of speech acquisition down to its dialectal nuances in the first 2 years of life, while the capacity for sustained phase-locked beat tracking takes another decade to be fully developed [8, 162] (see also Chap. 10 [161], the peer commentaries in Sects. 14.2.1 and 14.2.6 and my responses to them). As already mentioned, leisurely progress from the time-keeping deficiencies of infants and toddlers to the enthusiastic expertise, even virtuosity, of a teenager suggests that we owe our capacity for timekeeping to a mode of selection for which puberty is the terminus, namely, sexual selection.

Finally, Patel's invocation of neural mechanisms to support his rVLH [209] is vitiated by his arbitrary restriction—by stipulation—of rhythmic synchronization to that subset of such behaviour that presupposes beat extraction from complex stimulus sequences. This tends to confound sustained phase-locked synchrony with advanced auditory perceptual/cognitive capacity and accordingly skews his neural arguments toward advanced audio-motor circuitry instantiated in the brains of vocal learning birds and humans. This naturally increases the likelihood of finding correlations—possibly spurious ones—with vocal production learning.[18] Without that

[17]Another individual's voice (or other environmental sound source) does of course play a central role in vocal learning, but then typically as sensory source of the vocal pattern the learner eventually will produce, after passively (non-interactively) listening to a "tutor" or environmental sound, committing it to memory, and long after, following protracted solo practice, eventually duplicates. Even in cases where learners respond to tutor signals, as may happen when teaching a parrot to talk, their responses follow those signals rather than coincide with them.

[18]For example, when factors other than synchrony itself were controlled for in the meta-analysis [278], the involvement of the dorsal premotor area, a prominent member of the constellation of brain areas identified in imaging studies of rhythmic synchrony, did not reach significance.

confound we know that the fraction of a milligram of total brain tissue possessed by *Pteroptyx malaccae* suffices to orchestrate the exquisitely precise and sustained phase-locked group synchrony of these tropical "fireflies." They do so over a range of tempos by correcting both the phase and the period of their isochronous output, the very two functional parameters that are operative in human timekeeping.

In sum my answer to Patel's valiant defence of his rVLH is that sustained phase-locked entrainment to an external time-giver is dissociable from vocal production learning on each of Tinbergen's four dimensions of analysis, namely, function, phylogeny, mechanism, and ontogeny. As far as I can see, the problems that encumber each of the three composite strands of rVLH reviewed above prevent them from being joined into a coherent account of the dance-mimicry exhibited by the parrots of 2009 along the lines attempted by Patel, let alone of the nature and origin of the synchronizing capacity we exercise in music, dance, and drill. Alternatives therefore need to be explored, alternatives that do not arbitrarily exclude nature's foremost non-human group synchronizers, and with due cognizance of the distinctive feature of human rhythmic synchrony outlined in my introductory remarks. One such alternative has been sketched by me in earlier work [8, 66, 111, 112, 271] and is summarized in Sect. 14.1 as well.

Tudor Popescu and Tecumseh Fitch (Sect. 14.2.6) provide a commendably concise gist of that proposal in their commentary, and I thank them for the accuracy with which they have done so, and also for their good words concerning my efforts in the field of our shared interests. I am grateful as well for the way they underline the power of the "analogy" or "convergent evolution" leg of the comparative method when it comes to suggesting and testing adaptive hypotheses.

Since that is how my origins scenario for human rhythmic synchrony originated and is framed, Popescu and Fitch have no trouble finding their way to a potentially fatal flaw in my conjecture. How can selection *on males*, an unequivocal feature of the "beacon hypothesis" of the arthropod analogy, account for rhythmic synchrony in a species in which both sexes are equipped with the synchronizing capacity and—beyond its narrow compass—are equally musical?

Ever since Darwin discussed the conundrum of cases of secondary sexual characteristics and ornamentation in the female of species in which selection for the trait was focused on males [135], the default answer to this question has been that in such cases, patterns of trait inheritance constrain selection on one sex to be expressed in both.[19] Darwin made his proposal long before the modern synthesis of Darwin and Mendel, but it is now known as a correlated genetic response to selection on males [279], for which there is both theoretical [280] and empirical support [281].

[19] "The laws of inheritance, irrespectively of selection, appear to have determined whether the characters acquired by the males for the sake of ornament, for producing various sounds, and for fighting together, have been transmitted to the males alone or to both sexes [...] Why various characters should sometimes have been transmitted in one way and sometimes in another is, in most cases, not known" [135 vol. 2, p. 235].

Given the early stage of mapping the genetics of music at which we find ourselves [282], it will presumably be some time till we know whether our beat synchronization capacity is thus constrained. Its already known multiple pleiotropies [283] may, however, be preliminary evidence for barriers against selection for synchrony on males alone. Be that as it may, there is an additional factor to consider: Over the past few decades, the erstwhile focus of sexual selection theory on competing males and choosy females has been broadened to recognize sexual selection as a form of social selection operating across diverse mating systems [284]. As such it is shaped by many interacting ecological and social variables that need to be considered in addressing any given case. A crucial such variable in my origins scenario is the behavioural ecology context proposed to have framed our path to synchrony.

Unlike arthropod synchronizers, the common ancestor we share with chimpanzees lived permanently in mixed-sex groups. Present-day female chimpanzees may join pant-hoot choruses, though they do so less frequently than males [285]. There is no reason to assume that matters stood differently in the ancestral condition, which would mean that females were part of whatever changes vocal chorusing underwent to regularize its pacing and interindividual synchrony. Neither males nor females were presumably cognizant of the function of that synchrony, the group to which they belonged simply had its own "style" of hooting bouts, a style that ended up attracting migrating females to the synchronizing group at the expense of non-synchronizing ones through the wider reach of a synchronized signal.[20]

This marginal addition of females to synchronizing groups would ease somewhat the reproductive bottleneck of a K-selected taxon with long birth spacing [77, 270]. That advantage, so goes my proposal, was a first step on the synchronizers' path to populating the earth, but a first step only. It is meant as an account of no more than the synchronizing capacity itself, expressed in group chanting of something like the bonobo "staccato hoot," presumably to stomping in place as an aid to keeping the synchronizing beat. Utterly simple as such a performance is, it nevertheless provides the obligatory foundation for all rhythmic music, however structurally complex.

If we count that elementary beginning as an incipient form of singing and dancing, that would amount to the birth of music, but music devoid of almost all the structural attributes of what we today know as music. The elaborations came later: Some near at hand [8, 111, 112] including metrical grouping mentioned in Sect. 14.2.6), some more distant [66, 77, 271]. It would be a delight to join Popescu and Fitch in exploring the nature of these developments, broached in their commentary, but for now I rest content with having tried to meet their principal objection to my conjectural origins scenario.

Regarding their additional point about ontogeny, it is important to distinguish perceptual beat extraction from the capacity for behavioural synchrony with an external time giver. That capacity is absent in early infancy [8, 286] and emerges

[20] Such a group signal may also, of course, double as a deterrent (see my response to Herzing commentary in 14.2.2), as in the "beacon-oriented alternative hypothesis" of Popescu and Fitch (Sect. 14.2.6) or the "coalition signaling" of Hagen and Bryant [39, 263].

only gradually with development [215, 272, 287, 288] (see also Chap. 10 [161]). It appears not to be fully developed until puberty [215, 288, 289], in good agreement with the proposed origins scenario, assuming correlated selection in the two sexes (see also Chap. 10 and Sects. 14.2.1 and 14.2.5 on the protracted human path to full synchrony).

Finally, I note that the complex precisely coordinated multi-part duetting of plain-tailed wrens documented by Nigel Mann and colleagues does not require synchronization to a shared beat. I have proposed that each bird learns the entire stereotyped pattern as a single sequence of which it only contributes its own segment during performance (in hocket-like manner: [8] p. 5, fn. 1; see also Sect. 1.2.3.2. in Chap. 1 [290]). This conjecture received subsequent empirical support by direct recording of neural activity in wrens exposed to duet stimuli [291]. With such a memory template for performance, all that remains for a bird to render a seamless contribution to the duet is to determine "when to start." Because the wrens do not start their duet sequence in unison, the first song element of the bird that starts provides an adequate time cue for the rest of the participants to do so, given their wholistic, real-time, template.

Let me conclude this response to my commentators by thanking them all for the opportunity they have provided me to explicate a number of issues in the comparative study of human musical rhythmicity that I think are essential for progress toward understanding its nature and origin. I have followed with mounting apprehension developments triggered by the 2009 publication in *Current Biology* of papers on that topic by Patel et al. [68] and Schachner et al. [69], culminating in the introduction and rapid adoption of a local measure of isochrony claimed to be relevant to human musical rhythmicity by Roeske at al [253]. and those who have adopted their measure. It is my hope that the present peer commentary treatment has helped clarify the conceptual and empirical issues involved, and thereby will assist, in some small measure, recovery from errors made in the course of these developments.

Acknowledgments A. Patel thanks the organizers of the Nature Beats conference for the opportunity to write this commentary, Bjorn Merker for his interesting proposal, and Adena Schachner, Laura Cirelli, and Mimi Kao for helpful discussions. B. Merker thanks the volume editors for the opportunity to cast his chapter in peer commentary format, and for the assistance they rendered him in its implementation.

References

1. Grillner S, El Manira A (2020) Current principles of motor control, with special reference to vertebrate locomotion. Physiol Rev 100:271–320. https://doi.org/10.1152/physrev.00015.2019
2. Katz PS (2016) Evolution of central pattern generators and rhythmic behaviours. Philos Trans R Soc B 371:20150057. https://doi.org/10.1098/rstb.2015.0057
3. Prinz AA (2018) Rhythmic pattern generation in invertebrates. In: Byrne JH (ed) The Oxford handbook of invertebrate neurobiology, 1st edn. Oxford University Press, pp 391–400

4. Prochazka A, Yakovenko S (2007) The neuromechanical tuning hypothesis. Prog Brain Res 165:255–265. https://doi.org/10.1016/s0079-6123(06)65016-4
5. Sauciuc G-A, Schneider JN, Hattori Y (forthcoming) Chapter 9: neural and mechanical building blocks of rhythmic communication and how these constrain rhythm production and perception. In: Nature beats: the what, why and how of animal rhythmic behaviour. Springer
6. Greenfield MD (1994) Synchronous and alternating choruses in insects and anurans: common mechanisms and diverse functions. Am Zool 34:605–615. https://doi.org/10.1093/icb/34.6.605
7. Greenfield MD (2005) Mechanisms and evolution of communal sexual displays in arthropods and anurans. In: Advances in the study of behavior. Elsevier, pp 1–62
8. Merker BH, Madison GS, Eckerdal P (2009) On the role and origin of isochrony in human rhythmic entrainment. Cortex 45:4–17. https://doi.org/10.1016/j.cortex.2008.06.011
9. Jacoby N, Polak R, London J (2021) Extreme precision in rhythmic interaction is enabled by role-optimized sensorimotor coupling: analysis and modelling of West African drum ensemble music. Philos Trans R Soc B 376:20200331. https://doi.org/10.1098/rstb.2020.0331
10. Greenfield M, Merker B (2025) Chapter 5: endogenous rhythm, isochrony and the path to perfect synchrony. In: Nature beats: the what, why and how of animal rhythmic behaviour. Springer Nature
11. Yoshida S, Okanoya K (2005) Evolution of turn-taking: a bio-cognitive perspective. In: Nature beats: the what, why and how of animal rhythmic behaviour. Springer Nature
12. Greenfield MD, Aihara I, Amichay G et al (2021) Rhythm interaction in animal groups: selective attention in communication networks. Philos Trans R Soc B 376:20200338. https://doi.org/10.1098/rstb.2020.0338
13. Vanderhoff EN, Bernal Hoverud N (2022) Perspectives on antiphonal calling, Duetting and counter-singing in non-primate mammals: an overview with notes on the coordinated vocalizations of bamboo rats (*Dactylomys* spp., Rodentia: Echimyidae). Front Ecol Evol 10:906546. https://doi.org/10.3389/fevo.2022.906546
14. Arom S (1991) African polyphony and polyrhythm: musical structure and methodology, 1st edn. Cambridge University Press
15. Merker B (2014) Groove or swing as distributed rhythmic consonance: introducing the groove matrix. Front Hum Neurosci 8:454. https://doi.org/10.3389/fnhum.2014.00454
16. Polak R, London J, Jacoby N (2016) Both isochronous and non-isochronous metrical subdivision afford precise and stable ensemble entrainment: a corpus study of Malian Jembe drumming. Front Neurosci 10:285. https://doi.org/10.3389/fnins.2016.00285
17. Madison G, Merker B (2002) On the limits of anisochrony in pulse attribution. Psychol Res 66:201–207. https://doi.org/10.1007/s00426-001-0085-y
18. Fraisse P (1982) Rhythm and tempo. In: Psychology of music. Elsevier, pp 149–180
19. Madison GS (2000) On the nature of variability in isochronous serial interval production. In: Rhythm perception and production. Swets and Zeitlinger, Lisse, pp 95–113
20. Repp BH (2005) Sensorimotor synchronization: a review of the tapping literature. Psychon Bull Rev 12:969–992. https://doi.org/10.3758/bf03206433
21. Repp BH, Su Y-H (2013) Sensorimotor synchronization: a review of recent research (2006–2012). Psychon Bull Rev 20:403–452. https://doi.org/10.3758/s13423-012-0371-2
22. Ermentrout B (1991) An adaptive model for synchrony in the firefly *Pteroptyx malaccae*. J Math Biol 29:571–585. https://doi.org/10.1007/BF00164052
23. Repp BH, Keller PE (2008) Sensorimotor synchronization with adaptively timed sequences. Hum Mov Sci 27:423–456. https://doi.org/10.1016/j.humov.2008.02.016
24. Burchardt LS, Briefer EF, Knörnschild M (2021) Novel ideas to further expand the applicability of rhythm analysis. Ecol Evol 11:18229–18237. https://doi.org/10.1002/ece3.8417
25. Semjen A, Vorberg D, Schulze H-H (1998) Getting synchronized with the metronome: comparisons between phase and period correction. Psychol Res 61:44–55. https://doi.org/10.1007/s004260050012
26. Repp BH (2001) Processes underlying adaptation to tempo changes in sensorimotor synchronization. Hum Mov Sci 20:277–312. https://doi.org/10.1016/s0167-9457(01)00049-5

27. Repp BH (2001) Phase correction, phase resetting, and phase shifts after subliminal timing perturbations in sensorimotor synchronization. J Exp Psychol Hum Percept Perform 27:600–621. https://doi.org/10.1037/0096-1523.27.3.600
28. Hary D, Moore GP (1987) Synchronizing human movement with an external clock source. Biol Cybern 56:305–311. https://doi.org/10.1007/bf00319511
29. Hary D, Moore GP (1987) On the performance and stability of human metronome-synchronization strategies. Br J Math Stat Psychol 40:109–124. https://doi.org/10.1111/j.2044-8317.1987.tb00872.x
30. Miall RC, Weir DJ, Wolpert DM, Stein JF (1993) Is the cerebellum a smith predictor? J Mot Behav 25:203–216. https://doi.org/10.1080/00222895.1993.9942050
31. Lewis PA, Miall RC (2003) Distinct systems for automatic and cognitively controlled time measurement: evidence from neuroimaging. Curr Opin Neurobiol 13:250–255. https://doi.org/10.1016/s0959-4388(03)00036-9
32. Todd NPM, Seiss E (2004) Electrophysiological correlates of beat induction as internally- and externally-guided action. In: Lipscomb SD, Ashley R, Gjerdingen RO, Webster P (eds) Proceedings of the 8th international conference on music perception and cognition. Causal Productions, Adelaide, pp 212–218
33. Teki S, Grube M, Kumar S, Griffiths TD (2011) Distinct neural substrates of duration-based and beat-based auditory timing. J Neurosci 31:3805–3812. https://doi.org/10.1523/jneurosci.5561-10.2011
34. Grahn JA, Rowe JB (2013) Finding and feeling the musical beat: striatal dissociations between detection and prediction of regularity. Cereb Cortex 23:913–921. https://doi.org/10.1093/cercor/bhs083
35. Repp BH (2000) Compensation for subliminal timing perturbations in perceptual-motor synchronization. Psychological Research 63:106–128. https://doi.org/10.1007/PL00008170
36. Thaut MH, Tian B, Azimi-Sadjadi MR (1998) Rhythmic finger tapping to cosine-wave modulated metronome sequences: evidence of subliminal entrainment. Hum Mov Sci 17:839–863. https://doi.org/10.1016/s0167-9457(98)00031-1
37. Madison G, Merker B (2004) Human sensorimotor tracking of continuous subliminal deviations from isochrony. Neurosci Lett 370:69–73. https://doi.org/10.1016/j.neulet.2004.07.094
38. Merker B (2019) When extravagance impresses: recasting esthetics in evolutionary terms. In: Thaut MH, Hodges DA (eds) The Oxford handbook of music and the brain. Oxford University Press, pp 65–86
39. Mehr SA, Krasnow MM, Bryant GA, Hagen EH (2021) Origins of music in credible signaling. Behav Brain Sci 44:e60. https://doi.org/10.1017/s0140525x20000345
40. Savage PE, Loui P, Tarr B et al (2021) Music as a coevolved system for social bonding. Behav Brain Sci 44:e59. https://doi.org/10.1017/s0140525x20000333
41. Mora C, Tittensor DP, Adl S et al (2011) How many species are there on earth and in the ocean? PLoS Biol 9:e1001127. https://doi.org/10.1371/journal.pbio.1001127
42. Ødegaard F (2000) How many species of arthropods? Erwin's estimate revised. Biol J Linn Soc 71:583–597. https://doi.org/10.1111/j.1095-8312.2000.tb01279.x
43. Sauciuc G-A, Schneider JJN, Burchardt LS et al (2025) Chapter 3: the functional and structural diversity of rhythmic communication across species. In: Sauciuc G-A, Lameira AR, Persson T (eds) Nature beats: the what, why and how of animal rhythmic behaviours. Springer Nature
44. Todd NPM, Lee CS (2015) The sensory-motor theory of rhythm and beat induction 20 years on: a new synthesis and future perspectives. Front Hum Neurosci 9:444. https://doi.org/10.3389/fnhum.2015.00444
45. Ten Cate C, Spierings M, Hubert J, Honing H (2016) Can birds perceive rhythmic patterns? A review and experiments on a songbird and a parrot species. Front Psychol 7:730. https://doi.org/10.3389/fpsyg.2016.00730
46. Norton P, Scharff C (2016) "Bird song Metronomics": isochronous Organization of Zebra Finch Song Rhythm. Front Neurosci 10:309. https://doi.org/10.3389/fnins.2016.00309

47. Merker B (2002) Music: the missing Humboldt system. Music Sci 6:3–21. https://doi.org/10.1177/102986490200600101
48. Lehiste I (1977) Isochrony reconsidered. J Phon 5:253–263. https://doi.org/10.1016/s0095-4470(19)31139-8
49. Dauer RM (1983) Stress-timing and syllable-timing reanalyzed. J Phon 11:51–62. https://doi.org/10.1016/s0095-4470(19)30776-4
50. Bertrán AP (1999) Prosodic typology: on the dichotomy between stress-timed and syllable-timed languages. Lang Des 2:103–130
51. Ramus F, Nespor M, Mehler J (1999) Correlates of linguistic rhythm in the speech signal. Cognition 73:265–292. https://doi.org/10.1016/s0010-0277(99)00058-x
52. Levinson SC (2016) Turn-taking in human communication—origins and implications for language processing. Trends Cogn Sci 20:6–14. https://doi.org/10.1016/j.tics.2015.10.010
53. Lameira AR, Eerola T, Ravignani A (2019) Coupled whole-body rhythmic entrainment between two chimpanzees. Sci Rep 9:18914. https://doi.org/10.1038/s41598-019-55360-y
54. https://www.youtube.com/watch?v=f2K3kjKGLgM
55. Skutch AF (1944) Life-history of the prong-billed barbet. Auk 61:61–88
56. Buck J (1988) Synchronous rhythmic flashing of fireflies. II Q Rev Biol 63:265–289. https://doi.org/10.1086/415929
57. Hanson FE (1978) Comparative studies of firefly pacemakers. Fed Proc 37:2158–2164
58. Hanson FE (1982) Pacemaker control of rhythmic flashing of fireflies. In: Cellular pacemakers. Wiley, New York, pp 81–100
59. Murphy MA, Thompson NL, Schul J (2016) Keeping up with the neighbor: a novel mechanism of call synchrony in *Neoconocephalus ensiger* katydids. J Comp Physiol A 202:225–234. https://doi.org/10.1007/s00359-016-1068-1
60. Nityananda V, Balakrishnan R (2009) Modeling the role of competition and cooperation in the evolution of katydid acoustic synchrony. Behav Ecol 20:484–489. https://doi.org/10.1093/beheco/arp022
61. Walker TJ (1969) Acoustic synchrony: two mechanisms in the snowy tree cricket. Science 166:891–894. https://doi.org/10.1126/science.166.3907.891
62. Madison G (2001) Variability in isochronous tapping: higher order dependencies as a function of intertap interval. J Exp Psychol Hum Percept Perform 27:411–422. https://doi.org/10.1037/0096-1523.27.2.411
63. Repp BH (2010) Self-generated interval subdivision reduces variability of synchronization with a very slow metronome. Music Percept 27:389–397. https://doi.org/10.1525/mp.2010.27.5.389
64. Moelants D (2002) Preferred tempo reconsidered. In: Stevens C, Urnham D, McPherson G et al (eds) Proceedings of the seventh international conference on music perception and cognition. Causal Productions, Adelaide, pp 580–583
65. MacDougall HG, Moore ST (2005) Marching to the beat of the same drummer: the spontaneous tempo of human locomotion. J Appl Physiol 99:1164–1173. https://doi.org/10.1152/japplphysiol.00138.2005
66. Merker B, Morley I, Zuidema W (2018) Five fundamental constraints on theories of the origins of music. In: Honing H (ed) The origins of musicality. The MIT Press, pp 49–80
67. Sismondo E (1990) Synchronous, alternating, and phase-locked stridulation by a tropical katydid. Science 249:55–58. https://doi.org/10.1126/science.249.4964.55
68. Patel AD, Iversen JR, Bregman MR, Schulz I (2009) Experimental evidence for synchronization to a musical beat in a nonhuman animal. Curr Biol 19:827–830. https://doi.org/10.1016/j.cub.2009.03.038
69. Schachner A, Brady TF, Pepperberg IM, Hauser MD (2009) Spontaneous motor entrainment to music in multiple vocal mimicking species. Curr Biol 19:831–836. https://doi.org/10.1016/j.cub.2009.03.061
70. Cook P, Rouse A, Wilson M, Reichmuth C (2013) A California Sea lion (Zalophus californianus) can keep the beat: motor entrainment to rhythmic auditory stimuli in a non vocal mimic. J Comp Psychol 127:412–427. https://doi.org/10.1037/a0032345

71. Patel AD (2006) Musical rhythm, linguistic rhythm, and human evolution. Music Percept 24:99–104. https://doi.org/10.1525/mp.2006.24.1.99
72. Piaget J (1962) Play, dreams and imitation in childhood. W.W. Norton & Co, New York
73. Meltzoff AN (1988) Infant imitation and memory: nine-month-olds in immediate and deferred tests. Child Dev 59:217–225. https://doi.org/10.1111/j.1467-8624.1988.tb03210.x
74. Courage ML, Howe ML (2002) From infant to child: the dynamics of cognitive change in the second year of life. Psychol Bull 128:250–277. https://doi.org/10.1037/0033-2909.128.2.250
75. Merker B (2005) The conformal motive in birdsong, music, and language: an introduction. Ann N Y Acad Sci 1060:17–28. https://doi.org/10.1196/annals.1360.003
76. Merker B (2008) Ritual foundations of human uniqueness. In: Malloch S, Trevarthen C (eds) Communicative musicality. Oxford University Press, Oxford, pp 45–59
77. Merker BH (2012) The vocal learning constellation: imitation, ritual culture, encephalization. In: Bannan H (ed) Music, language and human evolution. Oxford University Press, Oxford, pp 215–260
78. Farabaugh SM (1982) The ecological and social significance of duetting. In: Acoustic communication in birds. Academic, New York, pp 85–124
79. Hall ML (2009) Chapter 3 a review of vocal Duetting in birds. In: Advances in the study of behavior. Academic, pp 67–121
80. Elie JE, Mariette MM, Soula HA et al (2010) Vocal communication at the nest between mates in wild zebra finches: a private vocal duet? Anim Behav 80:597–605. https://doi.org/10.1016/j.anbehav.2010.06.003
81. Henry L, Craig AJFK, Lemasson A, Hausberger M (2015) Social coordination in animal vocal interactions. Is there any evidence of turn-taking? The starling as an animal model. Front Psychol 6:1416. https://doi.org/10.3389/fpsyg.2015.01416
82. Benichov JI, Benezra SE, Vallentin D et al (2016) The forebrain song system mediates predictive call timing in female and male zebra finches. Curr Biol 26:309–318. https://doi.org/10.1016/j.cub.2015.12.037
83. Mann NI, Dingess KA, Slater PJB (2006) Antiphonal four-part synchronized chorusing in a Neotropical wren. Biol Lett 2:1–4. https://doi.org/10.1098/rsbl.2005.0373
84. Brenowitz EA (2021) Taking turns: the neural control of birdsong duets. Proc Natl Acad Sci USA 118:e2108043118. https://doi.org/10.1073/pnas.2108043118
85. Coleman MJ, Day NF, Rivera-Parra P, Fortune ES (2021) Neurophysiological coordination of duet singing. Proc Natl Acad Sci USA 118:e2018188118. https://doi.org/10.1073/pnas.2018188118
86. Bögels S, Torreira F (2021) Turn-end estimation in conversational turn-taking: the roles of context and prosody. Discourse Process 58:903–924. https://doi.org/10.1080/0163853x.2021.1986664
87. Demartsev V, Strandburg-Peshkin A, Ruffner M, Manser M (2018) Vocal turn-taking in meerkat group calling sessions. Curr Biol 28:3661–3666.e3. https://doi.org/10.1016/j.cub.2018.09.065
88. Takahashi DY, Narayanan DZ, Ghazanfar AA (2013) Coupled oscillator dynamics of vocal turn-taking in monkeys. Curr Biol 23:2162–2168. https://doi.org/10.1016/j.cub.2013.09.005
89. Gamba M, Torti V, Estienne V et al (2016) The indris have got rhythm! Timing and pitch variation of a primate song examined between sexes and age classes. Front Neurosci 10:249. https://doi.org/10.3389/fnins.2016.00249
90. Zanoli A, De Gregorio C, Valente D et al (2020) Sexually dimorphic phrase organization in the song of the indris (*Indri indri*). Am J Primatol 82:e23132. https://doi.org/10.1002/ajp.23132
91. De Gregorio C, Valente D, Raimondi T et al (2021) Categorical rhythms in a singing primate. Curr Biol 31:R1379–R1380. https://doi.org/10.1016/j.cub.2021.09.032
92. Haimoff EH (1984) Acoustic and organisational features of gibbon song. In: The lesser apes: evolutionary and behavioural biology. University of Edinburgh Press., Edinburgh, pp 333–353

93. Merker B, Cox C (1999) Development of the female great call in *Hylobates gabriellae*: a case study. IJFP 70:97–106. https://doi.org/10.1159/000021680
94. Geissmann T (2000) Duet songs of the siamang (*Hylobates syndactylus*): I. Structure and organisation. Primate Rep 56:33–60
95. Aihara I, Takeda R, Mizumoto T et al (2011) Complex and transitive synchronization in a frustrated system of calling frogs. Phys Rev E 83:031913
96. Aihara I, Mizumoto T, Otsuka T et al (2014) Spatio-temporal dynamics in collective frog choruses examined by mathematical modeling and field observations. Sci Rep 4:3891. https://doi.org/10.1038/srep03891
97. Patel AD (2021) Vocal learning as a preadaptation for the evolution of human beat perception and synchronization. Philos Trans R Soc B 376:20200326. https://doi.org/10.1098/rstb.2020.0326
98. Chakraborty M, Walløe S, Nedergaard S et al (2015) Core and shell song systems unique to the parrot brain. PLoS One 10:e0118496. https://doi.org/10.1371/journal.pone.0118496
99. Baylis JR (1982) Avian vocal mimicry: its function and evolution. In: Acoustic communication in birds. Academic, New York, pp 51–83
100. Dowsett-Lemaire F (1979) The imitative range of the song of the Marsh Warbler *Acrocephalus palustris*, with special reference to imitations of African birds. Ibis 121:453–468. https://doi.org/10.1111/j.1474-919x.1979.tb06685.x
101. Serpell J (1981) Duets, greetings and triumph ceremonies: analogous displays in the parrot genus *Trichoglossus*. Z Tierpsychol 55:268–283. https://doi.org/10.1111/j.1439-0310.1981.tb01272.x
102. Ten Cate C, Vos DR (1999) Sexual imprinting and evolutionary processes in birds: a reassessment. In: Advances in the study of behavior. Elsevier, pp 1–31
103. Miklósi Á (1999) The ethological analysis of imitation. Biol Rev 74:347–374. https://doi.org/10.1017/s000632319900537x
104. Whiten A, Horner V, Litchfield CA, Marshall-Pescini S (2004) How do apes ape? Learn Behav 32:36–52. https://doi.org/10.3758/bf03196005
105. Tennie C, Call J, Tomasello M (2006) Push or pull: imitation vs. emulation in great apes and human children. Ethology 112:1159–1169. https://doi.org/10.1111/j.1439-0310.2006.01269.x
106. Large EW, Gray PM (2015) Spontaneous tempo and rhythmic entrainment in a bonobo (*Pan paniscus*). J Comp Psychol 129:317–328. https://doi.org/10.1037/com0000011
107. Madison G (2006) Experiencing groove induced by music: consistency and phenomenology. Music Percept 24:201–208. https://doi.org/10.1525/mp.2006.24.2.201
108. Janata P, Tomic ST, Haberman JM (2012) Sensorimotor coupling in music and the psychology of the groove. J Exp Psychol Gen 141:54–75. https://doi.org/10.1037/a0024208
109. Dufour V, Poulin N, Curé C, Sterck EHM (2015) Chimpanzee drumming: a spontaneous performance with characteristics of human musical drumming. Sci Rep 5:11320. https://doi.org/10.1038/srep11320
110. De Waal FBM (1988) The communicative repertoire of captive bonobos (Pan Paniscus), compared to that of chimpanzees. Behaviour 106:183–251. https://doi.org/10.1163/156853988x00269
111. Merker B (1999) Synchronous chorusing and the origins of music. Music Sci 3:59–73. https://doi.org/10.1177/10298649000030S105
112. Merker B (1999) Synchronous chorusing and human origins. In: Wallin NL, Merker B, Brown S (eds) The origins of music. The MIT Press, pp 315–328
113. Bermejo M, Omedes A (2000) Preliminary vocal repertoire and vocal communication of wild bonobos (*Pan paniscus*) at Lilungu (Democratic Republic of Congo). IJFP 70:328–357. https://doi.org/10.1159/000021717
114. Zarco W, Merchant H, Prado L, Mendez JC (2009) Subsecond timing in primates: comparison of interval production between human subjects and rhesus monkeys. J Neurophysiol 102:3191–3202. https://doi.org/10.1152/jn.00066.2009

115. Merchant H, Honing H (2014) Are non-human primates capable of rhythmic entrainment? Evidence for the gradual audiomotor evolution hypothesis. Front Neurosci 7:274. https://doi.org/10.3389/fnins.2013.00274
116. Proksch S, Honing H, Ravignani A, Merchant H (2025) Chapter 7: converging notions on the main hypotheses regarding the origins of beat perception and synchronization: the key role of audiomotor interactions. In: Nature beats: the what, why and how of animal rhythmic behaviour. Springer Nature
117. Kauranen K, Vanharanta H (1996) Influences of aging, gender, and handedness on motor performance of upper and lower extremities. Percept Mot Skills 82:515–525. https://doi.org/10.2466/pms.1996.82.2.515
118. Gámez J, Yc K, Ayala YA et al (2018) Predictive rhythmic tapping to isochronous and tempo changing metronomes in the nonhuman primate. Ann N Y Acad Sci 1423:396–414. https://doi.org/10.1111/nyas.13671
119. Castillo-Almazán A, Pérez O, Prado L et al (2025) Flexible tapping synchronization in macaques: dynamic switching of timing strategies within rhythmic sequences. J Neurophysiol 134:580–590. https://doi.org/10.1152/jn.00158.2025
120. Hasegawa A, Okanoya K, Hasegawa T, Seki Y (2011) Rhythmic synchronization tapping to an audio–visual metronome in budgerigars. Sci Rep 1:120. https://doi.org/10.1038/srep00120
121. Hattori Y, Tomonaga M, Matsuzawa T (2013) Spontaneous synchronized tapping to an auditory rhythm in a chimpanzee. Sci Rep 3:1566. https://doi.org/10.1038/srep01566
122. Rouse AA, Cook PF, Large EW, Reichmuth C (2016) Beat keeping in a sea lion as coupled oscillation: implications for comparative understanding of human rhythm. Front Neurosci 10:257. https://doi.org/10.3389/fnins.2016.00257
123. Buck J, Buck E, Case JF, Hanson FE (1981) Control of flashing in fireflies: V. Pacemaker synchronization in *Pteroptyx cribellata*. J Comp Physiol 144:287–298. https://doi.org/10.1007/BF00612560
124. Cook P, Reichmuth C, Hanke FD (2021) The mind of a sea lion. In: Ethology and behavioral ecology of marine mammals. Springer, Cham, pp 323–345
125. Charrier I, Ahonen H, Harcourt RG (2011) What makes an Australian sea lion (*Neophoca cinerea*) male's bark threatening? J Comp Psychol 125:385–392. https://doi.org/10.1037/a0024513
126. Schusterman RJ (1977) Temporal patterning in sea lion barking (*Zalophus californianus*). Behav Biol 20:404–408. https://doi.org/10.1016/S0091-6773(77)90964-6
127. Schneider J, Burchardt LS, Sauciuc G-A, Lameira AR (2025) Chapter 6: how environmental and social contexts shape and constrain rhythm structure and production. In: Sauciuc G-A, Lameira AR, Persson T (eds) Nature beats: the what, why and how of animal rhythmic behaviour. Springer
128. Gustison ML, Sauciuc G-A (2025) Chapter 11: moving the body and the mind – links between emotion and rhythmic communication and the underlying neurobiological mechanisms. In: Sauciuc G-A, Lameira AR, Persson T (eds) Nature beats: the what, why and how of animal rhythmic behaviours. Springer
129. Nagasaka Y, Chao ZC, Hasegawa N et al (2013) Spontaneous synchronization of arm motion between Japanese macaques. Sci Rep 3:1151. https://doi.org/10.1038/srep01151
130. Tyack PL (1997) Studying how cetaceans use sound to explore their environment. In: Owings DH, Beecher MD, Thompson NS (eds) Perspectives in ethology. Plenum Press, New York, pp 251–297
131. Janik VM, Simard P, Sayigh LS et al (2011) Chorussing in delphinids. J Acoust Soc Am 130:2322–2322. https://doi.org/10.1121/1.3654281
132. Herzing D (2015) Synchronous and rhythmic vocalizations and correlated underwater behavior of free-ranging Atlantic spotted dolphins (*Stenella frontalis*) and bottlenose dolphins (*Tursiops truncatus*) in The Bahamas. AB&C 2:14–29. https://doi.org/10.12966/abc.02.02.2015

133. Lopez Marulanda J, Adam O, Huetz C et al (2021) Acoustic behaviour of bottlenose dolphins under human care while performing synchronous aerial jumps. Behav Process 185:104357. https://doi.org/10.1016/j.beproc.2021.104357
134. Ravignani A, Fitch WT, Hanke FD et al (2016) What pinnipeds have to say about human speech, music, and the evolution of rhythm. Front Neurosci 10:274. https://doi.org/10.3389/fnins.2016.00274
135. Darwin CR (1871) The descent of man, and selection in relation to sex. John Murray, London
136. Fisher RA (1930) The genetical theory of natural selection. Clarendon Press, Oxford
137. Zahavi A (1975) Mate selection—a selection for a handicap. J Theor Biol 53:205–214. https://doi.org/10.1016/0022-5193(75)90111-3
138. Fedurek P, Machanda ZP, Schel AM, Slocombe KE (2013) Pant hoot chorusing and social bonds in male chimpanzees. Anim Behav 86:189–196. https://doi.org/10.1016/j.anbehav.2013.05.010
139. Fedurek P, Schel AM, Slocombe KE (2013) The acoustic structure of chimpanzee pant-hooting facilitates chorusing. Behav Ecol Sociobiol 67:1781–1789. https://doi.org/10.1007/s00265-013-1585-7
140. Mitani J, Gros-Louis J (1998) Chorusing and call convergence in chimpanzees: tests of three hypotheses. Behaviour 135:1041–1064. https://doi.org/10.1163/156853998792913483
141. Eleuteri V, Henderson M, Soldati A et al (2022) The form and function of chimpanzee buttress drumming. Anim Behav 192:189–205. https://doi.org/10.1016/j.anbehav.2022.07.013
142. Duda P, Zrzavý J (2013) Evolution of life history and behavior in Hominidae: towards phylogenetic reconstruction of the chimpanzee–human last common ancestor. J Hum Evol 65:424–446. https://doi.org/10.1016/j.jhevol.2013.07.009
143. Ember CR (1975) Residential variation among hunter-gatherers. Behav Sci Res 10:199–227. https://doi.org/10.1177/106939717501000302
144. Ember CR (1978) Myths about hunter-gatherers. Ethnology 17:439. https://doi.org/10.2307/3773193
145. Ghiglieri MP (1987) Sociobiology of the great apes and the hominid ancestor. J Hum Evol 16:319–357. https://doi.org/10.1016/0047-2484(87)90065-0
146. Lalueza-Fox C, Rosas A, Estalrrich A et al (2011) Genetic evidence for patrilocal mating behavior among Neandertal groups. Proc Natl Acad Sci USA 108:250–253. https://doi.org/10.1073/pnas.1011553108
147. Lovejoy CO (2009) Reexamining human origins in light of *Ardipithecus ramidus*. Science 326:74. https://doi.org/10.1126/science.1175834
148. Wrangham R (1979) On the evolution of ape social systems. Soc Sci Inf 18:336–368. https://doi.org/10.1177/053901847901800301
149. Pusey A (2007) Inter-community transfer of chimpanzees in Gombe National Park. In: Hamburg D, McCown E (eds) The great apes. Benjamin/Cummings, Menlo Park, pp 465–479
150. Pusey A, Williams J, Goodall J (1997) The influence of dominance rank on the reproductive success of female chimpanzees. Science 277:828–831. https://doi.org/10.1126/science.277.5327.828
151. Merker B (2025) Human vocal production learning: the nature and origin of our watershed adaptation. In: Delafield Butt J, Reddy V (eds) Intersubjective minds, 1st edn. Oxford University Press, Oxford, pp 373–398
152. Byrne RW (2000) Evolution of primate cognition. Cogn Sci 24:543–570. https://doi.org/10.1207/s15516709cog2403_8
153. Lockard JS (1980) The evolution of human social behavior. Elsevier, New York
154. Crook JH, Gartlan JS (1966) Evolution of primate societies. Nature 210:1200–1203. https://doi.org/10.1038/2101200a0
155. Crook JH (1970) Behaviour in birds and mam. In: Social behavior in birds and mammals: essays on the social ethology of animals and man. Academic, New York, pp 103–166
156. Cook PF, Hood C, Rouse A, Reichmuth C (2025) Sensorimotor synchronization to rhythm in an experienced sea lion rivals that of humans. Sci Rep 15:95279. https://doi.org/10.1038/s41598-025-95279-1

157. Fodor JA (1983) The modularity of mind: an essay on faculty psychology. MIT Press, Cambridge, MA
158. Drewing K, Aschersleben G, Li SC (2006) Sensorimotor synchronization across the life span. Int J Behav Dev 30:280–287. https://doi.org/10.1177/0165025406066764
159. Thompson EC, White-Schwoch T, Tierney A, Kraus N (2015) Beat synchronization across the lifespan: intersection of development and musical experience. PLoS One 10:1–13. https://doi.org/10.1371/journal.pone.0128839
160. Ireland K, Parker A, Foster N, Penhune V (2018) Rhythm and melody tasks for school-aged children with and without musical training: age-equivalent scores and reliability. Front Psychol 9:1–14. https://doi.org/10.3389/fpsyg.2018.00426
161. Sauciuc G-A, Raimondi T, De Gregorio C (2025) Chapter 10: the developmental trajectory of rhythmic abilities in humans and singing primates. In: Sauciuc G-A, Lameira AR, Persson T (eds) Nature beats: the what, why and how of animal rhythmic behaviour. Springer
162. Greenfield MD, Merker B (2023) Coordinated rhythms in animal species, including humans: entrainment from bushcricket chorusing to the philharmonic orchestra. Neurosci Biobehav Rev 153:105382. https://doi.org/10.1016/j.neubiorev.2023.105382
163. Wilson M, Cook PF (2016) Rhythmic entrainment: why humans want to, fireflies can't help it, pet birds try, and sea lions have to be bribed. Psychon Bull Rev 23:1647–1659. https://doi.org/10.3758/s13423-016-1013-x
164. Hannon EE, Vanden Bosch Der Nederlanden CM, Tichko P (2012) Effects of perceptual experience on children's and adults' perception of unfamiliar rhythms. Ann N Y Acad Sci 1252:92–99. https://doi.org/10.1111/j.1749-6632.2012.06466.x
165. Hannon EE, Soley G, Ullal S (2012) Familiarity overrides complexity in rhythm perception: a cross-cultural comparison of American and Turkish listeners. J Exp Psychol Hum Percept Perform 38:543–548. https://doi.org/10.1037/a0027225
166. Buzsáki G, Draguhn A (2004) Neuronal oscillations in cortical networks. Science 304:1926–1929. https://doi.org/10.1126/science.1099745
167. Large EW (2025) Chapter 8: neural resonance and comparative rhythm cognition. In: Sauciuc G-A, Lameira AR, Persson T (eds) Nature beats: the what, why and how of animal rhythmic behaviours. Springer
168. Canolty RT, Knight RT (2010) The functional role of cross-frequency coupling. Trends Cogn Sci 14:506–515. https://doi.org/10.1016/j.tics.2010.09.001
169. Henry MJ, Cook PF, De Reus K et al (2021) An ecological approach to measuring synchronization abilities across the animal kingdom. Philos Trans R Soc B 376:20200336. https://doi.org/10.1098/rstb.2020.0336
170. Lameira AR, De Gregorio C, Gamba M et al (2025) Chapter 4: uncharted rhythmic diversity across the animal world. In: Sauciuc G-A, Lameira AR, Persson T (eds) Nature beats: the what, why and how of animal rhythmic behaviour. Springer
171. Merchant H, Yarrow K (2016) How the motor system both encodes and influences our sense of time. Curr Opin Behav Sci 8:22–27. https://doi.org/10.1016/j.cobeha.2016.01.006
172. Patel AD, Iversen JR (2014) The evolutionary neuroscience of musical beat perception: the Action Simulation for Auditory Prediction (ASAP) hypothesis. Front Syst Neurosci 8:57. https://doi.org/10.3389/fnsys.2014.00057
173. Hoffman-Kuhnt M, Herzing D (2016) Whose line sound is it anyway? Identifying the vocalizer on underwater video by localizing with a hydrophone Array. AB&C 3:288–298. https://doi.org/10.12966/abc.07.11.2016
174. Lammers MO, Castellote M (2009) The beluga whale produces two pulses to form its sonar signal. Biol Lett 5:297–301. https://doi.org/10.1098/rsbl.2008.0782
175. Myers AJ, Herzing DL, Bjorklund DF (2017) Synchrony during aggression in adult male Atlantic spotted dolphins (*Stenella frontalis*). Acta Ethol 20:175–185. https://doi.org/10.1007/s10211-017-0262-7
176. Raji JI, Potter CJ (2021) The number of neurons in *Drosophila* and mosquito brains. PLoS One 16:e0250381. https://doi.org/10.1371/journal.pone.0250381

177. Azevedo FAC, Carvalho LRB, Grinberg LT et al (2009) Equal numbers of neuronal and nonneuronal cells make the human brain an isometrically scaled-up primate brain. J Comp Neurol 513:532–541. https://doi.org/10.1002/cne.21974
178. Fallon TR, Lower SE, Chang C-H et al (2018) Firefly genomes illuminate parallel origins of bioluminescence in beetles. elife 7:e36495. https://doi.org/10.7554/elife.36495
179. Evans SD, Hughes IV, Gehling JG, Droser ML (2020) Discovery of the oldest bilaterian from the Ediacaran of South Australia. Proc Natl Acad Sci USA 117:7845–7850. https://doi.org/10.1073/pnas.2001045117
180. Clutton-Brock TH, Lukas D (2012) The evolution of social philopatry and dispersal in female mammals. Mol Ecol 21:472–492. https://doi.org/10.1111/j.1365-294x.2011.05232.x
181. Milich KM (2023) Male-philopatric nonhuman primates and their potential role in understanding the evolution of human sociality. Evol Anthropol 33:e22014. https://doi.org/10.1002/evan.22014
182. Renault D (2020) A review of the phenotypic traits associated with insect dispersal polymorphism, and experimental designs for sorting out resident and disperser phenotypes. Insects 11:214. https://doi.org/10.3390/insects11040214
183. Maquitico Y, Vergara A, Villanueva I et al (2022) *Photuris lugubris* female fireflies hunt males of the synchronous firefly Photinus palaciosi (Coleoptera: Lampyridae). Insects 13:915. https://doi.org/10.3390/insects13100915
184. Schweinfurth MK, Baldridge DB, Finnerty K et al (2022) Inter-individual coordination in walking chimpanzees. Curr Biol 32:5138–5143.e3. https://doi.org/10.1016/j.cub.2022.09.059
185. Merker B, Okanoya K (2007) The natural history of human language: bridging the gaps without magic. In: Lyon C, Nehaniv C, Cangelosi A (eds) Emergence of communication and language. Springer, London, pp 403–420
186. Okanoya K, Merker B (2007) Neural substrates for string-context mutual segmentation: a path to human language. In: Lyon C, Nehaniv C, Cangelosi A (eds) Emergence of communication and language. Springer, London, pp 421–434
187. Abla D, Katahira K, Okanoya K (2008) On-line assessment of statistical learning by event-related potentials. J Cogn Neurosci 20:952–964. https://doi.org/10.1162/jocn.2008.20058
188. Kudo N, Nonaka Y, Mizuno N et al (2011) On-line statistical segmentation of a non-speech auditory stream in neonates as demonstrated by event-related brain potentials: segmentation of non-speech auditory stream. Dev Sci 14:1100–1106. https://doi.org/10.1111/j.1467-7687.2011.01056.x
189. Ono S, Okanoya K, Seki Y (2016) Hierarchical emergence of sequence sensitivity in the songbird auditory forebrain. J Comp Physiol A 202:163–183. https://doi.org/10.1007/s00359-016-1070-7
190. Katsu N, Yuki S, Okanoya K (2021) Production of regular rhythm induced by external stimuli in rats. Anim Cogn 24:1133–1141. https://doi.org/10.1007/s10071-021-01505-4
191. Yosida S, Kobayasi KI, Ikebuchi M et al (2007) Antiphonal vocalization of a subterranean rodent, the naked mole-rat (*Heterocephalus glaber*). Ethology 113:703–710. https://doi.org/10.1111/j.1439-0310.2007.01371.x
192. Jain A, Bansal R, Kumar A, Singh K (2015) A comparative study of visual and auditory reaction times on the basis of gender and physical activity levels of medical first year students. Int J App Basic Med Res 5:124. https://doi.org/10.4103/2229-516x.157168
193. Benichov JI, Vallentin D (2020) Inhibition within a premotor circuit controls the timing of vocal turn-taking in zebra finches. Nat Commun 11:221. https://doi.org/10.1038/s41467-019-13938-0
194. Patel AD, Iversen JR, Bregman MR, Schulz I (2009) Avian and human movement to music: two further parallels. Commun Integr Biol 2:485–488. https://doi.org/10.4161/cib.2.6.9373
195. Patel AD, Iversen JR, Bregman MR, Schulz I (2009) Studying synchronization to a musical beat in nonhuman animals. Ann N Y Acad Sci 1169:459–469. https://doi.org/10.1111/j.1749-6632.2009.04581.x

196. Patel AD, Iversen JR, Chen Y, Repp BH (2005) The influence of metricality and modality on synchronization with a beat. Exp Brain Res 163:226–238. https://doi.org/10.1007/s00221-004-2159-8
197. Brown S (2022) Group dancing as the evolutionary origin of rhythmic entrainment in humans. New Ideas Psychol 64:100902. https://doi.org/10.1016/j.newideapsych.2021.100902
198. Doupe AJ, Kuhl PK (1999) Birdsong and human speech: common themes and mechanisms. Annu Rev Neurosci 22:567–631. https://doi.org/10.1146/annurev.neuro.22.1.567
199. Vernes SC, Kriengwatana BP, Beeck VC et al (2021) The multi-dimensional nature of vocal learning. Philos Trans R Soc B 376:20200236. https://doi.org/10.1098/rstb.2020.0236
200. Shamma S, Patel P, Mukherjee S et al (2021) Learning speech production and perception through sensorimotor interactions. Cereb Cortex Commun 2:tgaa091. https://doi.org/10.1093/texcom/tgaa091
201. Barreda S (2020) Vowel normalization as perceptual constancy. Language 96:224–254. https://doi.org/10.1353/lan.2020.0018
202. Sakata JT, Brainard MS (2006) Real-time contributions of auditory feedback to avian vocal motor control. J Neurosci 26:9619–9628. https://doi.org/10.1523/jneurosci.2027-06.2006
203. Schachner A (2010) Auditory-motor entrainment in vocal mimicking species: additional ontogenetic and phylogenetic factors. Commun Integr Biol 3:290–293. https://doi.org/10.4161/cib.3.3.11708
204. Zentner M, Eerola T (2010) Rhythmic engagement with music in infancy. Proc Natl Acad Sci USA 107:5768–5773. https://doi.org/10.1073/pnas.1000121107
205. Cirelli LK, Trehub SE (2019) Dancing to Metallica and Dora: case study of a 19-month-old. Front Psychol 10:1073. https://doi.org/10.3389/fpsyg.2019.01073
206. Kragness HE, Ullah F, Chan E et al (2022) Tiny dancers: effects of musical familiarity and tempo on children's free dancing. Dev Psychol 58:1277–1285. https://doi.org/10.1037/dev0001363
207. Kragness HE, Anderson L, Chow E et al (2023) Musical groove shapes children's free dancing. Dev Sci 26:e13249. https://doi.org/10.1111/desc.13249
208. Sauciuc G-A, Persson T, Madsen E (2022) The natural history of musical rhythm: functional and mechanistic theories on the evolution of human rhythm cognition and the relevance of rhythmic animal behaviors. CCBR 17:33–87. https://doi.org/10.3819/CCBR.2022.170003
209. Patel AD (2024) Beat-based dancing to music has evolutionary foundations in advanced vocal learning. BMC Neurosci 25:12. https://doi.org/10.1186/s12868-024-00843-6
210. Patel AD (2024) Response to commentaries by Schmidt and Kaplan, Penhune, Hickok and Theofanopoulou on "Beat-based dancing to music has evolutionary foundations in advanced vocal learning.". BMC Neurosci 25:21. https://doi.org/10.1186/s12868-024-00853-4
211. Reed KE (1997) Early hominid evolution and ecological change through the African Plio-Pleistocene. J Hum Evol 32:289–322. https://doi.org/10.1006/jhev.1996.0106
212. Miller G (1999) Evolution of human music through sexual selection. In: Wallin NL, Merker B, Brown S (eds) The origins of music. The MIT Press, pp 329–360
213. Miller GF (2001) The mating mind: how sexual choice shaped the evolution of human nature., 1st Anchor books ed. Anchor Books, New York
214. Winkler I, Háden GP, Ladinig O et al (2009) Newborn infants detect the beat in music. Proc Natl Acad Sci USA 106:2468–2471. https://doi.org/10.1073/pnas.0809035106
215. McAuley JD, Jones MR, Holub S et al (2006) The time of our lives: life span development of timing and event tracking. J Exp Psychol Gen 135:348–367. https://doi.org/10.1037/0096-3445.135.3.348
216. Haimoff EH (1986) Convergence in the duetting of monogamous Old World primates. J Hum Evol 15:51–59. https://doi.org/10.1016/S0047-2484(86)80065-3
217. Hall ML (2004) A review of hypotheses for the functions of avian duetting. Behav Ecol Sociobiol 55:415–430. https://doi.org/10.1007/s00265-003-0741-x
218. Serpell JA (1981) Duetting in birds and primates: a question of function. Anim Behav 29:963–965. https://doi.org/10.1016/s0003-3472(81)80041-3

219. Thorpe WH, North MEW (1965) Origin and significance of the power of vocal imitation: with special reference to the antiphonal singing of birds. Nature 208:219–222. https://doi.org/10.1038/208219a0
220. Wickler W (2010) Vocal Dueting and the pair bond: I. Coyness and partner commitment. A hypothesis. Zeitschrift für Tierpsychologie 52:201–209. https://doi.org/10.1111/j.1439-0310.1980.tb00711.x
221. Odom KJ, Hall ML, Riebel K et al (2014) Female song is widespread and ancestral in songbirds. Nat Commun 5:3379. https://doi.org/10.1038/ncomms4379
222. Mebes H-D (1978) Pair-specific duetting in the peach-faced lovebird *Agapornis roseicollis*. Naturwissenschaften 65:66–67. https://doi.org/10.1007/bf00420646
223. Power DM (1966) Antiphonal dueting and evidence for auditory reaction time in the orange-chinned parakeet. Auk 83:314–319. https://doi.org/10.2307/4083033
224. Adret P, Dingess K, Caselli C et al (2018) Duetting patterns of Titi monkeys (primates, Pitheciidae: Callicebinae) and relationships with phylogeny. Animals 8:178. https://doi.org/10.3390/ani8100178
225. Mann NI, Dingess KA, Barker FK et al (2009) A comparative study of song form and duetting in Neotropical Thryothorus wrens. Behaviour 146:1–43
226. Mann NI, Marshall-Ball L, Slater PJB (2003) The complex song duet of the plain wren. Condor 105:672–682
227. Haiduk F, Fitch WT (2022) Understanding design features of music and language: the choric/dialogic distinction. Front Psychol 13:786899. https://doi.org/10.3389/fpsyg.2022.786899
228. Fitch WT (2011) The biology and evolution of rhythm: unravelling a paradox. In: Rebuschat P, Rohmeier M, Hawkins JA, Cross I (eds) Language and music as cognitive systems. Oxford University Press, pp 73–95
229. Fitch WT (2013) Rhythmic cognition in humans and animals: distinguishing meter and pulse perception. Front Syst Neurosci 7:68. https://doi.org/10.3389/fnsys.2013.00068
230. Fitch WT (2016) Dance, music, meter and groove: a forgotten partnership. Front Hum Neurosci 10:64. https://doi.org/10.3389/fnhum.2016.00064
231. Koehler O (1951) Der Vogelgesang als Vorstufe von Musik und Sprache. J Ornithol 93:3–20. https://doi.org/10.1007/bf02007606
232. Richman B (1993) On the evolution of speech: singing as the middle term. Curr Anthropol 34:721–722. https://doi.org/10.1086/204217
233. Brown SM (2000) The "Musilanguage" model of music evolution. In: Wallin NL, Merker BH, Brown S (eds) The origins of music. MIT Press, Cambridge, pp 271–300
234. Mithen SJ (2005) The singing neanderthals: the origins of music, language, mind and body. Weidenfeld & Nicolson, London
235. Fitch WT (2013) Musical protolanguage: Darwin's theory of language evolution revisited. In: Rebuschat P, Rohmeier M, Hawkins A, Cross I (eds) Birdsong, speech, and language. The MIT Press, pp 489–504
236. Wagner B, Mann DC, Afroozeh S et al (2019) Octave equivalence perception is not linked to vocal mimicry: budgerigars fail standardized operant tests for octave equivalence. Behaviour 156:479–504. https://doi.org/10.1163/1568539x-00003538
237. Wagner B, Šlipogor V, Oh J et al (2023) A comparison between common marmosets (*Callithrix jacchus*) and human infants sheds light on traits proposed to be at the root of human octave equivalence. Dev Sci 26:e13395. https://doi.org/10.1111/desc.13395
238. Bannan N, Dunbar R, Bamford J (2024) The evolution of gender dimorphism in the human voice: the role of octave equivalence. Curr Anthropol 65:503–527. https://doi.org/10.1086/730147
239. Wagner B, Hoeschele M (2022) The links between pitch, timbre, musicality, and social bonding from cross-species research. CCBR 17:13–32. https://doi.org/10.3819/ccbr.2022.170002
240. Bowling DL, Gahr J, Ancochea PG et al (2022) Endogenous oxytocin, cortisol, and testosterone in response to group singing. Horm Behav 139:105105. https://doi.org/10.1016/j.yhbeh.2021.105105

241. Weinstein D, Launay J, Pearce E et al (2016) Singing and social bonding: changes in connectivity and pain threshold as a function of group size. Evol Hum Behav 37:152–158. https://doi.org/10.1016/j.evolhumbehav.2015.10.002
242. Bowling DL, Purves D, Gill KZ (2018) Vocal similarity predicts the relative attraction of musical chords. Proc Natl Acad Sci USA 115:216–221. https://doi.org/10.1073/pnas.1713206115
243. McDermott JH, Schultz AF, Undurraga EA, Godoy RA (2016) Indifference to dissonance in native Amazonians reveals cultural variation in music perception. Nature 535:547–550. https://doi.org/10.1038/nature18635
244. McPherson MJ, Dolan SE, Durango A et al (2020) Perceptual fusion of musical notes by native Amazonians suggests universal representations of musical intervals. Nat Commun 11:2786. https://doi.org/10.1038/s41467-020-16448-6
245. Popescu T, Neuser MP, Neuwirth M et al (2019) The pleasantness of sensory dissonance is mediated by musical style and expertise. Sci Rep 9:1070. https://doi.org/10.1038/s41598-018-35873-8
246. Curio E, Ernst U, Vieth W (1978) The adaptive significance of avian mobbing: II. Cultural transmission of enemy recognition in blackbirds: effectiveness and some constraints. Z Tierpsychol 48:184–202. https://doi.org/10.1111/j.1439-0310.1978.tb00255.x
247. Boesch C (1991) The effects of leopard predation on grouping patterns in Forest chimpanzees. Behav 117:220–241. https://doi.org/10.1163/156853991X00544
248. Werdelin L, Lewis ME (2005) Plio-Pleistocene Carnivora of eastern Africa: species richness and turnover patterns. Zool J Linnean Soc 144:121–144. https://doi.org/10.1111/j.1096-3642.2005.00165.x
249. Sauciuc G-A, Schneider JN (2025) Chapter 2: how animals produce rhythm—an overview of ethological mechanisms for rhythmic communication. In: Sauciuc G-A, Lameira AR, Persson T (eds) Nature beats: the what, why and how of animal rhythmic behaviours. Springer Nature
250. Fraisse P (1980) Les synchronizations sensori-motrices aux rhythmes. In: Anticipation et comportement. Editions de C.N.R.S, Paris, pp 233–257
251. McAuley JD (2010) Tempo and rhythm. In: Riess Jones M, Fay RR, Popper AN (eds) Music perception. Springer, New York, NY, pp 165–199
252. Best RE (2007) Phase-locked loops: design, simulation, and applications, 6th edn. McGraw-Hill, New York
253. Roeske TC, Tchernichovski O, Poeppel D, Jacoby N (2020) Categorical rhythms are shared between songbirds and humans. Curr Biol 30:3699. https://doi.org/10.1016/j.cub.2020.08.026
254. Fitch WT (2009) Biology of music: another one bites the dust. Curr Biol 19:R403–R404. https://doi.org/10.1016/j.cub.2009.04.004
255. Schusterman RJ (2010) Historical perspectives. Aquatic Mammals 36:84–110. https://doi.org/10.1578/am.36.1.2010.84
256. Merker B (2013) Cortical gamma oscillations: the functional key is activation, not cognition. Neurosci Biobehav Rev 37:401–417. https://doi.org/10.1016/j.neubiorev.2013.01.013
257. Merker BH (2016) Cortical gamma oscillations: details of their genesis preclude a role in cognition. Front Comput Neurosci 10:78. https://doi.org/10.3389/fncom.2016.00078
258. Lloyd JE (1973) Fireflies of Melanesia: bioluminescence, mating behavior, and synchronous flashing (Coleoptera: Lampyridae). Environ Entomol 2:991–1008. https://doi.org/10.1093/ee/2.6.991
259. Greenfield MD (1994) Cooperation and conflict in the evolution of signal interactions. Annu Rev Ecol Syst 25:97–126. https://doi.org/10.1146/annurev.es.25.110194.000525
260. Greenfield MD, Schul J (2008) Mechanisms and evolution of synchronous chorusing: emergent properties and adaptive functions in *Neoconocephalus katydids* (Orthoptera: Tettigoniidae). J Comp Psychol 122:289–297. https://doi.org/10.1037/0735-7036.122.3.289
261. Moiseff A, Copeland J (2010) Firefly synchrony: a behavioral strategy to minimize visual clutter. Science 329:181–181. https://doi.org/10.1126/science.1190421
262. Greenfield MD, Marin-Cudraz T, Party V (2017) Evolution of synchronies in insect choruses. Biol J Linn Soc 122:487–504. https://doi.org/10.1093/biolinnean/blx096

263. Hagen EH, Bryant GA (2003) Music and dance as a coalition signaling system. Hum Nat 14:21–51. https://doi.org/10.1007/s12110-003-1015-z
264. Connor RC, Smolker RA, Richards AF (1992) Two levels of alliance formation among male bottlenose dolphins (*Tursiops* sp.). Proc Natl Acad Sci USA 89:987–990. https://doi.org/10.1073/pnas.89.3.987
265. Sharpe F (2001) Social foraging of the southeast Alaskan humpback whale *Megaptera novaeangliae*. Doctoral thesis,. Simon Fraser University
266. Catchpole C, Slater PJB (1995) Bird song: biological themes and variations. Cambridge University Press, Cambridge, UK
267. Andersson MB (1994) Sexual selection. Princeton University Press, Princeton
268. Bradbury JW, Vehrencamp SL (2011) Principles of animal communication, 2nd edn. Sinauer Associates, Sunderland
269. Harcourt AH (1998) Does primate socioecology need nonprimate socioecology? Evol Anthropol 7:3–7. https://doi.org/10.1002/(SICI)1520-6505(1998)7:1<3::AID-EVAN2>3.0.CO;2-H
270. Kennedy G (2005) From the ape's dilemma to the weanling's dilemma: early weaning and its evolutionary context. J Hum Evol 48:123–145. https://doi.org/10.1016/j.jhevol.2004.09.005
271. Merker B (2025) Human vocal production learning: the nature and origin of our watershed adaptation. In: Delafield-butt J, Reddy V (eds) Intersubjective minds, 1st edn. Oxford University Press, Oxford, pp 373–398
272. Provasi J, Bobin-Bègue A (2003) Spontaneous motor tempo and rhythmical synchronisation in 2½- and 4-year-old children. Int J Behav Dev 27:220–231. https://doi.org/10.1080/01650250244000290
273. Zhao Z, Teoh HK, Carpenter J et al (2023) Anterior forebrain pathway in parrots is necessary for producing learned vocalizations with individual signatures. Curr Biol 33:5415–5426.e4. https://doi.org/10.1016/j.cub.2023.11.014
274. Feenders G, Liedvogel M, Rivas M et al (2008) Molecular mapping of movement-associated areas in the avian brain: a motor theory for vocal learning origin. PLoS One 3:e1768. https://doi.org/10.1371/journal.pone.0001768
275. Thorpe CW (1961) Bird-song: the biology of vocal communication and expression in birds, Cambridge monographs in experimental biology no. 12. Cambridge University Press, Cambridge, UK
276. Konishi M (2004) The role of auditory feedback in birdsong. Ann N Y Acad Sci 1016:463–475. https://doi.org/10.1196/annals.1298.010
277. Theunissen FE, Amin N, Shaevitz SS et al (2004) Song selectivity in the song system and in the auditory forebrain. Ann N Y Acad Sci 1016:222–245. https://doi.org/10.1196/annals.1298.023
278. Kasdan AV, Burgess AN, Pizzagalli F et al (2022) Identifying a brain network for musical rhythm: a functional neuroimaging meta-analysis and systematic review. Neurosci Biobehav Rev 136:104588. https://doi.org/10.1016/j.neubiorev.2022.104588
279. Lande R (1987) Genetic correlations between the sexes in the evolution of sexual dimorphism and mating preferences. In: Bradbury JW, Anderson MB (eds) Sexual selection: testing the alternatives. Wiley, pp 83–94
280. Bonduriansky R, Chenoweth SF (2009) Intralocus sexual conflict. Trends Ecol Evol 24:280–288. https://doi.org/10.1016/j.tree.2008.12.005
281. Poissant J, Wilson AJ, Coltman DW (2010) Sex-specific genetic variance and the evolution of sexual dimorphism: a systematic review of cross-sex genetic correlations. Evolution 64:97–107. https://doi.org/10.1111/j.1558-5646.2009.00793.x
282. Wesseldijk LW, Ullén F, Mosing MA (2023) Music and genetics. Neurosci Biobehav Rev 152:105302. https://doi.org/10.1016/j.neubiorev.2023.105302
283. Niarchou M, Gustavson DE, Sathirapongsasuti JF et al (2022) Genome-wide association study of musical beat synchronization demonstrates high polygenicity. Nat Hum Behav 6:1292–1309. https://doi.org/10.1038/s41562-022-01359-x
284. Clutton-Brock TH, Huchard E (2013) Social competition and selection in males and females. Philos Trans R Soc B 368:20130074. https://doi.org/10.1098/rstb.2013.0074

285. Soldati A (2023) The ontogeny of pant hoot vocalisations and social awareness in wild chimpanzees. Doctoral thesis,. University of Neuchâtel
286. Longhi E (2003) The temporal structure of mother–infant interactions in musical contexts. Doctoral thesis
287. Yu L, Todoriki K, Myowa M (2022) From spontaneous rhythmic engagement to joint drumming: a gradual development of flexible coordination at approximately 24 months of age. Front Psychol 13:907834. https://doi.org/10.3389/fpsyg.2022.907834
288. Carrer LRJ, Pompéia S, Miranda MC (2023) Sensorimotor synchronization with music and metronome in school-aged children. Psychol Music 51:523–540. https://doi.org/10.1177/03057356221100286
289. Hugardt A (2001) Individuality and development in children's spontaneous tempo and synchronization. PhD thesis. Gothenburg University
290. Burchardt LS, Hoeschele M, Honing H et al (2025) Chapter 1: what is rhythm and how can we study it? In: Sauciuc G-A, Lameira AR, Persson T (eds) Nature beats: the what, why and how of animal rhythmic behaviours. Springer
291. Fortune ES, Rodríguez C, Li D et al (2011) Neural mechanisms for the coordination of duet singing in wrens. Science 334:666–670. https://doi.org/10.1126/science.1209867

15 The Next Frontier in Research on Rhythm Cognition and Rhythmic Communication

Nature Beats, Lara S. Burchardt, Peter F. Cook, Chiara De Gregorio, Tecumseh W. Fitch, Marco Gamba, Michael D. Greenfield, Morgan L. Gustison, Yuko Hattori, Denise Herzing, Gerlinde Höbel, Marisa Hoeschele, Henkjan Honing, Adriano R. Lameira, Edward W. Large, Hugo Merchant, Bjorn Merker, Kazuo Okanoya, Nao Ota, Aniruddh Patel, Marcus Perlman, Tomas Persson, Tudor Popescu, Shannon Proksch, Teresa Raimondi, Andrea Ravignani, Stephan A. Reber, Colleen Reichmuth, Gabriela-Alina Sauciuc, and Jennifer N. Schneider

Abstract

In this concluding chapter, we focus on outlining the next frontier in research on rhythm cognition and rhythmic communication (RCC), synthesising conceptual, methodological, comparative, and applied directions, as they emerge from the

Disclaimer. This chapter does not represent a unified position or consensus among the contributing authors. Rather, it provides a brief overview of the current state of the field and outlines possible directions for future research. The field remains fragmented, with differing perspectives on what should be studied and how. The chapter is intended to be inclusive of the diverse frameworks, hypotheses, and viewpoints represented in this volume.

Nature Beats

L. S. Burchardt
Leibniz Zentrum für Allgemeine Sprachwissenschaften Berlin, Berlin, Germany

Institute for Theoretical Biology, Humboldt-Universität zu Berlin, Berlin, Germany

Donders Institute for Brain, Cognition, and Behaviour, Radboud University, Nijmegen, Netherlands
e-mail: lara.sophie.burchardt@hu-berlin.de

G.-A. Sauciuc et al. (eds.), *Nature Beats*, Fascinating Life Sciences,
https://doi.org/10.1007/978-3-032-24250-1_15

contributions to the *Nature Beats volume*. While consensus has grown regarding the composite nature of rhythmic competence and the need for integrated quantitative approaches, the field remains marked by conceptual heterogeneity and

P. F. Cook
Marine Mammal Science/Neuroscience, New College of Florida, Sarasota, FL, USA

Institute for Marine Sciences, University of California Santa Cruz, Santa Cruz, CA, USA
e-mail: pcook@ncf.edu

C. De Gregorio · A. R. Lameira
Department of Psychology, University of Warwick, Coventry, UK
e-mail: Chiara.De-Gregorio@warwick.ac.uk; Adriano.Lameira@warwick.ac.uk

T. W. Fitch
Department of Behavioural and Cognitive Biology, University of Vienna, Vienna, Austria
e-mail: tecumseh.fitch@univie.ac.at

M. Gamba
Department of Life Sciences and Systems Biology, University of Turin, Turin, Italy
e-mail: marco.gamba@unito.it

M. D. Greenfield
ENES Bioacoustics Research Lab, CRNL, University of Saint-Etienne, CNRS, Inserm, Saint-Etienne, France

Department of Ecology and Evolutionary Biology, University of Kansas, Lawrence, KS, USA
e-mail: michael.greenfield@univ-tours.fr

M. L. Gustison
Department of Psychology, Western University, London, ON, Canada
e-mail: mgustiso@uwo.ca

Y. Hattori
Institute for the Evolutionary Origins of Human Behavior (EHUB), Kyoto University, Inuyama, Japan

D. Herzing
Department of Biological Science, Florida Atlantic University, Boca Raton, FL, USA

G. Höbel
Behavioural and Molecular Ecology Group, Department of Biological Sciences, University of Wisconsin–Milwaukee, Milwaukee, WI, USA
e-mail: hoebel@uwm.edu

M. Hoeschele
Acoustics Research Institute, Austrian Academy of Sciences, Vienna, Austria
e-mail: marisa.hoeschele@univie.ac.at

H. Honing
Music Cognition Group, Institute for Logic, Language and Computation, University of Amsterdam, Amsterdam, The Netherlands
e-mail: H.J.Honing@uva.nl

E. W. Large
Department of Psychological Sciences, University of Connecticut, Storrs, CT, USA
e-mail: edward.large@uconn.edu

H. Merchant
Department of Neurophysiology, Instituto de Neurobiología, Universidad Nacional Autónoma de México, Campus Juriquilla, Santiago de Querétaro, Mexico
e-mail: hugomerchant@unam.mx

inconsistent operational criteria. To advance RCC research, the chapter identifies pressing needs for conceptual clarity, methodological standardisation, and ecologically grounded definitions of rhythmic phenomena across modalities and species. We also highlight future research trajectories for major theoretical

B. Merker
Independent Scholar, Kristianstad, Sweden

K. Okanoya
Advanced Comprehensive Research Organization (ACRO), Teikyo University, Tokyo, Japan

N. Ota
Museum of Nature and Human Activities, Hyogo, Sanda, Japan
e-mail: n-ota@hitohaku.jp

A. Patel
Department of Psychology, Tufts University, Medford, MA, USA

CIFAR Program in Brain Mind and Consciousness, MaRS, Toronto, ON, Canada
e-mail: a.patel@tufts.edu

M. Perlman
Department of Linguistics and Communication, University of Birmingham, Birmingham, UK
e-mail: m.perlman@bham.ac.uk

T. Persson · S. A. Reber · G.-A. Sauciuc (✉)
Department of Philosophy/Cognitive Science, Lund University, Lund, Sweden
e-mail: Tomas.Persson@lucs.lu.se; Stephan.Reber@lucs.lu.se;
Gabriela-Alina.Sauciuc@lucs.lu.se

T. Popescu
Department of Cognition, Emotion, and Methods in Psychology, University of Vienna, Vienna, Austria

S. Proksch
Department of Psychology, Augustana University, Sioux Falls, SD, USA

Program in Neuroscience, Center for Interdisciplinary Studies, Augustana University, Sioux Falls, SD, USA
e-mail: Shannon.Proksch@augie.edu

T. Raimondi
Department of Life Sciences and Systems Biology, University of Turin, Torino, Italy

Department of Human Neurosciences, Sapienza University of Rome, Rome, Italy
e-mail: teresa.raimondi@unito.it

A. Ravignani
Department of Human Neurosciences, Sapienza University of Rome, Rome, Italy

Center for Music in the Brain, Department of Clinical Medicine, Aarhus University, Aarhus, Denmark

Istituto di Scienze e Tecnologie della Cognizione, CNR, Rome, Italy
e-mail: andrea.ravignani@uniroma1.it

C. Reichmuth
Institute for Marine Sciences, University of California Santa Cruz, Santa Cruz, CA, USA
e-mail: coll@ucsc.edu

J. N. Schneider
Department of Psychology, LCC International University, Klaipeda, Lithuania
e-mail: jschneider@lcc.lt

frameworks, including the Vocal Learning–Beat Perception and Synchronisation hypothesis, Gradual Audiomotor Evolution hypothesis, Neural Resonance Theory, socio-ecological hypotheses of synchrony, and evolutionary accounts linking rhythm to social bonding and stress regulation. A bottom-up, species-specific, and comparative approach is advocated to capture the full structural and functional diversity of RCC and to evaluate mechanisms spanning neural oscillations and functional connectivity, biomechanics, development, reward systems, and environmental constraints. Finally, the chapter emphasises the translational value of RCC research. Rhythm-based interventions hold promise for educational, clinical, welfare, conservation, and human-animal interaction contexts. By integrating mechanistic, evolutionary, and applied perspectives, future research can uncover the biological principles underlying rhythmic behaviour and its central role in communication, cognition, and sociality across the animal kingdom.

15.1 Introduction

Rhythm cognition and rhythmic communication (RCC) have become an effervescent area of research. Research on these topics has gained significant momentum as a way of tracking the evolutionary roots of human musicality, following the discovery that some nonhuman mammals and birds are capable of synchronising movements with an external rhythm. Recently, however, research on RCC has branched out to also include other aspects of RCC, besides rhythmic synchronisation.

The aim of this book was to capture the full span of this expanding area of research, starting with its conceptual, terminological, and methodological debates (Chap. 1 [1]), and continuing with forms of RCC in humans and the rest of the animal kingdom (Chaps. 2 [2], 3 [3], 4 [4], 5 [5], 6 [6], 10 [7], 12 [8], 13 [9], Chap. 14 [10] and peer commentaries therein), as well as underlying and modulating mechanisms (Chaps. 1 [1], 2 [2], 5 [5], 6 [6], 7 [11], 8 [12], 9 [13], 10 [7], 11 [14], 13 [9], and 14 [10]). On the backdrop of this broad thematic span, we sought to emphasise both the intrinsic and the translational value of comparative research, while also highlighting a comprehensive need for both meticulous analyses with a narrow focus and increasingly integrative approaches.

We dedicate this last chapter to summarising current gaps in RCC research to set a—hopefully collaborative—research agenda for the next two decades, starting with demands for conceptual and methodological development (Sect. 15.2). We then refine and further expand on this agenda in sections that contemplate the next frontier in RCC for the theories represented in the book (Sect. 15.3), as well as in terms of exploratory (Sect. 15.4) and applied work (Sect. 15.5).

15.2 Conceptual and Methodological Development

Currently, there is a broad consensus regarding the composite nature of rhythmic competence, where RCC can be studied from multiple perspectives reflecting increasingly complex rhythmic abilities, from the physical level of the signal to music and group interactions (Chap. 1 [1]). Even within each of these perspectives, the complexity of the analysed phenomenon can vary. For example, from the physical signal perspective, structural complexity can vary from metronomic isochrony to hierarchically stratified rhythms. On the methodological front, there is emerging consensus that studies of RCC should report durational, density, and variance measures and that RCC requires combined approaches integrating local and global metrics.

Beyond such points of convergence, however, there is a comprehensive need to work towards increased conceptual consistency, methodological development, and standardisation. Conceptual stances diverge even for well-researched topics such as isochrony or synchronisation, where we lack accepted operational criteria such as thresholds for minimal sequence duration or maximal variance.

This ambiguity is also reflected in debates regarding adequate quantification tools or the very topics that should constitute the field and be worthy of scientific investment. In a human-centric perspective, studies of nonhuman RCC are framed as having scientific value as a way of gaining insight into the evolution of human musicality and/or speech. Within this perspective, however, opinions vary as to whether only sustained rhythmic synchronisation (Chaps. 5 [5] and 14 [10]) is relevant to this goal or RCC more broadly, i.e. also encompassing the production and perception of solo rhythms (Chaps. 3 [3], 4 [4], 6 [6], 7 [11], 8 [12], 9 [13], 10 [7], 11 [14], 12 [8], and 13 [9]). An additional view that is increasingly gaining support emphasises the intrinsic value of investigating the temporal organisation of animal communication and behaviour regardless of what we can learn about the evolution of musicality (see also Sect. 15.5). Ultimately, these perspectives form a continuum, with findings from humans guiding research towards new insight about RCC in other species and findings from nonhuman animals inspiring new directions of research in humans.

On a conceptual level, there is a need to acknowledge—and indeed discover—additional components of rhythmic competence beyond commonly studied notions such as beat perception and motor entrainment. For example, rhythm cognition also involves interval estimation, tempo perception, pattern recognition, statistical learning, chunking, categorical perception, a preference for integer ratios and more (a point to which we return in Sect. 15.4.1). Rhythmic complexity itself remains poorly conceptualised, and current data highlight both the promise and the gaps in current understanding (Sect. 15.4.1).

Methodologically, there is a growing need to refine experimental and observational paradigms to better study rhythm perception, including its automatic versus intentional dimensions and potential biological significance across species. This requires accounting for neuromuscular constraints and socio-cognitive influences to avoid confounds in cross-species comparisons (as detailed in Sects. 15.3 and 15.4).

15.3 The Next Frontier for Current RCC Theories

Having identified the key coordinates of conceptual and methodological needs (Sect. 15.2), we turn to the Next Frontier for the key theories and hypotheses that have been represented in the contributions to this book, focusing on empirical research designed to test and further develop them. Rather than revisiting these frameworks in detail (for which we refer to the relevant chapters), our goal here is to suggest concrete avenues for future studies that can help validate or expand current theoretical proposals.

15.3.1 The Vocal Learning-Beat Perception and Synchronisation Hypothesis

One of the most influential frameworks in the study of RCC is Aniruddh Patel's Vocal Learning Hypothesis, which posits a close connection between *advanced* vocal learning (VL) capacities and beat perception and synchronisation (BPS) (for reviews and critical discussions, see Chaps. 7 [11] and 14 [10]). According to the VL-BPS hypothesis, the strong auditory-motor connections required for VL also support the neural architecture necessary for BPS. The dorsal auditory pathways are central to this framework, with VL-BPS emphasising their role in auditory-motor integration.

As an adaptation, vocal learning is likely reinforced by reward mechanisms. Thus, a corollary of VL-BPS is that vocal learning species exhibit enhanced motivation to attend to complex auditory structure and spontaneously synchronise with external rhythms (Chap. 7 [11]). The VL-BPS hypothesis has inspired a wave of comparative studies in both vocal learning species—such as parrots, songbirds, and seals—and vocal non-learning species, including sea lions and primates (Chaps. 3 [3], 7 [11], and 14 [10]), in particular, the peer commentaries by Cook and Reichmuth (Sect. 14.2.1) and Okanoya (Sect. 14.2.4).

The next frontier for research driven by the VL-BPS hypothesis spans several objectives. In humans, it involves deepening our understanding of the cognitive and neural relationships between beat-based rhythmic processing and aspects of language in both adult and developmental populations. In nonhuman species, the next frontier involves continued exploration of the neural mechanisms of temporal prediction in beat-based auditory rhythms and investigating how these mechanisms compare to those used by humans. A particularly intriguing question is whether any animals besides parrots can spontaneously develop predictive and tempo-flexible synchronisation to musical beats through social interaction with humans, without formal training. Horses, for example, could be a promising candidate species for such investigations. Finally, another important future direction involves understanding the developmental and neurobiological factors that explain why some pet parrots develop dancing behaviour in response to musical rhythms while others do not.

15.3.2 The Gradual Audiomotor Evolution Hypothesis

The Gradual Audiomotor Evolution (GAE) hypothesis proposes that the neural and behavioural capacities for beat perception and synchronisation (BPS) evolved incrementally within primates rather than emerging suddenly in humans (Chap. 7 [11]). GAE views human rhythmic and musical behaviour as the outcome of a stepwise strengthening of audiomotor connectivity—linking auditory cortices, motor planning circuits, and subcortical timing structures. This contrasts with the VL-BPS hypothesis (Sect. 15.3.1), which ties rhythmic synchronisation more tightly to the evolution of complex vocal learning. Although early studies suggested that only vocal-learning species could synchronise predictively to a beat, recent behavioural and neural evidence has reshaped this view.

Rhesus macaques (*Macaca mulatta*), for example, can synchronise predictively to auditory and visual metronomes when motivated by reward, showing tempo flexibility, error correction, and continuation tapping (Chap. 7 [11]). Neural recordings reveal cyclical population dynamics in the medial premotor cortex that encode tempo and beat timing—evidence for an internal rhythmic clock. These findings suggest that primate brains already possess the computational substrate for beat-based timing, even if rhythmic behaviour is not intrinsically rewarding as it is for humans.

Recent extensions of GAE, such as the BPS Components Continuum (Chap. 7 [11]), frame rhythmic ability as a graded spectrum from basic interval timing to predictive beat entrainment. Future work should test this continuum using comparative behavioural paradigms and multimodal neuroimaging to quantify how cortical–subcortical connectivity, particularly within the medial cortical–basal ganglia–thalamo (mCBGT) circuit, scales with rhythmic proficiency. Another key question concerns the evolution of intrinsic reward: while macaques rely on external reinforcement, humans find synchronisation pleasurable, perhaps due to dopaminergic coupling between motor and auditory systems.

Expanding research to lemurs, gibbons, and chimpanzees—species showing proto-rhythmic behaviours such as duetting, drumming, swaying, etc.—will further test GAE. Integrating field data, rhythm analytics, and cross-species training will help link neural circuitry, motivation, and ecology in the evolution of rhythmic behaviour.

15.3.3 The Neural Resonance Hypothesis

The Neural Resonance Theory (NRT) proposed by Ed Large (Chap. 8 [12]) posits that rhythmic perception and synchronisation emerge from oscillatory neural dynamics that may be conserved across species. According to NRT, intrinsic neural oscillations entrain to external rhythmic stimuli, forming the foundation for beat perception and temporal prediction.

A central challenge for future NRT research is to uncover the evolutionary origins of these resonance mechanisms through cross-species, comparative work. Key milestones include determining whether nonhuman species capable of rhythmic synchronisation can also perceive complex rhythmic structures such as pulse and meter, including "missing pulse" rhythm. To address these questions, experimental tasks involving tempo perturbations and syncopated rhythms will be essential. Additionally, invasive neural recordings in animal models, combined with dynamical systems modelling, will permit identifying conserved oscillatory mechanisms.

Developmental studies comparing human infants and trained animals may illuminate how rhythmic attunement emerges and unfolds over time. Understanding how intrinsic frequencies adapt and stabilise through Hebbian learning could clarify the role of plasticity in rhythm acquisition. Moreover, examining integer-ratio preferences in human development and across cultures and by comparison with animal vocalisations may reveal evolutionary constraints on rhythm production.

Taken together, cross-species and developmental comparisons of rhythmic learning and synchronisation driven by NRT could help distinguish between convergent evolution and shared ancestry. Ultimately, integrating behavioural, neural, and computational approaches will be essential to trace the evolutionary trajectory of musical rhythm cognition.

15.3.4 Rhythmic Synchronisation as a Beacon Display Evolved in the Context of Male Territoriality and Female Exogamy

In a provocative hypothesis (Chap. 14 [10] but see also Chap [5]) based on striking examples of precise inter-individual group synchrony in arthropods and anurans, Björn Merker has conjectured that human rhythmic synchronisation evolved in ancestral socio-ecological conditions characterised by male territoriality and female exogamy. Analogous to the socio-ecology of clumped males and migrating females that is the setting for precise synchrony in the animal examples, such circumstances create a selection pressure for conspicuous, long-range signalling aimed at attracting dispersing females. In such a setting, rhythmic synchronisation of, say, sound signals would amplify their environmental reach by superposition of their discrete elements, allowing the signal to reach more female ears compared to unsynchronised signalling.

Looking ahead to research challenges posed by this hypothesis, three principal issues may prove fruitful. Not all species that feature clumped males and migrating females exhibit group synchrony in their signalling. That is, in fact, what is expected if signal synchrony happens to be a factor in a speciation event: the non-synchronising subpopulation remains behind a reproductive isolating barrier relative to the synchronising subpopulation, and the two subpopulations split into separate evolutionary lineages. Since both humans and chimpanzees exhibit male territoriality and female exogamy, while only humans synchronise their signalling, Merker has suggested that signal synchrony played a role in the human-chimpanzee split. Close scrutiny of related species that share similar ecological circumstances but differ in signal synchrony may disclose the more particular factors at play in such splits.

A further New Frontier issue concerns the fact that in arthropods and anurans, only males engage in synchronous chorusing, while in humans, both sexes are enthusiastic synchronisers (as noted by the peer commentary of Popescu and Fitch in Chap. 14 [10]). As suggested in Merker's response, genome-based studies that target components of human musicality may eventually be able to tell us whether genetic constraints on the human capacity for inter-individual synchrony promote female synchrony as a correlated genetic response to selection on males. And finally, do bonobos in the wild exhibit the vocal behaviour which has been named "staccato hooting" in captive bonobos (as described in Chap. 14 [10])? It is said to exhibit inter-individual synchrony and needs to be confirmed as more than a chance occurrence. Should it occur with synchrony in the wild, this might mean that the conjectural scenario for a human–chimpanzee split was repeated in a subsequent bonobo–chimpanzee split.

15.3.5 The Social Bonding Hypothesis of Music Evolution

The social bonding hypothesis proposes that music evolved primarily to strengthen social cohesion. Among others, this proposal is based on data showing that joint music-making—even just joint listening—promotes group unity, effectively acting as a powerful social glue. For instance, group singing has been shown to promote interpersonal closeness, cooperation, and a rapid sense of group identity (as discussed in Chap. 11 [14]). One particularly important mechanism enabling this effect is rhythm. By getting people to anticipate one another's actions and synchronising them in time, shared rhythm and coordinated movement are also in a position to align emotional states, in turn enhancing trust and affiliation within the group (Chap. 11 [14]).

To further elucidate the validity of the social bonding hypothesis of music evolution, future research should continue on its way to a more fine-grained comparative approach. Such data has indeed already provided a rich source of insights that have led to the initial formulation of this hypothesis. In particular, more systematic work is needed on nonhuman primates and other (vertebrate and nonvertebrate) species that engage in coordinated vocal behaviour, such as duetting and chorusing (Chaps. 3 [3], 5 [5], 10 [7], 12 [8], and 14 [10]). These natural phenomena offer a unique opportunity to test whether rhythmic coordination serves similar affiliative functions outside our species. Empirical efforts should prioritise field-based observations complemented by controlled experiments to evaluate how rhythmic synchrony influences cooperation, tolerance, and behavioural alignment among individuals. Developing quantitative approaches to examine the relationship between rhythmic regularity and behavioural coordination in animal communication will be key to understanding how the temporal structure of calls supports social functions.

Ultimately, investigating the role of rhythm in nonhuman vocal interactions may clarify whether the mechanisms that support musical bonding in humans are unique evolutionary innovations or reflect continuities in primate social behaviour, or even more deeply rooted traces of convergent evolution.

15.3.6 The Stress Relief and Social Bonding Hypothesis of Dance Evolution

Tandem-walking or "conga-line" behaviour in traumatised great apes—first documented in chimpanzees and later confirmed independently in multiple sanctuary populations (Lameira, personal observations)—appears to function as a rhythmic co-stereotypy that stabilises dysregulated autonomic systems. Through synchronous, full-body entrainment, individuals may achieve stress relief and regain the social buffering lost through maternal or early-life deprivation (discussed in Chap. 12 [8] and Chap. 13 [9]). In this form, tandem-walking offers a plausible precursory mechanism for the stress-modulating and bond-reinforcing functions that many theories attribute to early human dance.

Building from this hypothesis, a central priority for future research is to expand the empirical record across species and populations. Although the initial evidence emerged from a single chimpanzee dyad, subsequent reports from American zoos and African sanctuaries, suggest that the phenomenon arises repeatedly and convergently. Systematic data mining of online video repositories (e.g. YouTube) could identify additional occurrences in chimpanzees, gorillas, bonobos, and orangutans. Ideally, these efforts should be complemented by direct, high-frame-rate recordings in sanctuaries and rehabilitation centres where orphaned or socially deprived apes are housed. Such work would allow the collection of standardised data on bout duration, tempo, phase-locking accuracy, initiation cues, tactile dynamics, and social context, ultimately clarifying whether tandem-walking reflects a generalised great-ape stress-coping strategy or a phenomenon tied to specific ontogenetic conditions.

Current advances in AI-based pose-estimation technology make this expansion especially timely. Next-generation primate-specific models—such as OpenMonkeyStudio, PrimatE-Pose, LeapTrack, or dedicated DeepLabCut models fine-tuned for bipedal synchrony—can now extract highly precise pose and gait metrics even from low-quality footage. Transformer-based pose–time-series architectures can detect micro-lags, leader–follower reversals, and continuous mutual adjustments between partners. Likewise, graph neural network motion encoders are well suited to modelling synchrony as a multi-body coupling system rather than as two independent trajectories. When combined with 3D reconstructions (or monocular depth inference), these tools can quantify interpersonal distance regulation, hand placement, and the geometry of tactile coupling, thus revealing the underlying sensorimotor mechanisms that maintain synchrony.

Future research should also integrate behavioural analyses with physiological, affective, and social data. Non-invasive endocrine sampling or monitoring (e.g. using cine-infrared-thermography) could test whether tandem-walking aligns with known biological markers of stress relief and social bonding. Where possible, lightweight sensors or video-based techniques (e.g. remote photoplethysmography) could capture heart rate and HRV changes during synchrony, while sociometric mapping could assess whether tandem-walking alters affiliation networks or reduces aggression. Together, these measures would help determine whether the behaviour primarily functions for autonomic regulation, social cohesion, or both.

To move beyond correlation, carefully designed semi-experimental paradigms will be needed. Mild environmental manipulations—such as temporary separation, the introduction of new individuals, or controlled interruptions of ongoing bouts—could reveal the triggers and reinstatement dynamics of tandem-walking. Rhythmic versus arrhythmic auditory cues could test whether external beats influence synchrony stability. Observations of whether apes spontaneously resume phase-locked motion after pauses would clarify whether they maintain an internal representation of a joint rhythm. Comparisons between strongly bonded and weakly bonded dyads, or across different ages and developmental backgrounds, may illuminate how social and affective factors shape the behaviour.

Finally, comparative work across species remains essential. If tandem-walking consistently emerges in chimpanzees, gorillas, bonobos, and possibly orangutans under similar developmental conditions—especially in populations of orphans or socially disrupted individuals—this would point to a shared hominid predisposition for using rhythmic synchrony as a means of stress regulation and affiliative repair. Such evidence would strengthen the argument that early human dance could have originated as a small-scale, dyadic mechanism for co-regulation, later elaborated culturally into collective ritual and aesthetic expression. Expanding the search for this behaviour to new populations, both through digital repositories and direct fieldwork, will be critical for understanding how rhythmic synchrony functions in great apes today and how it may have shaped the earliest stages of dance evolution in our own lineage.

15.4 Exploring Evolutionary Trends with Bottom-Up Research

Scientific interest in rhythm as a dimension of animal communication has been mainly driven by human-centric models, using musicality and speech as benchmarks. Such top-down approaches may obscure the broader evolutionary and biological significance of RCC, which, as demonstrated by this book, is widespread across the animal kingdom. In light of the research presented in previous chapters, there is a pressing need to embrace bottom-up approaches that prioritise species-specific multidisciplinary investigations. Such research could account for ecological, bio-cognitive, and developmental constraints, offering a more nuanced and inclusive framework for rhythm science. This reframing also challenges the notion of rhythm as primarily a cognitive and/or cultural construct, instead positioning it as a biological phenomenon deeply embedded in the neuromechanical and ecological realities of diverse species, humans included.

By focusing on how RCC is shaped by organism–environment interactions, bottom-up research can uncover the evolutionary logic behind rhythmic behaviours and their communicative functions. It can also help identify which traits—e.g. vocal learning, signal complexity, audiomotor coupling, group living, imitation, bipedal locomotion, and more—predict which rhythmic capacities. Moreover, bottom-up research is essential to address the field's empirical fragmentation, conceptual heterogeneity, and lack of methodological standardisation. To systematically explore the milestones of this new frontier, we examine research gaps through the lens of the book's guiding questions: *What* should be studied as part of RCC research? *How* is RCC generated and modulated? *Why* is rhythm pervasive in communication?

15.4.1 The WHAT Question

To be able to uncover evolutionary trends in RCC, the scope of inquiry has to be broadened in terms of *species, modalities* of production and perception, aspects of *rhythmic structure*, and the range of *rhythmic abilities*. Regarding species, an obvious focus of expansion is the primate lineage, given its shared ancestry with humans (Chaps. 3 [3], 7 [11], 10 [7], 12 [8], and 14 [10], in particular Sects. 14.2.3 and 14.2.5). This will enable a *homology*-based perspective on rhythmic communication broadly, and on the evolution of human speech, music, and dance in particular. The (socio)-cognitive complexity of primates, along with neuroanatomical and functional homologies relevant to rhythm perception and motor control, further strengthen the case for a systematic investigation of RCC in primates.

Beyond our closest relatives, this expansion could follow a taxonomic approach: building outward from species that have been heavily studied—e.g. laboratory model organisms—to gradually incorporate more species within the same order, and eventually more orders within the same family (Chaps. 3 [3] and 14 [10]). This strategy would support comparative rigor while avoiding overgeneralisations and would help establish a broader, ecologically valid understanding of RCC across taxa.

It is important to note that some of these model species (e.g. insects, anurans, zebra finches, budgerigars, bats, singing mice) have attracted the interest of RCC researchers due to exhibiting RCC-relevant traits, such as vocal learning, the capacity to produce complex vocalisations, rhythmic synchronisation, etc. (Chaps. 3 [3], 5 [5], 7 [11], 9 [13], 10 [7], and 14 [10]). Thus, for such cases, a *convergent evolution* perspective may offer insights that complement (or even surpass) those gained from homology. In other words, the RCC traits that these taxa share with humans may reflect shared mechanistic affordances and/or socio-ecological selective pressures that in our species have favoured the evolutionary emergence of musicality (including dance) and speech. This is one reason why songbirds have been extensively studied, but there is a strong case for expanding research to other groups that share relevant abilities yet remain underexplored.

Additionally, rhythm must be studied as a multimodal phenomenon. Acoustic rhythms dominate current research (Chap. 3 [3]). Yet, visual, tactile, and vibrational modalities may be equally widespread and important (Chaps. 2 [2], 3 [3], 4 [4], 5 [5], 6 [6], 11 [14], 12 [8], and 13 [9]), especially in species with limited auditory capabilities (Chap. 3 [3]) or in environments where sound transmission is compromised (Chap. 6 [6]). These alternative rhythmic channels represent an area of study that is not only empirically underrepresented but also methodologically underdeveloped. A key target for future research is to investigate the interplay of different modalities in RCC, and how this interplay contributes to communicative effectiveness (Chaps. 2 [2], 3 [3], and 9 [13]).

With respect to rhythmic structure and rhythmic abilities, isochrony production and synchronisation have dominated both human and comparative research, while isochrony perception and beat perception have been mainly studied in humans. Hence, most conceptual and methodological developments have been tailored to these topics, although even in this case, the progress of comparative research is hampered by a lack of consensus (discussed in Sect. 15.2).

An important aspect in these debates is the need for defining and operationalising rhythmic phenomena from a functional—rather than formal—perspective. Such a shift is essential to the reliability of comparative research, in particular to the comparison of rhythmic signals that unfold at considerably different timescales across species. For example, fruit flies and many fish species produce acoustic signals that are tens of times faster than human music, dance, or speech (Chaps. 3 [3] and 9 [13]). Even the rhythmic signals of other mammalian species (e.g. singing mice, bats, canids, felids, fossorial rodents, and insectivores) can be several times faster than those of humans (Chaps. 3 [3], 4 [4], and 9 [13]). Conversely, perceptual capacities in these taxa are optimised for species-relevant rhythmic properties (Chaps. 3 [3] and 9 [13]). These examples highlight the inadequacy of applying human-centric operational criteria - i.e. derived from human research and, thus, constrained by human-specific neurophysiology and anatomy - and the need for developing ecologically valid measures (Chaps. 9 [13] and 14 [10], in particular, the commentary of Cook and Reichmuth in Sect. 14.2.1). In turn, this exposes the need for research aimed at evaluating RCC in terms of flexibility, motivation, and naturalistic timing ranges in order to permit a more accurate and inclusive understanding of rhythmic abilities across species.

Besides addressing these inconsistencies, another milestone for future research is to develop conceptual and methodological approaches for studying non-isochronous and complex rhythms in a comparative perspective, including turn-taking and antiphonal calling, that is, a phase-shifted synchronisation (as detailed in the commentaries of Okanoya in Chap. 14, Sect. 14.2.4 and Popescu and Fitch in Sect. 14.2.6). To advance this frontier, inspiration can be drawn from musicological and speech analysis, and from recent efforts to deal with this topic in nonhuman communication (Chaps. 3 [3], 9 [13], 10 [7], and 12 [8]). Given the vast diversity of rhythmic signals and our limited understanding regarding their possible patterns, exploratory, bottom-up research—inspired by the actual signals—will be needed for methodological development.

Research on the developmental precursors of human rhythmic synchronisation offers valuable insights (Chaps. 10 [7] and 14 [10], in particular, the peer commentaries of Cook and Reichmuth and Patel), providing a more adequate framework for interpreting findings in nonhuman species. From this perspective, future research should address forms of rhythmic engagement and temporal alignment, along with other insights provided by data on the maturational curves of rhythm perception and spontaneous rhythm production in human infants (Chaps. 10 [7] and 14 [10], in particular, Sects. 14.2.1 and 14.2.5).

On the cognitive front, the next research frontier lies in developing a holistic understanding of rhythm cognition, extending beyond commonly studied aspects like predictive timing and beat perception. Rhythm cognition also encompasses temporal interval estimation, tempo perception, pattern recognition, sequence learning, chunking, categorical perception, preference for integer ratios, and recursion, among others (Chaps. 7 [11], 8 [12], 9 [13], 11 [14], 12 [8], and 14 [10], in particular Sects. 14.2.4 and 14.2.6). In animals, even the more familiar components of rhythm cognition remain minimally explored, highlighting a clear need for empirical expansion.

15.4.2 The HOW Question: Mechanisms and Constraints

The chapters of this book reveal a myriad of pieces in the complex mechanistic puzzle that underlies RCC, yet only a few are captured by current hypotheses (Sect. 15.3). Indeed, research on the role these play in RCC is still in an early phase, highlighting a comprehensive need for scientific exploration. While our understanding of the neurobiology and biomechanics of non-communicative rhythms (e.g. locomotion, breathing) has advanced considerably over the past decades, progress in the study of RCC has been limited. In this domain, thus, there is a pressing need to deepen our understanding of how low- and high-level mechanisms—from the muscles to motoneurons to central pattern generators to subcortico-cortical circuits—contribute to the production of communicative rhythms and how they enable or constrain rhythm perception (Chaps. 7 [11], 8 [12], 9 [13], 11 [14], 13 [9], and 14 [10], in particular Sects. 14.2.3, 14.2.4, and 14.2.5).

Given the overlapping substrates of communicative and non-communicative rhythms (Chaps. 2 [2] and 9 [13]), a first step in this direction could be to compare these two categories of rhythmic output to identify similarities and differences. Another promising research avenue is to examine the coupling and decoupling between physiological (breathing, locomotion) and communicative rhythms (percussion, stridulation, vocalisation, etc., Chap. 9 [13]). However, such advances will not be possible without detailed examination of the individual components, and a deeper understanding of how variations in circuitry or organ function—across species or between different organs within the same species—constrain RCC (Chap. 9 [13]).

Taken together, the data generated from these research directions will deepen our understanding of species-specific optimisation across the mechanisms that underpin RCC. This, in turn, will enable better-tailored experimental tasks and more accurate interpretations of cross-species performance differences, clarifying whether these stem from cognitive or other mechanistic factors (as argued in Chap. 9 [13] and the commentary of Cook and Reichmuth in Chap. 14 [10]). This will allow the field to grapple with the diversity of motor behaviours used in entrainment studies. The intrinsic frequency and volitional control of different effectors—from head bobs to foot taps—vary across species and must be accounted for in experimental design. Additionally, such mechanistic data will provide empirical foundations for advancing our understanding of ritualisation, a process widely regarded as a driving force in the evolution of communication.

Understanding RCC in a comparative framework also requires close attention to its developmental trajectories, both in humans and in nonhuman species. Human data show that rhythm cognition and rhythmic communication develop at the cross-roads between bio-cognitive abilities that follow certain maturational curves and environmental influences. Different components of RCC follow distinct ontogenetic trajectories, with early-emerging abilities scaffolding later ones. These insights could inform research with nonhuman species, where developmental studies remain scarce (for some exceptions, see Chaps. 3 [3] and 10 [7]).

While existing data demonstrate the plasticity of rhythmic abilities, research on the socio-ecological scaffolding of RCC during development remains limited—both in humans and nonhuman species. Socio-cognitive modulators, such as social motivation for synchronised behaviour, appear to play a crucial role (Chaps. 7 [11], 11 [14], and 14 [10], in particular, the peer commentary of Cook and Reichmuth), yet they are still underexplored. Understanding how social rewards influence rhythmic learning, how caregivers scaffold rhythmic development, and how early exposure shapes rhythmic competence is essential for tracing the ontogeny of rhythm. Expanding research to include non-Western musical stimuli and contexts, as well as ethological data from diverse species, will further contextualise rhythm within broader cultural and ecological frameworks.

Overall, there is a pressing need for longitudinal studies to track how RCC components emerge, stabilise, and diversify over time. Such studies can identify critical periods for rhythmic learning and clarify how social contexts influence development—insights essential for reconstructing the evolutionary pathways of RCC and for refining our methods.

Finally, another underdeveloped but promising area for understanding evolutionary trends in RCC is the role of environmental constraints. As discussed in Chap. 6, factors such as habitat acoustics, population density, and substrate properties can shape the structure and efficacy of rhythmic signals. Over evolutionary timescales, these constraints may drive signal divergence and even speciation. Additionally, anthropogenic changes are altering the ecological contexts in which RCC occurs, impacting signal transmission, timing, and reception. These disruptions can have serious consequences for reproductive success and species survival, making rhythm increasingly relevant to conservation biology (see Chap. 6 [6] and Sect. 15.5.3). Despite its significance, research on this broad set of mechanisms remains limited and fragmented.

15.4.3 The WHY Question

While more traditional functional hypotheses (discussed in Sect. 15.3) focus on rhythmic synchronisation, this volume expands the scope of RCC, offering several additional hypotheses. First, rhythmic patterning—through its inherent repetitiveness—may provide a robust structural solution to communicative challenges across diverse ecological contexts (Chaps. 3 [3] and 6 [6]). Additionally, the flexibility of rhythmic patterning may also support speciation by catering to demands for species recognition, as well as individual recognition (Chap. 3 [3]). Moreover, because RCC is tightly constrained by neuromuscular substrates (Chaps. 2 [2] and 9 [13]), rhythmic patterning can convey individual fitness. Differences in rhythmic stability, pattern accuracy, and frequency rate may reflect motor precision, stamina, and cognitive control, which would align with theories of sexual selection and social signalling (Chaps. 3 [3] and 9 [13]). The intimate link between rhythm and emotion (Chap. 11 [14]) further suggests the role of RCC in managing social interactions, by conveying arousal and valence and influencing emotional states in both conspecifics and allo-specifics.

To test some of these functional hypotheses about RCC, research must address within-species and within-individual variability in rhythmic competence. Systematic studies of rhythmic stability, pattern accuracy, and perceptual acuity could uncover neurobiological, developmental, social, and ecological influences—thereby illuminating the selective pressures that have shaped rhythm as an adaptive communicative strategy. Complementary studies could assess signal transmission in noisy environments or evaluate cross-habitat resilience of rhythmic versus arrhythmic sequences.

However, to validate functional hypotheses about RCC, perception studies are a necessary complement to the meticulous analyses mentioned above. This remains a major gap in RCC, and we still know very little about the biological significance of the pervasive rhythmicity that patterns animal communication. It is, for example, unclear whether animals use rhythmic patterning, rather than other cues, for conspecific recognition, individual identification, or when reacting to signals in context-adaptive ways. Without an ability for perceiving rhythm, the rhythmic sequences produced by individuals may not necessarily be meaningful (Chaps. 1 [1], 3 [3], 9 [13], and 14 [10]). They may simply reflect mechanistic constraints rather than functional communicative roles. Clarifying this distinction is crucial for understanding the evolutionary and biological relevance of rhythmicity in animal signalling.

For complex rhythms with hierarchical organisation, it is important to assess whether animals perceive these stratified structures or focus on specific levels. A related question is whether particular levels—or the entire hierarchical arrangement—convey biologically significant information (Chap. 3 [3]). Understanding which aspects of rhythm animals attend to, and how these influence behaviour, is essential.

Historically, it has been difficult to demonstrate that nonhuman species—even those with advanced vocal abilities like songbirds—attend to rhythm rather than single intervals. Recent studies (Chaps. 1 [1], 7 [11], 9 [13], and 14 [10], in particular, the commentaries of Cook and Reichmuth in Sect. 14.2.1, Patel in Sect. 14.2.5, and Okanoya in Sect. 14.2.4) have begun to provide positive evidence, but this gap must be narrowed to validate functional hypotheses about the evolution of rhythmic communication. To increase the efficacy of experimental tasks, the field will have to acknowledge—and capitalise on—the species-specific mechanistic peculiarities that shape RCC at all levels of the neuroethological architecture (see Chap. 9 [13]) and the peer commentary of Cook and Reichmuth in Chap. 14 [10] and Sect. 15.4.2.

15.5 The Applied Relevance of RCC Research in Humans and Nonhuman Species

15.5.1 The Relevance of RCC to Clinical and Educational Contexts in Humans

Rhythm-based interventions hold significant promise across clinical and educational contexts throughout the human lifespan. The mechanisms underlying rhythmic communication and coordination (RCC) are closely linked to a wide range of cognitive and socio-emotional processes, including attention, working memory, impulse control, temporal sequencing, arousal regulation, and social bonding (Chaps. 7 [11], 8 [12], 9 [13], and 11 [14]). In educational settings, knowledge about the role of rhythm in these processes can be exploited to inform curriculum design, teaching methods and after-school activities. Preliminary work substantiates this applied potential, showing that music-based interventions improve reading fluency, executive function, classroom engagement, and cooperation.

In early development, rhythm plays a foundational role. One of the factors that puts prematurely born infants at risk for developmental delays and deficits regarding sensorimotor, respiratory, and cognitive capacities is the abrupt interruption of the multi-layered rhythmic stimulation experienced in the womb (Chap. 10 [7]). At later developmental stages, deficits of rhythm perception and motor control have been linked to disorders that affect (socio-) cognitive abilities implicated in communication, cooperation, executive control, and emotion regulation. Virtually, all the major developmental disorders are in this list: dyslexia, stuttering, developmental coordination disorder, self-regulation disorder, ADHD, and autism (Chap. 10 [7]).

Rhythm-based interventions capitalising on these insights have recently begun to emerge. For example, interventions leveraging rhythmic stimulation in the tactile, vestibular, and auditory modalities—stroking, carrying, water mattresses, rockers, maternal heartbeat, and music—have been devised and implemented in the preterm neonatal care with positive outcomes regarding emotion regulation, motor and respiratory development (Chap. 10 [7]).

In older children, different forms of music-based interventions have been found to enhance communicative skills, emotional regulation, and executive control across several disorders. For children with developmental language problems, the efficiency of music-based interventions could be anchored in the link between vocal learning and rhythmic competence theorised by the VL-BPS (see Chap. 7 [11] and the commentary of Patel in Chap. 14 [10]), the timing mechanisms emphasised by GAE (Chap. 7 [11]) and/or the neural resonance mechanisms posited by NRT (Chap. 8 [12]). If so, designing further interventions inspired by these frameworks could be beneficial. In this domain, rhythm-based approaches may improve verbal abilities by enhancing audiomotor integration, the brain's sensitivity to temporal patterns, and phonological awareness.

For children with autism spectrum diagnoses, rhythm-based interventions could improve social and communicative skills. Rhythmic movement provides predictable sensory input, which can help regulate the nervous system by reducing sensory overload. This is known to be an issue in these populations, which pushes individuals towards avoidance, withdrawal, and stimming behaviours, which are essentially rhythmic. Hence, dance routines, which involve repetition and structure, are comforting and engaging for autistic individuals.

In ADHD conditions, rhythm-based procedures have been explored both as diagnostic and therapeutic tools. ADHD is characterised by underactive dopaminergic (DA) pathways, which affect timing, working memory, and sequencing capacities. Thus, in line with mechanisms of rhythm genesis and rhythm tracking discussed in (Chaps. 7 [11], 8 [12], 9 [13], and 11 [14]), rhythm-based interventions have the potential to ameliorate ADHD deficits through DA upregulation. In turn, this could boost temporal and sequence processing, as well as executive control. In concert with effects exerted on other neurochemical mechanisms, rhythmic movement could also contribute to regulating arousal, impulsivity, and hyperactivity.

RCC research is also increasingly relevant for conditions diagnosed later in life, including neurodegenerative diseases and psychiatric disorders (as mentioned in Chaps. 7 [11], 9 [13], 11 [14], and 13 [9]). In Parkinson's disease, rhythmic stimulation can bypass impaired basal ganglia circuits to improve motor control (Chaps. 7 [11] and 9 [13]), capitalising, for example, on neural resonance mechanisms (Chap. 8 [12]). In Alzheimer's disease and other forms of dementia, rhythmic stimulation may have beneficial effects by inducing neural entrainment and engaging biochemical pathways that support the brain's natural cleanrance of harmful agents. The same mechanisms may also contribute to improving emotional, attention, and memory.

Cognitive timing is impaired in schizophrenia and other psychotic disorders. This also affects the processing of event sequencing and duration, making it difficult to align behaviour to external timing cues. This type of impairment has been linked to altered coupling of neural oscillations between slow and fast bands—a mechanism that is crucial for coordinating brain activity across regions (as discussed in Chaps. 8 [12], 9 [13], and 13 [9]). In this context, rhythmic interventions could help restore timing mechanisms and improve attention, memory, and executive function by resynchronising relevant brain networks and boosting predictive timing.

Finally, rhythmic stimulation—in the form of controlled breathing, exposure to rhythmic music, chanting, music-making, or rhythmic movement—is increasingly used in the management of personality and mood disorders (Chaps. 11 [14] and 13 [9]). In these contexts, the benefits of rhythmic engagement are likely mediated by neurophysiological and neurochemical mechanisms of emotion regulation, self-awareness, and interpersonal skills that are activated by rhythmic stimuli. For example, rhythmic activities can engage mechanisms of emotion regulation to balance arousal, facilitate emotional release, or promote relaxation (Chaps. 11 [14] and 13 [9]). In addition, rhythmic stimulation is known to foster social engagement and social bonding, which provides another psychobiological path towards enhanced well-being (Chap. 11 [14]).

While rhythmic practices are increasingly adopted to manage neurodegenerative and psychiatric disorders, systematic basic RCC research is required in order to tailor these practices to the treatment needs of the specific syndrome and the sensitivities of the individual. Research that improves our understanding of why, how, and when rhythmic intervention helps will pave the way towards evidence-based guidelines and practices.

15.5.2 The Applied Relevance of RCC Research to Animal Welfare

Studying rhythmic patterns in animal communicative signals can be a valuable tool for understanding emotional and social states (Chaps. 3 [3], 12 [7], and 11 [14]). In turn, this has direct applications for animal welfare assessment and husbandry practices in all managed care environments, from pet animals to zoos, sanctuaries, farms, and laboratories.

Rhythmicity—whether expressed through sound, substrate-borne vibrations, movement, or interactive behaviours—is a fundamental component of animal communication. As shown in Chaps. 3 [3], 9 [13], and 11 [14], variations in the rhythmic properties of communicative signals may convey information regarding fitness, sex, age, readiness to mate, arousal, etc., providing cues about the physiological and emotional states of the individuals. In addition, these can modulate the emotional states of receivers (Chap. 11 [14]). Basic research focused on systematically mapping animal RCC will deepen our understanding of the emotion–rhythm link in animal communication, having tremendous applied potential to inform behavioural indicators of well-being. In managed care settings, rhythmic behaviours can be systematically observed and analysed to assess welfare, helping to identify signs of boredom, anxiety, or social isolation. This research will also have the potential to support the development of evidence-based enrichment programs that align with species-specific needs. To design such interventions, however, we first need to understand the species' capacities for rhythm perception and production, and how these are linked to social communication, emotional regulation, and behavioural expression.

15.5.3 The Applied Relevance of RCC Research to Conservation Efforts

As shown in Chap. 6, basic research on RCC can play a crucial role in conservation efforts, particularly when considering the anthropogenic transformation of the natural world. Ecosystems undergo rapid and profound alterations due to human activities, ranging from climate change and habitat fragmentation to noise and light pollution. These changes cause disruptions that affect whether and how animals produce and perceive rhythmic signals, since the signals are influenced by environmental rhythms and their communicative efficacy depends on adaptations moulded by habitat properties. Because rhythmic signals are essential to multiple survival functions (Chap. 3 [3])—from mating and offspring viability, to navigation, feeding, or defence—these disruptions can have deleterious effects on the conservation status of species (Chap. 6 [6]). Understanding these disruptions can inform conservation efforts and habitat restoration.

For example, artificial light pollution interferes with natural light cycles, affecting neurochemical and endocrine processes that regulate RCC. This can lead to mistimed behaviours like premature birdsong, reducing mating success, and increasing predation risk. Climate-driven temperature changes impact rhythm-based communication in ectothermic species (e.g. frogs, fish, spiders), decoupling signal production and perception, and threatening reproductive success. Anthropogenic noise masks natural sounds, forcing animals to alter their calls. These adjustments can reduce signal clarity and attractiveness, with energetic and survival costs. Habitat changes and invasive species introduce acoustic competition, compelling native species to modify their rhythms, potentially affecting reproduction (Chap. 6 [6]).

Understanding these rhythm-related dynamics provides conservationists with tools to assess species welfare in the wild and design targeted interventions. By monitoring rhythmic behaviours and their disruptions, researchers can identify stressors, evaluate habitat quality, and implement enrichment strategies that support species-specific communication needs. Ultimately, integrating basic RCC research into conservation science enhances our ability to mitigate the impacts of environmental change and promote the resilience of animal populations in managed and wild ecosystems.

15.5.4 The Applied Relevance of RCC to Human–Animal Interaction

Rhythm plays a subtle yet powerful role in shaping human–animal interactions, influencing behaviour, emotional responses, and the quality of interspecies relationships. In everyday contexts such as training and caregiving, rhythmic cues—whether in voice, movement, or routine—can help regulate animal behaviour and foster emotional stability (Chap. 11 [14]). The predictability of human rhythmic actions may contribute to stress reduction, enhancing trust and learning outcomes.

In therapeutic settings, rhythmic synchrony between humans and animals has shown promise in enhancing bonding and therapeutic outcomes. Equine-assisted therapy, for instance, often involves coordinated movement between horse and rider, where shared rhythm may facilitate emotional attunement and physiological regulation. Similar effects have been observed in canine-assisted interventions, where rhythmic walking or breathing patterns between handler and dog appear to support emotional co-regulation.

Despite these promising observations, the mechanisms underlying rhythm in human–animal interaction remain underexplored. Future research should investigate how different types of rhythmic signals—acoustic, tactile, visual, and movement-based—contribute to interspecies communication and emotional synchrony. Studies could examine whether rhythmic entrainment between humans and animals enhances learning, reduces anxiety, or strengthens social bonds, and how these effects vary across species and contexts.

Moreover, comparative research across domesticated and wild animals could reveal whether sensitivity to human rhythm is a product of domestication or a broader feature of animal cognition. Investigating rhythmic dynamics in animal-assisted therapy could also inform clinical practices, helping to optimise intervention protocols based on rhythmic compatibility or entrainment potential.

As rhythmic communication is increasingly recognised as a cross-species phenomenon, integrating it into the study of human–animal relationships provides a rich avenue for interdisciplinary research—bridging behavioural science, neuroscience, ethology, and therapeutic practice.

15.6 Conclusions

To truly advance rhythm science, the field must embrace conceptual clarity and methodological rigor. At present, conceptual, terminological, and methodological heterogeneity hampers data comparability and undermines the comparative project. Harmonising research protocols and developing standardised metrics for rhythmic analysis will be essential.

Future RCC research will have to expand beyond acoustic rhythms to also investigate visual, tactile, and vibrational modalities. It will also have to embrace evolutionary and mechanistic depth. Combining field observations with lab experiments will help distinguish homology from convergence and identify selective forces shaping rhythmic capacities—shedding light on the adaptive functions of rhythmicity and on the evolution of human musicality and speech. Additionally, by bridging mechanistic and functional perspectives—e.g. linking neural oscillations, motor constraints, reward systems, etc. with the ecological and social significance of RCC—will enable explanatory models connecting proximate mechanisms to ultimate causes.

Studying these interactions and dynamics demands innovative methods and cross-disciplinary input from neuroscience, ethology, musicology, and computational modelling. Only through such integrative efforts can we begin to reveal the principles that govern rhythmic behaviour across species and contexts and understand its role in the broader tapestry of communication, cognition, and evolution. Finally, RCC research has broad applications—from clinical and educational interventions to animal welfare and conservation. Thus, rhythm science occupies a unique position at the intersection of fundamental and applied research, offering insights into one of nature's most pervasive organising principles.

References

1. Burchardt LS, Hoeschele M, Honing H et al (2026) Chapter 1: What is rhythm and how can we study it? In: Nature beats: the what, why and how of animal rhythmic behaviours. Springer Nature Switzerland, Cham
2. Sauciuc G-A, Schneider JN (2026) Chapter 2: How animals produce rhythm – an overview of ethological mechanisms for rhythmic communication. In: Nature beats: the what, why and how of animal rhythmic behaviours. Springer Nature Switzerland, Cham
3. Sauciuc G-A, Schneider JN, Burchardt LS et al (2026) Chapter 3: The functional and structural diversity of rhythmic communication across species. In: Sauciuc G-A, Lameira AR, Persson T (eds) Nature beats: the what, why and how of animal rhythmic behaviours. Springer Nature Switzerland, Cham
4. Lameira AR, De Gregorio C, Gamba M et al (2026) Chapter 4: Uncharted rhythmic diversity across the animal world. In: Sauciuc G-A, Lameira AR, Persson T (eds) Nature beats: the what, why and how of animal rhythmic behaviour. Springer Nature Switzerland, Cham
5. Greenfield M, Merker B (2026) Chapter 5: Endogenous rhythm, isochrony and the path to perfect synchrony. In: Nature beats: the what, why and how of animal rhythmic behaviour. Springer Nature Switzerland, Cham
6. Schneider JN, Burchardt LS, Sauciuc G-A, Lameira AR (2026) Chapter 6: How environmental and social contexts shape and constrain rhythm structure and production. In: Sauciuc G-A, Lameira AR, Persson T (eds) Nature beats: the what, why and how of animal rhythmic behaviour. Springer Nature Switzerland, Cham
7. Sauciuc G-A, Raimondi T, De Gregorio C (2026) Chapter 10: The developmental trajectory of rhythmic abilities in humans and singing primates. In: Sauciuc G-A, Lameira AR, Persson T (eds) Nature beats: the what, why and how of animal rhythmic behaviour. Springer Nature Switzerland, Cham
8. Sauciuc G-A, Persson T, Hattori Y, Lameira AR (2026) Chapter 12: The significance of the great ape heritage in the evolution of human rhythm in speech, music and dance. In: Sauciuc G-A, Lameira AR, Persson T (eds) Nature beats: the what, why and how of animal rhythmic behaviour. Springer Nature Switzerland, Cham
9. Lameira AR, Perlman M, Sauciuc G-A, Persson T (2026) Chapter 13: Rhythm-induced altered states and the evolution of human dance and music. In: Sauciuc G-A, Lameira AR, Persson T (eds) Nature beats: the what, why and how of animal rhythmic behaviour. Springer Nature Switzerland, Cham
10. Merker B, Cook PF, Reichmuth C et al (2026) Chapter 14: Human group synchrony in light of the comparative method: what we know and would like to know. A peer commentary treatment. In: Sauciuc G-A, Lameira AR, Persson T (eds) Nature beats: the what, why and how of animal rhythmic behaviour. Springer Nature Switzerland, Cham
11. Proksch S, Honing H, Ravignani A, Merchant H (2026) Chapter 7: Converging notions on the main hypotheses regarding the origins of beat perception and synchronization: the key role of audiomotor interactions. In: Nature beats: the what, why and how of animal rhythmic behaviour. Springer Nature Switzerland, Cham
12. Large EW (2026) Chapter 8: Neural resonance and comparative rhythm cognition. In: Sauciuc G-A, Lameira AR, Persson T (eds) Nature beats: the what, why and how of animal rhythmic behaviours. Springer Nature Switzerland, Cham
13. Sauciuc G-A, Schneider JN, Hattori Y (2026) Chapter 9: The neuroethology of rhythmic communication: neural and mechanical building blocks and constraints on rhythm production and perception. In: Sauciuc G-A, Lameira AR, Persson T (eds) Nature beats: the what, why and how of animal rhythmic behaviours. Springer Nature Switzerland, Cham
14. Gustison ML, Sauciuc G-A (2026) Chapter 11: Moving the body and the mind – links between emotion and rhythmic communication and the underlying neurobiological mechanisms. In: Sauciuc G-A, Lameira AR, Persson T (eds) Nature beats: the what, why and how of animal rhythmic behaviours. Springer Nature Switzerland, Cham

GPSR Compliance

The European Union's (EU) General Product Safety Regulation (GPSR) is a set of rules that requires consumer products to be safe and our obligations to ensure this.

If you have any concerns about our products, you can contact us on ProductSafety@springernature.com

In case Publisher is established outside the EU, the EU authorized representative is:

Springer Nature Customer Service Center GmbH
Europaplatz 3
69115 Heidelberg, Germany

Batch number: 10789951

Printed by Printforce, the Netherlands